P9-ECS-143

WITHDRAWN

Awaken students to science in their world!

EARTH SCIENCE

**Presenting current,
accurate science content**
■ ■ ■ ■ ■ ■ ■ ■ ■

**Incorporating the excitement
and relevance of science**
■ ■ ■ ■ ■ ■ ■ ■ ■

Featuring the most effective learning strategies

NASA

Current content that promotes interest and involvement in science

Earth Science incorporates the latest research and information in science and technology—such as wind and geothermal energy, landsat photography, and radiotelescopes.

Geothermal Power   In some regions, such as in Italy, New Zealand, and Iceland, hot igneous rocks produce usable energy within the earth. Rainwater penetrates porous rocks near the heat source and is converted to steam. The steam may surface through natural vents, or it can be extracted by drilling. The most famous steam field is at Larderello in Tuscany, Italy. Steam from Larderello is used to drive turbines that produce electricity. This steam power from the earth is called **geothermal power.** Geothermal power accounts for only 0.4 percent of the power used in Italy, but in New Zealand geothermal power accounts for 5 percent of the power produced.

Hot water and steam from geothermal fields are also used as a source of direct heat for buildings. The entire city of Reykjavik, Iceland, is heated in this way.

Solar Power   On a cold day, you have probably tried to stay in the sun to keep warm. The power produced by energy from the sun, called **solar power,** can be used directly as a source of heat or to produce electricity. The most common use of direct solar energy is for heating water. An array of dark-colored pipes placed on the roof of a building will provide hot water for personal use, and it may provide some heat for the building as well. Why would the pipes need to be dark in color?

Figure 17–22. Geothermal energy is widely used in Icel (right). In California some electricity is produced from steam fields such as the one shown here (left).

Figure 17–23. Solar energy be collected passively to hea water (right) or entire homes (left).

24.9 Observing the Earth

The value of observing the earth from space became clear when the first piloted flights began. Photographs taken by early astronauts and cosmonauts showed features of the earth that were not clear from the ground. The first satellite specifically designed to photograph the earth was *ERTS 1,* launched into a polar orbit on July 23, 1972. *ERTS 1,* which stands for *Earth Resources Technology Satellite 1,* was later renamed *Landsat 1.* Five Landsats were launched between 1972 and 1984.

The first application of Landsats was in cartography—the process of making maps. In addition, Landsat pictures of crop areas have made it possible to forecast production and to establish a worldwide food watch. Landsat photos are also important in monitoring dust storms, forest fires, and air pollution.

Landsats have also been used to locate mineral resources. The structure of the upper crust of the earth is studied as a means of identifying subsurface mineral deposits. The color of bare rocks and the pattern of vegetation where bedrock is covered often reveal the structure and mineral composition of the rocks. The changing snow cover in middle and high latitudes is also monitored by Landsats as a way of forecasting spring floods. The ice cover in Antarctica, Greenland, and many mountainous areas is monitored as a way of predicting possible sea-level changes. Several Landsat photographs, similar to the ones in Figure 24–21, have been used in this textbook to illustrate Earth features that standard aerial photography could not show nearly as well.

Figure 24–21. Landsat photographs of New York (left), Chicago (center), and Los Angeles (right) are shown here.

537

Sample pages are reduced. Actual sizes are 8″x10″.

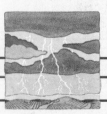

Applications that make science real and learning active

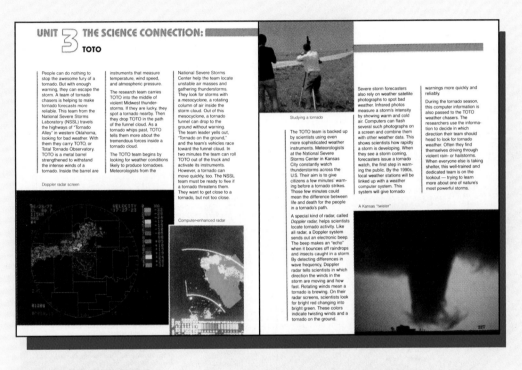

UNIT 3 — THE SCIENCE CONNECTION: TOTO

People can do nothing to stop the awesome fury of a tornado. But with enough warning, they can escape the storm. A team of tornado chasers is helping to make tornado forecasts more reliable. This team from the National Severe Storms Laboratory (NSSL) travels the highways of "Tornado Alley" in western Oklahoma, looking for bad weather. With them they carry TOTO, or Total Tornado Observatory. TOTO is a metal barrel strengthened to withstand the intense winds of a tornado. Inside the barrel are

Doppler radar screen

instruments that measure temperature, wind speed, and atmospheric pressure.

The research team carries TOTO into the middle of violent Midwest thunderstorms. If they are lucky, they spot a tornado nearby. Then they drop TOTO in the path of the funnel cloud. As a tornado whips past, TOTO tells them more about the tremendous forces inside a tornado cloud.

The TOTO team begins by looking for weather conditions likely to produce tornadoes. Meteorologists from the

Computer-enhanced radar

National Severe Storms Center help the team locate unstable air masses and gathering thunderstorms. They look for storms with a mesocyclone, a rotating column of air inside the storm cloud. Out of this mesocyclone, a tornado funnel can drop to the ground without warning. The team leader yells out, "Tornado on the ground," and the team's vehicles race toward the funnel cloud. In two minutes the team can roll TOTO out of the truck and activate its instruments. However, a tornado can move quickly, too. The NSSL team must be ready to flee if a tornado threatens them. They want to get close to a tornado, but not too close.

Studying a tornado

The TOTO team is backed up by scientists using even more sophisticated weather instruments. Meteorologists at the National Severe Storms Center in Kansas City constantly watch thunderstorms across the U.S. Their aim is to give citizens a few minutes' warning before a tornado strikes. Those few minutes could mean the difference between life and death for the people in a tornado's path.

A special kind of radar, called *Doppler radar*, helps scientists locate tornado activity. Like all radar, a Doppler system sends out an electronic beep. The beep makes an "echo" when it bounces off raindrops and insects caught in a storm. By detecting differences in wave frequency, Doppler radar tells scientists in which direction the winds in the storm are moving and how fast. Rotating winds mean a tornado is brewing. On their radar screens, scientists look for bright red changing into bright green. These colors indicate twisting winds and a tornado on the ground.

Severe storm forecasters also rely on weather satellite photographs to spot bad weather. Infrared photos measure a storm's intensity by showing warm and cold air. Computers can flash several such photographs on a screen and combine them with other weather data. This shows scientists how rapidly a storm is developing. When they see a storm coming, forecasters issue a tornado watch, the first step in warning the public. By the 1990s, local weather stations will be linked up with a weather computer system. This system will give tornado

A Kansas "twister"

warnings more quickly and reliably.

During the tornado season, this computer information is also passed to the TOTO weather chasers. The researchers use the information to decide in which direction their team should head to look for tornado weather. Often they find themselves driving through violent rain- or hailstorms. When everyone else is taking shelter, this well-trained and dedicated team is on the lookout — trying to learn more about one of nature's most powerful storms.

To conclude each unit, **The Science Connection**—in a high-interest magazine format— explores a unit topic and helps students see the relationship among concepts and apply science to their own lives.

Discover features in each section enable students to experience science through interesting activities that require a minimum of material and preparation time.

DISCOVER

Making Oxygen

Light a small candle and cover it with a glass jar. **CAUTION: Be careful with the flame.** Time how long the candle will burn before it goes out. Why does the candle go out?

Now place a small green plant next to the candle and light the candle again. Cover both the plant and the candle with a glass jar. Time how long the candle burns before it goes out. Does the candle burn for the same time? Explain any difference in time.

The **Activity** in each chapter helps students visualize scientific concepts and apply them to everyday situations.

ACTIVITY: Measuring Evaporation

How can you show that evaporation is a cooling process?

MATERIALS (per group of 2 or 3)

thermometers in notched corks (2); ring stand; clamps (2); cotton or gauze; rubber bands; beakers, 100 mL (2); rubbing alcohol; water; wax pencil

PROCEDURE

1. Using rubber bands, carefully fasten a piece of cotton or gauze to the bulb of each thermometer.
2. Attach the clamps to the ring stand, 10–20 cm above the table. Insert the corks into the clamps and tighten.
3. Fill one beaker with rubbing alcohol and the other with water. Label each beaker with a wax pencil.
4. Read and record the temperature on each thermometer.
5. Immerse one thermometer in the alcohol and the other thermometer in the water. Be sure to saturate the cotton.
6. At 30-second intervals, record the temperature on each thermometer.

CONCLUSIONS/APPLICATIONS

1. After the cotton was saturated with water or alcohol, did either thermometer show a change in temperature?
2. If so, which thermometer showed a faster change in temperature? Explain.

An instructional design that ensures success in learning

Highly readable text develops understanding by moving from the simple and familiar to the new and more complex. Analogies and models help students understand and visualize scientific principles, while outstanding charts, graphs, and diagrams clarify concepts and reveal science in real-life situations!

UNIT **5**

EARTH HISTORY

In 1977 geologist Walter Alvarez set out to study a thin layer of clay rock about 65 million years old. He concluded that the clay had been deposited when a giant meteor struck the earth. Alvarez knew that at about that same time nearly all the dinosaurs had vanished in a mysterious mass extinction. Alvarez and his scientist father, Luis, put these facts together and stated a new theory to explain the extinction of the dinosaurs. This theory started a scientific controversy. The Alvarezes claimed that the dinosaurs died when a cloud of meteor dust covered the earth.

- How can fossils be dated to show the evolution of animal and plant life?
- What may have caused the evolution and extinction of animal species?
- How will studying Earth's past enable humans to prepare for the future?

By reading the chapters in this unit, you will begin to learn the answers to these questions. You will also begin to develop an understanding of concepts that will help you to answer many of your own questions about the history of the earth.

Fossil dinosaur bones

321

Each unit opens with intriguing photographs and thought-provoking questions that set the stage for learning.

Sample pages are reduced. Actual sizes are 8″ × 10″.

CHAPTER 23

Observing the Universe

Stars, gas clouds, novae, and neutron stars all emit energy. Cosmic rays, gamma rays, X rays, ultraviolet rays, visible light rays, infrared rays, and radio waves all travel through space at the speed of light. How are these rays detected? What instruments do astronomers use?

Kitt Peak Observatory, in Arizona

SECTION 1
Optical Astronomy
23.1 Lenses and Light
23.2 Refracting Telescopes
23.3 Reflecting Telescopes

SECTION 2
Radio Astronomy
23.4 Radio Waves
23.5 Radiotelescopes

502

SECTION
1 Optical Astronomy

SECTION OBJECTIVES
After completing this section, you should be able to:
■ **Discuss** the importance of telescopes.
■ **Explain** what refraction and reflection are.
■ **Define** what is meant by *focus* and *focal length*.
■ **Describe** the difference between refractors and reflectors.

NEW SCIENCE TERMS
refraction
reflection
lens
focal length
telescope
magnification
spectrograph

23.1 Lenses and Light

Have you ever looked at the millions of stars in the night sky? The stars shine because of the energy they give off. The energy you see is in the form of light. You may recall from Chapter 8 that light travels through space as particles called *photons*.

Visible light is a very small part of all the energy in the universe. Some other forms of energy include radio waves, X rays, infrared radiation, and ultraviolet radiation. Visible light, however, is the easiest for people to detect. Light travels through space at the constant speed of 299 792 458 m/s (3.0 × 10⁸ m/s). Light also travels through transparent matter but at slower rates of speed. Table 23–1 shows the speed of light in various substances. Notice that the speed of light in air is similar to the speed of light in space.

Figure 23–1. Observing the stars is not simply a matter of looking through a telescope. Modern astronomical telescopes are controlled by computers.

TABLE 23–1: THE SPEED OF LIGHT IN DIFFERENT MEDIA

Substance	Speed of Light (m/s)
Empty space	299 792 458 (3.0 × 10⁸)
Air	299 702 547 (3.0 × 10⁸)
Water	225 407 863 (2.2 × 10⁸)
Glass	197 231 880 (2.0 × 10⁸)
Diamond	123 881 181 (1.2 × 10⁸)

Refraction Light behaves in certain predictable ways. A

Chapter and section openers spark interest, preview organization, and set goals for learning.

To monitor and reinforce learning, frequent review opportunities include brief questions at the end of each subsection to check understanding and encourage students to think about what they have just read.

Additional review features include:
- The **Section Review** to assess comprehension of section content
- The **Chapter Review,** which summarizes chapter concepts, reviews science vocabulary, and includes a quiz, extension activities, critical-thinking questions, and a list of related reading materials

---

DISCOVER

Observing Metamorphic Rocks

Study pieces of marble, slate, and quartzite. What sedimentary rocks might each of these have come from? On what observations did you base your answers?

of magma within the earth. A body of magma can cause changes in rocks for hundreds of kilometers around it.

○ *How might heat and pressure affect the minerals in a rock?*
○ *Name two sources of heat for the formation of metamorphic rocks.*

4.10 Classification of Metamorphic Rocks

Metamorphic rocks are divided into two groups: foliated and nonfoliated. **Foliated rocks** have obvious layers. Foliated rocks are formed by pressure from one direction. The layers often appear shiny, due to the new mineral crystals that have formed. Some foliated rocks show a pattern of bands. The amount of foliation is used to classify metamorphic rocks. For example, *slate, phyllite, schist,* and *gneiss* (NYS) show increasing amounts of foliation. **Nonfoliated rocks** have no visible layers. This is because pressure during formation comes from several directions. *Marble,* for instance, is nonfoliated.

Identifying the minerals in the original rocks helps to classify metamorphic rocks. For example, when limestone is changed to a metamorphic rock, the original limestone becomes more compact. However, the mineral can still be identified as calcite. The resulting rock is marble. Several kinds of metamorphic rocks are shown in Figure 4–25.

Another example of the metamorphic process occurs when bituminous coal is put under intense heat and pressure. This soft coal becomes a harder, more compact form of coal called *anthracite.*

○ *How can foliated rocks be identified?*
○ *What is a nonfoliated rock?*

Figure 4–25. Marble (below), quartzite (bottom left), and slate (bottom right) are all metamorphic rocks.

92   Chapter 4 Rocks

P 5

Strategies that help students build thinking and process skills

13.3 Sand Dunes

Sand dunes are the best known form of wind deposit. Dunes are characterized by gently sloping sides, called the *windward side*, facing into the wind. The steep side, facing away from the wind, is called the *leeward side*. Dunes move, or migrate, as sand is rolled up the gentle slope and deposited over the crest onto the other side.

The most common type of dune is the **barchan** (BAHR kahn) **dune**. A barchan is a crescent-shaped dune with the bulging side facing into the wind. This type of dune forms in areas that have a limited supply of sand. As sand is separated from the larger pebbles and rocks in an area, the rocks that are left are called *desert pavement*. In areas with a large supply of sand or with variable winds, different kinds of dunes form. Some of these different dunes are shown in Table 13–1.

○ *What is a barchan dune?*

Figure 13–9. When the supply of sand is plentiful and the wind blows constantly from the same direction, giant ridges of sand (right) can be produced. The area from which the sand is removed is often left barren, resulting in a hard surface called *desert pavement* (left).

Section Review

READING CRITICALLY

1. List four factors that form arid lands.
2. Explain how a dune migrates.

THINKING CRITICALLY

3. How can a coastal environment be very much like a desert?
4. Explain why the term *shadow* is especially useful in picturing the concept of a rain shadow.

280   Chapter 13 Wind and Ice

Concluding each section, **Reading Critically** helps students check understanding of what they have read, while **Thinking Critically** encourages students to analyze and apply what they have read. At the end of each chapter, **Writing Critically** helps students use thinking and writing skills to answer challenging essay questions.

Writing Critically

23. Explain how the production of specialty steels could be interrupted by international disagreements.

24. In what ways are the making of specialty glasses similar to the making of specialty steels?

25. Why are peat, lignite, and coal considered fossil fuels, but not hydrocarbons?

At the end of each chapter, **Application/Critical Thinking** encourages students to think critically and inventively and to apply science to everyday life.

Challenge Your Thinking presents intriguing photographs and questions that encourage students to apply what they have learned in the chapter to other scientific phenomena.

APPLICATION/CRITICAL THINKING

1. Explain why meander cutoffs form a single river channel rather than several small channels.

2. A special type of drainage pattern develops over an area underlain by limestone. This type of pattern is called *centripetal*, or circular. Using your knowledge of karst topography, describe what you think the characteristics of centripetal drainage would be.

3. Many large farms are found in the hilly regions of the northeastern United States and in the foothills of the eastern mountains. Explain how modern contour plowing could decrease the amount of runoff into streams of these areas.

FOR FURTHER READING

Bain, I. *Water on the Land.* New York: The Bookwright Press, 1984. Major concepts about surface and ground water are clearly presented in this book.

Gunston, B. *Water.* Morristown, N.J.: Silver Burdett, 1980. This classic book presents a thorough treatment of water characteristics, water activity on land and in the atmosphere, and water uses by living things and industry.

Maltby, E. *Waterlogged Wealth: Why Waste the Earth's Wet Places?* Washington, D.C.: International Institute for Environment and Development, 1986. An excellent review of wetlands of the world.

Challenge Your Thinking

Flowing through the Grand Canyon, the Colorado River provides a puzzle as to its stage of development. The river shows the wide, looping meanders characteristic of an old-age river; however, the valley walls are steep and V-shaped, characteristics of a youthful river. How might this apparent conflict of characteristics be explained?

Chapter 12 Review   **273**

A **Skill Activity** in every chapter helps students develop and use the skills of a scientist—including making measurements, interpreting data, drawing conclusions, and using inductive and deductive reasoning.

SKILL ACTIVITY: Drawing Profiles

BACKGROUND

When analyzing the geomorphology of an area, you need to be able to interpret both the rock structure underlying the area and the resulting shape of the land surface. Sometimes geologists draw a diagram called a *profile* to show accurately the shape of the land surface. A profile is a representation of an object as seen from the side. A profile of an area of land is like looking at hills outlined against the horizon. The contour lines on a topographic map are used to draw a profile.

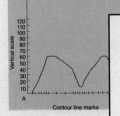

PROCEDURE

1. Identify the area to be profiled. The first step in drawing a profile is to draw a line across the map to show where the profile is to be taken. On the map, the line marked AB is drawn so that it crosses the peaks of two hills. Near the center of a piece of graph paper, draw a horizontal line. Fold your paper along this line, lay the fold along line AB on the map, and mark the ends of the line on your graph paper with dots labeled A and B.
2. Determine the vertical and horizontal scales. The horizontal scale is determined by the scale of the map. The scale of the map in this exercise is 1:20 000. The vertical scale is your choice; however, if you use a vertical scale that is the same as the horizontal scale, the slopes in your profile will be proportional to

the actual slopes of the land. On your graph paper, draw a vertical line upward from dot A. Make each line along the vertical a 20-foot contour interval starting with 0.
3. To mark the positions of the contour lines, place your line AB along line AB on the map. Mark on the horizontal where each contour line crosses line AB. Carefully label each mark with the elevation listed below the line.
4. Draw the profile. Find the mark that represents the position where the 20-foot contour line crossed your line AB. Directly above this mark, make a dot on the 20-foot line of your vertical scale. Mark the elevations of other contour lines in the same fashion. Connect the points on your graph paper with a smooth, curving line. You have now drawn a profile along line AB.

Vertical scale: 120, 110, 100, 90, 80, 70, 60, 50, 40, 30, 20, 10

A · Contour line marks

APPLICATION

Follow the same procedure and mak[e] the area between points C and A.

USING WHAT YOU HAVE LEARNE[D]

Compare your profile to the contour on the map.

1. Does it look as you expected? Ex[plain.]
2. Are the slopes steepest where th[e] closest together? Explain.

In each chapter, an **Investigation** helps students develop laboratory skills and encourages them to think like scientists. Step-by-step instructions and helpful illustrations clarify each investigation.

INVESTIGATION 11: Weathering by Carbonation

PURPOSE

To examine the effects of carbonic acid on the weathering of different rocks

MATERIALS (per group of 3 or 4)

Safety goggles
Laboratory apron
Laboratory balance
Small food jars with lids (4)
Tap water
Limestone
Carbonated water
Granite

PROCEDURE

1. Copy the chart shown.
2. **CAUTION: Wear safety goggles and a laboratory apron during this investigation.** Using the balance, determine the mass of the limestone specimen.
3. Fill a small jar half full of tap water, and add the limestone to the jar. Put the cap on the jar and gently shake the jar 200 times.
4. Remove the limestone from the jar, dry it, and determine its mass.
5. Repeat steps 3 and 4. After shaking the jar 400 times, record the mass.

6. Repeat steps 3 and 4 for a third time, shaking the jar 600 times. Record the results.
7. Compute the percentage mass change, using the following formula:
$$\frac{\text{Initial Mass} - \text{New Mass}}{\text{Initial Mass}}$$
8. Repeat steps 2 through 7, using limestone in carbonated water. Record your results.
9. Repeat steps 2 through 7, using granite in both tap water and carbonated water. Record your results.

ANALYSES AND CONCLUSIONS

1. Which combination of liquid and rock produced the greatest change in mass?
2. What kind(s) of weathering occurred in this investigation? List and describe the weathering in as much detail as possible.
3. In what way is the weathering process in this investigation similar to weathering in nature? In what way is it different?

APPLICATION

Using what you have learned in this investigation, predict two types of rock that would probably not weather very much. Predict one type of rock that would probably weather a lot. State the reasons for your predictions.

TABLE 1: EFFECTS OF WEATHERING				
	Limestone		Granite	
	Water	Carbonated Water	Water	Carbonated Water
Initial mass				
Mass after 200 shakes				
Mass after 400 shakes				
Mass after 600 shakes				

Sample pages are reduced. Actual sizes are 8″ × 10″.

Captivating features that complement instruction

To illustrate practical applications of scientific knowledge, **Technology** explores current issues in science and provides up-to-date information on technological advances.

TECHNOLOGY: Modern Farming

Most people in the United States live in urban areas. They get their food, wrapped in plastic and paper containers, from a supermarket. With each new day the demands on the land increase; more people must be fed, more trees must be cut for paper and building materials, and more land must be turned into areas where people can live.

Total land area is being carefully evaluated to make the best use of every available hectare.

Plants called legumes are being planted on marginally fertile soil because they produce nitrogen compounds that other plants can use. The contour of the land is studied, and crops are planted on terraces, or flat steps, rather than on slopes, to prevent loss of soil due to excess runoff.

Many farmers are now using a practice called minimum till farming. In this program, the soil is protected in the winter by the stubble of the previous crop. New crops are planted in narrow strips between the stubble to reduce the exposure of the soil to the drying wind.

New techniques in irrigation are turning semiarid parts of the world into productive cropland. In some places, water is dripped onto individual plants instead of being sprayed over entire fields, where m...

Modern harvesting equipment

Contour plowing and drip irrigation (inset)

BIOGRAPHIES: Then and Now

JOHANNES KEPLER (1571–1630)

Johannes Kepler was born on December 27, 1571. In his youth, he studied theology, mathematics, and philosophy at the University of Tübingen, in Germany. In the sixteenth century many people believed that the earth was the center of the solar system. In that model, the sun and all the planets revolved around the earth. Kepler became a strong defender of another theory, which had been proposed by Nicolaus Copernicus in 1540. That theory states that the sun, and not the earth, is the center of the solar system. In his mid-twenties Kepler published a work entitled Mysterium Cosmographicum. In the book, he strongly put forth this theory.

He later worked with the accomplished astronomer Tycho Brahe who invited him to become an assistant. When Tycho Brahe died a year later, Kepler took Brahe's position at court and inherited all of the famed astronomer's notes and records. Brahe's notes included much important information about the position and movement of the planets.

Kepler made many of his own observations and also used Brahe's work to discover several laws about the motion of the planets. In 1609 he wrote Astronomia Nova (New Astronomy). In that book he proposed that the planets did not move in circles as was previously thought but instead moved in elliptical orbits around the sun. Kepler's laws of planetary motion served as the basis for Sir Isaac Newton's laws of gravity.

IRENE DUHART LONG (1951–)

Irene Duhart Long was born in Cleveland, Ohio, in 1951. She became interested in flight at a very young age, when she accompanied her father on his flying lessons. By the age of nine, she had decided that she wanted to work for NASA.

She began her college education at Northwestern University, in Evanston, Illinois, and received her medical degree from the St. Louis University School of M... aerospace medicine. Sh...

In 1982 Dr. Long wa... residency at the Kenned... Medical and Environ... NASA. She is currently... tions. Dr. Long and her... how the free-fall envi... affects the health of as... vides medical care for th... of an emergency and c... tions of the human bo... group is particularly con... space travel on the hea... system. The informatio... to develop methods to... space travel on the hum...

One of the highes... Kennedy Space Center,... to be named Chief of... next goal is to travel in s... ical officer on a spacefli... lect in-flight informatio... space travel first hand.

Section 2 The Inner Planets a...

Biographies: Then and Now presents profiles of people in science—both historical and contemporary. Students gain new appreciation for the growth and vitality of science.

CAREERS

LAPIDARY

A lapidary is a person who cuts rough mineral samples into beautiful gemstones. A single cut can be the difference between a gem worth thousands of dollars and one that is worthless. Lapidaries must understand the optical effects of light and the structure of minerals.

People who would like to learn the art of gem cutting usually begin as apprentices.

For Additional Information
Gemological Institute of America
1660 Stewart Street
Santa Monica, CA 90406

MINING GEOLOGIST

Mining geologists play a major role in finding minerals and getting them out of the ground. They are also involved in the important task of restoring the environment in mined areas.

Most jobs in mining geology are through private employers in the industry. A bachelor's degree is required, and an advanced degree is often recommended. People who have gained experience as technical assistants to mining geologists can obtain positions by passing a professional licensing test.

For Additional Information
American Geological Institute
5205 Leesburg Pike
Falls Church, VA 22041

STONEMASON

Although most large buildings are now built with steel, stonemasons still play an important part in construction work. Many buildings have stone decorations, window sills, door frames, and exteriors. These stones must be cut, fitted, and secured with precision. Most stonemasons learn their skills through apprenticeship programs. Sometimes they take classes in general construction skills.

For Additional Information
Associated General
    Contractors of America, Inc.
1957 E Street, NW
Washington, DC 20006

58   Chapter 3   Minerals

Careers explores occupations related to science and includes names and addresses of professional organizations for more information.

Sample pages are reduced. Actual sizes are 8″ × 10″.

A MATTER OF FACT

In 1815 the Tambora volcano in Indonesia killed 12 000 people when it erupted. The eruption sent over 80 km^3 of ash into the atmosphere, blocking out so much of the sun that there was a general cooling of the entire earth, making 1816 the "year without a summer."

Throughout the textbook, **A Matter of Fact** provides intriguing information about discoveries and people in science.

Extension activities encourage independent research and apply science to other curriculum areas.

EXTENSION

1. Write a report that traces the evolution of the transistor from vacuum tube to silicon chip. Be sure to describe the materials that were, and currently are, used in the manufacture of each device.

2. Research and report to the class on the technological and environmental problems associated with the surface mining of coal.

3. Construct a model of an oil rig from simple materials, such as popsicle sticks and plastic straws, and then use your model to explain the drilling process to your class.

For Further Reading lists and briefly summarizes magazine articles and books for further exploration of chapter topics.

FOR FURTHER READING

Bartusiak, M. "Megascope." *Omni* 8(July 1986): 16. This article is a discussion of how improved technology may once again make arrays of reflecting telescopes valuable to astronomers.

Bartusiak, M. "Ultimate Catalog." *Omni* 8(February 1986): 26. This article is a summary of how scientists use complex computers to catalog over 20 million celestial objects.

Berry, R. *Build Your Own Telescope: Complete Plans for Five High-Quality Telescopes That Anyone Can Build.* New York: Scribner's, 1985. This is a step-by-step manual that shows anyone with basic carpentry skills how to build a telescope.

Kerrod, R. *The All Color Book of Space.* New York: Arco, 1985. This is a series of essays that relates astronomy to space exploration. The author describes early telescopes, space exploration technology, and space pioneers.

Renner-Smith, S. "New Satellite Antennas." *Popular Science* 227(December 1985): 69. This article gives pointers on how to select inexpensive, state-of-the-art television receivers.

Loaded with helpful resources, the comprehensive **Reference Section** (not shown) includes a complete **Glossary** and a cross-referenced **Index,** plus:

- Safety Guidelines
- Laboratory Procedures
- Common Laboratory Equipment
- Key Discoveries in Earth Science
- Periodic Table

- Suggested Science Projects
- Scientific Notation
- Significant Digits
- Building a Science Vocabulary
- Mineral Classification Key
- Rock Classification Key

- Weather and Climatic Charts
- Geologic Time Line
- Earth Science Atlas

A powerful Teacher's Edition for total support

For ease in lesson planning and teaching, the *Annotated Teacher's Edition* provides valuable suggestions and information—conveniently located before each unit and chapter:

- Unit and chapter overviews

- Advance preparation

- Bulletin board suggestions

- Issues in earth science

- Suggested projects

- Teacher resources

- Charts for planning and organizing chapters and sections

- Teaching suggestions, including class activities, field trips, and outside speakers

Helpful teaching annotations at point of use include:

1. Questions that prompt classroom discussion and promote thinking and process skills

2. Suggestions for extending section content

3. Background information

4. Ideas for teacher demonstrations

5. Answers to all questions in the student book

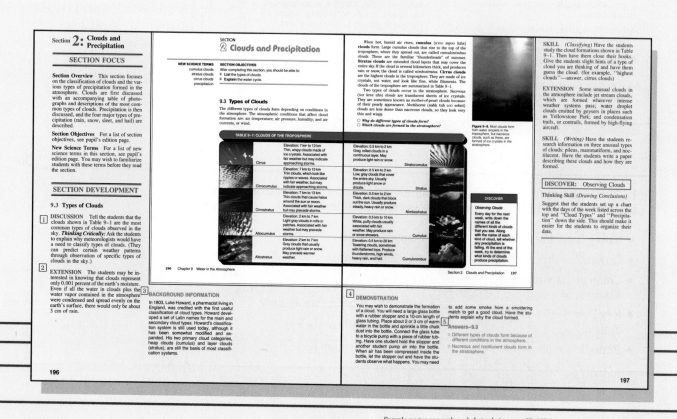

P 10

Sample pages are reduced. Actual sizes are 9"x10" (*Annotated Teacher's Edition*).

Exciting resources that make teaching a pleasure

Teacher's ResourceBank™

For rich and varied classroom instruction, the *Teacher's ResourceBank*™ includes a wealth of helpful copying masters with answer keys:

- Vocabulary Activities
- Reading for Content Activities
- Critical-Thinking Activities
- Cross-Discipline Activities

- Concept-Extension Activities
- Laboratory Investigations
- Student Record Book: Textbook Investigations

- Unit and Chapter Tests
- Science Sites: Connecting Science, Geography, and History
- Science Project Guide

For complete teaching support, the *Teacher's ResourceBank*™ also contains sample *Teaching Transparencies*, Classroom Reference Posters, and a demonstration disk for the *Earth Science Computer Test Bank*. A tabbed, three-ring binder provides organized storage for teaching resources.

Additional components

Laboratory Manual/Study Guide
The *Laboratory Manual/Study Guide* contains two additional laboratory investigations for each chapter, as well as Vocabulary Activities, Critical-Thinking Activities, and Concept-Extension Activities. The *Laboratory Manual/Study Guide, Annotated Teacher's Edition*, provides answers and annotations.

Teaching Transparencies
Teaching Transparencies of selected labeled illustrations in the textbook facilitate classroom instruction. Accompanying copying masters provide illustrations, without labels, that teachers may use as quizzes or tests.

Classroom Instructional Posters
To reinforce learning, *Classroom Instructional Posters* highlight major unit topics.

Earth Science Computer Test Bank
The *Earth Science Computer Test Bank* provides a complete testing program and also allows teachers to create their own tests.

Annotated Teacher's Edition

Earth Science

Annotated Teacher's Edition

EARTH SCIENCE

Cesare Emiliani

Professor and Chairperson,
Department of Geological Sciences
University of Miami
Coral Gables, Florida

Linda B. Knight

Earth Science Teacher
Paul Revere Middle School
Houston, Texas

Mark Handwerker

Instructional Support Teacher for Science
Temecula Middle School
Temecula, California

HBJ **Harcourt Brace Jovanovich, Publishers**

Orlando   San Diego   Chicago   Dallas

Copyright © 1989 by Harcourt Brace Jovanovich, Inc.

All rights reserved. No part of this publication may be reproduced or transmitted in any form or by any means, electronic or mechanical, including photocopy, recording, or any information storage and retrieval system, without permission in writing from the publisher.

Requests for permission to make copies of any part of the work should be mailed to: Permissions, Harcourt Brace Jovanovich, Publishers, Orlando, Florida 32887

Printed in the United States of America

ISBN 0-15-361454-4

Cesare Emiliani

Dr. Emiliani earned a D.Sc. in Geology from the University of Bologna and a Ph.D. in Geology from the University of Chicago. He is the Chairperson of the Department of Geological Sciences and Director of the Marine Science/ Marine Affairs Program at the University of Miami. Dr. Emiliani is the author of more than 100 technical papers and three books in the geological sciences, and is a contributor to the *McGraw-Hill Encyclopedia of Science and Technology*. Dr. Emiliani is a Fellow of the American Geophysical Union and a Fellow of the American Association for the Advancement of Science, and has received the Swedish Gold Vega Medal for achievement in Geological Sciences.

Linda B. Knight

Dr. Knight received a B.A. in English Literature from Knox College in Galesburg, Illinois; an M.A.T. in earth science and an Ed.D. in science education from Indiana University, Bloomington, Indiana. Dr. Knight is presently teaching earth science at Paul Revere Middle School in Houston, Texas and is President-elect of the Texas Earth Science Teacher's Association. She has also taught earth science at Bloomington High School North, Bloomington, Indiana. In addition, she has taught science methods classes to both elementary and secondary teacher candidates. She has served on the Midwest Space Shuttle Project Review Committee to review science projects for the shuttle program. She was also instrumental in the development of an expanded middle school earth science curriculum for the gifted and talented in the Houston Independent School District, and has served as co-author of several articles for professional journals.

Mark J. Handwerker

Dr. Handwerker received a B.S. in Psychology and Biology from the City College of the City University of New York, and a Ph.D. in biological sciences from the University of California, Irvine, California. Dr. Handwerker is presently science department chairperson, science teacher, and mentor teacher at Carnegie Junior High School in Los Angeles, California. In addition, he has been a lecturer and laboratory instructor at the University of Puerto Rico, and a laboratory instructor at the University of California, Irvine. Dr. Handwerker has also been a research associate and has served as co-author of several scientific books.

T4

ACKNOWLEDGMENTS

SENIOR EDITORIAL ADVISOR

Lee Suttner, Ph.D.
Professor of Geology
Indiana University
Bloomington, Indiana

CONTENT ADVISORS

George Greenstein, Ph.D.
Professor of Astronomy
Amherst College
Amherst, Massachusetts

Brian Skinner, Ph.D.
Professor of Geology and Geophysics
Yale University
New Haven, Connecticut

Carolyn Sumners, Ed.D.
Director of Astronomy and Physics
Houston Museum of Natural Science
Houston, Texas

CURRICULUM ADVISORS

Ronald E. Charlton, Ph.D.
Science Coordinator
Mt. Lebanon School District
Pittsburgh, Pennsylvania

Dee Drake*
Earth Science Teacher
Huron High School
Ann Arbor, Michigan

Larry Enochs, Ed.D.
Professor of Curriculum and Instruction
Kansas State University
Manhattan, Kansas

Robert Frank*
Earth Science and Geography Teacher,
  Science Coordinator K-12
Jefferson Junior High School
Caldwell, Idaho

Nancy Hampton Johnson
Earth Science Teacher
Haggard Middle School
Plano, Texas

Loistene Harrell
Earth Science Teacher,
  Department Chairperson
Lake Braddock Secondary School
Fairfax County, Virginia

Tom Howick*
Earth Science Teacher
Woodward Academy
College Park, Georgia

Bobbie G. Reed
Department Chairperson
Crestwood Middle School
Baton Rouge, Louisiana

David Sorenson
Science Teacher
Southwest High School
Green Bay, Wisconsin

Doris Tucker*
Earth Science Teacher
Rutherford County Schools
Spindale, North Carolina

Lou Travelsted
Science Teacher
Franklin-Simpson Middle School
Franklin, Kentucky

READING/ LANGUAGE ADVISORS

Patricia Bowers, Ph.D.
Mathematics and Science
  Coordinator
Division of Curriculum and
  Instruction
Chapel Hill-Carrboro City Schools
Chapel Hill, North Carolina

Sue Porro
English Teacher
Dr. Phillips High School
Orlando, Florida

Karen Rugerio
Chairperson, Fine Arts Department
Dr. Phillips High School
Orlando, Florida

*Outstanding Earth Science Teacher, as awarded by the National Association of Geology Teachers

CONTENTS

TEACHER'S BACKGROUND AND RESOURCE INFORMATION

TEACHER'S UNIT AND CHAPTER GUIDES

USING THE PROGRAM

Nowhere has the impact of modern technology on human lives and the explosion of knowledge been more apparent than in the sciences. New discoveries are announced with regular and increasing frequency. The advances in science require today's students to prepare themselves to make informed decisions on such questions as pollution control, alternative energy resources, and conservation of natural resources. In light of these advances and decisions, the students' success in an earth science course is especially important. What they learn, and the science-related skills they develop, will prepare them for science courses they will be taking in the future.

The general organization of **EARTH SCIENCE** proceeds from the concrete to the abstract. Unit 1 provides basic skills and information needed for succeeding chapters. Chapter 1, for example, sets the stage for all later discussions by explaining scientific methods and their reliance on a working model or hypothesis. Chapter 2 deals with maps and the use of models for studying the surface of the earth and the processes that create the earth's features.

The remainder of **EARTH SCIENCE** is divided into six main areas of study. Unit 2 introduces geology, the study of rocks and minerals and the processes that produce mountains, volcanoes, and earthquakes. Also in Unit 2 is a chapter dealing with Earth's resources. Unit 3 discusses meteorology, the study of weather and climate and the structure of the atmosphere. Weathering and erosion, the processes that shape the earth's surface, are discussed in Unit 4. Unit 5, Earth History, includes the study of Earth's past and future. In Unit 6, Oceanography, the study of the oceans and ocean resources is discussed. Unit 7 covers astronomy, the study of the solar system, the universe, and space exploration.

The reading level of the text has been carefully analyzed using computer-applied readability formulas. A basic vocabulary of earth science terms is identified through the use of boldface and by inclusion in the end-of-chapter list of science terms and in the Glossary. Many terms are pronounced, using a simple phonetic respelling, and all are defined in context when first used.

Equally important as the readability scores is the comprehensibility of a text. The comprehensibility of **EARTH SCIENCE** is enhanced by the graphic design, the lucid narrative style, the logical organization of topics, and the careful transitions between concepts. In addition, the use of analogies throughout the textbook provides the students with easy-to-understand explanations of difficult concepts. Practical applications of these concepts allow the students to incorporate what they learn into their daily lives.

Many features of the *Annotated Teacher's Edition* are designed to facilitate the teacher's task. The wrap-around design of the *Annotated Teacher's Edition* provides the teacher with a great variety of support material at point of use. For example, all questions in the pupil's edition, including essay-type questions, are answered in the margin adjacent to the question. Additional demonstrations, activities, suggestions for outside speakers, and suggestions for field trips are provided to help the teacher better plan the course of study.

The visual components of **EARTH SCIENCE** are also integral parts of the teaching program. Landsat photographs, computer-enhanced photographs, and topographic maps employ the latest scientific technology. The use of multiple images further serves to clarify concepts. The illustrations of processes and geologic events are as accurate and instructive as they are beautiful.

Every person in today's complex society needs a knowledge of science and its applications to function effectively. **EARTH SCIENCE** provides coverage of all phases of earth science, including geology, astronomy, meteorology, and oceanography topics. **EARTH SCIENCE** emphasizes thinking skills, an appreciation of technology, and the practical application of scientific principles.

Planning and implementing an earth-science course require careful thought and development. A number of fundamental issues must be addressed before the teaching program can be developed. These issues include the ability level of the students, the teacher's individual teaching style, how the textbook can be used most effectively, time constraints, and areas of content to be emphasized. This part of the *Annotated Teacher's Edition* is designed to assist you in resolving such issues.

BECOMING FAMILIAR WITH THE TEXTBOOK

EARTH SCIENCE is designed to aid the students in learning and the teacher in teaching. For the student, each component of the textbook is an important resource designed to facilitate learning. Many students will not take advantage of the various learning resources unless encouraged to do so. Therefore, it is important that you familiarize the students with the learning devices incorporated in the textbook.

Before making the first reading assignment, you may wish to have the students thumb through a unit of the textbook. The consistency of unit and chapter features makes it easy for your students to find the features in the textbook. Discuss the unit opener photograph and look at the questions asked beside the photograph. Tell the students that while they may be able to answer some of these questions before they study the unit, they will be able to provide more thorough and thought-out answers after the unit has been completed. Next, have the students look at the chapter opener photograph and read the introduction. The introduction is designed to provide motivation to the students. The discussion and preview can also provide an informal assessment of any background knowledge the students bring to each unit and chapter. This assessment will help you fine-tune your planning for the chapters.

Look through the first chapter with the class, locating such features as the introduction, chapter outline, *Section Objectives,* and headings and subheadings. Discuss various methods of organizing information that can be used as the students read and study the chapter. If you encourage your students to outline material, they may use the chapter outline as a basis for the development of a more detailed outline of chapter material.

Encourage your students to look at the science terms, which are listed at the beginning of each section and highlighted in **boldface** type throughout the textbook. Point out that the study of science requires the mastery of many new terms, including some familiar terms that may have a different, specialized meaning when used in scientific contexts. For example, ask the class for a definition of the term *law.* Then, direct their attention to the use of the word on page 8, where it is printed in boldface type. Point out that the definition of the term *theory* is also discussed on the same page. Next, have the students turn to the *Chapter Review* on pages 21–23, where the word is listed among the chapter's *Science Terms,* and its usage tested in the science quiz questions that follow. Finally, have the students find the word in the *Glossary,* which begins on page 594.

While reviewing how the textbook is organized, alert the students to the regular features located in the page margins: *Discover* and *A Matter of Fact. Discover* provides an exercise that involves either a brief experiment, research, or other hands-on activity. *A Matter of Fact* presents interesting facts related to the topic being discussed.

Other features can be used to expand on material presented in the textbook. The *Skill Activity* in each chapter focuses on one or more process skills, such as classifying, interpreting, and communication. The *Activity* is a short laboratory investigation about a topic in the chapter. The *Investigation* provides a more involved laboratory experience that carefully follows traditional scientific method.

In addition to the features that appear in each chapter, a triad of features—*Careers, Biographies: Then and Now,* and *Technology*—are also found in the chapters. One of these three features appears in each chapter. Each career feature provides information on three different careers in a field related to earth science. Differing levels of education are required for the careers presented. *Biographies: Then and Now* presents two biographies, one historical and one modern, of people who have done work in areas related to topics in the chapter. *Technology* gives information on developments that have caused or may cause tremendous advances in the earth sciences and that are or will be impacting people's lives.

The Science Connection, found at the end of each unit, is a two-page feature of up-to-date information on such topics as Landsat imaging of Earth (pages 46–47), theories on how the dinosaurs died (pages 386–387), and the use of tidal power to produce electricity (pages 450–451). *The Science Connection* provides closure for the unit, relating the unit-opener paragraph and the chapters in the unit to the specific topic being discussed. After having read the chapters in the unit and *The Science Connection,* the students should be able to answer the questions posed in the unit opener.

USING PERFORMANCE OBJECTIVES

EARTH SCIENCE contains *Section Objectives* written in terms of student performance. These performance objectives are one of the most useful tools available to teachers and students alike. Students should be encouraged to use the objectives when they begin, and again when they complete each section. The objectives will provide them with a clear goal for their reading.

The *Section Review* questions reflect the objectives and, therefore, provide an effective means of self-assessment. Some of these questions are designed to test the students' ability to recall certain facts within the section. However, other questions in this feature require higher levels of thinking. The students will find both *Reading Critically* and *Thinking Critically* questions in this feature.

Reading Critically is the first part of a triad of features that are found throughout the textbook and are designed to foster and reinforce thinking skills. *Reading Critically* asks the students to recall and apply information presented in the textbook. The second part of the triad is *Thinking Critically*. These questions require higher-level thinking. They ask the students to take the information they have learned in the section and apply it to new situations. These questions are vital in assessing student understanding of important concepts.

Performance objectives are often misused by students who regard them merely as guides to material that will someday appear on a test. Such students often approach learning by searching through the text for the answers to each objective. Not only is this inefficient, but it results at best in the acquisition of numerous disjointed facts. This approach is easy to correct through a discussion of how the *Section Objectives, Section Review* questions, and the *Chapter Review* materials can be used together for self-evaluation.

Within the *Chapter Review,* you will find the third part of the critical-thinking triad—*Writing Critically*. This feature is designed to help students develop higher-level thinking skills and write their responses clearly and concisely.

TEACHING PROCESS SKILLS

Scientific knowledge is generally viewed as the product of scientific acts, or processes. Therefore, process skills must be integral to science instruction and learning.

The "science process skills," referred to by some people as thinking skills or skills used in a scientific method, are defined as a set of broadly transferable skills representative of the behaviors of scientists. These skills, which are so important to the scientist, are crucial to the student of science as well. They enable the student to explore and manipulate the environment, to hypothesize and reason from data, to formulate explanations, and to communicate these explanations to others. In fact, process skills are prerequisites to success not only in science but also in other subjects and in the community and workplace, too.

The acquisition of process skills is a key goal in **EARTH SCIENCE.** To accomplish this goal, the program provides numerous opportunities for skills development through engaging narrative style and thought-provoking questions. In addition, the program provides direct instruction in developing and applying process skills through the pupil's edition feature *Skill Activity*. Since research suggests that skills taught in isolation from subject matter are not likely to transfer readily or become functional, skills in **EARTH SCIENCE** are taught in context, drawing from information directly related to chapter content. Other pupil's edition features—including *Investigation, Activity,* and *Discover*—offer opportunities for students to practice the skills in additional contexts.

The common process skills emphasized in **EARTH SCIENCE** are listed and described below:

- *Communicating*—conveying information verbally in oral and written forms as well as visually through models, charts, tables, graphs, diagrams, and maps.
- *Comparing*—identifying common and distinguishing characteristics among objects or events.
- *Drawing conclusions*—making general statements about particulars based on reasoning.
- *Experimenting*—designing data-gathering procedures to test hypotheses under conditions in which variables are both controlled and manipulated.
- *Formulating hypotheses*—stating assumptions that can be tested.
- *Inferring*—constructing explanations of observations based on limited information.
- *Interpreting*—explaining the meaning of organized data.
- *Measuring*—making quantitative observations using both conventional and nonconventional standards.
- *Observing*—using the senses to gather information about objects or events.
- *Organizing*—grouping information into categories based on criteria.
- *Predicting*—anticipating outcomes of future events based on patterns of experience.
- *Problem solving*—identifying and solving problems by applying systematic sets of procedures.
- *Relating*—recognizing time, space, and cause-and-effect relationships among objects and phenomena.
- *Sequencing*—grouping or ordering processes in chronological order.

HANDS-ON EXPERIENCES

EARTH SCIENCE provides a wide variety of hands-on experiences—*Investigations, Activities,* and *Discovers.* When planning the *Investigations* or *Activities,* be sure to utilize the Safety Guidelines for Teachers on pages T34–T35, the Materials List on pages T36–T37, Laboratory Suppliers on pages T45–T46, and the materials list immediately preceding each chapter. All experiences that require planning in advance are noted on the interleaved pages immediately preceding each unit.

Using the Investigations

The *Investigation* found in each chapter of **EARTH SCI-ENCE** applies chapter content while teaching the students basic laboratory skills within the context of a scientific method. You are encouraged to augment the textbook investigations with the investigations found in the companion **Laboratory Manual/Study Guide.**

Using the Activities

There is one *Activity* in each chapter of **EARTH SCI-ENCE.** These activities apply new concepts by having the students use basic laboratory skills. The *Activities* mirror the more complex *Investigations* in style. However, the Activities take a shorter amount of class time to complete and usually require simpler materials.

Using the Discovers

There is one *Discover* per section in each chapter of **EARTH SCIENCE.** The *Discovers* are short activities that apply new concepts in the chapter to everyday life. The *Discovers* may require laboratory work, library research, or manipulative skills. They may be done either in the classroom or at home. Most *Discovers* require simple, easy-to-obtain materials.

USING THE REFERENCE SECTION

While helping the students familiarize themselves with the textbook, do not overlook the extensive *Reference Section*, which begins on page 546. The *Reference Section* includes Safety Guidelines for the classroom and the laboratory, Laboratory Procedures used throughout the year, Common Laboratory Equipment, a Metric Conversion Table, Key Discoveries in Earth Science, a special section on Building a Science Vocabulary, The Periodic Table, Science Project list, information on Scientific Notation and Significant Digits, and maps and charts. The *Reference Section* also contains an illustrated Glossary and a fully cross-referenced Index.

MAKING FULL USE OF THE *ANNOTATED TEACHER'S EDITION*

The *Annotated Teacher's Edition* is designed pedagogically to provide all the information you need to teach **EARTH SCIENCE.** It features comprehensive strategies for teaching units, chapters, and chapter sections. The textual material presents class-discussion questions, teacher demonstrations, background information, skill activities, answers to all questions, and other useful information.

The basic design of the *Annotated Teacher's Edition* includes reduced student's pages bordered by margin columns, which provide a wealth of information for experienced and less-experienced teachers alike. All pertinent teacher information is located at point of use for ease and convenience. Most notably, the notes pertaining to a Pupil's Edition page will be found in the margins directly adjacent to the page. The margin notes include a complete plan for teaching each section and subsection and for custom tailoring your teaching style to the ability levels of your students.

Unit and Chapter Teaching Strategies

Each unit begins with several motivating paragraphs and questions. Teacher's notes in the margin offer suggestions for using the unit opener to set the stage for the chapters that follow.

Each chapter contains a comprehensive teaching plan located in the side columns. The *Chapter Overview* lists and describes the major concepts of the chapter. In addition, a *Chapter Motivating Activity* is provided to introduce the students to a major concept in the chapter.

There are three basic parts to the teaching plan for each lesson: SECTION FOCUS, SECTION DEVELOPMENT, and SECTION REVIEW. The Section Focus contains a short overview of the section. A suggested activity is presented for introducing the new science terms.

The core of the teaching plan is found in the Section Development, which features notes for *Discussion, Skill, Thinking Skill,* and *Extension.* The *Discussion* is useful for a general class discussion, calling students' attention to illustrations, photography, tables, and other information on the student's page. Within the *Discussion* a *Thinking Critically* question is frequently used to assess the students' understanding of the concept or skills taught. A suggested answer is provided for each *Thinking Critically* question.

The *Skill* is similar to the *Discussion* except it focuses on a measurable or manipulative skill. For example, the students may be asked to measure the area of a room, the volume of a container, or the energy used by a toaster-oven. The *Thinking Skill* is similar to the *Skill,* but it focuses on higher-level skills that are not manipulative. For example, the students may be asked to interpret a graph, to apply a concept, or to sequence steps in a process. The teacher can expand on the topic covered in the text by using the *Extension.* It is much like background information, but it is intended to be shared with the students if the teacher so desires. The *Extension* can be used with honor students. Some *Extensions* are related to other sciences, as well as to mathematics, social studies, language arts, and so on. *Thinking Critically* questions appear in the *Skill* and *Extension* notes.

In the margins below the student's pages are additional notes for background information, demonstrations, or references to demonstrations in the interleaved pages. This area is also reserved for all the answers to review questions. The answers are printed in red to easily distinguish them from other notes.

Also included in the margin columns are instructional notes for the special features found in the Pupil's Edition. For features such as the *Discover, Activity, Skill Activity,* and *Investigation,* there are suggestions for acquiring and preparing materials. Suggested answers to any questions are provided, as well as hints to head students in the right direction of a research-type activity. There are suggestions for teachers in guiding students through the activity. Boldface caution statements are provided when special care is to be taken, especially in regard to hazardous materials or procedures. For the *Technology, Carreers,* and *Biography* features there are suggestions for using the feature, as well as for students to do additional research.

After all of the subsections within a section are developed, the section teaching plan ends with Section Review. This part consists of two subsections: Summary and Reinforcement. These are suggestions for "wrapping up" the section and preparing for the Section Review in the textbook. These subsections provide good opportunities for cooperative learning activities.

The chapter ends with a three-page review. Suggestions for using the Summary and Science Terms and answers to all questions are included in the margins of these three pages.

Each unit ends with a two-page feature called *The Science Connection.* This feature ties together the various concepts presented in the unit through a specific application. In the side column there are additional notes for background information and discussions, as well as information about the feature.

Unit Interleaf

The *Annotated Teacher's Edition* includes two interleaved pages before each unit. These pages include a Unit Overview, Advance Preparation, Bulletin Board Suggestions, Issues in Earth Science, Suggested Projects, and Teacher Resources.

Chapter Interleaf

Immediately preceding each chapter are four interleaved pages of useful information. These pages include three tables: Planning the Chapter; Chapter Concepts, Objectives, and Terms; and Chapter Materials.

Planning the Chapter includes all chapter sections, chapter features, and program resources that supplement each section, all with appropriate page references. Chapter Concepts, Objectives, and Terms provides an immediate overview of the chapter—section by section—and the content for which the students will be responsible.

Chapter Materials lists all materials necessary to perform Investigations, Activities, Discovers, and Skill Activities. The remaining interleaved pages include suggestions for Class Activities, Outside Speakers, Field Trips, and Demonstrations keyed to each section.

USING ANCILLARY MATERIALS

A complete program of supplementary materials accompanies **EARTH SCIENCE.** The consumable **Laboratory Manual/Study Guide** includes two investigations keyed to each chapter of the textbook. The *Investigations* are designed to help students apply textbook concepts to everyday experiences. The **Laboratory Manual/Study Guide** also includes three worksheets per chapter—one designed to review vocabulary, one designed to improve students' reading strategies, and one designed to extend concepts. The

Annotated Teacher's Edition for the **Laboratory Manual/Study Guide** contains overprinted answers to all questions, directions for preparing materials, and a complete materials list.

The **Teacher's ResourceBank**™ contains ten copying-master components. This resource contains the *Investigations* and *Activities* included in the **Laboratory Manual/Study Guide** plus two additional types of activities—*Critical Thinking Activities* and *Cross-Discipline Activities.* The **Teacher's ResourceBank**™ also contains a complete testing program consisting of *Unit and Chapter Tests.* The *Science Project Sheets* contained in the **Teacher's ResourceBank**™ describe how to do a science project in a step-by-step manner, from formulating an idea to constructing the final display. A unique feature in the **Teacher's ResourceBank**™ is *Science Sites, Connecting Science, Geography, and History,* maps of continents and states that show sites of scientific interest and ask critical-thinking questions about the sites. Finally, the **Teacher's ResourceBank**™ also contains *Student Record Sheets,* recording devices to be used in conjunction with the textbook *Investigations.*

The above-mentioned copying masters are housed in a convenient three-ring, tabbed binder. The **Teacher's ResourceBank**™ also contains *Teaching Transparency* samples, *Classroom Reference Posters,* and the *Computer Test Bank* demonstration diskette.

Other supplementary materials include *Teaching Transparencies* and *Classroom Instructional Posters.*

Teaching Transparencies are a separately boxed set of 28 four-color transparencies, with accompanying copying masters and *Answer Key.* The copying masters of the *Teaching Transparencies* are designed for use as instructional or evaluative tools. As such, the labels found on the four-color *Teaching Transparencies* have been replaced with numbers on the illustrations. The *Classroom Instructional Posters* are designed to supplement instruction of each of the seven units in the textbook.

A separate *Computer Test Bank* provides a complete testing program compatible with the Apple IIe® and IIc® and IBM computers. This test bank will enable teachers to test

TEACHING THREE LEVELS OF STUDENTS

The **EARTH SCIENCE** program is designed to assist the teacher in meeting the needs of three different levels of students. One level may be composed of basic students; a second, of students of average ability; and a third may be composed of honors students.

The basic student may be defined as the student who is of lesser ability or slightly below average. Such students may be able to read at or slightly below grade level, but they may take slightly more time to comprehend concepts. The interest level of basic students, however, may be a grade level or slightly higher than that of their peers. The **EARTH SCI-**

ENCE course can be modified to meet the needs of basic students by omitting certain chapters or sections of chapters. Such omissions would not diminish the efficacy of the course, but would allow the basic student to achieve at his or her level and to attain a level of understanding commensurate with established curriculum guidelines. The chapters and sections appropriate for the basic student are indicated in the Pacing Chart found on pages T16–T20. In addition, the Planning the Chapter charts found immediately preceding each chapter indicate sections, subsections, and ancillary materials appropriate for the basic student.

The average student reads on grade level and understands and retains material at a pace expected of his or her grade or age. The average student should be able to handle most of the material presented in the textbook, except for a few selected sections of some chapters and certain ancillary materials. Sections, subsections, and ancillary materials appropriate for the average student may also be found in the Pacing Chart and the Planning the Chapter charts.

The honors student reads on or above grade level, reads and comprehends quickly, and needs a challenge. Thus, the honors student should be able to complete the entire textbook and all ancillary materials within the school year. Honors students may be especially challenged by certain ancillary materials, including *Concept Extension Activities* and *Critical Thinking Activities*.

It is likely that there will be three levels of students within one class. It is possible to meet the needs of all students by carefully selecting sections of the chapter and altering assignments as appropriate to each level. Some suggestions for scheduling follow.

SCHEDULING

Each teaching situation is unique. Therefore, each teacher is the most appropriate person to determine the best schedule for his or her earth-science course. In addition, local, county, and state requirements vary. Often, predetermined curricula dictate the selection and the sequence of topics. The teacher can arrive at an appropriate course schedule by considering the ability level of the students, their previous science experience, teacher interests, teaching style, local science-related resources, and institutional requirements. The length of the class sessions and the length of the school year must also be considered.

Establishing a schedule for the school year can be an important aid in making arrangements for audiovisual aids and for guest speakers. Planning a yearly schedule will also help maintain the pace that you feel is appropriate to your class.

While **EARTH SCIENCE** is designed with the intention that all the main topics be covered during the typical one-year earth-science course, the individual teacher can adapt the way in which he or she uses the textbook to the needs of the local teaching situation. No teacher should feel compelled to teach every chapter or to give equal emphasis to all chapters of the textbook. The teacher and the school system should dictate what constitutes appropriate course content.

The Pacing Chart beginning on page T16 is designed to assist teachers in making decisions regarding scheduling, course content, and emphasis. The recommended coverage of a particular chapter or section for a basic, average, or honors course is indicated by a ■. The number of recommended class sessions to be devoted to each chapter and section is also indicated. When an entire section is not recommended, the recommended numbered subsections are indicated, for example 1.1, 1.2, and so on. The days indicated in the Pacing Chart allow for laboratory work related to the pertinent section. Time for chapter and unit tests is also considered. Use of the Pacing Chart may be supplemented by the Planning the Chapter chart found on the interleaved pages immediately preceding each chapter. The teacher is cautioned to use these planning recommendations merely as a starting point from which to build an individual planning guide that meets his or her needs.

PACING CHART

Unit	Chapter	Section Number and Title	Basic Course	days	Average Course	days	Honors Course	days
1 INTRODUCTION TO EARTH SCIENCE				23		15		15
	1 Studying the Earth			10		7		7
		1 What is Earth Science?	■	1	■	1	■	1
		2 Methods of Scientific Study	■	3	■	2	■	2
		3 Measurements in Science	■	5	■	3	■	3
	Chapter Review and Test		■	1	■	1	■	1
	2 Earth Science Models			12		7		7
		1 Models of the Earth	■	3	■	1	■	1
		2 World Maps	■	3	■	2	■	2
		3 Topographic Maps	■	5	■	3	■	3
	Chapter Review and Test		■	1	■	1	■	1
The Science Connection and Unit Test			■	1	■	1	■	1
2 GEOLOGY				41		39		38
	3 Minerals			6		7		7
		1 Atomic Structure					■	1
		2 Chemical Compounds					■	1
		3 Kinds of Minerals			■	2	■	1
		4 Physical Properties of Minerals	■	5	■	4	■	3
	Chapter Review and Test		■	1	■	1	■	1
	4 Rocks			5		9		9
		1 Studying Rocks	4.1	1	■	1	■	1
		2 Igneous Rocks	4.5	1	■	3	■	3
		3 Sedimentary Rocks	4.8	1	■	2	■	2
		4 Metamorphic Rocks	4.11	1	■	2	■	2
	Chapter Review and Test		■	1	■	1	■	1

■ indicates section is to be taught

Unit	Chapter	Section Number and Title	Basic Course	days	Average Course	days	Honors Course	days
	5	**Earth's Resources**		10		6		6
		1 Metals and Nonmetals	■	5	■	3	■	3
		2 Fossil Fuels	■	4	■	2	■	2
		Chapter Review and Test	■	1	■	1	■	1
	6	**Structural Geology**		6		7		7
		1 Inside the Earth			■	2	■	2
		2 Crustal Adjustments	■	5	■	4	■	4
		Chapter Review and Test	■	1	■	1	■	1
	7	**The Dynamic Earth**		13		9		8
		1 Plate Tectonics	■	5	■	3	■	2.5
		2 Volcanoes	■	3	■	2	■	2
		3 Earthquakes	■	4	■	3	■	2.5
		Chapter Review and Test	■	1	■	1	■	1
		The Science Connection and Unit Test	■	1	■	1	■	1
3		**METEOROLOGY**		16		23		21
	8	**The Atmosphere**				10		9
		1 Composition of the Atmosphere		8.2		1	■	1.5
		2 Solar Radiation		8.4		2	■	2.5
		3 Structure of the Atmosphere			■	3	■	2
		4 Motions of the Atmosphere			■	3	■	2
		Chapter Review and Test			■	1	■	1
	9	**Water in the Atmosphere**		10		6		5
		1 Evaporation and Condensation	■	5	■	3	■	2.5
		2 Clouds and Precipitation	■	4	■	2	■	1.5
		Chapter Review and Test	■	1	■	1	■	1
	10	**Weather**		5		6		6
		1 Frontal Weather			■	3	■	3
		2 Violent Weather	■	4	■	2	■	2
		Chapter Review and Test	■	1	■	1	■	1
		The Science Connection and Unit Test	■	1	■	1	■	1

Unit	Chapter	Section Number and Title	Basic Course	days	Average Course	days	Honors Course	days
4 WEATHERING AND EROSION				37		31		30
	11 Weathering and Soil Formation			9		8		8
		1 Weathering	■	5	■	3	■	3
		2 Mass Movement and Landforms	■	3	■	2	■	2
		3 Soils			■	2	■	2
	Chapter Review and Test		■	1	■	1	■	1
	12 Running Water			12		7		7
		1 Surface Water	■	3	■	1	■	1
		2 Erosion and Deposition	■	4	■	2	■	2
		3 Ground Water	■	4	■	3	■	3
	Chapter Review and Test		■	1	■	1	■	1
	13 Wind and Ice			9		8		8
		1 Wind	■	4	■	2	■	2
		2 Ice	■	4	■	2	■	2
		3 Glacial Landforms			■	3	■	3
	Chapter Review and Test		■	1	■	1	■	1
	14 Geomorphology of North America			6		7		6
		1 Geomorphic Regions			■	2	■	1.5
		2 Provinces of North America	■	5	■	4	■	3.5
	Chapter Review and Test		■	1	■	1	■	1
The Science Connection and Unit Test			■	1	■	1	■	1
5 EARTH HISTORY				16		22		24
	15 Historical Geology			6		5		7
		1 The Record in Rock	■	4	■	2	■	2
		2 The Geologic Clock	15.5	1	15.5	2	■	4
	Chapter Review and Test		■	1	■	1	■	1
	16 Earth's Past						8		8
		1 The Cryptozoic Eon			■	4	■	4
		2 The Phanerozoic Eon			■	3	■	3
	Chapter Review and Test				■	1	■	1

Unit	Chapter	Section Number and Title	Basic Course	days	Average Course	days	Honors Course	days
	17	**Earth's Future**		9		8		8
		1 Managing Natural Resources	■	6	■	4	■	4
		2 Managing Energy Resources	17.6, 17.7	2	17.6, 17.7	3	■	3
		Chapter Review and Test	■	1	■	1	■	1
		The Science Connection and Unit Test	■	1	■	1	■	1
6	**OCEANOGRAPHY**			25		22		22
	18	**The Water Planet**		9		7		7
		1 Bodies of Water	■	4	■	3	■	3
		2 Coasts	■	4	■	3	■	3
		Chapter Review and Test	■	1	■	1	■	1
	19	**Ocean Waters**		6		8		8
		1 Characteristics of Ocean Waters			■	4	■	4
		2 Motions of the Ocean	■	5	■	3	■	3
		Chapter Review and Test	■	1	■	1	■	1
	20	**Ocean Resources**		9		6		6
		1 Ocean Sediments	■	4	■	2.5	■	2.5
		2 Ocean Life	■	4	■	2.5	■	2.5
		Chapter Review and Test	■	1	■	1	■	1
		The Science Connection and Unit Test	■	1	■	1	■	1
7	**ASTRONOMY**			22		29		31
	21	**The Solar System**		11		9		9
		1 Motions of Earth and the Moon			■	2	■	2
		2 The Inner Planets and the Asteroids	■	5	■	3	■	3
		3 The Outer Planets and the Comets	■	5	■	3	■	3
		Chapter Review and Test	■	1	■	1	■	1

Unit / Chapter	Section Number and Title	Basic Course	days	Average Course	days	Honors Course	days
22 The Universe			2		5		7
	1 Stars	22.1, 22.2	2	■	3	■	3
	2 The Sun			■	1	■	1
	3 Galaxies					■	2
	Chapter Review and Test			■	1	■	1
23 Observing the Universe					5		6
	1 Optical Astronomy			■	4	■	4
	2 Radio Astronomy					■	1
	Chapter Review and Test			■	1	■	1
24 Exploring Space			8		9		8
	1 The Road to Space	24.2, 24.3	3	■	4	■	4
	2 Space Travel	■	4	■	2	■	1.5
	3 Space Science			■	2	■	1.5
	Chapter Review and Test	■	1	■	1	■	1
The Science Connection and Unit Test		■	1	■	1	■	1

SCIENCE IN A TECHNOLOGICAL SOCIETY

For students, the present and the future loom larger and are more important than the past. Students approach the future with a sense of awe and, in many cases, anxiety as they question what their lives will be like in the years to come.

The problems of dwindling energy resources and the search for alternate energy supplies are addressed almost daily in newspapers and on television. The impact of computers and related technology on our daily lives has been significant. Social and ethical issues are constantly raised as technology progresses at a rapid pace in the fields of medicine, nuclear power, and genetic engineering.

In order to reduce anxiety about the future, to understand the role of technology, and to make informed decisions about the social implications of new scientific technologies, students first need a strong foundation in the principles and processes of science. The more a student knows about the inner workings of a computer, the more likely he or she may be to incorporate computers into daily routines. Students must do more than merely gain scientific knowledge, however. They must also develop an accurate image of the nature of science and the usefulness of science in solving problems. Finally, it is also important for students to gain confidence in their ability to identify science-related social issues and to use their own scientific knowledge to resolve these problems.

Teachers can play a vital role in helping students gain the knowledge and the skills necessary to make responsible decisions about social issues related to science and technology. The sections that follow provide suggestions to help teachers in this role, using features from **EARTH SCIENCE** in conjunction with a variety of classroom strategies and outside resources.

STRATEGIES USING EARTH SCIENCE

EARTH SCIENCE provides students with opportunities to analyze science-related and technology-related issues such as fusion power, modern farming, and radiotelescopes. A discussion of these issues can provide excellent additional opportunities for the development of thinking skills. Additional issues for discussion or debate may be found on the interleaved pages immediately preceding each unit.

EARTH SCIENCE contains a wealth of information and features that can be invaluable in teaching students about the relationship of science and technology to society.

Chapter 1 provides a basis for the understanding of the nature of science, scientific methods, and the metric system. This chapter is crucial for developing an understanding of what science is and how it is used to solve problems.

Other chapters in the book address the role of technology in almost all aspects of the students' lives. Chapter 17 provides an overview of energy resources, geothermal power, solar power, and fusion reactions. A comparison is also made between the relative benefits and the potential hazards of fission and fusion as possible sources of energy in Chapter 17.

Chapter 11 presents basic information about weathering and the importance of soil conservation. Chapter 5 presents the full impact of how dramatically the use of quartz has affected our lives. This can be demonstrated by asking the students how their lives would be different without silicon chips.

"What is a silicon chip?" and "How does it work?" are two questions often asked by students. Both of these questions are answered in Chapter 5.

These are just a few examples of how earth science is related to the lives of the students. You will find many opportunities to encourage the students to think about how science affects their lives.

OTHER CLASSROOM STRATEGIES

One popular way to introduce social issues into the science classroom is to have students bring in articles on science and technology from newspapers and magazines. Class time should be set aside on a regular basis to discuss the articles and the students' views. The teacher's role is vital in these discussions. However, the teacher must remember not to make judgments or to reject or praise student responses, but rather to accept all possible answers, ideas, and positions.

Debates are another natural way of approaching the issues of science, technology, and society. For this strategy to be useful, however, students must do more than repeat the opinions of others. They must be given adequate time and direction to research the issues, collect data, formulate their own opinions, and support their positions. Students might develop a questionnaire to gather information from their peers, interested adults, the scientific community, or other groups. Students can analyze items to be included in the questionnaire, predict possible responses, and then determine the best way to administer the questionnaire. The data collected may represent a type of observation unfamiliar to many students. However, using and trying to make sense of such observations exemplify the process of science.

The study of **EARTH SCIENCE** may introduce students to such topics as the use of nuclear energy and the disposal of organic wastes, about which there may be several points of view. A discussion of these topics can provide excellent learning opportunities, especially for the development of higher-level thinking skills. Suggested topics for discussion and debate may be found on the interleaved pages immediately preceding each unit. When such topics are encountered in the classroom, several strategies are possible. Students may also be encouraged to gather information from references. Once information has been gathered, a positive learning environment can be created if students are taught to respect the right of others to express their views.

Microcomputers are powerful teaching tools. They can store, manipulate, and interpret large quantities of information. They can produce graphics, charts, and other visual materials to help the students comprehend important concepts. Computers are being used in science classrooms to enhance instruction, organize and manipulate data, help students prepare papers, and assist teachers with classroom preparation and organization. If you have access to only one computer, you can attach it to a large-screen monitor or television set, and your class can work through computer programs. Interested students can work independently at the computer.

There are many types of computer programs that can be used in the earth-science classroom. Computer simulations are designed to give the students the opportunity to experience a computer version of some real event or situation. This can give the students a chance to learn about some aspect of the world in which they would normally not have experience. The students can change some of the variables in a simulation to see how these changes affect the program.

Interactive tutorials, a second type of software, help the students by providing appropriate question-and-answer techniques. The more effective tutorials provide help to the students (as necessary), assess the students' understanding of the material before permitting them to proceed, and generally allow the students to learn material at their own pace.

A third type of software is drill and practice software. Effective drill and practice software can help the students review difficult concepts. It can give the students an opportunity to learn terminology, classifications, and computation and problem-solving skills.

Some types of software also can be used effectively in helping the students collect laboratory data during class experiments or while working on individual research projects. Some software also can help the students record and analyze data collected in laboratory investigations. This is an excellent use of the computer because it shows the students how to use the computer as a tool.

The microcomputer can perform time-consuming but necessary calculations quickly. Using specific programs, the computer can do tasks that otherwise would be tedious, time consuming, and perhaps frustrating to the students.

Students can also use the computer as a tool in preparing and presenting laboratory reports and term papers. Those who have access to word-processing software, either at home or at school, will have an opportunity to edit their work more efficiently and to gain some computer-literacy skills.

Microcomputers can be a valuable aid in developing lessons and in classroom administration. They can also be used in many aspects of classroom management such as maintaining an inventory of science equipment and supplies. You can use some types of software to write lessons or to customize tests to suit your needs. The *Earth Science Computer Test Bank,* available for Apple II® and IBM computers, is a software package you can use to create your own tests.

With minor modifications, some word-processing programs can be used to keep track of grades and attendance to calculate a student's total points and final grade. While it does take some investment of time to become proficient in using word-processing software, word processing will eventually save you many hours of work and enable you to produce better materials at the same time.

As software and hardware continue to improve, the computer will play an increasingly important role in the classroom.

A list of software suppliers can be found on page T43. In addition, specific software resources are listed on the interleaved pages immediately preceding each unit.

CLASSROOM TECHNIQUES

The textbook is the students' primary source of information. By utilizing prereading strategies and guiding students in their reading assignments, teachers can significantly increase comprehension and make the task of reading the textbook easier for the students. Writing assignments also help develop language skills and reinforce or integrate students' learning.

DEVELOPING PREREADING STRATEGIES

Several research studies have shown that the amount of prior knowledge that students have about a topic directly influences their comprehension when reading about that topic. For example, the more students know about energy—including concepts, vocabulary terms, functions, relationships, and dangers—the easier it is for students to understand new information about energy. Conversely, when students know very little about energy, they find it much more difficult to read and comprehend textbook information about the topic. Because some of the information presented in the textbook is unfamiliar, students have no prior knowledge structures to link with the new information.

Other related research concerns the misconceptions students may have about a topic. This research shows that students tend to maintain their misconceptions about a topic when reading textbook information that deals with those misconceptions. In other words, students have difficulty grasping textbook information that conflicts with the ideas they already have about a topic, such as energy. They may ignore the textbook because the new information does not fit with their existing information. Students are able to discard or modify their misconceptions, however, when the teacher provides direct instruction that corrects the misinformation.

Helping students identify the information they already know about a topic prior to reading about that topic aids the students in at least three ways.

1. Students are able to build a framework of concepts related to the topic. As a result, students can relate the new information from the textbook to existing knowledge structures.
2. Teachers are able to identify the general awareness level of the class about a particular topic. The teacher can then plan instructional time accordingly.
3. The teacher has the opportunity to identify any misconceptions students may have about the topic. Then teachers can plan discussions, demonstrations, or other experiences to help students confront and correct their misconceptions prior to reading the information in the textbook.

ACCESSING PRIOR KNOWLEDGE

One effective strategy for accessing prior knowledge is to brainstorm with the class as a whole to determine what the group already knows about a topic. Write student responses on the chalkboard or on an overhead-projector transparency. Ask key questions to probe relationships among the concepts the students have identified. Ask how related terms are alike and how they are different. Answer your own questions if no one knows the answers, and correct any misinformation that turns up. Using the ideas from the brainstorming, make a diagram or outline that shows visually how the terms and concepts of the chapter are related.

Use class discussions to build a common vocabulary, using the list of new science terms at the beginning of each section. Many terms used in science have multiple scientific and nonscientific meanings. Providing a context for terms will help the students arrive at accurate definitions of science terms.

The interleaved pages immediately preceding each chapter and the margin notes on the chapter-opening pages include many other specific ideas for establishing students' prior knowledge about a topic. Motivating activities, demonstrations, and teaching suggestions not only stimulate interest and curiosity but also prepare students to understand what they are about to read. You may wish to vary the methods that are used from chapter to chapter.

ESTABLISHING A PURPOSE FOR READING

Research on reading shows that students comprehend better when they have a purpose for reading. Making a study guide of questions that students can answer as they read is one way of establishing a purpose. You can make questions out of the *Section Objectives* at the beginning of each section or use the subsection questions and *Section Review* questions as a prereading guide. You can also make questions out of chapter headings, captions, and *New Science Terms*. Once you have provided students with a model, they can be directed to make their own study guides for subsequent chapters.

PROVIDING EFFECTIVE READING STRATEGIES

Many students have no real strategies for reading textbook materials other than simply rereading difficult sections of a chapter. Students do not generally realize there are things they can do to enhance their comprehension of textbook materials. Students often have little awareness of higher-level cognitive strategies such as organizing, reflecting upon, or evaluating what they read. Good readers are not much different from poor readers in the type of strategies they use, although they do differ in the number of strategies employed.

Science teachers can use three major steps to help students learn and practice effective reading strategies.

1. Help students understand how science books organize information in general. Usually, each section of a chapter contains one main idea, stated first or last. Supporting details follow or precede the main idea in a recognizable pattern. Students should look for key phrases that mark the specific type of relationship of the information being presented. For example, the phrase "there are five traits" indicates an enumerative passage; terms such as "like" and "in contrast to" indicate comparisons and contrasts.

2. Point out the basic organizational features of this textbook. Help students understand that features such as the chapter outline, *Section Objectives, New Science Terms,* and *Section Review* questions can guide their reading. Important terms are printed in **boldface** type. Important concepts are illustrated by charts, labeled drawings, or photographs with captions.

3. Model effective reading and study practices for students. Students should be taught to preview a chapter. Previewing involves looking at all the chapter headings, illustrative material, and boldfaced terms page by page prior to actually reading the text. This technique helps students to gain a "feel" for the chapter and to build a basic structure of information. Later, when they read the chapter, students can fill in the details on the basic structures they have built.

USING THE TEXTBOOK TO STUDY

Students also should be taught how to study from the textbook. If they can pick out the important information from a textbook, they can put it on cards, make study guides, write questions, or orally rehearse what they have learned. All of these methods are more effective study strategies than simply rereading a chapter.

Suggest, for example, that students put a key concept, the five steps of a process, or a difficult vocabulary term on one side of a note card. On the other side, have them write a sentence in which the concept is used, the steps of the process, or the definition of the term. Encourage students to put on the cards not only the boldfaced science terms but other important and unfamiliar terms also. As they study these, have the students keep sorting the cards into two piles, one for terms they know and one for terms they do not know.

WRITING A RESEARCH REPORT

Research shows that writing is an effective way of improving reading. As students write, they are creating a text for others to read. The more they write, the more they are able to recognize compositional devices they encounter in the textbook.

A common problem that students meet in writing a science research report is a lack of practice in locating reference and resource materials. Many students also have insufficient knowledge of what a science report should contain. Some suggested topics for written reports may be found under Extension in each chapter review. (For a discussion of writing a laboratory report, see the **Laboratory Manual/Study Guide** that accompanies this textbook.)

The teacher can guide the students' writing most efficiently by using a series of checkpoints. Direct the students to follow a uniform sequence of steps, such as those given below, and to seek the teacher's approval before proceeding beyond the checkpoints marked with an asterisk.

1. Choose a broad topic area.
*2. List several specific questions about the topic.
3. Do some preliminary reading about the topic in an encyclopedia or in general reference books.
4. Take notes.
*5. Narrow the focus of your topic.
6. Gather additional information, using books, magazines, filmstrips, or vertical files.
7. Take more notes on your sources. (Some teachers require a certain number of note cards to be written, according to the length or complexity of the topic.)
*8. Make an outline for your paper:
   a. Introduction, telling what the topic is and what kind of information will be found in the remaining sections of the paper.
   b. Three (or more) paragraphs, each telling interesting or important things about the topic.
   c. Summary/Commentary, telling why you found this topic interesting, why it is important, what you learned, and what questions you still have.
*9. Write the first draft of your paper. (Commenting on the students' first draft is essential, as it gives the students the opportunity to revise their work according to specific guidelines.)
10. Revise your paper.

MEETING THE NEEDS OF EXCEPTIONAL STUDENTS

The science class offers a variety of students important information that will help them function in an increasingly scientific and technological society. Guidance counselors, special-education teachers, and the school nurse may be consulted to help work out the best learning environment as well as realistic goals for each student.

Since HB 94:142 requires that there be an individual education plan (IEP) on file for every special-education student, teachers who instruct these students in a mainstream setting should request access to those IEPs. The information contained may be explained by a counselor or a special-education teacher if the IEP itself is not provided. Content teachers may expect to learn that their special-education students are to be provided specific testing conditions, adaptations in presentations, or adjustments in grading expectations.

The following recommendations apply to specific types of special-education students.

LEARNING DISABLED

- Allow the student access to the special-education instructor as needed or as recommended on the IEP.
- Where feasible, make use of teaching helps available from such companies as National Teaching Aids, NASCO, and others. These include helps such as kits, models, and rubber stamps to reproduce diagrams.
- Allow for group work during oral assignments.
- Provide for simplified rephrasing of concepts, tests, and reviews.
- Make use of oral examinations.
- Make certain that easy-to-read science reference material is available.
- Provide a daily, unvarying routine so your expectations are clear.
- Establish special teaching procedures to take into account a short attention span and restlessness.
- Make certain that instructions are understood before the student starts work.
- Seat the student where classroom distractions are minimized.
- Allow the student to express ideas with drawings or models if the disability permits. Dyslexic students will be able to develop drawings of their own, while dysgraphic students will be more successful at labeling figures that have been prepared for them.
- Use tape recorders where appropriate.
- Within reasonable expectations, evaluate the student's grasp of the concept, not of spelling, punctuation, or intricate sentence structure. Provide spelling lists to which the students may refer when writing. Accept the use of printing when cursive writing presents too great a challenge.

MENTALLY RETARDED

- Stress critical attributes when presenting concepts. Use of examples and nonexamples may help clarify the concepts.
- Be selective about which concepts the special-education student should master.
- Provide for simplified rephrasing of concepts, tests, and reviews.
- Make use of oral examinations.
- Allow for group work during oral assignments.
- Make certain that easy-to-read science reference material is available.
- Give simple, clear directions.
- Encourage repeated efforts.
- Within reasonable expectations, evaluate the student's grasp of the concept, not of spelling, punctuation, or intricate sentence structure. Provide spelling lists to which the students may refer when writing. Accept the use of printing when cursive writing presents too great a challenge.

HEARING IMPAIRED

- Allow the student access to your lecture notes if he or she has serious difficulty in learning auditorily.
- Avoid speaking while facing the chalkboard.
- Provide seating where the student can hear best or lip-read most easily.
- Allow for group work during oral assignments.
- Obtain close-captioned films for the deaf.
- Provide a classroom partner who knows signing.
- Rephrase instructions. Some sounds may be heard better than others.

VISUALLY IMPAIRED

- Seat the student close to the chalkboard.
- Stand facing the windows to avoid putting a demonstration into shadow.
- Encourage the handling of materials before or after a demonstration.
- Assign a student to make copies of notes.
- Use verbal cues rather than nods or facial expressions.
- Provide a sighted student guide.
- Provide high-contrast copies of worksheets or chalkboard diagrams.
- Check with the special-education instructor or with local or state organizations regarding the availability of large-print textbooks.

HEALTH IMPAIRED

- Students who have diabetes, asthma, a heart condition, or other general health impairments may vary considerably in their degree of impairment. The school nurse and special-education teacher can acquaint you with each student's limits.
- Become acquainted with the various symptoms of any health emergencies that might occur in the classroom.
- Obtain training in first-aid procedures to be used in the event of a health emergency.

ORTHOPEDICALLY IMPAIRED

- Arrange for seating that is comfortable for the student.
- Provide rest breaks.
- Cover the student's desk with felt so that materials do not slip.

THE STUDENT WITH LIMITED ENGLISH PROFICIENCY

The study of science can be especially challenging and frustrating to the student who is limited in speaking, reading, writing, or understanding English. Technical science terms and definitions can be especially difficult. Many school systems and districts have teachers whose main responsibility is to instruct the limited-English-proficient (LEP) student. Work with this teacher as a resource person in designing lessons plans for the LEP student. If your school system or district has no such resource person, the following suggestions may be of help in dealing with the LEP student.

- Use pictures, diagrams, models, and other props as often as possible.
- Read the captions of illustrations and photographs aloud to the student, simplifying the language if necessary.
- Use video tapes or films with accompanying audio to help clarify concepts.
- Prepare read-along tapes for each lesson.
- Provide a word list of key terms (in addition to New Science Terms) that may come up often in the discussion of a particular topic.
- Allow the LEP student to work with a partner or in a group whenever feasible.
- Simplify language in descriptions and discussions whenever possible.
- Speak slowly and enunciate clearly. Restate sentences and phrases if necessary.
- Provide for oral testing with oral or pictorial responses from the student. Exercise patience while waiting for a response.

Allowing students to share their learning with one another is a strategy that has been used since the day of the one-room schoolhouse. Currently the concept has been refined and its merits clarified by such educators as Marilyn Burns, William Glasser, and the team of David and Roger Johnson. Many professional journals have carried articles encouraging team effort in learning.

COOPERATIVE LEARNING GROUPS

Grouping students for cooperative learning gives them the opportunity to work toward both group and individual goals. The basic elements of a cooperative learning group include: positive interdependence, face-to-face interaction, individual accountability, and interpersonal and small-group cooperative skills.

Positive Interdependence

A learning activity becomes cooperative only when the group members realize that no one member can be successful unless all members of the group are successful. Therefore, students must work with and depend on each other to achieve the group goal. Positive interdependence may be achieved through a division of labor, resources, and roles in the group. Rewards for the group, such as the same grade for every member of the group, may be included.

Face-to-Face Interaction

In cooperative learning, it is important that all group members provide support, encouragement, and help to each other. Members should be encouraged to share and discuss their ideas. For cooperative learning to be successful, students must interact with each other.

Individual Accountability

Individual accountability means that each group member is responsible for knowing the assigned material. In addition, each group member is responsible for the learning of other group members.

Interpersonal and Small Group Cooperative Skills

An important part of cooperative learning is teaching students the skills that are necessary for effective collaboration. Teachers need to specify interpersonal and small group cooperative skills such as "staying with your group," "taking turns," "looking at the person who is talking," and "checking to make sure that other members understand and agree with the answers." It is important for students to learn what a cooperative relationship should look and sound like. It is also important to allow students the time to analyze (process) how well their groups are cooperating.

It is important to provide students with opportunities to work as members of both small and large groups to encourage sharing, acceptance of responsibility, and decision-making. Successfully guiding cooperative learning activities in the classroom requires a complete understanding of cooperative learning strategies. For further information, you may wish to refer to *Circles of Learning: Cooperation in the Classroom,* Revised (Johnson, Johnson, and Houlubec, 1986).

THE TEACHER'S ROLE IN COOPERATIVE LEARNING

In cooperative learning situations, the teacher should identify the group goal; decide on the group size, group make-up, room arrangement, materials needed for the activity; and, sometimes, assign specific roles for group members. The teacher will also need to explain the task, structure the activity, monitor the interpersonal and small-group cooperative skills, and evaluate the product as well as the cooperative skills.

Identify the Group Goal

Students need to understand what is expected of them. Identify the group goal, whether it be to master specific objectives or to create a product such as a chart, report, or booklet. In addition, you should identify and explain the specific cooperative skills for each activity.

Decide Group Structure and Arrangement

For each activity, you will need to make some basic decisions. These decisions include group size, group make-up, room arrangement, materials needed for the activity, and, sometimes, specific role assignments for each group member.

Group Size Cooperative learning groups may consist of two to six members. If you are using cooperative learning activities for the first time, however, you may find it easier to keep groups to two or three members. Smaller groups require fewer cooperative skills of students than do larger groups.

Group Make-Up Use heterogeneous grouping of members when possible. Heterogeneous groups are those that include students with high, average, and low ability levels. This type of grouping encourages greater diversity of thinking. Use homogeneous grouping (grouping of students with similar levels of ability) when mastery of specific skills is the goal.

Room Arrangement Have each cooperative learning group sit in a circle so that every member can see every other member.

Materials Encourage interdependence by providing only one set of the materials to the group. By doing this, you can help members learn to work together to be successful. You may also give different materials to each group member. The completion of the task then depends on how well members work together.

Role Assignment For some cooperative learning activities, you may want to assign a *recorder,* who writes down the group's decisions and edits reports; a *summarizer-checker,* who makes sure all group members understand the material; a *researcher-runner,* who communicates with other groups and gets materials; a *reader,* who reads the directions and questions; and an *observer,* who keeps a record of how well the students perform the required cooperative skills.

Explain the Task

For each activity, make sure that all group members understand the task. To ensure understanding, you may want to follow these suggestions:

Give clear and specific instructions.

Explain lesson objectives.

Help students see the relationship of the concepts to past experiences and prior learning.

Define concepts and provide models of what students must do to finish a task.

Ask questions to be sure that students understand the task.

Structure Positive Goal Interdependence

In cooperative learning, it is essential that students work together. You should stress the "sink or swim together"

relationship that is necessary to achieve the group goal. To facilitate positive goal interdependence, you can structure activities in the following ways:

Have the group produce one product (answer), which each group member signs. Each member must know the reason for the agreed-upon product or answer and be able to explain it.

Give a group grade. You may evaluate members individually, but reward the group based on the total achievement of the group.

Evaluate often by giving practice tests, by randomly choosing group members to explain answers or to read papers, and by having members edit each other's work.

Encourage groups to help other groups by rewarding the entire class if a certain criterion is met.

Monitor Cooperative Skills

As the teacher, you have an important role in cooperative learning. While students are working together, you can pick up vital information on the task because, in a cooperative setting, learning takes place "out loud." You can actually hear the learning while it is occurring. This is also an opportune time to observe systematically how well students work together, especially on the skills that you specified before the activity started. You can count and record the number of times appropriate skills occur and provide the group with feedback. Student observers may also be used to record some cooperative skills.

Evaluation

It is important that you evaluate the product that the group creates. You may also want to evaluate the group on how well they worked together.

SCIENCE IN THE LABORATORY AND IN THE FIELD

SAFETY GUIDELINES FOR TEACHERS

The science laboratory will be a safe and productive working environment only if safety standards are established at the outset and their importance stressed to the students. When working in the laboratory, the students will need frequent reminders that safe practices are to be followed at all times. You can shape and reinforce proper student attitudes toward laboratory safety by setting a good example.

Many states have school laboratory regulations covering such topics as eye and body protection, storage of combustible materials, fire protection, and the availability of first-aid supplies. Check with your state department of education. Even in the absence of local regulations, the following safety precautions should be routine procedure in any laboratory.

EQUIPMENT

- Microscopes, hot plates, and other electrical equipment should be kept in good working order. Three-prong plugs must always be plugged into compatible outlets, or an adapter must be used. Never remove the grounding prong.
- Before each laboratory session, examine glassware for cracks and ask the students to do the same. Never use cracked or chipped glassware. Glassware used for heating should be made of heat-resistant material.
- Broken glassware should be swept up immediately, never picked up with the fingers. Broken glassware should be disposed of in a container specifically denoted for this purpose. If such a container is not available, broken glass should be adequately wrapped in paper and the paper secured. Alert the maintenance staff that the package contains broken glass.
- Never permit students to operate an autoclave or a pressure cooker for sterilization. Conduct the sterilization yourself.
- Mechanical equipment should be set up according to instructions. Students should read all instructions for use before handling mechanical equipment.

MATERIALS

- Volatile liquids such as alcohol should be used only in ventilated areas and never near an open flame. To heat such substances, warm them on a hot plate or in a water bath.
- For the correct procedures to be used by students when handling or heating chemicals, see pages 548–549 in the Reference Section.
- Used chemicals should be disposed of in a manner that does not pollute the environment. Check on school and state guidelines for proper chemical disposal.

- When teaching large classes, you may wish to set up materials stations in several parts of the room. The students can obtain chemicals and other supplies there, thus minimizing the distance they must carry supplies to their work areas.
- Chemicals should be stored in a well-ventilated area that is kept locked at all times. Flammable chemicals should be stored in a fire-resistant cabinet. Never store such chemicals in a refrigerator, unless the refrigerator is specifically marked *explosion proof*.
- Chemical storage areas should be kept clean, orderly, and well lighted.
- Dissection of preserved specimens should be conducted in well-ventilated areas. If fresh specimens are to be used for longer than one day, they should be preserved or kept under refrigeration. Specimens preserved in formaldehyde should not be used.
- **Pathogenic bacteria should never be used in the laboratory.**
- Petri dishes containing bacterial cultures should be sealed with tape.
- Before washing Petri dishes, the cultures should be killed by heating in an autoclave or a pressure cooker or by applying alcohol or a strong disinfectant.
- Wire loops used to transfer microorganisms should be flamed before and after the organisms are transferred.

LIVE ANIMALS

When live animals are introduced into the laboratory for observation and experimentation, a double safety standard must be maintained. The safety of the students is one objective; the humane treatment of the animals is the other. In addition to the following guidelines, consult the publication "Guidelines for the Use of Live Animals at the Preuniversity Level," which is available from the National Association of Biology Teachers, 11250 Roger Bacon Drive, Reston, VA 22090.

- Before introducing animals into the laboratory, complete all plans for their care and feeding, including their maintenance over weekends and school holidays.
- Be sure that all mammals used in a school laboratory have been inoculated for rabies unless they have been purchased from a reliable biological supply house.
- Wild animals should never be brought into the classroom.
- Animals should not be teased or subjected to unnecessary handling.
- Experiments with animals should not involve the use of drugs, toxic products, anesthetics, surgery, carcinogens, or radiation.
- Make certain that any student who is scratched or bitten by an animal receives immediate attention by the school nurse or physician.

- Remind the students to wash their hands thoroughly after handling animals.
- Animals taken from their natural environment should be returned there when no longer needed.

PLANTS

Using plants in the laboratory does not require elaborate caution. However, many plants, or parts of plants, including some common house plants, are poisonous.
- Caution the students never to put any part of a plant in their mouths or near their eyes.
- The students should always wash their hands thoroughly after handling plants.

GENERAL

- Always perform an experiment yourself before asking students to try it. Cautionary statements, which are printed in boldface type, are included in the procedures given for all laboratory investigations. Refer to those statements when preparing for the experiments. In addition, safety symbols alert the students to procedures that require special care. These symbols and their explanations follow.
- Do not permit the students to work in the laboratory without your supervision.
- Do not allow the students to conduct unauthorized experiments.
- When an investigation has been completed, insist that the students clean their work areas; wash and store all materials and equipment; and turn off all water, gas, and electrical appliances.

SAFETY EQUIPMENT

Safety equipment commonly found in school laboratories should include fire extinguishers, fire blankets, sand buckets, eyewash fountains, emergency showers, safety goggles, laboratory aprons, gloves, tongs, respirators, and a first-aid kit.
- Note the location of each piece of safety equipment in your laboratory, learn to use it correctly, and teach the students to do the same.
- Learn how to use the first-aid equipment. Call the school nurse or a physician immediately in case of serious injury.
- Post the telephone number of a poison-control center in your area.

SAFETY SYMBOLS

The instructions for all laboratory investigations will include cautionary statements where necessary. In addition, the following safety symbols will appear whenever a procedure requires extra caution:

 Wear safety goggles

 Wear a laboratory apron

 Sharp/pointed object

 Biohazard/disease-causing organisms

 Flame/heat

 Dangerous chemical/poison

 Electrical hazard

 Radioactive material

INVESTIGATIONS AND ACTIVITIES MATERIALS LIST

The following materials list has been compiled to help the teacher order supplies for the textbook Investigations and Activities. The items are keyed to the Investigation(I) and Activity(A) in each chapter in which the item is needed. The numeral following the letter indicates the chapter number. Additional information regarding amounts and preparation of materials may be found on the interleaved pages immediately preceding each chapter.

Apparatus and Equipment	Investigations and Activities
anemometer	A10
balance	I1, I11, I19
barometer, aneroid	A10
Bunsen burner	A8, A19, A20
clamps	A9
clock with second hand	A15
compass, magnetic	I17, A10, A22
crucible	I10
hand lens	I4, I16, A4
hose and clamps, fitted to tube	I12
hot plate	I5
laboratory apron	I4, I10, I11, A8, A19
lamp	I8
magnet, bar	I1, I17
microscope, binocular	A17
ring stand	I10, I24, A9, A24
ring stand and clamp	I20, A20
rubber tubing	A20
safety goggles	I4, I10, I11, A8, A13, A19, A20
spring balance	I24
streak plate	I3
test-tube clamps	I10
test-tube holder	A19
tripod	A8
vise	I15
wire gauze	I10, A8

Chemicals and Reagents	Investigations and Activities
alcohol, rubbing	A9
copper sulfate	I5
ferrous sulfate	I5
hydrochloric acid	I4
mineral oil	A5
sodium chloride	I5

Glassware	Investigations and Activities
aquarium	I18
beakers	
100 mL	I5, I10, I19, A9, A17
150 mL	I20
250 mL	I19
500 mL	I12, I20, A19
1000 mL	I10
1500 mL	I18
convex lenses	I23
fish tank, small	I10
glass blocks	A23
glass plate	I3, A3
glass rod	I1
glass tubing	A20
graduates	
10 mL	I19
50 mL	I12, A1
100 mL	I1
medicine dropper	A5, A17
Petri dish	I5, I10, I23, A5, A17
stirring rod	I5, I19
test tubes	I19, A19
thermometers, Celsius	I8, I9, I10, I19, A9, A10

Local Supply	Investigations and Activities
balloons	I24, A24
bell wire, insulated	I17
bleach (containing 5% sodium hypochlorite)	A17
bones, chicken or lamb	I15
bowl	A8
bucket	I18
can, empty soda	A8
can, empty soup	I9, I20
candle, small	I10, I23
cardboard	I10
charcoal, crushed	I20
clay, modeling	A5
clay, soil	I12
compass, drawing	I22, A22
cork	A1, A9
cotton or gauze	A9
flashlight	I9, A23
food coloring, red and blue	A19
food jars with lids, small	I11
gloves, heat proof	A8
glue	A6
gravel	I12, I20
hair dryer	A13
hammer	I15, I20
hole punch	I17

Local Supply	Investigations and Activities	Local Supply	Investigations and Activities
ice cubes	I9, I10	string	I1, I5, I21, I24, A1, A20, A21, A24
ice, crushed	I19, A19, A20		
ice-filled tray	I5	table of tangents	I22
index card	A21	tape	I8, I10, I21, I22, I24, A16 A21, A24
insect pins	I22		
maps, local topographic and geologic	A14		
markers	I21, A16, A18		
matches	I10, I23		
metal washer	A21	tape, electrical	I17
meter stick	I23	thumbtacks	I17
nail, finishing	I20	tray	I15, A13
nail, steel	I3, A3	water, carbonated	I11
pan, shallow	A18	water, muddy	I20
paper towels	I1, I10	wax pencil	I19, A9
paper, construction	I21, I22, A16, A22	wooden block (5 cm × 10 cm × 10 cm)	I17
black	I8, A23		
white	I8, I23, A23	**Rocks and Minerals**	**Investigations and Activities**
paper, graph	I7		
paper, legal size	A6	calcite	I1, I3, A1
paper, tracing	A6	chalcopyrite	I3
pencils, colored	A6, A14, A22	fossils, set of 10 numbered	I16
pennies	I3, A3, A15	galena	I1, I3, A1
petroleum jelly	I15	granite	I1, I11, A1
plaster of Paris	I15	hematite	I3
plastic bag	A20	igneous rock set	A4
plastic lid from a margarine container	I17	limestone	I11, A5
plastic straw	I24, A18, A21, A24	limonite	I3
		magnetite	I3
plastic tube	I12	mineral samples	A3
pliers, needle-nosed	I15	pyrite	I3
pond water (containing microscopic organisms)	A17	quartz	I3
		rock specimens	I4
protractor	I22, A21, A22	sandstone	A5
razor blade	A18	sediments	A18
rubber bands	A9	shale	A5
ruler, metric	I1, I18, I21, I22, I24, A1, A14, A16, A22	sphalerite	I3
		talc	I3
salt	I19		
sand	I12, I20, A13		
saw	I15		
scissors	I8, I10, I22, I23, A6, A16		
shoe box	I8, A15, A21		
snail shells or sea shells	I15		
soil or sand, coarse	I18		
stopwatch	I8, I12, I19, I24		

FIELD TRIPS

Field trips are a desirable part of any science program. Field trips serve to remind students that the real world exists outside the classroom. It is only in the field, for example, that organisms and physical phenomena can be observed and studied under natural conditions.

Whether you go to a museum, a local industry, or to an outdoor site, a truly successful field trip should add a new dimension to the students' grasp of science. It should make classroom work more meaningful and more enjoyable. Three things are required of the teacher in order to make any field trip an enriching experience: proper safety precautions, setting and evaluating goals, and careful trip planning.

ESTABLISHING SAFETY GUIDELINES

Some preparation in advance of the field trip may eliminate potential safety problems. The following suggestions will help ensure a safe and enriching trip.

- When planning a trip to an outdoor site, visit the site in advance and note any potential safety hazards. These include bodies of water, poisonous plants, venomous insects and snakes, areas where falls might occur, and electrical or mechanical hazards.
- When planning a trip to a museum, local laboratory, or government agency, visit the location and meet with institution staff prior to the trip. Ask what precautions students need to be given.
- Discuss necessary safety measures with the students in advance. Include warnings of any hazards discovered during your advance visit.
- Arrange for additional adult supervision, whenever feasible. A good rule of thumb is to have one teacher or parent for every ten students.
- If any part of the field trip is to take place on private property, get written permission in advance from the landowner.
- Make certain that any consent forms required of either parents or guardians or of school officials are drawn up, signed, and filed in advance.
- Caution the students to dress in a manner that will keep them comfortable, warm, and dry.
- On any trip to a large body of water, be sure to include an adult skilled in water safety and CPR.
- Pack a basic first-aid kit. Depending on the type of trip, area, and season, consider including an insect repellant.
- At the beginning of each trip, insist that students travel in pairs. Make each member of the pair responsible for his or her partner.
- Caution students to report any injury immediately to an adult supervisor.

SETTING AND EVALUATING GOALS

Assessing the value of a field trip means first setting goals so that you have some standard against which to measure student accomplishments. Well-defined goals, and the obligation to meet them, must then be transmitted to the students. This will encourage the students to regard the field trip as a learning experience rather than as simply time away from school.

Use the following guidelines to set and evaluate realistic goals:

- Make sure that students understand how the field trip relates to the appropriate textbook chapter.
- Provide students in advance with an outline of what they will experience on the field trip. If the trip involves an on-site guide, such as a museum curator or an animal breeder, try to get an advance summary of the material he or she plans to cover.
- Give students a short list of questions to be answered by them after completing the field trip. This will give the students a concrete goal to work toward and will provide you with a means of evaluating the field trip.
- On trips to natural habitats, encourage the students to make sketches of specimens, such as insects and leaves, rather than to collect specimens.
- If an on-site guide is involved, encourage the students to ask questions.
- Follow up the field trip with a class discussion. This discussion will provide you with an opportunity to evaluate the students' answers to the questions you have assigned. Encourage the students to ask additional questions they may have at this time.
- Send personal or student-written thank-you letters to on-site guides and institution or company representatives for their help in planning and conducting the field trip.

FIELD TRIP CHECKLIST

The following checklist is designed to help you with the details of field trip planning. Make copies of the checklist and use one each time you plan a trip.

Person to Contact at Field Trip Site
Name _____
Address _____
Telephone _____

Preliminary Visit to Trip Site
_____ Safety hazards checked out
_____ Meeting held with trip contact
_____ Contact informed of trip goal

Permission for Trip Granted by
_____ Appropriate school official
_____ Contact person at trip site
_____ Landowner (of private property)
_____ Parent or guardian

Information for Parents or Guardians
_____ Purpose of trip
_____ Site of trip
_____ Times of departure and return
_____ Cost, if any
_____ Equipment needed, if any
_____ Type of transportation provided

Confirmation of Trip with Site Contact
_____ Number of students
_____ Grade level
_____ Times of arrival and departure
_____ Reminder of trip objective
_____ Special arrangements, if any

Safety Precautions
_____ First-aid kit
_____ Safety personnel, if needed

Food, Transportation, and Supplies
_____ Transportation arranged
_____ Lunch, if any, provided for
_____ Special equipment acquired

Student Orientation
_____ Field-trip goals clarified
_____ Field-trip questions assigned
_____ Safety guidelines discussed
_____ Warning issued on site hazards
_____ Appropriate clothing suggested

ADDITIONAL RESOURCES FOR TEACHERS

The audiovisuals listed preceding each unit are available from the suppliers listed below. In addition, you may wish to write to the suppliers for their catalogs from which you may select additional audiovisual materials.

Agency for Instructional Television
Box A, 111 W. 17th Street
Bloomington, IN 47402

AIMS Media
6901 Woodley Avenue
Van Nuys, CA 91406

Barr Films
P.O. Box 5667
Pasadena, CA 91107

Carolina Biological Supply Company
2700 York Road
Burlington, NC 27215
or
Box 187
Gladstone, OR 97027

Center for Humanities, Inc.
Communications Park
Box 1000
Mt. Kisco, NY 10549

Charles Clark Company, Inc.
170 Keyland Court
Bohemia, NY 11716

Churchill Films
662 N. Robertson Boulevard
Los Angeles, CA 90069

Clearvue, Inc.
5711 N. Milwaukee Avenue
Chicago, IL 60646

Coronet/MTI Film and Video
108 Wilmot Road
Deerfield, IL 60015

CRM/McGraw-Hill Films
110 15th Street
Del Mar, CA 92014

Educational Images, Ltd.
P.O. Box 3456, West Side
Elmira, NY 14905

Educational Services, Inc.
1730 I Street, N.W.
Washington, D.C. 20006

Encyclopedia Britannica Educational Corporation
310 S. Michigan Avenue
Chicago, IL 60604

Films for the Humanities, Inc.
Box 2053
Princeton, NJ 08540

Geoscience Resources
2990 Anthony Road
P.O. Box 2096
Burlington, NC 27216-2096

Human Relations Media
175 Tompkins Avenue
Pleasantville, NY 10570

International Film Bureau
332 S. Michigan Avenue
Chicago, IL 60604

Journal Company
930 Pitner
Evanston, IL 60202

National Geographic Society
Educational Services, Department 84
Washington, D.C. 20036

National Teaching Aids, Inc.
1845 Highland Avenue
New Hyde Park, NY 11040

Phillips Petroleum Company
Advertising Department
310 W. 5th Street
Bartlesville, OK 74003

Prentice-Hall Media
150 White Plains Road
Tarrytown, NY 10591

Psychology Today/Educational Services
1725 K Street N.W., Suite 408
Washington, D.C. 20006

Sunburst Communications
39 Washington Avenue
Pleasantville, NY 10570

Twyman Films
329 Salem Avenue
Box 605
Dayton, OH 45401

Walt Disney Educational Media Co.
500 S. Buena Vista Street
Burbank, CA 91521

Ward's Natural Science Establishment, Inc.
5100 West Henrietta Road
Rochester, NY 14692-9012
or
11850 E. Florence Drive
Santa Fe Springs, CA 90670

West Wind Productions
P.O. Box 3532
Boulder, CO 80303

COMPUTER SOFTWARE SUPPLIERS

The following is a list of companies that produce computer software. You may wish to write to the companies for their catalogs of educational products.

A-Squared Software
P.O. Box 1828
Riverton, WY 82501

Austin Computer Workshop
100 West 22d Street
Austin, TX 78705

Bergwall Educational Software, Incorporated
106 Charles Lindbergh
   Boulevard
Uniondale, NY 11553-3695

Berkshire Scientific Software Concepts Associates
Astor Square—
   19 US Route 9
Rhinebeck, NY 12572

BioLearning Systems
Route 106
Jericho, NY 11753

Carolina Biological Supply Company
2700 York Road
Burlington, NC 27215

Central Scientific Company
11222 Melrose Avenue
Franklin Park, IL 60131

CONDUIT
The University of Iowa
Oakdale Campus
Iowa City, IA 52242

Connecticut Valley Biological Supply Company
82 Valley Road, Box 326
Southampton, MA 01073

Create A Test
80 Tilley Drive
Scarborough, Ontario,
Canada M1C 2G4

Creative Technology, Incorporated
P.O. Box 1009
Carlisle, PA 17013

Cross Educational Software
1802 North Trenton Street
P.O. Box 1536
Ruston, LA 71270

Datatech Software Systems, Incorporated
19312 East Eldorado Drive
Aurora, CO 80013

Diversified Educational Enterprises
725 Main Street
Lafayette, IN 47901

Educational Activities, Incorporated
P.O. Box 392
Freeport, NY 11520

Educational Computing
10661 John Ayres Drive
Fairfax, VA 22032

EduTech
1927 Culver Road
Rochester, NY 14609

Encyclopedia Britannica Educational Corporation
425 North Michigan Avenue
Chicago, IL 60611

Fisher Scientific
4901 West LeMoyne Street
Chicago, IL 60651

Frey Scientific Company
905 Hickory Lane
Mansfield, OH 44905

Kemtec Educational Corporation
P.O. Box 57
Kensington, MD 20895

Kons Scientific Company, Incorporated
P.O. Box 3
Germantown, WI
53022-0003

Merlan Scientific
247 Armstrong Avenue
Georgetown, Ontario,
Canada L7G 4X6

Micro Learningware
Route #1, Box 162
Amboy, MN 56010-9762

Nasco
901 Janesville Avenue
Fort Atkinson, WI 53538

Nasco West Incorporated
P.O. Box 3837
Modesto, CA 95352

Sargent-Welch Scientific Company
7300 North Linder Avenue
Skokie, IL 60077

Scholastic Software
730 Broadway
New York, NY 10003

Science Kit and Boreal Laboratories
777 East Park Drive
Tonawanda, NY 14150

Sunburst Communications
39 Washington Avenue
Pleasantville, NY
10570-9971

Videodiscovery, Incorporated
1515 Dexter Avenue, #200
Seattle, WA 98109

DIRECTORY OF ORGANIZATIONS

Many government agencies, business and professional organizations, and voluntary groups are useful sources for pamphlets and other teaching resources. In addition to those listed below, you will find other resources noted in the Career features in the *Pupil's Edition*. Check your telephone directory for local or regional offices of the organizations listed.

American Association for the Advancement of Science

1333 H Street, N.W.
Washington, DC 20005

American Geological Institute

4220 King Street
Alexandria, VA 22302

American Institute of Chemists

7315 Wisconsin Avenue, N.W.
Bethesda, MD 20814

American Institute of Professional Geologists

7828 Vance Drive
Suite 103
Arvada, CO 80003

Institute for Chemical Education

Department of Chemistry
University of Wisconsin
Madison, WI 53706

Institute for Earth Education

P.O. Box 368
Lawerence, KS 66044

Narcotics Anonymous

P.O. Box 9999
Van Nuys, CA 91409

National Academy of Sciences

2101 Constitution Avenue, N.W.
Washington, DC 20418

National Science Foundation

1880 G Street, N.W.
Washington, DC 20550

National Science Teacher's Association

1742 Connecticut Avenue, N.W.
Washington, DC 20009

Society of Independent Professional Earth Scientists

4925 Greenville Avenue
Suite 170
Dallas, TX 75206

Superintendent of Documents

U.S. Government Printing Office
710 N. Capitol Street, N.W.
Washington, DC 20401

U.S. Environmental Protection Agency

Office of Public Affairs
Washington, DC 20460

U.S. Public Health Service

Department of Health and
Human Services
200 Independence Avenue, S.W.
Washington, DC 20201

American Optical Company

Instrument Division
Eggert and Sugar Roads
Buffalo, NY 14215

Ann Arbor Biologicals

6780 Jackson Road
Ann Arbor, MI 48103

Baltimore Biological Laboratory

1640 Gorsuch Avenue
Baltimore, MD 21218

Bausch & Lomb

Scientific Optical Products Division
Rochester, NY 14602

Bico Scientific Company

2325 South Michigan Avenue
Chicago, IL 60616

Calbiochem

3625 Medford Street
Los Angeles, CA 90054

Carolina Biological Supply Company

2700 York Road
Burlington, NC 27215
or
Powell Laboratories Division
Gladstone, OR 97027

Central Scientific Company

2600 South Kostner Avenue
Chicago, IL 60623

Clinton Misco Corporation

P.O. Box 1005
Ann Arbor, MI 48106

Coe-Palm Biological Supply House

1130 North Milwaukee Avenue
Chicago, IL 60622

Connecticut Valley Biological Supply Company

82 Valley Road
Southampton, MA 01073

Dale Scientific Company

P.O. Box 1721
Ann Arbor, MI 48106

Damon

Instructional Division
80 Wilson Way
Westwood, MA 02090

Difco Laboratories

209 Henry Street
Detroit, MI 48104

Eastman Organic Chemicals

343 State Street
Rochester, NY 14650

Edmund Scientific Company

103 Gloucester Pike
Barrington, NJ 08007

Fisher Scientific Company

Stansi Educational Materials Division
4901 West LeMoyne Avenue
Chicago, IL 60651

Forestry Suppliers, Inc.

Box 8397
Jackson, MS 39204

Hach Chemical Company

P.O. Box 907
Ames, IA 50010

Harvard Apparatus Company

150 Dover Road
Mills, MA 02054

Hubbard Scientific Company

2855 Shermer Road
Northbrook, IL 60062

J. R. Schettle Biologicals

P.O. Box 184
Stillwater, MN 55082

La Pine Scientific Company

6001 S. Knox Avenue
Chicago, IL 60629

MacAlaster Scientific Company

Route 111 and Everett Turnpike
Nashua, NH 03060

Mogul-Ed

P.O. Box 2482
Oshkosh, WI 54901

Nasco

901 Janesville Avenue
Fort Atkinson, WI 53538

Nasco West

1524 Princeton Avenue
Modesto, CA 95352

Northern Biological Supply

P.O. Box 222
New Richmond, WI 54017

Owens-Illinois Scientific Sales

1290 Bayshore Boulevard
Burlington, CA 94010

Parco Scientific

P.O. Box 595
Vienna, OH 44473

Sargent-Welch Scientific Company

7300 North Linder Avenue
Skokie, IL 60076

Science Kit Inc.

777 East Park Drive
Tonawanda, NY 14150

Scientific Products Division

American Hospital Supply
1210 Leon Place
Evanston, IL 60201

Sherwin Scientific Company

N. 1112 Ruby Street
Spokane, WA 99202

Swift Instruments, Inc.

San Jose, CA 95106

Triarch, Inc.

P.O. Box 98
Ripon, WI 54971

Turtox, Inc.

500 West 128th Place
Chicago, IL 60658

U.S. Geological Survey

Branch District
1200 South Eads Street
Arlington, VA 22202

Van Waters & Rogers

3745 Bayshore Boulevard
Brisbane, CA 94005

Ward's Natural Science Establishment, Inc.

5100 West Henrietta Road
Rochester, NY 14692-9012
or
11850 E. Florence Drive
Santa Fe Springs, CA 90670

Wilkens-Anderson Co.

4525 West Division Street
Chicago, IL 60651

Will Scientific, Inc.

Box 1050
Rochester, NY 14603

EARTH SCIENCE

EARTH SCIENCE

Cesare Emiliani

Professor and Chairperson,
Department of Geological Sciences
University of Miami
Coral Gables, Florida

Linda B. Knight

Earth Science Teacher
Paul Revere Middle School
Houston, Texas

Mark Handwerker

Instructional Support Teacher for Science
Temecula Middle School
Temecula, California

 Harcourt Brace Jovanovich, Publishers

Orlando   San Diego   Chicago   Dallas

Copyright © 1989 by Harcourt Brace Jovanovich, Inc.

All rights reserved. No part of this publication may be reproduced or transmitted in any form or by any means, electronic or mechanical, including photocopy, recording, or any information storage and retrieval system, without permission in writing from the publisher.

Requests for permission to make copies of any part of the work should be mailed to: Permissions, Harcourt Brace Jovanovich, Publishers, Orlando, Florida 32887

Printed in the United States of America

ISBN 0-15-361451-X

ACKNOWLEDGMENTS

SENIOR EDITORIAL ADVISOR

Lee Suttner, Ph.D.
Professor of Geology
Indiana University
Bloomington, Indiana

CONTENT ADVISORS

George Greenstein, Ph.D.
Professor of Astronomy
Amherst College
Amherst, Massachusetts

Brian Skinner, Ph.D.
Professor of Geology and
    Geophysics
Yale University
New Haven, Connecticut

Carolyn Sumners, Ed.D.
Director of Astronomy and
    Physics
Houston Museum
    of Natural Science
Houston, Texas

CURRICULUM ADVISORS

Ronald E. Charlton, Ph.D.
Science Coordinator
Mt. Lebanon School District
Pittsburgh, Pennsylvania

Dee Drake *
Earth Science Teacher
Huron High School
Ann Arbor, Michigan

Larry Enochs, Ed.D.
Professor of Curriculum and
    Instruction
Kansas State University
Manhattan, Kansas

Robert Frank *
Earth Science and Geography
    Teacher, Science
    Coordinator K–12
Jefferson Junior High School
Caldwell, Idaho

Loistene Harrell
Earth Science Teacher,
    Department Chairperson
Lake Braddock Secondary
    School
Fairfax County, Virginia

Tom Howick *
Earth Science Teacher
Woodward Academy
College Park, Georgia

Nancy Hampton Johnson
Earth Science Teacher
Haggard Middle School
Plano, Texas

Bobbie G. Reed
Department Chairperson
Crestwood Middle School
Baton Rouge, Louisiana

David Sorenson
Science Teacher
Southwest High School
Green Bay, Wisconsin

Lou Travelsted
Science Teacher
Franklin-Simpson Middle
    School
Franklin, Kentucky

Doris Tucker *
Earth Science Teacher
Rutherford County Schools
Spindale, North Carolina

* Outstanding Earth Science Teacher,
    as awarded by the National
    Association of Geology Teachers

READING/LANGUAGE ADVISORS

Patricia S. Bowers, Ph.D.
Science Reading Coordinator
Division of Curriculum and
    Instruction
Chapel Hill–Carrboro City
    Schools
Chapel Hill, North Carolina

Sue Porro
English Teacher
Dr. Phillips High School
Orlando, Florida

Karen Rugerio
Chairperson, Fine Arts
   Department
Dr. Phillips High School
Orlando, Florida

**SERIES FIELD TEST
TEACHERS**

John Benning
Morse Middle School
Milwaukee, Wisconsin

Jerry East
Nimitz Middle School
Tulsa, Oklahoma

Freddie Fight
Bartow Junior High School
Bartow, Florida

Corinne Fish
Morse Middle School
Milwaukee, Wisconsin

Larry French
Deland Junior High School
Deland, Florida

Rick Herbert
Mac Arthur Junior High School
Lawton, Oklahoma

Stan Hitomi
Monte Vista High School
Danville, California

Jennifer Jones
Coosa Middle School
Rome, Georgia

Ellen McCullough
Edwards Middle School
Conyers, Georgia

Carl Nebelsky
Bloomfield Junior High School
Bloomfield, Connecticut

Anna Rice
Bloomfield Junior High School
Bloomfield, Connecticut

John Richardson
Letha Raney Junior High
   School
Corona, California

Bernard Sanner
South Division High School
Milwaukee, Wisconsin

Mary Satterwaite
Morse Middle School
Milwaukee, Wisconsin

Charles Siebert
South Division High School
Milwaukee, Wisconsin

Peggy Stewart
Glasgow Middle School
Baton Rouge, Louisiana

CONTENTS

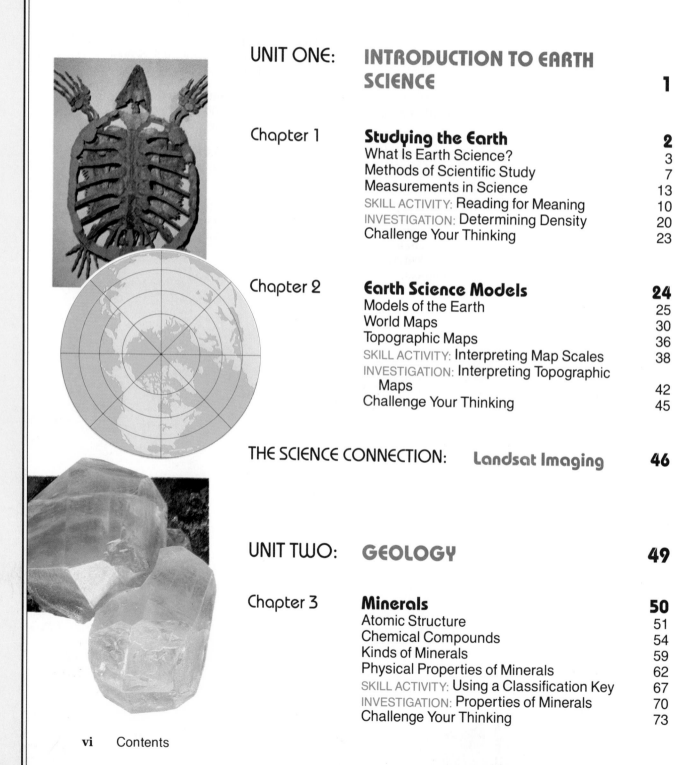

Contents   **vii**

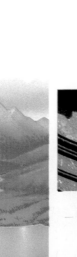

Contents   **ix**

Contents   **xi**

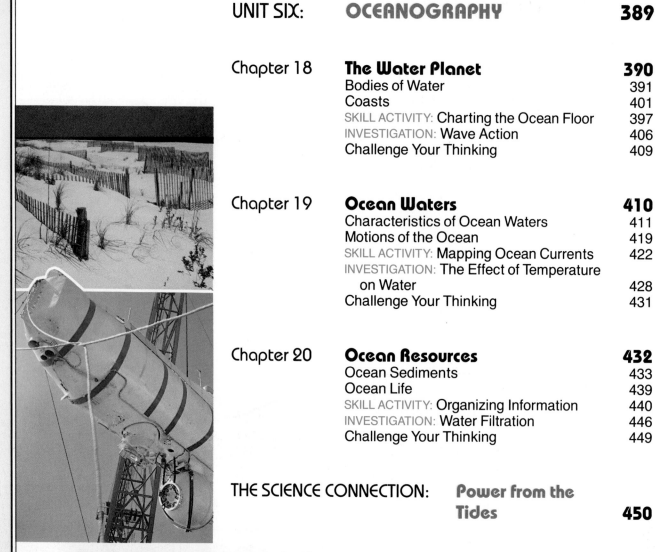

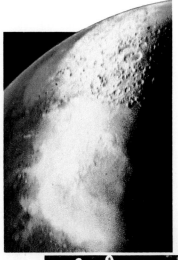

Contents   xiii

REFERENCE SECTION 547

SKILL ACTIVITIES

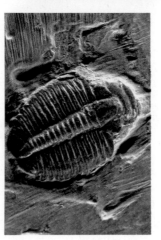

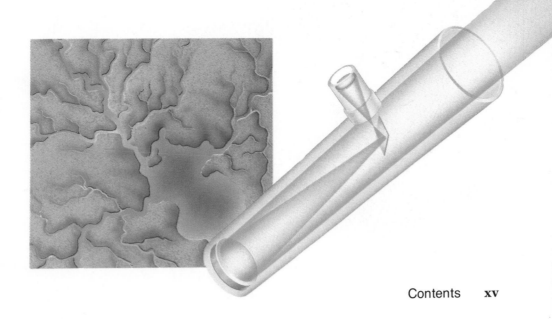

INVESTIGATIONS

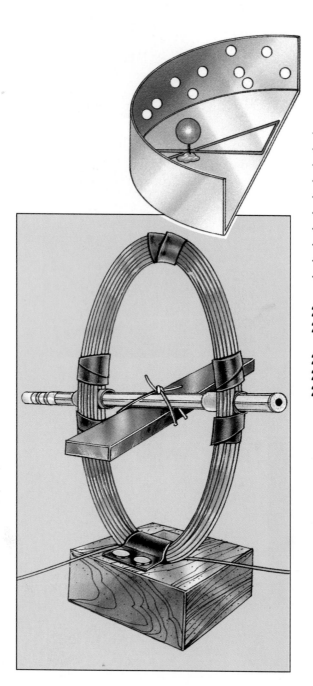

xvi   Contents

BIOGRAPHIES: THEN AND NOW

CAREERS

Contents **xvii**

TECHNOLOGY

ACTIVITIES

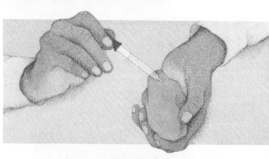

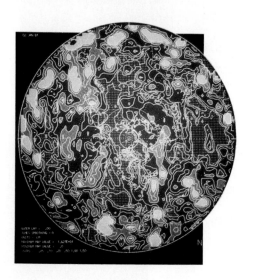

DISCOVERS

A Message to Students About
Earth Science

How often is your life touched by some aspect of earth science? You might quickly answer "never" or "not very often." However, if you stop and think for a few minutes, you might be surprised as to just how important earth science is to your life.

Do you know what the weather forecast is for today? If you do, you probably received your information from a newspaper or a radio or television broadcast. In turn, these media might have received their weather information from NOAA, the National Oceanic and Atmospheric Administration. NOAA is a department of the federal government that supplies weather data and forecasts for the United States. NOAA employs many meteorologists—earth scientists who specialize in the study of the weather.

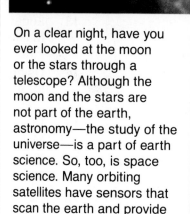

On a clear night, have you ever looked at the moon or the stars through a telescope? Although the moon and the stars are not part of the earth, astronomy—the study of the universe—is a part of earth science. So, too, is space science. Many orbiting satellites have sensors that scan the earth and provide

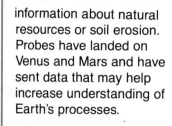

information about natural resources or soil erosion. Probes have landed on Venus and Mars and have sent data that may help increase understanding of Earth's processes.

Do you live on a mountain, on a plateau, or on a coastal plain? Have you ever seen a volcano, felt an earthquake, or heard thunder? Have you ever built a sand castle, tasted salt, or smelled rain? All these things are part of earth science.

You may be taking this course because you are interested in studying the earth or perhaps because it is required by your school. Whatever the reason you are taking this course, you should prepare yourself to make a discovery. In your study of *Earth Science,* you will learn many interesting and exciting things. Allow yourself to explore *Earth Science* and to discover the ideas, information, and beauty within its pages. You will find that learning about earth science is enjoyable as well as beneficial. You will learn many things that will help you understand the processes of the earth. You may even develop an interest in earth science as a career.

In addition to learning scientific concepts, with *Earth Science* you will develop other skills that will help you now and in the future. For example, you will learn how to improve your reading comprehension, how to make and read tables and graphs, and how to understand diagrams and photographs. As you begin your study of *Earth Science,* you also will start developing your skills as a trained observer. Much of science is based on precise observations of nature. From their observations, scientists form questions and develop experiments to expand on their observations. From these close and precise observations, scientists try to answer their questions and form conclusions.

Answers to questions and records of observations have allowed earth scientists to create a body of knowledge that serves as the primary resource for engineers and technicians. This body of knowledge is the basis of technological growth. Computers, radiotelescopes, fusion reactors, submersibles, and the space shuttle are just a few applications of scientific knowledge. These devices, which are related to earth science, help improve our standard of living.

Technology is the method of putting science to work. For example, through the study of scientific principles, Thomas Edison discovered

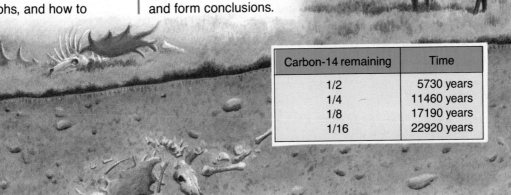

Carbon-14 remaining	Time
1/2	5730 years
1/4	11460 years
1/8	17190 years
1/16	22920 years

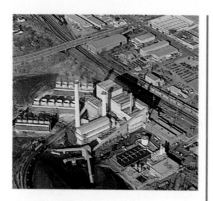

that electricity flowing through a thin wire would make the wire glow. As a result of this knowledge, Edison was able to create the light bulb. The invention of the light bulb allowed people to put electricity to work to improve their lives.

Further scientific discoveries created an understanding of how to generate the quantities of electricity needed to light every home. This information led to the development of fossil-fuel power plants that produce electricity by burning coal and oil. The need for large quantities of fossil fuels led to further studies of the earth and its resources. A lack of fossil fuels led to the discovery of alternative energy sources. In the 1950s,

nuclear energy was first used to generate electricity commercially. Nuclear power may help overcome a shortage of fossil fuels in the immediate future, but the 1986 accident at Chernobyl, in the Soviet Union, has shown that nuclear power is not the final answer to our need for electricity. In the future, technology may make solar energy and fusion energy practical and safe alternatives.

In *Earth Science,* you will learn more about some topics with which you may already be familiar. For example, you will review the use of a scientific method for solving problems. You will also practice measuring, using SI. You will learn how maps are made and about the different kinds of maps an earth scientist uses.

The structure of the earth and the composition of the atmosphere are described in great detail in *Earth Science.*

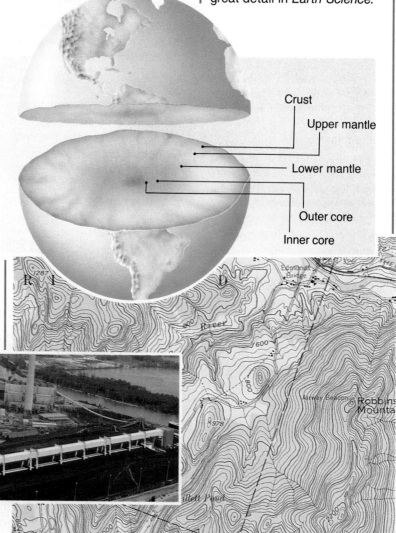

Crust

Upper mantle

Lower mantle

Outer core

Inner core

The processes that build and erode the earth are discussed. *Earth Science* presents the history of our planet and projects its future as well. Finally, in *Earth Science* you will learn about our great frontiers—the oceans and space.

The study of *Earth Science* will give you a new perspective on why and how things happen around you. You will begin to ask your own questions and supply your own answers. *Earth Science* includes the ongoing processes of asking and answering questions about the earth. Curiosity and imagination are guiding lights for earth scientists. By virtue of this curiosity and imagination, earth scientists are able to create a never-ending list of questions about the universe in which we live.

Many of the questions asked by today's earth scientists will not be answered in their lifetimes. Their observations and research, however, will help future scientists achieve the answers. By pursuing a career in earth science, you may become a beneficiary of the knowledge, the dreams, and the aspirations of the scientists who lived before you. Your work in earth science may help answer your own questions as well as questions posed by the earth scientists of the past.

The authors of *Earth Science* hope that you will enjoy reading and learning about the earth, and that perhaps one day you may help others discover the importance of earth science in their lives.

Unit 1: INTRODUCTION TO EARTH SCIENCE

UNIT OVERVIEW

This unit provides an introduction to the study of earth science. Scientific problem solving, the use of models in science, and measuring in SI are included. The development of map projections and the use of topographic maps are also presented.

Chapter 1: Studying the Earth, page 2

In this chapter, the divisions of earth science are listed and defined. The relationship of earth science to other fields of science, such as biology, chemistry, and physics, is discussed. Next, the methods of scientific studies are outlined, and the International System of Units is then introduced.

Chapter 2: Earth Science Models, page 24

This chapter defines maps as models of the earth. The history of mapmaking, or cartography, is discussed. Topographic, geologic, and hydrologic maps are identified. Latitude and longitude are explained, and map projections are described. Additional detail on topographic maps includes discussion of map scales and contour lines.

The Science Connection: Landsat Imaging, page 46

The Science Connection describes the use of Landsat satellites to produce images of the earth's continents. A description of the imaging process and of the many uses for Landsat data are included.

ADVANCE PREPARATION

Chapter 1

For the Discover on page 5, you will need several college catalogs. For more information, see page 5.

Chapter 2

For the Discover on page 39, you will need a large amount of modeling clay for each group of students. For more information, see page 39.

BULLETIN BOARD SUGGESTIONS

Chapter 1

Have the students design a Mystery Metric Measurement Board. Assign them the task of determining the distance, volume, and mass of various objects in the school. They can then design flip cards with a question on one side of the card and the answer on the flip side. Questions should be challenging, such as ''What is the volume of Mrs. Jones' desk?'' or ''What is the length of the principal's tie?''

Chapter 2

Make a matching board with photographs or pictures and topographic maps. Display pictures of mountains, plains, hills, or other common landforms. Label these with letters. Also display small sections of topographic maps that are labeled with numbers. Have the students match the letter on the picture to the number on the map that shows the type of terrain in the picture.

ISSUES IN EARTH SCIENCE

Inform the students that the United States is the only industrialized country that does not use SI for measurements. Discuss the advantages and disadvantages of having the United States convert to SI.

Have the students research the history of using satellite imaging to improve the accuracy of North American maps. Then debate the pros and cons of spending large sums of money to accurately map the rest of the continents and oceanic islands.

SUGGESTED PROJECTS

Have groups of students use sheets of cardboard to represent 10-m contour intervals. The students will cut out landform shapes on the cardboard and stack them to produce raised relief maps. If several groups work on different landforms, a large map suitable for school display can be produced.

Have the students bring in as many different kinds of maps as they can. Then create bulletin board displays by grouping similar types of maps together or grouping different maps of the same area together.

Have groups of students survey their neighborhoods to get opinions about the United States converting to SI. As an alternate activity some groups could interview local business persons or industrial managers to see how, or if, they use SI in their businesses.

TEACHER RESOURCES

Readings

Chapter 1

Experimentation and Measurement. Washington, D.C.: U.S. Government Printing Office, 1985. This book provides an excellent introduction to the realistic consideration of errors in measurement.

Funk, et al. *Learning Science Process Skills*. New York: Kendall Hunt, 1979. This book contains activity ideas for teaching the process skills of measurement and experimental design.

Chapter 2

Hamblin, W. K., and J. D. Howard. *Exercises in Physical Geology*. Edina, MN: Burgess International Group, Inc., 1980. Contains clear and concise explanations and applications of topographic maps. Also serves as an excellent reference for geologic maps and many other topics in geology.

Map Reading. Washington, D.C.: U.S. Government Printing Office, 1969, reprint 1983. This book provides detailed information on the use of maps, aerial photographs, and more.

Audiovisual/Software

Chapter 1

Scientific Method. Filmstrip. Nasco #SB7017M The film illustrates problem solving using a scientific method.

Chapter 2

U.S.G.S. Topographic Maps and Map Symbols Brochure, which may be obtained from:

U.S. Geological Survey
Box 25286, DFC
Denver, CO 80225

Other Materials

Chapter 1

A variety of objects for students to practice measuring in metric units.

Chapter 2

Raised relief topographic maps
Contour Models
Available from Geoscience Resources

PLANNING THE CHAPTER

Chapter Sections	Page	Chapter Features	Page	Program Resources	Page
Section 1: What Is Earth Science?	3			Reading for Content: *Applying What You Read* **(B)**	**TRB 1** **LM 173**
1.1 The Earth Sciences **(B)**	3				
1.2 Related Sciences **(B)**	5	**Discover:** Studying Science **(B)**	5		
		Section Review	6		
Section 2: Methods of Scientific Study	7			Critical Thinking **(H)**	**TRB 1**
				Investigation 1.2: *Variables and Controls* **(A)**	**TRB 3** **LM 5**
1.3 Scientific Method **(B)**	7	**Discover:** Using a Scientific Method **(B)**	9		
		Skill Activity: Reading for Meaning **(B)**	10		
1.4 Models in Science **(B)**	11	A Matter of Fact	11		
		Section Review	12		
Section 3: Measurements in Science	13			Cross-Discipline: Science and Mathematics, *Using Units and Formulas in Science* **(A)**	**TRB 1**
1.5 The International System of Units **(B)**	13	**Discover:** Making SI Measurements **(B)**	15	Investigation 1.1: *Measuring in SI* **(B)**	**TRB 1** **LM 3**
1.6 Measuring Volume **(B)**	15	**Activity:** Finding Volume **(B)**	16	Concept Extension: *Metric Measurements*	**TRB 1** **LM 223**
1.7 Comparing Mass, Weight, and Density **(B)**	17	A Matter of Fact	17		
		A Matter of Fact	18		
		Section Review	18		
		Biographies: Then and Now James Hutton, Luis Alvarez	19		
		Investigation 1: Determining Density **(A)**	20	Student Record Book: Textbook Investigations **(A)**	**TRB 1**
Chapter 1 Review	21			Vocabulary **(A)**	**TRB 1** **LM 123**
				Tests **(A)** Computer Test Bank	**TRB 1**

(LM) Laboratory Manual/Study Guide, **(TRB)** Teacher's ResourceBank™
B = Basic   **A** = Average   **H** = Honors
The coding Basic, Average, and Honors indicates sections or subsections that might be appropriate for different levels of learners. For additional suggestions regarding choice of topic and depth of coverage, see the Pacing Chart on pages T16–T20.

CHAPTER CONCEPTS, OBJECTIVES, AND TERMS

Section	Concepts	Objectives	Science Terms
Section 1: What Is Earth Science?	■ The major fields studied in earth science are geology, hydrology, oceanography, meteorology, and astronomy. **(1.1)** ■ Geology is the study of minerals and rocks and of the processes that shape the earth. **(1.1)** ■ Hydrology is the study of the underground and surface waters of lakes and streams. **(1.1)** ■ Oceanography is the study of the oceans. **(1.1)** ■ Meteorology is the study of the atmosphere and the weather of the earth. **(1.1)** ■ Astronomy is the study of the planets, the sun, and the stars. **(1.1)** ■ The earth sciences apply principles from the basic scientific fields of biology, physics, and chemistry. **(1.2)** ■ Biology is the study of living organisms. **(1.2)** ■ Chemistry is the study of the composition of matter. **(1.2)** ■ Physics is the study of matter and energy. **(1.2)**	■ **Describe** the five scientific fields that are studied in earth science. ■ **Explain** how the fields of biology, physics, and chemistry relate to the study of earth science.	geology hydrology oceanography meteorology astronomy science fossils
Section 2: Methods of Scientific Study	■ Problems can be studied by using the investigative procedure called a scientific method. **(1.3)** ■ Scientists often present theories in the form of models. **(1.4)**	■ **Explain** the steps in a scientific method. ■ **Compare** and **contrast** a hypothesis and a theory. ■ **Discuss** the value of models to the study of science.	data conclusions hypothesis theory experiment law variable scientific method control model
Section 3: Measurements in Science	■ Scientific work uses the International System of Units. **(1.5)** ■ The space that an object occupies is the object's volume. The volume of an irregularly shaped object is measured by displacement. **(1.6)** ■ Mass is the amount of material or matter in an object. Gravitational attraction affects an object's weight but not its mass. **(1.7)** ■ The density of an object depends on the relationship between the object's mass and volume. **(1.7)**	■ **Measure** length, volume, mass, and weight using SI. ■ **Calculate** the density of an object.	volume mass weight density

Title	Page	Materials
Discover: Studying Science	5	*(per student)* college catalog
Discover: Using a Scientific Method	9	*(per student)* paper, pencil
Skill Activity: Reading for Meaning	10	*(per student)* paper, pencil
Discover: Making SI Measurements	15	*(per student)* measuring tools, objects to measure, paper, pencil
Activity: Finding Volume	16	*(per group of 3 or 4)* graduate, 50 mL; water; string; galena; calcite; granite; cork; metric ruler
Investigation 1: Determining Density	20	*(per group of 3 or 4)* balance; galena; calcite; granite; bar magnet; glass rod; metric ruler; string; graduate, 100 mL; paper towels

TEACHING SUGGESTIONS

Section 1: What Is Earth Science?

Outside Speaker

Have an earth scientist come in and speak to the class about what he or she does.

Section 2: Methods of Scientific Study

Field Trip

Visit a research laboratory and have a research assistant explain how a scientific method is used to solve a problem.

Section 3: Measurements in Science

Demonstration: Measuring in Metric

Purpose

To practice metric measurements

Background

Students often do not understand either customary or metric measures. To help them with metric units, measure a series of common objects and write the measurements on the board.

Materials: As many of the following as possible

Meter stick
Centimeter ruler
Balance
Metric (kg) bathroom scale
Graduates of various sizes

Various objects for determination of mass, dimensions, and volume

Procedure

Using one or more of the measuring instruments, measure as many dimensions of the selected objects as possible. A chart of width, height, breadth, diameter, mass, and so on, can be maintained on the chalkboard. Do not hesitate to measure some physical dimensions of students who volunteer. You can measure cap size, mass, height, ring size, or shoe size.

Questions to Ask the Students

1. Why are metric units easier to use than customary units? (*All measurements are in base 10 units.*)
2. What metric units are used to measure distance, mass, and volume? (*distance–meters, mass–grams, volume–milliliters*)

Class Activity

The students can demonstrate differences in density by pouring water, cooking oil, and automobile oil into a graduate. Have the students explain why the three substances do not mix. Relate the concept of density to the fact that oil rises to the top in certain layers of rocks. Salt domes are created because salt has a very low density. This causes it to rise and force its way into domes or pockets of concentration.

Class Activity

Have the students practice reading a graduate. First, direct them to examine the scale of the graduate to determine the interval between each line. Explain that most liquids form a concave surface due to surface tension. The only exception is mercury. Caution students to position themselves at eye level with the curve and to read the volume from the bottom of the curve. Have the students fill the graduate to a specified reading.

UNIT 1

INTRODUCTION TO EARTH SCIENCE

In 1972 earth science took a giant leap forward with the launch of the first Landsat satellite. In the years that followed, four more Landsat satellites were launched and are still circling the earth. These satellites have given scientists more than a million pictures of the earth. Each multi-colored photograph shows an area about 185 km wide. Details as small as a football field can be seen.

- Why do you think different colors are used in Landsat photographs?

- How has the exploration of space helped to solve some problems on Earth?

- How might a mapmaker use a tool like Landsat to create more accurate maps?

By reading the chapters in this unit, you will learn the answers to these questions. You will also develop an understanding of concepts that will allow you to answer many of your own questions about the study of the earth.

Satellite photographing Earth

1

INTRODUCING THE UNIT

Have the students look at the Unit Opener photograph on these pages. Point out that Landsat is an earth-orbiting observational spacecraft that is 4.5 m long by 2.1 m wide and has a mass of 2000 kg. It circles the earth every 99 minutes in a polar orbit 700 m above the earth's surface, collecting data on such things as the atmosphere, pollution of the earth's environment, vegetation growth, and mineral resources hidden deep within the earth. Read the text and use the questions to guide a class discussion. Do not expect the students to be able to answer the questions now. The questions are motivational, and the students should be aware that they will be able to answer them after completing the Unit.

Discussion From ancient times, maps have been drawn to help explain the world in which we live. As travellers discovered and explored unknown regions and then returned to tell of their adventures, maps were revised and made more accurate. This updating of maps continues today. Ask the students why the perspective of Landsat photographs provides data about the shape of land masses that was previously unavailable to mapmakers. Ask the students why information about the location of vegetation, mineral resources, water and air pollution, depth and clarity of water, and changes in the atmosphere is important to scientists who are concerned about the future of our planet.

Extension You can extend the discussion by bringing in other Unit themes. Do maps serve different functions? Are there different kinds of maps? Where is "north" located on maps? Why is "north" always located in the same position on maps? What do the terms *longitude* and *latitude* mean? You may wish to record the answers and review them at the end of the Unit.

CHAPTER 1

CHAPTER OVERVIEW

In this chapter, the divisions of earth science are listed and defined. The relationship of earth science to other fields of science, such as biology, chemistry, and physics, is discussed. Next, the methods of scientific studies are outlined, and the International System of Units is then introduced.

Section 1: What Is Earth Science? The divisions of earth science are geology, hydrology, oceanography, meteorology, and astronomy. These are defined and the relationships among all fields of science are explained.

Section 2: Methods of Scientific Study The plan, or steps, that a scientist uses to reach the goal of a project is called a scientific method. This section outlines a scientific method and discusses its use in problem solving. The place of models in scientific study is also explained

Section 3: Measurements in Science In this section, the International System of Units (SI) is outlined. The need for this system and the use of appropriate units of measure are explained. Finding the volumes of regularly and irregularly shaped objects is presented. Finally, the concepts of mass, weight, and density are defined and compared.

Studying the Earth

Few people can look at this photograph and not admire the beauty of Earth. The cloud patterns in the air contrast with the land masses below. The large bodies of water provide the rich blue color that makes Earth so different from other planets. Earth is a fascinating place, and its study reveals many wonders.

Earth, as it appears from space

2

CHAPTER MOTIVATING ACTIVITY

Distribute four 10 cm x 10 cm squares of aluminum foil, a spiral birthday candle, a magnifying glass, and a ruler to each lab group. Have the students make a list of at least 10 observations about the candle. Encourage them to include both descriptive and measurable observations. Then allow them to light the candle and make further observations. **CAUTION: Make sure the students hold the candle horizontally, so the melted wax will drip onto the foil square.** Discuss the types of observations they were able to make.

Students could also make volume and mass measurements before and after burning the candle.

You might point out that the candle is made of paraffin, a hydrocarbon. The carbon can be collected by carefully holding a foil square above the flame.

1 What Is Earth Science?

SECTION OBJECTIVES

After completing this section, you should be able to:

- **Describe** the five scientific fields that are studied in earth science.
- **Explain** how the fields of biology, physics, and chemistry relate to earth science.

NEW SCIENCE TERMS

geology
hydrology
oceanography
meteorology
astronomy
science
fossils

1.1 The Earth Sciences

Have you ever taken something apart to see how it works? In order to really understand how something works, you must examine all its parts. Complicated things seem so much simpler if you can study one part at a time. The same is true when you study the earth. In order to understand how the earth works, you must look at it from the point of view of scientists who specialize in studying its parts.

Scientists are constantly striving to understand the earth and its secrets more fully. To do this, scientists observe, classify, and collect information about the earth.

Earth science can be divided into the areas of *geology, hydrology, oceanography, meteorology,* and *astronomy.* Although you can get a better understanding of parts of the earth through these specialties, you must study all of them to really understand how the earth works.

Figure 1–1. The oceans, the land, the clouds, and space are all subjects of study in earth science.

Section 1   What Is Earth Science?   **3**

BACKGROUND INFORMATION

Before the 1950s the branches of earth science remained rather distinct. However, as people became more concerned about environmental issues, specialization in the earth sciences became a necessity. As pollution became a greater threat, earth science and the study of our air and water became more important to the quality of life.

Section **1**: What Is Earth Science?

SECTION FOCUS

Section Overview   This section defines earth science and divides it into the areas of geology, hydrology, oceanography, meteorology, and astronomy. The interdependence of all fields of science is then discussed.

Section Objectives   For a list of section objectives, see pupil's edition page.

New Science Terms   For a list of new science terms in this section, see pupil's edition page. You may wish to have the students read the terms aloud.

SECTION DEVELOPMENT

1.1 The Earth Sciences

DISCUSSION   Have the students compare the photograph in the chapter opener and Figure 1–1. ***Thinking Critically:*** Ask them to name the features in the photographs that could be studied. (Some answers might include land, water, rivers, oceans, and clouds.) After naming the five areas of earth science, discuss how the photo features fall into the five areas.

DISCUSSION Ask the students to look at the photographs on pages 4 and 5 and tell in which area each scientist is working. Then have them read the captions. You may wish to have students suggest other occupations in each area.

EXTENSION (Tie-in/Language Arts) Have the students look up the names of the five areas of earth science in a dictionary and find the origin of each word.

SKILLS *(Making a Chart, Classifying Data)* Have the students set up a chart with the five areas of earth science as headings. List several terms associated with each area on the chalkboard and have the students write the terms on their charts under the proper headings. For example, rocks, minerals, sand, jewels, and earthquakes would fall under the geology heading. Hydrology would include ponds, wells, riverbeds, and fresh-water pollution. Oceanography would include water, tsunamis, clams, seaweed, and sharks. Rain, hurricanes, air pollution, oxygen, and wind would fall under meteorology. The study of astronomy might include words such as *Halley's comet, eclipse, mass and sunspot.*

DISCUSSION After the students have read about the five areas of earth science, ask them to think about how the areas are related. ***Thinking Critically:*** Ask the students how meteorology and hydrology are related. (The earth's water supply is greatly affected by the weather.) Ask how astronomy and meteorology are related. (The sun's energy affects the weather.) Then you might ask how an astronomer could help a hydrologist. (An astronomer could give a hydrologist information about the sun's energy that could help the hydrologist understand the distribution of water on Earth.)

Figure 1–2. From a small rock sample, a geologist can learn many things about the earth. The presence of fossils in the sample may even help to determine the rock's age. What else might a geologist learn from a rock sample? **1**
1 the rock's composition
Figure 1–3. These oceanographers (right) are studying the organisms living in a 1-m² section of the ocean floor. The hydrologist (below) is determining the amount of water running off a bare field. She might compare this to the amount of water flowing from a planted field of the same size.

Geology The study of the solid earth is called **geology** (jee AHL uh jee). A geologist studies minerals, rocks, and fossils, as well as the processes that change the earth. Studying volcanoes, mountain building, and erosion is also part of geology.

Hydrology and Oceanography The study of all the water found on land is called **hydrology** (hy DRAHL uh jee). A hydrologist studies the water under the ground, as well as the water in streams and lakes. Hydrologists are concerned with the movement of ground water and the quality of fresh water supplies.

The oceans are studied in **oceanography** (oh shuh NAHG ruh fee). In this field, all aspects of the oceans are studied. Oceanographers study the shape and composition of the ocean floor, the chemical composition of ocean water, the movements of tides and currents, and the plants and animals that live in the oceans.

Meteorology The study of the earth's atmosphere is called **meteorology** (meet ee uh RAHL uh jee). Some meteorologists analyze the composition of the gases that surround the earth. They also study the motions of the atmosphere and the interaction of the atmospheric gases with the liquid and solid surface of the earth. Such interactions affect weather and climate. Other meteorologists predict the weather for airlines, governments, newspapers, and radio and TV stations.

4   Chapter 1   Studying the Earth

Astronomy   The study of the planets, the sun, and the stars is called **astronomy** (uh STRAHN uh mee). Astronomers probe the universe trying to answer questions about the origin of the earth, the solar system, and the universe.

○ *What are the five specialties of earth science?*
○ *Why is the study of earth science divided into these parts?*

Figure 1–4. This weather forecaster (left) has studied current weather conditions, including satellite photographs, and is now predicting tomorrow's weather for his TV audience. The astronomer (right) is studying more distant sources of information—stars that are billions of kilometers from Earth.

1.2 Related Sciences

Science is an organized body of knowledge that has developed through observation and experimentation. Earth science is just one branch of science. The five divisions of earth science use the same basic principles and methods as the other fields of science. The scientific fields of biology, chemistry, and physics are each important to an understanding of the earth. All fields of science are related to one another.

Biology is the study of living things, their environment, and their behaviors within their environment. *Chemistry* is the study of the composition and characteristics of matter. The science of *physics* deals with the study of matter and energy. How do you think each of these areas of study relates to earth science? **1**

To see the relationships among the sciences, consider an example from geology—the search for oil. This search might actually begin with the study of fossils. **Fossils** are the remains of once-living organisms, preserved in the earth. *Paleontology* (pay lee uhn TAHL uh jee), the study of fossils, is an area in which scientists need a thorough knowledge of biology.

1 Biology is important for understanding fossils. Chemistry is important for studying mineral compositions. Physics is important for understanding energy resources.

Section 1   What is Earth Science?   **5**

DISCOVER

Studying Science

In a college catalog, look at the courses required for a major in earth science. Decide how each course is related to the five earth science areas.

1.2 Related Sciences

DISCUSSION   You may wish to have the students share what they know about biology, chemistry, and physics. Ask them to share any experiences they might have had in those fields. (Some students may have chemistry sets; some may have grown plants; some may have raised animals for 4-H projects.) Stress the importance of mathematics and language arts in the successful pursuit of the sciences.

EXTENSION   Putting a human into space involves many areas of science. ***Thinking Critically:*** Have the students name the various phases of a successful space flight and explain how different branches of science helped make the flight a success. (Some answers might include developing a material for a spaceship that can withstand launch and re-entry—chemistry and physics; providing an environment for a human to live in space—biology; providing food for a person in space—biology and chemistry.)

DISCOVER: Studying Science

Thinking Skill *(Drawing Conclusions)*

You may obtain catalogs from a high school library, a public library, or a high school guidance counselor. You may also write to the admissions offices of colleges and request current catalogs. Students' answers will vary, but they should show an understanding of the various areas of science.

Answers–1.1

○ The five specialties of earth science are geology, hydrology, oceanography, meteorology, and astronomy.

○ These five areas cover all aspects of the earth and its position in space.

SECTION REVIEW

DISCUSSION You may wish to direct the students' attention to the photographs and related captions. If possible, have some fossil samples in class for the students to examine and identify. *Thinking Critically:* Ask the students how fossils may have been preserved in the earth. (A body of water flooded the land and the object became buried under mud; resin, or sap, from trees trapped tiny insects or other animals; animals or plant objects fell into cracks in ice and froze.)

Figure 1–5. These fossils are all that remain of organisms that once were common on the earth. What can scientists learn about the earth's past by studying fossils? **1**

1 Fossils may reveal characteristics of the environment of these once-living organisms.

Figure 1–6. The people working on this oil platform are trying to recover the remains of other fossil organisms. The oil that they seek is called a *fossil fuel*.

Sometimes paleontologists are hired by oil companies to help locate rock layers in which oil might be found. Paleontologists exploring for oil must work closely with chemists who analyze oil.

You can see from this example that earth science problems require the experience of specialists. No one scientist could have all the knowledge necessary to solve complicated problems.

○ *What is science?*
○ *How do biology, chemistry, and physics help in the study of earth science?*

Section Review

READING CRITICALLY

1. Which one of the fields of earth science would include the study of glaciers?
2. What type of work would a physicist do?

THINKING CRITICALLY

3. Why would a geologist studying the way in which ground water dissolves limestone rock need a strong background in chemistry?
4. After the discovery of an oil deposit, what science specialties might be needed to recover, transport, and refine the oil? Explain your answer.

6   Chapter 1   Studying the Earth

SECTION REVIEW

Summary   Have the students list the five areas of earth science and briefly explain each.

Reinforcement   Have each student make up two more "Reading Critically" type questions. Have them write each question on a slip of paper with the answer on the back of the paper. Then ask groups of four or five students to pool their questions and take turns answering questions.

Answers–1.2

○ Science is an organized body of knowledge developed through careful observation and experimentation.

○ Earth scientists need to apply the basic concepts about matter, energy, and life forms to processes and events observed on the earth.

Answers to Section Review

Reading Critically

1. Geology would include the study of glaciers.
2. Answers will vary. For example, physicists study the release of energy on the earth and the motions of the earth, such as earthquakes.

continues

2 Methods of Scientific Study

SECTION OBJECTIVES

After completing this section, you should be able to:

- **Explain** the steps in a scientific method.
- **Compare** and **contrast** a hypothesis and a theory.
- **Discuss** the value of models to the study of science.

NEW SCIENCE TERMS

data
hypothesis
experiment
variable
control
conclusions
theory
law
scientific method
model

1.3 Scientific Method

What is your goal in life? Do you have any idea what you want to do after high school? Maybe you want to go to college or trade school, join the military, or begin working. Everyone needs to set goals for the completion of a task. Often it is much easier to decide what needs to be done when you have a goal to reach. In a similar manner, scientific problem solving requires goals.

Before scientists begin working on a project, they must determine the goal of the project. Whether the goal is to discover more oil or to predict the location of an earthquake, the whole scientific team must work toward that goal. What might happen if one person on a team were working toward a different goal from the rest of the team? **1**

Making Observations   Often the goal that scientists hope to reach is the answer to a question or the solution to a problem. In order to reach their goal, scientists must make predictions and decisions based on data (DAY tuh). **Data** is the information gathered by observation or investigation. For instance, observing that it takes the earth about $365\frac{1}{4}$ days to orbit the sun provides data about the earth's orbit.

1 The goal might not be reached, or someone might be injured.

Figure 1–7. Just as a mining team must work together for mutual safety, scientists must also work as a team to solve problems.

SECTION FOCUS

Section Overview   This section explains a scientific method that a scientist might use to reach a goal. The terms *hypothesis, theory,* and *law* are compared. The use and value of models in science are then discussed.

Section Objectives   For a list of section objectives, see pupil's edition page.

New Science Terms   For a list of new science terms in this section, see pupil's edition page. You may wish to call the students' attention to terms with which they may not be familiar, such as *hypothesis*.

SECTION DEVELOPMENT

1.3 Scientific Method

DISCUSSION   You may wish to ask the students about goals they have worked toward. (Goals may include buying a new bike, finding a lost object, making a cake for the first time, or winning a game.) They may then compare these goals with the goals of a scientist and define the role of data. ***Thinking Critically:*** Ask the students to name some ways data is collected. (Examples may include weighing, measuring, cutting apart or breaking apart, observing or looking at, heating, and cooling.)

Thinking Critically

3. The geologist who studies how limestone dissolves would need to understand how the chemicals that make up limestone react with water.
4. Engineering and physics would be involved in plans for removing and transporting the oil. Chemistry would determine how to refine the oil.

DISCUSSION   Point out that the headings on this page are steps in a scientific method. *Thinking Critically:* Ask the students if the steps could be performed in any other order. (Usually the hypothesis is stated first so that experiments can be performed to test the hypothesis. Then conclusion(s) are drawn from the data resulting from the experiments. Sometimes a scientist could observe some unexpected data from an experiment and state a new hypothesis to be tested.) You may wish to mention that a hypothesis is not a wild guess; it is based on observation. Scientists, however, do follow their imaginations and hunches when working on problems. Also, note that a well-designed experiment only changes or tests one variable at a time so that any changes in the data can be credited to the one variable.

[plural, *hypotheses*]

Figure 1–8. These students are being careful not to introduce unwanted variables into their experiment.

Forming Hypotheses   Have you ever predicted that something was going to happen and had your prediction come true? Maybe your prediction was not just a wild guess, but based on careful observations over a long period.

Scientists often begin their investigations by making an educated guess about what they think will occur. This scientific guess is called a **hypothesis** (hy PAHTH uh sihs). Once a hypothesis is stated, it becomes the goal the scientists try to reach.

Conducting Experiments   Scientists work carefully collecting data to test a hypothesis. This data is usually gathered through experiments. An **experiment** is a test designed to give a scientist information under carefully controlled conditions.

Scientists know that reliable data can come only from well-designed experiments. Any well-designed experiment tests a variable. In a scientific experiment, a **variable** is something that can be changed. Characteristics such as temperature, volume, time, and color are examples of the variables in an experiment. The part of an experiment that does not change is called the **control,** or the *controlled variable.*

In the experiment shown in Figure 1–8, the students try to find out if water will move more rapidly through gravel or sand. They think that the water will pass through the gravel faster. The variable in this experiment is time. In order to test their hypothesis, the students must control all other variables that might affect their experiment. For example, they must be sure they pour the same amount of water into each tube. What other variables do you think they need to control?

Drawing Conclusions   Scientists analyze the data they have collected to reach a conclusion. **Conclusions** are statements about the original hypothesis, based on all the information that has been gathered. Sometimes the conclusion confirms the hypothesis. In most cases, however, the conclusion creates more questions than it answers.

Scientists usually conduct many experiments to test a single hypothesis, so that they can be more sure about their conclusions. When new, conflicting information becomes available, the conclusion must be changed. Once scientists have collected data from many experiments, they may find that the data consistently supports their hypothesis. The hypothesis may then become a theory or a scientific law.

A **theory** is a statement that explains why things happen the way that they do. A scientific law is different from a theory. A **law** describes what happens in a given situation.

BACKGROUND INFORMATION

An example of a scientific law in earth science is the law of horizontality. The law states that sedimentary rocks are deposited in horizontal layers. An example of a theory in earth science is the theory of plate tectonics. This theory states that the earth is divided into a number of plates that move and interact with each other.

A scientific law applies to many situations and explains what has occurred. A law does not, however, explain why events occur. For example, the law of gravity allows you to predict that a book will hit the floor if you drop it from your desk. However, a theory would help you explain to your teacher why the book is on the floor.

Using a Scientific Method The steps that scientists use to answer questions are called a **scientific method.** Scientists do not always use these steps in the same order. The order chosen depends on the goal of the scientists.

Scientists are not the only people who use a scientific method. A scientific method is also used by historians, police detectives, and even football coaches. Examine Figure 1–9 to see how a scientific method might be used to plan strategies for a high-school football game. Using the steps of a scientific method, try to think of labels for each of the drawings.

The football coach in the illustration found that his hypothesis did not lead to a successful game. Scientists experience similar frustrations. Often experiments fail to support the scientists' hypotheses. Scientists know that they must conduct many experiments. They may revise a hypothesis several times before they completely understand the problem they are studying. This may take years of work and experimentation.

○ *What is a hypothesis?*
○ *How is data usually obtained?*

DISCOVER

Using a Scientific Method

Think of something that you do every day, such as getting from your home to your school. Now pretend that you have never done this before. Write out the steps that you would use to solve this "problem" by using a scientific method.

Figure 1–9. How does this situation illustrate a scientific method of problem solving? **1**

1 The coach has formed a hypothesis, his game plan, and the team has experimented by carrying out the plan. Data collected, the score, shows that the hypothesis needs to be modified because it was not successful.

Section 2 Methods of Scientific Study 9

DISCOVER:
Using a Scientific Method

Thinking Skill *(Problem Solving)*

Answers will vary, but the students should have a goal, conduct experiments, gather data, and then draw a conclusion; they should show understanding of a scientific method.

___

DISCUSSION Have the students study the sequence of pictures in Figure 1–9 and relate them to a scientific method. ***Thinking Critically:*** Ask the students to suggest what the coach might do since his hypothesis did not lead to a successful conclusion. (He should form one or two new hypotheses and test them until his results are successful.) Science is not a static field; it is always changing.

SKILL *(Collecting Data)* Give small groups of students wooden building blocks, about 4 cm to 8 cm long. Have each group measure, as accurately as possible, all dimensions in centimeters and then in millimeters. Then have the students compare their measurements. (They should find that millimeter measurements are more accurate.) ***Thinking Critically:*** Ask the students why results may vary with the one doing the measuring. (Because some measurement is approximation, and the smaller the unit of measure, the greater the accuracy will be.)

Answers–1.3

○ A hypothesis is a prediction, based upon previous observations, about what will happen in a certain situation.

○ Data is gathered through observations using the senses.

Objectives

- Recognize definitions of new terms.
- Write summary sentences for concepts.
- Use summary sentences to outline a chapter.

Discussion Reading science can be difficult because new terms and new concepts are used frequently. Have the students read the text on page 17 and pick out sentences that contain definitions. (They can define *mass, balance, weight,* and *newtons,* for example.) Ask the students for the main concepts presented on page 17. (mass and weight) ***Thinking Critically:*** Ask students why it is helpful to be able to recognize definitions and main concepts. (It makes the text easier to understand, and it is easier to learn the new concepts when you can recognize them easily.)

Answers to Application

1. Boldface type; use of the term *called*
2. Answers will vary, but the words for Section 1 include *hydrology, oceanography, meteorology, astronomy, science,* and *fossils.*

BACKGROUND

Reading science books and magazines is often a challenge because each paragraph may contain both new concepts and new vocabulary terms. The chapters in this textbook are designed with special features to help you learn vocabulary and understand concepts easily and quickly.

PROCEDURE

1. VOCABULARY New vocabulary terms are highlighted in several ways in this textbook. A list of terms is presented at the beginning of each section as an introduction to vocabulary that might be new to you. When the term is defined in the chapter, it is in **boldface type.** The name of an unfamiliar object or process is in *italic type.*

   Sometimes there are signals that a definition is coming. Definitions are often written using defining verbs like *is* and *are.* Words such as *called, means,* and *explained as* are also good clues. The definition may appear before or after the term in the sentence. For example, the definition of geology might be written:

   A. "**Geology** is the study of the solid earth."

   or

   B. "The study of the solid earth is called **geology.**"

   To help you even more, new science terms are listed at the beginning of each section and at the end of the chapter. The page number on which the term is defined follows each term. How many new terms are there in Section 1?

2. CONCEPTS Each chapter develops a group of related ideas, or concepts. You can use the headings in each chapter to help you understand the concepts presented. The headings are presented in outline form on the first page of each chapter.

   The technique of writing summary sentences for concepts may help you understand the material being presented. These summary sentences provide you with an expanded outline of the chapter. Often the statements in your expanded outline include definitions of new terms. Use the summary statements in the Chapter Review to check your outline.

APPLICATION

1. Find the definition of *geology* in Section 1. What clues were used to help identify the new term?
2. Locate three other definitions in Chapter 1. Copy these terms and their definitions. Then rewrite each sentence. Make sure each term is still defined.
3. Look at the first section of Chapter 1. An expanded outline of Section 1 follows.

A. What Is Earth Science?

 1. The Earth Sciences

   a. The study of the solid earth is called geology. Some of the topics geologists study are minerals, rocks, fossils, and processes that change the earth.
   b. Hydrology is the study of the water on land.
   c. Oceanography is the study of the oceans.
   d. Meteorology is the study of the earth's atmosphere.
   e. Astronomy is the study of the planets, the sun, and the stars.

 2. Related Sciences

   a. Biology is the study of living things.
   b. Chemistry is the study of the composition of matter.
   c. Physics is the study of matter and energy.

---

4. Practice outlining by completing the outline for Chapter 1.

Figure 1–10. Some models, unlike this ship, do not look like the real thing. In science, it is more important that models act like the real thing. Why do you think that this is so? **1**

1 So that data about real situations can be collected.

1.4 Models in Science

If you have ever built a model plane, boat, or car, you know that a lot of time is involved. You want your model to be perfect! Sometimes your goal is not just to have your model look like the real thing, but also to have it work in the same way as the real thing. A **model** is something scientists create to help them understand how things work. With these models, scientists can study objects and events that would be difficult to study otherwise.

Scientists often use a theory to create a model. Models show the relationships that scientists observe as they gather data. Such models are useful for explaining structures and processes, especially if an object is too large or too small to be observed directly. Models are continually refined and changed as more information is gathered. A good model allows scientists to predict what will happen in experiments that have not yet been performed. Today scientists use many computer models.

Consider how the model of the solar system has changed. In about 150 B.C., scientists constructed the first mathematical model of the solar system with the earth in the center and the sun orbiting around the earth. Since then, the model of the solar system has been refined many times as more information has been obtained.

> **A MATTER OF FACT**
>
> Sometimes models require special units of measurement. For example one astronomical unit is equal to the distance from the earth to the sun, or about 150 000 000 km, or 150 million km.

1.4 Models in Science

DISCUSSION Have the students discuss how a model differs from the real object. (size, materials, location) *Thinking Critically:* Ask what must be the same about a model and the real thing. (The relative sizes of the parts must be the same; that is, if part A is twice as large as part B in the real object, then part A must be twice as large as part B in the model. The operation must also be the same, or cause and effect must also be the same.) You may wish to point out that computers have increased the ability to construct and test models without the expense of building them.

EXTENSION (Tie-in/Mathematics) The packaging box for a model usually states the scale to which it has been made, 1/264 or 1/60, for example. Maps and an architect's plans are drawn "to scale." *Thinking Critically:* Ask the students to calculate the distance from Boston to Indianapolis if the distance on the map is 29 cm and the map is 1 cm = 50 km. (1450 km) Then ask them to calculate how far apart on the map two cities would be if they are actually 900 km apart. (18 cm)

DISCUSSION Have the students study the models of the solar system. ***Thinking Critically:*** Ask the students what kind of data scientists had to collect before making this model. (the size of the planets, the number of planets, the shape of the orbits)

EXTENSION Scientists may use models as part of a scientific method. ***Thinking Critically:*** Have the students decide when a scientist might make a model as part of an investigation and how it could help. (Since models show relationships that scientists observe as they collect data, it would be made after an experiment. It could help the scientists to understand the results that were observed and to possibly make new predictions.)

Figure 1–11. Early models of the solar system were Earth centered; that is, Earth was believed to be the center of the solar system. Because of the work of scientists during the sixteenth century, the earth-centered model was replaced by the sun-centered model of the solar system.

The present model of the solar system places the sun in the center with the planets revolving around the sun. Space travel has added more data to refine this model.

○ ***Why are models used in science?***
○ ***Why do models change?***

Section Review

READING CRITICALLY

1. What is the difference between a theory and a law?
2. Why is it necessary for scientists to do more than one experiment to form a valid conclusion?

THINKING CRITICALLY

3. Think of a problem, and tell how you might solve it using a scientific method.
4. List the control and variables for an experiment you might conduct to answer a question about temperature.

12   Chapter 1   Studying the Earth

SECTION REVIEW

Summary Have the students name the parts of a scientific method and explain each one. Ask them to explain how a scientist uses a model.

Reinforcement Have the students find definitions in the section of the "New Science Terms" on page 7.

Answers—1.4

○ Models are used to help explain what has been observed or to present relationships between observed data.

○ Models represent the data observed. If data is gathered that does not fit the model, then the model must be revised to accommodate the additional data.

Answers to Section Review

Reading Critically

1. A law explains what has happened; a theory explains why events happen.
2. A series of experiments will eliminate most errors due to chance alone. This will allow more confidence when drawing conclusions.

continues

3 Measurements in Science

SECTION OBJECTIVES
After completing this section, you should be able to:
- **Measure** length, volume, mass, and weight using SI.
- **Calculate** the density of an object.

NEW SCIENCE TERMS
volume
mass
weight
density

1.5 The International System of Units

As scientists gather information in support of a hypothesis, they collect two types of data. The first type of data is descriptive. This data is based on observations made directly by the senses of sight, touch, smell, hearing, and taste.

The other type of data is measurable. As scientists study a problem, they may take measurements of distance, volume, time, or mass. The quality of any scientific effort depends on the correctness of both types of data.

Although there are several different systems of measurement, scientists around the world have agreed to use one system. This system is called the *International System of Units (SI)*. Using SI allows scientists to relate their research results in measurements understood by scientists everywhere. What do you think would happen if scientists in every country used a different system for measuring common objects?**1** Some of the SI units are shown in Table 1–1.

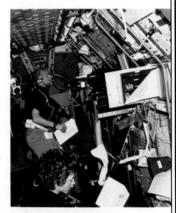

Figure 1–12. These scientists are collecting data from an experiment on board the space shuttle. They are recording the data in SI units. Why do scientists use SI units to record data? **2**

TABLE 1–1: SOME SI UNITS		
SI Unit	**Symbol**	**Property Measured**
meter	m	length or distance
liter	L	liquid volume
cubic meter	m^3	volume
kilogram	kg	mass
newton	N	weight
kelvin degrees Celsius	K °C	temperature

Figure 1–13. For most laboratory work, temperature is measured in degrees Celsius.

1 They could not communicate data to one another.
2 so that everyone can understand their data measurements

SECTION FOCUS

Section Overview The International System (SI) of units is presented in this section. The multiples and fractions of the basic units are also outlined. The methods for calculating the volume of a regularly and an irregularly shaped object are presented. The concepts of mass, weight, and density are compared.

Section Objectives For a list of section objectives, see pupil's edition page.

New Science Terms For a list of new science terms in this section, see pupil's edition page. You may wish to tell students that they will be learning the scientific meanings for terms that may already be familiar to them.

SECTION DEVELOPMENT

1.5 The International System of Units

DISCUSSION Direct the students' attention to Table 1–1. Have the students study the table of SI units so that they will know the symbol for each unit and the property measured for each unit. ***Thinking Critically:*** Have the students suggest situations when these units might be used. (Some answers might be: meter—buying a rug for a particular room; liter—preparing a punch recipe; cubic meter—making concrete for a patio.)

Thinking Critically

3. Answers will vary but should include the following steps: identifying a problem, stating a hypothesis, gathering data, and drawing a conclusion.
4. Answers will vary, but one variable might be the amount of heat that is applied to the object, while a control might be using the same type of material throughout the experiment.

BACKGROUND INFORMATION

"SI" stands for *Systeme Internationale,* which is French for "International System." The French initials have been used ever since the system was proposed in France in 1670. SI will be used exclusively in this text. Topographic maps are the only exception in which English measurements will be used.

DISCUSSION The students should be familiar with the basics of the decimal system from math class. Read through Table 1-2 with the students so they can pronounce all the prefixes. Explain to the students that in everyday life, they will probably only use *kilo-, centi-,* and *milli-* with the basic unit names. Ask the students what unit names are made by adding those three prefixes to *liter, meter,* and *gram.* (kiloliter, centiliter, and milliliter; kilometer, centimeter, and millimeter; and kilogram, centigram, and milligram)

THINKING SKILL (*Drawing Conclusions*) Have the students carefully study the top half of the left column of Table 1-2. Point out that the small raised numerals are called exponents; that is, in 10^2, 2 is the exponent. Ask how the exponent for each 10 is related to the number of zeros to the left of the equal sign. (The number of zeros is equal to the value of the exponent.) Ask what exponent could be used in 10 = 10. (1) Following the pattern, have the students decide what the next line in the table should be. Ask what the negative exponent means. (The negative exponent shows the number of places to the right of the decimal point.) You may wish to refer students to the article on scientific notation on page 560 of the Reference Section.

Figure 1-14. SI units are easier to imagine if you think of common objects. For example, a softball bat is about 1 m long and most soft drinks come in 2-L bottles.

The basic unit is not always appropriate for common measurements. For example, distance measurements in SI are in meters. A softball bat is nearly a meter long. However, many measurements must be made of distances much shorter than or much longer than a meter. To measure a distance shorter than a meter, you start with the basic unit, the meter. Then you add a prefix to indicate the fraction of the meter needed.

SI is based on the decimal system, or units of ten. Each prefix indicates some fraction or multiple of ten. Some common prefixes used in SI are shown in Table 1-2.

TABLE 1-2: SI UNIT PREFIXES				
Multiplication Factor	**Prefix**	**Symbol**	**Pronunciation**	**Term**
$1\ 000\ 000\ 000\ 000\ 000\ 000 = 10^{18}$	exa	E	EHKS uh	one quintillion
$1\ 000\ 000\ 000\ 000\ 000 = 10^{15}$	peta	P	PEHT uh	one quadrillion
$1\ 000\ 000\ 000\ 000 = 10^{12}$	tera	T	TEHR uh	one trillion
$1\ 000\ 000\ 000 = 10^{9}$	giga	G	JIHG uh	one billion
$1\ 000\ 000 = 10^{6}$	mega	M	MEHG uh	one million
$1\ 000 = 10^{3}$	kilo	k	KIHL uh	one thousand
$100 = 10^{2}$	hecto	h	HEHK toh	one hundred
$10 = 10^{1}$	deka	da	DEHK uh	ten
$0.1 = 10^{-1}$	deci	d	DEHS uh	one tenth
$0.01 = 10^{-2}$	centi	c	SEHN tuh	one hundredth
$0.001 = 10^{-3}$	milli	m	MIHL uh	one thousandth
$0.000\ 001 = 10^{-6}$	micro	μ	MY kroh	one millionth
$0.000\ 000\ 001 = 10^{-9}$	nano	n	NAN oh	one billionth
$0.000\ 000\ 000\ 001 = 10^{-12}$	pico	p	PEEK oh	one trillionth
$0.000\ 000\ 000\ 000\ 001 = 10^{-15}$	femto	f	FEHM toh	one quadrillionth
$0.000\ 000\ 000\ 000\ 000\ 001 = 10^{-18}$	atto	a	AT oh	one quintillionth

14   Chapter 1   Studying the Earth

DEMONSTRATION

For a demonstration of measuring in metric, see page xxiiie preceding this chapter.

For instance, if you needed to measure the length of your little finger, you would certainly need a unit shorter than a meter. Even a decimeter, which is one-tenth of a meter, would be too long. You would probably need to use the unit for one-hundredth of a meter (0.01 m)—the centimeter (cm). What unit would you use to measure the total length of interstate highways in your state? Why? **1**

These same prefixes are also used for measurements of volume and mass. For instance, you might need less than a liter of water to conduct an experiment. The best unit might be the milliliter—one-thousandth of a liter. What unit would you use to measure the mass of a blue whale? Why? **2**

○ *Why do all scientists use the International System of Units?*
○ *What units would you use to measure your textbook? Why?*

1.6 Measuring Volume

All objects take up space. The total space that an object occupies is its **volume.** Consider the volume of a shoe box. A shoe box has height, length, and width. Since a shoe box is rectangular in shape, its volume could be calculated by multiplying its dimensions.

Sample Problem

If you measured a shoe box and found the dimensions to be 30 cm, 15 cm, and 10 cm, what would the volume be?

Write the equation: Volume = length × width × height
$$V = l \times w \times h$$

Substitute the values of l, w, and h into the equation:
$$V = 30 \text{ cm} \times 15 \text{ cm} \times 10 \text{ cm}$$

Solve the equation: $V = 4500 \text{ cm}^3$

Notice that the unit for volume is cubic centimeters (cm^3). This is because the units are multiplied as well as the numbers.

The standard unit for measuring liquid volume is the liter. Since liters and cubic meters are both measurements of volume, their relationship can be calculated. A cube that is 1 cm × 1 cm × 1 cm (1 cm^3) will hold exactly one milliliter of liquid; therefore, one cubic centimeter equals one milliliter ($1 \text{ cm}^3 = 1 \text{ mL}$).

1 the kilometer; because it is used to measure large distances
2 the kilogram; because it is used to measure large masses

Section 3 Measurements in Science **15**

DISCOVER

Making SI Measurements

Pick out 10 objects to measure. For each object, pick a unit that you think would be appropriate and the tool that you would use to make the measurement. Now guess what you think the measurement will be. Then make the actual measurements, and check your estimates. Try the same thing with 10 new objects and see if you improve your skill at estimating.

DISCOVER:
Making SI Measurements

Thinking Skill *(Predicting)*

Ask the students when it would be acceptable to use estimated measurements. (When you are buying paint to paint a room, you can estimate the surface area of the walls; when you are buying food for a meal, you can estimate how much food each person will eat.) Discuss with the students that when precision is required, you must have measurements as exact as possible—when deciding if a tractor-trailer truck will clear a bridge, you must not be off by even a cm or the truck might hit the bridge. When measuring cross-beams for the roof a house, they must all be exactly the same.

DISCUSSION After reviewing the text with the students, give some other examples and let them choose the appropriate units. Some examples might be the width of a chair (cm), the contents of a jelly jar (mL), the mass of a cherry (g), the mass of a watermelon (kg), the length of the Mississippi River (km), and the contents of a picnic jug (L).

1.6 Measuring Volume

DISCUSSION Use three or four rectangular objects to review the dimensions of length, width, and height. Also use a cube to point out that the three dimensions are all the same. Direct the students' attention to the photograph and the sample problem. Then review the formula for finding volume. You may wish to give another example, such as a packing box with dimensions of 1.1 m, 0.7 m, and 2.1 m. ($V = 1.617 \text{ m}^3$) *Thinking Critically:* Ask the students what they would need to do before they could find the volume of a box that is 1m by 24 cm by 50 cm. (The units of measure would all have to be the same: 100 cm by 24 cm by 50 cm, or 1 m by 0.24 m by 0.5 m.)

Answers—1.5

○ Scientists can communicate their research to others easily because the same system of measurement is being used by all scientists.

○ cm—Since the measurement is linear, meter would be the basic unit. Due to the small size, cm would be the most reasonable subdivision to use.

DISCUSSION Review with the students the meaning of *displacement*—"being removed from the usual place, taking the place of something else." The shoes move an equal volume of water from its usual place, the shoes take the place of an equal volume of water. ***Thinking Critically:*** Have the students discuss why you cannot use $V = l \times w \times h$ to find the volume of shoes or sneakers or a rock or a toy bear. (You cannot get a single measure for length or width or height.)

ACTIVITY: Finding Volume

Skills   *(Measuring, Calculating)*

Preparation of Materials Water that has been dyed with vegetable coloring may be easier for the students to see.

Hint

Remind the students that their eyes should be level with the surface of the liquid when they are reading the measurement in a graduate. To measure the volume of the cork accurately, push it just below the surface of the liquid with the point of a pencil.

Answers to Conclusions/ Applications

1. Answers will vary depending on the size of the samples the teachers use.
2. An object will displace a volume of water equal to its own volume.

1 No, the volume of the box is greater.
2 You would have to use displacement, which would probably ruin the shoes.

Do you suppose the shoes that come in a box have the same volume as the box?**1** How could you find the volume of the shoes?**2** In order to calculate accurately the volume of an irregularly shaped object, like a pair of shoes or a rock, you use a method called *displacement*. The object is placed in a container with a known volume of water. The water level in the container will rise, or be displaced by the object. The amount of water displaced by the object will be equal to the volume of the object.

Figure 1–15. These two vessels, a flask (right) and a graduate (left), may be used to measure the volume of liquids.

○ *What is the formula for finding the volume of a regularly shaped object, such as a box?*
○ *What method is used to determine the volume of an irregularly shaped object?*

ACTIVITY: Finding Volume

How can you measure the volume of an irregularly shaped solid?

MATERIALS (per group of 3 or 4)

graduate, 50 mL; water; string; galena; calcite; granite; cork; metric ruler

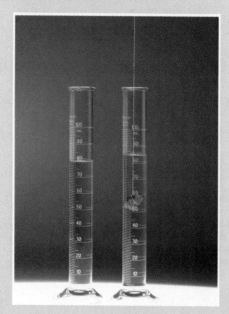

PROCEDURE

1. Copy the data table.

TABLE 1: VOLUME BY DISPLACEMENT

Object	Starting Volume	Final Volume	Volume of Object
Galena			
Calcite			
Granite			
Cork			

2. Fill the graduate about half full with water. Record the volume on your data table.
3. Tie a string around the galena and gently lower it into the water. Record the new volume of the water on your data table.
4. Subtract the starting volume from the final volume to find the volume of the galena.
5. Repeat steps 2, 3, and 4 using the other three specimens. Describe how you handled the cork.

CONCLUSIONS/APPLICATIONS

1. The faces of galena and calcite are nearly rectangular. Measure their dimensions with a ruler and calculate their volumes. How close are the measured volumes to the calculated volumes? Why don't they match exactly?
2. Explain how the displacement method works.

Answers—1.6

○ $V = l \times w \times h$

○ The displacement method is used to determine the volume of an irregularly shaped object.

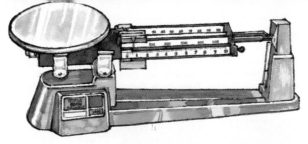

Figure 1–16. The balance (left) is used to measure mass, whereas the spring scale (right) is used to measure weight.

1.7 Comparing Mass, Weight, and Density

Have you ever heard of an object being described as "massive"? A mountain is a massive natural structure. The Pentagon in Washington, D.C., is a massive building. Jupiter is a massive planet.

Mass An object is massive because it has a lot of matter. **Mass** is the amount of matter an object contains. All objects, whether they are solids, liquids, or gases, have mass. In SI, mass is measured in kilograms. A balance is the instrument scientists use to measure mass. Figure 1–16 shows a common laboratory balance.

Weight Many people confuse mass with weight. **Weight** is the measure of the force, or pull, of gravity on the mass of an object. A spring scale, like the one shown in Figure 1–16, measures weight. The SI unit for weight is the newton. All forces are measured in newtons.

The earth's gravitational force decreases as you move away from the earth. Astronauts have traveled far enough from the earth to be almost weightless. On the moon the gravitational force is about one-sixth of the gravity on Earth. If an astronaut weighed 780 N on the earth, on the moon the same astronaut would weigh 130 N. Does this mean that the astronaut's body has lost mass? Explain your answer. **1**

1 The mass of the astronaut has not changed, but the force of gravity on the moon is one-sixth that of gravity on Earth. Therefore, the weight of the astronaut is less on the moon.

A MATTER OF FACT

The *standard mass* is a carefully machined one-kilogram cylinder of platinum-iridium alloy stored in a special vault in the International Laboratory of Weights and Measurements at Sèvres, near Paris, France.

Figure 1–17. These astronauts are experiencing free fall in a training airplane, while preparing for a space shuttle flight.

17

1.7 Comparing Mass, Weight, and Density

DISCUSSION The students have probably seen a photograph similar to Figure 1–17 showing astronauts drifting in free fall in the cabin of a space vehicle or in an experimental chamber. ***Thinking Critically:*** Ask the students why the astronauts can walk on Earth but float around in a space vehicle. (The gravitational force that holds them on Earth is not very strong in outer space.) Have the students look at the instruments in Figure 1–16 and explain the purposes of each.

SKILL *(Measuring)* Have the students find the mass of several objects on a laboratory balance. Some of the objects should be quite small, and some of the objects should be of different densities.

BACKGROUND INFORMATION

The difference between mass and weight is easier to understand if you think of an object being on the moon, where the pull of gravity is about one-sixth of what it is on Earth. For example, an object shown to have a mass of 6 kg on Earth would still have a mass of 6 kg on the moon. However, if the same object is put on a spring scale on Earth and shows a weight of 6 N, on the moon it would only show a weight of 1 N. The mass stays the same, but the measure of weight is relative to the pull of gravity on it.

A *newton* is equal to 100 000 dynes. A *dyne* is the unit of force that, if applied to a mass of 1 gram, would give it an acceleration of 1 centimeter per second for every second the force was applied.

DISCUSSION Have the students study the photograph and describe what it shows. (Two objects are on a balance scale; the object on the left has the greater mass, since that side of the scale is lower.) The caption of Figure 1–18 states that the two objects have the same volume. Point out to the students the definition of *density* in the first paragraph. Then ask which object has the greater density. (The object on the left has the greater density, since it has more mass per volume.)

EXTENSION You may wish to tell students the old joke/riddle: Which weighs more: a newton of feathers or a newton of lead? *Thinking Critically:* Have the students discuss why most people's first reaction is to say "a newton of lead." (Lead is denser than feathers, and people think of lead as weighing more than feathers. The trick is realizing that two objects each weighing a newton are exactly the same weight.)

SECTION REVIEW

Summary Review the basic units of the International System of Units, commonly called the metric system. Have the students name the property that each unit measures. Review the methods for finding the volumes of regularly and irregularly shaped objects. Have the students tell the difference between mass, weight, and density.

Reinforcement Have groups of four or five students write definitions of terms and properties to be measured on small sheets of paper and place them in a container. The groups can exchange containers and have a Jeopardy-like game, naming the defined term or required SI unit.

Figure 1–18. Both objects on the balance have the same volume; however, the lead has more mass than the wood.

Density Another property of matter related to mass is density. **Density** is the measure of the mass contained in a certain volume of matter. For example, if a brick and a block of wood are the same size, they would have the same volume. They would have different masses, however, so they would also have different densities.

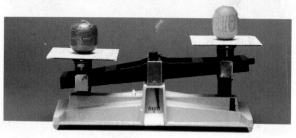

The density of a substance does not change. A large piece of copper will have the same density as a small piece. Figure 1–18 shows that the densities of different substances vary greatly. Since density is always the same for a substance, it is a characteristic property that can be used to identify the substance.

The greater the mass of an object, the greater the pull of gravity on that object. Suppose you have two objects with the same volume. The object that is more dense will be heavier than the less dense object. A bowling ball, for example, is heavier than a melon of the same size because the bowling ball is denser. Mass, weight, and density are all related quantities.

○ *What SI unit is used to measure mass?*
○ *What is density?*

A MATTER OF FACT

Early measurements were based upon crude standards. For instance, the span of a king's hand was once used as a standard for length.

Section Review

READING CRITICALLY

1. A scientist recorded a measurement of 10 g. Was this a measurement of mass, weight, or density?
2. Explain how you would find the volume of an irregularly shaped rock.

THINKING CRITICALLY

3. Compare a 1-cm^3 sample of lead and a 1-cm^3 sample of wood. Do they have the same mass? Which sample has the greater density? Explain.
4. Why is it important for scientists to use the same system of measurements?

18   Chapter 1   Studying the Earth

Answers—1.7

○ Kilograms or grams are used to measure mass.
○ Density is the measure of the amount of mass per a certain volume.

Answers to Section Review

Reading Critically

1. It is a measure of mass.
2. You would find the volume by displacement, that is, after lowering the rock into a container with a known volume of water, you record the volume of the water displaced by the rock.

Thinking Critically

3. The lead has a greater mass and a greater density. Since the volumes of the two substances are the same, the difference in their masses indicates that there will also be a difference in their densities.
4. Scientific knowledge has increased at such an amazing rate because scientists have shared their findings with each other. If they did not use the same standard measurements, their ability to share their research would be greatly limited.

BIOGRAPHIES: Then and Now

JAMES HUTTON (1726–1797)

James Hutton lived most of his life in Edinburgh, Scotland. His interest in science led him to investigate many parts of his environment. As a result, he made major contributions to the fields of geology, agriculture, chemistry, physics, and philosophy.

As a scientist, Hutton was a careful observer. He studied the land and drew conclusions about the origin of the rocks and the processes that changed them. Eventually he constructed his "theory of the earth." This theory proved to be so important to the development of the field of geology that Hutton is now called the founder of modern geology.

Hutton believed that the interior of the earth was composed of lava, with the crust acting as a container for the lava. Volcanoes acted as safety valves, allowing the molten rock to escape from the interior. According to Hutton, as this lava came to the surface, the overlying sedimentary rocks were tilted.

In his book *Theory of the Earth*, Hutton suggested a theory that the earth was much older than most scientists then thought. He believed that earth processes were continuous cycles. As support for this theory, he described the rock-forming processes that had occurred. He hypothesized that the same processes that can be observed today were working in the past and will continue to work in the future. Hutton's theory has been so thoroughly tested and so well supported by experimental data that it forms the basis for the science of geology.

LUIS ALVAREZ (1911–1988)

Luis Alvarez was an American physicist. He was awarded the 1968 Nobel Prize in physics. In 1980 he discovered that there was an unusually large amount of the element iridium in certain layers of rock in Italy. Iridium is very rare in rocks formed on Earth. This layer was deposited about 65 million years ago, at about the time of the disappearance of the dinosaurs.

Alvarez was surprised that such a concentration of iridium would appear in rocks of this age. Once he had observed this concentration in Italy, he hypothesized that there might be similar concentrations in rocks of the same age in other places. He discovered that there were other similar deposits.

This information led Alvarez to hypothesize that the dinosaurs, and other large animals, died because an asteroid collided with the earth. This event would have released so much dust into the atmosphere that the radiation from the sun would have been blocked for about three years. During this time, plant life would have been reduced and large animals would have starved. Although Alvarez's hypothesis seems reasonable, it is not accepted by all scientists.

An extension of the hypothesis proposed by Alvarez has recently been applied to the possibility that a similar situation migh develop if a large-scale nuclear war should occur. Dust and smoke produced by many nuclear explosions could block out the sun and lower the earth's temperature enough to cause the extinction of many plants and animals.

Section 3   Measurements in Science   **19**

BIOGRAPHIES: **Then and Now**

Discussion   Have the students read the two biographies and discuss how each person studied the earth to find out information about the past.

Thinking Skill   *(Generalizing)* After the students have read the two biographies, ask them why they think Hutton's theory became an important law in geology. Then ask why they think Alvarez's hypothesis is not accepted by all scientists.

Extension   Interested students might be encouraged to research further the life and work of Hutton or Alvarez or one of the scientists listed below.

Abraham Gottlob Werner

Sir James Hall

William Smith

Louis Agassiz

Charles Van Hise

Adolph Knopf

James D. Dana

INVESTIGATION 1:
Determining Density

Skills *(Estimating, Weighing, Measuring, Calculating)*

Preparation of Materials

In order for densities to be predictable, use pure mineral specimens.

Hints

The students can use the calculation method for galena, calcite, and the bar magnet. They will have to use the displacement method for granite and the glass rod. Advise students to be very careful in their measuring activities.

Answers to Procedure Questions

1. Students may not all choose the bar magnet as having the greatest density or the glass rod as having the lowest density. But previous experience should help them in forming hypotheses.
2. Students should complete the table.
3.–5. Answers will vary according to samples provided.

Answers to Analyses and Conclusions

1. glass rod
2. bar magnet
3. Answers will vary, but should show understanding of density.
4. No, the density will remain the same.
5. None of the objects float on water; anything less dense than water will float.

Answer to Application

Ice will float in a glass of water because frozen water is less dense than liquid water. (You may have to explain that the density of many substances changes as the temperature changes.) Since the students can see from the photograph that oil floats on water, they should conclude that oil is less dense than water. Therefore, ice sinks in the glass of oil because it is more dense than the oil.

INVESTIGATION 1: Determining Density

PURPOSE

To calculate the density of several objects

MATERIALS (per group of 3 or 4)

Balance
Galena
Calcite
Granite
Bar magnet
Glass rod
Metric ruler
String
Graduate, 100 mL
Paper towels

PROCEDURE

1. Look at, but do not pick up, the objects to be tested. Write a hypothesis stating which object you think will have the greatest density. Record your hypothesis on your paper. *Which object do you think will have the lowest density?*
2. Make a data table like the one shown.
3. Using a balance, determine the mass of each object. Record the masses in the data table.

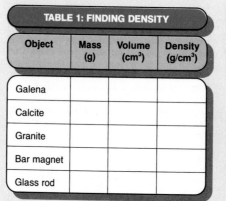

Object	Mass (g)	Volume (cm³)	Density (g/cm³)
Galena			
Calcite			
Granite			
Bar magnet			
Glass rod			

TABLE 1: FINDING DENSITY

4. Determine the volume of each object. For some objects you may measure the dimensions and multiply. For others you will have to determine the volume by the displacement method described in the activity on page 16.
5. Calculate the density of each object. (Density = mass ÷ volume.) Your answer will be in g/cm³ or g/mL. Since 1 mL equals 1 cm³, convert all answers to g/cm³ for easy comparison.

ANALYSES AND CONCLUSIONS

1. Which object has the lowest density?
2. Which object has the highest density?
3. Were your hypotheses correct? Explain. If you were incorrect, discuss what properties might have misled you.
4. If you doubled the size of the galena sample, would you double the density? Explain.
5. Explain how this activity proves that the density of water is less than that of calcite, granite, and galena.

APPLICATION

Look at the two photographs. Using your knowledge of density, explain why the ice floats in a glass of water, but sinks in a glass of oil.

GOING FURTHER

Make a list of five common materials of different densities found in your home or classroom. Before finding their densities, try to arrange them in order of increasing density. Then determine the density of each to see if you listed them in the correct order.

Answer to Going Further

Answers will vary according to the materials chosen, but answers should show that students have improved their knowledge of density. The students will probably "eyeball" the results by basing their predictions on how heavy the objects "feel." See if any other basis for predictions can be made.

1 REVIEW

SUMMARY

- The major fields studied in earth science are geology, hydrology, oceanography, meteorology, and astronomy. (1.1)
- Geology is the study of minerals and rocks, and of the processes that shape the earth. (1.1)
- Hydrology is the study of the underground and surface waters of lakes and streams. (1.1)
- Oceanography is the study of the oceans. (1.1)
- Meteorology is the study of the atmosphere and the weather of the earth. (1.1)
- Astronomy is the study of the planets, the sun, and the stars. (1.1)
- The earth sciences apply principles from the basic scientific fields of biology, physics, and chemistry. (1.2)
- Biology is the study of living organisms. (1.2)
- Chemistry is the study of the composition of matter. (1.2)
- Physics is the study of matter and energy. (1.2)
- Problems can be studied by using the investigative procedure called a scientific method. (1.3)
- Scientists often present theories in the form of models. (1.4)
- Scientific work uses the International System of Units. (1.5)
- The space that an object occupies is the object's volume. (1.6)
- The volume of an irregularly shaped object is measured by displacement. (1.6)
- Mass is the amount of material or matter in an object. (1.7)
- Gravitational attraction affects an object's weight but not its mass. (1.7)
- The density of an object depends on the relationship between the object's mass and volume. (1.7)

Write all answers on a separate sheet of paper.

SCIENCE TERMS

Correctly use each of the following terms in a sentence.

astronomy **(5)**
conclusions **(8)**
control **(8)**
data **(7)**
density **(18)**
experiment **(8)**
fossils **(5)**
geology **(4)**
hydrology **(4)**
hypothesis **(8)**
law **(8)**
mass **(17)**
meteorology **(4)**
model **(11)**
oceanography **(4)**
science **(5)**
scientific method **(9)**
theory **(8)**
variable **(8)**
volume **(15)**
weight **(17)**

SCIENCE QUIZ

Modified True-False

Mark each statement *true* or *false*. If the statement is false, change the underlined term to make the statement true.

1. The water in lakes is studied in the field of <u>geology</u>.
2. Scientists use a theory with more confidence than they use a <u>hypothesis</u>.
3. A kilogram equals 1000 <u>liters</u>.
4. The density of a sample of iron ore will be <u>less</u> on the moon than it is on Earth.
5. The International System of Units is based upon the <u>decimal</u> system.

continues

Chapter 1 Review   **21**

CHAPTER REVIEW

SUMMARY

The students may review the major concepts in the chapter by reading the summary statements. The statements are cross-referenced to the chapter sections to facilitate reinforcement of any concepts of which the students feel unsure. Encourage the students to work in pairs to quiz one another.

SCIENCE TERMS

The science terms to be reviewed are listed in alphabetical order, followed by the page reference. The sentence in which the science term is used should reflect an understanding of the definition of the term. Encourage the students to work in groups to compare the sentences they wrote and to revise their sentences if necessary to make them better.

SCIENCE QUIZ

Answers to Modified True-False

1. false, hydrology
2. true
3. false, grams
4. false, the same
5. true

continues

Answers to Multiple Choice

6. c
7. b
8. d
9. a
10. c

Answers to Completion

11. physics
12. conclusion
13. volume
14. mass
15. weight or force

Answers to Short Answer

16. Answers will vary but they should all be observations made by the senses rather than inferences or names of the parts of the object.
17. a. Fill a 100 mL graduate half full with water. Note the initial volume.
   b. Tie a string around the eraser and submerge it in the water.
   c. Measure the change in the volume of the water to find the volume of the eraser.
18. density = m/V = 1620 g/600 cm^3 = 2.7 g/cm^3

Answers to Writing Critically

19. A theory explains why things happen in a particular way. A scientific law applies to many different situations and explains what has occurred. Laws do not explain why events occur.
20. A meteorologist must explain how the exchange of energy and heat causes changes in weather patterns.

ANSWERS TO EXTENSION

1. The students' answers will vary. They might calculate the density of a block of wood and then cut the block in half. After calculating the density again, they should conclude that the size of the sample has nothing to do with the density.
2. Stories will vary but all should have the historian forming a hypothesis about the river pirate and then doing research to test the hypothesis.
3. Most materials in the building-supply store will have only common units. In a grocery store, products usually have both common and SI units. Conversions are difficult because

the common system is not based upon a uniform numbering system.
4. The reports should include at least the basic units for mass, volume, weight, length, and time.

SCIENCE QUIZ continued

Multiple Choice

Write the letter of the term that best answers the question or completes the statement.

6. A length that is 1/100 of a meter is a
   a) centigram.   b) millimeter.
   c) centimeter.   d) decimeter.

7. A kilogram is a measure of
   a) length.   b) mass.
   c) weight.   d) volume.

8. In order to compute the volume of an object, you must obtain measurements of
   a) length, height, and weight.
   b) mass and weight.
   c) mass, height, and length.
   d) length, width, and height.

9. The formation of clouds is studied in the field of
   a) meteorology.   b) geology.
   c) astronomy.   d) hydrology.

10. One liter is equal to
   a) 10 cm^3.   b) 100 cm^3.
   c) 1000 cm^3.   d) 10 000 cm^3.

Completion

Complete each statement by supplying the correct term.

11. Matter and energy are studied in the science of _____.
12. Usually, the step in a scientific method that follows experimentation is _____.
13. The cubic meter is a measure of _____.
14. The _____ of 1 cm^3 of water is 1 g.
15. A spring scale is used to measure _____.

Short Answer

16. Select a single object in your classroom and make five observations about it. Find out if your classmates can identify the object based on your list of observations.
17. Explain the procedure that you would follow to measure the volume of a heart-shaped eraser.

18. A block of metal 4 cm × 15 cm × 10 cm has a mass of 1620 g. What is the density of the block?

Writing Critically

19. Explain the difference between a scientific theory and a scientific law.
20. Why would a meteorologist need a strong background in physics?

EXTENSION

1. Design and build a model to demonstrate that the density of a substance will not change even if the amount of the substance is reduced.
2. Write a short story about a historian who has decided to research the life of a Mississippi River pirate who lived in the 1800s. Demonstrate in your story how the historian uses a scientific method in his or her effort to obtain material for the book he or she plans to write.
3. Visit a building-supply store and make a list of 20 items that are not measured by the use of SI. Visit a grocery store and note 20 products that use SI measurements. What difficulties might a person find in trying to convert the SI units to the system used in the United States?
4. Go to your school or public library and find out about the conference of international scientists that decided on the standard units of measurement for scientific research. In a written report, describe the basic units they decided on, and explain what the standard for each unit is and where it is located.

APPLICATION/CRITICAL THINKING

1. Many people are "afraid" to change to SI because they feel the conversion back and forth between SI and conventional measurements will be very difficult. Is this really a problem? Explain.

ANSWERS TO APPLICATION/ CRITICAL THINKING

1. Conversions will not be necessary. Once people use SI, or metric, units for a while, they will find the system easy and convenient to use.

2. Your uncle drives his restored 1957 car across the border into Canada. He immediately notices that the speed limits are in kilometers/hour (km/h). You know that there are 0.6 km/mile. Devise a simple chart that shows him the SI equivalents of driving 15 mph, 25 mph, 35 mph, 45 mph, 55 mph, and 65 mph.

FOR FURTHER READING

Harrison, J., ed. *Science Now.* New York: Arco Publishing, 1984. Sixty-five illustrated articles describe the way scientists have affected our lives and are shaping our future.

National Bureau of Standards, U.S. Department of Commerce. *The International System of Units (SI).* Special Publication 330, 1981. This publication provides descriptions and examples of using SI in everyday situations.

Trefil, J. *Meditations at Sunset.* New York: Scribner, 1987. In this book, the author explains that the laws of nature, on which scientists work for countless years in laboratories, can be seen all around us in the natural world. This book specifically explores the wonders of the atmosphere and explains many of the objects and phenomena seen in the night sky.

Challenge Your Thinking

This airplane is being built by a consortium, or group, of manufacturers from Britain, France, Germany, and Spain. How do you suppose they get all the parts to work together if they are made in four different countries? What problems in getting parts to fit might develop if the United States were part of this consortium?

Chapter 1 Review   **23**

ANSWER TO CHALLENGE YOUR THINKING

All four countries use the same specifications and the same units of measurement. Since the United States uses customary units of measurement, they would have to change to metric units or U.S.-made parts would not fit.

2. The numbers in the chart are approximations for convenient and easy conversions.

mph	15	25	35	45	55	65
km/h	25	40	60	75	90	110

PLANNING THE CHAPTER

Chapter Sections	Page	Chapter Features	Page	Program Resources	Page
Section 1: Models of the Earth	25				
2.1 Early Models **(B)**	25				
2.2 Modern Maps **(B)**	27	A Matter of Fact **Discover:** Making Maps **(B)** Section Review	27 28 29		
Section 2: World Maps	30			Critical Thinking **(H)** Cross-Discipline: Science and Language, *Obtaining a Map from the U.S. Geological Survey* **(B)**	TRB 3 TRB 3
2.3 Latitude and Longitude **(B)**	30	**Discover:** Finding Your Home Town **(B)** **Activity:** Finding Places on a World Map **(B)**	31 32		
2.4 Map Projections **(B)**	32	A Matter of Fact A Matter of Fact Section Review **Careers:** Surveyor, Geologist, Cartographer	32 34 34 35		
Section 3: Topographic Maps	36			Reading for Content: *Comparing and Contrasting to Understand What You Read* **(A)**	TRB 3 LM 175
2.5 Map Scales **(B)**	36	A Matter of Fact **Skill Activity:** Interpreting Map Scales **(B)**	37 38	Investigation 2.1: *Making a Topographic Map* **(B)**	TRB 5 LM 7
2.6 Contour Lines **(B)**	39	**Discover:** Drawing Contours **(B)** A Matter of Fact Section Review **Investigation 2:** Interpreting Topographic Maps **(B)**	39 40 41 42	Investigation 2.2: *Topographic Map Symbols* **(B)** Concept Extension: *Reading a Topographic Map* **(B)** Student Record Book: Textbook Investigations **(B)**	TRB 7 LM 9 TRB 3 LM 225 TRB 3
Chapter 2 Review	43			Vocabulary **(A)** Tests **(A)** Computer Test Bank 💻	TRB 2 LM 125 TRB 41

(LM) Laboratory Manual/Study Guide, **(TRB)** Teacher's ResourceBank™

B = Basic   **A** = Average   **H** = Honors

The coding Basic, Average, and Honors indicates sections or subsections that might be appropriate for different levels of learners. For additional suggestions regarding choice of topic and depth of coverage, see the Pacing Chart on pages T16–T20.

CHAPTER CONCEPTS, OBJECTIVES, AND ITEMS

Section	Concepts	Objectives	Science Terms
Section 1: Models of the Earth	■ As long ago as 2000 B.C., people constructed maps and models of the earth. **(2.1)** ■ Ptolemy was the first mapmaker to put north at the top of a map. **(2.1)** ■ Since 1807 the United States Coast and Geodetic Survey (now called the National Ocean Survey) has been making accurate maps. **(2.2)** ■ Bench marks are points where the exact elevation and location are known. **(2.2)**	■ **Describe** some maps used by early civilizations. ■ **Explain** how a map is a model. ■ **Identify** some modern techniques used in mapmaking.	cartography bench mark topographic maps geologic maps hydrologic maps
Section 2: World Maps	■ Latitude and longitude lines are used to indicate on maps the location of parts of the earth's surface. **(2.3)** ■ Lines of latitude are also known as parallels and lines of longitude are known as meridians. **(2.3)** ■ Map projections show a spherical Earth on a flat image. **(2.4)** ■ Many map projections show parts of the world in a distorted view. **(2.4)**	■ **Compare** and **contrast** latitude and longitude. ■ **Explain** the advantages and disadvantages of various map projections.	latitude longitude
Section 3: Topographic Maps	■ The representative fraction scale, the verbal scale, and the graphic scale are three ways to present the relationship between map distance and represented distance on the earth's surface. **(2.5)** ■ Although scientific work is in SI, many topographic maps are drawn with customary units of measurement. **(2.5)** ■ Contour lines show places of equal elevation above sea level. **(2.6)** ■ Topography is the shape of part of the earth's surface. **(2.6)**	■ **Relate** map scales to actual distances. ■ **Interpret** contour lines on a topographic map.	scale topography

Title	Page	Materials
Discover: Making Maps	28	*(per student)* paper, pencil, meter stick
Discover: Finding Your Home Town	31	*(per student)* paper, pencil
Activity: Finding Places on a World Map	32	*(per student)* paper, pencil
Skill Activity: Interpreting Map Scales	38	*(per student)* paper, pencil
Discover: Drawing Contours	39	*(per student)* modeling clay, paper, metric ruler
Investigation 2: Interpreting Topographic Maps	42	*(per group of 3 or 4)* topographic map of Boothbay quadrangle

TEACHING SUGGESTIONS

Section 1: Models of the Earth

Class Activity

Have the students talk about any kinds of models they collect and build. Ask them to describe the characteristics they look for in a model.

Outside Speaker

See if you can find a geologist or a surveyor to speak to your classes about his or her work.

Section 2: World Maps

Class Activity

After reading Section 2, have the students try to make map projections from a globe. If you have a clear globe with a light inside, the activity will be especially easy.

Field Trip

Visit a survey crew working in the field and have them explain and show how their work is used to improve the accuracy of maps.

Section 3: Topographic Maps

Demonstration: Making a Topographic Map

Purpose

To make a topographic map from a model

Background

A topographic map shows features of the surface of the earth. These features may be mountains, valleys, or rivers. The height of different features is shown by contour lines. In this demonstration you will make a topographic map of a model mountain.

Materials

Modeling clay
Food coloring
Meter stick
Aquarium and glass cover
Wax pencil

Procedure

1. Use the meter stick and the grease pencil to mark off centimeter lengths on one side of the aquarium. Start at the bottom and mark the centimeters to the top. The distance between marks represents the "contour interval" of the map. The contour interval is the difference in height of the points along two adjacent contour lines.
2. Make a mountain from modeling clay. Make an irregular shape, perhaps with two or more peaks of different heights. Make sure your model fits inside the aquarium.
3. Place the model inside the aquarium. Add water colored with food coloring until you fill the aquarium to the first centimeter mark near the bottom of the tank. Put a glass cover on the aquarium. Look straight down through the cover. With the grease pencil trace on the glass the outline of the mountain at the level of the water. This represents all points on the model at a height of 1 cm. The line might represent a contour line connecting all points 50 meters (m) above sea level. Now add more colored water to the aquarium until the level of the liquid is at 2 cm. Using the grease pencil, again trace a line on the glass, showing the contact points between the liquid and the clay model. Examine these two contour lines drawn on your map. If the lines are close together, the slope respresented between the lines is very steep. Add

the colored water until the aquarium is filled to the 3–cm mark. Then draw the third contour line.

4. Draw as many contour lines as possible for your mountain model. The bottom of the aquarium represents sea level. Number sea level zero. Then number the contour lines beginning with one. If your mountain has more than one peak, what is true of the contour lines? (The contour lines on different mountains will never connect.)

5. You have drawn a topographic map on the glass. Have the students examine your map. Ask them to examine each feature of the model and to explain how the topographic map identifies the features. Then have the students look at a commercial topographic map. How does it compare with the topographic map? See if the students can identify the features represented by the map. If the commercial topographic map is of your area, have students identify the earth structures with which they are familiar.

Questions to Ask the Students

1. What does a topographic map show? (*The features of the earth's surface, represented by lines indicating height above sea level.*)
2. Each line drawn was a contour line. What is a contour line? (*A contour line connects points of equal elevation.*)
3. What does each contour interval represent? (*A contour interval represents the difference in height between adjacent contour lines.*)
4. How do contour lines show that slopes are steep? (*Steep slopes are shown by contour lines drawn close together.*)
5. How can the map show several peaks of a mountain? (*Peaks are represented by close, concentric sets of contour lines.*)

CHAPTER 2

CHAPTER OVERVIEW

This chapter defines maps as models of the earth. The history of mapmaking, or cartography, is discussed. Topographic, geologic, and hydrologic maps are identified. Latitude and longitude are explained, and map projections are described. Additional detail on topographic maps includes discussion of map scales and contour lines.

Section 1: Models of the Earth In this section the history of cartography is outlined, including the contributions of Ptolemy. Modern mapmaking techniques are discussed, with emphasis on the work of the United States Geological Survey. The three basic types of maps used in earth science are described.

Section 2: World Maps This section details the use of lines of latitude and longitude and discusses mapmaking techniques for producing world maps. Map projections are introduced, with emphasis on the Mercator projection.

Section 3: Topographic Maps The widely used topographic map is used to develop an understanding of map scales and contour lines.

CHAPTER

Earth Science Models

For centuries, scientists and explorers have been gathering information about the earth's surface. Early civilizations considered the exploration of the earth's surface an exciting challenge. Even today, there are still unexplored regions of the earth that hold mysteries yet to be discovered.

The southern California coast

24

CHAPTER MOTIVATING ACTIVITY

To introduce some of the basic uses and characteristics of maps, bring to class a variety of maps to display. Inform the students that maps vary widely in their content. As some examples, display road maps, state maps, historical maps, weather maps, topographic maps, population maps, and even astronomical maps. Have the students discuss how to use a map and the details that are necessary to make a map usable. They should point out the need for a scale, names of locations, symbols, and compass directions. A point of reference is also necessary.

You may wish to follow this discussion with an activity that utilizes a map. Distribute regional or state road maps to groups of two or three students. Have the students write out sets of directions for driving between specified cities. They should mention route numbers and compass directions.

1 Models of the Earth

SECTION OBJECTIVES

After completing this section, you should be able to:

■ **Describe** some maps used by early civilizations.
■ **Explain** how a map is a model.
■ **Identify** some modern techniques used in mapmaking.

NEW SCIENCE TERMS

cartography
bench mark
topographic maps
geologic maps
hydrologic maps

2.1 Early Models

Try to imagine yourself as a Babylonian (bab uh LOH nee uhn) explorer in 750 B.C. You think that the earth is a flat disk completely surrounded by an ocean called *Bitter Waters*. You also believe that outside the disk are seven islands that link the world to an outer *Heavenly Ocean*, which is where the gods live.

The ancient map shown in Figure 2–1 is an example of one of the oldest surviving maps of the Mediterranean area. Even older maps were used to show land boundaries and personal property. These maps date back to 2200 B.C.

Early civilizations used a variety of mapmaking techniques. Inuits (Eskimos), for instance, cut shapes of coastal islands out of dark-colored animal skins. The shapes were then sewn onto a light-colored skin that represented the ocean. The Egyptians engraved maps on gold, silver, and copper plates to identify the

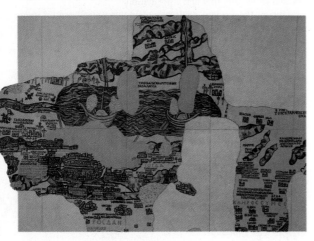

Figure 2–1. This is a photograph of one of the oldest known maps of the Mediterranean Sea.

Section 1   Models of the Earth   **25**

Section **1:** Models of the Earth

SECTION FOCUS

Section Overview This section describes early mapmaking and highlights the contributions of Ptolemy. It explains that a map is a model of the earth and discusses modern mapmaking techniques. The three basic types of maps are given, and the work of the United States Geological Survey (USGS) is identified.

Section Objectives For a list of section objectives, see pupil's edition page.

New Science Terms For a list of new science terms in this section, see pupil's edition page. You may wish to familiarize the students with these terms by having them read the terms aloud before they read the section.

SECTION DEVELOPMENT

2.1 Early Models

DISCUSSION Discuss some of the early world explorations and how little was known then about the size and physical characteristics of the world. Have the students study some ancient maps and try to identify the misinformation in them. *Thinking Critically:* Ask the students to suggest sources for this misinformation and to discuss the difficulties caused by the erroneous beliefs. (Answers will vary, but the students should report that earlier civilizations did not have the scientific tools and research to document their facts. Much of their knowledge was based on imagination and theory.)

DISCUSSION Have the students look at Figure 2–3 and locate the islands. *Thinking Critically:* Ask the students why it was important for people to make maps similar to this one. (Possible answers include for trading purposes, for exchanging ideas among the various island groups, and for having places to sail to in case of inclement weather or other natural disasters.) Review with the students Ptolemy's contribution to mapmaking.

Figure 2–2. The people of Micronesia made maps such as this to help them locate neighboring islands.

locations of gold mines and other valuable properties. The people of the Pacific Ocean islands made maps similar to the one in Figure 2–2 to indicate the location of nearby islands.

These ancient maps led to the development of **cartography** (kahr TAHG ruh fee), the science of mapmaking. In ancient times cartography was more of an art than a science. Mapmaking was based more on imagination, theory, and travelers' tales than on precise measurements. Ancient cartographers constructed the best models they could, based on the small amount of information available. Each time an adventurer returned from a trip, new information was added to the data already collected. Maps were revised to include this new information.

During the second century A.D., mapmaking became more precise. Ptolemy (TAHL uh mee), an Egyptian astronomer and geographer, made cartography into a science. Many of his contributions still influence mapmaking today.

Ptolemy was the first mapmaker to put north at the top of a map. Through his work in astronomy, he improved the understanding of the actual size of the earth and the distances between land masses by designing his maps based on a round Earth.

Figure 2–3. Ancient cartographers used the descriptions of explorers to make models of the earth.

Ptolemy's maps were so good that they were still being used in the 1400s. At that time Johann Gutenberg, a German printer, printed copies of Ptolemy's maps, making them readily available. About the same time, interest in geography blossomed as adventurers explored unknown parts of the world. With this new exploration, more was learned about the sizes and positions of the continents.

○ *Of what materials were some early maps made?*
○ *Who was Ptolemy?*

Answers–2.1

○ Some early maps were made of animal skins or metals like gold, silver, or copper.

○ Ptolemy was the Egyptian astronomer who made mapmaking into a science.

2.2 Modern Maps

Scientists are still involved in making maps. Since 1807 the United States Coast and Geodetic Survey (now called the *National Ocean Survey*) has been surveying much of the surface of the earth to improve the accuracy of maps.

Within the United States, each specific place that is surveyed is marked with a bronze marker called a **bench mark.** Perhaps you have noticed such markers while walking along a trail, in a field, or even in a city. Engraved on the bench mark are the point's elevation and exact location. Bench marks enable cartographers to show exact points on their maps. These points help land surveyors to set property boundaries and engineers to plan bridges and highways. In addition, bench marks allow construction crews to locate sites properly for buildings and dams.

During an average year, surveys covering about 64 000 km^2 of land are completed. These surveys establish about 3000 exact geographic positions. As a result of this effort, large parts of the United States have now been mapped. Greater accuracy and speed are now possible through the use of satellite and aerial photographs. However, it will still take many years to finish the job of accurately mapping the entire country.

Figure 2–4. A bench mark shows the exact location and elevation of this spot.

Figure 2–5. Satellite photographs help cartographers draw accurate representations of coastal areas.

A MATTER OF FACT

Colorado is the highest state, with an average elevation of about 2073 m.

DISCUSSION Bench marks are indicated on a topographic map with the symbol *B.M.* The elevation of the bench mark will also be written on the map. Have the students look at Figure 2–4 and tell where they may have seen bench marks locally. Also discuss various surveying instruments the students may have seen along highways being used by construction crews. ***Thinking Critically:*** Have the students identify recent changes in their own neighborhoods that would probably not show up on a current map. (Construction of new roads and streets may not show up on a current map.)

DISCUSSION Review with the students the three basic maps used in earth science: topographic maps, geologic maps, and hydrologic maps. Ask the students to describe each map and tell how it is used by scientists. (Topographic maps show the shape of the land; geologic maps show rock layers and types of rock; hydrologic maps show surface water and where underground water is located.) *Thinking Critically:* Ask the students which kind of map would be most useful for engineers and geologists to find new sources of water for human consumption. (Answers may include all three kinds of maps. However, the most important map would be the hydrologic map, which shows surface and underground water locations.)

EXTENSION Maps and charts have been made ever since people wanted to represent graphically the features of an area. Maps have been used as records of the features or as guides to follow when traveling. Ptolemy's maps were part of his eight-book *Geographia*. Sailors' charts, or *portolanos,* became a common guide to the coasts of the Mediterranean Sea in the fourteenth and fifteenth centuries. Even Christopher Columbus used *portolanos* to make maps. A map of the first explorations of Columbus exists in the Naval Museum in Madrid, Spain; it is dated 1500 and was made by Juan de la Cosa, a pilot for Columbus. The explorations of the 1500s led to great progress in mapmaking as cartographers incorporated into their own maps the findings of the mapmakers who were on almost every exploratory voyage.

DISCOVER: Making Maps

Skills *(Measuring, Diagramming)*

The students should measure carefully so that their maps are accurate representations of the room. You may wish to have them develop symbols for desks, windows, and other objects in the room. The students may find it easier to make their maps on graph paper. They could also work in groups.

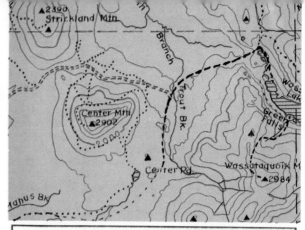

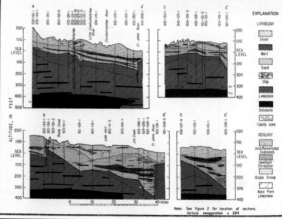

Figure 2–6. Shown here are examples of a topographic map (top), a hydrologic map (bottom), and a geologic profile map (center).

DISCOVER

Making Maps

Make a map of your classroom using a scale of 1 cm = 0.1 m. Draw the map with north at the top and measure to accurately locate doors, windows, desks, and other permanent features.

28

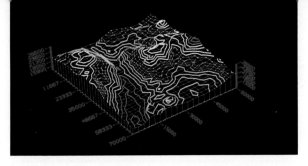

Figure 2–7. Computer simulations help cartographers visualize the shape of the land.

The survey work of the National Ocean Survey is very important to mapmaking efforts in the United States. However, the actual development of many maps is done by another agency of the government, called the *United States Geological Survey* (USGS). The USGS was founded in 1879 to prepare maps as well as to perform other, related tasks.

Maps are an important tool of the earth scientist. There are three basic types of maps used in earth science: *topographic maps, geologic maps,* and *hydrologic maps.*

Topographic maps show the shape of the land. These maps use fine lines drawn in patterns to indicate features such as mountains, valleys, and rivers. **Geologic maps** show different layers and types of rock by the use of different colors. Geologists use the patterns of colors to represent the age and structure of rock layers. **Hydrologic maps** show where surface and underground water are located. They show drainage patterns and areas where water has produced special landforms. Figure 2–6 shows examples of different types of maps.

○ *What government agency is responsible for surveying the United States?*

○ *What are three types of maps used by earth scientists?*

Section Review

READING CRITICALLY

1. Why are the works of Ptolemy so important to modern cartography?
2. How is data gathered for modern mapmaking?

THINKING CRITICALLY

3. How are modern maps similar to maps used by ancient civilizations? How are they different?
4. Why is the ancient map of the world in Figure 2–1 really a model of the world?

DISCUSSION   Have the students study the three maps in Figure 2–6. Have them point out the features that distinguish each type of map.

SKILL   *(Diagramming)* You may wish to have each of the students choose one of the three basic types of maps and make that type of map of an area in your locale.

SECTION REVIEW

Summary Have the students review each subsection and then complete the activities listed in the Section Objectives.

Reinforcement Encourage the students to quiz each other on the meanings of the new science terms.

Answers—2.2

○ The United States Geological Survey (USGS) is responsible for surveying the United States.

○ The three types of maps used by earth scientists are topographic maps, geologic maps, and hydrologic maps.

Answers to Section Review

Reading Critically

1. Ptolemy was the first person to map scientifically many of the Old World areas. Some of his maps were still in use in the fifteenth century.
2. Information is obtained mainly from satellites and aerial photographs.

Thinking Critically

3. They were models, based on the best information available of the real world. Modern maps are much more accurate.
4. The map was based on the best knowledge available at the time and provided useful information for the people.

Section **2: World Maps**

SECTION FOCUS

Section Overview Lines of latitude and longitude are introduced and defined, and the international date line and the prime meridian are explained. The problems involved in making accurate maps of the world are discussed. Various projection methods are outlined, with emphasis on the Mercator projection.

Section Objectives For a list of section objectives, see pupil's edition page.

New Science Terms For a list of new science terms in this section, see pupil's edition page. You may wish to write the terms on the chalkboard and read them aloud for the class.

SECTION DEVELOPMENT

2.3 Latitude and Longitude

DISCUSSION Some students may have trouble remembering which way the lines of latitude and longitude run. Imagining that lines of latitude lie FLAT on a globe and that lines of longitude lie the LONG way on the globe may help the students to distinguish between the two. Review the concepts of and the need for lines of latitude and longitude, emphasizing that they are imaginary lines. Have the students refer to Figure 2–8 or any globes that you may have in the classroom to identify the lines. *Thinking Critically:* Ask the students why locations must be given in terms of both latitude and longitude. (If a location were given only in terms of a parallel or a meridian, it could be anywhere on that line. A parallel and meridian together locate the intersection of two lines, and that is one, unique point.)

THINKING SKILL *(Relating Ideas)* The prime meridian was set by international agreement. Ask the students why it could have been anywhere in the world. (Any point on a circle can be a starting point. Any place on the earth could have been chosen as the point from which to mark off the 360° around the world.)

30

NEW SCIENCE TERMS
latitude
longitude

SECTION OBJECTIVES
After completing this section, you should be able to:
- **Compare** and **contrast** latitude and longitude.
- **Explain** the advantages and disadvantages of various map projections.

2.3 Latitude and Longitude

Imagine again that you are an early explorer about to set sail for some distant port. You have a map to help guide you from your home to your destination, but the map has few points of reference. You cannot tell which way to sail to reach your destination.

In order to locate places accurately on a map, a system of north-south and east-west lines was developed. The imaginary lines that run east-west around the earth are called lines of **latitude** (LAT uh tood). Latitude lines are parallel—they never cross. As you can see in Figure 2–8, the zero latitude line is the equator. All other lines of latitude are parallel to the equator and are measured from 0° to 90° from the equator. Another name for lines of latitude is *parallels*. The equator divides the earth into two equal halves—the Northern Hemisphere and the Southern Hemisphere. Therefore, a compass direction of north or south must always be included with the degrees from the equator when describing latitude—for example, 23°N or 45°S.

The imaginary lines that run from the North Pole to the South Pole are called lines of **longitude** (LAHN juh tood). Longitude lines are not parallel. They touch at the poles and are farthest apart at the equator. Lines of longitude cross parallels at right angles.

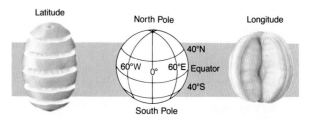

Figure 2–8. Lines of latitude are like slices of an orange (left). Lines of longitude are like sections of an orange (right).

30 Chapter 2 Earth Science Models

In 1884 a zero-degree longitude line was established by international agreement. This longitude line, called the *prime meridian*, passes through Greenwich, England. Lines of longitude, also called *meridians*, are measured, in degrees, either east or west of the prime meridian. As Figure 2–9 shows, the prime meridian and the 180° meridian divide the earth into two equal halves. The halves are called the *Eastern Hemisphere* and the *Western Hemisphere*. All longitude readings must include both the degrees and direction from the prime meridian—for example, 29°E or 41°W.

Any position on the earth can be located with latitude and longitude. Latitude and longitude can also be used to measure distance. A single degree of latitude is equal to about 111 km. Sometimes, however, maps must show distances of less than one degree. Each degree can be divided into 60 minutes (60′). Each minute is equal to 1.85 km and can be divided into 60 seconds (60″). Each second is equal to 0.03 km, or 30 m.

Meridians can serve as a measurement of time. The earth rotates through 15° of longitude each hour. One 24-hour period is a complete circle. The 180° meridian separates consecutive days—it is one day earlier east of this meridian. This meridian is called the *international date line*. If you stood with one leg on each side of it, each half of you would be in a different day.

○ **What are lines of latitude and longitude?**
○ **In addition to location, for what can latitude and longitude be used?**

DISCOVER

Finding Your Home Town

On a world map, find your home city or the city nearest to you and indicate its latitude and longitude.

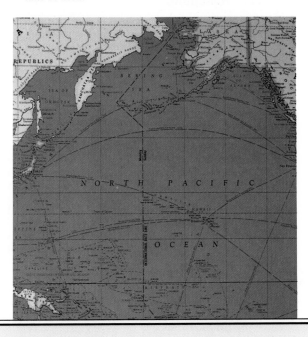

Figure 2–9. The international date line is the same as the 180° meridian, except in cases where the date line would split a nation.

Section 2   World Maps   **31**

DISCOVER:
Finding Your Home Town

Skill (*Reading a Map*)

After the students locate the city, they should give the latitude and longitude to the nearest whole degree. They may then wish to find the latitude and the longitude of their ancestors' home towns or those of some distant friends.

DISCUSSION   Point out to the students that these measurements of minutes and seconds are angle, not time, measurements. You may wish to locate two points, one directly above the other, on labeled parallels on a map and find the distance between the two points (1° of latitude equals 111 km). Then find the same distance using the scale on the map. Now refer the students to Figure 2–9. Decide on a time in hours. Start at the international date line and go east (to the right), counting off the hours (1 hour later) at each 15°. When you get back to the international date line, the time will be 24 hours later. This should help the students understand how it is one day on the east side of the international date line and the next day on the west side.

EXTENSION   You may wish to have the students look at the map of your state in the Science Sites booklet and add the appropriate lines of longitude and latitude to the map.

Answers—2.3

○ The latitude and longitude system is a grid system of imaginary E–W and N–S lines used to locate places on a map or globe.

○ Latitude and longitude can be used for a measurement of time, or as an indicator of general climatic conditions.

Skill *(Interpreting Maps)*

Preparation of Materials Have the students place a bookmark at the page with the world map for easy reference.

Hint The students can often correctly determine the coordinate readings but then attach the incorrect compass readings. Stress that they must always locate the zero line for either the equator or the prime meridian before determining compass directions.

Answers to Procedure

3. Asia, Africa, North America
4. North America, South America, Antarctica

Answers to Conclusions/Applications

1. Chicago
2. Cape Town
3. Calcutta
4. 31°N, 95°W
5. 41°N, 74°W
6. 23°N, 114°E
7. 34°S, 151°E
8. Latitude: 90°
   Longitude: 180°

2.4 Map Projections

DISCUSSION You may wish to have a small inflatable ball in the classroom. Ask the students to suggest ways to cut the ball so it can be laid flat. Use Figure 2–10 to discuss Mercator's map, relating the diagram on the left to the projection map on the right. Have a globe with Greenland's true proportions available. As the students compare the globe with the Mercator map, they can appreciate the amount of distortion.

ACTIVITY: Finding Places on a World Map

How can you use latitude and longitude to locate places on a map?

MATERIALS

A world map

PROCEDURE

1. Review the definitions of latitude and longitude lines.
2. Familiarize yourself with the map by locating the equator and the prime meridian. Determine the number of degrees between the latitude and longitude lines.
3. Locate the 30°N latitude line on the east and west borders of the map. List the continents that 30°N passes through.
4. Locate the 80°W longitude line on the north and south borders of the map. List the continents that 80°W passes through.

CONCLUSIONS/APPLICATIONS

1. What city is located at 42°N, 87°W?
2. What city is located at 34°S, 18°E?
3. What city is located at 22°N, 88°E?
4. What are the latitude and longitude of Houston, Texas?
5. What are the latitude and longitude of New York City?
6. What are the latitude and longitude of Hong Kong?
7. What are the latitude and longitude of Sydney, Australia?
8. What are the largest latitude and longitude readings possible?

2.4 Map Projections

Maps have many shortcomings as models of the earth. Maps are two-dimensional models of the earth's three-dimensional surface. Making a flat representation of a spherical surface causes distortions in the appearance of many of the earth's features. As a result, maps do not show totally accurate shapes of the continents and oceans.

In the 1500s, Gerardus Mercator (juhr AHR duhs muhr KAYT uhr), a Flemish cartographer, made a breakthrough in the art of mapmaking. The map that Mercator made showed all parallels and meridians at right angles to each other. This type of map is known as a *Mercator projection*.

The Mercator projection made it much easier to navigate using a compass and a map. For example, if you were sailing from New York to London, you would simply draw a straight line between the two cities. Then you would determine the angle of the line from any meridian and sail at that angle until reaching London. The Mercator projection widens and lengthens the areas at high latitudes. Greenland, for instance, appears almost as large as Africa. In fact, Africa is 15 times larger than Greenland.

A MATTER OF FACT

The earth is not perfectly round. The distance around the equator is over 100 km greater than the distance around the earth through the poles.

32 Chapter 2 Earth Science Models

DEMONSTRATION

Fill a 30-centimeter balloon with air just until it is inflated but not stretched. With a felt marker, draw circles and triangles on it, or you may even make some map-like outlines. Deflate the balloon. Cut it along what would be one meridian. Try to lay it flat. Make other cuts and again try to lay it flat, pointing out to the students how the drawings are stretched.

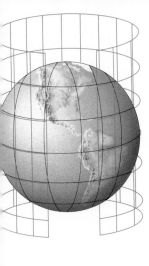

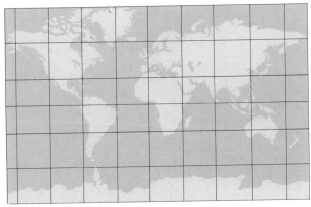

Figure 2–10. This illustration shows the development of a Mercator projection. Notice the distortion of the land areas at high latitudes.

There are other map projections that can be used to study specific parts of the world. If you wanted to study the countries around the Arctic Ocean, you might use a *polar projection*, such as the one shown in Figure 2–11. Polar projection maps are made as if the observer were looking down on the world from the North or South Pole. The polar projection is circular rather than rectangular like the Mercator projection. The pole is at the center of the map, and the equator forms the outer boundary. Land areas near the pole are shown in true proportions, but areas near the equator are greatly distorted.

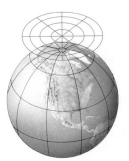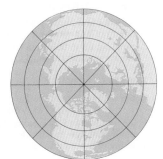

Figure 2–11. Polar projections, such as this, show areas at high latitudes without much distortion.

Section 2   World Maps   **33**

DISCUSSION Direct the students' attention to Figure 2–11. The diagram on the left shows how the image is projected onto the flat surface. Have the students compare the polar projection with a globe to appreciate the distortion. ***Thinking Critically:*** Ask the students why someone would want a polar projection map. (None of the other maps would help a person navigate around the poles because the land masses are always separated. In this case, a polar projection would be a useful map.)

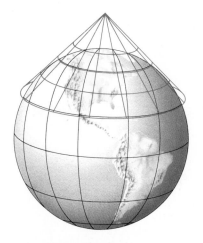

Figure 2–12. Polyconic projections are often used to make small area maps because such projections do not distort the shape of the land.

A MATTER OF FACT

Delaware has the lowest average elevation of any state, with an average of about 18 m.

The *polyconic projection*, shown in Figure 2–12, shows small areas of the world without distortion of any land masses. As you can see, however, these maps would not be very good for navigation because many different maps would have to be used. Polyconic projections are often used for road and topographic maps.

○ *Why is it hard to produce an accurate world map?*
○ *What one problem do most map projections share?*

Section Review

READING CRITICALLY

1. Describe how and why a Mercator projection changes the shape or size of land masses that are located near the poles.
2. Write out 23°15′30″W, using words instead of symbols. Is this a longitude reading or a latitude reading? Why?

THINKING CRITICALLY

3. Explain why 95°30′15″N and 190°30′15″E are impossible locations on a map of the world.
4. On Mercator and polar projections, extend the edge of a sheet of paper from New York City to Moscow. Why do these maps show totally different routes for travel between the same cities?

34   Chapter 2   Earth Science Models

DISCUSSION Have the students study the polyconic projection. Discuss how this map could be made from a globe. (You cut from the poles, about halfway to the equator, along the meridians every 30° or 40°.) *Thinking Critically:* Have the students compare this projection with the Mercator projection. (The polyconic projection shows small areas with very little distortion of the landmasses. But large areas are broken up, so it could not be used for traveling or as a distance guide. The Mercator projection shows the relative positions of countries and continents, but their sizes are very distorted.)

SECTION REVIEW

Summary Have the students summarize each subsection in a paragraph. (The first paragraph should include latitude and longitude with reference to the equator and the prime meridian, respectively. The second paragraph would describe the Mercator projection, polar projection, and polyconic projection maps, explaining the advantages and disadvantages of each.)

Reinforcement Have the students make up two questions for each subsection and form teams for a quiz.

Answers—2.4

○ Maps are two-dimensional representations of the earth's three-dimensional surface. This causes distortions.
○ They all distort some portion of the earth's surface.

Answers to Section Review

Reading Critically

1. Lines of longitude near the poles are very close together. On a Mercator projection, they are equidistant everywhere. Because of this, land area near the poles appears to be very large.
2. Twenty-three degrees, fifteen minutes, thirty seconds west. This is a longitude reading because the *W* for west signifies reference to the prime meridian.

Thinking Critically

3. The maximum latitude is 90 degrees (located at the poles). The maximum longitudinal reading is 180 degrees (located at the international date line).

continues

CAREERS

SURVEYOR

In the original American colonies, most boundaries were determined by natural features, such as rivers. Early *surveyors* were responsible for determining the boundaries of most of the states that were not part of the original 13 colonies. Surveyors today determine ownership of private property from fixed points. such as USGS bench marks.

Many surveyors work for state and local governments, where they establish right-of-ways for roads, waterlines, and government buildings.

To do the work of a surveyor, you must have a high school diploma.

For Additional Information
American Congress on
    Surveying and Mapping
210 Little Falls Road
Falls Church, VA 22046

GEOLOGIST

Studying rocks and the processes that change the surface of the earth is the job of a *geologist.* Most geologists have majored in the sciences and mathematics in college and have obtained an advanced degree in a specialized field of geology.

Some geologists who specialize in earth materials are economic geologists. Petroleum geologists specialize in the search for oil and natural gas. Other geologists study the earth's more violent changes, such as volcanoes, that sculpture the earth's surface.

For Additional Information
American Geological
    Institute
5205 Leesburg Pike
Falls Church, VA 22041

CARTOGRAPHER

Mapmaking, a combination of science and art, is the job of a *cartographer.* Cartographers use data from field geologists, surveyors, and other sources to make maps. This work takes imagination and a high mechanical-drawing aptitude. Computers are playing an increasingly important part in this field.

Part of cartography is the actual drafting of the maps from computer interpretation of data. This position requires two years of training after high school and some on-the-job training.

For Additional Information
American Congress on
    Surveying and Mapping
210 Little Falls Road
Falls Church, VA 22046

Section 2   World Maps   35

CAREERS

Discussion Have interested students write to one or all of the addresses provided for additional information. Have the students discuss the opportunities offered by the various careers.

Extension You may wish to hold a "Career Day" in the class during which the students report on the information they have obtained. If possible, invite a person in each career to speak to the class about why he or she chose the career, what specific training he or she received, and what a "typical" day at work involves. As an alternative to outside speakers, have one of the students assume the role of a person working in each career and make a presentation to the class covering the same topics mentioned above. The students may be interested in researching information on the following, additional careers:

Aerial Photographer

Draftsperson

Photogrammetrist

Geographer

---

4. The Mercator projection misrepresents distances at the poles; thus the shortest route—across the North Pole—is obscured.

SECTION FOCUS

Section Overview The importance of a map scale is explained, and the three kinds of map scales—verbal, representative fraction, and graphic—are described. Topography is defined, and the use of contour lines on topographic maps is discussed.

Section Objectives For a list of section objectives, see pupil's edition page.

New Science Terms For a list of new science terms in this section, see pupil's edition page. You may wish to familiarize the students with these terms before they read the section.

SECTION DEVELOPMENT

2.5 Map Scales

DISCUSSION You may wish to have the students describe instances when they have used maps. (planning trips, locating sites) Have the students describe why they would want to know the scale of a map. (They might want to know which of a number of places is closer; or they might want to know how far away a place is so that they can calculate how long it would take to get there.) Explain and compare the verbal scale and the representative fraction scale. Look at the map in Figure 2–14. Ask the students what type of scales are used on this map (an RF and a graphic scale). A scale of 1:62 500 is used for 1 in. = 1 mile instead of 1:63 360. This is a rounded figure and is easier to use.

NEW SCIENCE TERMS
scale
topography

SECTION OBJECTIVES
After completing this section, you should be able to:
- **Relate** map scales to actual distances.
- **Interpret** contour lines on a topographic map.

2.5 **Map Scales**

If you were planning a trip from your home to your state capital, you probably would not find a world map very helpful. Many times cartographers need to show more details than can be shown on a world map. To show more details of an area, cartographers change the scale of a map. The **scale** of a map is the relationship between a distance on the map and a distance on the earth. For example, a scale of 1 cm = 100 km means that 1 cm on the map represents 100 km on the earth's surface. What would 2 cm on the map represent with a scale of 1 cm = 1 km?

The example of 1 cm = 1 km is called a *verbal scale*. This type of scale equates two different units, centimeters and kilometers in this case. This type of scale is often found on road maps, because it is easy for people to understand.

Figure 2–13. Changing the scale of a map allows for greater detail (right) of a small portion of a larger area map (below).

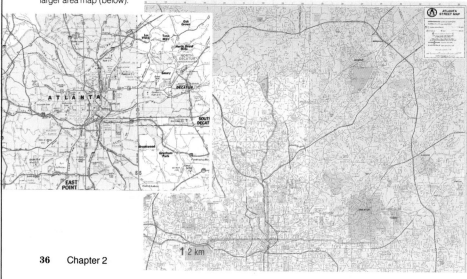

36   Chapter 2

DEMONSTRATION

For a demonstration of making a topographic map, see page 23c preceding this chapter.

Map scales are usually written as ratios. Such a scale is called a *representative fraction (RF)*. The RF shows the ratio of a represented distance on a map to the actual distance on the earth's surface. The two numbers must be in the same units. For instance, a verbal scale of 1 cm = 1 km would be written as 1:100 000. Since a ratio is not really a measurement, it is never written with units.

In addition to or in place of an RF, many maps show a *graphic scale*. A graphic scale consists of a measured line marked off in specific distances. To show the scale of 1:100 000, the line would be drawn 1 cm long. One end of the line would be marked 0 cm, while the other end would be marked 100 000 cm or 1 km. Look at Figure 2–14 to see an example of a graphic scale.

Although scientists have agreed to use SI measurements for all scientific work, most of the maps available in the United States are still drawn using customary units. This system of measurement uses inches and miles. Many of the maps available to you will use verbal scales of 1 in. = 1 mi., which is the same as the RF 1:62 500 (63 360).

○ *What is a map scale?*
○ *What are representative fractions and graphic scales?*

A MATTER OF FACT

The USGS distributes enough topographic maps each year to make a stack higher than the Empire State Building.

Figure 2–14. The key of this topographic map shows a graphic scale and a representative fraction.

DISCUSSION Explain the use of the graphic map scale. ***Thinking Critically:*** Have the students discuss why some people consider the graphic scale to be the most useful on a road map. (If you are traveling in a car, it is very easy to use the given distance to go along a road on the map and to determine a distance. If the width of a finger is used, for example, no calculations are involved.)

Answers–2.5

○ A map scale is the relationship between a distance on the map and a distance on the earth.

○ Representative fraction is the ratio of the distance on the map to the represented distance on the ground. A graphic scale is a measured line marked off in specific distances, showing the scale of the map.

Objectives

- Interpret map scales in both SI and customary units.
- Use map scales to determine distances.

Discussion   Often distances on maps will not measure out to be exactly the unit of the scale. For instance, High Island measures out to be a little longer than one mile. The most efficient method of using the subdivided section of the graphic scale is the following:

Answers to Application

1. Mark the distance between two locations on your paper as you have previously been instructed.
2. Hold one mark up to the zero mark on the scale. You will see that the second mark does not fall directly on a mile mark.
3. Slide the paper to the left until the right hand mark is exactly on a mile mark.
4. Now read the excess distance on the subdivided section of the scale and add this distance to the whole unit measured to the right of the zero line.

Answers to Using What You Have Learned

1. 7.6 km
2. 5 km
3. 5.8 km
4. 2.8 km

SKILL ACTIVITY: Interpreting Map Scales

BACKGROUND

Since maps are scale representations of actual locations, any interpretation of a map depends on an accurate understanding of the scale of the map. Although there are three different types of map scales, only the graphic scale and the representative fraction usually appear on the map. When people discuss the scale of a map, they frequently use the verbal scale.

Map scales can be presented in either the SI or the customary units of measurement. Most of the maps produced in the United States have been based on the customary units. These maps are still in use today, even though the scientific community works in SI units. As a student of maps, you will need to be able to interpret scales in both systems.

PROCEDURE

The graphic scale is especially handy to use because you do not need to convert from one unit to another. You can use the graphic scale of a map even if you do not have a ruler.

To use the graphic scale to measure a distance on a map, hold the edge of a sheet of paper along the distance you wish to measure. Mark both ends of the distance on your paper. Then hold the marked paper along the graphic scale of the map to read the distance.

APPLICATION

Using the Wet Fish Island map, complete the following:

1. Hold the side of a sheet of paper along an imaginary north-south line between the two coastlines.
2. Mark both ends of the island on the paper.
3. Place the paper along the graphic scale on the map, lining up one of the marks with the zero mark on the scale.
4. By comparing where the second mark on your paper falls along the scale, you can determine the distance between your two marks.

USING WHAT YOU HAVE LEARNED

Use the graphic scale to determine the following distances on the Wet Fish Island map.

1. What is the length of High Island?
2. What is the width of High Island?
3. What is the width of the map?
4. What is the width of Wet Fish Island?

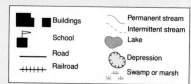

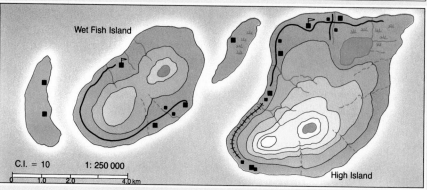

Wet Fish Island

High Island

C.I. = 10   1: 250 000

0   1.0   2.0   4.0 km

2.6 Contour Lines

If you were planning a hike through mountainous country, you would probably want to know the elevation of various places along your path. You might also want to know how steep certain trails were, so you could pick the easiest one to reach your destination. One of the most useful maps for these purposes is the topographic map. On a topographic map, a pattern of lines is used to show elevation. Streams, lakes, and other natural features are also represented on topographic maps, as well as land boundaries, towns, roads, and other structures.

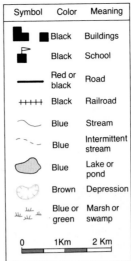

Symbol	Color	Meaning
▉	Black	Buildings
⚑	Black	School
—	Red or black	Road
+++++	Black	Railroad
~	Blue	Stream
- -	Blue	Intermittent stream
◯	Blue	Lake or pond
◯	Brown	Depression
⍑⍑	Blue or green	Marsh or swamp

0   1Km   2 Km

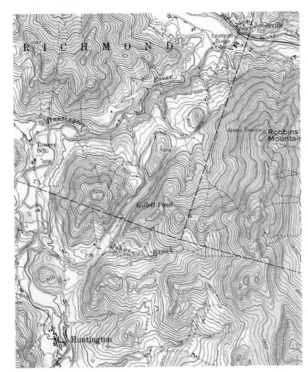

Figure 2–15. Contour lines connect places with the same elevation. Notice that the contour lines on this map are brown, and that some of them are labeled with the elevation. The labeled contour lines are called *index contours.* Special map features can be identified from the key above.

Topography is the contour, or shape, of the land surface. On a topographic map, topography is shown by using *contour lines.* Contour lines are lines on a map that connect points of equal height above sea level. A zero contour line is at sea level. Follow one of the contour lines in Figure 2–15. Notice that you stay at the same elevation.

DISCOVER

Drawing Contours

Take a large lump of modeling clay and form it into a hill-shaped mass. On a plain sheet of paper, draw contour lines to represent this hill. Use a contour interval of 1 cm and an RF of 1:1.

Section 3   Topographic Maps   **39**

2.6 Contour Lines

DISCUSSION   Have the students study the map shown and identify as many features as possible. *Thinking Critically:* Have the students imagine that they are planning to drive through an area they believe has flat terrain. Ask them how they might have to alter travel plans when they find out it is mountainous. (If they had allowed two hours to drive a distance of 75 flat kilometers, they might now have to allow four hours to drive up and down mountains.)

DISCOVER: Drawing Contours

Skill *(Representing Data)*

You may wish to have the students shape the clay twice, once with a cone-shaped mountain and once with an irregular-shaped (more natural) mountain. Drawing lines in the clay and then looking down on it may help the students visualize the lines.

DISCUSSION Steep slopes are shown with closely-spaced lines, gentle slopes with widely-spaced lines, and valleys with V-shaped or U-shaped lines. Have the students look for steep slopes and gentle slopes in Figure 2–16. *Thinking Critically:* Have the students discuss whether or not they could tell the steepness of the slopes in Figure 2–17 without the help of the contour interval. (Answers will vary but should show appreciation of the value of contour interval.) The contour interval for Figure 2–17 would probably need to be 10 or 20 feet in order to show elevation differences. In the mountainous area, the contour interval would probably be 100 feet. Topographic maps usually use the customary feet and miles instead of SI.

Figure 2–16. Shown here is a topographic map (right) and an aerial photograph (left) of Mount St. Helens, an active volcano in the state of Washington.

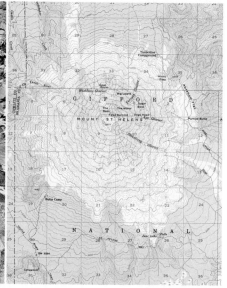

Look at the pattern on the map in Figure 2–16. This pattern represents a mountain. Anytime you see a pattern similar to this one on a topographic map, it represents a mountain. The difference in elevation between two consecutive contour lines is called the *contour interval*. For example, if the contour interval were 10 feet, the lines on either side of a line representing the 400-foot contour would represent 390 feet and 410 feet.

Many times a topographic map has so many contour lines that the elevation of a single line is difficult to determine. As an aid in reading the topographic map, every fifth contour line is drawn darker and thicker so that it stands out. Each of these darker lines, called *index contours*, is marked with the elevation. Locate an index contour line in Figure 2–16. What is its elevation?

The topographic map shown in Figure 2–16 is a map of the area also pictured in Figure 2–16. Notice the contour lines that show the hill. This map also shows special patterns for steep slopes, gentle slopes, and valleys. What is the pattern for each of these features?

Compare the topography in Figures 2–16 and 2–17. Why would you want to use a different contour interval for each area? 1

A MATTER OF FACT

The average elevation of the United States (excluding Alaska and Hawaii) is about 762 m above sea level.

1 To show the topography as accurately as possible.

By decreasing the contour intervals, mapmakers are able to show more details of flat surfaces. Larger contour intervals are used to show the features of mountainous areas.

Table 2–1 explains a few simple rules about contour lines. These rules should help you interpret topographic maps.

○ *What is topography?*
○ *How is elevation represented on a topographic map?*

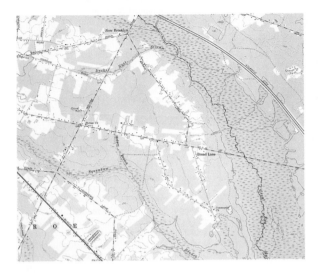

TABLE 2–1: INTERPRETING CONTOUR LINES

1. A contour line connects points of equal elevation.
2. A contour line is an endless line; it either closes upon itself on the map or at some point outside the map area.
3. Contour lines never branch or fork.
4. Contour lines never cross one another.
5. Closely spaced contour lines represent steep slopes; widely spaced contour lines represent gentle slopes.
6. Circular depressions in the earth's surface are shown by closed contours that have marks on the inside called *hachure marks*.
7. Contour lines point upstream when crossing a stream, river, or valley.

Figure 2–17. This topographic map is of a relatively flat coastal area of New England. Using the steps in Table 2–1 above, study the features of this map.

Section Review

READING CRITICALLY

1. Maps always use symbols to represent features. Study some of the features shown on the symbol sheet in Figure 2–15. Which ones can also be found in the map shown in Figure 2–17?
2. What is a topographic map?

THINKING CRITICALLY

3. Write a representative fraction and draw a graphic scale for the verbal scale of 1 cm = 100 km.
4. Examine the two road maps shown in Figure 2–13. They show some of the same places, although the scales of the two maps are different. What are some of the advantages of a small scale, such as 1:625 000, compared to a larger scale, such as 1:62 500? What are some of the advantages of the smaller scale?

Section 3   Topographic Maps   41

DISCUSSION   You may wish to have the students read the table and compare the items in Table 2–1 with Figure 2–17. They should look to see that contour lines never fork and never cross.

SECTION REVIEW

Summary   You may wish to have the students review the section by writing a sentence for each of the following: verbal scale, representative fraction, graphic scale, topographic map, contour line, and contour interval.

Reinforcement   Have the students complete the Section Review. The students may work in small groups, or answers can be compared when the students have completed the exercise, working individually.

Answers–2.6

○ Topography is the shape of the land.
○ Elevation is represented by contour intervals.

Answers to Section Review

Reading Critically

1. The symbols for contour lines, roads, rivers, and buildings are found on both topographic maps.
2. A topographic map is a map that shows the shape of the land by using patterns of contour intervals.

Thinking Critically

3. 1:10 000 000 cm

0   100 km   200 km   300 km

4. The larger scale allows an overall view of an area. The smaller scale allows for more detail.

INVESTIGATION 2:
Interpreting Topographic Maps

Skill (Interpreting Graphics)

Hint

This investigation is designed to reinforce the skills that the students need to read topographic maps. Encourage them to try to figure out the answers on their own. Allow them plenty of time before you offer assistance. You may need to remind them to locate the contour interval of the map or encourage them to count the lines between the unknown point and the closest index contour.

Answers to Procedure Questions

1. a). sea level
   b). 20 ft
   c). sea level
2. Outer Head, 43°47′N–69°44′W; Whittum Island, 43°51′N–69°44′W
3. a). Pumpkin Island, 43°45′N–69°32′W
   b). White Island, 43°46′N–69°31′W
   c). John's Island, 43°52′N–69°31′W
4. 20, 120, 200, 40, 180, 60, 260
5. less than 20 ft
6. 40 ft

Answers to Analyses and Conclusions

1. very low elevation
2. The eastside, because it is very steep. The contour lines are close together.
3. The profile would be very pointed, because the island is over 100 feet in elevation.
4. south; The elevation gets lower as you move to the south.
5. There will be no brown contour line on it.
6. There is a bench mark located there, with the exact elevation indicated.

Answer to Application

Answers will vary.

INVESTIGATION 2: Interpreting Topographic Maps

PURPOSE

To identify landforms and elevations from contour-line patterns

MATERIALS (per group of 3 or 4)

Topographic map of Boothbay quadrangle on page 581 of the Reference Section

PROCEDURE

Answer the following questions by using this map.

1. What is the maximum elevation of each of the following locations?
   a) Pumpkin Ledges   b) Pumpkin Island
   c) The Cuckolds

2. Name and locate (using latitude and longitude readings) two islands below 20 feet in elevation.
3. Name and locate one example of an island for each of the following elevations:
   a) More than 20 feet, less than 40 feet
   b) More than 40 feet, less than 60 feet
   c) More than 60 feet, less than 80 feet
4. Of the following elevations, which ones would you expect to be indicated with a contour line on this map? 20, 59, 120, 200, 40, 75, 150, 180, 60, 105, 260
5. What is the elevation of the house on Ram Island?
6. What is the elevation of Adams Pond?

ANALYSES AND CONCLUSIONS

1. Why do you think there are so many swamps in the southern part of Georgetown Island?
2. Locate Cushman Hill in the NW quarter of the map. Which side of the hill would prove to be the most difficult hike to the top? Explain how you reached your conclusion.
3. Describe the way High Island would look if you were approaching it from a distance. Sketch or describe its profile, or side view.
4. Locate the small stream that flows along the western side of Whaleback Ridge. In which direction does this stream flow? Explain how you reached your conclusion.
5. Explain how you can identify an island below 20 feet in elevation.
6. Locate Riggs Hill along the western boundary of the map. Its exact elevation is identified as 217 feet. Why was the number written on the map instead of just being indicated with the contour lines?

APPLICATION

On a topographic map of your own area, find your home, your school, or a nearby park, and several other natural features.

SUMMARY

- As long ago as 2000 B.C., people constructed maps and models of the earth. (2.1)

- Ptolemy was the first mapmaker to put north at the top of a map. (2.1)

- Since 1807 the United States Coast and Geodetic Survey (now called the *National Ocean Survey*) has been making accurate maps. (2.2)

- Bench marks are points where the exact elevation and location are known. (2.2)

- Latitude and longitude lines are used to indicate on maps the location of points on the earth's surface. (2.3)

- Lines of latitude are also known as parallels and lines of longitude are known as meridians. (2.3)

- Map projections show a spherical Earth on a flat page. (2.4)

- Many map projections show parts of the world in a distorted view. (2.4)

- The representative fraction scale, the verbal scale, and the graphic scale are three ways to present the relationship between map distance and represented distance on the earth's surface. (2.5)

- Although scientific work is in SI, many topographic maps are drawn with customary units of measurement. (2.5)

- Contour lines show places of equal elevation above sea level. (2.6)

- Topography is the shape of the earth's surface. (2.6)

Write all answers on a separate sheet of paper.

SCIENCE TERMS

Correctly use each of the following terms in a sentence.

bench mark **(27)**
cartography **(26)**
geologic maps **(29)**
hydrologic maps **(29)**
latitude **(30)**
longitude **(30)**
scale **(36)**
topographic maps **(29)**
topography **(39)**

SCIENCE QUIZ

Modified True-False

Mark each statement *true* or *false*. If a statement is false, change the underlined term to make the statement true.

1. The scale 1:100 000 is an example of a <u>verbal scale</u>.

2. The closer the contour lines, the <u>steeper</u> the slope that they represent.

3. The Mercator projection distorts landforms in the area around the <u>equator</u>.

4. A <u>verbal</u> scale never appears on a map.

5. All contour lines are measured in height above <u>sea level</u>.

6. A <u>polar projection</u> shows areas near the poles without distortion.

7. <u>Topographic</u> maps show rock structure.

8. <u>Bench marks</u> show exact elevation.

9. <u>Contour lines</u> connect points of equal elevation.

10. Most small area maps are <u>polyconic</u> projections.

continues

CHAPTER REVIEW

SUMMARY

The students may review the major concepts in the chapter by reading the summary statements. The statements are cross-referenced to the chapter to facilitate reinforcement of any concepts of which the students feel unsure. Encourage the students to work in groups to quiz one another.

SCIENCE TERMS

The sentence in which the science term is used should reflect an understanding of the definition of the term. Have the students quiz each other on the meaning of each term.

SCIENCE QUIZ

Answers to Modified True/False

1. false, representative fraction
2. true
3. false, the poles
4. true
5. true
6. true
7. false, Geologic
8. true
9. true
10. true

continues

Answers to Multiple Choice

11. b
12. b
13. c
14. a
15. d

Answers to Completion

16. cartographer
17. equator
18. polar
19. prime meridian
20. topographic

Answers to Short Answer

21. People used maps for purposes of travel, for marking property boundaries, and for determining taxes.
22. 2 A.M., October 15. As the earth rotates, the sun rises and sets at different times along different longitudes. By having different times zones, all places can organize their time so that the sun rises in the morning, or A.M. hours, and sets in the evening, or P.M. hours.
23. By using lines of latitude and longitude, individual locations can be pinpointed. The latitudinal reading provides the distance from the equator. The longitudinal reading provides the distance from the prime meridian.

Answers to Writing Critically

24. Curitiba, Brazil
25. Southern and Eastern hemispheres; continent of Australia

SCIENCE QUIZ continued

Multiple Choice

Write the letter of the choice that best answers the question or completes the statement. All of the questions refer to the topographic map of Clark's Falls.

11. In which direction does Green Fall River flow?
    a) north   b) south
    c) east   d) west

12. What is the contour interval of this map?
    a) 5 feet   b) 10 feet
    c) 15 feet   d) 20 feet

13. Which side of the hill northwest of Clark's Falls is the steepest?
    a) north   b) south
    c) east   d) west

14. What are the elevations of the two bench marks?
    a) 90 and 98   b) 150 and 190
    c) 199 and 255   d) 90 and 199

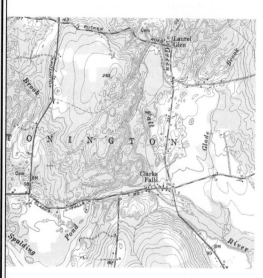

15. What is the elevation of Point D?
    a) 130 feet   b) 240 feet
    c) 200 feet   d) 170 feet

Completion

Complete each statement by supplying the correct term.

16. A person who makes maps is called a ____.

17. The zero latitude line is the ____.

18. The ____ projection is best when showing land at high latitudes.

19. Longitude lines are measured east and west of the ____.

20. A map that indicates elevation by the use of contour lines is called a ____ map.

Short Answers

21. Explain why even during the earliest civilizations, people needed to have maps.

22. Using the world map on page 592, determine what date and time it would be in Atlanta if it is 5:00 P.M., October 15, in Sydney, Australia. Explain why different time zones are necessary.

23. Describe how you could use latitude and longitude to locate any place on Earth to within 30 m.

Writing Critically

24. If you visited Los Angeles, you would be as far from the equator as what major city in the Southern Hemisphere?

25. If a city had a latitude reading of 25°S and a longitude reading of 130°E, in what hemisphere and on what continent would it be located? Explain how you arrived at your answer.

44   Chapter 2 Review

EXTENSION

1. Build a clay model of the Clark's Falls map. Using the point of a pencil or a straightened paper clip, carve the contour lines into the clay model.

2. Design your own topographic map showing an oval hill that is steepest on the west side and has an eastward-flowing stream with one branch, a lake, and a depression. The contour lines for each of these landforms must form a continuous sequence.

3. Research the development of Landsat maps of North America. Write a report describing the process for changing the satellite images into maps.

APPLICATION/CRITICAL THINKING

1. When comparing distances in Alaska with distances in Brazil, why would degrees of latitude be used instead of degrees of longitude to express the relationship between distance and degrees?

2. Explain why the international date line is not a straight line like all other meridians.

FOR FURTHER READING

Carey, H. H. *How to Use Maps and Globes.* New York: Franklin Watts, 1983. This book tells how to get information from maps and globes and how to draw maps.

Nordenskiold, A. E. *Facsimile Atlas to the Early History of Cartography.* Mineola, New York: Dover Publications, Inc., 1973. This atlas reproduces 169 of the most important maps printed before 1600.

Challenge Your Thinking

These photographs show a very complex system of levees, locks, and dams that protect much of the country of Holland. The components of these structures have been tested on a working scale-model. Of what value is a working model for predicting the consequences of various happenings?

Chapter 2 Review   45

ANSWERS TO APPLICATION/ CRITICAL THINKING

1. Parallels are always the same distance apart on flat maps, whereas lines of longitude tend to be exaggerated in high latitudes.
2. The international date line "bends" around island groups so as not to split some nations into different days.

ANSWER TO CHALLENGE YOUR THINKING

A model would make it easy to study variations in the system or consequences of storms without endangering the land or the people.

ANSWERS TO EXTENSION

1. Models will vary, but you may wish to check for closed contour lines and an accurate representation of the topographic map.
2. Maps will vary, but you may wish to check them for closed contour lines and accurate representations of the various land features on the topographic map.
3. There are several books available using Landsat photography to show geologic and environmental topics.

Discussion This *Science Connection* is about Landsat imaging. Some of the students may have seen Landsat photographs of the earth, as well as photographs of the earth taken by astronauts during space flights. Discuss the reasons for the difference in the appearance of conventional photographs and Landsat photographs. Ask the students why the color imaging used by Landsat helps scientists see more things than conventional photography reveals.

Discussion Landsat photography detects energy reflected from the earth's surface. Desert areas reflect more of the sun's light than do water, vegetation, and soil, and thus desert areas appear brighter (white) in these photographs. Scientists study Landsat photographs of desert areas to learn more about the movement of sand dunes, the drying up of water resources, and soil erosion in order to predict where deserts might be located in the future. Ask the students why this information is important, especially to those who are trying to prevent famines.

Discussion Scientists have been measuring the direction and speed of glaciers in Antarctica by using Landsat photography, which allows them to track the movements of surface features of the glaciers. They are trying to learn if the giant ice sheets are shrinking, growing, or remaining the same size. Ask the students why it is important to know this. Point out that more than 90 percent of the earth's fresh water is ice at the two poles, and any change in the ocean currents, atmospheric temperature, or levels of the oceans would severely affect the earth's climate.

Extension After completing the Unit and reading and discussing the *Science Connection*, have the students turn back to the Unit Opener on page 1 and answer the questions. When the class discussion has ended, you may wish to have interested students write to NASA's Goddard Space Flight Center in Landover, Maryland, for further information. These students can then present their findings in a group report.

UNIT 1 THE SCIENCE CONNECTION:

Landsat Imaging

Landsat satellites circle at altitudes as high as 920 km. They scan the earth with electronic sensors that find parts of the world not seen by the human eye.

The Landsat satellites gather information through two sets of sensors. A multispectral scanner (MSS) senses objects on the earth in great detail. The scanner can see in the visible light spectrum as well as in the infrared, or heat, spectrum.

The second type of sensor, the thematic mapper (TM), can record radiation from the surface of the earth. The TM sensor is responsible for the colors in the Landsat pictures. The TM records seven different types of radiation, or bands, as different colors. Colors are assigned to the bands to make it easier to analyze a region to be mapped. These computer-produced pictures are called *false-color images*.

The data for each of the seven bands can be made into separate black and white images. A computer combines the bands to make a colored image like that of the Texas coast shown below.

Landsat photograph of the Texas coast

Preparing maps from a Landsat image

Information provided by Landsat images has caused earth scientists to change some of their ideas about the earth and its resources. Through analysis of Landsat images, cartographers have surveyed regions in less developed countries where accurate maps have never been made. Using Landsat images, hydrologists have been able to locate unmapped lakes and other waterways. Landsat images have aided geologists in the discovery of oil in the Sudan, tin in Brazil, and copper in Mexico. The Landsat images have also been used to find uranium, zinc, copper, and nickel deposits in the United States.

A combination of bands can be selected to highlight certain features, such as a forest or river.

There are filters in the sensors that isolate the colors blue, green, and red. In addition, several types of radiation that cannot be seen by the human eye can be isolated.

Each band shows something different. The blue band shows areas where there are plants. This allows scientists to distinguish between bare soil and planted fields. The blue band can also detect clear, shallow water and show its depth. The green band shows healthy vegetation and can detect cloudy water. The red band separates different crops and shows other features such as cities and highways. Some of the other bands highlight fast-growing plants, show surface temperatures, and reveal sources of heat.

Testing a Landsat before launch

47

Unit 2: GEOLOGY

UNIT OVERVIEW

The composition of the earth's crust, including rocks, minerals, and natural resources is detailed in this unit. Also included are discussions of faulting and folding, which change the earth's surface, and the dynamic processes of plate tectonics, volcanoes, and earthquakes.

Chapter 3:   Minerals, page 50

This chapter introduces the structure of atoms and molecules. It presents the periodic table of elements and the bonding of atoms. The grouping of minerals into families and mineral identification is discussed.

Chapter 4:   Rocks, page 74

This chapter presents the three rock families that occur naturally on Earth. The rock cycle is discussed, as are the formations, characteristics, and possible uses of each family of rock.

Chapter 5:   Earth's Resources, page 98

This chapter presents an overview of some of the valuable resources that are found within the earth. Included are descriptions of specific metals and nonmetals and their various practical uses in industry. The earth's fossil fuels are then described, with an emphasis on the importance of these resources in today's world.

Chapter 6:   Structural Geology, page 116

This chapter presents the characteristics of the earth's complex layers. The results of forces acting on the earth's surface—faulting and folding—are discussed. The students are introduced to the theory of continental drift.

Chapter 7:   The Dynamic Earth, page 136

This chapter presents the theory of plate tectonics to explain the formation of crustal features. The activity at plate boundaries is related to dramatic geologic events, such as volcanoes and earthquakes. The formation and structure of volcanoes are explained. The causes of earthquakes are discussed, and the behavior and effects of the P, S, and surface waves that they generate are explained.

The Science Connection: Diamonds, page 160

The Science Connection describes the formation of diamonds inside igneous rock structures known as diamond pipes. The locations of large diamond deposits are detailed, and many of the industrial uses of diamonds are described.

ADVANCE PREPARATION

Chapter 3

For the Activity on page 64, you will need pennies, steel nails, and glass plates. For more information, see page 64.

Chapter 5

For the Discover on page 110, the students will need to record the distance driven and the amount of gasoline used in a one-week period. For more information, see page 110.

For Investigation 5 on page 112, you may want to prepare the saturated solutions of copper sulfate, ferrous sulfate, and sodium chloride ahead of time. For more information, see page 112.

Chapter 6

For the Discover on page 120, you will need corn syrup and food coloring. For more information, see page 120.

For the Discover on page 129, you or the students will need to bring in an old jigsaw puzzle. For more information, see page 129.

Chapter 7

For the Discover on page 146, you will need modeling clay and small plastic foam beads. For more information, see page 146.

For the Discover on page 154, you will need an old phonograph turntable and speaker. For more information, see page 154.

BULLETIN BOARD SUGGESTIONS

Chapter 3

Design a quiz board with pictures of minerals identified by name and separate cards listing various characteristics (i.e., hardness = 7, glassy luster, curved fracture, color: white, clear, pink, gray, purple). The students would match the characteristics cards to the picture of quartz.

Chapter 4

Make a model of the rock cycle using diagrams or humorous pictures to represent the various rock-forming processes (i.e., a muscle man squeezing rocks to form metamorphic rocks, a sieve to show sediment sorting and a bottle of school glue to show the process of cementation of sedimentary rocks, a bubbling pot of soup to represent magma for igneous rocks).

Chapter 5

Prepare a timeline illustrating the use of raw materials throughout the ages. Where applicable, be sure to include the inventions and inventors who created the processes that led to further technological innovations. A separate timeline may be required to illustrate the early years of the Industrial Revolution.

Chapter 6

Make a matching quiz problem for the layers of the earth. Design a model of the earth with differently colored concentric circles. Use zig zag tape to demonstrate how earthquake awareness. The questions and results can be displayed on the bulletin board. Some questions might be "Have you ever felt an earthquake?"; "Do you or your family have an emergency kit in case of an earthquake?";

Chapter 7

Have the students design and conduct a survey about earthquake awareness. The questions and results can be displayed on the bulletin board. Some questions might be: "Have you ever felt an earthquake?"; "Do you or your family have an emergency kit in case of an earthquake?"; "What should be in such an earthquake emergency kit?"

ISSUES IN EARTH SCIENCE

Over the past few decades, there have been several pendulum swings involving the conservation of natural resources. Discuss a plan for managing worldwide resources that will ensure long-term availability without drastically changing our lifestyle.

In California and in other places, nuclear power plants are built in seismically active areas, and in some cases, directly over tectonic faults. Have the students discuss the potential dangers of operating nuclear power plants in earthquake-prone areas.

SUGGESTED PROJECTS

Students can obtain crystal-growing kits from hobby shops and grow a variety of mineral crystals.

Have the students read and report on the advantages of growing crystals in zero gravity, such as in orbiting space stations.

TEACHER RESOURCES

Readings

Chapter 3

Desautels, P.E. *The Mineral Kingdom*. New York: Grosset & Dunlap, Inc., 1968. A beautifully illustrated introduction to gems, minerals, and mineral collecting.

Chapter 4

MacFall, R. P. *Rock Hunter's Guide*. New York: Crowell, 1980. This is a general geology book that is a useful tool for the collection and identification of rocks.

Chapter 5

"Materials for Economic Growth," *Scientific American* 255 (October, 1986) p. 50. This theme issue is a compilation of articles that range from innovations in advanced metals, ceramics, and polymers to their impact on present and future world economies.

Chapter 6

Hamblin, W. K., and J. D. Howard. *Physical Geology*. New York: Burgess Publishing Co., 1980. This book describes the geology of North America.

Miller, Russell. *Continents in Collision*. Alexandria, Virginia: Time-Life Books, 1983. This book discusses plate tectonics.

Chapter 7

McAlester, Lee. *The Earth, An Introduction to the Geological and Geophysical Sciences*. Englewood Cliffs, N.J.: Prentice-Hall, Inc., 1973. This classic covers the basics of geology.

Audiovisual/Software

Chapter 3

A computer program, "Mineralogy Laboratory," is available for IMB-pc computers from Geoscience Resources.

Chapter 4

Four films are available from Britannica Films: "The Rock Cycle," "Rocks That Form the Earth's Surface," "Rocks That Originate Underground," and "Minerals and Rocks."

Chapter 5

A computer program, "Rocks and Minerals Identification," is available for all Apple computers from Nasco.

Chapter 6

A computer program, "Introduction to Crystallography," is available for Apple II and IBM-pc computers from Geoscience Resources.

A videotape, "Aerial Photo Interpretations of Geologic Resources," is also available from Geoscience Resources.

Chapter 7

"Volcanoes and Volcanic Activity," 35 mm color, set of 43 slides, Wards.

A computer program, "Plate Tectonics," is available for Apple II computers from Geoscience Resources.

PLANNING THE CHAPTER

Chapter Sections	Page	Chapter Features	Page	Program Resources	Page
Section 1: Atomic Structure 3.1 Atoms **(H)** 3.2 Isotopes and Ions **(H)**	51 51 53	**Discover:** Making Atomic Models **(H)** Section Review	52 53		
Section 2: Chemical Compounds 3.3 Molecules and Bonds **(H)** 3.4 The Periodic Table **(H)**	54 54 56	**Discover:** Naming Simple Compounds **(H)** Section Review **Careers:** Lapidary, Mining Geologist, Stonemason	57 57 58	Investigation 3.2: *Comparing Compounds* **(H)**	**TRB 11** **LM 13**
Section 3: Kinds of Minerals 3.5 Silicates, Carbonates, and Sulfates **(A)** 3.6 Oxides, Halides, Sulfides, and Native Elements **(A)**	59 59 61	**Discover:** Observing Silicates **(A)** Section Review	60 61	Cross-Discipline: Science and Social Studies, *Understanding Maps, Gemstones, and World Trade* **(A)**	**TRB 5**
Section 4: Physical Properties of Minerals 3.7 Simple Mineral Properties **(B)** 3.8 Crystal Structure **(B)**	62 62 69	A Matter of Fact **Activity:** Testing a Mineral's Hardness **(B)** **Skill Activity:** Using a Classification Key **(A)** **Discover:** Breaking Crystals **(B)** Section Review **Investigation 3:** Properties of Minerals **(B)**	62 64 67 69 69 70	Critical Thinking **(H)** Investigation 3.1: *Specific Gravity* **(H)** Concept Extension: *Crystal Shapes* **(A)** Reading for Content: *Using Mapping to Understand What You Read* **(B)** Student Record Book: Textbook Investigations **(B)**	**TRB 5** **TRB 9** **LM 11** **TRB 5** **LM 227** **TRB 5** **LM 177** **TRB 5**
Chapter 3 Review	71			Vocabulary **(A)** Tests **(A)** Computer Test Bank	**TRB 3** **LM 127** **TRB 12**

(LM) Laboratory Manual/Study Guide,   **(TRB)** Teacher's ResourceBank™

B = Basic   **A** = Average   **H** = Honors

The coding Basic, Average, and Honors indicates sections or subsections that might be appropriate for different levels of learners. For additional suggestions regarding choice of topic and depth of coverage, see the Pacing Chart on pages T16–T20.

CHAPTER CONCEPTS, OBJECTIVES, AND TERMS

Section	Concepts	Objectives	Science Terms
Section 1: Atomic Structure	■ All things are made of atoms. Atoms are made of protons, electrons, and neutrons. **(3.1)** ■ There are 90 different kinds of atoms, called elements, that occur naturally. **(3.1)** ■ Isotopes of an element have the same number of protons but a different number of neutrons in their nuclei. **(3.2)** ■ An ion is an atom with an electrical charge. **(3.2)**	■ **Identify** the parts of the atom. ■ **Compare** atoms, isotopes, and ions.	atoms elements protons electrons neutrons nucleus isotopes ion
Section 2: Chemical Compounds	■ Atoms are often found joined together as molecules. **(3.3)** ■ Compounds are held together by two types of bonds: ionic and covalent. **(3.3)** ■ To help scientists and students organize information about each element, all the elements have been placed in a periodic table. **(3.4)** ■ The elements in the periodic table are listed in order of increasing proton number. The number of protons in an element is its atomic number. **(3.4)**	■ **Describe** two kinds of chemical bonds. ■ **Identify** an element's properties by studying the periodic table.	molecule compound ionic bond covalent bond atomic number
Section 3: Kinds of Minerals	■ Minerals are grouped into families based on their chemical composition. **(3.5)** ■ Families of minerals include silicates, carbonates, sulfates, oxides, halides, and sulfides. **(3.5, 3.6)** ■ Minerals formed of only one element are native elements. **(3.6)**	■ **Define** the term *mineral.* ■ **Classify** minerals into families based on their chemical formulas. ■ **Compare** native elements to other mineral families.	mineral tetrahedron
Section 4: Physical Properties of Minerals	■ Minerals can be identified by common physical properties. **(3.7)** ■ Some minerals have special properties that aid in their identification. **(3.7)** ■ Most minerals build crystal formations. **(3.8)** ■ A gem is a rare mineral with the special characteristics of beauty, durability, and value. **(3.8)**	■ **Describe** the five major mineral tests. ■ **List** a mineral example for each of the crystal forms. ■ **Explain** what qualities make a mineral a gem.	streak luster cleavage fracture crystal gem

CHAPTER MATERIALS

Title	Page	Materials
Discover: Making Atomic Models	52	*(per student)* colored paper, scissors
Discover: Naming Simple Compounds	57	*(per student)* paper, pencil
Discover: Observing Silicates	60	*(per student)* quartz, plagioclase, orthoclase feldspar
Activity: Testing a Mineral's Hardness	64	*(per group of 3 or 4)* paper, pen or pencil, penny, steel nail, glass plate, mineral samples (5)
Skill Activity: Using a Classification Key	67	*(per student)* galena, mineral samples (4), mineral classification key, reference section page 566
Discover: Breaking Crystals	69	*(per student)* table salt, dark paper, magnifying glass, pencil
Investigation 3: Properties of Minerals	70	*(per group of 3 or 4)* penny, steel nail, glass plate, streak plate, minerals: pyrite, chalcopyrite, sphalerite, galena, magnetite, quartz, hematite, talc, calcite, limonite

TEACHING SUGGESTIONS

Section 1: Atomic Structure

Demonstration: Producing Three Elements

Purpose
To produce three common elements

Background
Most of the substances that make up the earth are compounds. Water, one of the most familiar of these compounds, is made up of the elements hydrogen and oxygen. Hydrogen and oxygen, then, might be called important building blocks of the earth. Metals are building blocks, too. Many rocks and minerals are composed of metal compounds. In this demonstration you will break down some compounds and find out what their "building blocks" are.

Materials
Lead oxide
Carbon
Crucibles (2)
Copper oxide
Clay triangle or wire gauze
Bunsen burner
Zinc
Test tube
Dilute hydrochloric acid
Wood splints
Ring stand and ring
Test-tube holder

Procedure
A. Separating Metals from Compounds
1. Thoroughly mix about 3 parts lead oxide with 1 part carbon. Put the mixture in a crucible. Do the same with 3 parts of copper oxide and 1 part carbon. Then place this mixture in another crucible.
2. Place each crucible, in turn, on the clay triangle (or wire gauze) and heat in the flame of the Bunsen burner. When a definite change has taken place inside the crucible, remove the flame.
3. Have students look carefully at the crucibles. Draw a table on the chalkboard. Record student observations in the first two columns of the table.
4. A gas is given off when each crucible is heated. Test the gas with a glowing splint for the presence of oxygen or carbon dioxide. What gas was given off during heating? *(Carbon dioxide)* Add this information to the table.

Oxide Used	Appearance After Heating	Metal Formed	Gas Produced
Lead oxide	*silver-gray*	*lead*	*carbon dioxide*
Copper oxide	*reddish-brown*	*copper*	*carbon dioxide*

5. Put a piece of zinc the size of a pea into a Pyrex or Kimax test tube. Just cover the zinc with dilute hydrochloric acid. Have the students make a prediction about the reaction that is taking place. (*Answers will vary.*)
6. Test the gas given off by the reaction. First test for oxygen and carbon dioxide. What are your results? (*Neither oxygen nor carbon dioxide is given off.*)
7. Now test for the presence of hydrogen. Bring a burning splint near the mouth of the test tube. A loud "pop" shows that the hydrogen gas is being given off. What do you think is the other product of the reaction? (*Zinc chloride*)

Questions to Ask the Students

You have observed three reactions in this demonstration. Complete word equations for these reactions.
copper oxide + carbon → *copper + carbon dioxide*
lead oxide + carbon → *lead + carbon dioxide*
zinc + hydrochloric acid → *zinc chloride + hydrogen*

Section 2: Chemical Compounds

Outside Speaker

You could invite a hobbyist to talk to the class about mineral-collecting sites, and to show and describe some of his or her collection.

Section 3: Kinds of Minerals

Field Trip

This unit provides an ideal opportunity for a field trip either to the local museum or to a place where students can collect rocks and minerals. Many communities have rock and mineral clubs.

Section 4: Physical Properties of Minerals

Class Activity

The process of testing a mineral's hardness should be demonstrated. Students often fail to understand that the test can be made either by scratching the mineral with the testing object or by scratching the testing object with the mineral. Demonstrate how to scratch the mineral's surface and then verify the scratch by trying to wipe the scratch off.

Class Activity

Cleavage and fracture also should be demonstrated. Use mica to demonstrate basal (one-directional) cleavage. Then break calcite or halite to demonstrate cleavages in more than one direction. Use pyrite, quartz, and magnetite to demonstrate fracture.

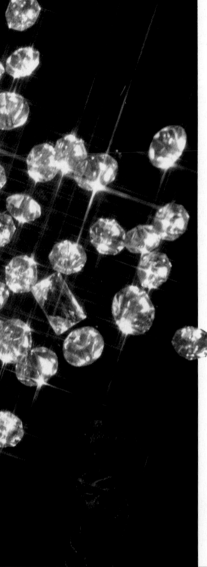

GEOLOGY

The formation of diamonds has been studied for a long time, but much of the process remains a mystery. Scientists are relatively certain that these pure-carbon gems begin to form at depths of about 150 km beneath the earth's surface. These diamond pipes begin in the upper mantle. In this region, there are many molten, or plasticlike, rocks.

- Is a diamond a mineral or a rock?
- What is fool's gold, and how is it different from real gold?
- Why are the rocks of the continental crust lighter than the rocks of the oceanic crust?
- What causes some of the rocks of the earth's interior to be plasticlike?

By reading the chapters in this unit, you will learn the answers to these questions. You will also develop an understanding of concepts that will allow you to answer many of your own questions about the study of the earth.

Cut diamonds

49

INTRODUCING THE UNIT

Have the students look at the Unit Opener photograph on these pages. Point out that these are cut diamonds. Read the text and use the questions to guide a class discussion. Do not expect the students to be able to answer the questions now. The questions are motivational and the students should be aware that they will be able to answer them when they have finished the Unit.

Discussion   Although the students will be familiar with diamond jewelry (rings, watches, necklaces, and so on), they may not be aware that only about 20 percent of the diamonds mined each year are gem quality. The great majority are used in industry for cutting tools. Ask the students to name the quality of a diamond that makes it suitable for use in industry. (Diamonds are extremely hard.)

Extension   You can extend the discussion by bringing in other Unit themes. What are minerals and how are they formed? What are the three types of rocks and how do they differ? How is a metal different from a nonmetal? What materials make up the earth's crust? You may wish to record student responses and review them at the end of the Unit.

CHAPTER 3

CHAPTER OVERVIEW

This chapter introduces the structure of atoms and molecules. It presents the periodic table of elements and the bonding of atoms. The grouping of minerals into families and mineral identification are discussed.

Section 1: Atomic Structure In this section, the concept and structure of atoms are presented. Ions and isotopes are defined and compared.

Section 2: Chemical Compounds The concept of the molecule is discussed in this section. Two types of chemical bonds, ionic and covalent, are presented and compared. The periodic table and families of elements are introduced.

Section 3: Kinds of Minerals A mineral is defined in this section. The classification of minerals into families is discussed, and several families are compared. Finally, native element is defined.

Section 4: Physical Properties of Minerals Some properties of minerals are explained: color, streak, luster, hardness, cleavage, and fracture. Mineral crystals are discussed, and qualities of gems are presented.

Minerals

How many different objects are pictured here? Do you think it is possible that all these objects are the same? Although their colors are different, these objects have many characteristics in common. What similar characteristics can you see? What other characteristics about these objects might help you decide if they are the same? 1

Various crystal minerals

1 Similar characteristics might be color and luster. Other characteristics might be crystal shape and hardness.

50

CHAPTER MOTIVATING ACTIVITY

Challenge the students to see who can make the longest list of objects, events, or concepts related to atoms. Examples might include atom smashers and nuclear power plants.

Use mineral and crystal samples to spark the students' interest. Pass a few specimens around the class. Samples of graphite, fluorite, pyrite, galena, muscovite, talc, and quartz are rather inexpensive and may help arouse interest.

1 Atomic Structure

SECTION OBJECTIVES

After completing this section, you should be able to:

- **Identify** the parts of the atom.
- **Compare** atoms, isotopes, and ions.

NEW SCIENCE TERMS

atoms
elements
protons
electrons
neutrons
nucleus
isotopes
ion

3.1 Atoms

Look around your classroom. How many different objects do you see? What do you think these objects have in common? People, tables, chairs, books, and even the air are all made of very small particles. These particles are called atoms. **Atoms** are the building blocks of matter.

Each kind of atom forms an element. **Elements** are substances made of only one kind of atom. There are 90 different elements that occur naturally on Earth. In addition to the 90 natural elements, 19 other elements are known. If they do not occur naturally, where do you think they come from?[1]

The tiniest speck of dust and the largest mountain are both made of atoms. A speck of dust is very small, but it contains thousands of atoms. Try to imagine how small an atom is. Take a piece of paper and tear it in half. Now tear the half in half, and continue this until you cannot tear the paper anymore. You now have the smallest part of a piece of paper that still has all the characteristics of paper, but it is still far bigger than an atom. Imagine taking that smallest piece of paper and dividing it 1000 more times. You would now be approaching something close to the size of an atom.

There are many different kinds of atoms. Figure 3–1 shows a substance made of two kinds of atoms: fluorine (F) and calcium (Ca). Atoms cannot be broken down into smaller parts without changing their characteristics.

Fluorine and calcium combine to form *fluorite*. Different atoms can combine to make many different substances. Fluorite can be separated into atoms of fluorine and calcium.

Atoms are small, but they are not the smallest particles that scientists know about. Atoms are made up of subatomic particles called *protons, electrons,* and *neutrons*. These subatomic particles do not have the characteristics of the elements from which they come. The protons, neutrons, and electrons of fluorine are the same as the protons, neutrons, and electrons of calcium.

[1] They are made in laboratories.

Figure 3–1. Calcium is a solid and fluorine is a gas. They combine to form the mineral fluorite, shown here.

Section 1   Atomic Structure   **51**

SECTION FOCUS

Section Overview   This section concentrates on atoms and the structure of atoms. Protons, neutrons, and electrons are discussed. Isotopes and ions are described.

Section Objectives   For a list of section objectives, see pupil's edition page.

New Science Terms   For a list of new science terms in this section, see pupil's edition page. You may wish to read over the terms with the students.

SECTION DEVELOPMENT

3.1 Atoms

DISCUSSION   You may wish to have one paper-tearing activity for the class or to have small groups do it. This activity should help them understand how small an atom is. Have the students study Figure 3–1. *Thinking Critically:* Ask the students why fluorine and calcium are important to humans. (Calcium forms strong bones and teeth; fluorine helps prevent tooth decay.) Point out the terms proton, electron, and neutron, which will be defined on the next page.

BACKGROUND INFORMATION

The Greek philosopher Democritus first put forth an atomic theory around 400 B.C. He proposed that if matter were divided over and over again, eventually a particle would be reached that could be divided no further. This particle he called an atom, from the Greek word meaning "not cuttable."

No advances were made in atomic theory until 1803, when John Dalton of England proposed that all elements are made of atoms that could not be split, nor could they be made or destroyed.

Sir Joseph Thomson discovered electrons in the late 1890s. Ernest Rutherford published his nuclear theory in 1911. And in 1913, Niels Bohr suggested a model for the atom with electrons arranged in shells surrounding a nucleus. In 1932, James Chadwick discovered neutrons. Lise Meitner and Otto Frisch in 1939 used the term *fission* for the splitting of an atom.

DEMONSTRATION

For a demonstration of producing three elements see page 49e preceding this chapter.

Thinking Skill *(Drawing Conclusions)*

The students' models should show a nucleus of protons and the corresponding number of electrons in orbits around the nucleus. If there are too many electrons in an energy level, they would repel each other and possibly tear the atom apart.

DISCUSSION Because the word *orbit* is also used in explaining the atom, review the structure of the solar system with the students. Emphasize that the planets travel in orbits around the sun. Direct the students' attention to the models of atoms, having them identify the nucleus and electrons.

THINKING SKILL *(Inferring)* As calcium and fluorine combine to form fluorite, the atoms of many other elements combine to form other substances. After the students have studied the models of atoms, ask why they think atoms combine. (The students might suggest that the orbits could link into one another, that the tiny particles of one atom could jump to another, or that the positive protons may attract the negative electrons.) If the students have access to an encyclopedia, you may wish to have them check their hypotheses by looking at the entry on ''atom.''

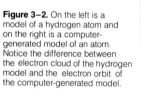

DISCOVER

Making Atomic Models

Cut out circles of colored paper to represent protons, electrons, and neutrons. Try to create flat models of atoms with one, five, and ten protons. In real atoms, the first energy level can have only one or two electrons, and the second level no more than eight electrons. Why do you think there are limits to the number of electrons in each level?

You may find it confusing that the same subatomic particles can form so many different kinds of atoms. Think of a common substance like wood. How many different things can be made of wood? Name at least ten. If wood is combined with two other substances, such as steel and concrete, how many things can be made? You can probably see now that making 109 different atoms from protons, neutrons, and electrons is not too difficult.

Protons are subatomic particles that have a positive electrical charge. **Electrons** are subatomic particles that have a negative charge. These opposite charges attract each other and hold atoms together, just as the opposite ends of two magnets attract and hold the magnets together. Protons repel protons, and electrons repel electrons. If you try holding like ends of two magnets together, they will repel each other.

Neutrons are subatomic particles that have no electrical charge. Protons and neutrons are clustered together to form the core of an atom. This core is called the **nucleus.** The neutrons keep the protons away from each other.

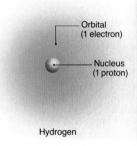

Orbital
(1 electron)

Nucleus
(1 proton)

Hydrogen

Figure 3–2. On the left is a model of a hydrogen atom and on the right is a computer-generated model of an atom. Notice the difference between the electron cloud of the hydrogen model and the electron orbit of the computer-generated model.

For hundreds of years, scientists have been trying to decide what atoms would look like if they could be seen. As a result, many models have been developed to help explain the structure of atoms. One of these models was developed by the Danish physicist Niels Bohr. The model Bohr proposed looks very much like a model of the solar system. At the center of the atomic model is the nucleus, with the electrons orbiting it—much as planets orbit the sun. The orbits in which the electrons move are called *energy levels.* The modern model of an atom is somewhat different from Bohr's model. The electron orbits are shown as a cloud surrounding the nucleus. Bohr's model is still used, however, because it shows the nucleus and electrons in a simplified manner.

○ *What are the three subatomic particles that make up atoms?*
○ *What is the electrical charge of each of these particles?*

52   Chapter 3   Minerals

Answers–3.1

○ Protons, neutrons, and electrons make up an atom.

○ Protons are positive; neutrons are neutral; and electrons are negative.

3.2 Isotopes and Ions

When there are equal numbers of electrons and protons, an atom is electrically neutral. The number of neutrons, however, may or may not be the same as the number of protons and electrons.

Suppose you have three different chocolate chip cookies. One has six chips, another has seven chips, and the third has eight chips. You know that they are not all exactly the same; however, they are all chocolate chip cookies. Similarly, every carbon atom has six protons and six electrons, but a carbon atom may have six, seven, or even eight neutrons. They are all carbon atoms, but they are not exactly the same. Atoms of an element that have different numbers of neutrons are called **isotopes.**

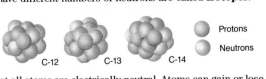

Protons

Neutrons

Not all atoms are electrically neutral. Atoms can gain or lose electrons. When an atom gains or loses electrons, the electrical charge is no longer neutral. If an atom gains electrons, it has more electrons than protons. Therefore, the atom has a negative charge. If an atom loses electrons, it then has more protons than electrons. What kind of charge does it have? [1] An atom that has gained or lost electrons and therefore has an electrical charge is called an **ion.** The opposite electrical charges of some ions pull them together. When ions are attracted to each other, they may combine, forming new substances.

○ *How many neutrons do different carbon isotopes have?*
○ *Why are ions attracted to each other?*

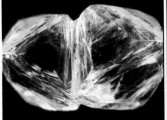

Figure 3–3. Although these forms of carbon appear to be very different, both the graphite and the diamond are composed mostly of Carbon-12. The other common carbon isotopes are shown at left.

Section Review

READING CRITICALLY

1. An atom of the element hydrogen has one proton. How many electrons would there be in a positive hydrogen ion?
2. Since carbon has six protons, how many electrons would you expect to find in an electrically neutral carbon atom? Why?

THINKING CRITICALLY

3. If two fluorine atoms were to lose one electron each, would they likely be attracted to each other? Explain your answer.
4. Suppose that there are two isotopes of an element with eight protons and eight electrons. How many neutrons might each of these isotopes have?

1 positive

DISCUSSION The students may have learned enough about signed numbers to know that, for example, $-3 + 3 = 0$, zero being neutral. ***Thinking Critically:*** Ask the students to think of other things that are the same but different. Some suggestions might be identical twins; two watermelons, one with seeds and one without; or two daisies, one may have 100 petals and the other 96. Have the students look at Figure 3–3 and discuss the isotopes of carbon. The students should be able to understand the attraction of ions by relating it to the attraction of a magnet.

SECTION REVIEW

Summary Review the structure of the atom and the names of the components of the atom. Compare isotopes and ions. Ask the students if they can now define all the new science terms in this section.

Reinforcement You may wish to draw and label simple diagrams of atoms with a specified number of protons, neutrons and electrons: for example, 2 protons, 2 neutrons, and 2 electrons; or 5 protons, 5 neutrons, and 5 electrons.

Answers to Section Review

Reading Critically

1. A hydrogen ion has no electrons.
2. An electrically neutral carbon atom has six electrons. The number of protons and electrons is the same in an electrically neutral atom.

Thinking Critically

3. No, they would both have a positive charge and therefore would not attract.
4. They might have 7, 8, or 9 neutrons.

Answers–3.2

○ Different carbon isotopes have 6, 7, or 8 neutrons.
○ Each ion has an unequal number of protons and electrons; thus, each carries a charge. Oppositely charged ions will attract each other.

SECTION FOCUS

Section Overview In this section, a molecule is defined in terms of atoms. Two basic chemical bonds—ionic and covalent—are discussed. The periodic table of elements is presented and element families are read from the table.

Section Objectives For a list of section objectives, see pupil's edition page.

New Science Terms For a list of new science terms in this section, see pupil's edition page. You may wish to mention to the students that some words have more than one meaning, for example, bond and compound. They will learn the scientific meaning in this section.

SECTION DEVELOPMENT

3.3 Molecules and Bonds

DISCUSSION Most students have probably heard water referred to as "H₂0." Remind them that this means water is a compound made of hydrogen and oxygen atoms. If you could divide a drop of water over and over again into smaller drops until you reached the smallest one possible, it would be a molecule of water. If it were divided any more, it would then be back to its components of hydrogen and oxygen atoms. *Thinking Critically:* Have the students discuss how molecules and compounds are alike. (Neither one is a single thing. A molecule is two or more atoms and a compound is two or more elements.) Making a simple diagram of a sodium atom (11 electrons) and a chlorine atom (17 electrons) might help the discussion of the ionic bond. Tell the students that elements with 10 electrons (sodium has 11) and 18 electrons (chlorine has 17) are chemically inactive.

NEW SCIENCE TERMS
molecule
compound
ionic bond
covalent bond
atomic number

SECTION OBJECTIVES
After completing this section, you should be able to:
- **Describe** two kinds of chemical bonds.
- **Identify** an element's properties by studying the periodic table.

3.3 Molecules and Bonds

You probably have friends that you like to spend time with. Maybe you spend time together at the library, in the lunchroom, or at the mall. Your group might even have a special name. Similarly, atoms are usually found together in groups called molecules. A **molecule** is a group of two or more atoms chemically combined. The atoms may be alike or they may be different.

Some molecules have only two atoms and are so small that millions of them could fit on the head of a pin. Some large molecules also exist. Many of the molecules in your body contain thousands of atoms; only a few hundred of these would fit on that same pinhead.

Atoms of different elements combine, or bond, to form compounds. A **compound** is a combination of two or more different kinds of elements.

There are two basic types of chemical bonds: *ionic* and *covalent*. An **ionic bond** forms when electrons move from one atom to another. Table salt, sodium chloride, has ionic bonds. When a sodium atom gives one of its electrons to a chlorine atom, the sodium becomes a positive ion, and the chlorine a negative ion. What do you think happens to these two ions? **1**

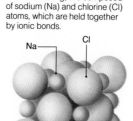

Figure 3–4. These halite crystals, shown in both the photograph and the drawing, are composed of sodium (Na) and chlorine (Cl) atoms, which are held together by ionic bonds.

Na   Cl

Halite

54   Chapter 3   Minerals

1 They are attracted to each other.

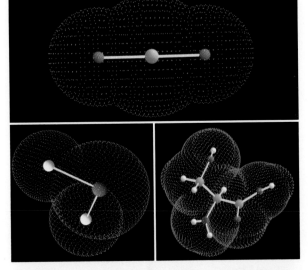

Figure 3–5. The computer-generated models show molecules that contain covalent bonds. The chemical formulas for carbon dioxiode (top), water (left), and serine (right) are shown in Table 3–1.

TABLE 3–1: COVALENT COMPOUNDS	
Covalent Compound	**Chemical Formula**
Water	H_2O
Carbon dioxide	CO_2
Serine	$CH_2CHOHNH_2COOH$

A **covalent bond** is formed when atoms share electrons. Water molecules form covalent bonds. In a single water molecule, two hydrogen atoms and one oxygen atom share electrons. Sharing electrons produces strong bonds between atoms, just as sharing experiences produces strong bonds between friends. Table 3–1 shows the chemical and structural formulas of several covalent compounds.

○ *What is the difference between an ionic bond and a covalent bond?*

○ *Give an example of each type of bond.*

Section 2   Chemical Compounds   **55**

DISCUSSION Making a simple diagram on the chalkboard of two hydrogen atoms (1 electron each) and one oxygen atom (8 electrons) might help the discussion of the covalent bond. Direct the students' attention to Figure 3–5. Tell the students to look at the photographs of the computer generated molecules.

THINKING SKILL *(Interpreting a Diagram)* Ask the students how the structural formula reflects the true shape of a molecule. (The structural formula shows, in a 2-dimensional way, the location of the atoms in the molecule.)

Answers–3.3

○ In an ionic bond, the electron(s) actually move from one atom to another atom. In a covalent bond, electrons are shared between two atoms.

○ Salt has ionic bonds. Water has covalent bonds.

55

3.4 The Periodic Table

DISCUSSION Before beginning the discussion, you may wish to have the students study the complete periodic table on pages 558–559 of the Reference Section. They should be informed that elements having similar properties are grouped together in the table and that elements are arranged horizontally by atomic number (the number of protons in the atom). Explain that the abbreviations for some of the elements come from Greek or Latin terms for the element. For example, Fe comes from *ferrum*, the Latin word for "iron."

SKILLS *(Reading a Table, Classifying)*
Elements are listed in the periodic table. Classify the following as elements or compounds.

tin (element)
pewter (compound)
brass (compound)
sulfur (element)
oxygen (element)
zinc (element)
cobalt (element)
carbon dioxide (compound)
glass (compound)
carbon (element)
neon (element)
salt (compound)
iron (element)
sugar (compound)
silver (element)
uranium (element)

3.4 The Periodic Table

In addition to the 90 naturally occurring elements, scientists have produced many other elements in the laboratory. To remember the characteristics of all these elements would be very difficult. To help scientists and students organize information about each element, all the elements have been placed in a table. This table, called the *periodic table*, may be found in the Reference Section on pages 558–559. Shown in Figure 3–6 are the 15 elements commonly found on Earth. The ten shown in red are common in the solid earth, and the five shown in blue are common in water and air.

The order of the elements in the periodic table is based on the properties of each element. The elements with similar properties are listed in groups. Each group is in order of increasing number of protons. The number of protons an atom contains is its **atomic number.** For example, helium has two protons and therefore an atomic number of two. Oxygen has eight protons. What is the atomic number of oxygen? **1**

Figure 3–6. Highlighted in this version of the periodic table are the most common elements found on Earth.

PERIODIC TABLE OF ELEMENTS

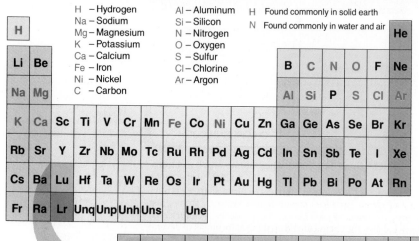

1 eight

BACKGROUND INFORMATION

Some of the elements, such as gold, arsenic, iron, lead, and mercury, were known to the ancient Greeks. The first "modern" element, platinum, was discovered by Julius Scaliger in 1557. Hennig Brand discovered phosphorus in 1669. Seventy-four other elements were discovered from the time that Georg Brandt discovered cobalt in 1737 until rhenium was discovered in 1925.

Beginning in the mid-1800s, chemists started to organize the known elements into a periodic table, listing elements in order of increasing atomic weight. The weight of an atom was compared with an atom of carbon 12, which had been assigned the weight of 12. Atomic mass is the second number, usually a decimal number, for each atom on the periodic chart. The number reflects the average unit mass of all the protons and neutrons in the atom's nucleus. The mass of the atom's electrons is almost negligible in this total. The mass number is an average because most elements exist as various isotopes.

As the atomic number increases, atoms may become less stable. Notice that the most common elements on Earth are elements with lighter nuclei.

The members of your family probably have many characteristics in common. Maybe all of you have dark hair and brown eyes. Each family of elements also has similar characteristics. Elements found in the same vertical column in the periodic table make up a family, or group. For example, the elements in Group 1 are all soft metals that are easily melted.

Elements in the same group sometimes substitute for each other in the formation of compounds. For instance, *feldspars*, a group of minerals, may contain potassium, sodium, or calcium. Although the metals affect the mineral's color, its other characteristics are not changed.

○ **What is meant by the atomic number of an atom?**
○ **What is the periodic table?**

Section Review

READING CRITICALLY

1. One neutral silicon atom has 14 protons and 14 neutrons. How many electrons does this silicon atom have? What is its atomic number?

2. Which elements are most commonly found in solid earth? Which are most common in water and air?

THINKING CRITICALLY

3. What elements could probably replace carbon in chemical compounds?

4. If there were an electrically neutral element with the atomic number of 110, how many protons and electrons would it have? How many neutrons might it have?

Figure 3–7. The two minerals shown at left are feldspar. The difference in color is due to the presence of potassium in one sample and sodium in the other sample. Sodium and potassium are shown above.

DISCOVER

Naming Simple Compounds

Look at the periodic table on pages 558–559. All the elements in Group 1 easily lose one electron, and all the elements in Group 17 easily gain electrons. See how many different compounds you can create by combining elements from these two columns. Using "sodium chloride" as an example, try naming the compounds as you form them.

Section 2   Chemical Compounds   **57**

DISCUSSION The vertical aspect of the grouping in the periodic table is emphasized. Call to the students' attention the fact that elements in the same vertical column sometimes substitute for each other in compounds. ***Thinking Critically:*** Ask the students what elements might substitute for cadmium in compounds. (zinc and mercury)

DISCOVER:
Naming Simple Compounds

Thinking Skill (*Interpreting Data*)

Using the data given, the students should discover that they use the element name from Group 1 with the Group 17 name, changing the -ine to -ide. Some examples might be sodium iodide, hydrogen chloride, potassium fluoride, and lithium bromide.

SECTION REVIEW

Summary You may wish to have the students name the two ways in which molecules bond, and compare the two types of bonds. Review the arrangement of the elements in the periodic table and have them identify and explain the atomic number of a few elements.

Reinforcement Have the students pair up with the person in front of or behind them and quiz each other on the new science terms on page 54. They could make a supplementary list of any other terms they think important.

Answers—3.4

○ The atomic number of an atom is the number of protons the atom contains.

○ The periodic table is a table displaying all the elements arranged horizontally by their atomic numbers and vertically by groups.

Answers to Section Review

Reading Critically

1. The silicon atom has 14 protons; its atomic number is 14.

2. Most common in solid earth: oxygen, sodium, magnesium, aluminum, silicon, sulfur, potassium, calcium, iron, and nickel. Most common in water and air: hydrogen, carbon, nitrogen, chlorine, and argon.

Thinking Critically

3. Silicon, germanium, tin, and lead could probably replace carbon in many chemical compounds.

4. It would have 110 protons and 110 electrons, and perhaps as many as 160 neutrons.

Discussion After the students have read the articles, you may wish to ask them if they think there are opportunities for the different careers in their locales. Interested students can write to one or all the addresses to get additional information. When they receive the information, the class should have a discussion on the career opportunities.

Extension You may wish to have a "Career Day" in the class. Students can report on information they have received. People involved in the particular careers could be invited to speak on their work and why they chose it. Pertinent films or videotapes might be available from guidance offices. Students might be interested in researching information on the following additional careers:

Jeweler

Mineralogist

Engraver

CAREERS

LAPIDARY

A *lapidary* is a person who cuts rough mineral samples into beautiful gemstones. A single cut can be the difference between a gem worth thousands of dollars and one that is worthless. Lapidaries must understand the optical effects of light and the structure of minerals.

People who would like to learn the art of gem cutting usually begin as apprentices.

For Additional Information
Gemological Institute of America
1660 Stewart Street
Santa Monica, CA 90406

MINING GEOLOGIST

Mining geologists play a major role in finding minerals and getting them out of the ground. They are also involved in the important task of restoring the environment in mined areas.

Most jobs in mining geology are through private employers in the industry. A bachelor's degree is required, and an advanced degree is often recommended. People who have gained experience as technical assistants to mining geologists can obtain positions by passing a professional licensing test.

For Additional Information
American Geological Institute
5205 Leesburg Pike
Falls Church, VA 22041

STONEMASON

Although most large buildings are now built with steel, *stonemasons* still play an important part in construction work. Many buildings have stone decorations, window sills, door frames, and exteriors. These stones must be cut, fitted, and secured with precision. Most stonemasons learn their skills through apprenticeship programs. Sometimes they take classes in general construction skills.

For Additional Information
Associated General
   Contractors of America, Inc.
1957 E Street, NW
Washington, DC 20006

58   Chapter 3   Minerals

Section 3: Kinds of Minerals

SECTION OBJECTIVES

After completing this section, you should be able to:
- **Define** the term *mineral*.
- **Classify** minerals into families based on their chemical formulas.
- **Compare** native elements to other mineral families.

NEW SCIENCE TERMS

mineral
tetrahedron

3.5 Silicates, Carbonates, and Sulfates

Have you seen the commercials on TV for products that contain a whole day's supply of vitamins and minerals? What is a day's supply of minerals? Where do these minerals come from?

A **mineral** is a naturally occurring substance consisting of a single element or compound. Most minerals come from nonliving things, such as the earth itself. In addition, most minerals are solids, formed from common elements.

Silicates   Silicon and oxygen are two of the most common elements in the earth's crust. Therefore, it is not surprising that many minerals contain these two elements. Minerals that contain silicon and oxygen are called *silicates* (SIHL uh kayts). The silicate minerals make up about 92 percent of the earth's crust.

One common silicate mineral is formed from a single silicon atom bonded to four oxygen atoms. These atoms form a network called a **tetrahedron,** or a four-sided structure. Silicon-oxygen tetrahedrons can bond together to form chains or sheets. These chains and sheets form a three-dimensional network that produces a very hard mineral called *quartz* (KWAHRTS).

Figure 3–8. This beautiful mineral is one form of the element sulfur.

Figure 3–9. Quartz is a silicate mineral formed of silicon-oxygen tetrahedrons.

59

SECTION FOCUS

Section Overview   In this section the term *mineral* is defined. Minerals are classified into families based on their chemical formulas. Native elements are distinguished from other minerals.

Section Objectives   For a list of section objectives, see pupil's edition page.

New Science Terms   For a list of new science terms in this section, see pupil's edition page. Students may need help in pronouncing *tetrahedron*. (teht ruh HEE druhn)

SECTION DEVELOPMENT

3.5 Silicates, Carbonates, and Sulfates

DISCUSSION   After answering the questions in the first paragraph, you may ask the students to describe a mineral. Discuss the minerals shown in Figures 3–8 and 3–9. You may wish to draw a tetrahedron, a triangular pyramid, on the chalkboard and supply the class with some tetrahedrons so they could try to arrange them in a chain on a flat surface.

EXTENSION   (Tie-in/Nutrition) Minerals are crucial to a person's well-being. Have lists of the recommended daily allowances of vitamins and minerals for each student or for small groups of students. ***Thinking Critically:*** Tell the students the names of the important minerals. (Iron, potassium, calcium, magnesium, iodine, sodium, and phosphorus are minerals necessary for good health. RDA lists may vary.) Have the students list foods that supply these minerals. (Dairy foods supply calcium, magnesium, and phosphorus. Seafood supplies iodine and sodium. Whole grains and meats supply iron and potassium.)

DISCOVER:
Observing Silicates

Thinking Skill (Comparing)

The students should note that the three silicates have similarities—texture and hardness. However, the three silicates may differ in color.

DISCUSSION You may wish to make a chart on the chalkboard, listing the three mineral families and the basic elements that are contained in each. Call the students' attention to Figure 3–10 and supply samples of as many of the minerals as possible. **Thinking Critically:** Ask why magnesium can substitute for some of the calcium in calcite and, thus, form dolomite, and why feldspar can contain sodium and potassium. (Because they are in the same columns on the periodic table.)

EXTENSION (Tie-in/Chemistry) If the students have access to a high school chemistry book or encyclopedia, they may wish to look up the chemical formulas for some of the minerals mentioned in the lesson. They can then identify the elements that form the compound. (quartz—SiO_2, silicon and oxygen; dolomite—$CaMgCO_3$, calcium, magnesium, carbon and oxygen; calcite—$CaCO_3$, calcium, carbon, and oxygen; feldspar—$KAlSi_3O_8$, potassium, aluminum, silicon, and oxygen)

DISCOVER

Observing Silicates

Look at a sample of quartz, plagioclase feldspar, and orthoclase feldspar. List as many similarities and differences as you can among these three silicates.

Feldspars are the largest group of silicate minerals. They are similar to quartz in that they are also silicates. Feldspars, however, contain calcium, potassium, and sodium atoms in addition to silicon atoms.

Just as paint may contain different-colored pigments, feldspars may contain other elements, such as iron and manganese. These different elements produce different-colored feldspars. However, they are all feldspars, just as all colors of paint have the basic characteristics of paint.

Carbonates   Only eight percent of the earth's crust is composed of nonsilicate minerals. One important nonsilicate family is the *carbonates*. Carbonate ions contain three oxygen atoms joined to a single carbon atom. A carbonate ion bonded to a calcium ion forms a mineral called *calcite*. If magnesium is substituted for half of the calcium, a different mineral, called *dolomite*, is formed. Dolomite is often used as a mineral supplement for people. It helps provide calcium and magnesium for preventing some bone diseases.

Sulfates   Another family of nonsilicate minerals is the *sulfates*. Sulfates have four oxygen atoms combined with one sulfur atom. If a sulfate ion bonds with calcium and water, a mineral called *gypsum* is formed. Gypsum, the most common sulfate mineral, is used to make drywall for buildings.

○ *What is the most common family of minerals?*
○ *What elements do sulfates contain?*

Figure 3–10. Many carbonate minerals, such as calcite (above), have commercial value. Gypsum, a sulfate mineral, can be made into drywall (right), a type of plasterboard used in the construction of many homes.

60   Chapter 3   Minerals

Answers–3.5

○ Silicates are the most common family of minerals.

○ Sulfates contain oxygen and sulfur atoms.

3.6 Oxides, Halides, Sulfides, and Native Elements

There are several other mineral families. Although these mineral families make up only a small percentage of the earth's crust, they are important as sources from which metals may be obtained.

Oxides In the *oxides*, oxygen is bonded to another element, such as iron or aluminum. Iron oxides include several different minerals; all are mined as sources of iron. The most common source of aluminum is also an oxide.

Halides and Sulfides Two other important mineral families are the *halides* and the *sulfides*. These families of minerals contain no oxygen. Sodium chloride is a halide called *halite*. *Pyrite*, also known as "fool's gold," is a sulfide of iron. Pyrite is known as fool's gold because it looks much like real gold, although it is nearly worthless. Look at Figure 3–12. Can you tell the fool's gold from the real thing? Some sulfides also have value as sources of metals such as lead, copper, and zinc.

Native Elements Some minerals contain only one kind of element. These rare minerals are called *native elements*. Many of the minerals in this group are valuable metals. Some examples are gold, silver, and copper. Uncombined nickel and iron are believed to exist in large amounts deep within the interior of the earth. However, uncombined nickel and iron are very rare in the earth's crust. Diamond and graphite are also native elements; they are pure carbon. They differ only in the way the carbon atoms are arranged.

○ *Why are oxides important?*
○ *Name two native elements.*

Figure 3–11. Oxides such as those shown above are commercial sources of iron, aluminum, and other metals.

Figure 3–12. Gold is a native element, while "fool's gold" is a compound called *iron pyrite*. The real gold is on the right in this sample.

Section Review

READING CRITICALLY

1. What are the major groups of minerals that make up the earth's crust?
2. What is the difference between native elements and other minerals?

THINKING CRITICALLY

3. Why does the silicate family of minerals make up such a large part of the earth's crust?
4. Why can magnesium replace calcium in the mineral dolomite?

3.6 Oxides, Halides, Sulfides, and Native Elements

DISCUSSION Few metals are found in the earth's crust in an uncombined state. Instead, they are obtained from ores, natural substances containing the metals. Discuss the importance of the oxides as ores. You may wish to point out that hematite, bauxite, and magnetite are oxides. Bauxite is an aluminum oxide; magnetite and hematite are iron oxides. *Thinking Critically:* Discuss why native elements are the most valuable metals. (Most metals are not found as pure metals, and the native elements are hard to find.)

EXTENSION (Tie-in/Economics) Twenty years ago, the price of gold was under $100 an ounce. Now it is about $400 an ounce. Have the students discuss what would account for such a price rise. (One reason is the general inflation of the world's economy—the price of everything has gone up in recent years. Another reason could be the law of "supply and demand": if there is a great demand for something available in small supply, then the price goes up.)

EXTENSION You may wish to have the students look at the map of Montana in the Science Sites booklet to see the location of Anaconda Reduction Works. This is one of the largest copper smelters in the world.

Answers–3.6

○ Oxides are sources for both iron and aluminum. These two elements are very important to the manufacturing industry.

○ Several answers are possible. Gold, silver, and copper are examples of native elements.

Answers to Section Review
Reading Critically

1. The major mineral groups that make up the earth's crust are silicates, carbonates, sulfates, oxides, halides, and sulfides.
2. Native elements are formed from only one type of element, such as gold or silver. The other mineral groups are formed from molecules made of combinations of different elements.

Thinking Critically

3. Silicon and oxygen are very abundant.
4. Both have the same number of electrons in their outer shells.

Section 4: Physical Properties of Minerals

SECTION FOCUS

Section Overview Some properties by which minerals can be classified without laboratory analysis are discussed in this section. Mineral crystals and the distinction of some crystals as gems are presented.

Section Objectives For a list of section objectives, see pupil's edition page.

New Science Terms For a list of new science terms in this section, see pupil's edition page. Students will be familiar with several of the words, and you can remind them that they will be learning the definitions as they apply to minerals.

SECTION DEVELOPMENT

3.7 Simple Mineral Properties

DISCUSSION As discussed earlier, gold miners in the nineteenth century were often deceived by "fool's gold," or pyrite. It helps to have some properties by which minerals can be quickly classified or identified. Discuss how impurities and tarnish affect the property of color. ***Thinking Critically:*** Ask the students how the old saying, "You can't judge a book by its cover," is true for minerals. (The saying means that you should not judge an object, or even a person, by outward appearances. For minerals, exposure to air and weather may dull and change their colors so that they cannot be readily identified. Many minerals take on various colors due to impurities, so color alone may not give you positive identification of a mineral.)

NEW SCIENCE TERMS
streak
luster
cleavage
fracture
crystal
gem

SECTION OBJECTIVES
After completing this section, you should be able to:
- **Describe** the five major mineral tests.
- **List** a mineral example for each of the crystal forms.
- **Explain** what qualities make a mineral a gem.

3.7 Simple Mineral Properties

Do you collect stamps, coins, or baseball cards? If you have a lot of them, you probably have them classified in some way, such as by postmark, date, or team. The minerals found on Earth can be classified, too.

A detailed analysis of minerals can be done only in a laboratory with special equipment. However, there are some simple physical properties that you can use to quickly classify a mineral sample.

A MATTER OF FACT

Wayne F. Downey, Jr., a high-school student in Harrisburg, Pennsylvania, discovered a new mineral in a burning coal bed. This oxide of selenium was named Downeyite to honor his discovery.

Color Probably the first characteristic that you will notice about a mineral is its color. Unfortunately, this is one of the least dependable characteristics you can use to classify a mineral. Color may change with even a small amount of impurity. For instance, pure quartz is clear, but with impurities it can be pink, tan, red, black, or purple.

The surfaces of some minerals change color when exposed to air. The mineral *chalcopyrite* (kal koh PY ryt), which contains copper, iron, and sulfur, can tarnish. The copper in the mineral reacts with the air just as the copper in a penny does, making it dark and dull.

Figure 3–13. These minerals are all samples of quartz. Pure quartz is colorless; the colors of these samples are due to a variety of impurities.

BACKGROUND INFORMATION

There may be several different colors of the same minerals, and they are usually given special names. Quartz is a good example. Red quartz is jasper; tan quartz is chalcedony; black quartz is flint; pink quartz is known as rose quartz; and purple quartz is amethyst. The varying colors in a mineral are caused by impurities. Copper will make a mineral green, mercury will make it red, and sulfur will make it yellow.

Streak If you rub a piece of chalk on a sidewalk, it leaves a mark on the cement. The color of the powder that is left when a mineral is rubbed on a rough surface is called the mineral's **streak.** Sometimes the color of the streak is different from the color of the mineral sample. For instance, the streak of golden-colored pyrite is greenish black, and the streak of silver-colored hematite is brick red.

Luster Have you ever admired the appearance of a shiny new car? What you are admiring is the luster of the surface. The **luster** of a mineral refers to the appearance of its surface in reflected light. There are two types of luster: metallic and nonmetallic. If the mineral shines like gold, copper, or silver, then the luster is metallic. If the mineral does not shine like a metal, then its luster is nonmetallic. A nonmetallic luster can be further described using terms such as waxy, pearly, glassy, dull, or brilliant. What common objects can be considered as examples of each of these lusters? **1**

1 Answers will vary but some common examples are: a waxy candle, a pearly button, a glassy water surface, dull dirt, and a brilliant diamond.

Figure 3–14. The girl on the left is preparing a mineral streak. Two mineral streaks are shown on the right. Streak is not influenced by such things as tarnishing and impurities. Therefore, streak is a better characteristic for identifying minerals than color is.

Figure 3–15. Metals, such as gold and silver, have a luster that can be easily recognized. Other types of mineral luster are named for common substances, such as pearls and earth (soil).

63

DISCUSSION Call the students' attention to Figure 3–14. After they identify what the student is doing, they should be able to explain why mineral streaks are better than color for identification. You may point out that scientists rub the minerals on a rough tile called a "streak plate." Have the students study the minerals in Figure 3–15 and discuss how they would apply the luster test to them.

SKILL (*Constructing a Chart*) Supply small groups of students with a streak plate and six to ten labeled minerals. Have them list the mineral, the color of the mineral itself, and the color of the streak from each mineral. Have them make charts listing their findings. (The charts will vary. The information could be displayed horizontally or vertically, but each should display the information in an orderly fashion, with appropriate headings and a title.)

DISCUSSION Have the students complete the introductory activity for hardness. Suggest that they use mineral (not organic) substances and that they write 1 to 10 next to the listed objects, with 1 being the softest and 10 the hardest. If possible, have the ten minerals listed in 2 so that the hardness test can be demonstrated. Students should understand that either the mineral can be scratched by the test object or the mineral can scratch the test object.

ACTIVITY:
Testing a Mineral's Hardness

Skill (Observing)

Preparation of Materials For each group of 3 or 4 students, have a set of the five minerals labeled with the proper number: orthoclase feldspar—mineral # 1, chalcopyrite—mineral # 2, galena—mineral # 3, talc—mineral # 4, and quartz—mineral # 5. You can often obtain free pieces of glass from hardware or window glass stores. Cut these pieces into 10 cm × 10 cm squares, using a glass cutter. Then sand the edges with a steel file to remove the glass slivers.

CAUTION: Have the pupils test the minerals against the glass by placing the glass on a flat table surface to prevent breakage.

Answers to Conclusions/Applications

1. #4 (talc)
2. # 3 (galena)
3. # 2 (chalcopyrite)
4. # 5 (quartz) and # 1 (feldspar)
5. # 1=6; # 2=3.5 to 4.0; # 3=2.5; # 4 =1; # 5=7

Hardness List ten different objects and rank them from the softest to the hardest. If you included a diamond, then it should be at the bottom of your list, because a diamond is the hardest substance on Earth. At the other extreme is the mineral *talc*, from which talcum powder is made. All minerals fall somewhere between these two extremes.

A set of ten standard minerals is used to make a scale to measure the hardness of all minerals. This scale, called *Mohs' Scale of Mineral Hardness*, is shown in Table 3–2. The number 1 is assigned to talc, and the number 10 is assigned to diamond. A mineral that can scratch calcite (hardness of 3) but cannot scratch fluorite (hardness of 4) would have a hardness between 3 and 4.

Frequently the minerals of the Mohs' Scale are not available for comparison, so geologists have created a set of more common objects that can be used. For instance, if a certain mineral can scratch a penny, its hardness is greater than 3. If a steel nail can scratch that same mineral, you know its hardness is less than 4.5, or somewhere between 3 and 4.5.

ACTIVITY: Testing a Mineral's Hardness

How can you use common materials to determine the hardness of mineral specimens?

MATERIALS (per group of 3 or 4)

penny, steel nail, glass plate, mineral samples (5)

TABLE 1: A SCALE OF HARDNESS

| Mineral | Softer than | | | | Mineral hardness |
	Finger-nail	Penny	Steel nail	Glass plate	
1					
2					
3					
4					
5					

PROCEDURE

1. Make a data table like the one shown.
2. Assign numbers to each of your minerals, using the following guide: 1 = pink or tan color, 2 = gold color, 3 = silver color, 4 = white, pearly color, 5 = clear to white color, and glassy.
3. Scratch mineral 1 with your fingernail, the penny, the steel nail, and the glass plate. Record your results in the data table.
4. Test the other samples in the same way. Record your results in the data table.

CONCLUSIONS/APPLICATIONS

1. Which mineral was softer than your fingernail?
2. Which mineral was harder than your fingernail but softer than the penny?
3. Which mineral was harder than the penny but softer than the steel nail?
4. Which minerals were harder than the glass?
5. Review the hardness of the glass, nail, and penny from page 65, then assign a hardness to each mineral.

64 Chapter 3 Minerals

TABLE 3–2: MOHS' SCALE OF MINERAL HARDNESS	
Hardness — mineral	**Common material — hardness**
1 — Talc	
2 — Gypsum	
3 — Calcite	Fingernail—2.5
4 — Fluorite	Penny—3.5
5 — Apatite	Steel nail—4.5
6 — Feldspar	Glass plate—5.5
7 — Quartz	Steel file—6.5
8 — Topaz	
9 — Corundum	
10 — Diamond	

DISCUSSION Advise the students that hardness of a mineral does not indicate strength. A diamond is very hard, but it can chip, and if hit with a hammer, it could shatter completely. A steel nail can be hit repeatedly with a hammer and not crack or break, but it may bend. The minerals should be treated with care. Have the students study Mohs' Scale of Hardness. Give individual students and groups several opportunities to classify mineral samples or common objects according to hardness, using the minerals from Mohs' scale, if possible.

SKILLS (*Classifying, Ordering*) The students can take several mineral samples and group together those of similar hardness. They can also take several samples and order them from softest to hardest or from hardest to softest.

THINKING SKILL (*Drawing Conclusions*) Glass is often engraved with a monogram or with a picture or design; plaques are engraved with data and names; jewelry is usually engraved with initials. Engraving is usually done with an electric engraver, a rapidly vibrating tip that is similar to a drill bit. The engraving tip can be made of various minerals. Have the students discuss what minerals would make good engraving tips and what could be engraved. (Answers will vary, but the students should show an understanding of the fact that the engraving tip would have to be harder than the object being engraved. A diamond tip could engrave anything, but a diamond could not be engraved by any common substances.)

EXTENSION The hardness of pennies changed in 1980 when the Federal Government began to mint zinc pennies with a copper coating. Prior to 1980 pennies had a hardness of 3.

BACKGROUND INFORMATION

Mohs' Scale of Hardness is the name given to a method for comparing the hardness of various of minerals. Friedrich Mohs (1773–1839), a German mineralogist, invented the method in 1822. Mohs chose ten minerals as standard and assigned the numbers 1–10 to them, 1 denoting the softest mineral and 10 the hardest. The hardness of other minerals is judged by whether the mineral can scratch or can be scratched by the Mohs standard minerals.

DISCUSSION Have the students study the various minerals, looking for cleavage and fracture. Galena and pyrite, as well as halite, break into cubic pieces. Mica, graphite, and talc have cleavage in one direction; they split into thin sheets. If a mineral does not have cleavage, then it has fracture. Most fractures have rough, irregular surfaces. The broken pieces of quartz and obsidian have smooth, spiral surfaces; this is called a conchoidal fracture. Have a lodestone available so the students can see its magnetic property. You may wish to discuss other special properties like the double refraction of Iceland spar (calcite variety), fluorescence, and other optical properties. **Caution students about performing any chemical reactions on their own.**

SKILLS *(Identifying, Sorting)* Have several minerals displayed that show cleavage or fracture. Have the class identify the property shown and sort the minerals into the two categories.

Figure 3–16. Halite (top left) will always break into cubes, while the cleavage of calcite (top center) is diagonal. Mica (top right) forms thin, flat sheets. The glass (below top) and asbestos (below bottom) fracture when they break.

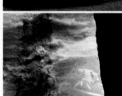

Cleavage and Fracture The terms *cleavage* and *fracture* describe how minerals break. Minerals that always break along flat surfaces have **cleavage.** Halite, for example, always breaks into little cubes. A mineral with cleavage will break where its chemical bonds are weakest.

If a mineral breaks along curved or irregular surfaces, it has **fracture.** Minerals can show fracture in several different ways. Some fractures have a curved surface similar to the inside of an eggshell. Another type of fracture produces long splinters or fibers. Fractures can also have jagged surfaces or sharp edges.

Special Properties Some minerals show special properties that can be used to identify them. Lodestone, for example, is a type of iron oxide that is magnetic. A piece of lodestone will attract iron filings just as a magnet does.

Calcite can be identified by its chemical reaction with weak acids. If a weak solution of hydrochloric acid is dropped on calcite, the acid begins to bubble. Many other minerals can be identified by special properties as well.

○ *What is meant by a mineral's streak?*
○ *What scale is used to rate the hardness of minerals?*

Figure 3–17. The lodestone shown here is magnetic and will attract a variety of iron objects.

66 Chapter 3 Minerals

Answers–3.7

○ Streak is the color of the powdered material left when a mineral is rubbed on a rough surface.

○ Mohs' Scale of Hardness is used to rate the hardness of minerals.

BACKGROUND INFORMATION

Many minerals have cleavage, or split apart neatly and clearly with smooth flat surfaces, because the molecules that make up the minerals form regular patterns. The joint where ten molecules meet is a cleavage zone, and when tapped, the mineral will split more easily along a cleavage zone. Because the molecules in each mineral form a distinctive pattern, each mineral has distinctive cleavage zones. A diamond cutter

continues

66

SKILL ACTIVITY: Using a Classification Key

BACKGROUND

Classifying things into groups in order to identify them is a skill that scientists find very useful. There are thousands of different minerals on the earth. To speed up the identification process, the testing information is often organized into a *classification key*. At each step in the key, one property is identified, and minerals not fitting the property are eliminated. This continues until only one mineral remains.

PROCEDURE

A mineral classification key is shown on page 566 in the Reference Section. Look at the first category of the key. *What property is tested?* Follow the sample that is given, and at each step, see if you can decide why that choice was made. Copy the table and fill it in as you classify your mineral samples.

1. Obtain a sample of galena from your teacher. Try to determine its luster. You should con-

clude that it fits in Category I because of its metallic luster.
2. Determine its streak. Galena's black streak puts it into Category A.
3. Test its hardness. Galena can be scratched by a penny, so its hardness is less than 3.5. It belongs in Group 1.

APPLICATION

Obtain four unknown samples of minerals from your teacher and classify them down to their group, using the key. Check your answers with your teacher when you have classified the samples.

USING WHAT YOU HAVE LEARNED

1. Did you classify your minerals correctly? If not, what problems did you have?
2. What are the advantages and disadvantages of this type of classification key?
3. Why is color an unreliable characteristic for classification?

TABLE 1: MINERAL CLASSIFICATION

Sample	Luster	Streak	Hardness	Color
1. Galena				
2.				
3.				
4.				
5.				

SKILL ACTIVITY:
Using a Classification Key

Objectives

- To test mineral properties.
- To use a mineral classification key.
- To classify sample minerals.

Discussion Familiarize the students with the Mineral Classification Key on pages 566–567 of the Reference Section. Have a sample of galena and four other minerals for each student to classify. The procedure can be a class activity with the galena to ensure that everyone understands the procedure. Students who finish quickly may enjoy doing more than four minerals. ***Thinking Critically:*** Have the class discuss why the classification process is made easier by a key. (The testing process is presented in an orderly way, and the scientist does not have to research the steps; the steps are already listed.)

Answer to Application

Answers will vary depending on the minerals used.

Answers to Using What You Have Learned

1. Answers will vary depending on samples used, but they should show that students used the process correctly.
2. Using a classification key allows an orderly sequence of tests and comparisons instead of a haphazard trial of various tests.
3. Color is an unreliable characteristic because impurities can cause a mineral to appear in many colors.

---

must be aware of a diamond's cleavage zones if he or she wants a large diamond split into several valuable diamonds and not shattered into a pile of small diamonds of much less value.

SKILL *(Observing)* Refer the students to Table 3–3: Crystal Shapes. Have the students observe the photographs of each crystal carefully. ***Thinking Critically:*** Ask the students to describe the physical properties of a crystal in their own words. (The students should respond by stating that a crystal is a (three-dimensional) solid with a definite shape, and a smooth plane surface called a face.)

TABLE 3–3: CRYSTAL SHAPES

Mineral	Geometry	Axes
Galena	Cubic	Three axes — all equal length and all right angles
Chalcopyrite	Tetragonal	Three axes—two equal length and all right angles
Quartz	Hexagonal	Four axes—three equal length and one at right angle to the others
Olivine	Orthorhombic	Three axes—all different lengths and all at right angles
Gypsum	Monoclinic	Three axes—length varies and two at right angles
Microcline	Triclinic	Three axes—all different lengths and no right angles

3.8 Crystal Structure

You may have seen the beauty of minerals in a jewelry store or a museum. Such minerals are not only beautiful, they usually exhibit interesting geometric patterns as well. These naturally shaped minerals are crystals.

A **crystal** is a three-dimensional structure with each face, or surface, having a definite shape and orientation. When minerals have enough space to grow, they form crystals. Each mineral has a characteristic crystal form that is useful in its identification. The basic crystal forms are shown in Table 3–3. Crystal forms should not be confused with cleavage or fracture surfaces. When a mineral fractures, the crystal form is destroyed. For instance, a quartz crystal is a beautiful six-sided column, topped by a pyramid. When quartz breaks, it forms surfaces with curved fractures.

Some minerals are classified as *gems*. A **gem** is a mineral that has beauty, durability, and value. The beauty of a gem is based on its richness of color, purity of composition, and brilliance of sparkle. Because gems have durability, they last when worn as jewelry. Since gems are rare and unusual, they usually have high monetary value. Sometimes jewelry is made from solids, such as glass or opal, which have no crystal structure. These stones are called *amorphous*, which means without regular form.

In some cases, only certain varieties of a mineral are considered to be gems. For instance, a ruby is a rare, dark-red variety of the mineral *corundum*. Emeralds and aquamarines are special varieties of *beryl*. These rare-colored gems are produced by impurities in the minerals.

○ *What is a crystal?*
○ *Why are gems valuable?*

Figure 3–18. These cut minerals have qualities that make them attractive as gems.

DISCOVER

Breaking Crystals

Pour some ordinary table salt onto a dark piece of paper. Using a magnifying glass, look at the shape of the crystals. What shape are the crystals?

With the point of a sharp pencil, try to break some of the crystals. Is the shape of the broken crystals the same as the shape of the unbroken crystals? Explain.

Section Review

READING CRITICALLY

1. Describe what is meant by the properties of streak and luster.
2. Explain why geologists might have to scratch or break a mineral to discover its true color.

THINKING CRITICALLY

3. Amethyst, a variety of quartz, is considered to be a semiprecious gem. Why do you think it is not as highly valued as emeralds and rubies are?
4. What do calcite and dolomite have in common that make them react to hydrochloric acid?

3.8 Crystal Structure

DISCUSSION Have the students study the crystal forms shown. Be sure they realize the distinction between the crystals and the pieces formed by a mineral that has cleavage. Photographs or slides of famous jewels, like those in the Tower of London, would be interesting for the students to see. Gems are minerals that have been cut to enhance their beauty. ***Thinking Critically:*** Discuss why one gem of the same type and size as another can be more valuable than the other gem. (The more valuable gem will be clearer, have a more desirable color, and be absolutely free of imperfections.) Some uncut gems, or photos of some, would help students understand the skill that a lapidary must have.

EXTENSION The students can grow salt crystals that can be seen with the naked eye. Have the students dissolve regular table salt in a small dish of water, stirring the salt solution and adding salt until no more salt will dissolve. You can point out that this means the solution is saturated. Leave the dish where the water can gradually evaporate. The more slowly the water evaporates, the larger the resulting crystals will be.

DISCOVER:
Breaking Crystals

Thinking Skill *(Observing)*

Different brands of salt have different-sized crystals. Samples from various brands might add more interest to the activity. If the pencil point does not break the crystals, perhaps the end of a metal rod could be used. The crystals are tiny cubes. The broken crystals are also cubes.

Answers–3.8

○ A crystal is a three-dimensional structure with each face having a definite geometric shape.

○ Gems are valuable because they are rare and unusual.

Answers to Section Review

Reading Critically

1. Streak is the color of the mineral powder. Luster is the way a mineral's surface appears in reflected light.

2. The mineral's surface might be tarnished or weathered.

Thinking Critically

3. Amethyst is more common than emeralds and rubies.

4. They both have a carbonate molecule that reacts with HCl, hydrocholoric acid.

SECTION REVIEW

Summary Have the students list and define the mineral properties and describe the test for each. Have them explain what makes a crystal a gem.

Reinforcement Encourage each student to each write a two-column matching test on the section. Have the students trade tests and quiz each other.

Skill (Observing)

Preparation of Materials

Caution the students to treat the mineral samples, glass plate, and streak plate carefully, keeping the two plates down on a flat surface. If you need to substitute any minerals for those in the chart, be sure students make the corresponding changes.

Hint

Determine in advance if any students are color-blind or have difficulty distinguishing colors so they can work with a noncolor-blind partner.

Answers to Procedure Questions

1.

Mineral	Hardness	Color	Streak	Luster
Pyrite	6–6½	brass-yellow	greenish or brownish black	metallic, shiny
Chalcopyrite	3½–4	brass-yellow	greenish black	metallic
Galena	2½	lead-gray	lead-gray	bright metallic
Magnetite	6	iron-black	black	metallic
Talc	1	apple-green; gray; white	white	pearly
Quartz	7	white		glassy
Calcite	3	white	white	glassy to dull
Sphalerite	3½–4.1	white or green	white to yellow brown	non-metallic
Limonite	5–5½	yellow brown	yellow brown	adamantine
Hematite	5½–6½	red brown to black	Red	metallic

Answers to Analyses and Conclusions

1. color and hardness
2. hardness
3. hardness and color
4. luster, streak, and hardness
5. hardness
6. streak and color
7. Minerals with metallic luster commonly have dark-colored streaks.

Answer to Application

Completing these tests should identify the unknown mineral sample.

INVESTIGATION 3: Properties of Minerals

PURPOSE

To use the properties of hardness, streak, color, and luster to identify minerals

MATERIALS (per group of 3 or 4)

Penny
Steel nail
Glass plate
Streak plate
Minerals: pyrite, chalcopyrite, sphalerite, galena, magnetite, quartz, hematite, talc, calcite, limonite

TABLE 1: CHARACTERISTICS OF COMMON MINERALS

Mineral	Hardness	Color	Streak	Luster
Pyrite				
Chalcopyrite				
Galena				
Magnetite				
Talc				
Quartz				
Calcite				
Sphalerite				
Limonite				
Hematite				

PROCEDURE

1. Make a data table like the one shown. Fill in the table as you complete the investigation.
2. Using the penny, nail, and glass plate, test the hardness of each mineral. Record your results in the data table.
3. Record the color of each mineral in the data table.
4. Using the streak plate, determine the streak of each mineral. Record your results in the data table.
5. Describe the luster of each mineral using the terms *metallic, glassy, pearly,* or *waxy.* Record your results in the data table.

ANALYSES AND CONCLUSIONS

1. Which two properties would be most helpful in distinguishing between galena and chalcopyrite?
2. Which one property would be helpful in telling calcite from quartz?
3. Which two properties would be helpful in telling pyrite from chalcopyrite?
4. Which three properties would be helpful in telling sphalerite from pyrite?
5. Which one property would be helpful in telling calcite from talc?
6. Which two properties would be helpful in telling hematite from limonite?
7. Compare the streaks of the minerals with metallic luster to those with nonmetallic luster. What general statement can you make about the relationship between luster and streak?

APPLICATION

If you had an unknown mineral sample, what tests would you use to determine the mineral family to which the unknown sample belongs?

SUMMARY

- All things are made of atoms. Atoms are made of protons, electrons, and neutrons. (3.1)

- There are 90 different kinds of atoms, called elements, that occur naturally. (3.1)

- Isotopes of an element have the same number of protons but a different number of neutrons in their nuclei. (3.2)

- An ion is an atom with an electrical charge. (3.2)

- Atoms are often joined together as molecules. (3.3)

- Compounds are held together by two types of bonds: ionic and covalent. (3.3)

- To help scientists and students organize information about each element, all the elements have been placed in a periodic table. (3.4)

- The elements in the periodic table are listed in order of increasing proton number. The number of protons in an element is its atomic number. (3.4)

- Minerals are grouped into families based on their chemical composition. (3.5)

- Families of minerals include silicates, carbonates, sulfates, oxides, halides, and sulfides. (3.5, 3.6)

- Minerals formed of only one element are native elements. (3.6)

- Minerals can be identified by common physical properties. (3.7)

- Some minerals have special properties that aid in their identification. (3.7)

- Most minerals build crystal formations. (3.8)

- A gem is a rare mineral with the special characteristics of beauty, durability, and value. (3.8)

Write all answers on a separate sheet of paper.

SCIENCE TERMS

Correctly use each of the following terms in a sentence.

atomic number **(56)**
atoms **(51)**
cleavage **(66)**
compound **(54)**
covalent bond **(55)**
crystal **(69)**
electrons **(52)**
elements **(51)**
fracture **(66)**
gem **(69)**
ion **(53)**
ionic bond **(54)**
isotopes **(53)**
luster **(63)**
mineral **(59)**
molecule **(54)**
neutrons **(52)**
nucleus **(52)**
protons **(52)**
streak **(63)**
tetrahedron **(59)**

SCIENCE QUIZ

Modified True-False

Mark each statement *true* or *false.* If a statement is false, change the underlined term to make the statement true.

1. A silicate is composed of silicon and <u>calcium.</u>

2. Compounds can have either ionic bonds or <u>electrical</u> bonds.

3. A carbon isotope is an atom of carbon that has the number of protons characteristic of carbon but a different number of <u>neutrons.</u>

4. Of all the properties of minerals, <u>color</u> is the least dependable to use for identification.

continues

Chapter 3 Review   **71**

CHAPTER REVIEW

SUMMARY

The students may review the major concepts in the chapter by reading the summary statements. The statements are cross-referenced to the chapter to facilitate reinforcement of any concepts of which the students feel unsure. Encourage the students to work in groups to quiz one another.

SCIENCE TERMS

The sentence in which the science term is used should reflect an understanding of the definition of the term. You may wish to have students exchange their lists of sentences and quiz one another.

SCIENCE QUIZ

Answers to Modified True-False

1. false, oxygen
2. false, covalent
3. true
4. true

continues

Answers to Modified True-False, (continued)

5. true
6. true

Answers to Multiple Choice

7. b
8. a
9. a
10. d
11. d
12. d

Answers to Completion

13. 90
14. oxygen and silicon
15. tetrahedron
16. talc
17. gems
18. luster

Answers to Short Answer

19. Fracture and cleavage are two properties that describe how minerals break. They differ in that cleavage refers to flat, clean surfaces, while fracture refers to irregular, rough surfaces. They also differ because cleavage directions are predictable and those of fracture are not.
20. Some specimens are exceptionally beautiful and rare. For a specimen to be considered a gem, it must be relatively hard and durable. It must also have exceptional color and brilliance or sparkle. Minerals with these characteristics are quite rare, and so are highly valued.
21. Try scratching the mineral with samples of a known hardness.
22. Elements are placed in order of increasing proton number. This is also their atomic number.

Answers to Writing Critically

23. Ice is similar to minerals in that it does have a fixed crystalline structure and is inorganic. It is also similar to minerals because it is made of a specific compound. It is different from minerals, however, because it is not a solid at normal surface temperatures. Students should decide if the melting

SCIENCE QUIZ continued

5. Gold and silver are native elements.
6. Elements in the periodic table are listed in order by increasing number of protons.

Multiple Choice

Write the letter of the choice that best answers the question or completes the statement.

7. The smallest particle of an element is
   a) a neutron.   b) an atom.
   c) an electron.   d) a proton.
8. The atomic number of an element is equal to its number of
   a) protons.   b) electrons.
   c) neutrons.   d) atomic particles.
9. Quartz belongs to the group of minerals called
   a) silicates.   b) feldspars.
   c) carbonates.   d) sulfides.
10. All of the following occur as native elements except
    a) graphite.   b) silver.
    c) gold.   d) quartz.
11. Fluorite has a mineral hardness of
    a) 1.   b) 2.
    c) 3.   d) 4.
12. Feldspar is a
    a) carbonate.   b) sulfate.
    c) halide.   d) silicate.

Completion

Complete each statement by supplying the correct term or phrase.

13. There are _____ naturally occurring elements.
14. Two of the most common elements in the earth's crust are _____.
15. The _____ is the basic structure of quartz.
16. The softest mineral on Mohs' Scale of Mineral Hardness is _____.

17. Diamonds and emeralds are minerals that are called _____.
18. The appearance of the surface of a mineral is known as its _____.

Short Answer

19. Compare and contrast the properties of fracture and cleavage.
20. Explain why some mineral specimens are considered to be gems.
21. Describe the steps to estimate the hardness of a mineral if Mohs' Scale of Mineral Hardness is not available.
22. Describe the placement of elements in the periodic table, and explain the relationship between atomic number and proton number.

Writing Critically

23. Considering the definition of a mineral, discuss why ice should or should not be considered a mineral.
24. Which two mineral identification tests do you think are the most useful? Explain why you chose these two.
25. Compare and contrast an atom, an element, a molecule, and a compound. Explain the fact that most minerals are compounds, while others are elements. Also explain why it is possible to have a single atom of some minerals, while for most minerals a molecule is the smallest unit possible.

EXTENSION

1. Gold is not used just in making jewelry. It has been and is still being used in the treatment of some diseases. Write a report about the medical uses of gold in both the past and the present.
2. Research the story of the Hope diamond. Be sure to include information about the people who have owned this diamond.

point is relevant in the classification of ice as a mineral.
24. Students may choose several different combinations of tests to answer this question. They should not choose the property of color, since it varies so much.
25. Atoms are the smallest units of elements, molecules, and combinations of atoms. Compounds are combinations of two or more different elements. Most minerals are compounds, so it would not be possible to have an atom of quartz, for example.

ANSWERS TO EXTENSION

1. Obvious would be the use of gold in repairing teeth, but gold has many other medical uses as well. Gold injections are sometimes given for arthritis.
2. Reports should include both the people who have owned the diamond and the legends associated with it.

APPLICATION/CRITICAL THINKING

1. Suppose you found a beautiful white mineral on the roadside while on vacation in Colorado, and you wanted to identify it. What tests would you perform on it? If the mineral you found was soft and showed three directions of cleavage, what might it be? Explain your answer.

2. Special organisms that live in the sea can take minerals from the water to build structures called reefs. The reefs are similar in composition to certain types of materials used for building human structures. Explain how artificial reefs might be constructed from building wastes. Describe the problems that might occur from using waste materials.

FOR FURTHER READING

Bains, R. *Rocks and Minerals.* Mahwah, New Jersey: Troll, 1985. This book simplifies the description of mineral composition for anyone looking for a clear explanation.

Cheney, G. A. *Mineral Resources.* New York: Watts, 1985. This book is a survey of America's most important minerals. It includes information on their composition, location, mining, and use.

Cotterill, R. *The Cambridge Guide to the Material World.* New York: Cambridge University Press, 1985. The book is well illustrated, and topics include crystals, minerals, metals, ceramics, and glass.

O'Neil, P. *Gemstones. Time-Life Planet Earth Series.* Alexandria, Virginia: Time-Life, 1983. This book presents extensive information on sources and processing of gems.

Challenge Your Thinking

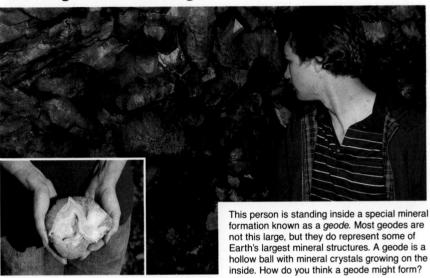

This person is standing inside a special mineral formation known as a *geode*. Most geodes are not this large, but they do represent some of Earth's largest mineral structures. A geode is a hollow ball with mineral crystals growing on the inside. How do you think a geode might form?

Chapter 3 Review   **73**

ANSWERS TO APPLICATION/ CRITICAL THINKING

1. You could test for luster, hardness, and cleavage very easily. It could be halite or calcite, depending on the angles of cleavage.

2. Building waste material could provide support and shelter for marine organisms, but some wastes may contain chemicals that would cause pollution.

ANSWER TO CHALLENGE YOUR THINKING

The ball was filled at one time. As water removed the contents, mineral crystals began to grow to fill the space.

PLANNING THE CHAPTER

Chapter Sections	Page	Chapter Features	Page	Program Resources	Page
Section 1: Studying Rocks 4.1 Rock Families **(B)** 4.2 The Rock Cycle **(A)**	75 75 77	A Matter of Fact **Discover:** Studying the Rock Cycle **(A)** Section Review	77 78 78	Concept Extension: *Rock Sections* **(A)**	**TRB 9** **LM 231**
Section 2: Igneous Rocks 4.3 Formation of Igneous Rocks **(A)** 4.4 Classification of Igneous Rocks **(A)** 4.5 Uses of Igneous Rocks **(B)**	79 79 79 82	**Activity:** Classifying Igneous Rocks **(A)** **Discover:** Floating Igneous Rocks **(B)** Section Review **Skill Activity:** Designing Tables **(A)**	81 82 82 83		
Section 3: Sedimentary Rocks 4.6 Formation of Sedimentary Rocks **(A)** 4.7 Classification of Sedimentary Rocks **(A)** 4.8 Uses of Sedimentary Rocks **(B)**	84 84 86 89	A Matter of Fact A Matter of Fact **Discover:** Finding Fossils in Sedimentary Rocks **(B)** A Matter of Fact Section Review **Biographies:** Then and Now René Just Haüy, Ronald B. Parker	86 87 88 89 89 90	Investigation 4.1: *Classifying Sedimentary Rocks* **(A)** Investigation 4.2: *Chemical Sedimentary Rocks* **(A)**	**TRB 13** **LM 15** **TRB 15** **LM 17**
Section 4: Metamorphic Rocks 4.9 Formation of Metamorphic Rocks **(A)** 4.10 Classification of Metamorphic Rocks **(A)** 4.11 Uses of Metamorphic Rocks **(B)**	91 91 92 93	**Discover:** Observing Metamorphic Rocks **(A)** A Matter of Fact Section Review **Investigation 4:** Classifying Rocks **(A)**	92 93 93 94	Critical Thinking **(H)** Cross-Discipline: Science and Language, *Writing the Story of a Rock* **(A)** Reading for Content: *Determining Cause and Effect* **(A)** Student Record Book: Textbook Investigations **(A)**	**TRB 7** **TRB 7** **TRB 7** **LM 179** **TRB 7**
Chapter 4 Review	95			Vocabulary **(A)** Tests **(A)** Computer Test Bank	**TRB 4** **LM 129** **TRB 15**

(LM) Laboratory Manual/Study Guide, **(TRB)** Teacher's ResourceBank™
B = Basic **A** = Average **H** = Honors
The coding Basic, Average, and Honors indicates sections or subsections that might be appropriate for different levels of learners. For additional suggestions regarding choice of topic and depth of coverage, see the Pacing Chart on pages T16–T20.

CHAPTER CONCEPTS, OBJECTIVES, AND TERMS

Section	Concepts	Objectives	Science Terms
Section 1: Studying Rocks	■ Rocks are naturally occurring minerals or combinations of minerals. **(4.1)** ■ Rocks are classified into three families: igneous, sedimentary, and metamorphic. **(4.1)** ■ Rocks are classified by their origin and the types of minerals they contain. **(4.1)** ■ The continuous process of change that affects all rocks is called the rock cycle. **(4.2)** ■ The rock cycle is not one way. Any rock may be changed into any other type of rock. **(4.2)**	■ **List** the three rock families and the general characteristics used to classify rocks into these families. ■ **Describe** the rock cycle and **relate** it to the formation of the three families of rocks.	rocks magma lava igneous rocks sedimentary rocks metamorphic rocks rock cycle
Section 2: Igneous Rocks	■ Igneous rocks are formed as magma or lava cools and crystallizes. **(4.3)** ■ The crystal size in igneous rocks is determined by the rate of cooling. **(4.4)** ■ Many igneous rocks are used in buildings because of their strength. **(4.5)**	■ **Discuss** the process that forms igneous rocks. ■ **Classify** common igneous rocks by their characteristics. ■ **List** several uses for igneous rocks.	texture intrusive rocks extrusive rocks
Section 3: Sedimentary Rocks	■ Sedimentary rocks are formed from the cemented fragments of other rocks. **(4.6)** ■ Sedimentary rocks, which are classified by the way the sediments form, are called clastic, chemical, or organic. **(4.7)** ■ Sedimentary rocks are used in many industrial processes. **(4.8)**	■ **Describe** three types of sedimentary rocks. ■ **Trace** the formation of compact limestone. ■ **Identify** common sedimentary rocks by their characteristics.	cementation clastic sediments chemical sediments organic sediments evaporites precipitates
Section 4: Metamorphic Rocks	■ Metamorphic rocks are formed when heat or pressure is applied to other rocks. **(4.9)** ■ Metamorphic rocks are classified as foliated or nonfoliated depending on whether or not mineral layers are visible. **(4.10)** ■ Many metamorphic rocks are very useful as building materials. **(4.11)**	■ **Explain** how metamorphic rocks form. ■ **Define** the terms *foliated rocks* and *nonfoliated rocks*. ■ **Identify** some common metamorphic rocks by their characteristics.	foliated rocks nonfoliated rocks

Title	Page	Materials
Discover: Studying the Rock Cycle	78	*(per student)* granite, sandstone, quartzite
Activity: Classifying Igneous Rocks	81	*(per group of 3 or 4)* paper, pencil, igneous rock set, hand lens
Discover: Floating Igneous Rocks	82	*(per student)* pumice stone, cork, dish, water
Skill Activity: Designing Tables	83	*(per student)* paper, pencil
Discover: Finding Fossils in Sedimentary Rocks	88	*(per student)* shale with visible fossils
Discover: Observing Metamorphic Rocks	92	*(per student)* marble, slate, quartzite
Investigation 4: Classifying Rocks	94	*(per group of 3 or 4)* safety goggles (1 per student); laboratory apron; unlabeled rock specimens; rock classification key, page 568; hydrochloric acid, 0.5 molar; hand lens

TEACHING SUGGESTIONS

Section 1: Studying Rocks

Demonstration: Identifying Rocks

Purpose

To identify rocks

Materials

Safety goggles
Laboratory apron
Newspaper
Hand lens
Rock samples
Cloth
Small hammer
Medicine dropper
Dilute hydrochloric acid

Procedure

1. Pick a steady surface to work on. Cover the surface with several layers of newspaper.
2. Have the students look at each sample with a hand lens. Record the color of each sample. As a class decide on the texture of the sample. Is it fine-grained? Coarse-grained? Is it glassy? These are characteristics of igneous rock. Is the rock made up of tiny well-rounded grains? This is a characteristic of sedimentary rock. Does the rock have colored bands? Does it have flat layers of crystals? These are characteristics of metamorphic rock. Have each student record observations in table.

3. Try to break small pieces off the rock sample. **CAUTION: Wrap the sample in cloth first. Then hit the sample with a small hammer. Rocks that break off in sheets or layers may be metamorphic.**
4. Separate any samples tht you think may be limestone (sedimentary) or marble (metamorphic). **CAUTION: Handle the hydrochloric acid carefully. It can cause serious burns. If any acid touches your skin, flood the affected area with water.** With a medicine dropper, place a few drops of hydrochloric acid on the sample. If bubbles of carbon dioxide form on the sample, the rock is made up of the mineral calcite. (calcium carbonate) This mineral is the main component of both limestone and marble.
5. Finally, have the students compare the rocks with the samples shown in the textbook. When the students have enough information to identify the sample, have them write the name of the sample in their tables. Put the table on the chalkboard.

Sample	Observations	Name of Sample
1		
2		
3		
4		

Questions to Ask the Students

1. Name some characteristics of each type of rock. (*Igneous rocks are made of molten minerals; sedimentary*

rocks are made of pieces of other rocks; and metamorphic rocks are changed by heat and pressure.)

2. Which observations were most helpful in identifying the rock samples? (*texture, hardness of minerals; not color*)

3. Is it possible to identify rocks by color only? By texture only? Explain. (*Rocks composed of different minerals and formed by different processes may have similar colors or textures.*)

Field Trip

Visit an industry in which rocks are used.

Outside Speaker

Invite a spokesperson from a quarry, gravel yard, mine, or other industry that uses rocks to explain how rocks are used in his or her particular industry.

Section 2: Igneous Rocks

Class Activity

Fill a 500-mL beaker nearly full with water. Add two drops of black ink. Have the students note that the water becomes uniformly black in a short time. Discuss with the students the fact that a very small amount of an impurity will color a large amount of a substance. Relate this to the black color of obsidian, which is due to a small amount of an impurity.

Class Activity

Have the students examine both pumice and scoria. Discuss how the holes originate in these rocks as gases escaping from the lava are trapped. Demonstrate how pumice floats in water. Try floating other light rocks.

Section 3: Sedimentary Rocks

Class Activity

Using a medicine dropper, demonstrate the differences in porosity of sandstone, shale, granite, and pumice. Encourage the students to hypothesize what differences in porosity mean when it rains in areas with underlying sandstone, compared to areas with less porous rocks.

Section 4: Metamorphic Rocks

Class Activity

Have the students place colored candle pieces in a beaker and gently heat the beaker until the the candle pieces begin to melt. As the colors begin to mix, remove the beaker from the heat and allow to cool. The students should see a metamorphic candle with a mix of colors.

CHAPTER 4

CHAPTER OVERVIEW

This chapter presents the three rock families that occur naturally on Earth. The rock cycle is discussed, as are the formations, characteristics, and possible uses of each family of rock.

Section 1: Studying Rocks In this section, rocks are defined as naturally occurring combinations of minerals. The three rock families—igneous, sedimentary, and metamorphic—are discussed. The rock cycle is then defined as the process by which rocks are continuously being changed from one family to another.

Section 2: Igneous Rocks The most common rocks in Earth's crust are the igneous rocks, formed when molten material cools. Crystal size and color of igneous rocks are presented in this section, and the effect of the rate of cooling on the mineral content and texture is discussed. This discussion leads to further classification of igneous rocks.

Section 3: Sedimentary Rocks Most rocks found on Earth's surface are sedimentary. The three types of sedimentary rocks—clastic, chemical, and organic—are named for the type of sediments that form them. The formation of sedimentary rocks and their uses are discussed.

Section 4: Metamorphic Rocks The formation of metamorphic rocks and their division into foliated and nonfoliated rocks are explained. The characteristics and uses of metamorphic rocks are then discussed.

CHAPTER

Rocks

Earth is a special planet. It has air, water, and a solid surface. This solid surface, or crust, is composed of rocks. As this picture shows, rock formations are often spectacular. As well as being spectacular, rocks are also important scientifically and economically. The rocks of Earth contain the history and most of the riches of our planet.

Long columns of igneous rocks

74

CHAPTER MOTIVATING ACTIVITY

Provide samples of igneous, sedimentary, and metamorphic rocks to small groups of students. Have the students try to group the rocks and make a list of their characteristics. Write the lists on the chalkboard and compare those chosen by each group.

Ask the students if they have ever seen rocks similar to the ones displayed in the classroom. Allow the students to share with the class descriptions of unusual or special rocks they have seen while traveling. Encourage them to think about buildings, monuments, or other structures in which rocks are used.

1 Studying Rocks

SECTION OBJECTIVES

After completing this section, you should be able to:

■ **List** the three rock families and the general characteristics used to classify rocks into these families.

■ **Describe** the rock cycle and **relate** it to the formation of the three families of rocks.

NEW SCIENCE TERMS

rocks
magma
lava
igneous rocks
sedimentary rocks
metamorphic rocks
rock cycle

4.1 Rock Families

What makes you a person? Is it your hair, or your face, or perhaps the fact that you have two arms and two legs? Whatever it is, you have definite characteristics that make you recognizable as a person. Still, you are different from every other person. You are unique.

Rocks have definite characteristics also. Look at the photographs in Figure 4–1. You can probably tell that the objects in the photographs are *rocks*. All rocks share some characteristics that make them rocks, but they also have many differences that make them unique. For example, the rocks in the photographs are different colors. One rock has large crystals, and the other appears to have no crystals at all. Rocks are like people; they are alike, yet each is unique.

In Chapter 3, minerals are defined as naturally occurring substances made of specific elements or compounds. **Rocks** are naturally occurring combinations of minerals. A few rocks consist of only one type of mineral. However, most rocks consist of two or more different minerals. Although hundreds of different kinds of minerals are found in rocks, only about 20 minerals are common. In Chapter 3, you practiced classifying minerals; rocks can be classified in much the same way.

Geologists classify rocks in several ways. One method of classification is based on the origin of the rocks; that is, on how

Figure 4–1. Although these rocks look different, they belong to the same rock family. As with brothers and sisters of human families, members of the same rock family share many characteristics.

Section 1 Studying Rocks **75**

BACKGROUND INFORMATION

Until the late eighteenth century, scientists believed that the earth had been fully formed by 4000 B.C. and only catastrophes caused further changes. In 1795, the Scottish geologist James Hutton became the father of modern geology by explaining two theories: 1) that rocks had formed from molten materials, not from minerals crystalized by water as had been previously thought; and 2) that the surface features of the earth had been formed gradually over millions of years and that they are still changing.

Section 1 : Studying Rocks

SECTION FOCUS

Section Overview This section presents the characteristics that distinguish the three rock families. The formation of each rock family and the rock cycle are discussed.

Section Objectives For a list of section objectives, see pupil's edition page.

New Science Terms For a list of new science terms in this section, see pupil's edition page. You may wish to read aloud the names of the rock families so that the students are familiar with the pronunciations.

SECTION DEVELOPMENT

4.1 Rock Families

DISCUSSION Ask the students to imagine that they are digging in the ground with a metal shovel, and that the shovel hits something hard. *Thinking Critically:* Ask the students how would they decide if the object they had hit was a rock. (Some possible ideas might be that the shovel would make a ringing sound if it had hit a rock; if the hard thing was not a rock, it might be broken up easily.)

DISCUSSION Snow could be used as a model for learning how rocks form. *Thinking Critically:* Ask the students to use an analogy to describe how the formation of snow and what happens to it on the ground parallels the formation of sedimentary and igneous rocks. Make sure the students understand that this description is only an analogy and that igneous rock is formed from heat, not cold. (Some answers might be that water is like magma. It cools into snow or ice; this would compare to the igneous rocks. The snowflakes could be thought of as sediment, and the snowflakes packed into a hard snowball would be like sedimentary rock.) Students may be able to think of other examples. Be sure to emphasize the origin of each rock family name.

SKILL (*Observing*) Have the students use a magnifying glass to examine the rock samples they previously examined without the aid of magnification. Have them list any additional characteristics they now observe.

EXTENSION At this point, you may wish to take a large poster board and begin a "rocks" chart. Separate the poster board into three sections and write the three rock families for headings. Add to the chart during the chapter as further classifications and characteristics are introduced.

DISCUSSION Ask the students to continue the snow example for metamorphic rocks. (When the sun shines on snow, it may not totally melt, but it does cause some melting and the snowflakes compact into an almost solid mass.) Have the students observe the rocks in the photos and discuss the changes.

they were formed. All rocks may be classified into one of three families according to the way in which they were formed.

Understanding how rocks form can be somewhat confusing. You can observe similar processes by using simple materials. Suppose your teacher takes some old candles and melts them in a pan, and then carefully pours the melted wax into a mold for you. If you allow the wax to cool completely, it will harden. You now have a solid object formed from liquid, or *molten*, material.

Rocks are formed in a similar manner inside the earth. Where temperature is very high, rocks melt. Molten rock within the earth is called **magma**. Many pockets of magma are found in the earth's interior.

Molten rock that reaches the surface of the earth is called **lava**. You may have heard the term *lava* used while listening to news reports about volcanoes. As magma or lava cools, it hardens, and solid rocks form. The rocks that form from magma or lava are known as **igneous** (IHG nee uhs) **rocks**. The word *igneous* comes from the Latin word *ignis*, meaning "fire."

Now suppose you take your wax block and carefully scrape off some of the wax with the edge of a knife. When you have made a small pile of wax shavings, you cover it with a piece of paper and a heavy book and press it as hard as you can, the wax shavings stick together. A second type of rock forms from small rock fragments in much the same way.

Rocks exposed to the weather break into small pieces, which form a sediment (SEHD uh muhnt). Sediments can be transported by water or wind to new locations. Eventually the sediments are deposited in layers. As layers of sediments are compressed and cemented, they stick together. Compressed sediments form **sedimentary** (sehd uh MEHN tuhr ee) **rocks**. The word *sedimentary* also comes from Latin, from the word *sedimentum*, meaning "that which has settled."

Figure 4–2. Molten rocks from within the earth often come to the surface as lava from a volcano. As the lava cools, igneous rocks form.

Figure 4–3. The mud of this dry lake bed may eventually become sedimentary rock. Some sedimentary rocks show mud cracks such as these or ripple marks from shallow water.

76 Chapter 4 Rocks

DEMONSTRATION

For a demonstration of identifying rocks see page 73c preceding this chaper.

Sandstone   Quartzite

Granite   Gneiss

Limestone   Marble

Figure 4–4. Shown here are three examples of metamorphic rocks. In each case the sample on the right is the metamorphic rock, and the sample on the left is the parent rock from which it formed.

Finally, if you warm the remaining wax slightly, you should be able to bend it and squeeze it without having to melt it. In a similar fashion, heat and pressure change igneous and sedimentary rocks.

Heat and pressure cause the minerals in rocks to form new combinations, or new crystals. These changes form rocks called **metamorphic** (meht uh MOR fihk) **rocks.** The word *metamorphic* comes from a Greek word, *metamorphosis*, which means "change."

○ *Name the three rock families.*
○ *How are rock families named?*

4.2 The Rock Cycle

People who don't know you may still be able to tell that you belong to a certain family because you have characteristics that identify you with that family. The family to which a rock belongs can usually be determined by studying certain characteristics of the rock. For instance, rocks can be identified by their mineral composition.

Some minerals occur in only a few types of rocks. The mineral calcite, for example, is rarely found in igneous rocks, but it is common in sedimentary rocks. Mica and feldspar usually occur in igneous rocks. Another characteristic of rocks is the layering of minerals, which often occurs in sedimentary and metamorphic rocks.

In our society, families change because of death, divorce, adoption, and marriage. The rock families change because of weather, heat, and pressure. New rocks are constantly being formed from old rock material. This continuous process of

A MATTER OF FACT
After the 1883 explosion of the volcano Krakatau in Indonesia, sailors pushed through 3 km of floating masses of pumice to get from their ships to the shore.

Section 1   Studying Rocks   **77**

4.2 The Rock Cycle

DISCUSSION   Ask the students to suggest ways that weather, heat, and pressure change rocks. Accept all logical suggestions at this time. Have the students make lists of their suggestions. After the discussion of the rock cycle has been completed, have the students return to their lists and make necessary modifications.

DISCUSSION   You may wish to remind the students that similar cycles, such as the nitrogen cycle and a food web, are found in biology.

Answers–4.1

○ Igneous, sedimentary, and metamorphic are the three rock families.

○ Rock families are named by the processes that form them.

EXTENSION (Tie-in/Social Studies) Rocks, rock formations, and mountains can affect how and where people live, what kind of homes they build, and how their towns look. ***Thinking Critically:*** Ask the students to discuss how rocks and their uses affect people's lives. (Some answers might be that mountains sometimes dictate where roads are built, which might affect the growth of towns; some rocks such as coal are natural resources—towns sometimes grow because there are natural resources nearby; and farms are not located in areas that have rocky soil.)

DISCOVER:
Studying the Rock Cycle

Thinking Skill (*Drawing Conclusions*)

Rock collections and kits may be obtained from any science supply company. The students should classify granite as an igneous rock, sandstone as a sedimentary rock, and quartzite as a metamorphic rock. Granite probably formed first because Earth was originally a molten mass.

SECTION REVIEW

Summary Have the students make a review chart, naming the three families of rocks and the method of formation of each.

Reinforcement Have each student make up three true-false questions on this section. The students can then exchange questions and answer them.

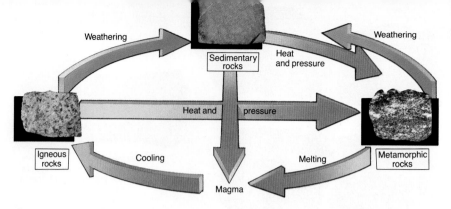

Figure 4–5. The rock cycle is not a one-way process. Rocks may enter the cycle at any place and be changed into any other type of rock.

DISCOVER

Studying the Rock Cycle

Look at samples of granite, sandstone, and quartzite. Try to decide which rock belongs in each family. Make a sketch of the rock cycle with these three samples. Which of the three rocks probably formed first? Why?

---

change within rock families is called the **rock cycle**. A simplified version of the rock cycle is shown in Figure 4–5.

All rocks go through this cycle. For example, as magma cools it may form *granite*, an igneous rock. Granite can be broken by sea waves into sand-sized sediments. The sand can then be deposited and compacted to become *sandstone*, a sedimentary rock. If the sandstone is exposed to heat and pressure, it can change to *quartzite*, a metamorphic rock. Quartzite may later be forced deep within the earth and remelted into magma. The cycle may repeat itself many times.

The rock cycle is not a one-way process. While it is possible for igneous rocks to be changed into sedimentary rocks, it is also possible for metamorphic rocks to be changed into sedimentary rocks. Granite may be changed into sandstone by weather and water. In fact, any rock may change from any one of the rock families to any one of the others.

○ ***What is the rock cycle?***
○ ***Into what kinds of rocks can igneous rocks change?***

Section Review

READING CRITICALLY

1. What are the origins of each of the rock families?
2. Explain why the rock cycle is not a one-way process.

THINKING CRITICALLY

3. What kind of rocks must have formed first on Earth? Explain your answer.
4. Starting with sandstone, a sedimentary rock, give several possible steps of a rock cycle that would end with granite, an igneous rock.

Answers—4.2

○ The rock cycle is the process of old rock material being formed into new rocks.

○ Igneous rocks can change into metamorphic rocks, sedimentary rocks, or other igneous rocks.

Answers to Section Review

Reading Critically

1. Igneous rocks are formed from lava or magma. Sedimentary rocks are formed from the compression of rock segments. Metamorphic rocks are formed when heat and pressure cause changes in existing rocks.
2. Depending on which process is applied, a rock can change into any of the other rock families.

continues

SECTION OBJECTIVES

After completing this section, you should be able to:

- **Discuss** the process that forms igneous rocks.
- **Classify** common igneous rocks by their characteristics.
- **List** several uses for igneous rocks.

NEW SCIENCE TERMS

texture
intrusive rocks
extrusive rocks

4.3 Formation of Igneous Rocks

What is the most common family name on Earth? What is the most common name in North America, or in your school? Which of your classmates has the largest family? You could ask similar types of questions about rocks.

Igneous rocks are by far the most common rock family in the earth's crust. Igneous rocks make up about two-thirds of the earth's crust. The type of igneous rock that forms depends mainly on two conditions: the chemicals within the magma or lava and the rate at which the magma or lava cools. If, for example, a large amount of a dark element such as iron is present in magma, dark-colored rocks, such as *basalt* (buh SAWLT), form.

As magma cools, ions bond to form mineral crystals. Magma that cools slowly forms visible crystals. These crystals join in tightly interlocking units, producing very dense rocks. Remember that lava is found on the earth's surface, so lava cools much more quickly than magma. This rapid cooling prevents the formation of large crystals.

○ *Does slow cooling form large or small crystals in rocks? Explain your answer.*

4.4 Classification of Igneous Rocks

Igneous rocks are classified by texture and mineral content. **Texture** is the size of the crystals in the rocks. Mineral content determines a rock's density and color.

Texture Geologists divide igneous rocks into two main groups: those with fine texture, formed from lava, and those with coarse texture, formed from magma. How does the texture of a rock relate to the size of the mineral crystals in the rock? 1

Rocks with large mineral crystals are called **intrusive** (ihn TROO sihv) **rocks**. Individual crystals can easily be seen in intrusive rocks.

1 Large crystals produce a coarse texture and small crystals produce a fine texture.

Figure 4–6. Shown here are basalt (top) and porphyry (bottom). The porphyry cools unevenly so small and large mineral crystals form.

Section 2: Igneous Rocks

SECTION FOCUS

Section Overview This section presents igneous rocks in more detail. The texture and mineral content of igneous rocks are studied, and some special igneous rocks are identified. The students are then introduced to some uses for igneous rocks.

Section Objectives For a list of section objectives, see pupil's edition page.

New Science Terms For a list of new science terms in this section, see pupil's edition page. You may wish to pronounce the terms for the class.

SECTION DEVELOPMENT

4.3 Formation of Igneous Rocks

DISCUSSION The students should be able to understand how igneous rocks are formed if they have ever seen melted wax drip from a candle and then solidify. Have the students study the rocks in Figure 4-6 and observe the difference in crystal size. You may wish to tell the students that they will learn more about magma in Chapter 5. You may also wish to review the definition of ion, which was given in Chapter 3.

4.4 Classification of Igneous Rocks

DISCUSSION You may wish to have samples of igneous rocks of different textures, such as granite and rhyolite. If individual minerals are large enough to distinguish, the rock is considered coarse-textured. Have the students decide which rock in Figure 4-6 is intrusive. (It will have coarse texture.)

Thinking Critically

3. Igneous rocks must have formed first because the earth was originally a molten mass.
4. Sandstone under heat and pressure forms quartzite; with more heat and pressure, it melts and then cools into granite.

Answers–4.3

○ Large crystals; the longer the cooling time, the more time the elements have to move around and join into large crystals.

DISCUSSION Have the students classify some igneous rock samples as intrusive or extrusive. ***Thinking Critically:*** Ask the students why intrusive rocks are not formed from lava. (Lava pours from a volcano or from a fissure in the earth into air that is hundreds of degrees cooler than the lava itself. It quickly loses heat to the atmosphere and begins solidifying. Intrusive rocks are formed very slowly within the earth.) The relationship between cooling temperature and the formation of specific minerals can be seen in a complete version of Bowen's Reaction Series. Have examples of igneous rocks in class so the students can see the variety of colors. Many felsic rocks are formed from light-colored minerals. Most mafic rocks are formed from dark-colored minerals.

EXTENSION Felsic and mafic are acronyms. FELSIC: FE for feldspar, L for lenads, or feldspathoid, SIC for silica; MAFIC: MA for magnesium, FIC for ferric.

DISCUSSION Point out to the students that obsidian and pumice look different from the other igneous rocks because of the cooling process that formed them. Obsidian also owes its color to the cooling process. It is often black because the impurities that give it a dark color are spread throughout the rock rather than isolated in small areas within the rock. Have samples of these rocks for the students to look at and to feel.

EXTENSION You may wish to continue the chart begun in Section 1, adding the classifications of igneous rock to the chart along with several examples of each.

Figure 4–7. This is rhyolite, an extrusive rock. Notice the fine texture of this rock.

Figure 4–8. The dark rocks of the Pacific islands (left) are mafic rocks, while the light rocks of the continent (right) are felsic rocks.

Smaller, invisible mineral crystals form in lava. Lava quickly loses its heat to the atmosphere, preventing the formation of large crystals. The fine-grained rocks formed from lava are called **extrusive** (ehk STROOS ihv) **rocks.** *Rhyolite*, shown in Figure 4–7, is an example of extrusive rock.

Mineral Content The rate of cooling determines the mineral content of rocks. The first crystals to form are those of minerals that are stable at high temperatures. Many of these minerals are rich in iron and magnesium. Two types of high-temperature minerals are olivine and pyroxene. These dark crystals sink into the lower parts of the magma because of their high density. This leaves the remaining magma with a high percentage of less dense minerals.

Igneous rocks that contain dense minerals are called *mafic* (MAF ihk) *rocks.* The magma that forms mafic rocks has a temperature of 900°C to 1200°C. Mafic rocks are usually black or dark green. Most of the rocks of the oceanic crust and of the Pacific islands are mafic rocks. Two examples of mafic rocks are basalt and *gabbro*.

Igneous rocks containing low-density minerals are called *felsic* (FEHL sihk) *rocks.* The minerals in felsic rocks are light-colored crystals of quartz and some types of feldspar. The temperature of the magma that forms felsic rocks is about 800°C. Rhyolite and *andesite* are felsic rocks.

Special Igneous Rocks Some igneous rocks do not fit neatly into either the intrusive or extrusive category. These rocks usually have no visible crystals at all.

One special type is volcanic glass, or *obsidian*, shown in Figure 4–9. Obsidian is an igneous rock that does not have crystalline structure. The rock hardens so fast that no crystals form. Obsidian contains many of the same minerals as rhyolite, but it looks very different. Arrowheads of early Native Americans were often made from obsidian.

Pumice, shown in Figure 4–10, is a low-density, glassy rock. Pumice cools so rapidly that gases in the lava are trapped inside the rock. The gases escape after the rock cools, leaving many spaces, or pores. There are so many of these pores in pumice that it can float on water. *Scoria* also has pores, but it is too dense to float.

○ *How do intrusive rocks form?*
○ *Which igneous rocks have light-colored minerals?*

Figure 4–9. Obsidian, or volcanic glass, cools so quickly that no crystals form.

Figure 4–10. Pumice (left) is so porous that it will float. Although scoria (right) is also porous, it will not float.

ACTIVITY: Classifying Igneous Rocks

How do you classify igneous rocks?

MATERIALS (per group of 3 or 4)

pencil, paper, igneous rock set, hand lens

TABLE 1: ROCK COMPOSITION AND TEXTURE			
TEXTURE	ROCK COMPOSITION (COLOR)		
	Felsic (Light)	Intermediate (Medium Light)	Mafic (Dark)
Glassy			
Fine			
Coarse			

PROCEDURE

1. Make a table like the one shown. Fill in the chart as you complete the activity. You may not have a rock for each box in the chart.
2. Divide the igneous rocks in your set into three groups based on their composition.
3. Separate each group by texture. Place the names of the rocks in your chart to indicate the proper texture and composition for each rock. You may need to review the chapter for help in matching the rock name to the proper texture.

CONCLUSIONS/APPLICATIONS

1. How would you distinguish granite from rhyolite?
2. Why would you have a problem placing pumice on this simple classification chart?
3. How were the holes formed in scoria?
4. Write a thorough description of granite. Explain what the texture of granite tells you about its origin.

Section 2   Igneous Rocks   **81**

Answers—4.4

○ Intrusive rocks form from magma within the crust. Slow cooling allows for the formation of large crystals.

○ Felsic rocks are most often light-colored.

ACTIVITY:
Classifying Igneous Rocks

Thinking Skill   *(Organizing)*

Preparation of Materials   Use any five igneous rocks for each set. Be sure to vary the sets. Put together a master set to facilitate the students' efforts to name the various samples.

Answers to Conclusions/Applications

1. Texture. Granite has large, obvious crystals; rhyolite has small crystals.
2. Pumice has no visible crystals.
3. The rocks cooled so quickly that gases in the magma were trapped in the rock; they could not escape into the atmosphere.
4. Granite has large crystals; therefore, it must form in well-insulated locations. It is an intrusive rock. The minerals are quartz, feldspar, and micas.

4.5 Uses of Igneous Rock

DISCUSSION Discuss some other uses of igneous rock. Basalt is used with concrete as a filler for road beds; granite is a good building stone; and pumice is sometimes used with soap or hand cream as a hand cleaner. *Thinking Critically:* Ask the students to name some properties of igneous rocks that make them useful. (strength and durability)

DISCOVER:
Floating Igneous Rocks

Thinking Skill *(Drawing Conclusions)*

The cork floats higher. Students' explanations will vary but will probably mention that cork is less dense than pumice, even though the pumice is full of holes.

SECTION REVIEW

Summary Have each student begin a summary chart to be used for the three rock families as the chapter is completed. The headings across the top could be Igneous, Sedimentary, and Metamorphic; down the side could be Formations, Classifications, and Uses.

Reinforcement Have the students quiz each other on the questions at the end of each subsection.

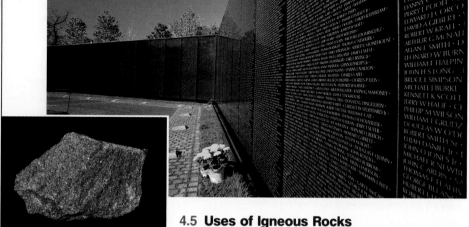

Figure 4–11. The black granite used to make this memorial will probably last for centuries.

DISCOVER

Floating Igneous Rocks

Float a piece of pumice stone in a dish with a piece of cork. Which floats higher in the water? Explain why you think this happens.

4.5 **Uses of Igneous Rocks**

Many igneous rocks are used in the construction industry because of their strength. These rocks can also be polished to shiny, smooth surfaces. Carvings and inscriptions in granite can withstand the effects of weather for centuries. The Vietnam Veterans' Memorial in Washington, D.C., is made of 150 panels of granite from Barre, Vermont.

Pumice is a valuable polishing material. Sometimes it is ground up and used as a scouring powder. Pumice is also used in a lump form called a *pumice stone*.

Igneous rocks are valuable sources for some minerals and metals. These rocks are often mined for uranium, iron, copper, and other metals.

○ *For what purpose is pumice used?*
○ *Why are igneous rocks used in monuments?*

Section Review

READING CRITICALLY

1. What are the two characteristics used to classify igneous rocks?
2. Explain the difference between felsic and mafic rocks and give examples of each.
3. What relationship exists between high-temperature stable crystals and density?

THINKING CRITICALLY

4. Give a possible origin of a rock that has large crystals of feldspar and very fine crystals of quartz.
5. What kinds of elements might be present in mafic rocks that make them dark and dense?

82   Chapter 4   Rocks

Answers–4.5

○ Pumice is used as a cleaner.

○ Igneous rocks are strong, durable, and polish well.

Answers to Section Review

Reading Critically

1. Texture and mineral content are used to classify igneous rocks.
2. Felsic rocks contain light-colored minerals such as quartz, muscovite, and light-colored feldspar. Mafic rocks contain dark-colored minerals such as olivine, magnetite, and dark-colored feldspar.
3. The minerals that form at higher temperatures are very dense.

continues

SKILL ACTIVITY: Designing Tables

BACKGROUND

As scientists study a particular topic, they gather large amounts of data. In order to make sense of all the information, they often organize the data into a table.

PROCEDURE

When you design a table, you must give the table a title. List all the information or relationships that you wish to record in your table. These are your headings. Headings may be listed either horizontally or vertically in the table.

For example, the activity on page 81 shows a data table for igneous rocks. In the table, the two major characteristics to be recorded are color and texture. Each of these characteristics is divided into smaller groups.

A different format is used for the table in Investigation 4, which is on page 94. One column lists the numbers of the rocks to be studied.

The other columns list categories of information about each rock.

Still another format can be seen in Table 1. This table has illustrations to show what the particles in different types of igneous rocks look like.

APPLICATION

Design a data table that shows the relationship between the degree of metamorphism and the metamorphic rocks formed from shale.

USING WHAT YOU HAVE LEARNED

1. Consider the characteristics of sedimentary rocks that are described in this chapter. Design a table to show the difference in sediment size of six sedimentary rocks.
2. Consider the origin of various metamorphic rocks and design a table to show the parent material of four different metamorphic rocks.

TABLE 1: IGNEOUS ROCK FORMATION

Rock Formed	Temperature of Magma	
	800°C	900–1200°C
Felsic rocks		
Mafic rocks		

Section 2   Igneous Rocks   83

Thinking Critically

4. The magma cools slowly at first. The first crystals to form are feldspar. They are large because of the slow cooling. As the magma begins to cool more rapidly, the remaining minerals form much smaller crystals.
5. Iron and magnesium are two elements that might make mafic rocks dark and dense.

SKILL ACTIVITY:
Designing Tables

Objectives

- Choose headings for the top and the side of a table.
- Communicate data in the form of a table.

Discussion

Have the students decide on a simple non-scientific table first. Some examples might be the favorite colors of the class or heights of people in the class. Then they should follow the Procedure step by step. When the students have this experience with choosing headings and deciding on format, they can complete the activity. Be sure they draw lines with a straight edge to make a neat table. *Thinking Critically:* Ask the students how tables can be useful to scientists. (When conducting experiments and otherwise collecting data, scientists are able to keep their data organized and are thus able to share it more easily with other scientists. Also, tables often make it easier to see trends and patterns in data collected.)

Answer to Application

Table design will vary, but the degree of metamorphism will be obvious if samples of shale, slate, schist, and gneiss are included. The students will be able to see the flat black minerals from the parent shale.

Answers to Using What You Have Learned

1. The table of sedimentary rocks should be constructed in order, from the smallest fragments to the largest or from the largest to the smallest.
2. The table of metamorphic rocks should show the parent rock and the metamorphic rock. For example:

granite	gneiss
sandstone	quartzite
shale	slate
limestone	marble

SECTION FOCUS

Section Overview   This section presents sedimentary rocks in more detail. The three types of sedimentary rocks are clastic, chemical, and organic. Some uses of sedimentary rocks are discussed.

Section Objectives   For a list of section objectives, see pupil's edition page.

New Science Terms   For a list of new science terms in this section, see pupil's edition page. Review the pronunciation of the terms with the class.

SECTION DEVELOPMENT

4.6 Formation of Sedimentary Rocks

DISCUSSION   Several models could help students understand the formation of sedimentary rocks. One is the wax-shavings model from the beginning of the chapter. Another example is salt in a shaker that becomes a lump in hot, humid weather.

EXTENSION   (Tie-in/Home Economics) A popular snack is made with crispy rice cereal, melted marshmallows, and shortening. ***Thinking Critically:*** Have the students describe cementation by relating it to the marshmallow/rice snack. (The crispy rice is like the grains of sand and pebbles. The melted marshmallows and shortening form the "cement" that makes the crispy rice stick together in a block, just as the natural cements bind sediments together.)

NEW SCIENCE TERMS
cementation
clastic sediments
chemical sediments
organic sediments
evaporites
precipitates

SECTION OBJECTIVES
After completing this section, you should be able to:
- **Describe** three types of sedimentary rocks.
- **Trace** the formation of compact limestone.
- **Identify** common sedimentary rocks by their characteristics.

4.6 **Formation of Sedimentary Rocks**

Even though igneous rocks make up most of the earth's crust, the rocks found on the earth's surface are mostly sedimentary rocks. You are probably unaware that sedimentary rocks are forming all around you. Fragments, or pieces of broken rocks, leaves falling from trees, and streams full of mud are all sources of sediments for new sedimentary rocks.

Recall the model sedimentary rock made by pressing together wax shavings. Fine sediments, such as clay and silt, also stick together if pressure is applied. They easily form solid rocks. Coarse sand and pebbles do not stick together unless something is present to hold them together.

The water that soaks through the earth carries with it natural cements. Some of these natural cements are silica (from quartz), calcium carbonate (from limestone), and iron oxides. They coat sediments, such as sand, and bind them together. This process is called **cementation.** The weight of additional sediments is usually enough to complete the formation of sedimentary rocks.

Natural cementation is really very similar to the process of making concrete. To make concrete for a driveway or sidewalk, you need three things: sand, water, and cement. The cement you

Figure 4–12. The fallen leaves on this pond are one source of sediment for the formation of sedimentary rocks.

Figure 4–13. The natural cement of sedimentary rocks is similar to the cement in the concrete these workers are pouring.

84   Chapter 4   Rocks

BACKGROUND INFORMATION

Although igneous rocks make up more than two-thirds of the earth's crust, most of the rocks on the earth's surface are sedimentary. A famous sedimentary rock is the Rock of Gibraltar, a huge chunk of limestone, that is about 426 m high and covers about 5.18 sq km. The White Cliffs of Dover in England are chalk formed from the shells of billions of microscopic sea organisms.

use in making concrete has the same function as the natural cement in sedimentary rocks; it holds the sand together. The water ensures even distribution of sand and cement.

Sand is only one kind of sediment that forms rocks. There are three types of materials that form sedimentary rocks: *clastic, chemical,* and *organic.*

Clastic sediments are pieces of other rocks. These sediments are produced as rocks are exposed to heat, plants, water, and air. For example, some rocks are broken into fragments when water collects in tiny cracks. The water freezes, and the ice expands and fractures the rock. The small rock pieces are then available for the formation of clastic sedimentary rocks.

Chemical sediments come from minerals dissolved in water. The salts in sea water can become chemical sediments. If the sea water evaporates, the chemicals collect on the bottom and form chemical sediments. Compaction turns the sediments into sedimentary rocks.

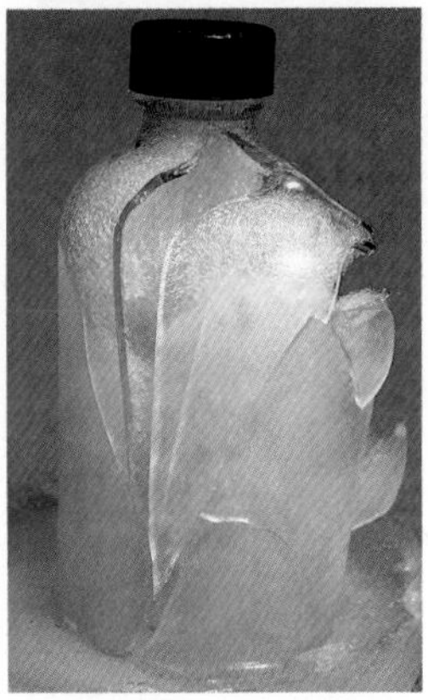

Organic sediments are the hard remains of once-living organisms, such as shells or the skeletons of corals. Many sea-floor sediments are organic. In some places these remains accumulate in thick layers, forming organic sedimentary rocks.

○ *Name three kinds of sediments.*
○ *What kind of sediments are the remains of once-living organisms?*

Figure 4–14. As the water in a bottle freezes, it expands and breaks the bottle. Water freezing in the cracks of a rock can break the rock.

Figure 4–15. The salt being loaded into these trucks is a chemical sediment.

Figure 4–16. Organic sedimentary rocks often contain fossils of the organisms from which they were formed.

85

DISCUSSION Have the students discuss the three types of sediments. *Thinking Critically:* Ask the students how the three types of sediments are alike and how they are different. (They are alike because they are all hard particles. They are different in their origin: fragments from other rocks, from evaporated water, and from organic remains.)

EXTENSION (Tie-in/Medicine) Ask the students if they have ever had to take a liquid medicine for a cough, an infection, or other ailment. Then ask if the liquid needed to be shaken. There will probably be both "yes" and "no" answers. *Thinking Critically:* Ask the students why only some medicines are shaken. (Because some medicines and liquids contain sediments that must be evenly distributed throughout the liquid to ensure proper dosage.) Cartoons sometimes show someone jumping up and down after taking medicine that was not shaken. Have the students discuss the futility of that activity. (The sediment is at the bottom of the medicine bottle, not the person's stomach.)

DEMONSTRATION

Put small amounts of silt, sand, and gravel into a jar of water. Cover the jar and shake it thoroughly. Point out the cloudiness of the water. Explain that this cloudiness is caused by the presence of sediments. Set the jar upright. The gravel will fall to the bottom; the sand will settle more slowly. After some time, the silt will also settle, leaving clear water at the top.

Answers—4.6

○ Sediments include clay, silt, and sand.

○ Organic sediments are the remains of once-living organisms.

4.7 Classification of Sedimentary Rocks

DISCUSSION Have the students discuss the examples of clastic rocks shown in the photographs. Have samples of clastic rocks for them to compare with the photos. ***Thinking Critically:*** Ask the students which type of clastic rock should have the smoothest surface and why. (Shale should have the smoothest surface because it is formed from microscopic grains and flakes.) The size specifications for clastic rocks are quite specific. This would be a good time to add to the rocks chart.

Figure 4–17. Coquina rock is a sedimentary rock formed from shell fragments.

Figure 4–18. Conglomerate (left), breccia (center), and sandstone (right) are all clastic sedimentary rocks.

4.7 Classification of Sedimentary Rocks

Remember that the classification of igneous rocks is based on whether the rock solidified from magma or lava. Sedimentary rocks are classified by the type of sediments present in the rocks. Therefore, the types of sedimentary rocks that form are the same as the types of sediments.

Clastic Sedimentary Rocks Clastic sedimentary rocks are formed from fragments of other rocks. Clastic rocks may be further classified by texture.

Coarse rocks are *conglomerates* (kuhn GLAHM uh rayts) and *breccias* (BREHSH ee uhs). Both of these rocks contain pebble-sized pieces with diameters of 2 mm or more. The difference between conglomerates and breccias is the shape of the rock pieces. Conglomerates have rounded pieces and breccias have sharp-edged pieces.

Sandstone is formed of sand-sized grains that are usually between 0.06 mm and 2.0 mm in diameter. The grains are usually quartz.

Clastic rocks formed from smaller sediments are *siltstone* (formed of grains 0.004 mm to 0.06 mm in diameter) and *shale* (formed of particles smaller than 0.004 mm). The minerals found in siltstone are quartz and feldspar, while those found in shale are clays.

A MATTER OF FACT

Sometimes special environmental records are preserved in sedimentary rocks. For example, ripple marks on wave-worked sands have been preserved with amazing detail.

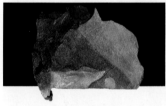

Figure 4–19. Shale (left) and siltstone (right) are composed of nearly microscopic particles.

BACKGROUND INFORMATION

Sedimentary rocks are less dense than igneous and metamorphic rocks and, to a geologist, they are softer. Geologists who study sedimentary rocks are called "soft-rockers." They can learn much about the past by studying the layers in clastic sedimentary rocks. Geologists who study igneous and metamorphic rocks are called "hard-rockers."

Clastic sedimentary rocks are deposited in layers. For example, in slow-moving water, only clay-sized sediments are carried by the water; larger, sand-sized sediments are deposited. If the water stops flowing altogether, even the clay-sized sediments are deposited. Two different types of sedimentary rocks are formed as a result—sandstone and shale. Each type of rock is in a separate layer.

Clastic sedimentary rocks can be identified by the layers they contain. Other types of sedimentary rocks do not show clearly defined layers.

Chemical Sedimentary Rocks

Chemical sedimentary rocks are made of substances dissolved in water. For chemical rocks to form, the dissolved substances must somehow be separated from the water.

Two chemical sedimentary rocks, evaporites and precipitates, are formed by different processes. **Evaporites** (ih VAP uh rytes) are chemical rocks that form when substances dissolved in water are left as the water evaporates. The two most common evaporites, shown in Figure 4–20, are *rock salt* and *rock gypsum*.

Precipitates (pruh SIHP uh tayts) are chemical rocks that form when dissolved substances precipitate, or fall out of solution. Precipitates are usually deposited on the sea floor. The most common precipitate is *compact limestone*, shown in Figure 4–21, which forms when calcite precipitates from sea water. Silica also precipitates out of sea water, forming some types of *chert*. Chert is a very dense rock formed of quartz.

Figure 4–20. Rock salt (top) and rock gypsum (bottom) are chemical sedimentary rocks.

Figure 4–21. Compact limestone (left) and chert (right) are both precipitated from sea water.

A MATTER OF FACT

Rock salt is mined from huge deposits found in several locations in the United States. For example, the salt beds near Detroit, Michigan, are about 300 m thick. The openings in the salt mines of Louisiana are more than 35 m high.

Section 3   Sedimentary Rocks   **87**

SKILL *(Making a Table)* Have the students make a table that gives an example of each type of classic rock. The table should also give the fragment size for each type.

EXTENSION Conglomerates consist of rounded, weather-worn pebbles cemented together, while the particles that make up breccias are broken and angular. Quartz is the most common mineral fragment in conglomerates. *Thinking Critically:* Ask the students why quartz might be the most common. (It is very hard, very common, and chemically stable.)

EXTENSION If possible, obtain some sea water and allow it to evaporate so the students can study what remains after evaporation. Otherwise allow a concentrated solution of table salt and water to evaporate. Have the students try adding more salt to the solution. Salt will precipitate out of water when the water becomes supersaturated, or overloaded, with the salt.

THINKING SKILL *(Drawing Conclusions)* The Bonneville Salt Flats were created when Lake Bonneville evaporated. The Great Salt Lake in Utah is all that is left of Lake Bonneville, which was many times larger than the Great Salt Lake. In Kansas and Michigan, there are beds of halite, or rock salt, that are more than 1000 m thick. Ask the students how they think those salt beds might have been formed. (It is believed that parts of the country were once covered by seawater. When the water evaporated, the salt beds were formed.)

DISCUSSION Organic sedimentary rocks are formed from the remains of once-living organisms. Organic sedimentary rocks are, therefore, a good source of fossils, such as those shown in Figure 4-22. Fossils are a fascinating subject for the students and they are a tangible record of the history of the earth. Try to have several fossils for the students to examine.

EXTENSION Compact limestones often contain large numbers of planktonic fossils. There may be microscopic fossils in chert because diatoms use silica in forming their shells. Gastropods (snails), pelecypods (clams), brachiopods, and bryozoa, along with plant fossils, are commonly found in sandstones and shales.

DISCOVER: Finding Fossils in Sedimentary Rocks

Thinking Skill (*Relating*)

Samples of shale are available from any geological supply company. The students should realize that they would not expect to find fossils in igneous rocks because these rocks form from magma and lava. Fossils can be found in organic sedimentary rocks because fossils are organic remains.

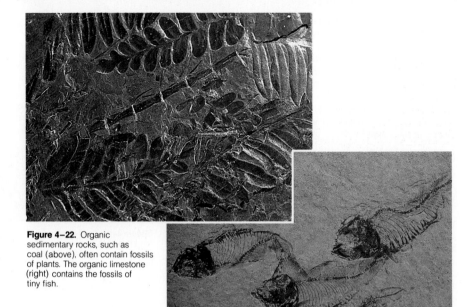

Figure 4–22. Organic sedimentary rocks, such as coal (above), often contain fossils of plants. The organic limestone (right) contains the fossils of tiny fish.

DISCOVER

Finding Fossils in Sedimentary Rocks

Look at a piece of shale that has fossils in it. See if you can recognize any of the fossils. Would you expect to find fossils in igneous or sedimentary rocks? Explain your answer.

Organic Sedimentary Rocks   Organic sedimentary rocks form from the remains of once-living organisms. Although limestone is a chemical rock, it can also be an organic rock. Organic limestone forms from the hard remains of animals such as corals. Organic limestone often contains fossils of animals trapped in the sediments, as shown in Figure 4–22. These fossils provide clues about life during the time the organisms lived.

Organic sedimentary rocks also form from compressed plant material. In a swamp, plants decay very slowly to form *peat*. Over years the buried peat becomes *lignite*, sometimes called brown coal. Brown coal is compacted, but it is not as hard as other coals. With more pressure and time, a harder coal called *bituminous* (bih TOO muh nuhs) *coal* forms. Bituminous coal is used primarily in the production of electricity. Millions of years are required for bituminous coal to be produced from plant material. Many good plant fossils are found in layers of bituminous coal, as Figure 4–22 shows. Coal often provides the only clues about ancient plant life, for soft plant parts do not easily form fossils.

○ *Give an example of an organic rock.*
○ *What are two types of chemical sedimentary rocks?*

Answers–4.7

○ Lignite, peat, bituminous coal, and organic limestone are all organic rocks.
○ Evaporites and precipitates are two types of chemical sedimentary rocks.

Answers–4.8 (p. 89)

○ The sandstone of the White House is painted to hide the damage caused by fire during the War of 1812.
○ Porcelain is a combination of clay, feldspar, and quartz.

4.8 Uses of Sedimentary Rocks

Many sedimentary rocks are used in industrial processes and construction projects. Shale and siltstone are used for manufacturing sewer pipe, drain tile, and chimney linings.

Many buildings are constructed of sandstone. Huge blocks are sawed out of thick sandstone beds. These blocks are then cut into smaller rectangular blocks to meet the size requirements of the construction industry. The White House in Washington, D.C., is built mostly of sandstone, painted white to hide the damage of a fire set by British soldiers during the War of 1812.

Rock gypsum is used by the building industry in the production of drywall and plaster. The quartz in sandstone is the basic mineral component of glass and porcelain. Porcelain—used in fine china, bathtubs, sinks, and toilets—is a combination of clay, feldspar, and quartz. Fiberglass insulation and the silicon chips used in computers are also made from quartz.

○ *Why is the sandstone of the White House painted?*
○ *What is porcelain?*

Section Review

READING CRITICALLY

1. What kind of rocks form from rounded pebble-sized sediments?
2. Why is rock salt considered to be an evaporite?

THINKING CRITICALLY

3. Why would silt be deposited by slow-moving water but not by more rapidly moving water?
4. Why might you expect to find fossils in organic limestone but not in chemical limestone?

Section 3   Sedimentary Rocks   **89**

Figure 4–23. The White House in Washington, D.C. is built of sandstone that is painted white. Sandstone is a clastic sedimentary rock.

A MATTER OF FACT

We have used as much coal since World War II as had previously been used throughout recorded time.

4.8 Uses of Sedimentary Rocks

DISCUSSION   Have the students make lists of materials used in building a house that are products of sedimentary rocks. Samples of drywall, plaster, and porcelain (a piece of broken sink or drywall) might be interesting to the students so they can see the texture of these products.

THINKING SKILL   (*Drawing Conclusions*) Bituminous coal is used primarily to power the generators that produce electricity. Coal is a fossil fuel, formed from the remains of ferns and other plants that thrived in giant swamps millions of years ago. Have the students discuss why coal is called a nonrenewable fuel. (Although coal is still being formed, it is being used faster than it can form.)

SECTION REVIEW

Summary   Have the students review the formation, classification, and use of sedimentary rocks by continuing the chart begun in the previous section.

Reinforcement   Have the students quiz each other on the facts they have recorded on their charts.

Answers to Section Review

Reading Critically

1. conglomerates
2. The salt is left behind when the water evaporates.

Thinking Critically

3. Rapidly moving water would keep the sand and silt in suspension, not allowing them to be deposited.
4. Chemical limestones are formed under conditions in which calcite precipitates out of seawater. These conditions often occur in areas not likely to support much life. Organic limestones form from hard remains where sea life is abundant.

BIOGRAPHIES: Then and Now

Discussion Have the students read the biographies and discuss the contribution each man made to the field of geology.

Thinking Skill *(Comparing)* After reading the two biographies, have the students compare how Haüy formed his hypothesis of crystal formation with the way Parker formed his Buffer Hypothesis.

Extension Interested students might be encouraged to research further the life and work of Haüy and Parker or one of the scientists listed below.

Norman L. Bown

James Hutton

BIOGRAPHIES: Then and Now

RENÉ JUST HAÜY (1743–1822)

René Haüy (ah YOO ee) was a French priest who developed an interest in science. While visiting a friend, he accidentally dropped a calcite specimen from his friend's mineral collection. The mineral broke into many small pieces. He was very embarrassed, of course, but as he picked up the mineral fragments, he noticed that all the broken pieces formed the same geometric shape. The shape was that of a rhombohedron, a slanted box. Later he tested more calcite and verified that calcite would always break into rhombohedral shapes.

Haüy then hypothesized that each mineral is built up of layers or sections of a simple geometric shape. He proposed that this geometric shape is related to the internal crystalline form of the mineral. That is, the form is based upon the molecular makeup of the mineral.

For example, crystals of halite are always cubic in shape. Crystals of quartz are always hexagonal (six-sided) with a pyramidal top.

This was the beginning of the science of crystallography. Haüy became a professor of mineralogy at the Museum of Natural History in Paris during the reign of Napoleon. While at the museum, he wrote the first important textbook on crystallography. Haüy's work with crystals has led to the development of many consumer products based on the characteristics of quartz.

RONALD B. PARKER (1932–)

Ronald Parker earned a Ph.D. in geology from the University of California at Berkeley. He has been teaching geology at the University of Wyoming for over seventeen years. Dr. Parker has written many articles and several books on various topics in geology. Through witty stories and simple language, he leads his readers into the complex study of the earth.

Parker has also done much research in geology and has proposed a hypothesis to explain why some geologic events occur as great bursts of energy rather than in gradual steps. He calls this the Buffer Hypothesis. In this hypothesis, Parker argues that many earth features, such as volcanoes, glaciers, mountain slopes, and geysers, store energy as a reservoir stores water. Eventually, so much energy is stored that the system reaches its energy capacity. At that point, the system is so out of balance that the energy is released in the form of a dramatic event such as a landslide, volcanic eruption, or burst of water from a geyser.

Although not proven, this hypothesis may explain why earthquakes and volcanic eruptions are so hard to predict. So far, scientists have not been able to measure the amount of energy that is built up in the earth. If a way could be found to measure this energy buildup, it might be possible to predict these natural disasters.

SECTION OBJECTIVES

After completing this section, you should be able to:

■ **Explain** how metamorphic rocks form.
■ **Define** the terms *foliated rocks* and *nonfoliated rocks*.
■ **Identify** some common metamorphic rocks by their characteristics.

NEW SCIENCE TERMS

foliated rocks
nonfoliated rocks

4.9 Formation of Metamorphic Rocks

Many animals go through a metamorphosis, or change, in their lives. Frogs, for example, start life as fishlike tadpoles. They gradually change their body form, by growing legs and lungs. Their life style changes also, as they become adults. Instead of living in water, they can live on land. Other animals, such as butterflies and moths, also undergo great changes in form.

Many rocks change physically, as well. Metamorphic rocks form when heat and pressure cause chemical and structural changes in existing rocks. During a structural change, heat and pressure rearrange the minerals within the rock. This change results in a more compact mineral pattern or in a different form of crystal. When a chemical change occurs, elements recombine to form different minerals from those that were in the original rocks.

The pressure that forms metamorphic rocks can come from mountain-building processes. Mountains, which are discussed more fully in Chapter 6, are built up as forces within the earth push and fold huge areas of land. A source of heat is large bodies

Figure 4–24. Heat and pressure can change the minerals in igneous and sedimentary rocks, forming metamorphic rocks.

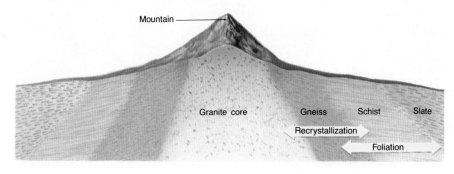

Mountain

Granite core   Gneiss   Schist   Slate

Recrystallization

Foliation

BACKGROUND INFORMATION

The verb metamorphose means "to change form." The noun metamorphosis means " a change of form." Metamorphic rocks are rocks that have undergone a change of form. Geologists have determined that heat and pressure cause the chemical and structural changes that transform one type of rock into another.

SECTION FOCUS

Section Overview   This section gives a detailed look at metamorphic rocks. The formation of metamorphic rocks is discussed and their classification as foliated and nonfoliated rocks is explained. The section concludes with a discussion of the uses of metamorphic rocks.

Section Objectives   For a list of section objectives, see pupil's edition page.

New Science Terms   For a list of new science terms in this section, see pupil's edition page. Review the pronunciation of foliated and nonfoliated.

SECTION DEVELOPMENT

4.9 Formation of Metamorphic Rocks

DISCUSSION   With a little pressure, a ball of new-fallen snow can transform into a lump of ice. Many students have probably had this experience and could relate it to the changes pressure can cause in rocks. Have the students discuss the formation of metamorphic rocks as shown in Figure 4-24.

EXTENSION   (Tie-in/Home Economics) Most students have probably seen cookies made "from scratch." Many recipes require butter and sugar to be combined. However, the crystals of sugar are still visible even after they have been combined with the butter. ***Thinking Critically:*** Ask the students how the action of heat on the cookie batter is similar to the action of heat in creating metamorphic rocks. (In the finished cookies, there are no separate sugar crystals or other separate ingredients. The heat has caused a chemical reaction in the ingredients so they are now just one substance—cookie.)

DISCUSSION   One source of heat that forms metamorphic rocks comes from magma. The changing of minerals near a large magma intrusion is called contact metamorphism.

DISCOVER:
Observing Metamorphic Rocks

Thinking Skills *(Comparing, Drawing Conclusions)*

Each student or small group of students should have samples of marble, slate, and quartzite. Have samples of other rocks available, including limestone, shale, and quartz sandstone, so the students can make comparisons and draw conclusions. Marble comes from limestone; slate comes from shale; and quartzite comes from quartz sandstone. The students should mention some characteristics of the various rocks.

4.10 Classification of Metamorphic Rocks

DISCUSSION   After the students learn the difference between foliated and non-foliated rocks, have them study the rocks in Figure 4-25 and classify them. You may wish to have other samples in class for more classification. One characteristic of a metamorphic rock is that it is harder than the rock from which it originated. ***Thinking Critically:*** Ask the students why they think marble is so often used by sculptors in making statues. (Answers might include hardness, color, durability, or because it is nonfoliated.)

DISCOVER

Observing Metamorphic Rocks

Study pieces of marble, slate, and quartzite. What sedimentary rocks might each of these have come from? On what observations did you base your answers?

Figure 4–25. Marble (below), quartzite (bottom left), and slate (bottom right) are all metamorphic rocks.

of magma within the earth. A body of magma can cause changes in rocks for hundreds of kilometers around it.

○ ***How might heat and pressure affect the minerals in a rock?***
○ ***Name two sources of heat for the formation of metamorphic rocks.***

4.10 Classification of Metamorphic Rocks

Metamorphic rocks are divided into two groups: foliated and nonfoliated. **Foliated rocks** have obvious layers. Foliated rocks are formed by pressure from one direction. The layers often appear shiny, due to the new mineral crystals that have formed. Some foliated rocks show a pattern of bands. The amount of foliation is used to classify metamorphic rocks. For example, *slate*, *phyllite*, *schist*, and *gneiss* (NYS) show increasing amounts of foliation. **Nonfoliated rocks** have no visible layers. This is because pressure during formation comes from several directions. *Marble*, for instance, is nonfoliated.

Identifying the minerals in the original rocks helps to classify metamorphic rocks. For example, when limestone is changed to a metamorphic rock, the original limestone becomes more compact. However, the mineral can still be identified as calcite. The resulting rock is marble. Several kinds of metamorphic rocks are shown in Figure 4–25.

Another example of the metamorphic process occurs when bituminous coal is put under intense heat and pressure. This soft coal becomes a harder, more compact form of coal called *anthracite*.

○ ***How can foliated rocks be identified?***
○ ***What is a nonfoliated rock?***

Answers—4.9

○ Heat and pressure may cause both chemical and structural changes in rocks.

○ Pressure from mountain-building forces and heat from large bodies of magma help to form metamorphic rocks.

Answers—4.10

○ Foliated rocks contain layers of minerals. These minerals are often shiny.

○ A nonfoliated rock does not contain layers.

4.11 Uses of Metamorphic Rocks

Metamorphic rocks have many common uses. Marble, for example, is extensively used as a building material because its surface polishes beautifully, displaying interesting patterns. Marble is also used for monuments and sculptures. The Lincoln Memorial in Washington, D.C., is made of polished marble.

Slate is used as roofing tile in some areas, because it easily breaks into thin slabs and it is waterproof. Slate was once widely used for gravestones. The inscriptions on these gravestones, some over 100 years old, are still readable. What characteristic of slate allows it to last so long? **1**

Anthracite coal is very hard and has few of the sulfur impurities found in bituminous coal. For these reasons, anthracite coal is a very good fuel for heating, but it is also very rare and expensive.

○ *Why is marble used for buildings and monuments?*
○ *What is slate used for?*

Figure 4–26. Because of its beauty and durability, many monuments, such as the Lincoln Memorial, are made of marble.

A MATTER OF FACT

Very pure, white marble is found in the mountains around Carrara, Italy. Mining of this marble has been going on for at least 2000 years. The interior of the Pantheon in Rome was built of this marble. It can be found in such different places as the Leaning Tower of Pisa, the pavement of St. Peter's Basilica in Rome, and in the Kennedy Center in Washington, D.C.

Section Review

READING CRITICALLY

1. What are metamorphic rocks?
2. What characteristic of a metamorphic rock determines whether it is classified as foliated or nonfoliated?

THINKING CRITICALLY

3. How can the alignment of crystals in a metamorphic rock provide information about the direction in which pressure was applied to the rock?
4. Why are mineral crystals not foliated in marble?

1 Water will not penetrate its surface.

4.11 Uses of Metamorphic Rocks

DISCUSSION The Washington Monument, the Jefferson Memorial, and the Lincoln Memorial in Washington, D.C., are all made of marble. ***Thinking Critically:*** Ask the students why they think metamorphic rocks are used so widely for outdoor objects. (Water has little effect on them, and they are very hard. They are very durable even when they are exposed to the weather.)

EXTENSION The relatively new phenomenon of acid rain is having bad effects on buildings, bridges, and monuments. Interested students may wish to research the effects of acid rain on metamorphic rocks.

SECTION REVIEW

Summary Have the students continue the summary chart begun in the second section in order to review the formation, classification, and uses of metamorphic rocks.

Reinforcement Students can quiz each other on the facts recorded on their chart.

Answers—4.11

○ Marble is often used for buildings and monuments because it has interesting patterns, and it polishes well.

○ Slate is used for roofing tile, chalkboards, and pool tables.

Answers to Section Review

Reading Critically

1. Metamorphic rocks are rocks formed from other rocks changed by pressure and heat.

2. Foliated rocks have layers or bands of minerals. Nonfoliated rocks do not show distinct bands.

Thinking Critically

3. The minerals will align perpendicular to the applied force.

4. Most types of limestone from which marble forms contain only one type of mineral—calcite.

INVESTIGATION 4:
Classifying Rocks

Skill *(Classifying)*

Preparation of Materials

For each group of students, have a sample of granite, breccia, and any six of the following: sandstone, rhyolite, pumice, limestone, marble, quartzite, basalt, shale, slate, gneiss, conglomerate, and coal. Make sure that different groups receive different sets of rocks, except for the granite and breccia.

Discussion CAUTION: Be sure to have the students put on safety goggles and laboratory aprons before doing this investigation. Introduce the rock key to the students by classifying two rock samples (granite and breccia are suggested) through the key. Then have the students do the rest of the classifications on their own. You may wish to allow two or three class periods for the students to practice classifying rocks. The rock key provides an opportunity to give a practical lab test.

Answers to Analyses and Conclusions

1. Answers will vary depending on the rocks selected.
2. Classification by rock color uses groups of minerals such as the groupings in granite. They are simply classified as light-colored or dark-colored, rather than by specific colors. This is less specific than color descriptions for minerals.
3. Answers will vary, but many students may choose marble because of its polished look and reaction with HCl.

Answer to Application

You could look for a compact sedimentary rock that does not show layering but that may contain some fossils. Acid will bubble vigorously as it is dropped on the rock.

INVESTIGATION 4: Classifying Rocks

PURPOSE

To classify igneous, sedimentary, and metamorphic rocks, using a rock classification key

MATERIALS (per group of 3 or 4)

Safety goggles
Laboratory apron
Unlabeled rock specimens
Rock Classification Key, page 568
Hydrochloric acid, 0.5 M
Hand lens

PROCEDURE

1. **CAUTION: Put on your safety goggles and laboratory apron and leave them on for the entire investigation.**
2. Make a table like the one shown. Fill in the chart as you complete the investigation.
3. Select one of the rock specimens. Classify it by using the key on page 568. This key is organized to lead you step by step, eliminating various possible rock types as you move through the key. The first category is sedimentary rocks. If a fresh surface of your rock specimen looks like fragments cemented together, then you should further explore the subdivisions in Category I.
4. If your rock does not look like a sedimentary rock, you should move through the key to Category II.
5. If Category II does not match, you should continue through the key until you find the description that matches your specimen.
6. When you decide the name of the rock, you should fill in all the information in your Rock Identification Table.
7. Follow this same procedure for all the specimens.

ANALYSES AND CONCLUSIONS

1. Which rock did you find to be the most difficult to classify? Explain why it was difficult.
2. Do you think the range of mineral colors would be less useful for classifying rock specimens than for classifying mineral specimens? Explain.
3. Of all the rock specimens discussed in this chapter, which one is the easiest to identify? Why?

APPLICATION

If you owned some property with a lot of rocks, how would you find out if the rocks were limestone and, therefore, good for building?

	TABLE 1: ROCK IDENTIFICATION			
Rock	Name	Mineral Content	Rock Family	Key Classification Characteristics
1				
2				
3				
4				
5				
6				
7				
8				

CHAPTER 4 REVIEW

SUMMARY

- Rocks are naturally occurring minerals or combinations of minerals. (4.1)

- Rocks are classified into three families: igneous, sedimentary, and metamorphic. (4.1)

- Rocks are classified by their origin and the types of minerals they contain. (4.1)

- The continuous process of change that affects all rocks is called the rock cycle. (4.2)

- The rock cycle is not one way. Any rock may be changed into any other type of rock. (4.2)

- Igneous rocks are formed as magma or lava cools and crystallizes. (4.3)

- The crystal size in igneous rocks is determined by the rate of cooling. (4.4)

- Many igneous rocks are used in buildings because of their strength. (4.5)

- Sedimentary rocks are formed from the cemented fragments of other rocks. (4.6)

- Sedimentary rocks, which are classified by the way the sediments form, are called clastic, chemical, or organic. (4.7)

- Sedimentary rocks are used in many industrial processes. (4.8)

- Metamorphic rocks are formed when heat or pressure is applied to other rocks. (4.9)

- Metamorphic rocks are classified as foliated or nonfoliated depending on whether or not mineral layers are visible. (4.10)

- Many metamorphic rocks are useful as building materials. (4.11)

Write all answers on a separate sheet of paper.

SCIENCE TERMS

Correctly use each of the following terms in a sentence.

cementation **(84)**
chemical sediments **(85)**
clastic sediments **(85)**
evaporites **(87)**
extrusive rocks **(80)**
foliated rocks **(92)**
igneous rocks **(76)**
intrusive rocks **(79)**
lava **(76)**
magma **(76)**
metamorphic rocks **(77)**
nonfoliated rocks **(92)**
organic sediments **(85)**
precipitates **(87)**
rock cycle **(78)**
rocks **(75)**
sedimentary rocks **(76)**
texture **(79)**

SCIENCE QUIZ

Modified True-False

Mark each statement *true* or *false.* If a statement is false, change the underlined term to make the statement true.

1. Granite is an example of an <u>igneous</u> rock.

2. When heat and pressure are applied to igneous rocks, <u>sedimentary</u> rocks are formed.

3. Rocks are classified by the properties of <u>density</u> and origin.

4. Rock salt is an example of an <u>evaporite.</u>

5. The term <u>foliated</u> refers to the layers of mineral crystals in metamorphic rocks.

6. The three main types of sedimentary rocks are <u>clastic</u>, organic, and chemical.

continues

Chapter 4 Review   **95**

CHAPTER REVIEW

SUMMARY

The students may review the major concepts in the chapter by reading the summary statements. The statements are cross-referenced to the chapter to facilitate reinforcement of any concepts of which the students feel unsure. Encourage the students to work in groups to quiz one another.

SCIENCE TERMS

The sentence in which the science term is used should reflect an understanding of the definition of the term. You may wish to review the terms and definitions in the form of a two-column matching test.

SCIENCE QUIZ

Answers to Modified True-False

1. true
2. false, metamorphic
3. false, mineral content
4. true
5. true
6. true

continues

Answers to Multiple Choice

7. a
8. a
9. a
10. d
11. b
12. a

Answers to Completion

13. three
14. sedimentary
15. mineral grains
16. slow
17. metamorphic
18. Coquina

Answers to Short Answer

19. Intrusive rocks cook while deeply buried. They cool slowly in this well-insulated condition, thus forming large mineral crystals. Extrusive rocks rise to the earth's surface. They cool very quickly, so large mineral crystals never have the time to form.
20. Anthracite coal is formed when layers of bituminous are pressed together for long periods of time.
21. Most sedimentary rocks are made of clastic materials that have been cemented together. The rocks are not very compact. Igneous rocks form from magma or lava. They are very crystalline and compact. Metamorphic rocks are also crystalline, often showing foliation. Nonfoliated metamorphic rocks are often a single, uniform color.
22. Because rocks may enter the cycle at any point and change in any direction.

Answers to Writing Critically

23. It would not be possible for fossils to be found in igneous rocks. Magma is so hot the fossils would melt. For this reason, fossils are never found in igneous rocks.
24. See Figure 4-5 for the rock cycle.
    Sandstone and stone → quartzite → granite → sandstone
    sandstone → quartzite → granite → sandstone
    limestone → marble
    shale → slate → basalt → shale
25. Because the pressure, which forms quartzite, covers any sandstone layering.

Multiple Choice

Write the letter of the choice that best answers the question or completes the statement.

7. Granite consists of feldspar, mica, and
   a) quartz.   b) olivine.
   c) shale.   d) magnetite.

8. What is the main difference between felsic and mafic rocks?
   a) the rate of cooling of the molten rock
   b) the degree of metamorphosis
   c) the color of the minerals in the rocks
   d) the depth of magma source

9. Bituminous coal is considered a sedimentary rock because it
   a) is formed from clastic material.
   b) hardens under great heat and pressure.
   c) has been compressed and made more compact.
   d) solidified from a molten body.

10. Which choice lists only metamorphic rocks?
    a) shale, gneiss, slate, schist
    b) slate, shale, granite, gneiss
    c) basalt, schist, slate, phyllite
    d) slate, phyllite, schist, gneiss

11. A rock formed from the cementing of fine clay minerals is
    a) sandstone.   b) shale.
    c) limestone.   d) marble.

12. Which of the following is not a type of clastic sedimentary rock?
    a) limestone   b) breccia
    c) sandstone   d) conglomerate

Completion

Complete each statement by supplying the correct term.

13. Rocks are classified into _____ families.

14. The term clastic refers to particles in _____ rocks.

15. The term texture refers to the overall appearance and feel of a rock due to the _____.

16. The _____ cooling of magma results in the formation of large mineral crystals.

17. When heat and pressure are applied to existing rocks, _____ rocks are formed.

18. Sedimentary rocks formed from the compressed, cemented fragments of shells are called _____.

Short Answer

19. Compare and contrast intrusive and extrusive rocks.

20. Describe how anthracite is formed.

21. Explain how rocks from each of the three rock families differ.

22. Explain why the rock cycle is not a one-way process.

Writing Critically

23. Would it be possible for fossils to be preserved in igneous rocks? Explain your answer.

24. Explain the rock cycle. Include changes that occur to at least three different rocks as they move through the rock cycle.

25. Why do the layers of sediment that are clearly visible in sandstone seem to be entirely absent in quartzite?

EXTENSION

1. Research the process that results in a rock formation called a *pegmatite*. Locate areas in the continental United States where pegmatites are found. Why are pegmatites of economic interest?

2. Find out what type of rocks are in your region of the country. Make a chart to classify the rocks.

3. Go to the library and read about *geodes*. Explain how scientists think geodes are formed and draw a map to show locations where geodes are found.

ANSWERS TO EXTENSION

1. Pegmatites often contain valuable minerals—even gem quality minerals.
2. Answers will vary depending on rocks in area.
3. Report should include information about minerals inside geodes dissolving away.

APPLICATION/CRITICAL THINKING

1. Oil is found in certain rocks. In what type of rocks could oil be found? Explain your answer.

2. Coquina is a rock made of pieces of broken shells. Describe the conditions in the environment that would allow such a rock to form.

3. Stone fireplaces are almost never constructed of sandstone because of the danger of fire spreading beyond the fireplace. What is there about the structure of sandstone that might cause sandstone fireplaces to be dangerous?

FOR FURTHER READING

MacFall, R. *Rock Hunters' Guide.* New York: Crowell, 1980. This book is a good introductory guide to finding and identifying collectible rocks as a hobby.

McGowen, T. *Album of Rocks and Minerals.* New York: Rand McNally, 1981. This book explains that rocks are more interesting than you might think. Their uses range from the practical to the destructive.

Selsam, M. and J. Hunt. *A First Look at Rocks.* New York: Walker and Company, 1984. This book relates some of the pleasures of looking for and at rocks.

Weiner, J. *Planet Earth.* New York: Bantam, 1986. This book provides the reader with many facts about the earth, including resources of the earth.

Challenge Your Thinking

Have you ever seen rocks that bend? Most people think of rocks as being very solid. This rock is a special type of sandstone called *flexible sandstone.* The rock can be bent without breaking. What characteristic of sedimentary rocks would allow them to bend in this way? Why are there no flexible igneous or metamorphic rocks?

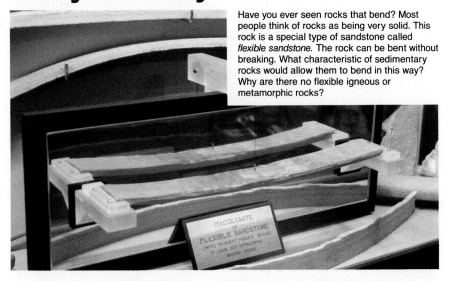

ANSWERS TO APPLICATION/ CRITICAL THINKING

1. Oil would be found in sedimentary rocks because the oil is formed in sedimentary rocks and is stored in porous rocks.
2. A beach environment where there are strong waves to break up the shells.
3. The pores of sandstone often contain moisture, which greatly expands upon heating. This may cause the rocks to explode.

ANSWER TO CHALLENGE YOUR THINKING

Incomplete cementing of the grains in sandstone would account for the flexibility. Igneous and metamorphic rocks could not be flexible because there are no pores in the rocks.

Chapter 5: EARTH'S RESOURCES

PLANNING THE CHAPTER

Chapter Sections	Page	Chapter Features	Page	Program Resources	Page
Section 1: Metals and Nonmetals	99			Investigation 5.2: *Extraction of Copper from its Ore* **(A)**	**TRB 19** **LM 21**
5.1 Metals **(B)**	99	**Discover:** Making Silver **(B)** **Skill Activity:** Writing a Report **(B)**	100 101		
5.2 Iron and Steel **(B)**	102	A Matter of Fact A Matter of Fact	102 103		
5.3 Quartz **(B)**	103	A Matter of Fact A Matter of Fact Section Review **Technology:** Computer Chips **Investigation 5:** Crystal Formation **(B)**	104 105 105 106 112	Student Record Book: Textbook Investigations **(B)**	**TRB 9**
Section 2: Fossil Fuels	107			Critical Thinking **(H)** Cross-Discipline: Science and Mathematics, *Examining Natural Resources in the United States and the World* **(A)**	**TRB 9** **TRB 9**
5.4 Hydrocarbons **(B)** 5.5 Oil and Gas **(B)**	107 108	**Activity:** Determining the Oil-Bearing Capacity of Rocks **(B)**	109	Investigation 5.1: *Fractional Distillation* **(H)** Concept Extension: *Coal Mining* **(B)**	**TRB 17** **LM 19** **TRB 11** **LM 233**
5.6 Coal **(B)**	110	**Discover:** Conserving Fossil Fuels **(B)** A Matter of Fact Section Review	110 111 111	Reading for Content: *Making Pie Graphs* **(B)**	**TRB 9** **LM 181**
Chapter 5 Review	113			Vocabulary **(A)** Tests **(A)** Computer Test Bank	**TRB 5** **LM 131** **TRB 18**

(LM) Laboratory Manual/Study Guide, **(TRB)** Teacher's ResourceBank™

B = Basic **A** = Average **H** = Honors

The coding Basic, Average, and Honors indicates sections or subsections that might be appropriate for different levels of learners. For additional suggestions regarding choice of topic and depth of coverage, see the Pacing Chart on pages T16–T20.

CHAPTER CONCEPTS, OBJECTIVES, AND TERMS

Section	Concepts	Objectives	Science Terms
Section 1: Metals and Nonmetals	■ Gold was probably the first metal used by humans. **(5.1)** ■ Copper was the first metal to be smelted from its ore. **(5.1)** ■ Bronze is an alloy of copper and tin. **(5.1)** ■ Iron is obtained by smelting iron ore in a forge. **(5.2)** ■ Steel is an alloy of iron and one percent carbon. Nickel and other metals may be added to produce specialty steels. **(5.2)** ■ Quartz, one of the most plentiful minerals, has many uses in the modern world. **(5.3)** ■ Colored glass is made by adding small amounts of specific metals or compounds to the molten glass. **(5.3)** ■ The property of piezoelectricity makes quartz important in the electronics industry. **(5.3)**	■ **List** the metals used by early humans. ■ **Discuss** the process of smelting metals. ■ **Explain** how iron and steel are produced. ■ **Describe** the uses of quartz.	ores smelting alloy forge
Section 2: Fossil Fuels	■ Hydrocarbons are chemical compounds of carbon and hydrogen. **(5.4)** ■ Most hydrocarbons burn easily. Since they are good sources of energy, they are known as fossil fuels. **(5.4)** ■ Petroleum is a mixture of liquid hydrocarbons; natural gas is a mixture of gaseous hydrocarbons; and asphalts are solid hydrocarbons. **(5.5)** ■ Oil and gas are produced in rock layers called *source rocks.* **(5.5)** ■ A drill is used to release oil and gas from reservoir rocks. **(5.5)** ■ North America has large deposits of hydrocarbons along the Pacific and Gulf coasts, and on the north slope of Alaska and Canada. **(5.5)** ■ Coal forms from plant matter that accumulates at the bottom of swamps. **(5.6)** ■ The United States has many deposits of coal in the eastern mountains and the western plains. **(5.6)**	■ **Describe** the chemical structure of simple hydrocarbons. ■ **Explain** how hydrocarbons form. ■ **List** the stages of coal development.	hydrocarbons

Title	Page	Materials
Discover: Making Silver	100	*(per student)* copper wire, silver nitrate solution, small beaker
Skill Activity: Writing a Report	101	*(per student)* index cards or notebook
Activity: Determining the Oil-Bearing Capacity of Rocks	109	*(per group of 3 or 4)* rock samples (sandstone, limestone, and shale); Petri dishes (3); mineral oil; medicine dropper
Discover: Conserving Fossil Fuels	110	*(per student)* paper, pencil
Investigation 5: Crystal Formation	112	*(per group of 3 or 4)* cotton string (30 cm), Petri dishes (3), water, hot plate, copper sulfate, ferrous sulfate, sodium chloride, beakers 100 mL (3), glass stirring rod, ice-filled tray

TEACHING SUGGESTIONS

Section 1: Metals and Nonmetals

Class Activity

Have the students select any single raw material mentioned in this chapter. Help each student to find the name of a large company that uses that raw material in the production of its major product. Suggest that each student write a letter to the chosen company requesting information about where and how they obtain the raw material and what happens to it during the manufacture of the company's product.

Field Trip

Take a field trip to an industrial plant that changes a mineral resource into a finished product.

Outside Speaker

Have someone from a steel mill or a computer chip factory come to the class to explain how his or her plant changes raw materials into finished products.

Section 2: Fossil Fuels

Demonstration: Burning Hydrocarbons

Purpose

To compare the amount of heat released from burning a variety of hydrocarbons

Background

Be sure to provide ventilation while performing this demonstration. The materials burned in this experiment may "pop and splatter" as they are burned, so take usual precautions to prevent fire or injury. The charcoal may have to be soaked with lighter fluid before igniting it. The students should be informed that lighter fluid (butane) is also a product of fossil fuels. The charcoal will release the most heat, followed by wood (a less efficient fuel source). The foodstuffs will, of course, release the least amount of heat.

Materials

Safety goggles
Laboratory apron
Calorimeter
Peanut, cracker, marshmallow, charcoal, wood
Laboratory burner
Laboratory balance

Procedure

1. Set up the calorimeter according to its instructions.
2. Mass a peanut and place it in the calorimeter.
3. Read and record the water temperature in the calorimeter using the thermometer and lift the thermometer out of the calorimeter.
4. Light the peanut with the match so that it begins to burn vigorously. The entire peanut must burn completely. If it does not, begin again with fresh water and a new peanut.
5. When the peanut is completely burned, lower the thermometer and read the water temperature again. Record the temperature.
6. Calculate the number of calories released by the burning peanut by using the following formula:
$$C = M \times (T_A - T_B)$$

where **C** is the number of kilocalories; **M** is the mass of the foodstuff; T_A is the temperature of the water after burning; and T_B is the temperature of the water before burning.

7. Put new water in the calorimeter and mass a cracker. Adjust the size of the cracker until it has the same mass as the peanut. Repeat steps 3–6 for the cracker.
8. Repeat the measuring procedure for the marshmallow, charcoal, and wood.

Questions to Ask the Students

1. Which "fuel" released the most energy? Explain. (*The peanut, because it contains oil.*)
2. How could this procedure be modified to find the heat released by the burning of liquid fuels? (*The "fuel" could be burned in a heat-resistant container such as a crucible.*)

CHAPTER 5

CHAPTER OVERVIEW

This chapter presents an overview of some of the valuable resources that are found within the earth. Included are descriptions of specific metals and nonmetals and their various practical uses in industry. The earth's fossil fuels are then described, with an emphasis on the importance of these resources in today's world.

Section 1: Metals and Nonmetals In this section, the earth's major metals and nonmetals, such as iron and quartz, are described. Included are discussions of historical discoveries of some of the most important metals, the past and present uses of these metals, and the processes required to manufacture the metals. Quartz, one of the most important nonmetallic minerals, is discussed in detail. Descriptions of some of the many uses of quartz, such as the manufacturing of glass and silicon chips, are included.

Section 2: Fossil Fuels In this section, the earth's fossil fuels are described. Hydrocarbons are defined, with an emphasis on their importance in today's economy. The formation of oil and gas in rocks is then explained, followed by a discussion of how these fossil fuels are obtained from the earth. The importance of coal is explained and a description of the forces forming coal within the earth is included.

Earth's Resources

What does this photograph show? You may think that this is an aerial view of a modern city. However, it is really a photograph of a computer chip from a personal computer. The chip is made from quartz. Quartz, a resource from the earth, has many unique properties that allow it to be used in a number of important industries.

A computer chip

98

CHAPTER MOTIVATING ACTIVITY

Obtain books with illustrations and photographs of people at work and at play throughout the ages. Display these illustrations to the class. For example, an illustration of the ancient Egyptians in the act of building the pyramids can be used to prompt a discussion of how their tools and construction methods were limited by available materials. A photograph of a family relaxing in front of a wood-burning stove in the early 1900s can prompt a discussion of how the use of resources today, such as oil and gas, has changed life-styles in the United States. A simple photograph of a scientist working at a microscope illustrates the importance of tools in the development of new scientific ideas. Remind the students that without innovations in techniques for working with glass, the microscope could not have been invented. Discuss the fact that raw materials and their use by humans shape the society in which they live.

1 Metals and Nonmetals

Section 1: Metals and Nonmetals

SECTION OBJECTIVES

After completing this section, you should be able to:

- **List** the metals used by early humans.
- **Discuss** the process of smelting metals.
- **Explain** how iron and steel are produced.
- **Describe** the uses of quartz.

NEW SCIENCE TERMS

ores
smelting
alloy
forge

5.1 Metals

You probably know that gold is one of the most valuable metals in the world. Part of its value is due to the fact that gold occurs as a native element—that is, it is not combined with other elements. Occasionally silver will occur as a native element, but it is found mainly as a compound with sulfur or chlorine. Most metals occur combined with other substances in economically important mineral compounds called **ores.**

Since gold occurs as a native element, it was probably the first metal used by early humans. Gold is soft and easily worked, so it is likely that it was used for making ornaments and jewelry. Gold has also been formed into coins and used as money since about 2000 B.C.

Copper, like silver, sometimes occurs as a native element also. Native copper was used 10 000 years ago to make ornaments, jewelry, and—since it is stronger than gold and silver—simple cutting tools. Unfortunately, copper is not often found as a native element. However, humans soon found that heating copper ore in air produced copper metal. The separation of metals from their ores by the use of heat is called **smelting.**

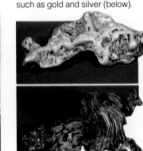
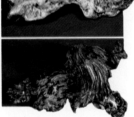

Figure 5–1. These coins (left) were made from native elements, such as gold and silver (below).

Section 1   Metals and Nonmetals   **99**

DEMONSTRATION

To show how metals can be extracted from ores, place powdered charcoal and a tiny amount of black copper oxide into a test tube. Heat the test tube gently for a few minutes, and then pour this mixture into a beaker of water. Students will soon observe that the copper will settle to the bottom of the water and the charcoal will float on the top of the water.

Section 1: Metals and Nonmetals

SECTION FOCUS

Section Overview   This section focuses on the earth's most valuable metals and nonmetals and their uses in today's world. Also included are brief histories of the development of some of the more important metals. The section concentrates on the manufacturing of specific metals and nonmetals and their practical applications in everyday use.

Section Objectives   For a list of section objectives, see pupil's edition page.

New Science Terms   For a list of new science terms in this section, see pupil's edition page. You may wish to have students define these terms before they read the section.

SECTION DEVELOPMENT

5.1 Metals

DISCUSSION   Explain that minerals are the raw materials from which metals and nonmetals can be obtained. Briefly review important characteristics of minerals with the students. Be sure they understand the definition of ore. ***Thinking Critically:*** Ask the students to think of some practical uses, other than jewelry and money, of gold and silver. (For example, gold and silver are used in dentistry; silver is used in photography.)

EXTENSION   Metals have properties that make them very useful, such as the ability to conduct electricity and heat. Most metals can also be hammered or rolled into thin sheets. This property is called malleability. An example of this is gold leaf. Some metals also have the ability to be pulled into thin strands without breaking. This property is called ductility. Copper in electric wires is a good example of this. ***Thinking Critically:*** Ask the students to think of other ways that these properties are useful to us. (Answers might include metal pots that conduct heat or aluminum that is shaped into foil and cans.)

EXTENSION (Tie-in/Mathematics)
Gold and silver have long been used as alloys rather than as pure metals. Less expensive metals are added to the gold and silver, reducing the cost of the precious metals, but still maintaining their valued appearance. Gold is commonly alloyed with silver, platinum, or copper. Explain to the students that the purity of gold is measured in karats. Twenty-four-karat gold is pure gold. Have students calculate the percentage of pure gold in an 18-karat gold ring (75 percent); a 14-karat ring (58 percent); and a 10-karat ring (42 percent).

DISCUSSION Explain to the students that the removal of a metal from its ore is called *smelting*. Ths usual source of energy for smelting is heat. Point out that an example of the type of furnace used to smelt ores is shown in Figure 5–2. One of the costs of producing metals depends on the cost of fuel in smelting the ore. *Thinking Critically:* Ask the students to name another important form of energy that is most likely used in smelting. (electricity)

THINKING SKILL (*Inferring*) Tell the students that new combinations and processes are always being researched for other practical applications of alloys. For example, superalloys have been developed that resist very high temperatures, oxidation, and internal pressures. Ask the students to infer what complex machinery today would require the use of superalloys. (spacecraft, jet engines)

DISCOVER: Making Silver

CAUTION: This should be a supervised activity.
Skill (Observing)

The students may research information about the chemical reaction they observed and present their findings to the class.

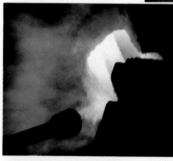

Figure 5–2. Copper (below) is separated from its ore in huge smelters (right).

DISCOVER

Making Silver

CAUTION: Silver nitrate will stain your hands.
Place a small piece of copper wire in a solution of silver nitrate (AgNO₃). In a few minutes you should begin to see pieces of pure silver forming on the wire.

Figure 5–3. Shown here are some tools of the Bronze Age.

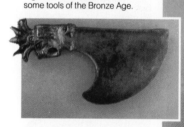

About 3000 B.C. it was discovered that copper could be made stronger by adding tin. This new material was an alloy called *bronze*. An **alloy** can be either a mixture of two or more metals or a mixture of metals and nonmetals.

For 2000 years bronze was commonly used to make knives, swords, helmets, shields, mirrors, combs, and measuring tools. Bronze was so important that this period of history is called the *Bronze Age.*

○ *What is an ore?*
○ *How is copper smelted from its ore?*

BACKGROUND INFORMATION

The Bronze Age, which followed the Stone Age, is not a specific period of time. Different areas in the world had their Bronze Ages at different times. The periods of duration also varied from region to region. Recent excavations in China indicate that bronze may have been made as early as 3600 B.C. Bronze was used by people until sometime between 1400 and 1000 B.C. Iron then became the common metal used for tools. In most areas, the Bronze Age and Iron Age overlapped.

Answers—5.1

○ An ore is a combination of metal and non-metal minerals.

○ The ore is heated with charcoal to melt and separate the copper.

SKILL ACTIVITY: Writing a Report

BACKGROUND

Humans first began using metals over 10 000 years ago. From copper and bronze to iron and steel, civilization has progressed through the use of metal alloys.

In the 19th Century, an English engineer, Henry Bessemer (1813–1898), invented an inexpensive way of making steel from iron. The method involved blowing air through the molten metal in a container called a Bessemer converter. This was done in order to remove impurities from the molten iron.

PROCEDURE

1. Use the card catalog of your school or local library to research the steel making process of Henry Bessemer. The words underlined in the background paragraphs should help you to locate the information you need.

2. After locating several sources of information, make notes on the process on index cards or in a notebook. Include sketches of the Bessemer converter in your notes.

3. Write a two-page report, complete with diagrams, that describes the Bessemer process of making steel. Be sure to include several paragraphs on how the process transforms iron into steel. You might also want to include some information about Henry Bessemer, and mention some of the advantages that steel has when compared to iron.

USING WHAT YOU HAVE LEARNED

After you have finished reading the chapter, pick another topic, perhaps the formation of coal. Use the card catalog to locate library sources, and write another report, including diagrams, on your chosen topic.

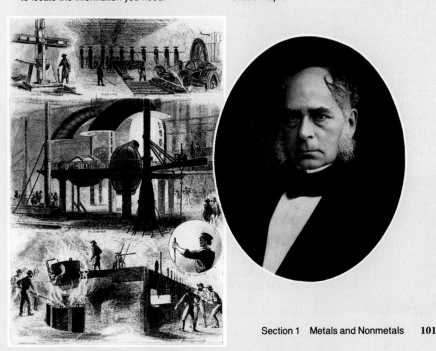

Section 1 Metals and Nonmetals 101

SKILL ACTIVITY:
Writing a Report

Objectives

- Communicate information in written form.
- Collect and organize data.
- Draw conclusions based on research.

Discussion Explain to the students that prior to the innovations in steelmaking made by Henry Bessemer, producing steel was a very costly and slow process. Emphasize that in addition to researching Bessemer and his steelmaking process, the students should also read about how steel was made during the early part of the Iron Age, in the Middle Ages, and up to the nineteenth century. This will help them understand why Bessemer's process revolutionized steelmaking.

Before the students begin their reports, emphasize that it is important to take clear, concise notes from various research sources.

5.2 Iron and Steel

EXTENSION (Tie-in/Social Studies) The early Sumerians are credited with the discovery of the reddish-brown iron ore. The first known word for the metal, however, was the Egyptian *ba-en-pet*, meaning "metal of the sky." This referred to the fact that the first iron was discovered in iron meteorites.

EXTENSION Hematite and magnetite are iron oxides (compounds of iron and oxygen) and contain about 70 percent iron. Magnetite is black and has magnetic properties. Hematite may occur as grainy rocks, shiny crystals, or as a loose, earthy material. It may be brownish-red, black, or dark red.

THINKING SKILL *(Inferring)* Have the students study the early iron tools shown in Figure 5–4. Ask the students to infer possible uses for these tools.

Figure 5–4. Iron for these tools (right) was made in a forge and then hammered into shape by a blacksmith (above).

A MATTER OF FACT

Most of the earth's iron-ore deposits were produced over 2 billion years ago. Oxygen-producing bacteria oxidized the iron in the earth's surface waters. The iron precipitated as iron oxide.

Figure 5–5. Steel, shown here being poured into a mold, has replaced iron as the most important metal in our society.

5.2 **Iron and Steel**

About 1100 B.C. a process for smelting iron was developed in Persia and Eastern Europe. The process for smelting iron caused a dramatic change in the ways in which humans worked and defended themselves.

Iron occurs most often as the iron oxides hematite (Fe_2O_3) and magnetite (Fe_3O_4). Producing metallic iron from these minerals is not as easy as producing copper from its ore.

To make iron, a forge is needed. A **forge** is a furnace through which a stream of air is forced. This produces a very hot flame—hot enough to force the iron oxide to react with the carbon in the charcoal or coal that is used for fuel. The reaction in the forge produces carbon dioxide and *bloom*, a spongy, gray mixture of metallic iron and iron carbide (Fe_3C). Blacksmiths, as iron workers were called, found that hammering red-hot bloom would remove most of the iron carbide, leaving fairly pure iron.

Blacksmiths also discovered that hot iron plunged into cold water hardened rapidly. The hardened iron would maintain a sharp edge longer than any known metal, even bronze. Good iron was so difficult to make that an iron sword or plow was considered a valuable possession. The Bronze Age had been replaced by the *Iron Age*.

Today most iron is made into steel. Although iron can be hardened, it remains brittle and can be broken easily. Iron is also hard to shape except when it is very hot. Steel, an alloy made of iron and about one percent carbon, is as hard as iron, but it is less brittle and can be shaped when cold.

Figure 5–6. Stainless steel has many uses because of its resistance to oxidation. The body of this sports car should last indefinitely.

You may recall from Chapter 3 that there are many metals in the center of the periodic table. Some of these metals, such as chromium, manganese, cobalt, and molybdenum, are added to steel. The addition of each of these metals improves certain, specific qualities of steel, producing alloys called *specialty steels*. One of the most important metals used in the production of specialty steels is nickel. Nickel occurs in igneous rocks and in some soils. Nickel gives steel added strength and improved electrical, thermal, and antirust properties.

The United States is the world's largest producer of specialty steels and the largest consumer of the metals needed for specialty steels. However, only iron and nickel occur in significant quantities in North America; the other metals must be imported from around the world.

○ *What is an alloy?*
○ *What two metals are alloyed most often in the production of steel?*

5.3 Quartz

Quartz, one of the most common nonmetallic minerals, is also one of the most important to the modern world. Quartz is used extensively in the production of common glass, fine crystal, high-temperature glass, precision optical glass, and electronic circuits.

A MATTER OF FACT

Tin neither rusts nor corrodes; therefore, it is ideal for coating the inside of steel food cans. Over 50 billion "tin cans" are made each year.

Figure 5–7. Glass, manufactured from quartz sand, can be made into many products.

103

DISCUSSION Explain to the students that iron is converted into liquid steel through a complex refinery process. (Refer the students back to Figure 5–5.) The liquid steel is then formed into beams, sheets, rods, wire, and other shapes used in making various products. Today, most steel mills perform all the steps necessary in the steelmaking process. *Thinking Critically:* Ask the students to think of a word that could be used instead of refining in the following sentence. The new word should help clarify the sentence: Steel is produced by refining iron. (purifying)

EXTENSION Explain to the students that there are many different kinds of iron and steel. However, iron can be classified as pig iron, cast iron, or wrought iron depending on how it is formed. Steel can be classified as carbon steel, alloy steel, stainless steel, or tool steel, depending on how it is produced. You may wish to have interested students research the main uses of these types of iron and steel.

THINKING SKILL (*Communicating*) Most students will be familiar with one of the most common types of steel, stainless steel. Explain that stainless steel resists corrosion better than any other type of steel. Have the students list as many products made from stainless steel as they can think of. (Some examples are knives, pots and pans, automobile parts, flatware, bathroom fixtures, and razor blades.)

5.3 Quartz

EXTENSION (Tie-in/Chemistry) Explain to the students that glass is composed primarily of silicon dioxide (SiO_2) but may contain varying percentages of other materials, such as sodium, potassium, calcium, aluminum, boron, and others. Remind the students that oxides were discussed in Chapter 3. *Thinking Critically:* Ask the students to name the metals that ferric oxide and cupric oxide are ores of. (iron and copper, respectively)

Answers–5.2

○ An alloy is a mixture of two or more metals.
○ Iron and nickel are alloyed most often in the production of steel.

EXTENSION (Tie-in/Social Studies) Refer the students to Figure 5–8. Tell them that the invention of the blowpipe occurred around 30 B.C. Prior to this, the manufacture of glass was costly and rare; only the very wealthy could afford glass products. With the advent of the blowpipe, glass became less of a luxury and more practical. In the eleventh through the thirteenth centuries, a guild of glassworkers developed that created a vast array of beautiful, stained-glass windows.

Figure 5–8. Although most glass products are now made in molds, the art of glassblowing is still practiced.

A MATTER OF FACT

Silicon makes up more than one-fourth of the earth's crust by mass.

Figure 5–9. Many specialty products, such as high-temperature glass (left) and fiberglass (right), can also be produced from quartz glass.

104

Have you ever seen anyone blowing glass as the person in Figure 5–8 is doing? At one time all glass was made this way. Today common glass is manufactured by heating quartz sand, sodium carbonate (Na_2CO_3), and calcium carbonate ($CaCO_3$) to a temperature of nearly 1000°C. At this temperature the minerals fuse, and molten glass is produced. The soft glass is removed from the furnace and poured into molds to produce the desired shape. Some fine crystal is still blown by hand.

TABLE 5–1: MATERIALS USED TO MAKE COLORED GLASS	
Addition to Glass	**Resulting Color**
Gold	Red
Selenium	Red
Ferric oxide	Brown
Silver	Yellow
Uranium oxide	Yellow
Ferrous oxide	Green
Cobalt oxide	Blue
Manganese oxide	Violet

Specialty glass is similar to specialty steel in that various elements and compounds are added to the molten glass to give it special properties. Table 5–1 shows the colors that result from the addition of metals and minerals to molten glass.

Most glass contains impurities. Glass made of pure silica (SiO_2) has a very high melting point (1713°C) and is used only for special purposes that require high temperatures.

Just as sugar can be spun into fibers of cotton candy, glass can be spun into *fiberglass*. In a loose condition, fiberglass can be used as a filtering material or as insulation for buildings. Mixed with chemicals called *resins*, fiberglass is used wherever a strong, lightweight substitute for metal is needed, such as in boat hulls or sports car bodies.

You have probably seen advertisements for quartz watches; in fact, you may even have one. You know that the watch is not made of quartz, but do you know why it is called a quartz watch? The reason is that these watches are controlled by a crystal of quartz. Quartz has a unique property that allows it to vibrate at a constant frequency, making a quartz watch very accurate.

Crystals of quartz can transform vibrations into electrical voltages, or conversely, transform electrical voltages into vibrations. Quartz has the property of *piezoelectricity* (pee AY zoh ih lehk TRIS uh tee)—that is, it can produce voltages. When pressure is applied to a quartz crystal, positive and negative charges develop on opposite sides of the crystal. If the pressure is alternately applied and relaxed, a current of electricity flows.

If subjected to a vibrating electric field, a quartz crystal will vibrate at a constant frequency. A typical quartz crystal may vibrate at a constant rate of 100 000 Hz, resulting in an extremely accurate timing device. Quartz watches work on this principle.

Silicon chips, made from quartz, have special properties that make them useful in electronic components. Miniature electronic circuits can be printed on them in much the same way that a diagram is printed on paper. This property makes silicon chips important in the production of microcircuits for computers and many other electronic devices.

○ *State five uses of quartz.*
○ *What is piezoelectricity?*

Figure 5–10. A quartz watch, such as the one shown here, is very accurate because of the special properties of the quartz crystal it contains.

A MATTER OF FACT

Substances called *electrides* may one day replace silicon in the manufacture of computer chips. Electrides are crystalline materials consisting of metals such as cesium and potassium.

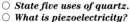
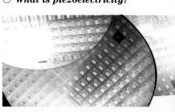

Figure 5–11. Most electronic products can now be made more economically because of silicon chips such as these.

Section Review

READING CRITICALLY

1. Why is iron more difficult to smelt than copper?
2. How is glass made?

THINKING CRITICALLY

3. Why was the production of bronze important to the development of civilization?
4. If quartz were as rare and expensive as gold, how would the electronics industry be affected?

EXTENSION   The students may be interested to know that quartz crystals are used in wave transmitters for televisions, radios, and most radars. In these transmitters, the electric signal produced is expanded and transformed into radio waves of a certain frequency.

SECTION REVIEW

Summary   Have the students list examples of metals and nonmetals. The students should then describe the values of these resources and the various products that the resources provide for human use.

Reinforcement   Have the students review the section concepts by completing the Section Review. Students who have difficulty with the Section Review questions should go over key concepts, preferably with a classmate who can quiz them.

Answers–5.3

○ Quartz is used in the manufacture of common glass, fine crystal, high-temperature glass, precision optical glass, and electronic circuits.

○ Piezoelectricity is the production of electricity in a substance by exposing it to heat or pressure.

Answers to Section Review

Reading Critically

1. More heat is required to separate iron from its ore. A forge is needed to accomplish this.
2. Quartz sand, sodium carbonate, and calcium carbonate are heated to 1000°C. At these temperatures the minerals fuse to produce glass.

Thinking Critically

3. The discovery that metals could be combined to produce alloys with more desirable characteristics was an innovation for civilized cultures. Bronze was the first alloy manufactured by humans.
4. Silicon chips would be very expensive. This would make computers expensive as well.

After the students have read this article, lead a class discussion on the benefits of computer chips. Also, discuss some of the considerations of computer-chip production, such as foreign competition and financial support for research and for developing new materials used in this technology.

TECHNOLOGY: Computer Chips

All computers rely on circuits to control the flow of electric current. In the late 1950s, a computer chip could hold only one circuit. Today the number of circuits molded onto a single chip can be greater than 1 million.

A printed circuit

Computer chips are made from a thin slice of silicon called a *wafer*. A single wafer may be used to make hundreds of separate computer chips.

Why is silicon used to make these chips? Silicon has several advantages that make it ideal for use in microcircuitry. For example, it is a semiconductor. That is, sometimes it can conduct electricity and sometimes it cannot. Silicon's properties as a semiconductor remain intact over a wide range of temperatures. Since silicon chips have a natural resistance to heat, they are more efficient semiconductors than other types of chips.

106

In most atoms the electrons in the outer energy level are ready to react with other atoms. In silicon, however, the outer electrons are bound tightly to each other. This situation can be changed by inserting materials that can "donate" or "steal" electrons from silicon atoms. The insertion of impurities can change the electrical conductivity of the chip. Doing this to the chip is called *doping the chip*. After doping, the flow of electric current in the wafer can be precisely controlled.

A data processing center

The manufacture of computer chips requires great precision. A procedure called *optical lithography* is used to "print" computer chips in much the same way that photographic plates are made for printing magazines. Computer chips are made under microscopes, since the circuits are microscopic in size.

With improved manufacturing technology, the size of computer chips continues to get smaller and smaller. In the 1960s the smallest circuits were about 25 micrometers across. During the 1980s, that size has been reduced to 2.5 micrometers. Some scientists predict that by the year 2000, the size could be 0.25 micrometers.

Of course, these advances in miniaturization will depend on the chemical and structural characteristics of the raw materials used and the photographic processes and materials used to make the chips.

A personal computer

SECTION

2 Fossil Fuels

SECTION OBJECTIVES

After completing this section, you should be able to:

- **Describe** the chemical structure of simple hydrocarbons.
- **Explain** how hydrocarbons form.
- **List** the stages of coal development.

NEW SCIENCE TERM

hydrocarbons

5.4 Hydrocarbons

What do you think is the single most important resource in the world? Do you think it might be gold or diamonds or iron? Would it help you to know that in the mid-1970s the United States' economy suffered severely as a result of less than a 20 percent reduction in the availability of this resource? You have probably guessed by now that the resource that is so important to the modern way of life is oil. No other resource is so directly involved in the way we work and play.

Oil belongs to a group of compounds known as hydrocarbons. **Hydrocarbons** are compounds containing only hydrogen and carbon. The simplest hydrocarbon, methane (CH_4), consists of four hydrogen atoms bonded to a single carbon atom.

Natural hydrocarbons are fossils that were once part of living organisms. Most hydrocarbons burn easily, so they are often used as sources of energy. For these reasons, hydrocarbons are called *fossil fuels.*

○ *What are hydrocarbons?*
○ *What are fossil fuels?*

Figure 5–12. In the mid-1970s, many oil refineries (left) had to close because of a lack of crude oil. This created a shortage of gasoline that resulted in many problems for motorists (above).

Section 2   Fossil Fuels   **107**

Answers—5.4

○ Hydrocarbons are compounds containing hydrogen and carbon.

○ Fossil fuels are hydrocarbons obtained from once-living organisms.

DEMONSTRATION

For a demonstration of burning hydrocarbons see page 97c preceding this chapter.

Section **2: Fossil Fuels**

SECTION FOCUS

Section Overview   This section focuses on the earth's fossil fuels, with an emphasis on their importance as world resources. The origins of the specific fossil fuels—petroleum, natural gas, and coal—are described. Important geographical locations of fossil fuels are also noted.

Section Objectives   For a list of section objectives, see pupil's edition page.

New Science Terms   For a list of new science terms in this section, see pupil's edition page. You may wish to have the students define new terms before they read the section.

SECTION DEVELOPMENT

5.4 Hydrocarbons

DISCUSSION   Discuss with the students the many ways in which our culture is dependent upon fossil fuels. Compare energy use in the United States today with energy use in our country two hundred years ago. You should note that most of our energy today comes from fossil fuels; two hundred years ago people burned wood to provide heat and energy. ***Thinking Critically:*** Ask the students to cite both advantages and disadvantages that fossil fuels have provided for the United States. (For example, an advantage is the use of modern conveniences; a disadvantage is the problem of air pollution.)

THINKING SKILL   (*Communicating*) Ask the students to describe how they think the oil crisis in the 1970s in our country affected people. (Gasoline shortages caused long lines and higher prices at gas stations; heating oil for homes was less plentiful; industries dependent on oil for production had to cut down usage, which affected workers.)

107

5.5 Oil and Gas

EXTENSION Explain to the students that exploring for and discovering oil deposits, extracting the crude oil from the ground, temporarily storing and then delivering the product to its consumers is not the end of the story. All hydrocarbons must undergo complex refining procedures before they are used by the average consumer. To give the students an overview of the refining process, you may wish to send for information from a major oil company.

THINKING SKILL Ask students to recall what other resources (discussed in Section 5.1) depend on a complex refining process to manufacture products. (iron and steel)

EXTENSION Many oil deposits are found offshore on the continental shelf where water depth rarely exceeds more than a hundred meters. Offshore oil rigs are positioned and anchored there by engineers working in underwater submersibles. Rig crews may spend many months on the platform, extracting oil from the shelf reservoirs.

THINKING SKILL (*Inferring*) Tell the students that natural gas and petroleum are quite often discovered in the same deposit. Ask the students to infer why the natural gas can usually be found above the petroleum. (Natural gas is lighter than petroleum, thus natural gas separates and remains on top.)

EXTENSION You may wish to have the students look at the maps of Louisiana, Texas, and Oklahoma in the Science Sites booklet for the location of major drilling areas. You may also wish to point out the Trans-Alaska pipeline on the map of Alaska.

Figure 5–13. As this diagram (right) shows, oil is produced in sedimentary deposits. To remove the oil, it is often necessary to drill into those deposits (above).

Figure 5–14. Oil under pressure often produces a gusher when the pressure is suddenly released.

5.5 Oil and Gas

A liquid mixture of hydrocarbons is called *oil* or *petroleum*, while a mixture of gaseous hydrocarbons is called *natural gas*. Solid hydrocarbons are *asphalts*.

Much oil and gas is found in rocks that are less than 100 million years old, and in rocks between 200 and 300 million years old. However, under favorable conditions, the physical and chemical processes that change organic matter into hydrocarbons may take as little as 1 million years.

Petroleum and natural gas form in fine-grained, sedimentary rocks called *source rocks*. Under the pressure of overlying rocks, the oil and gas are squeezed out of their source rocks and move up through whatever space exists. If the oil or gas reaches the surface, it is oxidized and lost. However, if on its way to the surface it becomes trapped beneath a layer of impermeable rock, the oil or gas forms a deposit.

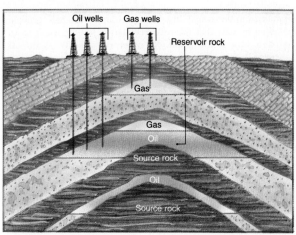

The rocks that contain large deposits of oil and gas are called *reservoir rocks*. Nearly 60 percent of the world's petroleum is found in sandstone reservoirs. The remainder is in limestone and other rocks. Petroleum is obtained by drilling into reservoir rocks and pumping out the oil; gas rushes out by itself.

Sometimes the oil is not concentrated enough to be pumped out directly, so water or steam is forced into the reservoir rock through a nearby well. On rare occasions the oil will be under such tremendous pressure that it rushes to the surface by itself, forming a gusher. Oil gushers waste much valuable oil, so modern wells are capped to prevent this from happening.

North America has large deposits of petroleum and natural gas along the Gulf and California coasts and on the north slope of Alaska and Canada. In spite of this, the United States still imports about 20 percent of its oil from the Middle East, South America, and Africa.

Many petroleum and natural gas deposits are found offshore, deep under ocean water. To drill at these locations, derricks are mounted on platforms, which are set on the sea floor. The largest platform—over 300 m tall—operates in the North Sea, between Britain and Scandinavia.

○ *What are source rocks and reservoir rocks?*

Figure 5–15. Oil deposited under the oceans requires the construction of special drilling platforms. Some of these platforms are constructed on the ocean bottom (above), while others are constructed on land, then towed to the drilling site (left).

ACTIVITY: Determining the Oil-Bearing Capacity of Rocks

How can you determine the oil-bearing capacity of different rocks?

MATERIALS (per group of 3 or 4)

rock samples (sandstone, limestone, and shale); plastic Petri dishes (3); mineral oil; medicine dropper

PROCEDURE

1. Place one of the rock samples into a Petri dish.
2. Repeat the procedure with the other samples, placing each sample in a separate dish. Be sure that the grain sizes of each sample are noted.
3. Fill the medicine dropper with mineral oil and place a drop of oil on top of each sample.

4. Observe and record the time required for the oil to be soaked up by each sample.

CONCLUSIONS/APPLICATIONS

1. Which sample soaked up the oil the fastest?
2. Do you think oil deposits can be found in most any rock? If not, which rocks could be oil reservoirs?

Section 2   Fossil Fuels   **109**

EXTENSION   Many types of products are derived from petroleum during the refinement process. You may wish to list these products and their percentages on the board. They include: gases for fuel, 10.9 percent; gasoline, 38.6 percent; jet fuel, 6.3 percent; heating oil and kerosene, 13.6 percent; industrial fuel, 5.8 percent; fuel for trains, ships, and diesel trucks, 5.6 percent; power plant fuel, 8.5 percent; lubricants, wax, asphalt, and petrochemical products, 8.4 percent.

SKILL   (*Writing*) Scientists predict that the United States could possibly run out of oil by the year 2060. Have students write a short report in which they analyze the effects that running out of oil would have in the United States. For example, students should consider the rising cost of energy, alternatives to oil, and other issues that would affect people's daily lives. Have them suggest ways that people can conserve energy today in preparation for the future depletion of oil.

ACTIVITY:
Determining the Oil-Bearing
Capacity of Rocks

Skill (*Observing*)

Preparation of Materials   For this activity, each of the rock samples should be approximately the same size. You may wish to prepare dropping bottles of mineral oil in advance.

Answers to Conclusions/Applications

1. Sandstone
2. No. Sandstone and limestone trap the world's oil.

Answers—5.5

○ Source rocks are fine-grained sedimentary rocks in which petroleum forms. Reservoir rocks are sandstone and limestone layers that trap oil.

5.6 Coal

5.6 Coal

EXTENSION From the early 1700s through the mid-1800s, coal was the major source of energy in the industrialized nations. It was used primarily in homes, offices, and stores as a source of heat. At the beginning of the 1900s, oil comprised less than 5 percent of the fossil fuels used in the United States. Today, however, oil has replaced coal as the primary fuel. Currently, coal satisfies less than 25 percent of our nation's needs. However, with our dependence on foreign sources for petroleum and the rapid rate at which petroleum sources are being depleted, coal may once again become a major source of fuel.

EXTENSION Lignite, often called "brown coal," is brownish black and burns with a smoky flame. Lignite deposits are found beneath the earth's surface. Bituminous, or soft, coal is a dark brown or black material with a high carbon content. Bituminous coal is the most abundant type of coal and also burns with a smoky flame. Most of the coal in the United States is bituminous. In places where formations of bituminous coal were tightly compressed and temperatures were very high, anthracite coal formed. Anthracite is a hard, black, and brittle form of coal. It gives off tremendous heat and burns with a blue, smokeless flame. Most anthracite coal is found in the Appalachian Mountains.

DISCOVER:
Conserving Fossil Fuels

Thinking Skill *(Problem Solving)*

Students should estimate the average number of kilometers per week that their parents drive, the kilometers driven per liter, and the cost of gasoline. (Answers will vary.)

Figure 5–16. Coal is used extensively for producing electricity (right). In many places, such as the Okefenokee Swamp in Georgia (above), coal is still being formed. However, reserves of coal are being used much faster than new deposits are forming.

DISCOVER

Conserving Fossil Fuels

Ask your parents or guardians to carefully record the distance they drive and the amount of gasoline they use in a week. Determine how much fuel could be saved in a year if they cut their driving by 10 percent.

Although the United States must import some percentage of the petroleum it uses, it has vast supplies of coal. Coal is no longer used much as a heating fuel, but it is very important for the production of electricity. Carbon from coal is also used in the steel industry.

Coal forms from decayed plants in areas that were once swamps. Fresh plant matter is about 80 percent water and 20 percent carbon. When plants die, they become waterlogged and sink to the bottom. In the first stage of coal development, the activity of bacteria and fungi transforms the plants into peat. Peat is a dark brown, mushy substance in which the carbon content is about 60 percent. Peat is now forming in many swamps, including the Dismal Swamp of Virginia and North Carolina and the Okefenokee Swamp of southeastern Georgia.

The second step in coal formation occurs as peat is buried under sediments and layers of additional peat. The pressure increases and the temperature rises. Water is expelled from the peat, and any oxygen present in the organic matter combines with carbon to form carbon dioxide. Next, the peat gradually becomes lignite, which is nearly 70 percent carbon. If the process continues, lignite becomes subbituminous coal, then bituminous coal, and eventually anthracite. Anthracite is over 90 percent carbon.

The major coal fields of the United States are in the Appalachian Mountains, extending from Pennsylvania to Alabama, and under the prairies of the Midwest. Coal is also found under the western plains from New Mexico to North Dakota.

DEMONSTRATION

Display samples of lignite, bituminous coal, and anthracite. After the students study the samples, break up the coal using a hammer. Have the students examine the pieces using hand lenses. They should look for any fossil fragments in the soft coal. Have them record their observations and explain why the anthracite coal does not contain fragments. (It has been compacted.)

Figure 5–17. In some areas, coal is mined by stripping away the overlying rocks and soil (left). In other areas, coal is removed from deep underground mines (right).

Coal beds range in thickness from a few centimeters to nearly 50 m in the Powder River Basin of Wyoming. Most Eastern coal is deep-mined from tunnels dug directly into the coal seams or stripped from the ground by removing the overlying rock layers. Western coal is almost exclusively stripped.

○ *What is peat?*
○ *Where are the major coal fields of the United States?*

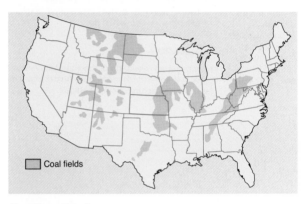

Coal fields

Figure 5–18. This map shows the location of major deposits of coal in the United States.

Section Review

READING CRITICALLY

1. How are hydrocarbons formed?
2. Why is coal important to the steel industry?

THINKING CRITICALLY

3. What are the dangers of the United States being dependent on imported oil?
4. How might the use of coal be expanded to take advantage of the vast reserves in the United States?

A MATTER OF FACT

By the year 2000 about 70 percent of the electric power in the United States will be generated in coal-fired power plants.

Section 2   Fossil Fuels   **111**

DISCUSSION (Tie-in/Ecology) Discuss with the students some of the major problems in using coal as a source of energy. Explain that the large amounts of sulfur contained in coal produce the gas sulfur dioxide when coal is burned. Sulfur dioxide is a major cause of air pollution and acid rain. Have the students research the current technology that is being employed to prevent air pollution. Tell the students they will learn more about the causes and effects of air pollution in Chapter 17. *Thinking Critically:* Ask the students to think of everyday examples of air pollution that they observe. Then ask for suggestions on preventing air pollution. (For example, an automobile with a defective pollution control system should be repaired immediately.)

SECTION REVIEW

Summary   Have the students name the fossil fuels and list ways in which they use these fuels every day. Answers may include transportation and home heating use. Briefly review the ways in which fossil fuels were formed in the earth.

Reinforcement   Encourage the students to make large tables listing the earth's resources that have been studied in this chapter. They can then classify and list important characteristics of the resources, such as physical descriptions, sources, and practical applications.

Answers–5.6

1. Peat is a dark brown substance in which the carbon content is 60 percent.
2. The major coal fields in the United States are in the Appalachian Mountains and in the west.

Answers to Section Review

Reading Critically

1. Hydrocarbons are formed from organic matter that has undergone chemical change due to heat and pressure.
2. Coal is used in forging the iron that is used to make steel.

Thinking Critically

3. Oil is a major source of fuel in the United States. If foreign countries control the sources of our fuel, they can also limit the amount of energy we have to function as a society.
4. Ways could be found to replace oil with coal products.

Investigation 5: Crystal Formation

Skill *(Observing)*

Preparation of Materials

If your school does not have copper sulfate, ferrous sulfate, or sodium chloride, they can be ordered from most scientific supply houses.

If the students are preparing their own solutions, caution them to wear safety goggles and laboratory aprons and to handle the solutions carefully. You may, however, prefer to prepare the solutions as a class demonstration.

During the procedure, have the students speculate on the necessity of having a carefully controlled environment for growing crystals. (The slightest impurities can deform the crystals.)

Answers to Analyses and Conclusions

1. As the solution evaporates, the crystal begins to grow slowly.
2. In order to make rock candy, begin with a "seed" of sugar prepared in the same manner as that used in this investigation. Immerse the seed attached to a cotton string in a saturated solution of common sugar and allow the liquid to evaporate.

Answer to Application

Understanding the growth of crystals might allow scientists to produce conditions necessary for growing even larger crystals.

INVESTIGATION 5: Crystal Formation

PURPOSE

To observe how crystals form

MATERIALS (per group of 3 or 4)

Cotton string (30 cm)
Petri dishes (3)
Water
Hot plate
Copper sulfate
Ferrous sulfate
Sodium chloride
Beakers, 100 mL (3)
Glass stirring rod
Ice-filled tray

PROCEDURE

1. Cut the string into three 10-cm lengths and lay each piece of string across a Petri dish.
2. Prepare a saturated solution of copper sulfate by heating water in the beaker on a hot plate and slowly adding copper sulfate until no more will dissolve.
3. Soak the string from the first petri dish with the copper sulfate solution.
4. Prepare saturated solutions of ferrous sulfate and sodium chloride in the same manner and soak the other strings in those solutions.
5. As the water evaporates from each string, observe the formation of crystals on the string.
6. When the strings are nearly dry, place the beakers into the ice-filled tray and then dip each string into the beaker containing its dissolved salt. Observe the results.

ANALYSES AND CONCLUSIONS

1. What happened when the "crystal-covered strings" were placed into the cooling solution of the same salt? Explain why this occurred.
2. Can you devise a method of making rock candy using a similar procedure? Explain exactly what you would do.

APPLICATION

How might a knowledge of the way that crystals grow help to produce large, pure quartz crystals for the manufacture of silicon chips?

SUMMARY

- Gold was probably the first metal used by humans. (5.1)

- Copper was the first metal to be smelted from its ore. (5.1)

- Bronze is an alloy of copper and tin. (5.1)

- Iron is obtained by smelting iron ore in a forge. (5.2)

- Steel is an alloy of iron and one percent carbon. Nickel and other metals may be added to produce specialty steels. (5.2)

- Quartz, one of the most plentiful minerals, has many uses in the modern world. (5.3)

- Colored glass is made by adding small amounts of specific metals or compounds to the molten glass. (5.3)

- The property of piezoelectricity makes quartz important in the electronics industry. (5.3)

- Hydrocarbons are chemical compounds of carbon and hydrogen. (5.4)

- Most hydrocarbons burn easily. Since they are good sources of energy, they are known as fossil fuels. (5.4)

- Petroleum is a mixture of liquid hydrocarbons; natural gas is a mixture of gaseous hydrocarbons; and asphalts are solid hydrocarbons. (5.5)

- Oil and gas are produced in rock layers called *source rocks*. (5.5)

- A drill is used to release oil and gas from reservoir rocks. (5.5)

- North America has large deposits of hydrocarbons along the Pacific and Gulf coasts, and on the north slope of Alaska and Canada. (5.5)

- Coal forms from plant matter that accumulates at the bottom of swamps. (5.6)

- The United States has many deposits of coal in the eastern mountains and the western plains. (5.6)

Write all answers on a separate sheet of paper.

SCIENCE TERMS

Correctly use each of the following terms in a sentence.

alloy **(100)**
forge **(102)**
hydrocarbons **(107)**
ores **(99)**
smelting **(99)**

SCIENCE QUIZ

Modified True-False

Mark each statement *true* or *false*. If a statement is false, change the underlined term to make the statement true.

1. One of the first metals used by humans was <u>gold</u>.

2. Native <u>copper</u> was used 10 000 years ago.

3. Steel is an <u>ore</u>.

4. The process of obtaining metals from ore is called <u>forging</u>.

5. <u>Petroleum</u> is a liquid, natural hydrocarbon.

6. <u>Coal</u> is a solid fossil fuel.

Multiple Choice

Write the letter of the choice that best answers the question or completes the statement.

7. Early metal tools were probably not made of
   a) gold.   b) copper.
   c) bronze.   d) iron.

8. Early blacksmiths hammered bloom into
   a) iron.   b) steel.
   c) iron carbide.   d) lead.

9. Some steel alloys contain
   a) nickel.   b) iron oxide.
   c) bloom.   d) carbon dioxide.

continues

CHAPTER REVIEW

SUMMARY

The students may review the major concepts in the chapter by reading the summary statements. The statements are cross-referenced to the chapter to facilitate reinforcement of any concepts of which the students feel unsure. Encourage the students to work in groups to quiz one another.

SCIENCE TERMS

The sentence in which the science term is used should reflect an understanding of the definition of the term. You may wish to orally quiz students to evaluate their comprehension of particularly difficult science terms.

SCIENCE QUIZ

Answers to Modified True-False

1. true
2. true
3. false, alloy
5. false, smelting
6. true

Answers to Multiple Choice

7. a
8. a
9. a

continues

Answers to Multiple Choice
(continued)

10. b
11. c
12. a

Answers to Completion

13. source rocks
14. alloys
15. soils
16. smelting
17. lignite
18. quartz

Answers to Short Answer

19. A stream of air is blown through a furnace, and the heat causes the iron oxide in the ore to react with carbon in the charcoal fuel. The reaction produces carbon dioxide and a mixture of metallic iron and iron carbide.

20. Long lengths of pipe with a drill bit at the bottom are used to drill a hole into the reservoir rock. A mixture of fine clay and water is forced through the hollow pipe to wash out loosened material. Water and steam are flushed into a nearby well, creating pressure that will force oil in the reservoir up the drill hole. Sometimes the oil will "gush up" by itself.

21. The Appalachian Mountains from Pennsylvania to Alabama are the primary coal ranges of the United States. New Mexico, North Dakota, and Wyoming also have large coal reserves.

22. Red glass contains gold or selenium. Yellow glass contains silver or uranium oxide. Blue glass contains cobalt oxide. Brown glass contains ferric oxide. Green glass contains ferrous oxide.

Answers to Writing Critically

23. Many of the metals used for specialty steels are imported from foreign countries. Wars, shortages, or boycotts could disrupt these supplies.

24. In both processes, materials are added to the glass or steel to produce special properties.

25. Solid fossil fuels contain carbon only, not hydrocarbon.

---

10. Which material can transform electrical voltage charges into vibrations?
a) iron ore   b) quartz
c) steel   d) copper

11. What are the elements in hydrocarbons?
a) carbon and oxygen
b) hydrogen and oxygen
c) carbon and hydrogen
d) methane and quartz

12. Hydrocarbons include all of the following except
a) coal.   b) oil.
c) gas.   d) asphalts.

Completion

Complete each statement by supplying the correct term or phrase.

13. Sedimentary rocks in which hydrocarbons form are called _____.

14. Combinations of either two or more metals or of metals and nonmetals are called _____.

15. Nickel is found in igneous rocks and some _____.

16. The _____ process is used to remove metals from ores.

17. Peat will become _____, and then coal.

18. Computer chips are made of silicon, which can be obtained from the mineral _____.

Short Answer

19. What is the procedure for making iron in a forge?

20. Describe the process for obtaining oil from reservoir rocks.

21. Which areas of the United States have large deposits of coal?

22. What must be added to molten glass to make it red? To make it yellow? To make it blue? To make it brown? To make it green?

114   Chapter 5 Review

Writing Critically

23. Explain how the production of specialty steels could be interrupted by international disagreements.

24. In what ways are the making of specialty glasses similar to the making of specialty steels?

25. Why are peat, lignite, and coal considered fossil fuels, but not hydrocarbons?

EXTENSION

1. Write a report that traces the evolution of the transistor from vacuum tube to silicon chip. Be sure to describe the materials that were, and currently are, used in the manufacture of each device.

2. Research and report to the class on the technological and environmental problems associated with the surface mining of coal.

3. Construct a model of an oil rig from simple materials, such as popsicle sticks and plastic straws, and then use your model to explain the drilling process to your class.

APPLICATION/CRITICAL THINKING

1. During the time of Napoleon, the element aluminum was considered more valuable than gold. Today aluminum is considerably less expensive than gold. What advances might have led to the decline in the value of aluminum?

2. How would your life be different if all fossil fuels suddenly disappeared?

3. This photograph shows one result of the energy crisis of the mid-1970s. What other changes resulted from this drive to conserve energy?

ANSWERS TO EXTENSION

1. The report should include the following pertinent facts: (1) the function of a transistor is to regulate the flow of electricity in a circuit; (2) vacuum tubes accomplish this by controlling the flow of current through two or more conducting grids inside an airless tube; (3) silicon chips perform the same function using semiconductors; and (4) silicon chips have lower power requirements and are more compact than vacuum tubes.

2. The report should include the impact of strip-mining on the land.

3. Models will vary.

ANSWERS TO APPLICATION/ CRITICAL THINKING

1. Students should be able to explain that the cost of a product (i.e., aluminum or gold) is dictated by the supply of the product and the demand for the product. During Napoleon's time, the cost of extracting aluminum

continues

FOR FURTHER READING

Bailey, D., and L. Castoro. *Careers in Computers.* New York: Julian Messner, 1985. If you are interested in a career working with computers, this book is a good guide to the various opportunities available in the field. It describes types of computer-related jobs in industry, business, government, education, and research.

Bass, G. "Splendors of the Bronze Age." *National Geographic* 172 (September 1987): 693. This article describes improvements in underwater research equipment which is expanding the science of marine archaeology. The article includes illustrations and photographs of the exploration of ancient sea-going vessels and their Bronze Age treasures.

Bell, P. and D. Wright. *Rocks and Minerals.* New York: Macmillan, 1985. This field book contains colorful photos and illustrations accompanied by fact-filled descriptions of the earth's many mineral resources.

Cotterill, R. *The Cambridge Guide to the Material World.* New York: Cambridge University Press, 1985. This is a beautifully illustrated and comprehensive description of many different types of materials, including metals, polymers, ceramics, and glass, as well as all the naturally occurring minerals.

Ward, F. "Jade: Stone of Heaven." *National Geographic* 172 (September 1987): 282. This article gives a fascinating history of jade, one of civilization's most prized precious stones.

Challenge Your Thinking

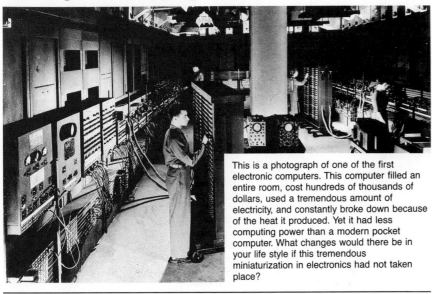

This is a photograph of one of the first electronic computers. This computer filled an entire room, cost hundreds of thousands of dollars, used a tremendous amount of electricity, and constantly broke down because of the heat it produced. Yet it had less computing power than a modern pocket computer. What changes would there be in your life style if this tremendous miniaturization in electronics had not taken place?

Chapter 5 Review   **115**

ANSWER TO CHALLENGE YOUR THINKING

There probably would not be small personal computers, radios, cassette recorders, televisions, or calculators.

from its ore was both difficult and very expensive. Thus, the pure substance was always in very short supply. That made the price of aluminum exceed that of gold. However, chemists perfected an electrolytic process for separating the element from its ore, and it soon became quite abundant and easily accessible to the masses.

2. Answers will vary. However, life-styles would tend to be more like those of the early nineteenth century when the major source of fuel was wood.

3. Answers will vary, but may include reducing the speed limit on major highways, setting thermostats higher in summer and lower in winter, developing of more energy-efficient cars, and increasing recycling programs.

PLANNING THE CHAPTER

Chapter Sections	Page	Chapter Features	Page	Program Resources	Page
Section 1: Inside the Earth	117			Reading for Content: *Finding Main Ideas* **(B)** Investigation 6.1: *Movement in Rock Layers* **(A)**	TRB 11 LM 183 TRB 21 LM 23
6.1 The Interior of the Earth **(A)**	117	A Matter of Fact	118		
6.2 Movement Within the Mantle **(A)**	119	**Discover:** Making Convection Cells **(A)**	120		
		Section Review	121		
Section 2: Crustal Adjustments	122			Critical Thinking **(H)** Cross-Discipline: Science and Language, *Explaining Continental Drift Theory* **(A)**	TRB 11 TRB 11
6.3 Folds **(B)**	122	A Matter of Fact	123	Investigation 6.2: *Drifting Continents* **(B)**	TRB 23 LM 25
6.4 Fractures and Faults **(B)**	124	A Matter of Fact	126	Concept Extension: *Crustal Movement* **(A)**	TRB 13 LM 235
6.5 Continental Drift **(B)**	129	**Skill Activity:** Interpreting Geologic Maps **(A)**	128		
		Discover: Solving the Puzzle **(B)**	129		
		Section Review	130		
		Activity: Studying Ancient Glaciers **(B)**	130		
		Careers: Seismologist, Scientific Translator, Oil-platform Worker	131		
		Investigation 6: Analyzing Structural History **(A)**	132	Student Record Book: Textbook Investigations **(A)**	TRB 11
Chapter 6 Review	133			Vocabulary **(A)**	TRB 6 LM 133
				Tests **(A)** Computer Test Bank	TRB 21

(LM) Laboratory Manual/Study Guide,   **(TRB)** Teacher's ResourceBank™

B = Basic   **A** = Average   **H** = Honors

The coding Basic, Average, and Honors indicates sections or subsections that might be appropriate for different levels of learners. For additional suggestions regarding choice of topic and depth of coverage, see the Pacing Chart on pages T16–T20.

CHAPTER CONCEPTS, OBJECTIVES, AND TERMS

Section	Concepts	Objectives	Science Terms
Section 1: Inside the Earth	■ The earth is made of several layers that are distinct in composition, density, and thickness. **(6.1)** ■ Beneath the crust are the mantle and the core of the earth. **(6.1)** ■ The heated material within the earth moves by convection. Most of the heat in the interior of the earth is supplied by the decay of radioactive materials. **(6.2)**	■ **Identify** the layers of the earth. ■ **Locate** the major discontinuities between layers. ■ **Explain** how convection currents may cause movements of the earth's crust.	crust mantle core lithosphere asthenosphere convection
Section 2: Crustal Adjustments	■ Forces within the earth's crust cause rocks to fold and fracture. Rocks fold into anticlines and synclines. **(6.3)** ■ Large scale folding produces mountain ranges by the process of orogeny. Plateaus may be formed by large-scale uplifting without folding. **(6.3)** ■ Landforms created by folding are often studied by using geologic maps. **(6.3)** ■ The position of rock formations can be described using strike and dip. **(6.3)** ■ Faults form because of tension, shear, or compression. **(6.4)** ■ Tension produces normal faults. Fault-block mountains may result from normal faults. Reverse faults are the result of compression. A low-angle reverse fault is a thrust fault. Lateral faults, like the San Andreas fault, are caused by shear. **(6.4)** ■ In 1915 Alfred Wegener proposed the theory of continental drift. **(6.5)** ■ The match of continental shapes, geological features, and fossil evidence are used to support the theory of continental drift. **(6.5)**	■ **Describe** the process that causes the crust to fold. ■ **Classify** faults. ■ **List** four kinds of evidence that support the theory of continental drift.	fold orogeny anticline syncline dip strike fault tension compression shear continental drift

Title	Page	Materials
Discover: Making Convection Cells	120	*(per student)* goggles, large laboratory beaker or glass baking dish, corn syrup, food coloring, hot plate or laboratory burner
Skill Activity: Interpreting Geologic Maps	128	*(per student)* geologic map
Discover: Solving the Puzzle	129	*(per student)* old jigsaw puzzle
Activity: Studying Ancient Glaciers	130	*(per group of 3 or 4)* pencils, tracing paper, textbook, scissors, colored pencils, glue, legal-size paper
Investigation 6: Analyzing Structural History	132	*(per student)* paper, pencil

TEACHING SUGGESTIONS

Section 1: **Inside the Earth**

Demonstration: Showing Magnetism

Purpose
To show the magnetism of the earth

Background
As you know, a study of magnetism has taught us much about changes that have taken place in the crust of the earth. You also know that the earth acts like a huge magnet with its magnetic field extending through the earth and far out into the atmosphere. In this demonstration you will make a model of the earth and its magnetic field. The model will help the students make predictions about a magnetic field.

Materials
Newspaper
Bar magnet
Large sheet of white paper
Pencil
Plate
Iron filings
Large sewing needles
Small cork
Nylon thread (20 cm long)

Procedure
1. Work on a sturdy table. Place a piece of newspaper on the table. Put a bar magnet on the newspaper so that the North Pole is nearest you and the South Pole points away from you. Now center a large piece of white paper over the magnet and press it against the edges of the magnet. An imprint of the magnet will be made in the paper. Remove the paper and outline the magnet with a pencil.

Get a plate that is slightly larger than the magnet. Place the plate upside down on the drawing of the magnet. Trace around the plate and then remove it.

2. This drawing now represents the earth with the magnet within it. The South Pole of the magnet is at the North Pole of the earth. And the North Pole of the magnet is at the South Pole of the earth. Mark the North and South poles of the magnet and mark the North and South Poles of the earth. Place the paper over the magnet so that the magnet imprint is directly over the magnet. Gently shake the iron filings over the paper. Tap the paper gently. The iron filings will outline the magnetic field of the magnet. This should give the students a very rough idea of the magnetic field that lies within and around the earth. Have the students use a pencil to draw lines representing the magnetic field indicated by the iron filings.

3. Examine the lines of force of the magnetic field drawn on the diagram of the earth. Have the students predict in which direction a compass would point at the North Pole.

4. Now make a compass free to move in any direction. Push a needle through a cork to make a hole through its center. Then remove the needle and magnetize it. To make the point of the needle the North Pole, pull the South Pole of a bar magnet along the needle from its eye to the point. Do this several times. Then carefully insert the needle into the hole in the cork. Tie one end of a thread around the cork. Tie the other end of the thread around a wooden support. Let the cork hang freely. Make sure the needle is balanced. The needle should act like a compass.

Questions to Ask the Students
1. Describe the magnetic field of the earth. (*Curved magnetic lines of force stretch around the earth from pole to pole.*)
2. In which direction would a compass point at the North Pole? (*It would point down.*)
3. In which direction would a compass point at the Equator? (*north or south*)

4. In which direction would a compass point if it were midway between the North Pole and the Equator? (*north*)
5. What observations did the free-swinging compass help you make? (*Answers will vary*.)

Class Activity

Demonstrate the motion of a convection current by using a large beaker, water, ring stand, candle or Bunsen burner, and grits or tapioca. Take care to wear safety goggles during this demonstration. Set up the beaker so that the heat source is directly heating only one side of the beaker. When the water begins to boil, sprinkle some of the grits into the beaker. Have the students explain what makes the grits follow a circular path. Have them then relate the motion of convection currents in the mantel.

Section 2: Crustal Adjustments

Class Activity

Use selected topographic maps for fault labs. Especially clear areas are Santaquin, Utah (block faults) and Bright Angel, Arizona (lateral fault).

Using 5 cm × 10 cm × 15 cm sanded blocks of wood, draw various layers of rocks to show tilted, faulted, and folded patterns. Leave one side or the top blank. Have the students sketch out the missing panel on their own paper.

Using a set of geological highway maps (available through Wards Science Supply), have students work in small groups to identify folds, faults, or tilted rocks on their maps. By referring to the cross-sectional views provided on each map, students should be able to locate and describe several structures from each map.

Field Trip

Take a trip to a spot where a highway cuts through a hillside. Look for evidence of folds and faults, synclines and anticlines, and strike and dip in the highway cut.

Outside Speaker

Have a geologist come to the class to explain how to figure out strike and dip and how to read geologic maps.

CHAPTER 6

CHAPTER OVERVIEW

This chapter presents the characteristics of the earth's complex layers. The results of forces acting on the earth's surface, faulting and folding, are discussed. Students are introduced to the theory of continental drift.

Section 1: Inside the Earth The layers of the earth are described, including their composition, density, and thickness. The boundaries that separate each of the layers are discussed. Convection of the material of the mantle is identified as a force that may cause the lithosphere to move.

Section 2: Crustal Adjustments In this section, it is explained that forces within the earth cause surface rocks to fold, fracture, and fault. Specific types of faults and folds are identified and described. Strike and dip are defined. Evidence supporting the theory of continental drift is presented. The theory itself is also briefly discussed.

Structural Geology

How was this beautiful landform created? What forces could have caused these rocks to be positioned like this? The formation of geologic features that may have taken millions of years to develop is difficult to imagine. Often only small pieces of landforms remain as evidence of past processes. An even greater challenge is trying to understand the internal processes of the earth.

Folded layers of rock

116

CHAPTER MOTIVATING ACTIVITY

Have the students respond to the question "What evidence is there that forces act on the earth's surface?" (Possible answers might include cracks in the earth's surface, earthquakes, volcanoes, and erosion.)

1 Inside the Earth

SECTION OBJECTIVES

After completing this section, you should be able to:
- **Identify** the layers of the earth.
- **Locate** the major discontinuities between layers.
- **Explain** how convection currents may cause movements of the earth's crust.

NEW SCIENCE TERMS

crust
mantle
core
lithosphere
asthenosphere
convection

6.1 The Interior of the Earth

Scientists know much about the surface of the earth, but no one has been able to take samples of the center of the earth for study. Using current technology, geologists can drill only about 13 km into the earth; the center of the earth is about 6400 km below the surface. However, information about the earth's center is available from the study of earthquakes. This information has helped scientists determine the structure and composition of the earth's interior.

The Crust   The earth is divided into three layers: the *crust*, the *mantle*, and the *core*. The **crust** is the thin, rocky layer on the surface of the earth.

Like the crust of a pie, the earth's crust is not the same thickness throughout. Under the oceans the crust is only about 7 km thick, and it is composed of dense mafic rock such as basalt. A thin layer of sediments lies on top of the basalt. The continental crust, which is much thicker than the oceanic crust, ranges from 25 km at the edges of the oceans to 70 km near the center of the

Figure 6–1. The continental crust, composed mostly of granite and sedimentary rocks, is much lighter and thicker than the oceanic crust, which is composed mostly of basalt and sediments. Therefore, the continental crust sits higher than the oceanic crust.

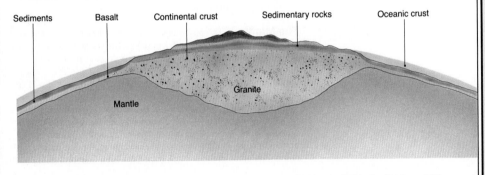

Sediments   Basalt   Continental crust   Sedimentary rocks   Oceanic crust

Mantle   Granite

DEMONSTRATION

For a demonstration of magnetism see page 115c preceding this chapter.

SECTION FOCUS

Section Overview   The layers that compose the earth, as well as the discontinuities that separate the layers, are discussed in detail. Earthquake waves are identified as the source of this information about the layers. The convection currents thought to exist in the mantle are described.

Section Objectives   For a list of section objectives, see pupil's edition page.

New Science Terms   For a list of new science terms in this section, see pupil's edition page. Because many of the terms in this section are unfamiliar to students, it may be helpful to read the terms aloud before you begin.

SECTION DEVELOPMENT

6.1 The Interior of the Earth

DISCUSSION   Explain to the students that as earthquake waves pass through the earth, they change speed and direction. These changes are due to the differing densities and compositions of the earth's layers. Tell the students that they will learn more about the way earthquake waves travel through the earth in Chapter 7. Although most students are familiar with the term, you may want to remind them that an earthquake is the trembling of the ground.

SKILL   *(Making Mathematical Calculations)* Ask the students to calculate what percentage of the earth's thickness from surface to center has been drilled so far. ($7.5/6400 = 0.00117$ or 0.117 percent) As a demonstration, you may want to slice an orange that has a radius of about 5 cm in half. Try to stick a straight pin into the rind about 0.06 mm, which is 0.117 percent of 5 cm. The students will see that this is nearly impossible and will recognize just how small 7.5 km is in comparison to the earth's total radius.

117

EXTENSION Have the students use clay, a polystyrene foam ball, or other materials to construct a model of the layers of the earth. They can use the diagram in Figure 6–2 as a guide. Be sure that the students use the proper proportions for each layer. Have the students label the layers and discontinuities. You may then wish to have the students make up a chart listing the discontinuities, the layers that they separate, and the depth at which they are found.

DISCUSSION You may want to tell the students the main elements that make up each layer of the earth. The crust consists of 46.6 percent oxygen, 27.7 percent silicon, and the remaining 25.3 percent is made up of aluminum, iron, calcium, sodium, potassium, magnesium, and various trace elements. The mantle probably consists mostly of silicon, oxygen, iron, and magnesium. The outer core is made of liquid iron and nickel and the inner core is made of iron and nickel. *Thinking Critically:* Ask the students what elements they think most rocks of the crust are made of. (silicon and oxygen) Point out that these elements are often found in the form of silicon dioxide (SiO_2), or as quartz.

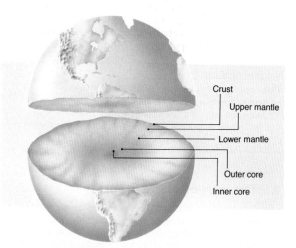

Figure 6–2. If the earth could be cut open, you would see several distinct layers. Some of the layers are solid, while others are pliable or even liquid.

Crust
Upper mantle
Lower mantle
Outer core
Inner core

A MATTER OF FACT

Mauna Kea in Hawaii is the tallest mountain in the world if measured from its base to its peak. Its peak is over 4200 m above sea level —but its base is about 6000 m below sea level.

continents. Continents are made of various igneous, sedimentary, and metamorphic rocks. Because of this mix of rock types, the continental crust is less dense than the oceanic crust. The continental crust is also much older than the oceanic crust. Some continental rocks may be more than 20 times older than the oldest rocks on the ocean floor.

There is a definite boundary to the earth's crust. Andrija Mohorovičić (ahn DREE jah moh hoh ROH vuh chihch), a Yugoslavian geologist, discovered the boundary between the crust and the rest of the interior by analyzing earthquake data. Mohorovičić noted that the waves produced by earthquakes did not pass directly through the earth. Instead, the waves were deflected at a point about 40 km below the surface. This boundary, or discontinuity, is called the *Mohorovičić discontinuity,* or *Moho,* in his honor.

The Mantle The Moho is about 7 km below the ocean floor and, on the average, about 35 km below the surface of the continents. The Moho separates the crust from the next layer. The **mantle** is the layer immediately beneath the Moho. The mantle, which is about 2900 km thick, is divided into two sections. The upper mantle and the lower mantle are separated by an unnamed boundary. This boundary marks the place where the rocks of the mantle become soft and pliable. The upper mantle is composed of rocks even more dense than the rocks of the oceanic crust. The lower mantle is made of similar rocks, but due to the great pressure, the rocks are even denser.

The Core Below the mantle lies the innermost layer of the earth, called the **core**. The core is separated from the mantle by a boundary called the *Gutenberg discontinuity.* This boundary is about 2900 km beneath the earth's surface.

Like the mantle, the core is also divided into two parts. The outer core is liquid and probably consists of iron. The inner core is solid and is probably composed of iron and nickel. Scientists believe the liquid outer core may be responsible for producing the magnetic fields of the earth.

The extreme pressure at this depth, due to the weight of the overlying rocks, probably keeps the inner core solid. Most elements expand when they melt, but because the pressure of the overlying layers is so great, the elements cannot expand or melt.

○ *Name the layers of the earth's interior.*
○ *What is a discontinuity?*

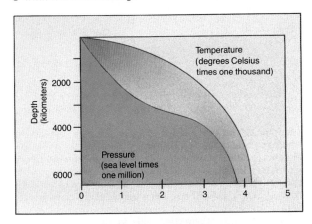

Figure 6–3. Temperature and pressure increase as you go deeper into the earth. At one point in the mantle, there is enough heat to overcome the pressure and to partially melt some of the rocks.

6.2 Movement Within the Mantle

The layers of the earth may also be grouped according to their consistency. The solid **lithosphere** (LIHTH uh sfihr), composed of crust and upper mantle, floats on the asthenosphere (as THEHN uh sfihr).

The **asthenosphere** is a 100-km-thick layer near the top of the mantle. In the asthenosphere high temperature and pressure combine to cause a small amount of melting of the rocks. This makes the asthenosphere a bit softer than the rest of the mantle.

Some of the heat still lingers from the earth's formation. However, scientists believe that most of the heat comes from the decay of radioactive elements. Radioactive decay is discussed in Chapter 15.

Section 1 Inside the Earth **119**

EXTENSION Based on new evidence, scientists now believe that the core of the earth is not completely round. Instead, it has large protrusions that reach far into the mantle.

THINKING SKILL *(Interpreting Graphs)* Have the students use Figure 6–3 to discuss the relationship between temperature, pressure, and depth within the earth. Next have the students make a to-scale drawing of the earth's interior, being careful to mark each discontinuity. Ask them to label the temperature and pressure at each discontinuity and at the earth's center.

6.2 Movement Within the Mantle

DISCUSSION The students may wonder why the crust floats rather than sinks into the asthenosphere. Explain that the crust is made of material that is less dense than the material of the mantle. You may want to demonstrate this point by showing the class how a piece of wood floats on water. Explain that thicker parts of the crust float lower on the mantle. Show this by using a thicker piece of wood.

Answers–6.1

○ The crust, mantle, and core are the three layers of the earth's interior.

○ A discontinuity is a boundary between layers of the earth.

DISCOVER:
Making Convection Cells

Thinking Skill *(Inferring Relationships)*

CAUTION: Goggles are required. The students should identify that warm syrup rises and cool syrup sinks.

Discussion The students should note the circular motion of the corn syrup when it is heated. ***Thinking Critically:*** Ask the students how this movement relates to the mantle. (The heating of the syrup near the bottom causes it to become less dense than the syrup at the top. This causes the warm syrup to rise and the cool syrup to sink. As the cool syrup sinks, it becomes heated and the process of rising and sinking is repeated again and again. This is very similar to the way the material of the mantle is heated by the core and moves in convection cells inside the earth.)

DISCUSSION Emphasize that the rock of the asthenosphere, while solid, has the ability to flow. This property is called *plasticity*. Explain that plastic rocks are somewhat like hot tar. To demonstrate plasticity, use a piece of plastic putty. Roll the putty into a ball and let the students hold it to determine that the putty is indeed solid. Leave the putty ball on a desk overnight. The next day, students will notice that the ball has slowly spread out overnight. This shows the plastic quality of a solid. Next place the plastic putty in a freezer. Once it has cooled, remove the putty from the freezer and throw it onto a desktop. The students will note that the putty is brittle and breaks. It does not behave like plastic. Hold the putty in your hand for a few minutes and show the students that it stretches again. ***Thinking Critically:*** Ask the students what the relationship is between temperature and plasticity. (Warm temperatures increase plasticity; cool temperatures decrease plasticity.)

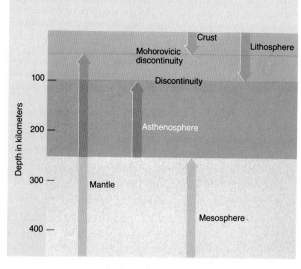

Figure 6–4. The lithosphere, composed of the crust and upper mantle, floats on the plasticlike asthenosphere.

DISCOVER

Making Convection Cells

CAUTION: Put on safety goggles and leave them on during this activity. Half fill a large laboratory beaker or a glass baking dish with corn syrup. Add several drops of food coloring, but do not stir the mixture. Place the container on a hotplate or over a very low burner flame. As the syrup heats, a convection cell should begin to form. Observe the flow of the convection cell by watching the movement of the food coloring.

Figure 6–5. Convection cells bring molten rock material to the earth's surface where new rock is added to the crust.

When rocks are heated, they become less dense and tend to rise through the mantle. Currents within the mantle carry warmer rocks up and cooler rocks down. These currents are very slow, moving the rocks only a few centimeters per year. The transfer of heat by the circulation or movement of solids, liquids, or gases is called **convection**. These currents create pockets of circulation, known as *convection cells*, throughout the mantle. Convection cells can be observed in a laboratory by carefully heating a thick, colored liquid in a beaker. However, movement within the earth is far more complicated than movement in a laboratory beaker. Convection cells in the earth may have eddies, or currents within currents.

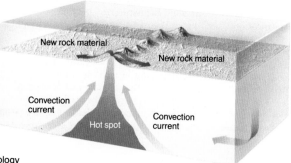

Answers–6.2 (p. 121)

○ Convection cells are small currents of heated mantle rock.

○ Heat to produce convection cells comes mainly from the decay of radioactive elements.

Since the lithosphere is floating on the asthenosphere, the movements within convection cells influence the entire lithosphere. There is much evidence that the lithosphere does not remain stationary. In fact, some evidence suggests that entire continents may be drifting on these convection currents.

○ *What are convection cells?*
○ *Where does the heat to produce convection cells come from?*

Figure 6–6. These photographs show evidence of movement of the earth's crust. For example, the Himalaya Mountains (right) grow several centimeters each year, the rift valley in Iceland (upper left) continues to expand, and these cracks in the earth continue to widen (lower left).

Section Review

READING CRITICALLY

1. Diagram the layers and discontinuities of the earth's interior.
2. Explain the difference between the lithosphere and the asthenosphere.

THINKING CRITICALLY

3. How might the study of earthquakes have been used to gain knowledge about the earth's interior?
4. Explain how convection cells could cause movement of the lithosphere.

Section 1   Inside the Earth   **121**

DISCUSSION   You may use Figure 6–6 to begin a discussion of the forces that shape the earth's crust. ***Thinking Critically:*** Ask the students what sort of force would form high mountains such as the Himalayas. (Answers will vary but the students may suggest "pushing" or "squeezing.") Then ask them what type of force may have formed this rift valley, a great crack in the crust. (Answers will vary but should include "pulling" or "stretching.") Emphasize the role of convection currents in this pushing and pulling.

SECTION REVIEW

Summary   Have the students describe the forces that cause folds and faults. Ask the students to classify and make diagrams of each type of fault. Also have them explain the many pieces of evidence that support the theory of continental drift.

Reinforcement   Have the students answer the questions at the end of each numbered section. Then let them compare and discuss their responses with their classmates.

Answers to Section Review

Reading Critically

1. Diagrams should include crust, mantle, core, Mohorovičić discontinuity, lithosphere, and asthenosphere.
2. The lithosphere is solid and is composed of the crust and upper mantle. The asthenosphere is putty-like and is composed of the remainder of the mantle.

Thinking Critically

3. As the energy from an earthquake travels through the layers of the earth, both the speed and direction of the waves are changed. This allows scientists to locate boundaries between layers and predict the composition and consistency of the layers.
4. As the putty-like material in the asthenosphere moves in circular currents, it drags the rigid lithosphere with it.

SECTION FOCUS

Section Overview This section describes the causes of folds in the earth's crust. The measuring of rock layer orientation using strike and dip is also explained. The types of faults and the associated forces that form them are classified. Evidence used by Alfred Wegener to arrive at his theory of continental drift is detailed.

Section Objectives For a list of section objectives, see pupil's edition page.

New Science Terms For a list of new science terms in this section, see pupil's edition page. Read the new terms aloud to familiarize students with their pronunciation. You may prefer to have the students read the terms aloud themselves.

SECTION DEVELOPMENT

6.3 Folds

EXTENSION Tell the students that formations, whether vertical, horizontal, or at any angle in between, are named according to their location and rock type. Often, the name comes from the place in which the rock layer was first identified; for example, Cody Shale for Cody, Wyoming, and Dakota Sandstone for Dakota County, Nebraska. The naming system makes it easy for geologists to identify and communicate about similar formations.

DISCUSSION At this time you may want to expand on the concepts of horizontal deposition, superposition, and deposition sequences by incorporating Section 15.1 pp. 323–324. You also may want to explain the formation of an uplift further by referring students to Section 14.3 on page 305 for a description of the formation of the Colorado Plateau.

NEW SCIENCE TERMS
fold
orogeny
anticline
syncline
dip
strike
fault
tension
compression
shear
continental drift

SECTION OBJECTIVES

After completing this section, you should be able to:

- **Describe** the process that causes the crust to fold.
- **Classify** faults.
- **List** four kinds of evidence that support the theory of continental drift.

6.3 Folds

Studying a map may give you the idea that the earth's surface is fixed, and cannot change. However, because of forces within the earth, the surface is slowly but constantly changing. Look at Figure 6–7. This is not a trick; these rocks really are positioned like this. Although sediments are deposited in horizontal layers, sedimentary rocks are not always horizontal. Some rock layers may even be vertical. Great forces within the earth have pushed these rocks into this position.

Sometimes these forces build up slowly, causing rocks to warp or fold. A **fold** is a bend in rocks caused by force. You can make a fold by pushing gently on one end of a sheet of paper while holding the other end in place. What happens to the paper? 1In the same way, slow, continuous force can build mountain ranges. This process of mountain building is called **orogeny** (oh RAHJ uh nee).

Figure 6–7. Forces within the earth can cause rocks, such as those shown here, to bend or break.

1 It folds up in the middle.

Mountain ranges, such as the Rocky Mountains, that are formed in this way are called *orogenic belts*. In addition, large areas can be lifted without breaking the rocks in the formation. Areas formed in this way are called *uplifts*. The Colorado Plateau, shown in Figure 6–8, is an example of a large-scale uplift.

Two factors that cause rock layers to bend are pressure and temperature. On or near the earth's surface, rocks fold if steady pressure is applied over a long time. Deep within the earth, high temperatures change rocks to a softer, plasticlike material that folds more easily.

Interesting landforms develop where folded rocks are exposed on the earth's surface. The landforms are most easily studied from a geologic map of the area. On a geologic map, each rock layer is represented by a color and a number. Folds often appear as U-shaped patterns on geologic maps. When formations are folded upward, an **anticline** is formed. A downward fold forms a **syncline**. Anticlines and synclines are shown in Figure 6–9.

Figure 6–8. The city of Denver, just east of the Rocky Mountains, is on a plateau, or an uplifted area.

A MATTER OF FACT

The highest elevations on the earth's surface are in Tibet. The average altitude of this area is about 4880 m above sea level.

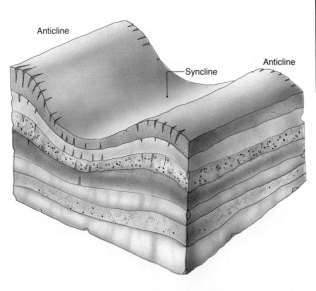

Figure 6–9. Anticlines and synclines form as rock layers are folded. Anticlines often form parallel ridges, with a syncline forming a valley between them.

Geologists have developed a system to describe synclines and anticlines and to show them on geologic maps. This system is called *strike and dip*. **Dip** is the angle at which the rock layer is tilted from the horizontal. Dip also includes the direction in

Section 2   Crustal Adjustments   **123**

EXTENSION   Show the students a geologic map of anticlines and synclines. Explain to the students that the U-shapes on the map represent plunging folds that have been eroded to expose the rock layers that were folded. A plunging fold is one that dips at an angle to a horizontal plane. Have the students make a three-dimensional drawing of a plunging anticline and plunging syncline, showing both the surface and a cross-sectional view. *Thinking Critically:* Ask the students where the youngest rocks appear in the anticline. Also ask the students where the oldest appear. (at the edge of the U; in the center) Ask the same questions about the syncline. (in the center; at the edge) Next explain that a nonplunging fold will appear as parallel formations on either side of the oldest layer (in an anticline) or youngest layer (in a syncline). The rocks become either progressively older or younger from the axis of the fold.

DISCUSSION   Emphasize to the students that although surface features often reflect the underlying rock structures, anticlines do not always form ridges and synclines do not always form valleys. It is the relative resistance of the rock layer's erosion that determines where hills and valleys form. An excellent example of synclinal ridges and anticlinal valleys can be seen in a cross section of parts of the Appalachian Mountains. Obtain such a cross section and display it as you conduct this discussion.

DEMONSTRATION

To demonstrate the system of strike and dip, hang a card with a compass direction on each wall of the classroom. Have a student hold a stiff cardboard panel as the horizontal plane. Use a second cardboard panel as a tilted rock layer. Draw a strike line on the rock layer with chalk. Draw the direction of dip perpendicular to the strike. Have the students describe the compass direction of the strike (for example, N or S) and dip (E or W). Tilt the rock layer in the opposite direction to show that the strike can stay the same but

the dip changes. Use a protractor to measure the angle of dip. Be sure to measure from the horizontal.

EXTENSION Strike and dip are represented on a map with a T-shaped symbol. The direction of strike is shown by the top of the T. The dip is shown as the direction in which the line extends away from the line of strike. A number representing the number of degrees of dip is written next to the T symbol. Rock layers are represented on maps with different colors and markings, though layers of the same rock type usually have the same markings.

6.4 Fractures and Faults

DISCUSSION Faulting, fracturing, and folding are all caused by forces that deform the crust. This deforming force is called *stress*. Stress slowly changes the shape and volume of rocks. The change in the size or volume is called *strain*. Explain to the students that they will learn about three types of stress: compression, tension, and shear. ***Thinking Critically:*** Ask the students what factors determine whether a rock layer faults or folds under stress. (A brittle rock layer is more likely to fracture than a softer, more pliable type of rock. Thicker layers tend to fracture, while thin layers tend to fold. Stress that is applied slowly and evenly will probably result in folded rock layers. Stress that is applied suddenly is more likely to cause rocks to fracture.)

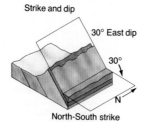

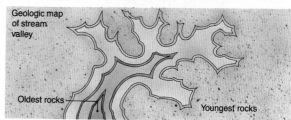

Figure 6–10. Geologic maps show rock layers, as well as special features such as strike and dip (left). Rock layers are usually shown in colors (right), while special features are shown with symbols.

which the layer is tilted. Thus, dip is described by both an angle and a compass direction. The angle tells to what degree the layer is tilted, and the compass reading tells the direction in which the layer is tilted. The rock layer shown in Figure 6–10 is dipping 30° east.

As the formation dips, it forms a straight line called the **strike** along the earth's surface. The direction of this line is always at right angles to the dip. If the dip is east, then the strike will be north-south. Therefore, the rock layer shown in Figure 6–10 would strike north-south and dip 30° east. Strike and dip are shown on geologic maps, along with individual formations.

○ *What are folds?*
○ *How are folds shown on a geologic map?*

6.4 Fractures and Faults

Sometimes the forces within the earth are strong enough to break, or fracture, rocks instead of folding them. An example of fracturing can be seen in the quartzite cliffs near Devil's Lake, Wisconsin.

Figure 6–11. Forces within the earth are sometimes strong enough to cause rock layers to fracture. The fractured rocks shown here are at Devil's Lake, Wisconsin.

124   Chapter 6   Structural Geology

Answers—6.3

○ Folds are bends in rock layers.
○ Some folds appear on geologic maps as U-shapes.

If there is no movement along the fracture, the fracture is called a *joint*. Joints can also form as igneous rocks cool and contract. These joints often form long columns in rocks, as shown in Figure 6–12.

If the forces are great enough, the fractured rocks may move, forming a fault. A **fault** is a rock fracture along which there is movement. All faults share some characteristics. For instance, all faults have two sides, the *footwall* and the *hanging wall*, as shown in Figure 6–13. All faults show movement between the walls, as well. Different types of faults result from movement in different directions. Earthquakes, which are discussed in the next chapter, often occur along faults.

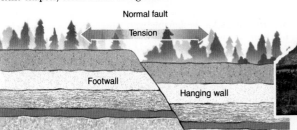

Normal fault

Tension

Footwall

Hanging wall

Normal Faults Movement along faults is caused by three different kinds of force—tension, compression, and shear. **Tension** occurs when the earth's crust is pulled apart. Take a block of modeling clay and pull it apart. Tension separates rocks in a similar manner, and the hanging wall moves down. The type of fault formed by tension is called a *normal fault*. Normal faults are seen in the Basin and Range area of the southern Rocky Mountains. In this area pieces of the earth's crust may move many kilometers. As crustal blocks move and tilt, they form *fault-block mountains*.

Figure 6–12. These rock joints are caused by fractures in the rocks. Where joints occur there is no movement along the fracture.

Figure 6–13. Fault-block mountains result from movement along normal faults. The fault-block mountains shown here are part of the Rocky Mountains.

Section 2   Crustal Adjustments   125

DISCUSSION Remind the students that a joint is a crack along which there is no appreciable movement. Explain that joints in an area often form in parallel sets. When several sets of joints intersect, they can break the rock into sheets or blocks. Joints that form in sedimentary rock are often the result of some type of stress. In igneous rocks, joints generally form as the melted rock cools and contracts. The type of joints shown in Figure 6–12, called *columnar joints*, were formed in this manner.

SKILL *(Interpreting Photographs)* Have the students trace the rocks in Figure 6–13. Have them label the footwall and hanging wall. Then ask them to draw arrows on the rocks showing the relative direction of movement. Lastly, have the students draw arrows showing the direction of the force that formed these fault blocks.

EXTENSION Obtain a set of geologic maps and topographic maps of the same area. Have the students work in small groups to identify folds, faults, and tilted strata on the geologic maps. By referring to the cross sections, the students should have little trouble locating and describing several structures on each map. Next have the students use the correlating topographic maps to describe the surface features and landforms that result from the underlying structures. You may want to use maps from some especially clear areas such as Santaquin, Utah (block faults) and Bright Angel, Arizona (lateral fault).

DEMONSTRATION

You may want to demonstrate the formation of fault-block mountains for the students. Place ten dominoes against one another on their sides lengthwise across a 30-cm long piece of wide elastic tape. Put small wooden blocks at each end of the dominoes to keep them upright. Slowly, stretch the elastic until the dominoes tilt. Have the students draw the resulting model block faults. Ask the students to classify the type of fault modeled. (normal fault) Also ask them to identify the type of force that formed the faults.

DISCUSSION Explain to the students that the block of rock that moves upward relative to a graben is called a *horst*. Tell them that the Great Rift Valley is still quite an active fault zone. Many scientists believe that the continued tension along the valley will cause it to continue opening and a new ocean basin will eventually form. Some scientists believe that the Basin and Range Province of the southwestern United States, an area broken up by many horsts and grabens, may also become a new ocean basin sometime in the far distant future.

EXTENSION A thrust fault is a reverse fault that dips at a very small angle. In such a fault, the hanging wall block is pushed almost horizontally on top of the footwall block. Rocks that have been thrust-faulted often are pushed several kilometers from their original location. Tell the students that the Lewis Thrust Fault in the northern Rocky Mountains has moved rocks as much as 48 km. Remind the students that older rock layers are usually found beneath younger rock layers. *Thinking Critically:* Ask the students how a geologist studying in an area of thrust-fault might become confused about the ages of the rock layers observed. (The geologist may believe that older rocks are actually younger. To avoid confusion, the geologist must recognize evidence that would indicate that thrust-faulting had occurred in the area.)

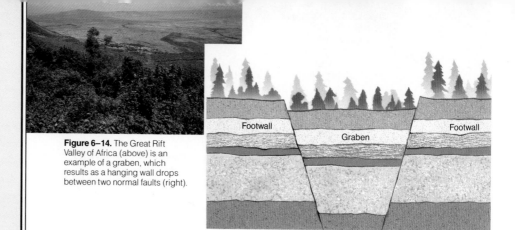

Figure 6–14. The Great Rift Valley of Africa (above) is an example of a graben, which results as a hanging wall drops between two normal faults (right).

Sometimes a large block falls between two normal faults. The sides of this block are the hanging walls of the two faults. This sunken area is called a *graben* (GRAH behn). The Great Rift Valley of Africa, shown in Figure 6–14, is an example of a graben.

A MATTER OF FACT

The Dead Sea lies about 395 m below the surface of the Mediterranean. This landlocked salt lake is really a northeastern extension of the East African Rift Valley.

Reverse Faults Instead of being pulled apart, sections of the crust may be pushed together. This pushing together of the crust is called **compression,** which is the second type of force. A fault formed by compression is called a *reverse fault.* In a reverse fault, the hanging wall moves up in relation to the footwall, as shown in Figure 6–15. If the angle of the reverse fault is very low, the fault is called a *thrust fault.* In some thrust faults, the hanging wall can slide hundreds of kilometers over the footwall.

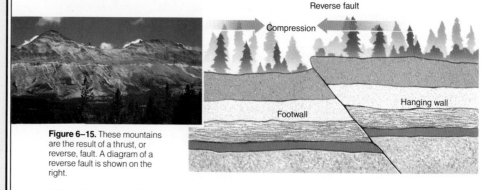

Figure 6–15. These mountains are the result of a thrust, or reverse, fault. A diagram of a reverse fault is shown on the right.

126   Chapter 6   Structural Geology

DEMONSTRATION

Obtain two 12-cm thick sheets of polystyrene foam. Cut one into two pieces at angles of 60 degrees, making a model footwall and hanging wall of a fault. Demonstrate the relative movements of a normal, reverse, and lateral fault. Using the second piece of foam, cut two faults. The one on your left should be cut at a 120° angle; the one on your right at 60°. Use these pieces of foam to demonstrate how a fault block slips down between two normal faults to form a graben. Post the compass directions in the room and have the students describe the strike and dip of the fault planes.

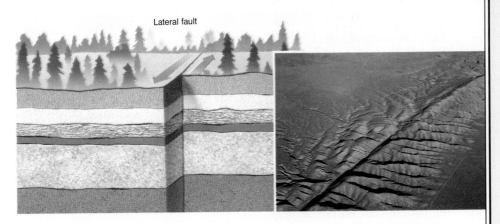
Lateral fault

Lateral Faults

The third type of force on pieces of the crust is parallel to the pieces. This force, called **shear**, occurs when pieces of the crust drag against each other in opposite directions. Try sliding two sheets of sandpaper past each other. The sheets of sandpaper resist sliding, but if enough force is applied, they may suddenly move.

This sudden movement, which is shear, produces a third type of fault, called a *lateral fault.* In a lateral fault the hanging wall moves parallel to the footwall but in the opposite direction. The famous San Andreas fault in California, shown in Figure 6–16, is an example of a lateral fault.

Like a fold, a faulted area also forms interesting patterns on geologic maps. A geologic map of a faulted area often indicates rock layers that are repeated or that are missing. These repeated or missing layers are caused by movements along faults. A formation may be displaced several hundred kilometers by faulting. Figure 6–17 shows a geologic map of an area that has been both folded and faulted.

○ *What is a fault?*
○ *Name the three forces that produce faults, and the type of fault each produces.*

Figure 6–16. The famous San Andreas fault in California (above) is a lateral fault. A diagram of a lateral fault is shown on the left.

Figure 6–17. A geologic map of an area that has been both folded and faulted can be very complex.

127

THINKING SKILL *(Inferring)* Have the students study Figure 6–16. Ask them if it would be possible to tell if a lateral fault were present in an area by looking at a cross section. Explain. (No. A lateral fault moves only horizontally, and thus the rocks layers still appear aligned when viewed from the side.)

DISCUSSION Have the students draw a cross-section view of the way a series of rock layers would look after the following events: Four horizontal sedimentary rock layers are laid down. These are deformed into anticlines and synclines. A normal fault forms through one of the anticlines. Erosion occurs. Two more sedimentary layers are laid down on the erosional surface. Finally, a fault-block mountain forms. Remind the students that a fault is always younger than the rocks through which it cuts. Although the students are all drawing the same geologic events, their pictures should be very different. Have the students compare their art with that of other class members. Discuss the various interpretations represented by the students' drawings.

Answers—6.4

○ A fault is a fracture along which there is movement.

○ Tension, compression, and shear are the three forces that produce faults. Tension produces a normal fault, compression produces a reverse fault, and shear produces a lateral fault.

Objectives

- Interpret a geologic map and cross section.
- Sequence chronologically rock layers on the map.
- Make inferences about the construction of geologic cross sections.

Discussion   You may want to point out to the students that along the line A B cross section, they can see that anticlines do not always form ridges and synclines do not always form valleys. By observing which layers are less eroded, the students can decide why the resistance of a particular layer to weathering is an important factor in the formation of surface features. *Thinking Critically:* Ask the students which rock layers appear most resistant (SL, DL). Then ask what type of rock these layers probably are. (igneous rocks)

Answers to Application

1. From A to C the rock layers are DM, DL, DM, DL, DM, DL, DM, and DU.
2. There is one anticline and one syncline.
3. The oldest layer should be DM, and the youngest, DL.

BACKGROUND

Interpreting a geologic map is often the first stage in understanding the geologic history of an area. Just as topographic maps represent land-scapes with symbols, geologic maps use symbols to show where different rock units appear on the earth's surface.

The rocks shown on a geologic map are called formations. On most geologic maps, graphic symbols are used to indicate more common rock types. Letters are also used to indicate the geologic age of a formation. Other symbols are used on geologic maps to show strike and dip, horizontal beds, and the axes of synclines and anticlines.

PROCEDURE

Study the geologic map symbols and the area of the map along the line marked AB. Along the line AB, locate different types of rock layers, anticlines and synclines, and locate the oldest and youngest rocks along the line.

APPLICATION

1. Along the line marked AC, name the different types of rocks.
2. How many anticlines and synclines are there along the line AC?
3. Name the oldest and youngest rock layers along the line AC.

USING WHAT YOU HAVE LEARNED

At the bottom of the map is a cross section of the area along the line AB. Study it to determine how it was constructed, and then try drawing a similar cross section for the area along the line AC.

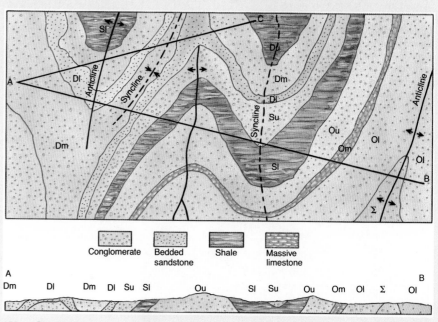

6.5 Continental Drift

The evidence presented so far has shown that small areas of the earth's crust move. There is, however, evidence that large portions of the crust, entire continents in fact, are also moving. This movement of large portions of the earth's crust is called **continental drift.**

In the early 1900s, a young German scientist, Alfred Wegener (VAY guh nuhr), became fascinated by the matching shapes of the coastlines of Africa and South America. In his book, *The Origin of Continents and Oceans*, published in 1915, Wegener claimed that these coastlines appeared similar because at one time they were joined. Wegener further proposed that the continents, because they are made of lighter rocks than the ocean basins, float on top of the oceanic crust.

When Wegener proposed his theory of continental drift, most scientists disagreed with him; he was even ridiculed by the scientific community. Like Wegener, many scientists had noticed that there is an amazing match in the coastlines of Africa and South America; however, most dismissed the similarity as mere coincidence. Careful study of the rocks along the edges of these two continents provided support for Wegener's theory. The rocks match in structure and type, indicating that the continents were once joined.

Other matches of similar rocks have been discovered. The Appalachian Mountain range, in the eastern United States, seems to disappear into the northern Atlantic Ocean. The rocks in this mountain range, however, are very similar to those found in Scotland's Grampian Mountains. This geologic evidence indicates that North America and Europe were once joined as well.

DISCOVER

Solving a Puzzle

On the back of an old jigsaw puzzle, sketch the rough outlines of Africa and South America joined along a common line. Now separate the puzzle pieces, and see if a classmate can reconstruct the joined supercontinent. How is solving this puzzle similar to what Wegener must have done in developing his theory of continental drift?

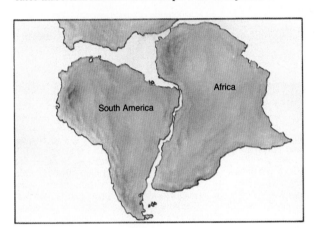

Figure 6–18. This diagram shows how the continents on either side of the Atlantic Ocean may have been arranged before they began drifting apart. The similarities of the coasts of Africa and South America support the theory of continental drift.

6.5 Continental Drift

DISCUSSION Discuss with the students the dilemma that the scientific community faced as they argued over the theory of continental drift. You may want to have the students find out about earlier beliefs concerning the shapes of the continents and the similarity of fossils and live organisms from continent to continent. Suggest that the students do research on people such as Charles Dwight, who believed the earth's features formed as the earth cooled and contracted, much in the way that an apple skin wrinkles when it dries out. Other scientists who could be studied are Eduard Suess and Alexander von Humboldt. Be sure the students describe thoroughly the reasons why the beliefs were accepted and why they were later rejected. If you prefer, you can have the students write reports on the scientists who gathered the data that was used by Wegener to arrive at his theory of continental drift. Some of these scientists are Sir George Airy, John Wesley Powell, and Clarence Dutton. Inform the students that they will learn more about continental drift and the more complicated theory of plate tectonics in Chapter 7.

DISCOVER: Solving a Puzzle

Thinking Skill (*Problem Solving*)

If the students do not have an old jigsaw puzzle, they may use a heavy piece of cardboard instead. Have the students outline the pieces of a puzzle on one side and draw South America and Africa on the other. The students should realize that they need to follow a scientific method in order to solve the puzzle, just as Wegener had to do in order to develop his theory. The students must observe the data, experiment with different configurations of the puzzle, and finally, reach the most reasonable solution to the puzzle.

EXTENSION You may want to point out that there are many other fossils found on the different continents. For example, *Kannemeyerid* (a small land-dwelling dinosaur) fossils are found in North America, South America, Africa, and Asia. *Kannemeyerids* could not swim and thus could not get from one continent to another.

THINKING SKILL *(Analyzing Maps)* Have the students study a world map to compare the coasts of the continents. Next have them study a map of the ocean basins so that they can compare the edges of the continental shelves. Ask the students if the view of the shelves helps to support or dispute the theory of continental drift. (Answers will vary, but the students should note that the fit of the continents is much more obvious when the shelves are used as the continental borders.)

SECTION REVIEW

Summary Have the students make diagrams of the three types of faults and an anticline and a syncline. Have the students describe the structures and the forces that formed each type of fault.

Reinforcement Have the students write the main ideas of each paragraph of this section. Then have the students compare their ideas with those of the other students.

ACTIVITY:
Studying Ancient Glaciers

Thinking Skill *(Inferring Relationships)*

Preparation of Materials You may want to prepare copies of a world map and distribute them to the class.

Answers to Conclusions/Applications

1. Continental shelves, though not visible on most maps, are part of the continents and must be considered.
2. yes
3. The massive continent must have been near the poles to support such a large glacier. The climate must have been very cold.

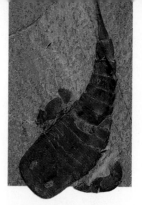

Figure 6–19. Fossils found on several continents, provide evidence for the theory of continental drift.

More evidence supporting the theory of continental drift comes from the comparison of fossils. Fossils of the same mammal-like reptiles were found in Antarctica, Africa, South America, and India. These ancient animals could not swim. Scientists wondered how these animals could have crossed hundreds of miles of open ocean to reach Antarctica from Africa or India. Each new bit of evidence added support to the theory of continental drift. However, more evidence was needed to convince scientists who found it difficult to believe that solid rock could flow.

○ *What is the theory of continental drift?*
○ *What evidence exists to prove this theory?*

Section Review

READING CRITICALLY

1. What are the three types of force and the faults they produce?
2. What conditions are needed to fold rocks?

THINKING CRITICALLY

3. Why would it be dangerous to construct a building on or near a fault?
4. If the theory of continental drift is correct, on what layer of the earth's interior could the continents actually be moving?

ACTIVITY: Studying Ancient Glaciers

How can data from ancient glaciers support the theory of continental drift?

MATERIALS (per group of 3 or 4)

pencils, tracing paper, textbook, scissors, colored pencils, glue, legal-size paper

PROCEDURE

1. Trace the continents of South America, Africa, India, and Australia from the world map in the reference section of pages 592–593. Cut out the tracings of the continents.
2. Using a colored pencil, color the areas where ancient glaciers might have been found. Add arrows to indicate the likely direction of movement of the glacier.

3. Position the tracings so that the glaciers appear to move from a central location outward onto the various continents. Glue the tracings into place on the legal-size paper to show the single continent.

CONCLUSIONS/APPLICATIONS

1. Why do the continents not fit together exactly?
2. If you consider the shape of the shallow shelves just offshore, does the fit improve?
3. If you assume that all the continents were together when this glacier covered parts of these continents, where on the globe would this continent have been? What type of climate existed in North America at that time?

130   Chapter 6   Structural Geology

Answers–6.5

○ Continental drift is the theory that sections of the earth's crust are moving.

○ The matching shape of the continental edges, similar mountain ranges, and fossils are all used as evidence to support the theory of continental drift.

Answers to Section Review

Reading Critically

1. Tension produces normal faults. Reverse faults are produced by compression, and shear produces lateral faults.
2. Pressure, high temperature, and time are conditions needed to fold rocks.

continues

CAREERS

SEISMOLOGIST

A seismologist studies waves in the earth produced by earthquakes. Seismologists not only record and study earthquake data, but they use the data to predict earthquakes.

Seismologists also use earthquake data to analyze the rock formations of an area. This is sometimes useful in the search for natural resources such as oil and gas.

Some seismologists work in laboratories interpreting seismic records, while others work in the field using lasers, computers, and other sophisticated equipment. A career in seismology requires a college degree in geology or physics.

For Additional Information
American Petroleum Institute
1220 L Street, N.W.
Washington, DC 20005

SCIENTIFIC TRANSLATOR

Research and scientific work are being done by scientists in many countries. There is a great need for qualified translators of scientific writing, both into English and from English into other languages.

Translators are often employed by international agencies like the World Health Organization. Scientific journals also use the services of translators.

There is no standard training program for translators. A basic understanding of science is necessary as well as reading and writing experience in one or more foreign languages.

For Additional Information
Contact your local community college or university.

OIL-PLATFORM WORKER

Much of the oil and gas needed to run an industrialized world is found deep below the earth's surface. Some of this oil is below the surface of the sea, many kilometers from shore.

The process of obtaining oil and gas from undersea wells requires both scientists and technicians. Most of the jobs on an oil platform are exciting and dangerous. Many require an ability to work with tools and machinery.

The technician's jobs require at least a high-school diploma; some require specialized training such as a trade school or technical school would offer.

The wages are high and there is opportunity for travel and advancement. However, the hours are long and the conditions in places such as the North Sea are unpleasant.

For Additional Information
American Geological Institute
5205 Leesburg Pike
Falls Church, VA 22041

Section 2   Crustal Adjustments   **131**

CAREERS

Discussion Have interested students write to one or all of the addresses provided for additional information. Have the students discuss the opportunities offered by the various careers.

Extension You may wish to hold a "Career Day" in the class during which students report on the information they have obtained. If possible, invite a person in each career to speak to the class about why he or she chose the career, what specific training he or she received, and what a typical work day involves. As an alternative to outside speakers, have students assume the role of the professionals listed below and make a presentation to the class covering the same topics mentioned above. The students may be interested in researching information on these related careers:

Geologic Field Assistant

Cartographer

Surveyor

Geomorphologist

Thinking Critically

3. When movement along the fault occurs, the structure could be damaged.
4. The continents as part of the lithosphere are floating on the asthenosphere.

Skill *(Reading Maps)*

Preparation of Materials

To familiarize the students with reading map cross sections, have them look at simplified examples of anticlines, synclines, and faults in individual cross sections. Also help the students use the map key to identify the rocks in the cross section. You may want to review the types of faults and the way to determine dip.

Answers to Procedure Questions

2. by a U-shaped valley; L, M, N, O, P, Q, R, S
3. A, B, C, D, E, T, U, V, W, and X
4. Thrust fault L, M, N, O, P, Q, R

Answers to Analyses and Conclusions

1. The fault probably formed last because there is an unconformity of the rocks in the syncline.
2. Layer X is the same as layer A.
3. Layer X dips east.
4. The fault dips west.
5. Layer Y is the youngest.
6. Layer K is the oldest.
7. It is a thrust fault.
8. L, M, N, O, P, Q, R
9. K–M, J–N, I–O, H–P, G–Q, F–R, T–S

Answer to Application

Each of the layers L–R would be extended through the syncline.

INVESTIGATION 6: Analyzing Structural History

PURPOSE

To analyze the history of an area using information from rock formations

MATERIALS

Paper and pencil

PROCEDURE

1. Study the cross section of rock layers shown. Examine it for rock type and structural deformations.
2. How is a syncline illustrated in the cross section? Identify a syncline in this cross section by listing the letters that represent the formation.
3. Locate an anticline in the cross section. List the letters of the formations involved.
4. Classify any faults found in this cross section. List the letters of the formations involved.

ANALYSES AND CONCLUSIONS

1. Which of the three structures in procedure steps 1, 2, and 3 was formed last? Explain your reasoning.
2. Which layer on the footwall side of the fault is the same as layer A?
3. In which direction does rock layer X dip?
4. In which direction does the fault dip?
5. Name the youngest rock layer.
6. Name the oldest rock layer.
7. Name the type of fault shown on this cross section.
8. List the rock layers that make up the footwall of the fault.
9. Match the rock layers of the footwall to the same layers in the hanging wall.

APPLICATION

What do you think the area east of the cross-section might look like? Draw an extension of the original cross-section showing how you think it might appear.

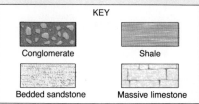

KEY

Conglomerate   Shale

Bedded sandstone   Massive limestone

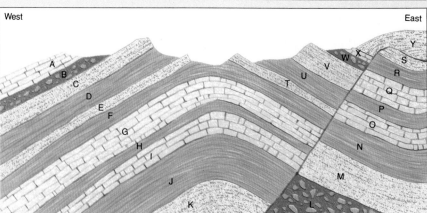

CHAPTER 6 REVIEW

SUMMARY

- The earth is made of several layers that are distinct in composition, density, and thickness. (6.1)

- Beneath the crust are the mantle and the core of the earth. (6.1)

- The heated material within the earth moves by convection. Most of the heat in the interior of the earth is supplied by the decay of radioactive elements. (6.2)

- Forces within the earth's crust cause rocks to fold and fracture. Rocks fold into anticlines and synclines. (6.3)

- Large-scale folding produces mountain ranges by the process of orogeny. Plateaus may be formed by large-scale uplifting without folding. (6.3)

- Landforms created by folding are often studied by using geologic maps. (6.3)

- The position of rock formations can be described using strike and dip. (6.3)

- Faults form because of tension, shear, or compression. (6.4)

- Tension produces normal faults. Fault-block mountains may result from normal faults. Reverse faults are the result of compression. A low-angle reverse fault is a thrust fault. Lateral faults, like the San Andreas fault, are caused by shear. (6.4)

- In 1915 Alfred Wegener proposed the theory of continental drift. (6.5)

- The match of continental shapes, geological features, and fossil evidence are used to support the theory of continental drift. (6.5)

Write all answers on a separate sheet of paper.

SCIENCE TERMS

Correctly use each of the following terms in a sentence.

anticline (123)
asthenosphere (119)
compression (126)
continental drift (129)
convection (120)
core (119)
crust (117)
dip (123)
fault (125)
fold (122)
lithosphere (119)
mantle (118)
orogeny (122)
shear (127)
strike (124)
syncline (123)
tension (125)

SCIENCE QUIZ

Modified True-False

Mark each statement *true* or *false*. If a statement is false, change the underlined term to make the statement true.

1. A <u>lateral fault</u> forms when the crust pulls apart.

2. According to the <u>theory of continental drift</u>, continents can move.

3. Hot molten material rises in the <u>mantle</u> due to convection currents.

4. <u>Grabens</u> are formed by compression.

5. <u>Mountains</u> may occur as a result of folds in the crust.

continues

Chapter 6 Review   133

CHAPTER REVIEW

SUMMARY

The students may review the major concepts in the chapter by reading the summary statements. The statements are cross-referenced to the chapter to facilitate reinforcement of any concepts of which the students feel unsure. Encourage the students to work in groups to quiz one another.

SCIENCE TERMS

The sentence in which the science term is used should reflect an understanding of the definition of the term. You may wish to stage a game of "Jeopardy" during which the definition is provided and the "contestants" must provide the correct term.

SCIENCE QUIZ

Answers to Modified True–False

1. false, normal fault
2. true
3. true
4. false, Reverse faults
5. true

continues

Answers to Multiple Choice

6. d
7. c
8. d
9. a
10. c

Answers to Completion

11. theory of continental drift
12. the Gutenberg discontinuity
13. joints
14. Antarctica and India
15. Mohorovičić discontinuity

Answer to Short Answer

16. These fossils are important in two ways. First, the animals could not have swum across large distances of open ocean. Second, as tropical animals, they could not have lived in arctic environments. Thus, Antarctica must have once been closer to the other continents and closer to the equator.
17. Compression forces form reverse or thrust faults. Tension forces form normal faults. Shear forces form lateral faults.
18. These two mountain ranges were formed as one continuous range when North America and Europe were connected as one continent.
19. See Figure 6–2.

Answers to Writing Critically

20. These discontinuities are caused by differences in density and composition.
21. The continental crust is made of less dense minerals than that in the oceanic crust; therefore, it floats higher.
22. The rising magma of convection cells separates the crust and spreads the continents.

ANSWERS TO EXTENSION

1. The students should show at least three different formations in order to fully develop a rock structure. For block-fault mountains, the students might research the mountains in the Basin and Range Province in the southern Rockies or the Teton range. For folded mountains, they might study the Valley and Ridge Province in the Appalachian mountains. You may wish to include domes such as the Black Hills or the central mineral regions of Texas.
2. Reports should include the type of bones found as well as a description of the type of rock in which the fossils were found. The environmental conditions could be interpreted from the rock type as well as from the other types of plants and animals found in the same rock formation. This activity could be extended by having the students research the fossil *glossopteris*.
3. Chief Mountain is located at the eastern edge of the Rocky Mountains in northwestern Montana. The entire mountain is a *klippe,* or erosional remnant of the Lewis Thrust sheet. It was once connected to the thin layer of Precambrian rock that was thrust on top of the Cretaceous formations during the building of the northern Rocky Mountains.

SCIENCE QUIZ **continued**

Multiple Choice

Write the letter of the choice that best answers the question or completes the statement.

6. The only portion of the earth thought to be entirely molten is
   a) the upper mantle.
   b) the inner core.
   c) the lower mantle.
   d) the outer core.

7. Pressure of overlying rocks affects the rocks below by
   a) lowering the melting point.
   b) altering the mineral content.
   c) raising the melting point.
   d) causing convection currents.

8. The lithosphere includes
   a) the crust and the mantle.
   b) the mantle and the upper core.
   c) the asthenosphere and the crust.
   d) the upper mantle and the crust.

9. Which type of fault is produced by shear?
   a) lateral
   b) normal
   c) reverse
   d) thrust

10. The evidence for continental drift comes from
    a) lateral faults.
    b) anticlines.
    c) matching rocks.
    d) radioactive material.

Completion

Complete each statement by supplying the correct term or phrase.

11. Wegener began a great scientific argument by proposing the _____.

12. The boundary between the core and the mantle is _____.

13. Fractures along which there is no movement are called _____.

14. Fossils of a land-dwelling African animal were found in _____.

15. The boundary between the crust and the mantle is called the _____.

Short Answer

16. Explain why the discovery of the same types of fossils in Africa and India was important in the development of the theory of continental drift.

17. What is the relationship between different forces and different types of faults?

18. How are the Appalachian Mountains similar to the Grampian Mountains? Why is this geologically important?

19. Draw a cross section of the earth's interior and label the discontinuities.

Writing Critically

20. What causes the discontinuities between the crust and the mantle, and between the mantle and the core?

21. Explain why the continents float higher than the ocean basin does.

22. Explain the relationship between spreading continents and convection cells.

EXTENSION

1. Build clay models of rock structures such as anticlines or fault blocks. Show the different landforms that are created and write a brief report identifying at least one location on Earth where such a landform exists.
2. Research *Lystrosaurus* fossils. Find out how they were preserved. Write a report about the type of environmental conditions that must have existed where they lived.
3. Find out what is unusual about Chief Mountain, Montana. Explain how it resulted from a thrust fault.

APPLICATION/CRITICAL THINKING

1. Design your own geologic map to show an anticline, a syncline, and a thrust fault.
2. Some geologists have said that the section of California west of the San Andreas fault will end up near Alaska someday. Using your knowledge of fault movement, explain why you agree or disagree.
3. Imagine two people walking out of a California travel agency. One character turns to the other and says, "The only way we'll ever get to Hawaii is by continental drift." Explain whether or not this would be possible.

FOR FURTHER READING

Lane, Frank W. *The Violent Earth.* Topsfield, Mass., 1986. This book offers an interesting discussion of violent actions on Earth, covering everything from earthquakes to avalanches.

McConnell, Anita. *The World Beneath Us.* New York: Facts on File, 1985. The author provides very interesting descriptions of the geologic time table, fossils, plate tectonics, earthquake prediction, and volcano activity.

Miller, R. *Continents in Collision.* Time-Life Books, 1983. This book traces the possible positions of the continents through the history of the earth and suggests possible movement in the future.

"Seafloor Spreading and Sinking." *Science News* 107 (March 22, 1986): 183. This article describes some of the sediments deposited in regions where there are plate boundaries.

Challenge Your Thinking

These two circular structures are related to synclines and anticlines. One is called a *basin* because of its bowl-like shape, and the other is called a *dome* because of its rounded shape. Basins and domes are circular, and both are folded formations. Using what you have learned in this chapter, explain why basins and domes might be considered as special types of synclines and anticlines.

Chapter 6 Review **135**

ANSWERS TO APPLICATION/ CRITICAL THINKING

1. The students may need to build a model of clay before making a map. Make certain that all the structures connect rather than stand isolated.
2. The San Andreas Fault is a right lateral fault. The western side is moving northwest and will eventually be moved closer to Alaska.
3. This is not possible because North America is not moving due west. Also, if it were moving west, the Pacific Plate would also be moving west.

ANSWER TO CHALLENGE YOUR THINKING

Basins are synclines in which the center has eroded, while domes are where the surrounding rock has eroded.

PLANNING THE CHAPTER

Chapter Sections	Page	Chapter Features	Page	Program Resources	Page
Section 1: Plate Tectonics	137			Investigation 7.1: *Plate Tectonics* **(A)**	**TRB 25** **LM 27**
7.1 Crustal Plates in Motion **(B)**	137	**Discover:** Shifting Plates **(A)**	138		
7.2 Plate Boundaries **(B)**	139	A Matter of Fact	139		
		Section Review	142		
		Activity: Examining Crustal Plates **(B)**	142		
Section 2: Volcanoes	143			Critical Thinking **(H)** Reading for Content: *Finding Main Ideas and Supporting Details* **(A)**	**TRB 13** **TRB 13** **LM 185**
7.3 Lava Flows **(B)**	143	A Matter of Fact	143	Investigation 7.2: *Cinder-Cone Volcanoes* **(A)**	**TRB 27** **LM 29**
7.4 The Structure of Volcanoes **(B)**	144	A Matter of Fact	145		
		Discover: Creating Composite Cones **(B)**	146		
7.5 Hot Spots **(B)**	146	A Matter of Fact	147		
7.6 Undersea Volcanoes **(B)**	147	Section Review	148		
		Skill Activity: Making a Graph **(B)**	149		
Section 3: Earthquakes	150			Cross-Discipline: Science and Mathematics, *Predicting the Next Earthquake at Pallett Creek* **(A)**	**TRB 13**
7.7 The Origin of Earthquakes **(B)**	150	A Matter of Fact	150	Concept Extension: *Predicting Earthquake Magnitude* **(A)**	**TRB 15** **LM 237**
7.8 Earthquake Waves **(B)**	151	A Matter of Fact	151		
		Discover: Making a Seismogram **(B)**	154		
		Section Review	154		
		Technology: Seismic Records	155		
		Investigation 7: Plotting Earthquake Data **(A)**	156	Student Record Book: Textbook Investigations **(A)**	**TRB 13**
Chapter 7 Review	157			Vocabulary **(A)**	**TRB 7** **LM 135**
				Tests **(A)** Computer Test Bank 💻	**TRB 24**

(LM) Laboratory Manual/Study Guide,   **(TRB)** Teacher's ResourceBank™

B = Basic   **A** = Average   **H** = Honors

The coding Basic, Average, and Honors indicates sections or subsections that might be appropriate for different levels of learners. For additional suggestions regarding choice of topic and depth of coverage, see the Pacing Chart on pages T16–T20.

CHAPTER CONCEPTS, OBJECTIVES, AND TERMS

Section	Concepts	Objectives	Science Terms
Section 1: Plate Tectonics	■ Alfred Wegener hypothesized that all the continents were once joined as the supercontinent Pangaea **(7.1)** ■ The theory of plate tectonics explains the movement of the earth's crust. **(7.1)** ■ There are three types of plate boundaries: divergent, convergent, and transform fault. **(7.2)** ■ Many geologic processes, such as subduction, folding, and faulting, occur along plate boundaries. **(7.2)**	■ **Explain** the theory of plate tectonics. ■ **Identify** the three types of crustal boundaries and **describe** their associated landforms. ■ **Predict** how the positions of the continents may change in the future.	theory of plate tectonics earthquakes volcanoes divergent boundary convergent boundary transform fault
Section 2: Volcanoes	■ Magma sometimes intrudes into other rock layers where it forms batholiths. Fingerlike projections between or across rock layers form sills or dikes. **(7.3)** ■ Lava flows have many characteristics that depend on the viscosity of the lava. **(7.3)** ■ Volcanoes are classified as shield volcanoes, cinder-cone volcanoes, and composite volcanoes. **(7.4)** ■ Hot spots form volcanic island chains in the middle of plates. Geysers form near hot magma. **(7.5)** ■ Seamounts and guyouts are undersea volcanoes. **(7.6)**	■ **Explain** why volcanoes usually form near plate boundaries. ■ **Classify** types of volcanoes. ■ **Describe** the relationship between hot spots in the crust and volcanic island chains.	batholith viscosity shield volcano cinder-cone volcano composite volcano caldera hot spot seamount
Section 3: Earthquake	■ The sudden release of energy along fault zones causes earthquakes. **(7.7)** ■ Earthquake waves are classified as P waves, S waves, and surface waves. **(7.8)** ■ Earthquake waves travel at different speeds. **(7.8)** ■ Earthquake magnitude is measured on a scale called the Richter scale. **(7.8)**	■ **Compare** and **contrast** the motions of P, S, and surface waves. ■ **Explain** how the epicenter of an earthquake is found. ■ **List** some factors that may be helpful in predicting earthquakes.	focus epicenter P waves S waves surface waves seismograph Richter scale

Title	Page	Materials
Discover: Shifting Plates	138	*(per student)* paper, pencil, scissors, thin sheets of polystyrene, glass baking dish, hot plate
Activity: Examining Crustal Plates	142	*(per group of 3 or 4)* paper, pencil
Discover: Creating Composite Cones	146	*(per student)* modeling clay, small plastic foam beads
Skill Activity: Making a Graph	149	*(per student)* pencil, graph paper
Discover: Making a Seismogram	154	*(per student)* stiff sheet of paper, tape, record, phonograph turntable
Investigation 7: Plotting Earthquake Data	156	*(per student)* textbook, graph paper, pencil

TEACHING SUGGESTIONS

Section 1: **Plate Tectonics**

Field Trip

If you know of a dormant or extinct volcano or of an active fault in your area, plan a visit to show the students what these features really look like.

Section 2: **Volcanoes**

Demonstration: Examining Volcanic Properties

Purpose

To examine the properties of some volcanic products

Background

There are several different kinds of volcanoes. Different types of volcanoes erupt in different ways. Some volcanoes erupt with one shuddering explosion. Others may have a series of smaller eruptions. Still others may shoot out solid fragments called *bombs*. In this investigation the students will examine the properties of some products of volcanic activity.

Materials

Safety goggles
Laboratory apron
Bottle of carbonated beverage
Bottle opener
Glass tubing, 10–cm
Bunsen burner
Ceramic square
Obsidian
Pumice

Procedure

1. Open a bottle of carbonated beverage. Have the students observe the liquid near the surface. They will see bubbles of carbon dioxide gas escaping from the liquid. Melted rock from a volcano may contain dissolved gases that escape from the liquid rock. The gases in the liquid rock are under pressure within the earth. As the pressure is released, the gases escape.

2. Hold the end of the 10–cm piece of glass tubing just below the tip of a Bunsen burner flame. This is the hottest part of the flame. Turn the tubing slowly so that it becomes heated uniformly on all sides. Have the students watch the end of the tubing. They should see the edges become rounded. When this happens, stop turning the tubing. Now the end of the rod will begin to soften and bend. Place the glass tubing on a piece of ceramic to cool. When the glass tubing is cool, examine the end that was molten. Compare it with a piece of obsidian. Obsidian is a glassy rock that has been produced by a volcano. Hold obsidian in front of a light source. Much obsidian is translucent and smoky. Have the students examine a piece that has been broken. Does the broken surface look like a piece of broken glass?

3. Place a piece of pumice in water. Pumice is a form of lava in which expanding gases have been trapped as the material hardened.

Questions to Ask the Students

1. Some lava contains dissolved gases that escape from the liquid when (*the pressure of overlying rocks is released*).
2. Obsidian is much like common glass. How is it formed? (*Obsidian forms as molten lava and cools rapidly, leaving no crystals.*)
3. What is an identifying characteristic of pumice? (*It has many pores and it floats.*)

Section 3: Earthquakes

Class Activity

To teach the concept of seismic gaps, use gelatin blocks. Materials for 8 lab groups: eight packages of unflavored gelatin, 2 liters hot water, two 20 cm × 20 cm baking pans, aluminum foil, 0.5 liter of loose sand, eight 10 cm × 10 cm sandpaper squares. To prepare, dissolve eight packages of unflavored gelatin in 2 liters of hot water. Add food coloring and pour into two baking pans lined with aluminum foil. When the gelatin has set, remove the gelatin and the foil from the pan. Cut the gelatin blocks and the foil into four squares. (Each 10-cm square can be used for a group of students.) Cut each square in half to make a "fault scarp." Place the gelatin square on a sandpaper sheet and sprinkle loose sand along the fault. Apply gentle but continuous pressure on one block of gelatin, forcing it to move along the fault line. Notice that the movement is not continuous. Some sections move; others resist movement initially and then spring to a new position.

Field Trip

Visit a seismic recording station in your area. Many of them are affiliated with universities.

Outside Speaker

Have a geologist visit the class to describe the operation of a seismograph and to explain the process of triangulation in locating earthquakes.

CHAPTER 7

CHAPTER OVERVIEW

This chapter presents the theory of plate tectonics to explain the formation of crustal features. The activity at plate boundaries is related to dramatic geologic events such as volcanoes and earthquakes. The formation and structure of volcanoes are explained. The causes of earthquakes are discussed, and the behavior and effects of the P, S, and surface waves that they generate are explained.

Section 1: Plate Tectonics This section uses the theory of plate tectonics to explain continental drift and how present-day continents were formed from the ancient continent, Pangaea. Geologic activities, such as volcanoes and earthquakes, that occur along boundaries are described, and the characteristic crustal features that can be found along each are presented.

Section 2: Volcanoes In this section the behavior of magma as it rises and cools to form intrusions or volcanoes is described. Shield, cinder-cone, and composite volcanoes are classified according to their composition, shape, and intensity of eruption. An explanation for the formation of midplate volcanoes over "hot spots" is suggested. The formation of undersea volcanoes is also discussed.

Section 3: Earthquakes In this section the forces that cause earthquakes are described. The degree of interaction between plate boundaries is related to the intensity of the earthquakes that result. This section also provides a detailed explanation of the properties and effects of the P and S waves that emanate from an earthquake's focus. A detailed description of the measurement of seismic activity and the cause of the range of earthquake magnitude as measured on the Richter scale are explained.

The Dynamic Earth

Erupting volcanoes are some of the most dramatic geologic events on Earth. They have always fascinated humans. Early civilizations explained these events by telling stories of angry gods and of animals with supernatural powers. You may find that the truth is even more fascinating! Geologists now know that volcanoes are produced as heated rocks are forced to the surface.

An active volcano erupting

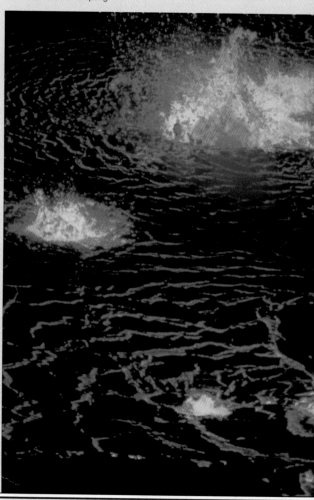

136

CHAPTER MOTIVATING ACTIVITY

Use the following Norse myth to introduce the students to the dramatic geologic events discussed in this chapter. Ask the students why ancient peoples made up myths about geologic events. Then tell them the story of Loki, a Norse god who was known as a clever prankster. The gods decided to punish Loki when they learned of his plot to kill the most powerful god, Baldur. Loki was chained in a deep cave, and an enormous snake was hung above him. Loki's sister protected him by catching the dripping poison in a cup. Sometimes, however, a drop of poison would splash Loki, causing him unbearable pain. At these times he would pull so violently on his chains that the earth would tremble.

1 Plate Tectonics

SECTION OBJECTIVES

After completing this section, you should be able to:

- **Explain** the theory of plate tectonics.
- **Identify** the three types of crustal boundaries and **describe** their associated landforms.
- **Predict** how the positions of the continents may change in the future.

NEW SCIENCE TERMS

theory of plate tectonics
earthquakes
volcanoes
divergent boundary
convergent boundary
transform fault

7.1 Crustal Plates in Motion

Imagine all the land areas of the world joined together in one supercontinent. This supercontinent would, of course, be surrounded by one superocean. Look at the shape of the continents on the world map in the Reference Section on page 592. Imagine cutting around the continents and joining the pieces together where there seems to be a natural fit. One fit, for example, would be the east coast of South America with the west coast of Africa.

Alfred Wegener hypothesized that all the continents were once joined together in one huge continent called *Pangaea* (pan GEE ah). Since the time of Pangaea, Wegener said, the continents have been drifting apart. Some scientists listened to Wegener's theory of continental drift, but many rejected his idea of a supercontinent. They could not believe that continents could drift through a solid crust.

Scientists now believe that about 550 million years ago two continental masses were drifting on the mantle. These early continents are called *Laurasia* and *Gondwana*. Evidence supports the belief that, about 300 million years ago, these two continents collided, forming Pangaea. Since that time the continents have been drifting apart, as Figure 7–1 shows.

Figure 7–1. This series of computer-generated diagrams shows the break-up of Pangaea about 200 million years ago, and the movement of the continents into their present positions.

137

Section **1**: Plate Tectonics

SECTION FOCUS

Section Overview   This section uses the plate tectonic theory to explain continental drift. Geologic activities characteristically found at convergent and divergent plate boundaries are described.

Section Objectives   For a list of section objectives, see pupil's edition page.

New Science Terms   For a list of new science terms in this section, see pupil's edition page. You may wish to introduce the terms after the concepts have been presented.

SECTION DEVELOPMENT

7.1 Crustal Plates in Motion

DISCUSSION   Have the students look at a map that shows present-day continents, and have them compare the continents with Pangaea as shown in Figure 7–1. Elicit from the students which present-day continents they think were once joined and how they can tell. (Outlines of present-day continents appear to match like pieces of a jigsaw puzzle.)

DISCOVER: Shifting Plates

Thinking Skill *(Inferring Relationships)*

You may wish to add food coloring to the liquid as you perform this activity. Add 2 or 3 drops as the water is being heated to make convection currents more visible to the students.

DISCUSSION Have the students study Figure 7–2. Some polystyrene foam cutouts might help them understand plate tectonics. Ask the students how far the North American plate moves in five years. (about 10 cm) In 20 years? (about 40 cm) Ask how far the Pacific plate has moved in five years. (65 cm) In 20 years? (260 cm) Have the distances marked off on the chalkboard so the students get a feel for the distance. ***Thinking Critically:*** Ask the students what produces the convection currents that cause plates to slide or float on the asthenosphere. (Molten rock in the mantle behaves as a semi-liquid. Magma rises, forming convection currents.)

THINKING SKILL *(Predicting)* Have the students look again at Figure 7–2 and at a present-day world map. Based on the directions of plate movement shown, ask them to predict where the continents will be located millions of years from now. (North and South America are moving toward Asia; the Pacific Ocean is getting smaller.)

DISCOVER

Shifting Plates

Using Figure 7–2, sketch the shape of several adjoining plates on a sheet of paper. Cut out the outlines, separate the plates, and transfer the patterns to thin sheets of polystyrene. Cut out the polystyrene plates, and float them in water in a glass baking dish. Gently warm the dish on a hot plate, and note the movement of the plates as convection cells begin to develop.

These conclusions led to the formulation of a new theory—that of plate tectonics. The **theory of plate tectonics** states that the lithosphere is divided into plates, each moving independently of the others.

Some of these plates consist of oceanic crust only, while other plates contain both continental and oceanic crust. These plates have an average thickness of about 70 km and extend to the top of the asthenosphere. These plates slide or float on the asthenosphere, driven by convection cells in the mantle.

The plates that form the lithosphere are moving at different speeds and in different directions. For example, the North American plate is moving toward the southwest at about 2 cm per year. The adjoining Pacific plate is moving to the northwest at about 13 cm per year. What might eventually happen if these plates continue to move as they do now? **1**

○ *What is the theory of plate tectonics?*
○ *What causes the individual plates to move?*

Figure 7–2. This map shows the location of the major plates of the world.

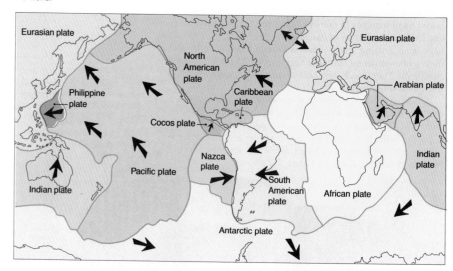

1 Part of California would continue moving to the northwest. In several million years Los Angeles and San Francisco would be next to each other and the Giants and the Dodgers would again be cross-town rivals.

138   Chapter 7   The Dynamic Earth

Answers–7.1

○ The theory of plate tectonics states that the crust is divided into several moving plates.

○ The plates probably move because of the currents produced by convection cells.

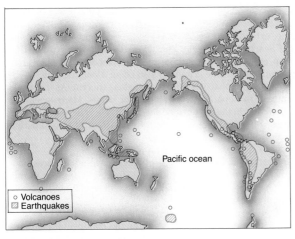

Pacific ocean

○ Volcanoes
▨ Earthquakes

7.2 Plate Boundaries

Exciting geologic events occur along the boundaries of plates. Although there are exceptions, most earthquakes and volcanoes occur near the edges of plates. **Earthquakes** are violent shakings of the earth's interior. **Volcanoes** are openings in the crust through which magma reaches the earth's surface.

So many earthquakes and volcanoes occur around the edge of the Pacific Ocean that this area is called the *Ring of Fire*. Earthquakes, volcanoes, mountains, and even ocean trenches occur near the boundaries between crustal plates. These earthquake- and volcano-prone areas are shown in Figure 7–3. You can see that these areas form a ring.

Divergent Boundaries Three different types of plate boundaries have been identified. Approximately in the middle of the Atlantic Ocean is an area where molten rock from the mantle rises to the earth's surface through cracks, or *rifts*, in the ocean floor. Here newly formed crust is adding width to an underwater mountain range called the *Mid-Atlantic Ridge*. An area of the earth where the crust is spreading is called a **divergent boundary.** The word *diverge* means "to move apart."

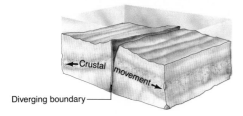

←Crustal movement→

Diverging boundary

A MATTER OF FACT

In 1883 there was a volcanic eruption on Krakatau in Indonesia. It was one of the world's worst disasters, killing nearly 36 000 people.

Figure 7–4. Along divergent plate boundaries (left) new material is added to the crust and the plates spread apart.

Section 1 Plate Tectonics **139**

7.2 Plate Boundaries

DISCUSSION As you progress through this section, remind the students that many of the geologic events that occur along plate boundaries are the result of hot magma rising from the asthenosphere. To better understand the formation of crustal features at each type of boundary, demonstrate to the students how the rising and cooling of magma force landmasses to change positions and shapes. You may wish to use construction paper to demonstrate the folding and faulting of the crustal plates. *Thinking Critically:* Ask the students to look at Figure 7–2. Ask them where they would expect to find frequent volcanic activity and earthquakes. (Students should name plates that share boundaries along which geologic activity would be expected.)

EXTENSION You may wish to introduce the term *isostasy*. Explain to the students that the continental crust is less dense than the oceanic crust and is therefore able to float higher than the oceanic crust. This floating is called isostasy. *Thinking Critically:* Ask the students why continents do not sink along subduction zones. (Due to isostasy, continental plates float higher than oceanic plates.)

DISCUSSION You may wish to show the students a map of Africa. Point out that many long lakes are situated along the Great Rift Valley between Ethiopia and Mozambique. A relief map will also show the mountainous regions bordering these bodies of water.

THINKING SKILL *(Drawing Conclusions)* Ask the students to compare the processes that form mountains along divergent and convergent boundaries. (Along divergent boundaries, magma rises through rifts; along convergent boundaries, folding and faulting occur as plates collide.)

Figure 7–5. The Arabian Sea is widening because of diverging plates. If the divergence continues, one day this sea may be known as the Arabian Ocean.

Divergent boundaries are also found in a few places on land, such as the Great Rift Valley of Africa. This huge valley, or rift zone, extends for over 4000 km, from the southern end of the Red Sea to Mozambique. The wide, flat valley is bordered by steep cliffs, over 600 m high in some places. Frequent earthquakes and volcanic eruptions also occur along the valley.

Africa is probably splitting apart along this rift. As the plates continue to separate, the valley floor drops. Many scientists believe that in the next few million years, the valley will become a sea, with the cliffs forming the coastlines of two separate continents.

Convergent Boundaries Where two plates press together a **convergent boundary** is formed. *Converge* means "to come together." Converging plates produce mountains, volcanoes, and ocean trenches.

If an oceanic plate collides with a continental plate, the denser oceanic plate is forced below the continental plate. An area where one plate is forced under another is called a *subduction zone*. As the oceanic plate is pushed under the continental plate and into the mantle, the rocks heat up and melt. Some of this molten rock rises again, forming volcanic mountains along the continental edge of the boundary. Volcanoes are discussed in Section 2.

During subduction the oceanic plate pushes downward and a trench forms between the two plates. Trenches occur in many places, although most are located along the edges of the Pacific Ocean. The greatest ocean depths are found in these trenches. The *Mariana Trench*, located off the Mariana Islands, is the deepest trench at about 11 000 m.

Figure 7–6. Along converging plate boundaries (right), part of the crust is subducted into the mantle. There it melts and forms pools of magma. This magma may again reach the surface as lava (left).

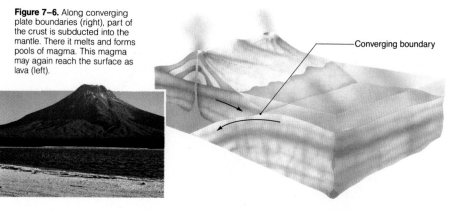

Converging boundary

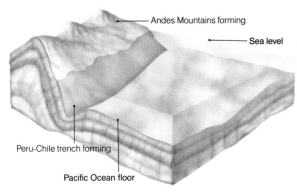

Andes Mountains forming

Sea level

Peru-Chile trench forming

Pacific Ocean floor

Figure 7–7. Where the Nazca Plate was forced under the South American Plate, a deep ocean trench (the Peru-Chile Trench) and a tall mountain range (the Andes Mountains) eventually formed.

On the western edge of the South American plate is another subduction zone and trench. Here, the Pacific plate is being forced below South America, forming a deep trench called the *Peru-Chile Trench*. The Andes Mountains are a result of this movement. Several trenches in the Pacific Ocean and in the Atlantic Ocean border chains of islands called *island arcs*. Look at the map on page 138. Where else might you find subduction zones?

When two continental plates converge, neither plate can be forced under the other. Instead, great mountain ranges are formed. The rocks fold and fracture, often rising to great heights. The Himalaya Mountains were formed when the Indian plate collided with the Eurasian plate. Mountains in this range continue to rise by about 5 cm each year. In the Western Hemisphere, the westward movement of North and South America formed the Cordilleras. This mountain system extends from Alaska to the southern tip of South America and includes the Rocky Mountains of the United States and Canada.

Figure 7–8. The Himalaya Mountains (left) were formed by the collision of continental plates. The Rocky Mountains (right) are fault-block mountains.

DISCUSSION You may wish to use the chalkboard to illustrate the subduction of oceanic plates beneath continental plates for the students. Remind the students that once the oceanic plate heats up as a result of its contact with the asthenosphere, rocks in the oceanic plate melt and rise.

THINKING SKILL *(Comparing Processes)* Ask the students to compare the causes of mountain formation along a subduction zone and along the boundaries of converging continental plates. (Along a subduction zone, melted rock from the oceanic plate rises to the surface; along the boundaries of colliding continental plates, compression causes rocks to fault and fold.)

EXTENSION The highest point on Earth is located in the Himalayan Mountains. Mt. Everest rises to 8848 m above sea level. The lowest point is in the Mariana Trench, 11 000 m below the ocean surface. There is a distance of approximately 20 km between these two points.

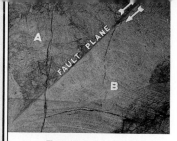

Transform Faults When two plates slide past each other in parallel but opposite directions, a type of lateral fault called a **transform fault** forms. The San Andreas fault in California is caused by the North American plate grinding past the Pacific plate. Although this type of boundary does not form mountains or trenches, there are many earthquakes associated with this movement.

Figure 7–9. Transform faults occur where adjoining plates slide past each other in parallel but opposite directions.

○ *What are the three types of plate boundaries?*
○ *Which type of boundary creates folded mountain ranges?*

Section Review

READING CRITICALLY

1. Compare and contrast the three types of plate boundaries.
2. How does the Ring of Fire provide evidence for the theory of plate tectonics?

THINKING CRITICALLY

3. Explain how the theory of plate tectonics is different from the theory of continental drift.
4. Describe the possible position of the continents fifty million years in the future. Support your answer with facts about present plate movement.

ACTIVITY: Examining Crustal Plates

How can you analyze information about crustal plates?

MATERIALS (per group of 3 or 4)

pencil, paper

PROCEDURE

1. Make a table similar to the one below, but allow room for 20 rows.
2. Carefully examine each plate shown on the map in Figure 7–2. On your data table, record the name of each plate, the name of any landmasses included on the plate, the type of boundary, and the direction of movement.

CONCLUSIONS/APPLICATIONS

1. How many different plates did you find?
2. How many plates contained landmasses?
3. Do you predict that any of the plates are going to eventually subduct under other plates? Explain your prediction.
4. Geologists have stated that it is relatively easy to identify a plate but often quite difficult to locate the plate boundaries exactly. Explain why this might be so.

TABLE 1: CRUSTAL PLATES				
Plate Name	Landmasses on Plate	Types of Plate Boundaries	Direction of Movement	Other Features

142 Chapter 7 The Dynamic Earth

SECTION REVIEW

Summary Have the students name crustal features associated with each of the three types of plate boundaries. (Divergent: rift valleys, cliffs; convergent: mountains, volcanoes, ocean trenches; transform faults: earthquakes.)

Reinforcement Have the students complete the Section Review. Allow the students to draw and label diagrams to illustrate their answers.

ACTIVITY:
Examining Crustal Plates

Skill (*Making Comparisons*)

Procedure

You may wish to provide the students with a relief map of the world to help them identify crustal features along plate boundaries. Direct the students to compare the map in Figure 7–2 with the relief map. Help them locate plates and the continents and oceans in which they occur.

Answers to Conclusions/ Applications

1. 7 major plates and 18 minor plates
2. Eight plates contain major landmasses.
3. The Pacific plate is subducting in many regions. It will become much smaller.
4. Plate boundaries are usually identified by trenches, volcanoes, or earthquakes. If these features are not found, the actual position of a boundary is inferred from other sources of data.

Answers—7.2

○ Divergent, convergent, and transform faults are three types of plate boundaries.

○ Convergent boundaries create folded mountain ranges.

Answers to Section Review

Reading Critically

1. Divergent boundaries occur where plates pull apart. Molten rock rises to form new crust. Convergent boundaries occur where plates collide. Oceanic plates subduct beneath continental plates. Subducted oceanic rock melts and rises to form volcanoes. Continental plates that converge form mountains by faulting and folding. Transform boundaries occur when plates slide past each other, often causing earthquakes.

2. Volcanoes and earthquakes along the Ring of Fire indicate plate boundaries.

continues

SECTION OBJECTIVES

After completing this section, you should be able to:

- **Explain** why volcanoes usually form near plate boundaries.
- **Classify** types of volcanoes.
- **Describe** the relationship between hot spots in the crust and volcanic island chains.

NEW SCIENCE TERMS

batholith
viscosity
shield volcanoes
cinder-cone volcano
composite volcano
caldera
hot spot
seamount

7.3 Lava Flows

One of the most dramatic effects of plate movement is the formation of volcanoes. Most volcanic activity occurs near the boundary of crustal plates. When plates collide, crustal material is subducted, or moves, into the asthenosphere. There an increase in pressure raises the temperature of the rocks. Eventually the rocks melt, forming magma. Where plates diverge, pressure is released, rocks melt, and magma rises through rifts, as in Iceland.

Often, rising magma does not reach the surface. When a body of magma cools beneath the surface, it forms an intrusion called a **batholith.** Smaller intrusive bodies are called *stocks.* The cores of many mountains are formed from batholiths. Sometimes thin fingers of magma extend from these intrusions. When magma squeezes between layers of rock, a *sill* is formed. When magma cuts across rock layers, it forms a *dike.*

If magma reaches the surface, it is called *lava.* Some magma rises slowly to the surface, creating lava flows that may stream great distances across the earth's surface. If the magma rises

> **A MATTER OF FACT**
>
> In 1815 the Tambora volcano in Indonesia killed 12 000 people when it erupted. The eruption sent over 80 km³ of ash into the atmosphere, blocking out so much of the sun that there was a general cooling of the entire earth, making 1816 the "year without a summer."

Figure 7–10. A lava flow, such as that shown, allows molten rock from within the earth to reach the surface.

Section 2   Volcanoes   **143**

Thinking Critically

3. In the theory of continental drift, continents were thought to float on a static crust. The theory of plate tectonics explains the movement of crustal plates above a soft asthenosphere in which convection currents cause motion.
4. North America and Asia will be closer. Western California will be near Alaska. A new ocean will form along Africa's Great Rift Valley.

DEMONSTRATION

For a demonstration of volcanic properties see page 135c preceding this chapter.

Section 2: Volcanoes

SECTION FOCUS

Section Overview This section focuses on the formation of volcanic eruptions along plate boundaries and above hot spots in the crust. A description of how intrusions form is also given. The three types of above-surface volcanoes—shield, cinder-cone, and composite—are described in terms of their composition and shape. Seamounts and guyots are also described.

Section Objectives For a list of section objectives, see pupil's edition page.

New Science Terms For a list of new science terms in this section, see pupil's edition page. You may wish to direct the students' attention to Figures 7–10 through 7–18 as the terms are presented. Remind the students that some of the terms in this section are derived from the characteristic shape associated with each type of volcano.

SECTION DEVELOPMENT

7.3 Lava Flows

DISCUSSION Students should keep in mind that volcanic activity occurs as a result of the building up of enormous pressure at plate boundaries. As plates diverge, magma is released and rises through rifts to form volcanoes.

EXTENSION Point out to the students that the Columbia Plateau in North America is an example of a lava plain. This plain formed as magma rose along cracks or fissures in the crust. Other such plains can be found on the continental landmasses.

DISCUSSION Remind the students that once magma rises to the surface it is called *lava.* Lava occurs in many forms. Small, rounded lava stones are called *lapilli.* Rocks with a crusty appearance, called *breadcrust bombs,* get their uneven surfaces as gases escape from cooling chunks of lava.

THINKING SKILL *(Applying Concepts)* You may wish to demonstrate the concept of viscosity by having the students describe observable differences as you pour water, oil, and molasses. Ask the students how the density of these materials affects their behavior. Ask the students to apply this concept to the relationship between the composition, density, and flow of lava. (Silica-rich magma is more dense and, therefore, flows more slowly. The more water present in lava, the less dense it is, and the faster its flow.)

7.4 The Structure of Volcanoes

DISCUSSION Draw the students' attention to Figures 7–12 through 7–14. Remind the students that each type of volcano has its characteristic shape due to the composition of the lava and the kind of eruption. *Thinking Critically:* Ask the students to name the volcano type they think has the densest lava. (Lava from silica-rich cinder-cone volcanoes is the densest.)

Figure 7–11. Pahoehoe (left) is smooth, ropy lava. Aa (center) is rough and blocky. Pillow lava (right) forms rounded lumps.

Figure 7–12. A shield volcano has gentle slopes and a wide base. Most of the volcanoes of the Pacific islands are shield volcanoes.

quickly to the surface, lava is thrown into the atmosphere in semisolid chunks. Most of this lava hardens as it falls to the ground, forming lava masses called *bombs* and *lapilli.* At times the force of a lava eruption is so violent that large chunks of old, hardened lava blocks, or cinders, are thrown into the air. Ash—particles less than 2 mm in diameter—is also produced by violent eruptions. Ash can form great clouds which, when they fall to Earth, cover everything.

Lava flow is affected by viscosity (vihs KAHS uh tee). **Viscosity** is a liquid's resistance to flow. For example, honey has high viscosity, so it flows very slowly. Water has low viscosity, so it flows easily. The viscosity of lava is determined by the amount of water and silica in the magma. Low-viscosity lava has little silica but a lot of water. High-viscosity lava is rich in silica but has little water. Low-viscosity lava moves rapidly, forming smooth ropy flows of basalt called *pahoehoe* (pah HOH ee hoh ee). High-viscosity lava moves more slowly and forms rough, blocky flows of andesite called *aa* (AH ah).

○ *What is viscosity?*
○ *What are the two types of lava flows?*

7.4 **The Structure of Volcanoes**

Many volcanoes begin with relatively peaceful lava flows and then erupt explosively. In other cases the first eruption is the most explosive. A volcano's shape depends on the type of lava and the force of the eruption. There are three kinds of volcanoes, which are classified by shape: *shield volcanoes, cinder-cone volcanoes,* and *composite volcanoes.*

Shield volcanoes are formed by quiet eruptions of basalt lava that has a low silica content. A shield volcano has a cone with gentle slopes and a base that covers a wide area. Shield volcanoes make up most of the Hawaiian Islands.

Silica-rich magma produces viscous lava. This magma traps gases inside the volcano until enough pressure builds up to push the magma out of the earth. If there is much water present, it

BACKGROUND INFORMATION

The shield volacanoes in Hawaii have been built upward from the ocean floor, which is about 7000 m below sea level. The sides of these volcanoes look ragged and twisted because of the numerous lava flows.

Answers–7.3

○ Viscosity is the resistance of a liquid to flow.

○ Pahoehoe and aa are the two types of lava flows.

quickly turns to steam. The pressure of the steam causes an explosion that may shoot gases, bombs, and ash several kilometers into the atmosphere. In contrast to shield volcanoes, this kind of eruption builds a steep-sided **cinder-cone volcano.** Cinder-cone volcanoes consist mainly of ash and *tuff.* Tuff is formed from compressed ash, cinders, and lapilli. Many of the volcanoes in Mexico and Central America are cinder cones.

If the lava flow changes, the nature of the eruption may change as well. A third type of volcano, formed from a series of alternating eruptions of different lavas, is a **composite volcano.** A composite volcano consists of alternating layers of ash, tuff, and andesite lava. The slope of a composite volcano is steeper than that of a shield volcano but not as steep as that of a cinder-cone volcano. Mount Fuji, in Japan, is a composite volcano.

Figure 7–13. A steep-sloped cinder-cone volcano (left) often erupts explosively (top). A great deal of ash is produced from a cinder-cone volcano (bottom).

A MATTER OF FACT

In 1902 Mount Pelee in Martinique erupted, leaving only two survivors in the town of St. Pierre. One of these was Auguste Ciparis, a criminal, who survived because he was protected by his thick-walled jail cell.

Figure 7–14. Japan's Mount Fuji, one of the world's most photographed mountains, is a composite volcano.

Section 2   Volcanoes   145

DISCUSSION   Point out to the students that often earthquake tremors are a warning signal that a volcano is about to erupt. When a composite volcano erupts, there may be a series of different kinds of eruptions that occur close together, with a quiet period between the eruptions.

EXTENSION   The smoke that rises from volcanoes consists mostly of steam and dust. Toxic gases, such as carbon dioxide, hydrochloric acid, and hydrogen sulfide, may also be released during eruptions.

EXTENSION   (Tie-in/Language Arts) Point out to the students that many of the science terms they have encountered in this section are names for the shapes of the features they describe. For instance, the term *caldera* is derived from the Latin word *caldaria,* which means kettle or pot.

EXTENSION Although volcanic activity is generally viewed as destructive, it also provides benefits. The ash of violent eruptions enriches soil. Some of the most fertile soil in the world can be found in areas that were once buried beneath volcanic ash.

DISCOVER:
Creating Composite Cones

Skill (*Modeling*)

You may wish to allow the students to use a spatula or dull blade to carve lava flows out of modeling clay. Then have them paint their models to show molten and solid lava color differences.

7.5 Hot Spots

DISCUSSION You may wish to initiate discussion about the formation of volcanic islands by performing the following demonstration. Construct clay models of the three types of volcanoes. Place the models on a piece of cardboard. Use an unlit alcohol burner to symbolize a hot spot beneath the crust. ***Thinking Critically:*** As you move the cardboard on which the models have been placed in a straight line above the burner, ask the students why they think volcanoes form over hot spots and then become dormant. (As the plate slides over the stationary hot spot, volcanoes form and then become dormant.)

DISCUSSION Point out to the students that the processes that cause volcanic eruptions at plate boundaries and over hot spots are similar; hot magma rises through rifts in the crust in both cases. However, in hot spots, different portions of the crust are melted as plates pass over the hot spots.

Figure 7–15. Old calderas often fill with water, forming beautiful lakes such as Crater Lake in Oregon.

DISCOVER

Creating Composite Cones

You can build a model of a composite-cone volcano by alternating thin layers of modeling clay with layers of small plastic foam beads pushed into the clay. After completing your model, try creating a caldera by removing the top quarter of your cone and molding the hollow interior into the shape of a bowl.

An explosive eruption of an old composite volcano may blow away entire sections of the cone wall. Large sections of the wall then collapse into the hollow center, forming a *caldera*. A **caldera** is a wide basin, usually more than a kilometer across, in the center of a volcanic cone. Crater Lake in Oregon is in an old caldera, while Mount St. Helens, a volcano in the state of Washington, formed a new caldera when it erupted in 1980.

○ ***What causes some volcanic eruptions to be explosive?***
○ ***Describe the three shapes of volcanoes.***

7.5 Hot Spots

Although most volcanoes are formed near plate boundaries, some seem to develop in the middle of plates. These mid-plate volcanoes often occur in long, straight island chains, such as the Hawaiian Islands. At one end of the island chain, the volcanoes are old and extinct. At the other end of the chain, the volcanoes are young and active, such as those on the island of Hawaii.

Figure 7–16. Active volcanoes on the island of Hawaii indicate that this island is located over a hot spot in the Pacific plate.

146

Answers–7.4

○ High water content and silica-rich magma produce explosive eruptions.

○ Shield volcanoes have gentle slopes and wide bases. Cinder-cone volcanoes have steep slopes made of cinder and ash. Composite volcanoes have moderately steep slopes made of alternating layers of lava and layers of ash and cinder.

The exact cause of these mid-plate volcanoes is unknown; however, geologists believe these volcanoes develop from columns of magma rising from hot spots in the mantle. A **hot spot** is an area of concentrated heat deep within the earth. Hot spots melt the rocks of the mantle with which they come in contact. The molten rock creates a column of magma that rises toward the surface. The magma then erupts through the lithosphere. As plates move across the rising column of magma, volcanoes are produced. Geologists have identified nearly 120 hot spots on the earth.

A MATTER OF FACT

In 1986 deadly gases escaped without warning from a lake in Cameroon, Africa. Over 1700 people died in this disaster.

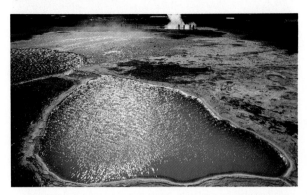

Figure 7–17. This steam field in Iceland (left) and Old Faithful geyser in Yellowstone National Park, Wyoming (below), are both located over hot spots.

Sometimes hot magma heats underground water, turning the water into steam. This steam is released through cracks in the earth's surface, producing steam vents called *geysers* (GY zuhrz). Yellowstone National Park, with its famous "Old Faithful" geyser, sits over a hot spot. Geologists predict that areas to the northeast of Yellowstone may soon begin showing signs of geyser activity, as well. Why would new geysers form to the northeast of Yellowstone National Park?

○ *What are hot spots?*
○ *What are geysers?*

7.6 Undersea Volcanoes

Volcanoes form not only on land but also under the oceans. In shallow water, volcanoes erupt violently, forming clouds of ash and steam. An underwater volcano is called a **seamount.** Seamounts look much like cinder-cone volcanoes on land.

There are more than 10 000 seamounts on the Pacific Ocean floor. However, many seamounts never reach the surface of the ocean. The Hawaiian Islands, for example, are a small part of an

EXTENSION Throughout history people have used hot springs for medicinal baths. The enormous pressure with which steam rises from cracks in the earth over hot spots has also been harnessed to provide heat and power to villages and factories in places such as Iceland.

7.6 Undersea Volcanoes

DISCUSSION Point out to the students that oceanic relief is similar in appearance to the continents in that it has mountain chains, valleys, and other geologic features. Many of these features are submerged, and only those features that have risen above the surface, such as the Hawaiian Islands, are visible to us without the aid of soundings and other technologies used to map the ocean floor.

Answers–7.5

○ Hot spots are areas of concentrated heat deep within the asthenosphere. The heat rises and melts the crust above it.

○ Geysers are heated underground water that turns to steam and rises through cracks to the earth's surface.

THINKING SKILL *(Identifying Similarities)* Ask the students how the processes that shape mountains on the continental crust are similar to the action of waves that flattens the tops of guyots. (Erosion in the form of rainwater and wind wears down the tops of continental mountains; weathering and wave action flatten guyots.)

SECTION REVIEW

Summary Have the students draw the three types of volcanoes and label them. (shield, cinder-cone, and composite) Ask the students to divide a piece of paper into two columns. Have them label one column *Similarities* and label the other column *Differences*. Ask them to list characteristics that are common to all three volcanoes in one column and characteristics that are distinctive to each in the other column. (Similarities: Shape depends on composition of lava; occur mostly along plate boundaries or hot spots. Differences: Composition, and therefore speed of lava flow, differs for each; shape; number and violence of eruptions.)

Reinforcement Have the students complete the Section Review. Allow them to work in pairs to quiz each other on new science terms.

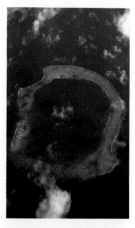

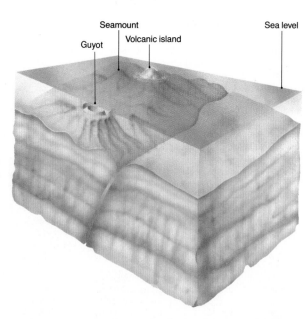

Figure 7–18. Seamounts are volcanoes that form under water. Guyots form as waves flatten the tops of seamounts near the water's surface (right). If the guyot is in a tropical ocean, a coral atoll may result (above).

underwater volcanic-mountain chain. Just eight of these volcanoes have become islands, while dozens remain submerged.

Another type of undersea volcano is called a *guyot* (gee OH). A guyot is an undersea volcano with a flat top. Geologists hypothesize that guyots were once seamounts that rose above the ocean's surface. There the tops of the volcanoes were flattened by weathering and wave action.

○ **What are seamounts?**

Section Review

READING CRITICALLY

1. Explain how a chain of volcanoes can form in the middle of a plate.
2. Why is most volcanic activity located near plate boundaries?

THINKING CRITICALLY

3. If guyots were flattened by weathering and the action of waves, why are most of them found far below the surface?
4. Where would you expect any new Hawaiian Island volcanoes to form? Explain.

Answer–7.6

○ Seamounts are undersea volcanoes.

Answers to Section Review

Reading Critically

1. A chain of volcanoes can form as a crustal plate slides over a hot spot.
2. As plates pull apart or push together, molten material rises to the surface from the asthenosphere.

Thinking Critically

3. The erosional effects of waves extend well below the surface of the ocean. These seamounts may have been at the ocean surface at one time.
4. The next island should form at the southeastern end. The chain gets progressively younger toward the southeast.

SKILL ACTIVITY: Making a Graph

BACKGROUND

Scientists often present information in the form of a table. Sometimes, however, this format may not be the best way to present the information. Often a better interpretation can be obtained by placing the data on a graph. The procedures for placing any type of data on a graph are the same.

PROCEDURE

1. Title your graph. Write the title on the top of your paper.
2. Draw two axes on a sheet of graph paper. The horizontal, or *X*, axis should be near the bottom of the paper. The vertical, or *Y*, axis should be near the left edge of the paper. Label both axes with the type of data you are going to plot.
3. Choose the scale for each axis. The scale should be such that the graph covers as much information as possible and is spaced evenly over the entire sheet of graph paper.
4. Plot the data on your graph.

APPLICATION

Below is a table giving the ages of several of the Hawaiian Islands. Use the information in the table to make a graph. On one axis, plot the longitude of each island, starting with the island of Hawaii and moving to the west. On the other axis, plot the age of each island from east to west, starting with Hawaii.

USING WHAT YOU HAVE LEARNED

Use your graph to answer the following questions:
1. Which is the youngest island?
2. Which is the oldest island?
3. What is the relationship between the nearness of the islands to Hawaii and their relative age?
4. Using the graph and a map, predict the longitude of the next volcanic island to form in this chain.
5. The age of one of the islands does not seem to fit the pattern formed by the others. Explain.

TABLE 1: THE HAWAIIAN ISLANDS AND SEAMOUNTS

Island (or reef)	Approximate Age (in millions of years)	Longitude
Hawaii	0	155°30′ W
Kanum	39.0	170° E
Kauai	4.1	159°30′ W
Maui	0.6	156°15′ W
Midway Island	18.0	177°30′ W
Molokai	1.8	157° W
Necker Reef	10.1	164°30′ W
Nihoa	no data	162° W
Oahu	3.1	158° W
Pearl Reef	20.1	176° W
Yuryaku	42.3	168°30′ E

Section 2 Volcanoes **149**

SKILL ACTIVITY: Making a Graph

Objectives

- Make a graph.
- Interpret data in a table.
- Interpret a graph.
- Correlate data from a graph and a map.
- Make predictions.

Discussion You may wish to review with the students how to locate lines of longitude on a map. Ask the students what lines of longitude are. (Lines measuring degrees of distance east and west from the prime meridian.) In preparation for the application of this activity, help the students find the longitude of the area in which they live.

Answer to Application

Point out to the students that Table 1 does not include data on the age of Nihoa. *Thinking Critically:* Ask the students if they can predict the age of Nihoa based on the data for other volcanic islands in the Hawaiian chain. (The students may infer that Nihoa may be between 3.1 and 10.1 million years in age. The reason may be that Nihoa (162° W) lies between the longitude of Oahu (158° W) and Necker Reef (164°30′ W). Please note that there are no data on Nihoa. Therefore, the students' ideas are inferences that need to be researched further for the approximate age of Nihoa.)

NOTE: Although the actual trend of the Hawaiian Island chain is NW to SE, latitude has been omitted from this activity to simplify the procedure. As an enrichment activity, you may wish to have the students determine the latitudinal positions and plot a graph showing latitude and age.

Answers to Using What You Have Learned

1. Hawaii
2. Yuryaku
3. The closer the island, the younger it is.
4. 152°–153°
5. Midway; students' answers may vary. Students may suggest that there was a pre-existing fracture in the crust, causing the course of magma flow to shift or that the plate shifted in direction.

149

SECTION FOCUS

Section Overview This section describes the events that cause earthquakes at plate boundaries. The properties of P, S, and surface waves and how their interactions are related to the intensity of an earthquake are discussed. A description of the measurement of seismic activity and earthquake magnitude, as measured on the Richter scale, is provided as well.

Section Objectives For a list of section objectives, see pupil's edition page.

New Science Terms For a list of new science terms in this section, see pupil's edition page.

SECTION DEVELOPMENT

7.7 The Origin of Earthquakes

DISCUSSION Point out to the students that some of the same forces that form mountains cause earthquakes including, the interaction of plates along plate boundaries.

SKILL *(Interpreting Maps)* Have the students locate the San Andreas fault on the map of California in the Science Sites booklet. ***Thinking Critically:*** Ask the students which areas are most likely to experience a great deal of damage should an earthquake occur. Then ask them what kind of damage they would expect to occur. (Densely populated areas, such as Los Angeles and San Francisco, are most susceptible. Buildings, major roadways, and industrial areas would experience the most damage.)

NEW SCIENCE TERMS

focus
epicenter
P waves
S waves
surface waves
seismograph
Richter scale

SECTION OBJECTIVES

After completing this section, you should be able to:
- **Compare** and **contrast** the motions of P, S, and surface waves.
- **Explain** how the epicenter of an earthquake is found.
- **List** some factors that may be helpful in predicting earthquakes.

7.7 The Origin of Earthquakes

You may recall from Section 1 that strong convection cells within the mantle are constantly moving the earth's plates. At the plate boundaries, pressure builds as the friction between the plates stops their movement. Finally, the plates slip, and the pressure is released in a series of waves. These waves are similar to the ripples that move away from a pebble dropped into a pond of water. However, instead of moving through water, the waves move through the solid earth. This movement shakes the earth, causing an earthquake.

Many of the earthquakes in the United States occur where the Pacific plate scrapes past the North American plate, along the San Andreas fault. Some sections along the fault move smoothly past each other, causing frequent, but minor, earthquakes. In other sections the plates stick to each other as they try to move. These stuck-together sections, called *seismic gaps*, are areas where minor earthquakes do not occur. However, seismic gaps have a high probability of producing major earthquakes. Both San Francisco and Los Angeles are located on seismic gaps of the San Andreas fault.

A MATTER OF FACT

Although earthquakes are less likely to occur in central North America, there have been a few very strong ones. Earthquakes at New Madrid, Missouri, which occurred in 1811 and 1812, were felt as far away as Canada.

Figure 7–19. The San Francisco earthquake of 1906 occurred along a seismic gap of the San Andreas fault.

150

Just as ripples in a pond move out from a falling pebble, earthquake waves move away from their source. The source of an earthquake is called the **focus**. The waves travel in all directions from the focus, which is located below the earth's surface. The area on the earth's surface directly above the focus is called the **epicenter** of the earthquake.

○ *What causes earthquakes?*
○ *What is the epicenter of an earthquake?*

7.8 Earthquake Waves

There are two types of waves that originate from the earthquake focus. The fastest-moving wave is a longitudinal wave. A longitudinal wave travels by compressing earth material in front of it and stretching material behind it. You can create longitudinal waves by moving a coil of wire back and forth.

Longitudinal waves are the fastest waves produced during an earthquake and, therefore, are the first waves that reach an earthquake recording station. For this reason, longitudinal waves are called *primary waves*, or **P waves**. P waves can travel through any type of material, even the dense center of the earth. A diagram of how P waves travel is shown in Figure 7–21.

The second type of earthquake waves is transverse waves, which move more slowly than P waves. These slower waves are called *secondary waves*, or **S waves**. They are similar to the movement of a rope shaken from side to side.

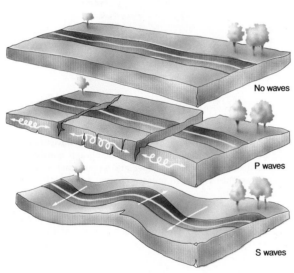

No waves

P waves

S waves

Figure 7–21. P waves and S waves, which originate at an earthquake's focus, cause only minor damage.

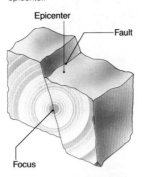

Figure 7–20. As this diagram shows, the focus of an earthquake is often far below the epicenter.

Epicenter

Fault

Focus

A MATTER OF FACT

Along the west coast of Mexico, the Cocos plate dives beneath the North American plate in a subduction zone. On September 19, 1985, the built-up pressure caused the Cocos plate to move 20 m below the surface. This caused a tremendous earthquake, releasing energy a thousand times greater than that of the atomic bomb dropped on Hiroshima, Japan, during World War II.

EXTENSION You may wish to have the students investigate new architectural designs that have been used in modern office buildings in places such as Los Angeles and San Francisco. These buildings have been designed to withstand the effects of moderate earthquakes.

EXTENSION Explain to the students that the safest place to be during an earthquake is in a doorway of a building. Falling bricks, electrical wires, telephone poles, and flying glass can cause injury to people caught outdoors. The greatest damage of earthquakes often occurs as a result of fires. In the great San Francisco earthquake of 1906, much more damage was caused by fires than by the direct effects of the earthquake.

7.8 Earthquake Waves

DISCUSSION To demonstrate the compressional movement of P waves you may wish to use a Slinky®. To demonstrate the movement of an S wave you may wish to use rope. ***Thinking Critically:*** After the students have observed the motions of P and S waves, ask them if the intensity of the wave is the same at a distance from the focus as it is at the focus. (No. Waves lose some energy as they travel through a solid material.)

EXTENSION Point out to the students that waves travel in all directions from the focus of an earthquake. The loud booming sounds heard after an earthquake are caused by the motion of air molecules as waves pass through the atmosphere.

Answers–7.7

○ Earthquakes occur as friction that builds up between plate boundaries is suddenly released.

○ The epicenter is the area on the earth's surface directly above the origin, or focus, of an earthquake.

DISCUSSION Point out to the students that longitudinal waves, which are also called compressional waves, are similar to sound waves. These waves cause rocks to vibrate in the direction of the motion. S waves, also known as shear waves, cause rocks to vibrate at right angles to the waves' direction of motion. Surface waves can be compared to the surface waves on an ocean because they only affect the surface layer of rocks.

THINKING SKILL *(Inferring Conclusions)* Ask the students how the type of rock in an area can be inferred from data about the characteristics of seismic waves traveling through the rock. (Waves change their characteristics as they pass through materials of different compositions and densities.)

EXTENSION In 1950 a major earthquake occurred in Asia, causing extensive damage in Tibet, India, and Burma. Five thousand deaths were reported. The most remarkable consequence of this earthquake is that Mt. Everest, the tallest mountain in the world, is said to have risen over 60 m.

EXTENSION You may wish to have the students look at the maps of Alaska and Tennessee in the Science Sites booklet. These maps show the location of the Queen Charlotte Fault near Anchorage and the New Madrid Fault near Memphis.

Figure 7–22. Surface waves, which originate at an earthquake's epicenter like ripples on a pond, cause severe damage because of their rolling action.

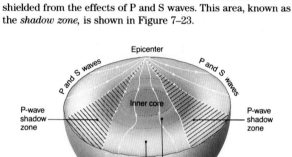

Surface waves

The interaction of P and S waves produces **surface waves** that cause the surface of the earth to shake and roll like a ship in a storm. These surface waves originate from the epicenter, not the focus, and cause more destruction than either P waves or S waves. P waves and S waves shake buildings back and forth or up and down, but the rolling action of surface waves causes many buildings to collapse.

The wave motion of earthquakes is very complex. As earthquake waves travel through the different layers of the earth, they tend to change characteristics because of the density differences of the layers. For example, P waves bend sharply as they enter the core. S waves, however, cannot pass through the liquid outer core. This creates an area on the earth's surface that is shielded from the effects of P and S waves. This area, known as the *shadow zone*, is shown in Figure 7–23.

Figure 7–23. The shadow zone is an area of the earth protected from the effects of P and S waves. However, this area is not protected from the effects of surface waves.

Epicenter

P and S waves   P and S waves

P-wave shadow zone   Inner core   P-wave shadow zone

Mantle   Outer core

The instrument used to record earthquake waves is called a **seismograph** (SYZ muh graf). A seismograph consists of a rotating drum wrapped with paper, and a pen attached to a suspended weight. The pen presses gently against the paper-wrapped drum. The structure that supports the drum is fastened to solid rock, so it will move up and down only if the rock moves. Any vibration of the rock will produce a zigzag line on the paper. The recording from a seismograph is called a *seismogram*.

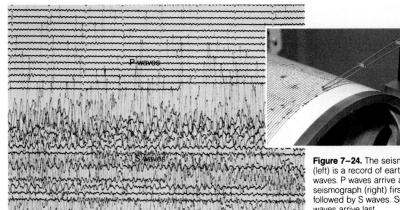

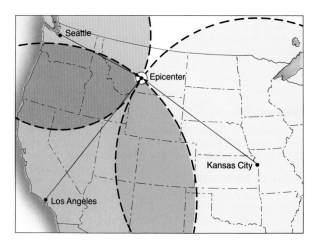

Figure 7–24. The seismogram (left) is a record of earthquake waves. P waves arrive at the seismograph (right) first, followed by S waves. Surface waves arrive last.

P waves travel fastest and are the first waves recorded on the seismogram. The P waves are recorded as a series of small zigzag lines. The S waves arrive after the P waves and appear as larger, more ragged lines. The surface waves arrive last and make the largest lines. The difference in travel time between the waves indicates the distance between the recording station and the epicenter. However, this is not enough information to locate the epicenter. The distance from at least three seismograph stations must be measured before the exact location of an earthquake epicenter can be determined.

Figure 7–25. Triangulation, diagrammed here, is the process used to determine the location of an earthquake from the recordings of three seismic reporting stations.

Section 3   Earthquakes   **153**

DISCUSSION   To explain the concept of lag time, tell the students to think of the two types of waves as though they are runners in a race. Runner P can run faster than runner S. They start off from the focus at the same time. Runner P reaches the 1 km marker 20 seconds before runner S. In other words, runner S lags behind by 20 seconds. Runner P reaches the 2 km marker 40 seconds before runner S. The farther the two runners travel, the greater the lag time will be.

Skill *(Constructing Models)*

To do this activity, use a phonograph turntable that has different speeds and a bass level. Use a variety of shapes and paper to get the best results.

DISCUSSION   Have the students study Table 7–1 to compare the magnitude and number of deaths listed. Point out that the magnitude of an earthquake is not a reliable indicator of the amount of damage it may cause. *Thinking Critically:* Ask the students what factors contribute to the amount of damage caused by an earthquake. (the material the crust is composed of in a particular area; the density of population)

SECTION REVIEW

Summary   Have the students create a table listing the features of P, S, and surface waves. Their tables should include descriptions of the movement and speed of each type of wave and a diagram illustrating relative wave lengths.

Reinforcement   Have the students complete the Section Review. Allow any students who need assistance to look up the answers or to work together in groups.

DISCOVER

Making a Seismogram

Roll a stiff sheet of paper into a cylinder, and tape the edges together. Place a record and then the paper cylinder on a phonograph turntable. If the speakers are separate from the turntable, carefully place the turntable on top of a speaker. Select the slowest speed available for the turntable, and turn the bass all the way up. Suspend a felt-tip pen from a string, and gently touch it to the paper cylinder. Create a base line for your graph by running the turntable without music for a few seconds, and then lower the needle onto the record to create the seismic recording.

The size, or *magnitude*, of an earthquake is reported using a special scale called the **Richter** (RIHK tur) **scale.** This scale measures the amount of energy released by an earthquake. Each increase of one unit on the Richter scale represents a tenfold increase in earthquake strength. For example, an earthquake magnitude of eight is ten times stronger than an earthquake magnitude of seven, and 100 times greater than a magnitude of six. Table 7–1 shows the magnitudes of some strong earthquakes rated on the Richter scale.

TABLE 7–1: MAJOR RECORDED EARTHQUAKES

Year	Location	Magnitude	Deaths
1556	China	unknown	830 000[1]
1906	San Francisco	8.3	700
1906	Colombia/Equador	8.9	unknown[2]
1923	Tokyo	8.3	100 000
1976	China	8.0	750 000
1976	Guatemala	8.1	22 000
1985	Mexico City	8.1	5 000

[1]Most deadly earthquake
[2]Strongest earthquake

○ **What is a seismograph?**
○ **What is an earthquake shadow zone?**

Section Review

READING CRITICALLY

1. Compare and contrast P waves, S waves, and surface waves.
2. Why are areas of seismic gaps likely to have stronger, but less frequent, earthquakes than other sections of a fault?

THINKING CRITICALLY

3. Explain why three different seismic readings are needed to find the epicenter of an earthquake.
4. Why is it easier to locate a potential earthquake area than to predict when an earthquake will occur?

154   Chapter 7   The Dynamic Earth

Answers–7.8

○ A seismograph is an instrument used to record earthquake waves.
○ The shadow zone is the area shielded from the P and S waves due to the bending of seismic waves as they cross Earth's layers.

Answers to Section Review

Reading Critically

1. P waves are the first waves, traveling faster than the other waves. These are smaller disturbances than the S waves. The stronger S waves come after the P waves. Surface waves are the last waves to arrive and cause the most damage.
2. Greater pressure is built up in seismic gaps. Thus, much energy is released by fault movement.

continues

TECHNOLOGY: Seismic Records

In the eighteenth century, an Italian geologist developed a pendulum-based instrument that recorded the strength of an earthquake. This instrument works because a heavy, suspended weight resists movement and remains in a fixed position while the earth moves below it.

Scientist at the National Earthquake Center

Later seismographs used a suspended pen, which lightly rests on a rotating, paper-covered drum. The drum is anchored in rock, so that any movement of the earth will cause the pen to make zigzag lines on the drum. This seismograph records vertical movement. In a second type of seismograph, the pen is attached to an arm that swings freely. This seismograph records horizontal movements. Both types are needed to accurately measure earthquake magnitude.

In the 1930s Victor Hugo Benioff designed a seismic instrument that was similar to a telephone. A continuous electrical current was passed through the instrument. When the earth vibrated in an earthquake, this instrument showed tiny variations in current, which could be recorded. Modern instru-

ments are based on this same principle.

Technological advances in the past 20 years have greatly improved the scientist's ability to interpret seismograms. The orientation of the fault plane, the type of movement, and even the direction of movement along the fault can be determined. Computers have also improved the speed and the accuracy of recording seismographic data.

Seismic recording stations are located around the world. All these stations report their data to the National Earthquake Information Service in Golden, Colorado. Each month approximately

A computer-enhanced seismogram

60 000 seismic records are fed into the computers at the center. This information is analyzed by scientists to obtain a better understanding of earthquakes and to aid in developing a method for predicting earthquakes in the future.

A remote seismic station

TECHNOLOGY: Seismic Records

Discussion After the students have read this article, you may wish to demonstrate the workings of a pendulum. Place a board on blocks. Then place a pendulum on the board. Simulate earthquake tremors by gently shaking the board. Have the students discuss the effect that varying degrees of motion has on the movement of the pendulum. *Thinking Critically:* Have the students compare the ability of a seismograph and a pendulum to accurately measure the magnitude of earthquakes. (The seismograph provides more accurate measurements; it is a more sensitive instrument.) If your school is located near an earthquake recording station, arrange for the class to visit the station to see the equipment there and how it is used. You may wish to have the class create a class chart listing earthquakes. Pairs of students may research particular earthquakes and contribute data on date, location, and magnitude. They may also provide any interesting data they find, such as the occurrence of fires, damage to property, or changes in terrain, that were caused by the earthquake. Compile all class data, and list it on a class chart either in order of magnitude or chronology.

Thinking Critically

3. If you think of a seismic reading as a circular distance from a recording station, one reading could locate the earthquake anywhere on the circle; two stations with overlapping circles have two points of intersection; three station circles only overlap in a single point. Thus, three stations are required to locate the epicenter.

4. Once an earthquake occurs, the epicenter can be located from three seismic recording locations. Even though scien-

tists can locate areas where seismic gaps exist, the ability to predict when the next earthquake will occur is beyond present technology.

Thinking Skill (Interpreting Data)

Discussion Use a world map to help the students locate lines of latitude and longitude. Have them match the coordinates of latitude and longitude to plates they learned about in the beginning of the chapter. Elicit observation of patterns of earthquakes in relation to plate boundaries.

Answers to Analyses and Conclusions

1. Sketches should show locations of earthquakes near Manila in the Philippines.
2. The graph makes a relatively straight line with a noticeable bend just below 300 km. The strongest earthquakes are nearer to the surface.

Answers to Application

1. Probably a convergent boundary. One plate is subducting below the other.
2. They are probably volcanic mountains because of the subduction and melting magma.

INVESTIGATION 7: Plotting Earthquake Data

PURPOSE

To use earthquake data to classify a plate boundary

MATERIALS (per student)

Textbook
Graph paper
Pencil

TABLE 1: EARTHQUAKE DATA

Latitude	Longitude	Depth of Quake	Magnitude of Quake
15° N	130° E	10 km	8.0
15° N	130° E	10 km	7.8
15° N	129° E	10 km	8.2
15° N	129° E	50 km	7.7
15° N	128° E	95 km	7.0
15° N	128° E	165 km	7.1
15° N	128° E	180 km	7.6
15° N	128° E	155 km	7.3
15° N	127° E	200 km	7.1
15° N	127° E	305 km	7.2
15° N	127° E	225 km	6.9
15° N	126° E	310 km	7.0
15° N	126° E	275 km	7.5
15° N	125° E	375 km	6.1
15° N	124° E	350 km	6.5
15° N	124° E	445 km	6.4
15° N	123° E	570 km	6.7
15° N	121° E	550 km	6.1
15° N	121° E	510 km	6.9
15° N	120° E	690 km	6.2

PROCEDURE

1. Examine the data presented in the Earthquake Data table. Note that the data was recorded along a single latitude line but at different longitude positions. Refer to the world map on page 592 of the Reference Section to locate the general area represented by this data.
2. Graph the information in the table. Since latitude is constant, graph longitude versus the depth of the earthquake. Longitude should be on the X, or horizontal, axis. Place the reading of 120° E at the intersection of the axes. Space your graph to fill the entire sheet of graph paper.
3. The Y, or vertical, axis should represent the depth of the earthquake. Place 700 km at the top and 0 km at the bottom.
4. Plot the earthquake data on the graph, using the following symbols to indicate the magnitude of the earthquake.

   + = 6.0 – 6.9
   × = 7.0 – 7.7
   O = 7.8 – 8.5

ANALYSES AND CONCLUSIONS

1. Sketch a map of the region represented by the earthquake data. On your sketch show where the earthquakes occurred.
2. Describe the pattern on your graph made by the different earthquake depths.

APPLICATION

1. What type of plate boundary do you think the earthquake data indicates? Explain what you think is happening at this boundary by discussing the plates involved and their direction of movement.
2. There are several mountains on the landmasses in this area. Based on your analysis of the type of plate boundary, what type of mountains would you expect to find? Explain how these mountains are related to the plate boundary.

SUMMARY

- Alfred Wegener hypothesized that all the continents were once joined as the supercontinent Pangaea. (7.1)

- The theory of plate tectonics explains the movement of the earth's crust. (7.1)

- There are three types of plate boundaries: divergent, convergent, and transform fault. (7.2)

- Many geologic processes, such as subduction, folding, and faulting, occur along plate boundaries. (7.2)

- Magma sometimes intrudes into other rock layers where it forms batholiths. Fingerlike projections between or across rock layers form sills or dikes. (7.3)

- Lava flows have many characteristics that depend on the viscosity of the lava. (7.3)

- Volcanoes are classified as shield volcanoes, cinder-cone volcanoes, and composite volcanoes. (7.4)

- Hot spots form volcanic island chains in the middle of plates. Geysers form near hot magma. (7.5)

- Seamounts and guyots are undersea volcanoes. (7.6)

- The sudden release of energy along faults causes earthquakes. (7.7)

- Earthquake waves are classified as P waves, S waves, and surface waves. (7.8)

- Earthquake waves travel at different speeds. (7.8)

- Earthquake magnitude is measured on a scale called the Richter scale. (7.8)

Write all answers on a separate sheet of paper.

SCIENCE TERMS

Correctly use each of the following terms in a sentence.

batholith **(143)**
caldera **(146)**
cinder-cone volcano **(145)**
composite volcano **(145)**
convergent boundary **(140)**
divergent boundary **(139)**
earthquakes **(139)**
epicenter **(151)**
focus **(151)**
hot spot **(147)**
P waves **(151)**
Richter scale **(154)**
seamount **(147)**
seismograph **(152)**
shield volcanoes **(144)**
S waves **(151)**
surface waves **(152)**
theory of plate tectonics **(138)**
transform fault **(142)**
viscosity **(144)**
volcanoes **(139)**

SCIENCE QUIZ

Modified True-False

Mark each statement *true* or *false.* If a statement is false, change the underlined term to make the statement true.

1. Aa is a type of <u>lava</u> with high viscosity.

2. Most earthquake activity occurs along <u>the center</u> of a plate.

3. <u>Hot spots</u> melt small areas of the mantle.

4. The distance between a recording station and an earthquake source is calculated by determining the difference in <u>arrival time</u> between different types of waves.

5. The <u>Ring of Fire</u> provides evidence in support of the theory of plate tectonics.

6. <u>Batholiths</u> are igneous intrusions.

continues

Chapter 7 Review   **157**

CHAPTER REVIEW

SUMMARY

Students may review the major concepts in the chapter by reading the summary statements. The statements are cross-referenced to the chapter to facilitate reinforcement of any concepts of which the students feel unsure. Encourage the students to work in groups to quiz one another.

SCIENCE TERMS

The sentence in which the science term is used should reflect an understanding of the definition of the term. You may wish to have the students play a game in which one group provides clues to another group that must determine which term is being described.

SCIENCE QUIZ

Answers to Modified True-False

1. true
2. false, the edge
3. true
4. true
5. true
6. true

continues

Answers to Multiple Choice

7. c
8. c
9. b
10. d
11. d
12. d

Answers to Completion

13. undersea
14. P wave
15. transform fault
16. seismic gaps
17. shield
18. Richter

Answers to Short Answer

19. The areas along a known fault that have not recently experienced an earthquake are considered to be seismic gaps. The longer the period of dormancy, the greater the pressure buildup. Eventually, pressure will be released by earthquake activity.
20. The bedrock will transmit accurate seismic waves from an earthquake. Loose rock would settle and vibrate as independent pieces of rock, causing the seismic waves to be distorted.
21. In a convergent boundary, two plates are pushing together. In a divergent boundary, two plates are pulling apart.
22. Triangulation involves the use of three seismic recording stations to locate an earthquake. Where the three circles overlap will be the earthquake's epicenter.

Answers to Writing Critically

23. The Hawaiian chain clearly shows the oldest cones in the northwest and the youngest in the southeast. If the Pacific plate continues to move in a northwesterly direction, the next volcano should form in the southeast.
24. A high silica content in lava makes the lava very viscous. This sticky lava makes very explosive volcanic eruptions. Water also makes a volcanic eruption explosive because the heated water turns into steam.

25. P waves are like the movement of a crawling earthworm. S waves are like a rope shaken from side to side. They move buildings back and forth. Surface waves are rolling waves like ocean waves. They cause buildings to move in a circle like the water in an ocean wave.

SCIENCE QUIZ continued

Multiple Choice

Write the letter of the choice that best answers the question or completes the statement.

7. The lava that forms a volcano comes from a _____ in the earth's crust.
   a) volcanic cone   b) seamount
   c) rift   d) caldera

8. Large, streamlined masses of lava are called
   a) tuff.   b) pahoehoe.
   c) bombs.   d) aa.

9. Which of the following is not a type of intrusion?
   a) sill   b) rift
   c) batholith   d) dike

10. Cinder cones form from lava that is
    a) low in water and low in silica content.
    b) high in water and low in silica content.
    c) low in water and high in silica content.
    d) high in water and high in silica content.

11. A convergent boundary may be identified by
    a) an ocean trench.
    b) volcanic activity.
    c) a mountain range.
    d) All choices are correct.

12. The magnitude of an earthquake depends on
    a) the speed of the P waves.
    b) the amount of energy that is released by the earth.
    c) the material at the epicenter.
    d) All choices are correct.

Completion

Complete each statement by supplying the correct term.

13. Guyots are _____ volcanoes.

14. The earthquake wave that is a longitudinal wave is a _____.

15. The boundary between the North American plate and the Pacific plate is a _____.

16. Areas where faults stick together are _____.

158   Chapter 7 Review

17. Lava that has a low silica content is likely to form a _____ volcano.

18. Earthquake magnitude is measured on the _____ scale.

Short Answer

19. How can scientists use seismic gaps to predict the location of future earthquakes?

20. Why do scientists anchor a seismograph into solid rock?

21. What is the difference between a convergent and a divergent plate boundary?

22. Explain the process of triangulation and relate it to determining the location of earthquake epicenters.

Writing Critically

23. Explain why scientists predict that any new Hawaiian islands that form will do so to the southeast of the present chain.

24. How do viscosity, silica content, and steam affect the eruption of volcanoes?

25. Describe the motions of P waves, S waves, and surface waves. Explain the effects that each one would have on buildings constructed of wood, stone, and steel. Why is the motion of surface waves more destructive than either P waves or S waves?

EXTENSION

1. Study the effects of the 1985 earthquake in Mexico City. Write a report describing how the ancient lake bed increased the effects of the wave motion.

2. Research the way in which geothermal energy is used as a source of electricity in certain areas of the world. Relate geothermal energy to magma. Present your findings to the class.

3. Find out where the closest seismic recording station is to your home. If possible arrange to visit the station.

ANSWERS TO EXTENSION

1. The report should contain information about the nature of the soil under Mexico City that contributed to the severity of the earthquake.
2. The hotspots where geothermal sources are located are near pools of magma, near the earth's surface.
3. Answers will vary.

APPLICATION/CRITICAL THINKING

1. Explain how a lava flow, such as the one that formed the Columbia River Plateau, might be related to the volcanic activity and geysers of a hot spot below Yellowstone National Park.

2. What clues would you look for in trying to determine whether a lava flow was composed of aa or pahoehoe?

3. After reviewing Table 7–1 on page 154, explain why the strongest earthquakes do not necessarily cause the most deaths. Also explain why an early warning system for strong earthquakes would be vital for heavily populated areas.

FOR FURTHER READING

Aylesworth, T. G., and V. L. Aylesworth. *The Mount St. Helens Disaster: What We've Learned.* Danbury, CT: Franklin Watts, 1983. This book offers a complete account of the 1980 eruption of Mount St. Helens.

McConnell, A. *The World Beneath Us.* New York: Facts on File, 1985. The author provides very interesting descriptions of the geologic time table, fossils, plate tectonics, earthquake prediction, and volcano activity.

Walker, B., ed. *Earthquake.* New York: Time-Life Books, 1982. This book is an outstanding reference on all aspects of earthquakes.

Challenge Your Thinking

These buildings were all in good condition before the 1985 earthquake in Mexico City. One building (right) was a solid old structure built with bricks and cement, while another (far left) was constructed of steel and glass. The newer building was built to withstand the effects of an earthquake, while the older buildings obviously were not. What kinds of things might have been built into the newer building to make it earthquake resistant?

Chapter 7 Review   **159**

ANSWERS TO APPLICATION/ CRITICAL THINKING

1. Lava could have flowed through rifts in the earth's crust, covering this flat land.
2. Lava that is smooth and ropey would be pahoehoe, while rough, blocky lava would be aa.
3. The strongest earthquakes may have occurred in a sparsely populated area, while some milder ones may have occurred in densely populated cities.

ANSWER TO CHALLENGE YOUR THINKING

The newer buildings were designed with flexible supports and perhaps even shock absorbers to lessen the effects of a damaging earthquake.

Discussion This *Science Connection* is about the diamond, the hardest substance on Earth. A diamond is the pure, crystal-line form of the element carbon. Graphite, a much softer mineral, is also pure carbon. Ask the students to discuss how the hardness of two minerals made from the same element can be so different. The students should realize that the difference results from the atomic structures of the two minerals. The carbon atoms in graphite are packed and bound loosely; in diamonds, they are packed and bound very tightly.

Discussion Diamonds are used widely in industry because they have the highest thermal conductivity of any known substance. Although diamonds are subjected to high temperatures when they are used to cut other objects, the heat is conducted through the diamond and the cutting edge stays cool. However, if diamonds are subjected to a sufficiently high temperature, they will burn. Ask the students to discuss why this is true. (The students should realize that diamonds are made of pure carbon, much like anthracite, or hard coal, which is almost pure carbon and is used as a fuel.)

Extension After completing the Unit and reading and discussing the *Science Connection,* have the students turn back to the Unit Opener on page 49 and answer the questions. When the class discussion has ended, you may wish to have the students write a report describing how diamonds are formed and how they are used in industry.

UNIT THE SCIENCE CONNECTION:
Diamonds

From laboratory experiments, scientists have concluded that diamonds are formed under very special circumstances. The temperature must be at least 1400°C and the pressure about 40 000 times the surface pressure on the earth in order to cause atoms of carbon to crystallize into diamonds. Diamonds appear to form inside structures called *diamond pipes,* which are slender, carrot-shaped holes filled with igneous rock.

The diamond-bearing rocks probably reached the surface through these pipes when ancient concentrations of carbon dioxide gas exploded. Apparently, no new diamond pipes have formed in the past 15 million years.

When the igneous rocks of the diamond pipes moved toward the surface, the magma mixed with bits of rock from the surrounding area. As it moved upward, the water in the ground also interacted with the magma. The resulting rock, called *kimberlite,* is a jumbled mixture of minerals, including diamonds. Kimberlite is found only in the oldest and most stable continental regions. Although kimberlites have been found in Africa, Brazil, India, Australia, North America, and Siberia, those in Africa provide 95 percent of the world's total diamond supply.

You may think of diamonds only as expensive gemstones. Diamonds are also highly valued in industry for their special properties. Diamonds are the hardest known substance on Earth. Many gems are cut using a paper-thin disk made of hardened bronze coated with diamond dust.

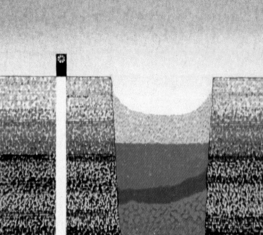

Diamond pipe formation

Sorting diamonds

Diamonds also have the highest thermal conductivity of any known substance. This means that they do not get hot when used as a cutting instrument. The heat that develops from friction is rapidly conducted through the diamond so that the cutting edge stays cool.

Surgeons use diamond-tipped scalpels for delicate operations such as eye surgery. Diamonds are also used in high-quality record-player needles and as heat conductors in miniature transmitters. The Pioneer space probe of Venus used a diamond wafer as a tiny window. The diamond wafer was used instead of glass or plastic because diamonds can withstand the extremes of heat and cold in outer space.

Grinding industrial diamonds

Diamonds are also used in drilling operations. Crystals of diamonds are used as an abrasive called *bort*. Bort is mounted on the end of a cylinder called a *bit*. When the rotating bit is forced against a rock, it cuts through the rock. The mining industry makes extensive use of the bort drill.

The greatest use of industrial diamonds is in the production of diamond grinding wheels. These wheels are especially hard because they have diamond dust embedded in them. Diamond wheels are able to shape and sharpen the new ultrahard tools that are being produced by modern industry.

Jewelry quality diamonds

161

Unit 3: METEOROLOGY

UNIT OVERVIEW

In this unit the structure and composition of the earth's atmosphere and the origin of planetary and local winds is described. The processes of evaporation and condensation, and the development of weather systems along the fronts between air masses are explained. Violent storms are also discussed.

Chapter 8: The Atmosphere, page 164

This chapter describes the main components of the atmosphere. The effect of the sun's radiation on the atmosphere is also explained. Details of the structure of the atmosphere are given. The motions of the atmosphere are discussed.

Chapter 9: Water in the Atmosphere, page 190

This chapter presents the various meteorological effects of water in the earth's atmosphere. Included are explanations of the terms *evaporation, humidity, condensation,* and *dew point.* The chapter concludes with descriptions of cloud types and the different types of precipitation formed in the atmosphere.

Chapter 10: Weather, page 206

This chapter introduces the students to the characteristics of fronts and the air masses that produce them. Also included are types of fronts and the weather they produce, including cyclones and anticyclones. Basic types of climate are discussed, and the causes of thunderstorms, tornadoes, and hurricanes are also discussed.

The Science Connection: TOTO, page 226

The Science Connection describes the work of the tornado chasers of the National Severe Storms Laboratory. TOTO, or the Total Tornado Observatory, is placed in the path of possible severe weather to record data about the development of tornadoes. Also described is the use of Doppler radar to study storms.

ADVANCE PREPARATION

Chapter 8

For the Discover on page 167, you will need a candle, a small green plant in a pot, and a large jar, with lid, into which both the plant and candle will fit. For more information, see page 167.

For the Discover on page 173, several students will need to bring in solar-powered calculators. You will need light bulbs of several different colors. For more information, see page 173.

For the Activity on page 177, the students will need to bring in empty aluminum soft-drink cans. For more information, see page 177.

For the Discover on page 183, you will need two small electric fans (nonoscillating). For more information, see page 183.

For Investigation 8 on page 186, the students will need to bring in shoeboxes with lids. For more information, see page 186.

Chapter 9

For the Discover on page 197, you will need to make observations of the clouds each day. For more information, see page 197.

For Investigation 9 on page 202, the students will need to bring in small soup cans. For more information, see page 202.

Chapter 10

For the Activity on page 218, you will need to make weather observations for five days. For more information, see page 218.

For the Discover on page 220, the students will need to bring in newspapers or magazine articles about severe storms. For more information, see page 220.

For Investigation 10 on page 222, you will need several small, rectangular fish tanks. For more information, see page 222.

BULLETIN BOARD SUGGESTIONS

Chapter 8

Have the students prepare a poster showing the layers of the atmosphere. Place the poster in the center of the bulletin board. Around the poster, the students could display magazine and newspaper articles that relate to the atmosphere. A red arrow should be drawn from specific articles to the layer of the atmosphere to which they correspond. For example, an article discussing the depletion of ozone in the atmosphere should point to the stratosphere on the poster.

Chapter 9

Have the students make models of clouds by gluing wads of cotton, shaped like various cloud types, onto blue paper on the bulletin board. The different "clouds" should be placed at the appropriate relative altitudes.

Chapter 10

Create tables and graphs to tabulate and plot the chronology of weather conditions for the duration of the unit. The bulletin board should contain information on local temperature, barometric pressure, relative humidity, and wind conditions. Rotate the responsibilities of groups of students so that each group has an opportunity to tabulate and graph each of the different quantities. Each temperature, barometric pressure, and relative humidity table should have a corresponding graph.

ISSUES IN EARTH SCIENCE

Have the students research the seemingly seasonal depletion of the ozone layer over Antarctica. This hole develops every fall and has been observed to be getting larger each year. Have the students report on the possible causes of ozone depletion and its effects on humans and other organisms.

In a few years it may be possible for technologically advanced countries to control the weather and cause rain and snow whenever or wherever they want. Have the students discuss the implications of this on world climate and for countries not as technologically advanced.

SUGGESTED PROJECTS

A group of students could keep and record daily high and low temperatures, using a minimum-maximum thermometer.

Students could determine the relative humidity each day, using a wet-dry bulb thermometer or sling psychrometer and a relative humidity table.

Students could construct and maintain a weather station for the school, including a minimum-maximum thermometer, a remote recording thermometer, a recording barometer, a recording hygrometer, and an anemometer.

TEACHER RESOURCES

Readings

Chapter 8

"The Biosphere." *Scientific American* (September 1970). This issue includes classic articles on the water, oxygen, carbon, and nitrogen cycles of our atmosphere.

Ehrlich, P., and C. Sagan. *The Cold and The Dark: The World after Nuclear War*. New York: W.W. Norton, 1984. The authors express the view that a limited nuclear war may drastically upset the radiation balance of the atmosphere.

Schubert, G., and C. Covey. "The Atmosphere of Venus." *Scientific American* 245 (July 1981) p. 66. The article

reviews the exploration by spacecraft of Earth's sister planet. Venus's atmosphere is set in stark contrast with that of Earth.

Chapter 9

Bowden, C. *Killing the Hidden Waters*. Austin: University of Texas Press, 1985. This book tells of the problems of using and abusing water resources that cannot be easily replaced by precipitation.

Chapter 10

Allen, O. *Atmosphere*. Chicago: Time-Life Books, 1983. This book will answer most student questions about the atmosphere, especially as it concerns weather conditions.

Audiovisual/Software

Chapter 8

A computer program, *Air Pollution*, that simulates atmospheric pollution in an urban environment is available from Nasco.

A film, *The Atmosphere in Motion*, dealing with the forces that keep the atmosphere in motion is available from Britannica Films and Video.

Chapter 9

A computer program, *The Weather Analyst*, is available from Geoscience Resources for Apple II, IBM-pc, TRS 80, and Atari computers.

Chapter 10

A computer series, *Water and Weather*, is available from Nasco for Apple II and Commodore 64 computers.

Films on weather and climate, including *What Makes Weather?*, *What Makes Rain?*, *The Atmosphere in Motion*, and *What Makes Clouds?*, are available from Encyclopedia Britannica Educational Corporation.

Other Materials

Chapter 8

Charts showing the layers of the atmosphere are available from most science supply houses.

Chapter 9

A set of illustrations of cloud types may be obtained from Geoscience Resources.

A water cycle simulator is also available from Geoscience Resources.

Chapter 10

Write your local office of the National Weather Service, which is a branch of the Federal Department of Commerce. Request pamphlets and booklets on the history and current policies of the National Weather Service.

Chapter 8: THE ATMOSPHERE

PLANNING THE CHAPTER

Chapter Sections	Page	Chapter Features	Page	Program Resources	Page
Section 1: Composition of the Atmosphere 8.1 The Primitive Atmosphere **(H)** 8.2 The Present Atmosphere **(A)**	165 165 167	A Matter of Fact **Discover:** Making Oxygen **(H)** Section Review **Skill Activity:** Reading for Meaning **(A)**	166 167 168 169		
Section 2: Solar Radiation 8.3 Electromagnetic Waves **(H)** 8.4 Radiation Balance **(A)**	170 170 171	A Matter of Fact **Discover:** Using Solar Power **(A)** A Matter of Fact Section Review **Investigation 8:** Solar Radiation **(A)**	170 173 174 174 186	Critical Thinking **(H)** Cross-Discipline: Science and Social Studies, *Collecting and Analyzing Survey Information* **(A)** Investigation 8.2: *Absorption by Soil and Water and Radiation of Energy* **(A)** Student Record Book: Textbook Investigations **(A)**	TRB 15 TRB 15 TRB 31 LM 33 TRB 15
Section 3: Structure of the Atmosphere 8.5 Atmospheric Pressure **(A)** 8.6 Layers of the Atmosphere **(A)**	175 175 177	**Careers:** Automotive Technician, Environmental Protection Agent, Atmospheric Chemist **Activity:** Demonstrating Atmospheric Pressure **(A)** A Matter of Fact **Discover:** Making Layers of Air **(A)** Section Review	176 177 177 179 180	Reading for Content: *Making Diagrams* **(A)** Investigation 8.1: *Investigating Air Pressure* **(A)**	TRB 15 LM 187 TRB 29 LM 31
Section 4: Motions of the Atmosphere 8.7 The Coriolis Effect **(A)** 8.8 Planetary Winds **(A)** 8.9 Local Winds **(A)**	181 181 183 184	A Matter of Fact **Discover:** Creating Wind Belts **(A)** A Matter of Fact Section Review	182 183 185 185	Concept Extension: *Weather Watch* **(A)**	TRB 17 LM 239
Chapter 8 Review	187			Vocabulary **(A)** Tests **(A)** Computer Test Bank 💻	TRB 8 LM 137 TRB 32

(LM) Laboratory Manual/Study Guide, **(TRB)** Teacher's ResourceBank™
B = Basic **A** = Average **H** = Honors
The coding Basic, Average, and Honors indicates sections or subsections that might be appropriate for different levels of learners. For additional suggestions regarding choice of topic and depth of coverage, see the Pacing Chart on pages T16–T20.

Section	Concepts	Objectives	Science Terms
Section 1: Composition of the Atmosphere	■ Carbon dioxide, nitrogen, and sulfur dioxide formed most of the primitive atmosphere of the earth. The removal of carbon dioxide from the primitive atmosphere left nitrogen as the most abundant atmospheric gas. Nitrogen, oxygen, and argon form 99 percent of the modern atmosphere. **(8.1)** ■ The gases of the earth's atmosphere that are heavier than helium do not escape into space because the velocity of their atoms and molecules is less than the escape velocity from the earth. **(8.2)** ■ Ozone in the stratosphere absorbs dangerous ultraviolet radiation from the sun. **(8.2)** ■ Fast protons from the sun heat up the upper part of the atmosphere. **(8.2)**	■ **Explain** why the earth has an atmosphere. ■ **Compare** the earth's early atmosphere with the present atmosphere. ■ **Describe** the earth's atmosphere and **explain** why it does not disappear into space.	atmosphere solar wind photosynthesis sunspots ozone
Section 2: Solar Radiation	■ Light consists of particles called photons, which travel from the sun to the earth as electromagnetic waves. The sun emits light of all colors. The human eye sees this combination of colors as white. **(8.3)** ■ The addition of carbon dioxide to the atmosphere may be enough to produce a strong greenhouse effect and melt the ice of Greenland and Antarctica. **(8.4)**	■ **Explain** the way in which Earth receives energy from the sun. ■ **Describe** the relationship of solar radiation to radiation balance. ■ **Discuss** how the addition of carbon dioxide to the atmosphere affects the radiation balance of the earth.	photons electromagnetic waves radiation balance greenhouse effect
Section 3: Structure of the Atmosphere	■ Atmospheric pressure is the weight of the atmosphere. **(8.5)** ■ The atmosphere is divided into layers, much like a multi-layered cake. From the earth outward, the layers of the atmosphere are troposphere, stratosphere, mesosphere, and thermosphere. **(8.6)**	■ **Describe** the cause of air pressure. ■ **List** the layers of the atmosphere and **discuss** their characteristics.	atmospheric pressure barometer Van Allen belts
Section 4: Motions of the Atmosphere	■ The Coriolis effect becomes evident when the Coriolis acceleration is inadequate. The Coriolis effect makes moving bodies veer to the right in the Northern Hemisphere and to the left in the Southern Hemisphere. **(8.7)** ■ The planetary winds form broad belts that surround the earth at various latitudes. **(8.8)** ■ Land and sea breezes are examples of local winds. **(8.9)**	■ **Discuss** how the Coriolis effect determines the circulation of winds. ■ **List** and **locate** the planetary wind belts. ■ **Describe** the cause of land and sea breezes.	Coriolis effect jet stream

Title	Page	Materials
Discover: Making Oxygen	167	*(per student)* small candle, glass jar, matches, small green plant
Skill Activity: Reading for Meaning	169	*(per student)* paper, pencil
Discover: Using Solar Power	173	*(per student)* solar-powered calculator, lamp with bulbs of different colors
Activity: Demonstrating Atmospheric Pressure	177	*(per group of 3 or 4)* safety goggles, laboratory aprons, gloves, water, empty soft-drink can, bowl, Bunsen burner or alcohol lamp, tripod, wire gauze
Discover: Making Layers of Air	179	*(per student)* small dish, ice, candle, matches, glass jar
Discover: Creating Wind Belts	183	*(per student)* fans (2), candle or match
Investigation 8: Solar Radiation	186	*(per group of 3 or 4)* scissors, shoe box with lid, black construction paper, white construction paper, tape, Celsius thermometers (2), lamp, stopwatch

TEACHING SUGGESTIONS

Section 1: Composition of the Atmosphere

Demonstration: Oxygen Content of the Air

Purpose

To demonstrate the approximate concentration of oxygen in the atmosphere

Background

The oxygen in the test tube will react slowly with the iron in the steel wool to form rust. The water level in the tube will rise to replace the lost gas. Ideally, the water level should rise one-fifth the length of the test tube, indicating a loss of 20 percent of the atmosphere in a day or two (possibly four to six days depending upon the brand of steel wool).

Materials

Steel wool
Test tube
Beaker, 100 mL
Ring stand with clamp

Procedure

1. Moisten the steel wool and push it down to the bottom of the test tube. It should remain loosely packed.
2. Fill the beaker nearly to the rim with water and place it at the foot of the stand under the clamp.
3. Invert the test tube into the water, well below the surface, and secure with the clamp.
4. Leave the setup undisturbed overnight.

Questions to Ask the Students

1. Does the changing color of the steel wool indicate a chemical reaction? (*yes*) If so, what chemical reaction might have taken place? (*oxidation of the iron*) What compounds may have been formed? (*iron oxide*)
2. Why did the water level in the test tube rise? (*The oxygen was removed by the chemical reaction.*)
3. How would you estimate the concentration of the "lost gas" in the atmosphere? (*The increase in the water level should be the same as the amount of gas lost.*)

Section 2: Solar Radiation

Class Activity

The students can contact the Environmental Protection Agency by phone and interview an agent.

Section 3: **Structure of the Atmosphere**

Outside Speaker

Contact the Environmental Protection Agency. Find out if they have any agents on assignment in the area. Invite one to come to class to discuss the work of the agency.

Section 4: **Motions of the Atmosphere**

Field Trip

Visit any local factory that is actively involved with controlling its potentially dangerous emissions.

METEOROLOGY

A tornado can toss a train as if it were a toy or drive a board through a concrete wall. Most people flee from this awesome power, but some scientists actually chase tornadoes. They measure a storm's power with a special package of instruments named TOTO, after Dorothy's dog in the movie *The Wizard of Oz.* With TOTO, scientists are learning more about what causes tornadoes and how they can be predicted.

- What kinds of weather conditions create tornadoes and other violent storms?
- How does weather radar show storms?
- Why does weather occur only in the lower layer of the atmosphere?
- How do the TOTO weather chasers know where to find tornadoes?

By reading the chapters in this unit, you will learn the answers to these questions. You will also develop an understanding of concepts that will allow you to answer many of your own questions about the study of weather and climate.

A rapidly moving tornado

163

INTRODUCING THE UNIT

Have the students look at the Unit Opener photograph on these pages. Point out that this is a rapidly moving tornado. Read the text and use the questions to guide a class discussion. Do not expect the students to be able to answer the questions now. The questions are motivational and the students should be aware that they will be able to answer them after they have finished the Unit.

Discussion Although they are among the smallest of all storms, tornadoes are the most violent storms, and can cause a great deal of damage in a very short period of time. Have the students recall the movie, *The Wizard of Oz,* and discuss the weather conditions that occurred and the damage the students witnessed as the tornado moved across the farm. Ask the students to share with the class any information they have about tornadoes and other forms of violent weather.

Extension You can extend the discussion of meteorology by bringing in other Unit themes. What is the atmosphere and how does it move? What is the role of water in the atmosphere? How do clouds and rain form? How do meteorologists predict the weather? You may wish to record student responses and review them at the end of the Unit.

CHAPTER 8

CHAPTER OVERVIEW

This chapter describes the main components of the atmosphere. The effect of the sun's radiation on the atmosphere is also explained. Details of the structure of the atmosphere are given. The motions of the atmosphere are discussed.

Section 1: Composition of the Atmosphere Carbon dioxide, nitrogen, and sulfur dioxide are identified as the main components of Earth's primitive atmosphere. The processes that transformed the primitive atmosphere into the present atmosphere are described. The composition of the present atmosphere is discussed.

Section 2: Solar Radiation Electromagnetic waves are described. The balance of incoming solar energy to the earth's atmosphere and outgoing heat energy from the earth's atmosphere is then discussed.

Section 3: Structure of the Atmosphere The characteristics of the four layers of the atmosphere are presented. Atmospheric pressure is described. The variations of pressure are explained.

Section 4: Motions of the Atmosphere The Coriolis effect is described. The planetary winds are defined as the broad belts that surround the earth at various latitudes. Each type of planetary wind is identified by its specific characteristics. The causes of land and sea breezes are described.

CHAPTER

The Atmosphere

If you live in the northern part of the United States, you may have seen lights like these in the sky. These lights are caused by charged particles from the sun interacting with the atmosphere. In the Northern Hemisphere, these lights are known as the *aurora borealis* and in the Southern Hemisphere as the *aurora australis*.

The aurora borealis, or northern lights

164

CHAPTER MOTIVATING ACTIVITY

Show a film or slide sequences of volcanic eruptions. Films on the effects of the Mt. St. Helens' eruption in Washington would be good to use. Ask the students if our planet has always had volcanoes. Describe how the earth might have appeared 4 billion years ago as it cooled from its original molten state and formed a solid, but heavily fractured, crust. Ask the students to explain why our atmosphere and climate would be different if there were more or fewer active volcanoes. The students should realize that the primitive atmosphere of Earth was in part a product of volcanic activity and that it was very different from the present atmosphere. Discuss how volcanic ash in the atmosphere might affect the absorption of solar radiation. Ask the students what other occurrences might have caused the content of the atmosphere to change.

1 Composition of the Atmosphere

SECTION OBJECTIVES

After completing this section, you should be able to:
- **Explain** why the earth has an atmosphere.
- **Compare** the earth's early atmosphere with the present atmosphere.
- **Describe** the earth's atmosphere and **explain** why it does not disappear into space.

NEW SCIENCE TERMS

atmosphere
solar wind
photosynthesis
sunspots
ozone

8.1 The Primitive Atmosphere

Air is just about the only thing that is free. Most people take for granted an unlimited supply of air. Why does Earth have air? Where did this air come from?

The **atmosphere** is the layer of gases that surrounds the earth. Earth's original atmosphere probably formed along with the planet. Scientists theorize that the sun and all the planets formed from a cloud of gas and dust about 4.5 billion years ago. The cloud collapsed into a central body, which formed the sun and ten rings. The rings eventually became the nine planets. One ring did not form a planet.

It is likely that high-energy radiation from the young sun, called the **solar wind,** scattered into space most of the gases that had originally collected around the planets closest to the sun, including Earth. Some of the more distant planets, such as Jupiter and Saturn, have retained their first atmospheres.

After the solar wind scattered Earth's original atmosphere, a new atmosphere formed from gases such as hydrogen, methane, and water vapor that were trapped inside the earth when it formed. Most of these gases escaped during volcanic eruptions. In fact, escaping gases are still adding to the atmosphere today.

Figure 8–1. Many of the gases of the primitive earth's atmosphere probably came from volcanoes.

165

DEMONSTRATION

For a demonstration of the oxygen content of the air, see page 163e preceding this chapter.

Section 1: Composition of the Atmosphere

SECTION FOCUS

Section Overview This section concentrates on the composition of the earth's primitive atmosphere and its transformation into Earth's present atmosphere. The relationship of the solar wind to the earth's atmosphere is discussed. The importance of the water vapor and ozone that are in the atmosphere is also described.

Section Objectives For a list of section objectives, see pupil's edition page.

New Science Terms For a list of new science terms in this section, see pupil's edition page. You may wish to familiarize the students with these terms by reading them aloud to the class.

Section Development

8.1 The Primitive Atmosphere

DISCUSSION Remind the students that the four gas giants—Jupiter, Saturn, Uranus, and Neptune—retain their original atmospheres. ***Thinking Critically:*** Ask the students to explain why they think these planets retain their gases. (These planets are much larger than the inner planets, so they have more gravity. The outer planets are better able to hold their atmospheres under the influence of the solar wind. Also, because the outer planets are farther from the sun, the solar wind is substantially weaker by the time it reaches those planets.)

THINKING SKILL *(Interpreting Photographs)* Direct the students' attention to Figure 8–1. Ask them to list some materials they think may be released by volcanic eruptions. (Accept all reasonable answers, including steam, carbon dioxide, and ash.)

DISCUSSION Some scientists have calculated that if the earth were 10 million kilometers closer to the sun than it is (an average of 140 million kilometers), its temperature would increase to the point that water vapor would not condense into liquid form. The oceans would never have formed, and carbon dioxide would not have been removed from the atmosphere. Under these conditions it is doubtful that life as we know it today would have evolved. *Thinking Critically:* Ask the students to explain what might have happened if the earth had been 10 million kilometers farther away from the sun. (Students will probably hypothesize that water would have been solid rather than liquid and that the oceans would not have formed.)

DISCUSSION Cyanobacteria, primitive blue-green algae, were the first living organisms to develop the process of photosynthesis. This enabled them to use sunlight to make their own food. *Thinking Critically:* Ask the students why cyanobacteria represent a very important stage in the development of life on Earth. (Cyanobacteria were partly responsible not only for the development of the present atmosphere from the primitive atmosphere, but perhaps also for the development of the higher life forms on Earth, all of whom require oxygen in order to live. Oxygen is a by-product of photosynthesis.)

EXTENSION Stanley Miller of the University of Chicago created a replica of what is believed to be the primitive atmosphere of Earth. Through this model atmosphere, Miller passed an electric current which he likened to lightning. Scientists believe that lightning was probably present in the primitive atmosphere. Many new compounds were created by the chemical reactions that occurred in Miller's experiment, including amino acids. Lightning in the primitive atmosphere may have been responsible for producing these building blocks of life.

A MATTER OF FACT

A little more than a century ago, most people, including scientists, believed that the earth's atmosphere was part of a substance that filled all space beyond the moon and planets of our solar system. The substance was called *ether.* In 1887, German-American physicist A. A. Michelson proved that this ether did not exist.

One abundant gas of this primitive atmosphere was water vapor. This water vapor produced rains that lasted for millions of years. These rains filled in the low places on the earth and formed the primitive ocean. Escaping hydrogen was lost into space. Other gases, such as carbon dioxide and sulfur dioxide, were created by chemical reaction and remained in the new atmosphere. This primitive atmosphere was very different from Earth's present atmosphere, as shown in Table 8-1.

TABLE 8–1: COMPARISON OF THE PRIMITIVE ATMOSPHERE AND THE PRESENT ATMOSPHERE

Gas	Percent of Molecules	
	Primitive Atmosphere	Present Atmosphere
Carbon dioxide (CO_2)	92.2	0.03
Nitrogen (N_2)	5.1	78.1
Sulfur dioxide (SO_2)	2.3	0.0
Hydrogen sulfide (H_2S)	0.2	0.0
Methane (CH_4)	0.1	0.0
Ammonia (NH_3)	0.1	0.0
Oxygen (O_2)	0.0	20.9
Argon (Ar)	0.0	0.9

The living organisms that evolved on Earth were probably responsible for the change from the primitive atmosphere to the present atmosphere. Bacteria, among the first living organisms on Earth, developed two processes that were fundamental to life. At first, they used ammonia from the primitive atmosphere to make the chemical compounds necessary for life. Later, the bacteria used sunlight as a source of energy to make their own food. These bacteria, called *cyanobacteria,* made their food through the process of photosynthesis (foht oh SIHN thuh sihs). During **photosynthesis,** water and carbon dioxide are combined chemically to make sugar.

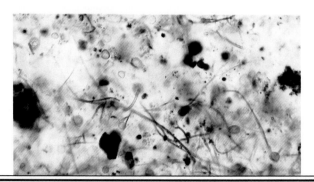

Figure 8–2. Shown here are fossils of cyanobacteria and algae. Cyanobacteria were the likely source of oxygen for the primitive earth's atmosphere.

Photosynthesis also releases oxygen as a byproduct. Other life forms needed this oxygen to survive. Today green plants, rather than cyanobacteria, produce most of the oxygen in the atmosphere.

Of the gases that were released into the primitive atmosphere, some, especially oxygen and carbon dioxide, dissolved in the ocean. In water, carbon dioxide reacts with rocks, forming calcium carbonate. By about 600 million years ago, enough oxygen had dissolved in the ocean to allow animal life to evolve there. Many of these early life forms had shells of calcium carbonate. Most of the carbon dioxide in the earth's atmosphere was used by these animals to form their shells. Only a trace of carbon dioxide remains in the atmosphere today.

Figure 8–3. Early corals and worms, from which these fossils formed, used much of the carbon dioxide dissolved in ocean water to make their skeletons and tubes.

With the removal of carbon dioxide from the primitive atmosphere, nitrogen became the most abundant gas. Oxygen, from photosynthesis, soon became the second most abundant gas. As oxygen increased, the sulfur dioxide in the primitive atmosphere was oxidized to sulfuric acid. The sulfuric acid then washed out of the atmosphere as "acid rain." The acid rain you read about today contains sulfuric and other acids, but these are from industrial pollution.

○ *What were the most common gases of the primitive atmosphere?*
○ *What caused the primitive atmosphere to change?*

8.2 The Present Atmosphere

You may wonder why the atmosphere does not disappear into space. What keeps it stuck to the earth? The answer is gravity.

Some gases, such as hydrogen and helium, the lightest gases, do escape because they tend to float to the top of the atmosphere. There they are scattered by the solar wind and escape into space.

DISCOVER

Making Oxygen

Light a small candle and cover it with a glass jar. **CAUTION: Be careful with the flame.** Time how long the candle will burn before it goes out. Why does the candle go out?

Now place a small green plant next to the candle and light the candle again. Cover both the plant and the candle with a glass jar. Time how long the candle burns before it goes out. Does the candle burn for the same time? Explain any difference in time.

Answers–8.1

○ The most common gases of the primitive atmosphere were carbon dioxide, nitrogen, sulfur dioxide, and water vapor.

○ The primitive atmosphere changed because early organisms produced oxygen and because much of the carbon dioxide dissolved in the ocean.

EXTENSION Since the beginning of the industrial revolution, great amounts of carbon dioxide have been added to the atmosphere. However, only about 50 percent of that amount remains in the atmosphere; the other 50 percent is absorbed by the oceans. An equilibrium is maintained because, as the percentage of carbon dioxide in the air increases, the ocean water dissolves more of the gas. However, in the future the ocean will not be able to absorb enough of the carbon dioxide to prevent a "greenhouse effect" from occurring. It is predicted that by the year 2000, carbon dioxide levels may reach nearly 400 parts per million, which is almost 90 parts per million higher than the levels in the mid-nineteenth century.

DISCUSSION Explain to the students that the balance between the numbers of plants and animals on the earth keeps the oxygen level of the atmosphere at about 20.9 percent. ***Thinking Critically:*** Ask the students to explain why the deforestation of the tropical rain forests of South America may have an effect on the survival of both plants and animals. (It would change the ratio of oxygen and carbon dioxide in the atmosphere. There may not be enough plants to produce sufficient oxygen for all the animals.)

8.2 The Present Atmosphere

DISCUSSION Ask the students to identify the force that holds them and all other objects on the earth. (gravity) Then review with them the laws of gravity. Explain that all matter is affected by gravity. ***Thinking Critically:*** Ask the students if air is matter. (Air is matter because it takes up space, has mass, and is affected by the force of gravity.)

DISCOVER:
Making Oxygen

Thinking Skill *(Analyzing Experiments)*

Students should realize that a flame can burn only in the presence of oxygen. The candle goes out when all the oxygen in the jar is used up. When the plant is present, oxygen is added as the plant photosynthesizes, giving off oxygen. Even though the burning candle uses oxygen, more oxygen is added to the jar.

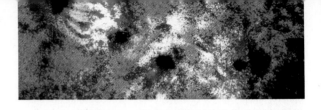

THINKING SKILL *(Communicating)*

Have the students use the information in this section to write an essay comparing the past and present atmospheres of Earth. Have the students who are interested include information on what our atmosphere may be like 2000 years into the future. They may want to do some library research on current trends in atmospheric changes to help provide material for their essays.

THINKING SKILL *(Inferring)* Emphasize that the air over tropical regions has a much higher water vapor content than does the air over deserts. Ask the students why this might be so. (In desert regions, there are rarely any large bodies of water from which water vapor can enter the air.)

EXTENSION When high-flying supersonic jets were first proposed, it was feared that they would cause pollution that would destroy the ozone layer. Today, the major threat to the ozone layer is from fluorocarbons that are used as propellents in aerosol sprays. Although they have been mostly banned in the United States, they are still used in other countries. Scientists have recently discovered a large hole in the ozone layer over Antarctica and a smaller hole over the Arctic Ocean. They believe that although these holes are seasonal, they are much larger than they should be because of damage by fluorocarbons.

SECTION REVIEW

Summary Have the students write a short paragraph describing each stage in the development of the atmosphere of the earth.

Reinforcement You may want to have the students review the section by having them quiz one another using the questions at the end of each numbered subsection.

Figure 8–4. Sunspots create strong magnetic fields that accelerate the solar wind.

Figure 8–5. Scientists have discovered a hole in the ozone layer of the upper atmosphere over Antarctica. If similar holes should develop over populated areas, large amounts of dangerous ultraviolet radiation could cause an increase in genetic disorders and skin cancers.

The particles of the solar wind, mostly protons, are accelerated by strong magnetic fields created by sunspots. **Sunspots** are enormous whorls of gases on the surface of the sun. The number of sunspots changes in a cycle that repeats itself every 11 years. When there are few sunspots, the solar wind is weak; when there are many sunspots, the solar wind is strong. The solar wind increases the speed of all gas atoms in the upper atmosphere.

In addition to the elements and compounds shown in Table 8–1, the present atmosphere also contains variable amounts of water vapor. The amount of water vapor ranges from almost zero over the deserts to 2.5 percent over the tropic regions near the equator.

Another small but very important component of the atmosphere is ozone. An **ozone** molecule consists of three atoms of oxygen. Ozone is extremely important to life on Earth because it absorbs dangerous ultraviolet radiation from the sun. Ultraviolet radiation causes sunburn, skin cancer, and possible genetic damage inside cells.

The ozone layer is in danger of being destroyed by chemical pollution, particularly from substances added to the atmosphere by industrial processes and by the propellants in aerosol cans. In order to save the ozone layer, laws have been passed to limit this kind of pollution.

○ *What are sunspots?*
○ *Why is the presence of ozone in the atmosphere important?*

Section Review

READING CRITICALLY

1. How did the early atmosphere differ from today's atmosphere?
2. Why do hydrogen and helium escape from the atmosphere?

THINKING CRITICALLY

3. Why has the level of oxygen in the atmosphere not continued to increase beyond its present percentage?
4. What might happen to the number of cases of skin cancer if the ozone level of the atmosphere is reduced?

168   Chapter 8   The Atmosphere

Answers–8.2

○ Sunspots are large whirls of gases on the sun's surface.

○ Ozone absorbs much of the dangerous ultraviolet radiation from the sun.

Answers to Section Review

Reading Critically

1. Today's atmosphere is a mixture of mostly nitrogen and oxygen. The early atmosphere of Earth consisted mostly of carbon dioxide, methane, hydrogen sulfide, sulfur dioxide, ammonia, and nitrogen.

2. Because these two elements are the lightest of all the elements, they reach the greatest velocities and are able to escape to higher altitudes.

Thinking Critically

3. The oxygen level has remained nearly constant because of the balance between plants and animals.

continues

BACKGROUND

In order for you to learn from textbooks, magazines, and newspapers, you must be able to understand what you read. One way to understand is to answer questions about what you have just read.

PROCEDURE

In this textbook, questions have been provided after sections to help you see if you have understood what you have read. When you read something that does not provide questions, you can ask yourself questions. Questioning is a very important technique in learning.

Of the questions you will answer about what you have read, some can be answered directly from the material and some you have to think about. If you can answer the ones you have to think about, you have really understood what you have read.

APPLICATION

Read the following selection and answer the questions that follow.

Atmospheric Gases

The many gases of the atmosphere are mixed in a delicate balance. Each gas, such as nitrogen or oxygen, is found in the atmosphere in very specific amounts. Each plays an important role in the maintenance of life on our planet.

The atoms and compounds that make up the gases circulate throughout the environment in very specific chemical pathways called *cycles*. There is a water cycle, an oxygen cycle, a carbon cycle, and a nitrogen cycle, which is shown below. Upsetting the flow of any of these cycles can have drastic consequences for our environment and the plants and animals that live in it.

Air pollution created by modern industries disturbs the balance of atmospheric substances. Industrial pollutants poured into the atmosphere can disturb natural chemical reactions that keep the cycles going. Although this problem has existed for several centuries, scientists have only, in the past few decades, begun to examine it very closely.

1. What is a cycle?
2. Why are cycles important?
3. Why would upsetting the way a cycle works cause problems for living organisms?
4. How does air pollution affect the cycles?

USING WHAT YOU HAVE LEARNED

Go to the library and find a newspaper or magazine article about air pollution. Write a brief report that answers the following questions:

1. What kind of pollution does the article discuss?
2. What specific pollutants are involved?
3. Which industry is mostly responsible for the pollution?

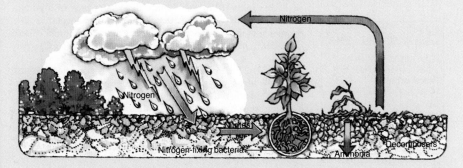

Nitrogen

Nitrogen

Nitrates

Nitrogen-fixing bacteria

Ammonia

Decomposers

Section 1   Composition of the Atmosphere   **169**

SKILL ACTIVITY: Reading for Meaning

Objectives

- To increase reading comprehension.
- To think critically while reading.

Discussion You may want to help the students decide which questions are important to ask while reading an article. Explain that questions that require the recall of facts from the article can be helpful. Point out, however, that questions about the article's general ideas require critical thinking and help increase knowledge. ***Thinking Critically:*** Ask the students why these types of questions are especially helpful in learning. (The students should suggest that by using critical thinking questions, they gain not only knowledge of the facts of an article but gain skills that they can employ when reading other articles and performing other tasks.)

Answers to Application

1. a pathway
2. They circulate gases throughout the environment.
3. Organisms depend on a flow of atmospheric gases.
4. Air pollution may disrupt the cycles.

Answers to Using What You Have Learned

1. Answers will vary.
2. Answers will vary.
3. Answers will vary.

4. The number of cases of skin cancer would probably increase significantly because more ultraviolet radiation would reach the earth.

Section 2: Solar Radiation

SECTION FOCUS

Section Overview The transport of energy from the sun to the earth by electromagnetic waves is described. Radiation balance is defined, and its effect on the earth's atmosphere is explained. The role of radiation balance in the climate of Earth is identified.

Section Objectives For a list of section objectives, see pupil's edition page.

New Science Terms For a list of new science terms in this section, see pupil's edition page. To familiarize the students with the new terms, have them read the terms aloud and discuss their meanings.

SECTION DEVELOPMENT

8.3 Electromagnetic Waves

DISCUSSION You may want to point out that the speed of light is the frequency times the wavelength, or $v = f \times w$. Emphasize that the speed of light is constant. *Thinking Critically:* Ask the students if a light wave with a frequency higher than another light wave will travel faster. (No. The speed of light, of any frequency, is always the same in the same medium. The speed of light in a vacuum is 3.0×10^8 m/s. Explain that because the speed of light is a known constant, the frequency of a wave can be determined if the wavelength is known, and vice versa, by using the formula presented earlier. You may want to have the students practice calculating the frequency or wavelength of a light wave by giving them a specific wavelength or frequency and the value of the speed of light.)

NEW SCIENCE TERMS

photons
electromagnetic waves
radiation balance
greenhouse effect

SECTION OBJECTIVES

After completing this section, you should be able to:
- **Explain** the way in which Earth receives energy from the sun.
- **Describe** the relationship of solar radiation to radiation balance.
- **Discuss** how the addition of carbon dioxide to the atmosphere affects the radiation balance of the earth.

8.3 Electromagnetic Waves

You might think that the use of solar energy is something new. The fact is, practically all the energy on Earth is received as *solar radiation*, or energy from the sun.

Light consists of particles called **photons** (FOH tahnz) that travel from the sun to the earth in 8.3 minutes. Photons travel as electromagnetic waves. **Electromagnetic waves** are energy waves that travel at the speed of light.

Figure 8–6. Volcanic dust in the atmosphere causes more red light waves to reach the earth (left). The structure of waves is shown in the diagram (right).

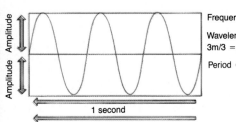

Frequency = 3 hertz

Wavelength = 3m/3 = 1m

Period = 0.33 sec.

1 second

3 m

Electromagnetic waves vibrate like the strings of a guitar. The number of vibrations that occur in one second is the *frequency* of the wave. Frequency is measured in SI units called *hertz* (Hz). One hertz is equal to one vibration, or one cycle, per second.

Consider, for example, the human heart. Your heart beats about 60 times per minute. The frequency of this heartbeat would be one hertz, or one beat per second.

Each wave has a high point, called a *crest*, and a low point, called a *trough*. The distance from crest to crest or from trough to trough is called the *wavelength*. The height of the wave, from the midpoint to the trough or from the midpoint to the crest, is the *amplitude*.

A MATTER OF FACT

The speed of photons in empty space is constant at 299 792 458 m/s and is called the *speed of light*. It would take 125 days to travel that distance in a car going 100 km/h.

170 Chapter 8 The Atmosphere

Electromagnetic waves range in frequency from less than 1 Hz to 100 000 billion billion (10^{23}) Hz. The higher the frequency of the wave, the shorter the wavelength. Figure 8–7 shows different wavelengths in the *electromagnetic spectrum*.

Human senses can detect only a portion of the electromagnetic spectrum. However, humans may be affected by wavelengths they cannot detect. Some waves that affect humans are ultraviolet rays, X rays, and gamma rays.

Notice that the range of visible light is a very small part of the electromagnetic spectrum. The sun emits light of all colors. The human eye sees the combination of these colors as white.

○ *What are electromagnetic waves?*
○ *What is frequency?*

Figure 8–7. White light contains all the colors of the spectrum (below). The electromagnetic spectrum is shown in the diagram (left).

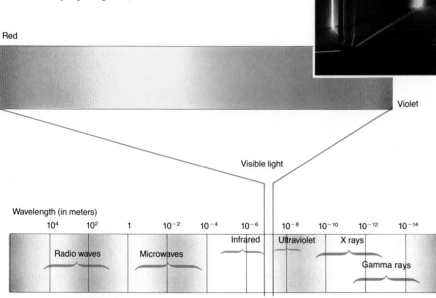

8.4 Radiation Balance

Have you ever noticed that you feel warmer on a sunny day if you are wearing something dark in color? This is because different surfaces absorb different amounts of solar radiation. A black surface absorbs nearly all the solar radiation that falls on it. Consequently, any dark-colored object set out in the sun becomes very hot. A white surface reflects almost all the solar radiation that falls on it. For this reason, white objects stay cooler in the sun than objects of any other color.

Section 2   Solar Radiation   **171**

Answers–8.3

○ Electromagnetic waves are energy waves that travel at the speed of light in empty space.

○ Frequency is the number of wave vibrations that pass a fixed point in one second.

DISCUSSION   Direct the students' attention to Figure 8–7. Explain that because the glass of the prism is of a higher density than air, electromagnetic waves are slowed and bent, or refracted, when they pass through the prism. The different colors of light have different wavelengths, and the light waves bend differently, separating the white light into its various colors. Ask the students to name the colors in the order in which they appear in the band of light called the *spectrum*. (red, orange, yellow, green, blue, violet) *Thinking Critically:* Ask the students if the order of colors in the spectrum will appear the same. (Yes. The wavelength of the waves of each color is always the same. The colors will always appear in the same order in the spectrum.)

DISCUSSION   Have the students study Figure 8–7 and answer the following question: What type of electromagnetic wave has the shortest wavelength? (gamma rays) the longest? (radiowaves) Which waves of visible light have the shortest wavelength? (violet) the longest? (red)

8.4 Radiation Balance

DISCUSSION   Within the troposphere, air temperature varies from region to region. The amount of solar energy absorbed by a given area of the earth's surface influences the air temperature above that area. The ratio of the amount of light and heat reflected by a surface to the total amount of light and heat received by that surface is called *albedo*. Snow, for example, has an albedo of 40–95 percent, while a forest has an albedo of 5–20 percent. In general, the air above light-colored surfaces will be cooler than the air above dark-colored surfaces.

EXTENSION Almost all of the shorter wavelengths of solar radiation are absorbed into the upper atmosphere. The ultraviolet rays that do reach the earth cause skin to burn when a person stays in the sun too long. Usually only the longer wavelength radiation, visible light, and infrared (heat) rays reach the earth's lower atmosphere. The density of this layer causes the waves to be bent, or scattered, as the light collides with particles of the atmosphere. Blue light is scattered more than the other wavelengths of light. *Thinking Critically:* Ask the students if they think that the scattering of blue light is related to the fact that the sky usually appears blue. (The other colors of light continue on to the earth's surface. Since the blue light is scattered within the atmosphere, the atmosphere appears blue.)

DISCUSSION Have the students refer to Figure 8–8. An average of about 20 percent of the sun's radiation is absorbed by the atmosphere or by clouds before reaching the earth's surface. Another 35 percent is reflected back into space or scattered within the atmosphere. The remaining 45 percent reaches the earth's surface where it is absorbed or reflected.

Most atmospheric gases let solar radiation through without reflecting or absorbing much of it. Clouds absorb or reflect about 25 percent of the solar radiation entering the atmosphere. About 30 percent of the solar radiation that does reach the earth is in the form of heat.

For the earth to remain livable, there must be a balance between the amount of solar radiation that is absorbed and the amount that is reflected into space. **Radiation balance** is the balance between solar radiation coming into Earth's atmosphere and radiation going out from Earth. Most of this outgoing radiation is in the form of heat. Heat radiation is in the infrared part of the electromagnetic spectrum.

Infrared radiation is reflected and radiated into space from the ground and from cloud tops. If too much heat is reflected into space, the atmosphere cools down. If too little heat escapes, the atmosphere heats up. Either case could upset the weather conditions that now exist on Earth. Radiation balance is maintained in delicate adjustment by many physical and environmental factors.

Figure 8–8. This diagram shows how the radiation balance of the earth is maintained.

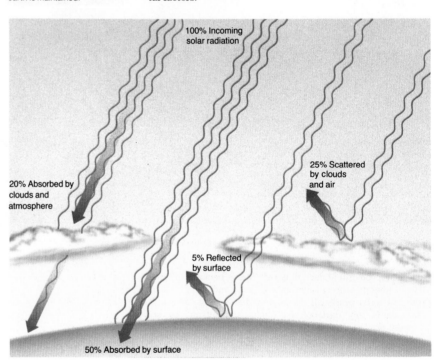

100% Incoming solar radiation

20% Absorbed by clouds and atmosphere

25% Scattered by clouds and air

5% Reflected by surface

50% Absorbed by surface

Figure 8–9. If the earth receives more heat than it loses, a warm climate results. If the earth receives less heat than it loses, a cold climate results.

The radiation balance of the earth has shifted many times throughout the earth's history. One extreme in atmospheric temperature occurs during times of extreme cold, when more of the earth's surface is covered with snow and ice than is now covered. These conditions cause more of the sun's radiation to be reflected into space. This higher rate of reflection causes the earth to be colder. Why would a covering of snow and ice cause the earth to reflect radiation?

The other extreme occurs when there is little or no ice on Earth, and the oceans extend over broad portions of the land. More solar energy is absorbed by the earth, and the earth is warmer. From one extreme to the other, however, the average temperature of the earth changes by only a few degrees Celsius. Temperature changes have never been large enough to endanger all life on the earth. However, some species, such as dinosaurs, may have become extinct because of temperature changes.

Unfortunately, activities of people worldwide threaten to change dangerously the radiation balance of the earth. Aerosols, chlorine-containing compounds, and carbon dioxide are being added continuously to the atmosphere. These substances may change the radiation balance enough to affect life on the planet.

Carbon dioxide, for example, traps infrared radiation that would normally be radiated back into space. This same effect can be seen in a greenhouse or in a car with the windows closed. Like carbon dioxide, glass lets sunlight through but traps

DISCOVER

Using Solar Power

Try operating a solar-powered calculator in various colors of light. Are there any colors that seem to work better than others? Are there any colors that will not work at all?

EXTENSION   The radiation that is absorbed by the earth's surface is reradiated as infrared radiation, or heat. This heat is distributed throughout the atmosphere in three ways. One way is by conduction, which is the direct transfer of heat from one substance to another. Air in direct contact with the warmed ground or warmed water is itself warmed. Another way is by convection, which is the transfer of heat through a fluid. (Air is a fluid.) Air near a warm surface becomes less dense and rises. As cooler air moves downward to replace the warm air, it is also warmed. The third way in which heat is transferred is by radiation. The students should, by this point, be able to define radiation. (It is the transfer of heat energy in the form of waves.)

EXTENSION   Invite an automotive technician or Environmental Protection Agency worker to lead a class discussion about the products and machinery that contribute to the pollution of the atmosphere. The automotive technician might show parts of cars such as the carburetor or fuel injector, catalytic converter, or muffler. He or she might explain how the parts prevent or control the introduction of harmful substances into the atmosphere. The EPA worker might discuss devices that can be used in the home and by industry to control pollution.

DISCOVER:
Using Solar Power

Skill *(Making Comparisons)*

Light waves on the left of the spectrum are lowest in energy. The calculator may not function in red light and may be slow in orange light. Its performance will improve in the light of the other colors because that light has greater energy. In addition, the solar cell may not be sensitive to all frequencies.

174

DISCUSSION Have the students list the different forms of electromagnetic radiation. Ask them to name some common uses of each of the nonvisible forms of radiation. (Some examples are using X rays for photographing bones, using microwaves for cooking, and using radar to detect moving objects.)

SECTION REVIEW

Summary Review electromagnetic waves with the students. Have them name the waves from longest to shortest. Also have them give a brief explanation of the way that radiation balance determines the temperature of the atmosphere.

Reinforcement Have the students answer the questions at the end of each numbered subsection. Let them compare and discuss their responses.

A MATTER OF FACT

For most of the history of the earth, the Arctic and the Antarctic have been free of ice.

---

the returning infrared radiation. The inside of the greenhouse or car warms up. The temperature rise caused by the addition of carbon dioxide to the atmosphere produces a **greenhouse effect.**

Some scientists believe that human activities may change the radiation balance of the earth enough to trigger another ice age or to melt the ice that now covers Antarctica and Greenland. What problems do you foresee if a new ice age were to begin? What would happen if all the polar ice were to melt and the level of the oceans were to rise?

○ *What is radiation balance?*
○ *What causes the greenhouse effect?*

Figure 8–10. A greenhouse traps heat, so the temperature remains high even on a cold day. The greenhouse effect would cause the average temperature of the earth to increase, possibly melting the polar ice.

Section Review

READING CRITICALLY

1. What is the effect of carbon dioxide on the radiation balance of the earth?
2. Why is the radiation balance of the earth important?

THINKING CRITICALLY

3. How could the fact that most of the sunlight received on Earth is in the blue-green wavelength affect the way life developed on Earth?
4. How can you explain the fact that some people think the earth's temperature may increase, while others think the temperature may decrease?

Answers—8.4

○ Radiation balance is the balance between solar radiation coming into the earth's atmosphere and radiation going out from the earth.

○ The greenhouse effect is the temperature rise caused by carbon dioxide in the atmosphere.

Answers to Section Review

Reading Critically

1. More carbon dioxide would mean an increase in the earth's overall temperature. Less carbon dioxide would mean a decrease in the earth's temperature.
2. An increase or decrease in radiation could change the temperature of the earth.

continues

SECTION OBJECTIVES

After completing this section, you should be able to:

■ **Describe** the cause of air pressure.

■ **List** the layers of the atmosphere and **discuss** their characteristics.

NEW SCIENCE TERMS

atmospheric pressure
barometer
Van Allen belts

8.5 Atmospheric Pressure

Have you ever noticed how hard it is to open a new jar of pickles? Many food products are vacuum packed—that is, packed without air. The atmosphere pushes down on the lid and makes it very difficult to open.

If you could measure a column of air 1 cm² across that reaches from sea level to outer space, you would find that the air has a total mass of 1.033 kg. This much mass exerts a considerable force, or weight, on objects on the earth. **Atmospheric pressure** is the weight of the atmosphere. Atmospheric pressure is determined by measuring the force the atmosphere exerts on a column of mercury. Mercury is 13.6 times denser than water, so atmospheric pressure that would raise a column of mercury 760 mm high would raise a column of water 10.33 m high.

The pressure exerted by the atmosphere decreases with altitude. Mount Everest is one of the highest mountains on Earth. Atmospheric pressure at the top of Mount Everest is less than one-third of the pressure at sea level. Mountain climbers trying to reach the top have to carry oxygen tanks and breathe through regulators. This is necessary because high on the mountain the atmosphere becomes so thin that there is not enough oxygen for the climbers to do the strenuous work of climbing.

One instrument used to measure atmospheric pressure is called a **barometer** (buh RAHM uh tuhr). The most commonly used type is the *aneroid* (AN uhr oihd) *barometer*, shown in Figure 8–11. An aneroid barometer consists of a thin-walled metal chamber from which some of the air has been removed. As atmospheric pressure changes, the shape of the chamber changes. The change in shape activates a pointer that indicates the atmospheric pressure on a gauge.

○ *Why does the atmosphere exert pressure?*
○ *What is atmospheric pressure?*

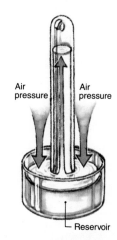

Air pressure   Air pressure

Reservoir

Figure 8–11. An early barometer (top) was made by an Italian scientist, Evangelista Toricelli. A modern aneroid barometer is also shown (bottom).

Section 3   Structure of the Atmosphere   **175**

SECTION FOCUS

Section Overview In this section, atmospheric pressure is defined. The layers of the earth's atmosphere are characterized and compared with one another.

Section Objectives For a list of section objectives, see pupil's edition page.

New Science Terms For a list of new science terms in this section, see pupil's edition page. You may want to read the new terms aloud to familiarize students with their pronunciations.

SECTION DEVELOPMENT

8.5 Atmospheric Pressure

DISCUSSION Explain to the students that air pressure varies, depending on altitude, temperature, and the amount of water vapor in the air. ***Thinking Critically:*** Ask the students to predict the effect of the three factors on the atmospheric pressure. (Pressure decreases with increasing altitude because there is less air. Higher temperatures result in lower air pressure because warm air expands and becomes less dense. The greater the amount of water vapor in the air, the lower the air pressure will be.)

EXTENSION There are quite a few units that have been used to measure air pressure. Standard atmospheric pressure, or pressure measured at sea level, is expressed as 30 in. of mercury, 760 mm of mercury, 760 Torr, 1013.25 millibar (mb), or 100 kiloPascals. Most weather maps use pressure expressed in millibars.

THINKING SKILL (*Analyzing Ideas*) Death Valley in California is 100 m below sea level. Ask the students what the effect on air pressure might be. (The air pressure will be greater than at sea level. There is more air above the ground at a lower altitude, and thus the weight is greater.)

Thinking Critically

3. Plants and animals adapt to their environments or die. Since most of the light received from the sun is in the green wavelength, organisms have evolved that make use of that range of energy.

4. Increased carbon dioxide may cause a greenhouse effect, which would increase Earth's temperature. An increase in solid pollutants could lower Earth's temperature by reducing the radiation that reaches it.

Answers—8.5

○ The atmosphere exerts pressure because of Earth's gravity.

○ Atmospheric pressure is the weight of the atmosphere.

Discussion Have interested students write to one or all of the addresses provided for additional information. Have the students discuss the opportunities offered by the various careers.

Extension You may wish to hold a "Career Day" in the class during which the students report on the information they have obtained. If possible, invite a person in each career to speak to the class about why he or she chose the career, what specific training he or she received, and what a "typical" day at work involves. As an alternative to outside speakers, have the students assume the role of a person in each career and make a presentation to the class covering the same topics mentioned above. The students may be interested in researching information on the following additional careers:

Meteorologist
Paleoclimatologist

CAREERS

AUTOMOTIVE TECHNICIAN

Since the 1960s, *automotive technicians* have played a very important role in trying to keep our air free of pollution. They are skilled at adjusting features of gasoline engines that improve their efficiency.

Most automotive technicians work at neighborhood garages. Some own or manage businesses that specialize in automobile maintenance, such as tune-ups. Others are employed by large companies that manufacture cars. All automotive technicians must have at least a high-school diploma, and many must pass state and local examinations that test their knowledge of engines and laws governing pollution-control standards.

For Additional Information
Society of Automotive Engineers
400 Commonwealth Drive
Warrendale, PA 15096

ENVIRONMENTAL PROTECTION AGENT

The Environmental Protection Agency is a government agency that works to improve the environment. *Environmental protection agents* are dedicated to the control of pollution caused by sewage, agricultural pesticides, factory noise and emissions, automobile emissions, and more.

The agency employs people of many educational backgrounds. However, most have college degrees. You do not need a degree in science to work for the EPA, but researchers and field agents usually have one.

For Additional Information
The Environmental Protection Agency
401 M Street, S.W.
Washington, DC 20460

ATMOSPHERIC CHEMIST

Atmospheric chemists conduct research on atmospheric pollution. They study the effects of modern industry on rain, snow, and cloud chemistry.

A new and exciting field for atmospheric chemists has grown with the success of the space program, that is, studying the atmospheres of other planets.

Atmospheric chemists must have an advanced degree in chemistry. They work for government agencies like the EPA or the National Aeronautics and Space Administration (NASA) or do research for universities.

For Additional Information
National Academy of Sciences
Office of Public Affairs
2101 Constitution Avenue
Washington, DC 20418

ACTIVITY: Demonstrating Atmospheric Pressure

Can the fact that the atmosphere exerts pressure be shown?

MATERIALS (per group of 3 or 4)

safety goggles, laboratory aprons, gloves, water, empty soda can, bowl, Bunsen burner or alcohol lamp, tripod, wire gauze

PROCEDURE

1. **CAUTION: Put on safety goggles, a laboratory apron, and gloves before you begin this experiment.**
2. Pour about 2 mL of water into the soda can.
3. Fill the bowl nearly to the brim with COLD water.
4. Put the burner or lamp on a low flame.
5. Set the can on the wire gauze atop the tripod over the Bunsen burner or lamp.
6. Count for 30 seconds after the water in the can gives off steam and boils steadily.
7. Remove the flame from under the can.
8. Quickly turn the can upside down into the bowl of water. Record your observations.

CONCLUSIONS / APPLICATIONS

1. What happened to the can when it was placed in the cold water? Explain.
2. What might have happened to the can above? Explain.
3. Draw a conclusion about the surrounding atmosphere based on your observations.

8.6 Layers of the Atmosphere

Imagine a wedding cake four layers high. This is a very unusual wedding cake, for the guests all have different tastes. The bottom, or first layer, is marbled. There is cool, dense, dark chocolate swirled with warm, light, milk chocolate.

The second layer is yellow cake, but since it has just come out of the oven, it is cooling at the bottom and still very warm at the top. The third layer is dull, plain white cake, warm at the bottom where it touches the second layer, but much cooler at the top.

The fourth layer is the most exciting to see. Near the bottom of this layer, the cake is greenish with touches of red, and it seems to glow with a light of its own. The top of this layer is angel food, all light and thin; it does not really end, it just sort of trails off into nothing.

The earth's atmosphere is much like this layered cake; each layer has its own characteristics of temperature and composition.

> **A MATTER OF FACT**
>
> If all the water were squeezed out of the atmosphere, the resulting rainfall would be only 0.3 cm. Yet without water vapor in the atmosphere, there would be no life on land.

Thinking Skill (*Drawing Conclusions*)

Preparation of Materials   CAUTION: Before beginning this activity, be sure that the students are wearing their safety goggles, laboratory aprons, and heat resistant gloves. Review the safety procedures for using a Bunsen burner or alcohol lamp. Removing air from the soda can allows the pressure in the can to decrease in relation to the surrounding atmosphere. Upon cooling, the can should collapse because of this difference in pressure. In some cases, however, the students may not heat the soda can sufficiently. This results in the atmospheric pressure pushing cold water up into the can. If this occurs, have the students repeat the procedure until they succeed in making the can collapse.

Answers to Conclusions/Applications

1. The can was crushed; it imploded. If the students answer that the water level in the bowl decreased as the can sucked in the water, have them repeat the activity as noted above.
2. Answers will vary depending on observed results. Heating the can pushed out some of the air. Cooling the can caused the remaining air to contract. The pressure of the surrounding atmosphere pushed the water from the bowl into the can.
3. The atmosphere exerts pressure.

8.6 Layers of the Atmosphere

DISCUSSION   Tell the students that, as explained earlier, gravity holds the atmosphere close to the earth. The pull of gravity causes the upper layers of the atmosphere to push down on the lower layers. *Thinking Critically:* Ask the students if they would expect the air pressure to be greater in the upper or lower atmosphere. (The air pressure is greater in the lower atmosphere because the pressure increases nearer to the earth's surface.)

EXTENSION Instruct the students to make a graph showing the relationship of temperature to altitude. Using the information given in Figure 8–12, plot the altitude along the x axis and the temperatures of the atmospheric layers along the y axis. Use the current temperature as the starting point for the troposphere. Have the students label each layer and each boundary.

DISCUSSION Have the students study the layers of the atmosphere shown in Figure 8–12. ***Thinking Critically:*** Have the students discuss the similarities and differences among the layers. Ask them to describe the various characteristics shown. (Answers will vary but should include differences in temperature, and so on.)

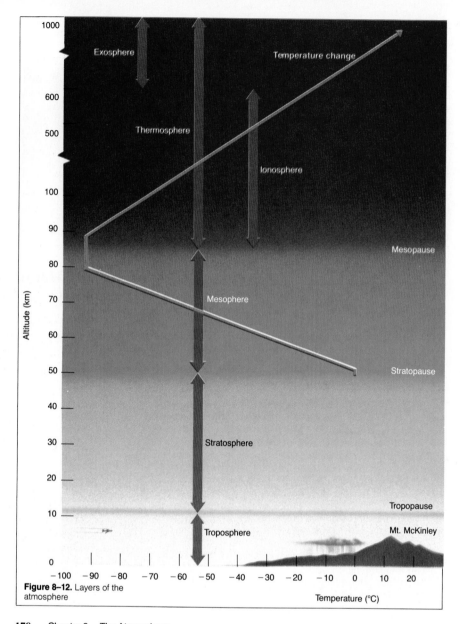

Figure 8–12. Layers of the atmosphere

Troposphere   The bottom layer of the atmosphere, the one nearest the earth, is called the *troposphere*. The troposphere is like the bottom layer of the wedding cake; it is here that much mixing of air occurs. Air near the ground is warmer and less dense than air above. Warm air rises and cold air sinks, forcing the troposphere to turn over. This overturning is a type of convection, which creates the weather systems of the world. Although it is not distinct, like the difference between the layers of a cake, the boundary between the troposphere and the next layer is called the *tropopause*.

Stratosphere   Above the tropopause is the second layer of the atmosphere, which is called the *stratosphere*. Like the second layer of the wedding cake, the stratosphere is cool on the bottom and warm at the top.

Most of the ozone of the atmosphere is located in the stratosphere. The concentration of ozone increases with altitude, reaching a maximum at 22 km. Ozone absorbs not only ultraviolet radiation but also some infrared and visible light. Most of the absorption of radiation occurs near the upper boundary of the stratosphere, called the *stratopause*.

Convection does not occur in the stratosphere because the less dense, warmer air is above the more dense, colder air. In fact, the stratosphere is named for its stable stratification, or layers.

Mesosphere   Above the stratopause lies the third layer of the atmosphere, which is called the *mesosphere*. Like the third layer of the cake, there is really nothing special about the mesosphere. The temperature drops constantly from the base of the mesosphere to the *mesopause*, at the top. This drop in temperature is due to a decrease in the concentration of ozone and an increase in distance from the warm surface of the earth.

Thermosphere   Above the mesopause lies the *thermosphere*, which goes all the way to the edge of the atmosphere. This is like the layer at the top of the cake.

Atmospheric gases are so thin in the thermosphere that the protons of the solar wind easily pass through. The atmospheric atoms and molecules hit by the protons move faster, causing a temperature rise. In addition, the solar wind removes electrons from atoms to form ions. This ionization occurs in the lower part of the thermosphere, called the *ionosphere*. Ionization is strongest at high latitudes because the ionizing particles are directed toward the poles by the earth's magnetic field. When the ions recapture their electrons, photons are released, and a colorful display known as an *aurora* is created. The upper

Section 3   Structure of the Atmosphere   **179**

DISCOVER

Making Layers of Air

Place a small dish of ice and a lighted candle under a glass jar. When the candle goes out, notice where the smoke collects inside the jar. Why do you think all the smoke collects in one layer?

Figure 8–13. Although the layers of the atmosphere are not visible, as this photograph seems to indicate, they can be distinguished by temperature differences.

DISCOVER:
Making Layers of Air

Thinking Skill *(Inferring)*

The students should realize that the bowl of ice cools the air just above the bowl, near the bottom of the jar. This is where the smoke collects.

EXTENSION   Emphasize that the upper layer of the stratosphere is the warmest because it is where the ozone is the densest and absorbs the most radiation. The upper mesosphere is the coldest layer of the atmosphere with temperatures of $-100°C$. Despite this coldness, the mesosphere is the layer in which most meteors burn up. ***Thinking Critically:*** Ask the students to explain why most meteors burn in the coldest layer of the atmosphere. (The temperature of the layer has little to do with the burning. The meteors burn because of the friction produced as they come into contact with the gas molecules of the atmosphere.)

DISCUSSION   Within the ionosphere there are several layers of ions. AM radio waves are bounced off the lowest of these layers in order to transfer them from station to station. In the evenings, however, the lowest layer dissipates and the waves bounce off the next highest layer. ***Thinking Critically:*** Ask the students to explain why it is possible to receive AM radio transmissions from very far away only in the evening. (The radio waves travel higher into the atmosphere in the evening, and thus can be transmitted farther.)

THINKING SKILL *(Applying Concepts)* An aurora is a curtain of colorful light seen at or just above the horizon. Ask the students why they think the aurora cannot be seen in the tropics. (The aurora cannot be seen there because the magnetic field necessary to create the ionization is very weak near the tropics. It is strongest near the poles.)

EXTENSION Have the students look up the meaning of the root *thermo* in a dictionary. (It means "heat.") The thermosphere is the layer of the atmosphere that has the highest temperature. However, a person exposed to the thermosphere would freeze to death. ***Thinking Critically:*** Ask the students to explain why a person would freeze in a layer of air having temperatures as high as 2000°C. Remind the students that temperature is a measure of the speed of a molecule. (The molecules of the thermosphere move very fast and thus are very hot. However, because the particles are spread so far apart, a person would not feel the heat.)

DISCUSSION The outer covering of a rocket must be constructed to withstand atmospheric conditions as it ascends into orbit through the layers of the atmosphere. ***Thinking Critically:*** Ask the students to describe how changes in temperature, chemical composition, and wind velocities might endanger the rocket. (Answers will vary but should be consistent with the students' knowledge of each layer of the atmosphere.)

EXTENSION The Van Allen belts are part of the earth's magnetosphere, which is the area just beyond the atmosphere. It is made up of positively and negatively charged particles that are concentrated within the earth's magnetic field.

SECTION REVIEW

Summary Have the students describe the characteristics of each layer of the atmosphere in the form of a table. Using separate columns, the students should list the temperature, relative pressure, and composition of each layer.

Reinforcement To summarize the main concepts of this section, have the students write a sentence or two on the main idea of each paragraph in the section.

portion of the thermosphere is called the *exosphere*. Here the atmospheric gases are so thin that a gas molecule travels an average of 650 km before hitting another molecule. At sea level a molecule can travel only 1/100 000 mm before hitting another molecule.

Van Allen Belts High-energy protons and electrons from beyond the solar system are trapped in a belt about 3200 km above the equator. A second belt, about 25 000 km above the equator, traps lower-energy protons and electrons from the sun. These belts are called the **Van Allen belts** after the American physicist, James Van Allen, who discovered them in 1958. These two belts act as holding areas for incoming protons and electrons and protect the earth from these high-energy particles.

○ *Name the layers of the atmosphere.*
○ *What characteristic is used to separate one atmospheric layer from another?*

Figure 8–14. The Van Allen radiation belts (right) were discovered by a Vanguard satellite, launched in 1958 (above).

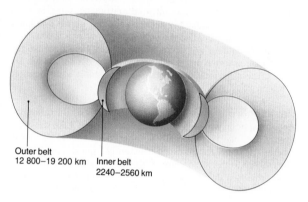

Outer belt
12 800–19 200 km   Inner belt
2240–2560 km

Section Review

READING CRITICALLY

1. What is a barometer and what does it measure?
2. Describe the temperature changes in each layer of the atmosphere.

THINKING CRITICALLY

3. If a barometer reads 760 mm at sea level and 660 mm at 1000 m above sea level, what might it read in Death Valley, which is nearly 100 m below sea level?
4. Why can auroras be seen only at high latitudes?

Answers–8.6

○ The layers of the atmosphere are the troposphere, the stratosphere, the mesosphere, and the thermosphere.

○ Temperature difference is the characteristic used to separate the layers of the atmosphere.

Answers to Section Review

Reading Critically

1. A barometer is a device used to measure atmospheric pressure.
2. In the troposphere, temperature decreases with altitude. In the stratosphere, the temperature increases with altitude. In the mesosphere, the temperature decreases with altitude. In the thermosphere, the temperature rises greatly.

continues

SECTION OBJECTIVES

After completing this section, you should be able to:

- **Discuss** how the Coriolis effect determines the circulation of winds.
- **List** and **locate** the planetary wind belts.
- **Describe** the cause of land and sea breezes.

NEW SCIENCE TERMS

Coriolis effect
jet stream

8.7 The Coriolis Effect

In the nineteenth century, Gaspard Gustaf de Coriolis, a French physicist, studied the path of bodies in motion on the earth's surface. His explanation of the force that influences these motions is complicated. However, it can be described with a simple analogy.

Suppose that the earth did not rotate and was a smooth sphere with a very icy surface. Now imagine that you are on ice skates at the North Pole and wish to go to New Orleans. A friend gives you a mighty push, and you move south toward New Orleans. The distance between the North Pole and New Orleans is 6673 km. If you were traveling at 1000 km/h, you would reach New Orleans in just over 6.5 hours. However, the earth does rotate. In 24 hours the earth makes one full turn. During the time that you are moving south, New Orleans is moving east. By the time you get to New Orleans, it is gone and you will find the Pacific Ocean in its place.

In the Northern Hemisphere, moving objects tend to move to the right because of what is called the **Coriolis** (cawhr ee OH lihs) **effect.** If you want to get to New Orleans, you cannot just

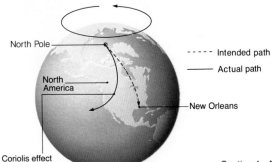

Figure 8–15. In the Northern Hemisphere, moving objects are deflected toward the right by the Coriolis effect.

SECTION FOCUS

Section Overview In this section, the relationship of the Coriolis effect to the motions of global winds is explained. The planetary winds are characterized. Also discussed are the forces behind land and sea breezes.

Section Objectives For a list of section objectives, see pupil's edition page.

New Science Terms For a list of new science terms in this section, see pupil's edition page. The students will probably be unfamiliar with these terms, but explain that they will learn the scientific definitions of the terms in this section.

SECTION DEVELOPMENT

8.7 The Coriolis Effect

EXTENSION (Tie-in/Physics) Explain to the students that the ice skater described in the text is assumed to be skating along without being affected by friction. The skater travels in a straight line heading south. *Thinking Critically:* Ask the students to relate the skater's assumed motion to Newton's law of inertia. (This law states that an object in motion will travel in a straight line unless acted upon by an outside force. With no outside forces acting upon the skater, he or she will continue south.)

Thinking Critically

3. If the barometer drops 100 mm while the altitude increases by 1000 m, then it should rise 10 mm if the altitude decreases by 100 m. The barometer should read 770 mm in Death Valley.

4. Auroras are colorful light displays that appear near the poles. They are caused by the interaction between the sun's charged particles and particles in the upper layers of the atmosphere. The magnetic fields near the poles cause much of the ionization of particles.

EXTENSION The students may have a very difficult time comprehending the Coriolis effect without a visual illustration. Secure a round piece of paper over the turntable of a record player, a Lazy Susan, or other freely rotating disk. Have one of the students attempt to draw a straight line from the center of the paper to its edge while the disk is rotating. Students will be able to observe that the line drawn is not straight but instead curves. Although the student's hand moves in a perfectly straight line relative to the rotating disk beneath the hand, the line becomes curved.

THINKING SKILL (*Applying Concepts*) Earlier, the students were asked to imagine the path a skater would take if the earth did not rotate. Ask the students to describe the movement of the earth's winds if the earth did not rotate. (The winds would flow in a straight line from the North Pole to the equator in the Northern Hemisphere and from the South Pole to the equator in the Southern Hemisphere.)

EXTENSION You may want to dispel a commonly held misconception about the Coriolis effect. It was once believed that water drained from basins in a counterclockwise direction in the Northern Hemisphere and in a clockwise direction in the Southern Hemisphere because of the Coriolis effect. The direction of drainage in both hemispheres is, in fact, arbitrary and has nothing to do with the Coriolis effect at all.

A MATTER OF FACT
Ocean currents are also affected by the Coriolis effect. In the Northern Hemisphere these currents are deflected to the right, the same as currents of air.

skate south. You have to go toward the east and chase after New Orleans. You have to continually change your direction. This change in direction is called *Coriolis acceleration*.

Now imagine you are at the South Pole and wish to skate to Rio de Janeiro. Your friend gives you another push, and you move north toward the equator. Traveling again at 1000 km/h, it takes you 7.5 hours to reach the latitude of Rio de Janeiro. By the time you get there, however, Rio de Janeiro is gone. In its place you find the South Pacific. Now, it seems you have veered to the left. In the Southern Hemisphere, moving objects tend to move to the left, again because of the Coriolis effect. If you wish to get to Rio de Janeiro, you will have to keep veering toward the east to follow the rotation of the earth.

The Coriolis effect occurs when Coriolis acceleration is insufficient or missing. A moving body is deflected to the right in the Northern Hemisphere and to the left in the Southern Hemisphere.

The Coriolis effect acts dramatically on the lower atmosphere. Since there is not enough friction between the air and the ground to force the air to follow the rotation of the earth, moving air masses are deflected to the right in the Northern Hemisphere and to the left in the Southern Hemisphere. These deflections cause the circular motions of the winds called *gyres* (JYRZ). Gyres are clockwise in the Northern Hemisphere and counterclockwise in the Southern Hemisphere.

○ *What is the Coriolis effect?*
○ *What causes the Coriolis effect?*

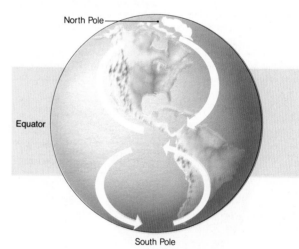

Figure 8–16. In the Northern Hemisphere, the gyres of the earth's winds are clockwise. These gyres are counterclockwise in the Southern Hemisphere.

Answers—8.7

○ The Coriolis effect is the apparent change of direction of moving objects on the earth's surface.

○ The Coriolis effect is caused by the rotation of the earth.

8.8 Planetary Winds

Earth's atmosphere is constantly in motion. This motion is due to heat energy. The source of the heat that keeps the atmosphere in motion is the sun, which warms up the low latitudes much more than the high latitudes. Since hot air is less dense than cold air, you would expect the heated air to rise at the equator, travel to the poles, sink there, and return to the equator along the ground. However, this is not what happens.

The motion of the earth also adds motion to the atmosphere in the form of the Coriolis effect. Because of the Coriolis effect, hot air rising at the equator cannot travel straight to the poles. By the time it reaches latitude 30° North, the air has been turned eastward. The air cools and sinks. As it sinks, some of the air continues toward the poles. Most of it, however, spreads out to the north and south along the ground. The air spreading toward the equator forms a planetary wind belt called the *trade winds*. The air spreading northward forms another planetary wind belt called the *westerlies*.

The polar areas are very cold. Cold, dense air sinks over the poles and streams out at ground level. The streaming is clockwise as seen looking down over the Arctic; it is counterclockwise as seen looking down over the Antarctic. As a result, the polar winds blow from east to west in both hemispheres, forming the *polar easterlies*. The planetary wind systems of the world are shown in Figure 8–17.

DISCOVER

Creating Wind Belts

Place two fans about 1 m apart, blowing toward each other. Light a candle or a match and then carefully blow it out, so that there is a lot of smoke. Direct the smoke into the area in which the air from the two fans meets. What happens to the smoke in this area? What might happen in the atmosphere where two wind systems meet?

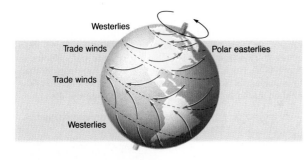

Westerlies

Trade winds

Polar easterlies

Trade winds

Westerlies

Figure 8–17. This diagram shows the planetary winds of the earth.

At latitudes below 40°, the tropopause is at an altitude of 16 km. Between latitudes 40° and 50°, the tropopause drops to an altitude of 10 km. There is a strong pressure difference across this drop. As air rushes toward the poles to balance the pressure, it is deflected eastward by the Coriolis effect. The result is a high-speed wind called the **jet stream**, circling each polar area at an altitude of 12 km.

Section 4   Motions of the Atmosphere   **183**

DISCOVER:
Creating Wind Belts

Thinking Skill *(Making Inferences)*

Using large wooden matches should produce a great deal of smoke. Students will notice that the smoke is drawn both up and down where the air from the two fans meet. The students should be able to infer that air in an area where two wind systems meet will either rise or sink.

8.8 Planetary Winds

EXTENSION   The circulating air of the atmosphere forms three major cells of winds in the Northern Hemisphere that are mirrored in the Southern Hemisphere. If these cells were visible, they would appear as six hollow doughnuts circling the earth. Between each cell is a belt of high or low pressure. Where the northern and southern trade winds converge at the equator, there is a belt of low air pressure called the doldrums. Where the trade winds and westerlies come together at 30°N and 30°S latitude, there is a high-pressure belt called the horse latitudes. Between the polar easterlies and the westerlies is the polar front, a belt of low pressure.

THINKING SKILL   *(Interpreting Diagrams)* Have the students study Figure 8–17. Ask them to explain the wind patterns as they appear in the Southern Hemisphere. (The winds move in a mirror image of those in the Northern Hemisphere due to the Coriolis effect.)

EXTENSION   Wind belts shift because of the changing position of the earth and the sun. In the Northern Hemisphere's summer, the belts shift about 10 degrees north, and in the winter the belts shift about 5 degrees to the south. The shift is not as great in the winter because the southern oceans tend to modify the weather conditions caused by the shift.

DISCUSSION A jet flying from Atlanta, Georgia, to Los Angeles, California, requires well over four hours to arrive at its destination. The return trip, however, takes less than four hours. *Thinking Critically:* Ask the students to explain the difference in travel time. (The westerlies flow from west to east, thus helping to speed the plane along on its return trip.)

EXTENSION Monsoons are most pronounced over the Indian subcontinent. During the spring, winds flow from the warm, dry interior of the Asian continent and bring long, dry spells and often drought. During the fall, winds shift and flow from the Indian Ocean toward the continent. These winds bring cooler, very moist air to India, as well as hundreds of centimeters of rain that often cause extensive flooding.

8.9 Local Winds

DISCUSSION Have the students imagine that they are going to a beach on a hot summer day. They arrive well before noon. *Thinking Critically:* Ask the students whether they would expect the water or the sand to be warmer. (The sand is warmer because it absorbs heat more quickly than water.) Next, have the students imagine that it is later in the evening, shortly after sunset. *Thinking Critically:* Ask the students which would cool faster, the sand or the water. (The sand would cool faster because water retains heat more efficiently.) Have the students relate this concept to the occurrence of land and sea breezes.

Figure 8–18. The monsoon winds of Asia bring seasons of torrential rain to India.

Figure 8–19. These diagrams show the cause of a land breeze (right) and a sea breeze (left).

The jet stream's speed averages 60 km/h in summer and 150 km/h in winter. The speed is greater in winter because the temperature contrast between the low latitudes and the high latitudes is much greater then. The jet stream does not circle the poles smoothly. Instead, it forms broad loops that may extend almost to the tropics. The jet stream strongly influences the weather at middle and high latitudes in both hemispheres.

Another planetary wind is the *monsoon*. A monsoon is characterized by a reversal of wind direction from summer to winter. Monsoons occur when air pressure over a large area reverses itself as the seasons change from summer to winter. The strongest monsoon is the Indian monsoon, which affects India and the northern Indian Ocean.

Monsoon circulation is not restricted to India, however. Weaker monsoons exist over the southwest United States, Australia, South America, and Africa.

The polar easterlies, the westerlies, the trade winds, and the monsoons form the planetary wind system. They are responsible for most of the weather in the middle latitudes and for distributing the heat of the tropics to the higher latitudes.

○ *What causes the planetary winds to curve?*
○ *Name the winds of the planetary wind system.*

8.9 **Local Winds**

In addition to the influences of planetary wind circulation, coastal areas often experience land breezes and sea breezes. During the day the land warms up more than the water. As the warmed air rises over the land, air from the ocean blows in to replace it. This flow of air is called a *sea breeze* and is shown in Figure 8–19. During the night the land cools more rapidly than the water, and a flow of air develops from land to sea. This flow is called a *land breeze* and is also shown in Figure 8–19.

Sea breeze — Warm air — Cool air

Land breeze — Cool air — Warm air

Answers—8.8

○ The Coriolis effect causes the planetary winds to curve.

○ Starting at the equator, and moving toward the poles, the planetary winds are the trade winds, the westerlies, and the polar easterlies.

○ Katabatic winds are warm, dry winds that blow down mountain slopes.

Answers—8.9 (p.185)

○ Land and sea breezes are caused by temperature differences between the land and the water.

Sea and land breezes are common in the tropics during most of the year and at middle latitudes during the summer. Sea breezes usually produce a line of clouds just inland along a tropical coastline or above tropical islands.

When air blows down a mountain slope, pressure and temperature rise, and the air is very dry. This type of wind is a *katabatic* (kat uh BAT ihk) *wind.* Katabatic comes from a Greek word that means "moving down."

Figure 8–20. Sea breeze creates puffy clouds over tropical islands (above). Cold, dry winds stream down from the Antarctic ice cap (left). Both are examples of local winds.

There are many regional examples of katabatic winds. The katabatic wind that blows down the eastern slope of the Rocky Mountains is called the *chinook.* Another katabatic wind is the *Santa Ana,* which blows down the coastal ranges of southern California. Katabatic winds may cause sudden snowmelts on the slopes down which they blow. These, in turn, may cause extensive flooding on the plains below. The strongest katabatic winds are those streaming down from the Antarctic icecap.

○ **What causes land and sea breezes?**
○ **What are katabatic winds?**

A MATTER OF FACT

The windiest place on Earth is Commonwealth Bay, Antarctica, where the south magnetic pole is located. Here, winds of more than 300 km/h are not uncommon.

Section Review

READING CRITICALLY

1. Why do winds turn to the right in the Northern Hemisphere and to the left in the Southern Hemisphere?
2. What causes the planetary winds to blow?

THINKING CRITICALLY

3. Why are monsoons part of the planetary wind system, while sea breezes are considered local winds?
4. If the earth did not rotate, what would the planetary wind system be like?

Section 4   Motions of the Atmosphere   **185**

DISCUSSION Explain that as air moves up the slope of a mountain most of the moisture in it is precipitated. Thus, the wind blowing down the other side of the mountain is very dry. The Santa Ana winds that blow down the Santa Ana Mountains around Los Angeles are responsible for increasing the danger of brush fires in the area. *Thinking Critically:* Ask the students to explain why this is so. (The wind is very dry, and thus it absorbs much of the moisture from the area on the leeward side of the mountain, leaving dangerously dry conditions there. Any small spark can start a raging fire.)

EXTENSION There are many katabatic winds around the world. Both the Santa Ana and the Chinook winds are very warm. The contraction of the air as it moves down the mountains causes the warming of the winds. Another warm katabatic wind is the forewind that blows down the southern Swiss Alps. However, some katabatic winds are so cold that the warming effect of contraction does not affect them. Some examples of these cold mountain winds are the winds that blow down the Alps toward the Mediterranean Sea and the bora winds that blow down the Yugoslavian mountains toward the Adriatic Sea. Have the students research these winds and write a short report describing the source region of the air and the effects of the winds on the regions through which they blow.

SECTION REVIEW

Summary Have the students write short paragraphs describing the characteristics of each of the global and local winds. They should include the location of the winds, the direction of flow, and so on.

Reinforcement Have the students quiz one another on the major concepts of this section. This can be done by using the Section Objectives.

Answers to Section Review

Reading Critically

1. The tendency for winds and ocean currents to turn to the right in the Northern Hemisphere and to the left in the Southern Hemisphere is due to the Coriolis effect.
2. Temperature differences, and therefore pressure differences, between the equator and the poles cause the planetary winds to blow.

Thinking Critically

3. The monsoons are created by large-scale pressure differences, while land and sea breezes are caused by local differences in air pressure.
4. If the earth did not rotate, the planetary winds would probably blow from the poles toward the equator because of differences in air pressure.

INVESTIGATION 8:
Solar Radiation

Skill (Making Inferences)

Preparation of Materials

It is best to obtain a gooseneck lamp for this investigation. The students may prepare the shoebox apparatus prior to the investigation so that the class period can be devoted solely to the experiment. Copies of the chart might be prepared in advance to hand out to the students.

Answers to Analyses and Conclusions

1. The temperature of the black side increased faster.
2. the black side
3. Dark surfaces absorb more light and heat than light surfaces do. The energy absorbed by the dark surface is transferred by convection and infrared radiation to the surrounding air.

Answer to Application

A light-colored car would be best because it reflects much of the solar radiation.

INVESTIGATION 8: Solar Radiation

PURPOSE

To examine the effect of solar radiation on light and dark surfaces

MATERIALS (per group of 3 or 4)

Scissors
Shoe box with lid
Black construction paper
White construction paper
Tape
Celsius thermometers (2)
Lamp
Stopwatch

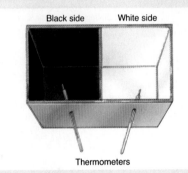

Black side   White side

Thermometers

PROCEDURE

1. Cut the lid of the shoe box as shown for use as a partition. Divide the shoe box in half with the partition.
2. Line one side of the box with black construction paper and the other side with white construction paper. Use tape to secure the linings.
3. Cut small holes in the box as shown and insert one thermometer in each side of the box.
4. Center the lamp directly above the box, about 3 cm above the rim.
5. Copy the chart shown.
6. Turn on the lamp and record the temperatures of both thermometers at time zero.

7. Record the temperatures again at one-minute intervals.
8. At the end of 10 minutes, turn off the lamp and continue recording until both thermometers show the same temperature as that recorded at time zero.

TABLE 1: RADIATION AND TEMPERATURE

Time	°C (dark)	°C (light)
0 min		
1 min		
2 min		
3 min		
4 min		
5 min		
6 min		
7 min		
8 min		
9 min		
10 min		

ANALYSES AND CONCLUSIONS

1. In which side of the shoe box did the temperature increase faster?
2. Which side of the shoe box was hotter at the end of ten minutes?
3. Compare the effects of white light on light and dark surfaces.

APPLICATION

Suppose you lived in a tropical area. What color car do you think would be the best to buy? Why?

SUMMARY

- Carbon dioxide, nitrogen, and sulfur dioxide formed most of the primitive atmosphere of the earth. (8.1)
- The removal of carbon dioxide from the primitive atmosphere left nitrogen as the most abundant atmospheric gas. (8.1)
- Nitrogen, oxygen, and argon form 99 percent of the modern atmosphere. (8.1)
- The gases of the earth's atmosphere that are heavier than helium do not escape into space because of gravity. (8.2)
- Ozone in the atmosphere absorbs dangerous ultraviolet radiation from the sun. (8.2)
- Fast protons from the sun heat up the upper part of the atmosphere. (8.2)
- Light consists of particles called photons, which travel from the sun to the earth as electromagnetic waves. (8.3)
- The sun emits light of all colors. The human eye sees this combination of colors as white. (8.3)
- The addition of carbon dioxide to the atmosphere may be enough to produce a strong greenhouse effect and melt the ice of Greenland and Antarctica. (8.4)
- Atmospheric pressure is the weight of the atmosphere. (8.5)
- The atmosphere is divided into layers, much like a multilayered cake. (8.6)
- From the earth outward, the layers of the atmosphere are troposphere, stratosphere, mesosphere, and thermosphere. (8.6)
- The Coriolis effect becomes evident when the Coriolis acceleration is inadequate. The Coriolis effect makes moving bodies veer to the right in the Northern Hemisphere and to the left in the Southern Hemisphere. (8.7)
- The planetary winds form broad belts that surround the earth at various latitudes. (8.8)
- Land and sea breezes are examples of local winds. (8.9)

Write all answers on a separate sheet of paper.

SCIENCE TERMS

Use each of the following terms correctly in a sentence.

atmosphere (165)
atmospheric pressure (175)
barometer (175)
Coriolis effect (181)
electromagnetic waves (170)
greenhouse effect (174)
jet stream (183)
ozone (168)
photons (170)
photosynthesis (166)
radiation balance (172)
solar wind (165)
sunspots (168)
Van Allen belts (180)

SCIENCE QUIZ

Modified True-False

Mark each statement *true* or *false*. If a statement is false, change the underlined term to make the statement true.

1. The atmosphere of the young earth consisted mainly of oxygen.

2. The distance between two crests of a wave is called the wavelength.

3. Infrared radiation has a longer wavelength than ultraviolet radiation.

4. Temperature decreases with altitude in the stratosphere.

5. The Coriolis effect causes bodies moving on the surface of the earth in the Northern Hemisphere to veer to the left.

6. The layer of the atmosphere in which all of the weather occurs is the troposphere.

continues

CHAPTER REVIEW

SUMMARY

The students may review the major concepts in the chapter by reading the summary statements. The statements are cross-referenced to the chapter to facilitate reinforcement of any concepts of which the students feel unsure. Encourage the students to work in groups to quiz one another.

SCIENCE TERMS

The sentence in which the science term is used should reflect an understanding of the definition of the term. You may wish to stage a game of "Password" in which teams of two students compete against one another. One student gives one-word clues while his or her partner tries to guess the term. The teams alternate after each clue is given.

SCIENCE QUIZ

Answers to Modified True-False

1. false, carbon dioxide
2. true
3. true
4. false, increases
5. false, right
6. true

continues

Answers to Modified True-False (continued)

7. false, local
8. false, ionosphere
9. false, cyanobacteria
10. true

Answers to Multiple Choice

11. b
12. d
13. b
14. c
15. c

Answers to Completion

16. oxygen and nitrogen
17. gravity
18. visible
19. radiation
20. troposphere

Answers to Short Answer

21. Green plants produced the oxygen. Most of the carbon dioxide was dissolved in the ocean.
22. If solar radiation increased, the earth's temperature would rise, the polar ice caps would melt, and the world's coastal areas would flood. If solar radiation decreased, the polar ice caps would grow, the earth's climate would be much cooler overall, and an ice age might begin.
23. Most of the warmth in the atmosphere is heat reflected from the earth's surface. As you ascend into the thermosphere, you travel farther from this heat source, and there is less atmosphere to absorb and hold this heat.

Answers to Writing Critically

24. If the earth were rotating faster, the Coriolis effect would be even more pronounced, so much so that wind and ocean currents might become circular cells or eddies rather than curving bands.
25. The overall temperature of the earth would probably increase. This could cause the polar ice caps to melt and sea level to rise significantly, flooding coastal areas around the world.

SCIENCE QUIZ continued

7. Land breezes and sea breezes are two kinds of planetary winds.
8. The aurora occurs in the stratosphere.
9. Oxygen in the early atmosphere was probably produced by green plants.
10. An increase in the temperature of the earth's atmosphere is called the greenhouse effect.

Multiple Choice

Write the letter of the choice that best answers the question or completes the statement.

11. The solar wind consists partially of
    a) dust.   b) protons.
    c) neutrons.   d) electrons.

12. The composition of the early atmosphere of the earth was changed into the modern one by the action of
    a) solar radiation.
    b) the earth's gravitational field.
    c) the earth's magnetic field.
    d) the activity of marine organisms.

13. The most abundant gases in the present atmosphere are
    a) hydrogen and helium.
    b) oxygen and nitrogen.
    c) argon and carbon dioxide.
    d) hydrogen and oxygen.

14. Light from the sun takes _____ minutes to reach the earth.
    a) 15.5   b) 30
    c) 8.3   d) 300

15. In the Northern Hemisphere, the trade winds blow from
    a) north to south.
    b) east to west.
    c) northeast to southwest.
    d) south to north.

Completion

Complete each statement by supplying the correct term or phrase.

16. The most abundant gases in the earth's atmosphere are _____ and _____.

17. The force that prevents the atmosphere from escaping into space is _____.
18. Most of the solar energy is emitted as _____ light.
19. The balance between incoming solar radiation and outgoing terrestrial radiation is called the _____ balance.
20. The lowest layer of the atmosphere is called the _____.

Short Answer

21. Why is the atmosphere of Earth now different from that of the primitive earth?
22. Why is it important not to disturb the earth's radiation balance?
23. Why does temperature decrease dramatically with altitude in the thermosphere?

Writing Critically

24. Describe how the Coriolis effect would affect the pattern of atmospheric and oceanic circulation if the earth were rotating faster than it does.
25. What would happen if the continued addition of industrial carbon dioxide to the atmosphere were to cause a large increase in the greenhouse effect? Explain.

EXTENSION

1. Research the atmospheric conditions of another planet. Find out what differences and similarities there are with Earth's atmosphere and report on these to the class.
2. Find out the direction of the predominant wind in your area and determine how this affects changes in your weather, both daily and long-range.

ANSWERS TO EXTENSION

1. NASA information could be helpful in this. Contact the education center at the following address: Kennedy Space Center, FL 32842.
2. Use the same source as above and also library information.

APPLICATION/CRITICAL THINKING

1. Why does the oxygen level in the earth's atmosphere remain nearly constant?

2. Describe what you think the Coriolis effect would be on Venus, which rotates very slowly, and in a direction opposite to that of Earth.

FOR FURTHER READING

Barth, M. and J. G. Titus, eds. *Greenhouse Effect* and *Sea Level Rise: A Challenge for This Generation.* New York: Van Nostrand Reinhold, 1984. This book discusses the possible global warming that could result from increased concentrations of carbon dioxide and other gases in the earth's atmosphere.

Heppenheimer, T. A. "Earth." *OMNI* (October 1985):18. This article presents an interesting comment on the marvels and inconveniences of wind turbines.

"Hole in the Sky." *Current Science* (December 1986):11. Scientists have detected a thinning of the ozone layer over the continent of Antarctica. The short article discusses the consequences of losing this layer.

Schneider, S. "Nuclear Winter: The Storm Builds." *Science Digest* 93(January 1985):48. The author examines the possibility that a nuclear war will plunge the earth into deadly cold by filling the atmosphere with smoke and dust.

Challenge Your Thinking

These athletes are extremely tired because of a lack of oxygen. The athletes at the 1968 Olympics suffered due to the thin atmosphere in 2240 m high Mexico City, yet many world records were set that year. Explain this apparent contradiction.

Chapter 8 Review   **189**

ANSWERS TO APPLICATION/ CRITICAL THINKING

1. There is a balance between the oxygen used by organisms for respiration and the amount that is replaced by photosynthesis.
2. The effect would be the reverse of that on Earth. Deflection on Venus would be to the left in the Northern Hemisphere and the right in the Southern Hemisphere.

ANSWER TO CHALLENGE YOUR THINKING

The thin atmosphere, although providing less oxygen for respiration, also provides less resistance, thus allowing many world records to be broken.

189

PLANNING THE CHAPTER

Chapter Sections	Page	Chapter Features	Page	Program Resources	Page
Section 1: Evaporation and Condensation	191			Cross-Discipline: Science and Language, *Writing About the Water Cycle* **(A)**	TRB 17
9.1 Evaporation and Humidity **(B)**	191	**Discover:** Demonstrating Evaporation **(B)**	192	Investigation 9.2: *Calculating Relative Humidity* **(A)**	TRB 35 LM 37
		Activity: Measuring Evaporation **(B)**	193		
9.2 Condensation and Dew Point **(B)**	194	A Matter of Fact	195		
		A Matter of Fact	195		
		Section Review	195		
		Investigation 9: Determining Dew Point **(B)**	202	Student Record Book: Textbook Investigations **(B)**	TRB 17
Section 2: Clouds and Precipitation	196			Critical Thinking **(H)** Reading for Content: *Comparing and Contrasting Types of Precipitation* **(A)**	TRB 16 TRB 16 LM 189
9.3 Types of Clouds **(B)**	196	**Discover:** Observing Clouds **(B)**	197	Investigation 9.1: *Creating Clouds* **(B)**	TRB 33 LM 35
		Skill Activity: Reading a Wind Chill Chart **(B)**	198	Concept Extension: *Cloud Identification* **(B)**	TRB 19 LM 241
9.4 Precipitation **(B)**	199	A Matter of Fact	199		
		A Matter of Fact	200		
		Section Review	200		
		Biographies: Then and Now Vilhelm Bjerknes, Warren M. Washington	201		
Chapter 9 Review	203			Vocabulary **(A)**	TRB 9 LM 139
				Tests **(A)** Computer Test Bank	TRB 35

(LM) Laboratory Manual/Study Guide,   **(TRB)** Teacher's ResourceBank™

B = Basic   **A** = Average   **H** = Honors

The coding Basic, Average, and Honors indicates sections or subsections that might be appropriate for different levels of learners. For additional suggestions regarding choice of topic and depth of coverage, see the Pacing Chart on pages T16–T20.

CHAPTER CONCEPTS, OBJECTIVES, AND TERMS

Section	Concepts	Objectives	Science Terms
Section 1: Evaporation and Condensation	■ The process of changing liquid water to water vapor is evaporation. **(9.1)** ■ When water evaporates, cooling occurs. **(9.1)** ■ Air is saturated when it can hold no more water vapor. **(9.1)** ■ Water vapor in the air is humidity. **(9.1)** ■ Relative humidity is the percentage of saturation of the air. **(9.1)** ■ The change from water vapor to liquid water is condensation. **(9.2)** ■ Clouds form when water vapor condenses around aerosol particles. **(9.2)** ■ Fog is similar to a cloud except that it forms near the ground and stays there. **(9.2)**	■ **Describe** how water evaporates from the ocean. ■ **Explain** how condensation produces clouds.	evaporation saturated air humidity condensation dew point aerosols cloud
Section 2: Clouds and Precipitation	■ Different types of clouds form, depending on conditions in the atmosphere. **(9.3)** ■ When hot, humid air rises, cumulus clouds form. Stratus clouds are extended cloud layers that cover the entire sky. Cirrus clouds are the highest clouds in the atmosphere. **(9.3)** ■ Water that evaporates from the ocean precipitates onto the land and the ocean. **(9.4)** ■ Rain, snow, sleet, and hail are all forms of precipitation. **(9.4)** ■ Raindrops form when cloud droplets fuse together. Snow forms when cloud temperatures are below 0°C. Sleet is rain that freezes as it falls through a layer of cold air near the ground. Hail forms as rain freezes in the clouds. **(9.4)**	■ **List** the types of clouds. ■ **Explain** the water cycle.	cumulus clouds stratus clouds cirrus clouds precipitation

Title	Page	Materials
Discover: Demonstrating Evaporation	192	*(per student)* thermometer, piece of cotton, water, pencil, paper
Activity: Measuring Evaporation	193	*(per group of 3 or 4)* thermometers in notched corks (2); ring stand; clamps (2); cotton or gauze; rubber bands; beakers, 100 mL (2); rubbing alcohol; water; wax pencil
Discover: Observing Clouds	197	*(per student)* paper, pencil
Skill Activity: Reading a Wind Chill Chart	198	*(per student)* paper, pencil
Investigation 9: Determining Dew Point	202	*(per group of 3 or 4)* small soup can, ice cubes, thermometer, flashlight

TEACHING SUGGESTIONS

Section 1: Evaporation and Condensation

Demonstration: The Water Cycle

Purpose

To demonstrate how water is recycled through the ocean and atmosphere

Background

The steam produced by the boiling water will condense on the cool inner surface of the metal bowl. After several minutes, the droplets will run down the sides of the bowl and "rain" around the beaker. Carefully control the flame of the burner so that the boiling in the beaker does not become too vigorous. Note that the "rain" is colorless.

Materials

Colored salt water
Beaker, 150 mL
Wire gauze
Tripod
Metal tray
Bunsen burner
Stainless steel mixing bowl
Ring stands with clamps (2)
Sponge
 (blue or green)

Procedure

1. Mix a solution of colored salt water and fill the beaker half full.

2. Place the beaker on the wire gauze on top of the tripod in the center of the metal tray.
3. Position the Bunsen burner under the beaker.
4. Secure the stainless steel bowl in an inverted position over the beaker (the top of the beaker slightly higher than the lip of the bowl) using the clamps and ring stands.
5. Wet the sponge with cold water and place it on top of the bowl to keep the metal cool.
6. Light the Bunsen burner and bring the water to a steady boil.

Questions to Ask the Students

1. Where did the "rain" come from?
   (*It condensed from the air under the bowl.*)
2. What was the purpose of the cool sponge?
   (*The cool sponge caused condensation.*)
3. Why was the rain colorless? tasteless?
   (*Only the water, not the color or the salt, evaporated from the beaker.*)

Class Activity

Have the students make posters showing the water cycle and how it can be disrupted by the activities of humans.

Section 2: Clouds and Precipitation

Field Trip

If possible, spend part of several class periods outside looking at different cloud formations.

CHAPTER 9

CHAPTER OVERVIEW

This chapter presents the various meteorological effects of water in the earth's atmosphere. Included are explanations of the terms, *evaporation, humidity, condensation,* and *dew point.* The chapter concludes with descriptions of cloud types and the different types of precipitation formed in the atmosphere.

Section 1: Evaporation and Condensation
This section focuses on the processes of evaporation and condensation in the earth's atmosphere. The causes of evaporation are discussed first with an emphasis on how temperature affects the rate of evaporation. Humidity is then defined, followed by an explanation of how humidity affects the evaporation rate. The section concludes with a discussion of condensation and dew point. Included are descriptions of aerosols and their effects on condensation.

Section 2: Clouds and Precipitation
This section describes the various types of clouds in the atmosphere, and includes a photographic table of the most important cloud types and characteristics for identification purposes. The section concludes with a discussion of precipitation, including specific descriptions of rain, snow, sleet, and hail.

Water in the Atmosphere

Almost everyone knows what causes a rainbow, or at least that sun and rain are needed for one to form. However, do you know why it rains or how water gets into the atmosphere? Besides the obvious one, what role does the sun play in forming this beautiful picture?

Rainbows form as the sun shines through water droplets in the air.

190

CHAPTER MOTIVATING ACTIVITY

Begin this chapter by showing a film or slides of different types of clouds. After the students have seen the presentation, have them draw pictures of the different types of clouds.

1 Evaporation and Condensation

SECTION OBJECTIVES

After completing this section, you should be able to:
- **Describe** how water evaporates from the ocean.
- **Explain** how condensation produces clouds.

NEW SCIENCE TERMS

evaporation
saturated air
humidity
condensation
dew point
aerosols
cloud

9.1 Evaporation and Humidity

Look closely at Figure 9–1. What causes the dominant blue color of the earth's surface, or the white streaks in the atmosphere?

Both the dominant blue of the surface and the white streaks in the atmosphere are water. There is a great deal of water on the earth and in the atmosphere. Before clouds can form, however, some of the surface water must become water vapor, a gas.

Evaporation The process of changing liquid water to water vapor is called **evaporation.** Most of the water in the atmosphere comes from the tropical oceans on both sides of the equator. The water temperature there is very warm, averaging 25°C. The warmer the water, the faster the water molecules move; the faster the molecules move, the more water evaporates. Why does the temperature affect the rate at which water molecules move? **1**

Figure 9–1. One of Earth's dominant features, as viewed from space, is the cloud cover of the planet (right). Much of the evaporation, which eventually produces clouds, occurs over warm, tropical oceans (left).

1 Heat adds energy to the water molecules, which speeds them up.

Section 1   Evaporation and Condensation   **191**

DEMONSTRATION

For a demonstration of the water cycle see page 189c preceding this chapter

Section **1**: Evaporation and Condensation

SECTION FOCUS

Section Overview This section describes the processes of evaporation and condensation in the earth's atmosphere. Saturated air, absolute humidity, and relative humidity are all explained in terms of their effects on evaporation rates. The process of condensation is described and includes an explanation of dew point.

Section Objectives For a list of section objectives, see pupil's edition page.

New Science Terms For a list of new science terms in this section, see pupil's edition page. You may wish to have the students define the terms before they read the section. You may also wish to read the terms aloud so the students are familiar with the pronunciation of each.

SECTION DEVELOPMENT

9.1 Evaporation and Humidity

DISCUSSION Have the students describe some examples of evaporation they may have noticed. (For example, streets eventually dry after a rain; wet clothes dry when put in a dryer or hung outside.) *Thinking Critically:* Ask the students to explain why they feel cold after they step out of a shower or a swimming pool. (When water evaporates from the skin, the body loses heat.)

EXTENSION Most of the water vapor in the atmosphere comes from the oceans. However, rivers, lakes, animals, and plants also release water vapor. Industrial plants are another source of water vapor in some parts of the world.

THINKING SKILL *(Inferring)* Point out to the students that on a warm, humid day with no breeze, very little sweat will actually evaporate from their bodies. Ask the students to infer which type of day would make them feel cooler—a calm, warm, humid day or a hot, dry, and windy day. (Hot, dry, and windy day because evaporation has a cooling effect and more sweat will evaporate from their bodies on this type of day.)

EXTENSION Explain that there is usually more water vapor in areas close to the earth's surface and particularly near the ocean. Water vapor enters the air by evaporation from sources such as bodies of water and masses of vegetation. Discuss with the students the fact that there is more water vapor in the air of warm regions than in cold regions since warm air can hold more water vapor than cold air. However, there are warm areas, such as the Sahara Desert, that are very dry. *Thinking Critically:* Ask the students to explain why warm desert regions are very dry, even though warm air holds more water vapor than cold air. (Desert regions are cut off from water supplies.)

DISCOVER:
Demonstrating Evaporation

Skill *(Recording Data)*

Use a rubber band to hold the piece of wet cotton around the thermometer bulb. The wet bulb thermometer should have a lower temperature reading because evaporation is a cooling process.

Figure 9–2. If you shake a box of plastic beads, some of the beads may have enough energy to escape the box. In a similar manner, water molecules, heated by the sun, may have enough energy to escape from the surface of the ocean.

DISCOVER

Demonstrating Evaporation

With a thermometer, record the air temperature of your classroom. Then take a piece of cotton, soak it in water, and wrap it around the thermometer bulb. Wait about 15 minutes and record the room temperature again. Is there a difference between the two temperature readings? Explain any difference.

Figure 9–3. Human bodies are cooled as sweat evaporates from the surface of the skin. A dog has no sweat glands, but water evaporating from its tongue as it pants has the same effect.

192   Chapter 9

Imagine that you have a shoe box half-full of small plastic beads. If you supply a little energy to the box by shaking it gently, the beads move around a little, but nothing much happens. Similarly, molecules in a cold body of water do not receive much energy, and the water molecules move very little.

Now imagine supplying that box of plastic beads with more energy by shaking it vigorously. The beads move around a lot; some may even escape from the box. If a body of water receives a lot of heat energy from the sun, the water temperature increases. Some of the faster molecules escape from the water's surface, and the molecules that remain in the liquid also speed up. The energy used to speed up these molecules causes a decrease in the total amount of energy in the water. The decrease causes a loss of heat and lowers the water temperature.

You have probably experienced the cooling effects of evaporation, after splashing yourself with water on a warm day. In the same way, evaporation of sweat helps keep your body cool. Dogs, however, do not sweat; to cool their bodies, they pant. Water evaporating from the mouth and throat helps reduce a dog's body temperature.

Under certain conditions evaporation stops. For example, when the air above a pond is saturated with water, the number of molecules that leave the surface of the pond equals the number of molecules that return to the pond. **Saturated air** is air that is filled with as much water as it can hold. When air is saturated, no more evaporation takes place. For the same reason, water in a capped bottle does not evaporate; the air above the water surface is saturated. If the water molecules in the air above the surface are removed, however, evaporation continues.

Most ocean evaporation takes place in the tropics. There, high temperatures speed up water molecules and increase the rate of evaporation, and winds blow the moisture away before the air becomes saturated.

DEMONSTRATION

Demonstrate to the students that evaporation will not occur when air is saturated. Obtain two wide-mouthed bottles of the same size. Fill both bottles to the same level with water and mark the level with a grease pencil. Use a stopper to seal the opening of one bottle, leaving the other open. Over a period of several days, the students will notice that the water level in the open bottle drops lower than the level in the sealed bottle. Explain that this occurs because the air in the sealed bottle quickly becomes saturated, and no more water will evaporate. The water in the open bottle will continue to evaporate.

ACTIVITY: Measuring Evaporation

How can you show that evaporation is a cooling process?

MATERIALS (per group of 2 or 3)

thermometers in notched corks (2); ring stand; clamps (2); cotton or gauze; rubber bands; beakers, 100 mL (2); rubbing alcohol; water; wax pencil

PROCEDURE

1. Using rubber bands, carefully fasten a piece of cotton or gauze to the bulb of each thermometer.
2. Attach the clamps to the ring stand, 10–20 cm above the table. Insert the corks into the clamps and tighten.
3. Fill one beaker with rubbing alcohol and the other with water. Label each beaker with a wax pencil.
4. Read and record the temperature on each thermometer.
5. Immerse one thermometer in the alcohol and the other thermometer in the water. Be sure to saturate the cotton.
6. At 30-second intervals, record the temperature on each thermometer.

CONCLUSIONS/APPLICATIONS

1. After the cotton was saturated with water or alcohol, did either thermometer show a change in temperature?
2. If so, which thermometer showed a faster change in temperature? Explain.

Humidity   Very seldom is the air saturated with water vapor. The amount of water vapor in the air is called **humidity.** The actual amount of water vapor per cubic meter of air is called *absolute humidity.* Humidity, however, is usually expressed as *relative humidity,* or the percentage of saturation. For example, air at 30°C can hold a maximum of 26 g of water per cubic meter. If the air actually contains this amount of water vapor, the air is saturated, and the relative humidity is 100 percent. If the same air contains only 13 g of water per cubic meter, the relative humidity is only 50 percent.

Relative humidity affects the rate of evaporation from a body of water or from a human body. The higher the relative humidity, the slower evaporation occurs. The combination of heat and high humidity can be very uncomfortable, even dangerous. If there is no evaporation of sweat from the surface of your skin, your body temperature goes up. A rise of only a few degrees produces heat exhaustion. A rise of a few more degrees causes collapse and possibly death.

Figure 9–4. If sweat cannot evaporate from a person's skin, the body temperature continues to rise. If the temperature rises high enough, the person may collapse or even die.

Section 1   Evaporation and Condensation   **193**

ACTIVITY:
Measuring Evaporation

Skill *(Measuring)*

Preparation of Materials   Before the students begin, you may wish to review the reading of a Celsius thermometer. Emphasize that the students should carefully observe the thermometer readings for accuracy. Remind the students that they should handle the thermometers with care.

Emphasize to students that the wet bulbs should be removed from the liquid and fanned before temperatures are read.

Answers to Conclusions/Applications

1. Both thermometers should show a temperature drop.
2. The alcohol thermometer should show a faster drop in temperature because alcohol evaporates more quickly than water.

EXTENSION   Explain to the students that temperature directly affects relative humidity. The warmer the air, the greater its capacity to hold water vapor and therefore the more moisture needed to generate high relative humidity. The opposite occurs if the air cools. The relative humidity increases, and the air nears the saturation point even though the absolute moisture content is unchanged. This explains why the relative humidity may be near 100 percent in a snowstorm, but the cold air holds less than half the moisture held in the same amount of hot desert air at the same level of absolute humidity.

SKILL   *(Calculating)* To help the students understand relative humidity, you may wish to give them the following practice problems: a) If $1m^3$ of air contains 3 g of water vapor but could hold 6 g, what is the relative humidity? (relative humidity = 3/6 = 0.5, or 50 percent) b) If $1m^3$ of air contains 9 g of water vapor but could hold 12 g, what is the relative humidity? (relative humidity = 9/12 = 0.75, or 75 percent)

EXTENSION Relative humidity is often measured with a psychrometer. The psychrometer has two thermometers, the wet-bulb thermometer and the dry-bulb thermometer. The wet-bulb thermometer is covered with a moist cloth. When air passes over the wet bulb, the water in the cloth evaporates. This evaporation cools the thermometer bulb. If humidity is low, evaporation will occur quickly and the temperature of the wet-bulb thermometer will drop rapidly. If humidity is high, however, evaporation will be slow and the temperature will remain steady. Meteorologists determine the relative humidity by calculating the difference between the dry-bulb and the wet-bulb temperature and reading a table such as the one on page 570 of the Reference Section.

9.2 Condensation and Dew Point

DISCUSSION Tell the students that dew forms best on still, clear nights. When the wind is blowing, the air cannot stay in contact with the cool objects long enough to cool to the dew point. ***Thinking Critically:*** Ask the students to explain why a clear sky would be a factor in the formation of dew. (The earth and its exposed objects lose heat more slowly when the sky is cloudy.)

The combined effect of heat and humidity, sometimes referred to as the "humiture," is called the *heat index*. The combined effect of cold and wind is called *wind chill*. These terms express what the temperature "feels like" to your body. High humidity can make a 35°C day feel warmer than a dry 40°C day. A 30 km/h wind can make 10°C feel like 0°C. A heat-index chart and a wind-chill chart are shown in the Reference Section on page 571.

○ ***What is evaporation?***
○ ***Why does evaporation stop when the air becomes saturated?***

9.2 Condensation and Dew Point

The warm air of day holds more water vapor than the cooler night air. As the temperature drops during the night, the relative humidity may reach 100 percent. If the temperature continues to fall, some of the water vapor in the air changes back to liquid. The change from water vapor to liquid water is called **condensation.**

When water condenses on the ground, it forms dew. The temperature at which dew forms is called the **dew point.** The dew point changes depending on the relative humidity of the air; humid air has a higher dew point than dry air. In the morning the temperature usually rises above the dew point and the dew evaporates. If the dew point is below 0°C, the dew freezes and becomes frost. As the temperature rises in the morning, the frost melts and then evaporates.

Have you ever seen a hot-air balloon floating gently in the cool morning breeze? The balloon rises because the warm air inside the balloon is less dense than the cool air outside the balloon.

The air in the balloon is warmed by a gas-fired burner. In nature, air is warmed by the ground, which is heated by the sun. As warm air rises, it cools, and its relative humidity increases. This causes the water vapor in the air to condense.

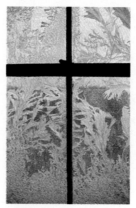

Figure 9–5. Water from the atmosphere often condenses on smooth surfaces, such as window panes. If the temperature is below 0°C, frost will form.

Figure 9–6. Hot-air balloons are usually flown early in the day because the air is cooler then and less heat is needed to make the balloons rise.

Answers–9.1

○ Evaporation is the process of a liquid changing to a gas.

○ Evaporation stops when the number of molecules leaving the surface of the liquid is equal to the number of molecules returning to the liquid.

BACKGROUND INFORMATION

In 1751, a French physician named Charles LeRoy made a breakthrough in the study of the behavior of air. Through experimentation, LeRoy discovered an essential principle: that any part of air has a certain temperature (the dew point) at which the water vapor it contains will condense to liquid form. This discovery helped other scientists, particularly John Dalton, develop the important theory that water vapor is one of several gases that form air. Prior to this, it was generally believed that water vapor was simply another form of air.

Condensation increases when microscopic solids, such as dust, smoke, salt crystals, and pollutants, are in the air. These solids in the air, called **aerosols,** act as nuclei for condensation, and water droplets grow around them. A **cloud** is a collection of water droplets in the atmosphere.

Fog is similar to a cloud except that it forms near the ground and stays there. If the temperature rises within a cloud or in an area where there is fog, the droplets evaporate and the cloud or the fog disappears.

A MATTER OF FACT

More than two-thirds of the fresh water on our planet is unavailable for drinking. This water is frozen in icecaps and glaciers.

Figure 9–7. As the morning sun warms the air, fog often lifts, forming clouds like those above. The fog lifts because the warmed air rises. Eventually the water droplets evaporate.

The droplets that form a cloud are so small that air currents keep them floating. Cloud droplets tend to combine with each other to form larger droplets and, eventually, raindrops. Raindrops are droplets that are too big to float. There are about one million cloud droplets in every raindrop.

○ *What is condensation?*
○ *Why does warm air rise?*

Section Review

READING CRITICALLY

1. Why do you feel chilly if, after a swim, you let your body dry in the wind?
2. What is the relationship between dew point and relative humidity?

THINKING CRITICALLY

3. How might an increase in pollution increase the amount of condensation in the atmosphere?
4. Why is frost more likely to occur on clear, calm nights than on cloudy, windy nights?

A MATTER OF FACT

More than two hundred years ago, the people of Allahabad, India, made more than 100 tons of ice without refrigeration. They ladled water into bowls and set the bowls deep in the ground. By the cooling of evaporation the water froze, even though the air temperature was well above 0°C.

Section 1 Evaporation and Condensation **195**

EXTENSION Tell the students that ice and aerosols are types of particles on which water molecules will condense. The largest of these particles are ice, but they are the smallest in number. Aerosols, or cloud condensation nuclei, are particles on which water molecules condense to form cloud droplets. These nuclei are submicroscopic solid or liquid particles usually produced by dust or smoke. Point out that the particles are very important in cloud formation. Without the particles, clouds would not form until the air became saturated. This would cause fog and desert conditions on Earth.

SECTION REVIEW

Summary Review the processes of evaporation and condensation with the students. Be sure they understand the difference between these two processes. Have the students explain the terms *saturated air* and *relative humidity*. Have the students explain the differences between fog and a cloud. (Fog is a cloud at ground level.)

Reinforcement You may wish to have the students quiz each other on the key processes discussed in this section.

Answers–9.2

○ Condensation is the change from a gas to a liquid.

○ Warm air rises because it is less dense than cool air.

Answers to Section Review

Reading Critically

1. The wind circulates and blows away the water molecules from the body's surface. Evaporation is a cooling process, so the skin is left feeling cool.

2. The dew point is the temperature at which the air is completely saturated, so at the dew point the relative humidity is 100 percent.

Thinking Critically

3. Water molecules are able to adhere to aerosols, such as dust. Other water particles will cohere to the water molecules already adhering to the aerosols, causing larger and larger droplets to condense out of the air.

4. Clouds tend to hold some of the heat of the day, keeping the temperature above the point at which frost will form. Wind tends to keep the air circulating, mixing warm air with cool, and keeping the temperature above freezing.

SECTION FOCUS

Section Overview This section focuses on the classification of clouds and the various types of precipitation formed in the atmosphere. Clouds are first discussed with an accompanying table of photographs and descriptions of the most common types of clouds. Precipitation is then discussed, and the four major types of precipitation (rain, snow, sleet, and hail) are described.

Section Objectives For a list of section objectives, see pupil's edition page.

New Science Terms For a list of new science terms in this section, see pupil's edition page. You may wish to familiarize students with these terms before they read the section.

SECTION DEVELOPMENT

9.3 Types of Clouds

DISCUSSION Tell the students that the clouds shown in Table 9–1 are the most common types of clouds observed in the sky. *Thinking Critically:* Ask the students to explain why meteorologists would have a need to classify types of clouds. (They can predict certain weather patterns through observation of specific types of clouds in the sky.)

EXTENSION The students may be interested in knowing that clouds represent only 0.001 percent of the earth's moisture. Even if all the water in clouds plus the water vapor contained in the atmosphere were condensed and spread evenly on the earth's surface, there would only be about 3 cm of rain.

NEW SCIENCE TERMS
cumulus clouds
stratus clouds
cirrus clouds
precipitation

SECTION OBJECTIVES
After completing this section, you should be able to:
- **List** the types of clouds.
- **Explain** the water cycle.

9.3 **Types of Clouds**

The different types of clouds form depending on conditions in the atmosphere. The atmospheric conditions that affect cloud formation are: air temperature; air pressure; humidity; and air currents, or wind.

TABLE 9–1: CLOUDS OF THE TROPOSPHERE

Type	Description
Cirrus	Elevation: 7 km to 13 km. Thin, wispy clouds made of ice crystals. Associated with fair weather but may indicate approaching storms.
Cirrocumulus	Elevation: 7 km to 13 km. Thin clouds, which look like ripples or waves. Associated with fair weather, but may indicate approaching storms.
Cirrostratus	Elevation: 7 km to 13 km. Thin clouds that cause halos around the sun or moon. Associated with fair weather but may precede storms.
Altocumulus	Elevation: 2 km to 7 km. Light gray clouds in rolls or patches. Associated with fair weather but may precede storms.
Altostratus	Elevation: 2 km to 7 km. Gray clouds that usually produce light rain or snow. May precede warmer weather.

196   Chapter 9   Water in the Atmosphere

BACKGROUND INFORMATION

In 1803, Luke Howard, a pharmacist living in England, was credited with the first useful classification of cloud types. Howard developed a set of Latin names for the main and secondary cloud types. Howard's classification system is still used today, although it has been somewhat modified and expanded. His two primary cloud categories, heap clouds (cumulus) and layer clouds (stratus), are still the basis of most classification systems.

When hot, humid air rises, **cumulus** (KYOO myoo luhs) **clouds** form. Large cumulus clouds that rise to the top of the troposphere, where they spread out, are called *cumulonimbus clouds*. These are the familiar "thunderheads" of summer. **Stratus clouds** are extended cloud layers that may cover the entire sky. If the cloud is several kilometers thick, and produces rain or snow, the cloud is called *nimbostratus*. **Cirrus clouds** are the highest clouds in the troposphere. They are made of ice crystals, not water, and look like fine, white filaments. The clouds of the troposphere are summarized in Table 9–1.

Two types of clouds occur in the stratosphere. *Nacreous* (NAY kree uhs) *clouds* are translucent sheets of ice crystals. They are sometimes known as mother-of-pearl clouds because of their pearly appearance. *Noctilucent* (nahk tuh LOO sehnt) *clouds* are less dense than nacreous clouds, so they look very thin and wispy.

○ *Why do different types of clouds form?*
○ *Which clouds are formed in the stratosphere?*

Figure 9–8. Most clouds form from water droplets in the troposphere, but nacreous clouds, such as these, are formed of ice crystals in the stratosphere.

Elevation: 0.5 km to 2 km Gray, rolled clouds in a continuous layer. May produce light rain or snow. **Stratocumulus**	
Elevation: 0.5 km to 2 km Low, gray clouds that cover the entire sky. Usually produce light snow or drizzle. **Stratus**	
Elevation: 0.5 km to 2 km Thick, dark clouds that block out the sun. Usually produce steady, heavy rain or snow. **Nimbostratus**	
Elevation: 0.5 km to 10 km White, puffy clouds usually associated with fair weather. May produce rain or snow showers. **Cumulus**	
Elevation: 0.5 km to 20 km Towering clouds, sometimes with flattened tops. Produce thunderstorms, high winds, heavy rain, and hail. **Cumulonimbus**	

DISCOVER

Observing Clouds

Every day for the next week, write down the names of all the different kinds of clouds that you see. Along with the name of each kind of cloud, tell whether any precipitation is falling. At the end of the week, try to determine what kinds of clouds produce precipitation.

Section 2   Clouds and Precipitation   **197**

SKILL *(Classifying)* Have the students study the cloud formations shown in Table 9–1. Then have them close their books. Give the students slight hints of a type of cloud you are thinking of and have them guess the cloud. (for example, "highest clouds"—answer, cirrus clouds)

EXTENSION Some unusual clouds in the atmosphere include jet stream clouds, which are formed wherever intense weather systems pass; water droplet clouds emitted by geysers in places such as Yellowstone Park; and condensation trails, or contrails, formed by high-flying aircraft.

SKILL *(Writing)* Have the students research information on three unusual types of clouds: pileus, mammatiform, and noctilucent. Have the students write a paper describing these clouds and how they are formed.

DISCOVER: Observing Clouds

Thinking Skill *(Drawing Conclusions)*

Suggest that the students set up a chart with the days of the week listed across the top and "Cloud Types" and "Precipitation" down the side. This should make it easier for the students to organize their data.

DEMONSTRATION

You may wish to demonstrate the formation of a cloud. You will need a large glass bottle with a rubber stopper and a 10-cm length of glass tubing. Place about 2 or 3 cm of warm water in the bottle and sprinkle a little chalk dust into the bottle. Connect the glass tube to a bicycle pump with a piece of rubber tubing. Have one student hold the stopper and another student pump air into the bottle. When air has been compressed inside the bottle, let the stopper out and have the students observe what happens. You may need to add some smoke from a smoldering match to get a good cloud. Have the students explain why the cloud formed.

Answers–9.3

○ Different types of clouds form because of different conditions in the atmosphere.

○ Nacreous and noctilucent clouds form in the stratosphere.

Objectives

- Interpret a weather chart.
- Find wind-chill temperatures for specific weather conditions.

Discussion   As with the heat index, wind chill is the combined effect of two weather conditions—temperature and wind speed. The combined effect makes it seem colder than it really is. The reason for this is that the wind removes heat from around the body. Review the process of evaporation with the students and ask them how evaporation causes the body to feel colder than it actually is outside. (Any sweat will quickly be evaporated by the usually dry, cold wind. Since evaporation is a cooling process, this will make the person feel colder as well.)

Background

In the United States, most wind-chill charts use degrees Farenheit and miles per hour.

Answers to Application

1. 0°F
2. −10°F
3. −15°F
4. −20°F
5. −25°F
6. −35°F
7. −40°F
8. −45°F
9. −50°F
10. −60°F

Answer to Using What You Have Learned

Answers will vary depending on the temperature selected.

BACKGROUND

Many times the easiest way to present a lot of information is by using a chart or table. The heat-index chart on page 571 of the Reference Section demonstrates this. Find the current temperature and the relative humidity and locate the point where these rows intersect. That point is the heat index.

There is a similar relationship between temperature and wind speed. Because moving air tends to speed up evaporation, and also to remove heat, a strong wind on a cool day makes it feel colder than it really is. This combination of temperature and wind is known as the wind chill.

PROCEDURE

Look at the part of a wind-chill chart presented here. The apparent wind-chill temperature is found by locating the column with the current temperature and then the row with the wind speed. Where the two intersect is the wind-chill temperature.

APPLICATION

Using the wind-chill chart, find the wind-chill temperature for each of the following conditions:

1. 15°F — 10 mph
2. 10°F — 10 mph
3. 5°F — 10 mph
4. 0°F — 10 mph
5. − 5°F — 10 mph
6. − 10°F — 10 mph
7. − 15°F — 10 mph
8. − 20°F — 10 mph
9. − 25°F — 10 mph
10. − 30°F — 10 mph

USING WHAT YOU HAVE LEARNED

Find a specific wind-chill temperature, such as − 25°F, and see how many combinations of wind speed and air temperature will produce this same apparent temperature. A complete wind-chill chart is shown on page 571 of the Reference Section.

WIND CHILL CHART (APPARENT TEMPERATURE)

Wind Speed		Cooling Power of Wind Expressed as "Equivalent Chill Temperature"															
Knots	MPH	Temperature (°F)															
Calm	Calm	40	35	30	25	20	15	10	5	0	−5	−10	−15	−20	−25	−30	−35	−4
		Equivalent Chill Temperature															
3	6	5	35	30	25	20	15	10	5	0	−5	−10	−15	−20	−25	−30	−35	−40	−
7	10	10	30	20	15	10	5	0	−10	−15	−20	−25	−35	−40	−45	−50	−60	−65	−7
11	15	15	25	15	10	0	−5	−10	−20	−25	−30	−40	−45	−50	−60	−65	−70	−80	−8
16	19	20	20	10	5	0	−10	−15	−25	−30	−35	−45	−50	−60	−65	−75	−80	−85	−9
20	23	25	15	10	0	−5	−15	−20	−30	−35	−45	−50	−60	−65	−75	−80	−90	−95	−10
24	28	30	10	5	0	−10	−20	−25	−30	−40	−50	−55	−65	−70	−80	−85	−95	−100	−11
				5									−75	−80	−90	−100	105	11

9.4 Precipitation

Rain, snow, sleet, and hail are all precipitation. **Precipitation** is solid or liquid water that falls to the earth's surface. This completes the water cycle that began with evaporation from the surface of the ocean.

Precipitation varies widely around the world. For example, in some areas of the Atacama Desert of northern Chile, no precipitation has ever been recorded. At the other extreme, on Mount Waialeale, Hawaii, the average yearly precipitation is 11.45 m. That is more than enough water to completely cover a three-story building.

Rain Cloud droplets fuse together to form raindrops. In calm air the drops are very small and are usually referred to as drizzle. Some raindrops, carried by air currents, travel up and down inside a cloud several times, getting larger with each trip. The maximum size of a raindrop is, however, limited by evaporation and friction. Figure 9–10 shows the relative sizes of droplets and a raindrop.

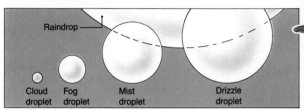

Raindrop

| Cloud droplet | Fog droplet | Mist droplet | Drizzle droplet |

Snow Ice crystals, which form at temperatures below 0°C, collect to form snowflakes. You may have heard that no two snowflakes are alike. That is probably not true, but there is a large variety of shapes possible among snowflakes.

Snow is much less dense than rain. One meter of snow equals only 10 to 15 cm of water. A heavy rainfall drops much more water than a heavy snowfall.

A MATTER OF FACT

The heaviest rainfall on record dropped almost 2 m of water in 24 hours on la Réunion, an island in the Indian Ocean, in 1952. The greatest 24-hour snowfall on record dropped almost 2 m of snow on Silver Lake, Colorado, in 1921.

Figure 9–10. Liquid precipitation, as the diagram shows, may range in size from microscopic cloud droplets to large rain drops that fall during thunderstorms. If condensation occurs at temperatures below 0°C, ice crystals or snowflakes form instead of rain.

Section 2 Clouds and Precipitation **199**

BACKGROUND INFORMATION

In the late 1800s, a young man named Wilson Bentley began viewing snowflakes through a microscope in his home in Vermont. He became so enthralled that over the next 40 years he photographed and catalogued thousands of snowflakes, using photography techniques that were very innovative for the time. In his thousands of microphotographs, Bentley never found two snowflakes that were identical.

9.4 Precipitation

DISCUSSION Tell the students that the first raindrops in a rain cloud that fall toward Earth pass through warmer, drier air below the cloud. Passing through this air causes many of the small drops to evaporate again. This results in a spattering of large raindrops on the ground. These falling drops then cool the ground and moisten the lower air and a downpour soon begins. ***Thinking Critically:*** Ask the students to explain why the air below a rain cloud causes raindrops to evaporate, based on the information in Section 1. (The air is unsaturated.)

THINKING SKILL *(Inferring)* Tell the students that snowflakes fall very slowly. Many atmospheric conditions, such as temperature and humidity, affect the descent of a snowflake. Ask the students to infer why it may not snow very heavily on extremely cold winter days. (Air that is very cold may not have enough moisture to form precipitation.)

EXTENSION Although no two identical snowflakes have ever been observed, the crystals can be widely categorized as either asymmetrical or symmetrical. It has been found that the shape of a snowflake is determined by the temperature and moisture condition of the cloud in which it forms.

Figure 9–11. Sleet (right) is rain that freezes as it falls. Hail (left) freezes in the clouds and then falls to the earth.

Sleet   Frozen rain falls when the air temperature near the ground is lower than the air temperature higher up. This frozen rain is called *sleet*, and it is different from snow. Sleet starts as rain; it then freezes before reaching the ground. Large accumulations of sleet are very rare. Usually either the temperature increases near the ground, changing the sleet to rain, or the temperature in the clouds decreases and the precipitation becomes snow.

Hail   In cumulonimbus clouds, ice crystals may grow into hailstones if the air in the lower part of a cloud is above 0°C. Ice crystals formed in the upper part of the cloud gather a layer of water as they fall. Strong updrafts carry the growing hailstones back up to where the temperature is low enough to freeze the accumulated water. Several round trips between the upper and lower parts of a cloud can produce hailstones the size of marbles. Baseball-sized hail is not uncommon. The largest hailstone on record fell in Kansas in 1970 and weighed 750 g. It was larger than a softball.

○ *What is precipitation?*
○ *What is the difference between sleet and hail?*

A MATTER OF FACT

Hail causes more economic damage by destroying crops in the United States than do tornadoes.

Section Review

READING CRITICALLY

1. What kinds of clouds usually produce precipitation?
2. Describe the process of hail formation.

THINKING CRITICALLY

3. What would happen if the water that evaporates from the ocean and falls on the ground as precipitation were not returned to the ocean? How would this affect the water cycle?
4. Evaporation and condensation are opposite processes. If evaporation is a cooling process, which absorbs energy, then describe the process of condensation.

200   Chapter 9   Water in the Atmosphere

SECTION REVIEW

Summary   Review the formation of a cloud with the students. Have them name some of the most common types of clouds in the atmosphere and describe some of the distinguishing features of these clouds. Review the four types of precipitation described in this section. Then have the students explain the formation of rain, snow, sleet, and hail in the atmosphere.

Reinforcement   Have the students observe and record the weather that they experience over a period of one week. Tell them to record types of clouds, any precipitation, dew, changes in humidity levels, and any other atmospheric changes they notice. Have them compare their notes and discuss what they have observed.

Answers—9.4

○ Precipitation is solid or liquid water that falls to the earth.

○ Sleet is rain that freezes as it falls to Earth. Hail is ice that travels up and down in clouds, gaining layers of water, which freeze in the cloud.

Answers to Section Review

Reading Critically

1. Nimbostratus and cumulonimbus clouds usually produce precipitation.
2. Ice crystals form in clouds when the temperature is below freezing and fall to lower levels, where they pick up a coating of water. They are then carried higher by updrafts in the cloud to where the water freezes. As they start falling again, they pick up another coating of water and

continues

BIOGRAPHIES: Then and Now

VILHELM BJERKNES (1862–1951)

Vilhelm Bjerknes was one of the world's first meteorologists. He was born in Oslo, Norway, on March 14, 1862. He spent most of his youth working with his father, who was a mathematics professor. Together they studied the motions and actions of liquids, such as water, under many conditions of temperature and pressure.

When Bjerknes was 38, he began to work with Heinrich Hertz at the University of Bonn in Germany. They studied the behavior of electrical discharges, such as lightning, through gases. Some of their findings were later used in the development of the radio.

In 1897, Bjerknes became a physics professor at the University of Stockholm in Sweden. Over the next several years, he developed theories about the large-scale movements of the oceans and atmosphere. He began to predict the weather based on the known principles of physics. He found that some of the movements of weather systems are due to the Coriolis effect. An understanding of this effect allows for many weather predictions.

As a teacher, he was greatly respected and admired by his students. In 1904, many of his assistants joined him in founding a program for weather prediction in Leipzig and Bergen. In 1919, his son Jakob was among several researchers who discovered how cyclones begin. Cyclones begin as waves along fronts—the boundaries between different air masses.

WARREN M. WASHINGTON (1936–)

Warren M. Washington was born on August 28, 1936, in Portland, Oregon. Washington showed an early interest in science, especially in physics and in the work of Albert Einstein.

Washington attended Oregon State University, where he graduated with a bachelor's degree in physics in 1958. During the summer after his graduation, Washington worked as a mathematician at Stanford University. His job was to solve mathematical problems relating to the atmosphere. Afterward, Washington returned to Oregon State to continue his education in physics. He received his master's degree in 1960.

Washington's work at Stanford had sparked his interest in the atmosphere. He continued his education at Pennsylvania State University, receiving a doctorate in meteorology in 1964. He had combined physics, mathematics, and computer science in his study of meterology.

After graduating from Penn State, Dr. Washington began work at the National Center for Atmospheric Research (NCAR) in Boulder, Colorado. In 1972 he was appointed by President Richard Nixon to serve on the National Advisory Committee on Oceans and Atmosphere.

Dr. Washington is currently the director of the Climate and Global Dynamics Division at NCAR. He is best known for creating computer models of the atmosphere and the oceans, which are used to study changes that occur there.

Section 2   Clouds and Precipitation   **201**

BIOGRAPHIES: Then and Now

Discussion   After the students have read the biographies, hold a class discussion on the importance of being able to predict the weather. Also discuss how computer modeling of the atmosphere helps us understand it better.

Extension   You may wish to invite some scientists with experience similar to that of the two men featured in this section to speak to the class. Examples might include an astronomer, oceanographer, mathematician, or meteorologist or one of the scientists listed below.

Jacques Cousteau

Geminus

Ptolemy

Carl Sagan

again are carried higher, where the water again freezes. This may continue several more times, until the updrafts can no longer overcome gravity and the hailstone falls.

Thinking Critically

3. If the water were not returned to the ocean, the water cycle would be broken. The cycle depends on the completion of all parts, including evaporation, precipitation, and runoff.

4. Condensation is a warming process. Energy, in the form of heat, is released into the atmosphere.

Skill *(Measuring)*

Preparation of Materials

Any shiny can will work for the investigation, although a smooth surface is preferable. Aluminum soft drink cans work well. In addition to using a flashlight, the students can run their fingers down the sides of the cans to detect moisture.

CAUTION: The students should not use the thermometers as stirring rods and should handle the thermometers carefully to prevent breakage. In step 4, the students should remove the thermometers before adding the ice cubes and stirring the water.

Answers to Analyses and Conclusions

1. Cold tap water may already be lower than the dew point, and condensation would have started at once.
2. Stirring keeps the water temperature even.
3. To record the exact temperature at that moment.
4. This is the dew point.
5. The outside air had a different amount of water vapor in it than did the air inside of the school.

Answers to Application

1. The relative humidity must be near 100 percent.
2. Frost would form.

INVESTIGATION 9: Determining Dew Point

PURPOSE

To determine the dew point and calculate the relative humidity

MATERIALS (per group of 3 or 4)

Small soup can
Ice cubes
Thermometer
Flashlight

PROCEDURE

1. Fill the can about half full of water and let it stand until the water reaches room temperature.
2. Record the air temperature.
3. Place the thermometer in the can and read and record the water temperature.
4. Add a few ice cubes to the water and stir carefully with a pencil.
5. The temperature of the ice-water mixture will drop and this, in turn, will lower the temperature of the can. Watch the outside of the can.
6. Have your lab partner observe the outside of the can so that he or she can be alert to when condensation first starts to appear. He or she should shine the flashlight on the can. This will assist in determining when the condensation first appears.
7. As soon as condensation is seen, your partner should say "Now!" At this instant the temperature of the water should be read and recorded. This temperature is the dew point.
8. If you have time, repeat the experiment outside the school building. Be sure that you are starting with water that is at air temperature.

ANALYSES AND CONCLUSIONS

1. Why did you have to let the water come to room temperature?
2. Why did the water and ice have to be constantly stirred?
3. Why was it important to notice the exact moment when condensation first appeared on the outside of the can?
4. Why did the temperature have to be read at the same moment as the first condensation appeared?
5. Why was the dew point different when you repeated the experiment outside?

APPLICATION

1. If the dew point were close to the air temperature, what could you say about the relative humidity?
2. If the dew point were below 0°C, what would form on the outside of the soup can?

9 REVIEW

SUMMARY

- The process of changing liquid water to water vapor is evaporation. (9.1)
- When water evaporates, cooling occurs. (9.1)
- Air is saturated when it can hold no more water vapor. (9.1)
- Water vapor in the air is humidity. (9.1)
- Relative humidity is the percentage of saturation of the air. (9.1)
- The change from water vapor to liquid water is condensation. (9.2)
- Clouds form when water vapor condenses around aerosol particles. (9.2)
- Fog is similar to a cloud except that it forms near the ground and stays there. (9.2)
- Different types of clouds form, depending on conditions in the atmosphere. (9.3)
- When hot, humid air rises, cumulus clouds form. Stratus clouds are extended cloud layers that cover the entire sky. Cirrus clouds are the highest clouds in the atmosphere. (9.3)
- Water that evaporates from the ocean precipitates onto the land and the ocean. (9.4)
- Rain, snow, sleet, and hail are all forms of precipitation. (9.4)
- Raindrops form when cloud droplets fuse together. Snow forms when cloud temperatures are below 0°C. Sleet is rain that freezes as it falls through a layer of cold air near the ground. Hail forms as rain freezes in the clouds. (9.4)

Write all answers on a separate sheet of paper.

SCIENCE TERMS

Correctly use each of the following terms in a sentence.

aerosols **(195)**
cirrus clouds **(197)**
cloud **(195)**
condensation **(194)**
cumulus clouds **(197)**
dew point **(194)**
evaporation **(191)**
humidity **(193)**
precipitation **(199)**
saturated air **(192)**
stratus clouds **(197)**

SCIENCE QUIZ

Modified True-False

Mark each statement *true* or *false*. If a statement is false, change the underlined word to make the statement true.

1. When water <u>condenses</u>, cooling occurs.

2. When the relative humidity of the cold night air reaches 100 percent, <u>dew</u> forms.

3. Water usually condenses around <u>aerosol</u> particles.

4. <u>Sleet</u> is frozen rain.

5. Snow forms when cloud temperatures are below <u>10°C</u>.

continues

CHAPTER REVIEW

SUMMARY

Students may review the major concepts in the chapter by reading the summary statements. The statements are cross-referenced to the chapter to facilitate reinforcement of any concepts of which the students feel unsure. Encourage the students to work in groups to quiz one another.

SCIENCE TERMS

The sentence in which the science term is used should reflect an understanding of the definition of each term. You may wish to quiz the students on these terms or stage a "science bee" in which two student teams quiz each other.

SCIENCE QUIZ

Answers to Modified True-False

1. false, evaporates
2. true
3. true
4. true
5. false, 0°C

continues

Answers to Multiple Choice

6. b
7. c
8. b
9. b
10. d

Answers to Completion

11. cooling
12. cirrus
13. snow
14. the same
15. aerosols

Answers to Short Answer

16. Condensation nuclei cause saturated air to condense and form cloud droplets.
17. Sleet is rain that freezes as it falls, while snow forms as ice crystals.
18. Humidity and wind might limit the size of raindrops.

Answers to Writing Critically

19. Water that evaporates into the atmosphere must condense and fall to Earth to complete the water cycle.
20. In the summer there is usually no cold layer of air near the earth to freeze rain. In the winter there is not enough convection to form large hailstones.

ANSWERS TO EXTENSION

1. Drawings will vary. The cycle is complete because all the precipitation eventually returns to the oceans, where the evaporation occurs.
2. Temperature, humidity, and type of glass may all affect patterns of frost.
3. Cirrus clouds are high clouds, stratus clouds are low clouds, and cumulus clouds develop vertically.

SCIENCE QUIZ continued

Multiple Choice

Write the letter of the choice that best answers the question or completes the statement.

6. Which of the following are types of frozen rain?
   a) sleet and snow   b) sleet and hail
   c) snow and hail   d) drizzle and sleet

7. Which type of cloud usually produces rain?
   a) cumulus   b) stratus
   c) nimbostratus   d) cirrus

8. As warm air rises, water vapor in the air will
   a) evaporate.   b) condense.
   c) rain.   d) precipitate.

9. Aerosols form nuclei for
   a) evaporation.   b) condensation.
   c) precipitation. d) dew.

10. Water that falls to the earth is called
    a) snow.   b) dew.
    c) condensation. d) precipitation.

Completion

Complete each statement by supplying the correct term.

11. Evaporation is a _____ process.

12. _____ clouds are high, thin clouds made of ice crystals.

13. Precipitation that forms from ice crystals is called _____.

14. The amount of water that evaporates from the earth's surface is _____ as the amount of precipitation that falls on the earth's surface.

15. Particles of dirt in the atmosphere are called _____.

Short Answer

16. Explain the process of cloud formation.

17. How do snow and sleet differ?

18. What factors might limit the size of raindrops?

Writing Critically

19. Explain why precipitation is necessary to complete the cycle of water.

20. Since hail and sleet are both frozen rain, why is there no sleet in the summer and no hail in the winter?

EXTENSION

1. Make a diagram of the water cycle and show, with arrows and labels, why this is a complete cycle.

2. Take some close-up photographs of frost on a pane of glass. Bring these photographs to school and discuss with your classmates some possible reasons for the various patterns of frost on glass.

3. Photograph or draw all the different types of clouds you see in a week. Classify your examples according to height of the clouds.

APPLICATION/CRITICAL THINKING

1. What happens to the energy released during condensation?

2. Consider the following situation. Near the ground, the temperature is $-10°C$. At 100 m, the temperature is $10°C$, and at 500 m, where condensation is taking place, the temperature is $-25°C$. Describe the precipitation that would form and any changes to it that would occur as it fell.

3. The cloud shown here has a flattened top, sometimes called an *anvil* because of its shape. What do you think causes this anvil shape?

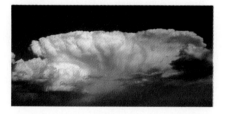

ANSWERS TO APPLICATION/ CRITICAL THINKING

1. The energy released during condensation is heat. This heat remains in the atmosphere.
2. The forming precipitation would be snow. As it fell, the snow would first melt into rain and then near the ground it would refreeze into sleet.
3. The upper level winds blow away the tops of the cloud, making them flat on top.

FOR FURTHER READING

Adler, D. *World of Weather.* Mahwah, New Jersey: Troll, 1984. This book provides basic weather information. The topic of evaporation and condensation is covered, as well as the topic of precipitation.

Cosgrove. M. *It's Snowing.* New York: Dodd, Mead, 1980. This well-illustrated book explains how ice crystals form in the atmosphere and grow into snowflakes. The book also describes some of the problems associated with heavy snowfalls in mountainous areas.

McFall, C. *Wonders of Dust.* New York: Dodd, Mead, 1980. This book provides many interesting descriptions of and facts about dust and its effect on the weather. Topics include dust storms, pollution, and a possible link between dust and hailstorms.

Schaefer, V. and J. Day. *A Field Guide to the Atmosphere.* Boston: Houghton Mifflin, 1981. This book describes the formation of clouds in the atmosphere and the types of precipitation they bring.

Vesilind, P. "Monsoons." *National Geographic* 166(December 1984): 712. This article is a brilliant photographic essay of the earth's most important storms.

Challenge Your Thinking

Recall from reading the chapter that hail grows in size by making several round trips between warm and cold layers inside a cloud. Most hailstorms reach an average size of 3 to 5 cm. Look at the number of hailstones in this picture. What must be happening inside the clouds to allow hail of this number to form?

Chapter 9 Review **205**

Strong updrafts inside the clouds prevent the hail from falling until the stones are quite numerous.

Chapter 10: WEATHER

PLANNING THE CHAPTER

Chapter Sections	Page	Chapter Features	Page	Program Resources	Page
Section 1: Frontal Weather	207			Critical Thinking **(H)** Cross-Discipline: Science and Graphic Arts, *Watching Weather on the Move* **(A)**	TRB 18 TRB 18
10.1 Air Masses and Fronts **(A)**	207			Reading for Content: *Asking Questions* **(A)**	TRB 18 LM 191
10.2 Cyclones **(A)**	209	**Discover:** Making a Water Cyclone **(A)**	209	Investigation 10.1: *Weather Forecasting* **(A)**	TRB 37 LM 39
		Skill Activity: Interpreting Weather Maps **(A)**	210		
10.3 Climate **(A)**	211	A Matter of Fact	211		
		Section Review	212		
		Technology: Using Computers to Model Climate	213		
		Investigation 10: Air Masses and Temperature **(A)**	222	Student Record Book: Textbook Investigations **(A)**	TRB 19
Section 2: Violent Weather	214			Investigation 10.2: *Tracking Hurricanes* **(B)**	TRB 39 LM 41
		A Matter of Fact	215	Concept Extension: *Tornado Safety* **(B)**	TRB 21 LM 243
10.4 Thunderstorms **(B)**	214	A Matter of Fact	217		
10.5 Tornadoes **(B)**	216	**Activity:** Making Weather Observations **(B)**	218		
10.6 Hurricanes **(B)**	219	A Matter of Fact	219		
		Discover: Reading About Storms **(B)**	220		
		Section Review	221		
Chapter 10 Review	223			Vocabulary **(A)**	TRB 10 LM 14
				Tests **(A)** Computer Test Bank 💻	TRB 38

(LM) Laboratory Manual/Study Guide,   **(TRB)** Teacher's ResourceBank™

B = Basic   **A** = Average   **H** = Honors

The coding Basic, Average, and Honors indicates sections or subsections that might be appropriate for different levels of learners. For additional suggestions regarding choice of topic and depth of coverage, see the Pacing Chart on pages T16–T20.

SECTION OBJECTIVES

After completing this section, you should be able to:

- **List** and **sketch** the different types of fronts.
- **Describe** air rotation around a cyclone in the Northern Hemisphere.
- **List** the major climatic zones of the earth.

NEW SCIENCE TERMS

air mass
front
cyclone
climate

10.1 Air Masses and Fronts

When you go home tonight, try to watch the local weather report on the evening news. See how many times the weather reporter mentions the terms *air mass* and *front*. These are terms that are used frequently in weather forecasting. What exactly are fronts? What influence do air masses have on weather?

An **air mass** is a large body of air with nearly uniform temperature and humidity. The characteristics of an air mass are determined by the nature of the surface beneath it. For example, a cool, moist air mass will form over a cold ocean, and a warm, dry air mass will form over a desert.

When different air masses meet, the surface between them is called a **front.** Perhaps, while studying history, you have heard the term *front* used in connection with war. The area where two opposing armies clash is referred to as a front. The term *front* has a similar meaning in meteorology, because there is often a clash between air masses along a front. For example, a front occurs where polar air meets tropical air.

Figure 10–1. This weather map shows several fronts, which are the boundaries between air masses.

207

Section 1: Frontal Weather

SECTION FOCUS

Section Overview This section concentrates on air masses, the fronts that separate them, and the cyclones and anticyclones that form because of the interaction of these air masses. The different climatic zones of the world are introduced, as are the factors that control these zones.

Section Objectives For a list of section objectives, see pupil's edition page.

New Science Terms For a list of new science terms in this section, see pupil's edition page. You may wish to familiarize the students with these terms before they read the section. You may wish to read the terms aloud and have the students repeat them after you so they are familiar with the pronunciation of each term.

SECTION DEVELOPMENT

10.1 Air Masses and Fronts

DISCUSSION Direct the students' attention to Figure 10–1. Have them locate the four low pressure areas shown on the map. ***Thinking Critically:*** Ask the students how many air masses are located over the United States. (Two; they are labeled mild and cool.)

THINKING SKILL *(Inferring)* Before the students read the text, ask them why they think the areas labeled mild and cool are air masses. (They have generally uniform characteristics and are separated by a front.)

EXTENSION The stability of an air mass is important in determining an area's weather. Warm, moist air masses are said to be unstable. Because it is lighter, the warm air at the surface will tend to rise, producing clouds and precipitation. The terms *cold* or *warm* refer to the temperature relative to the ground. A "cold" air mass is colder than the ground it is passing over.

DISCUSSION Direct the students' attention to Figure 10–2 and have them locate the diagrams of the cold and warm fronts. *Thinking Critically:* Ask the students to determine in which front they think air is being pushed up more rapidly. Have them explain their answers. (Air is pushed up more rapidly in a cold front.)

THINKING SKILLS *(Interpreting Diagrams/Inferring)* Direct the students' attention to the diagrams of the warm and cold fronts shown in Figure 10–2. Ask them which front they think will provide precipitation for the longest period of time. (The warm front, since the air is pushed up gradually over the "back" of a colder air mass.)

EXTENSION You may wish to review the symbols used on a daily weather map to show stationary and occupied fronts, and compare these symbols with those that represent warm and cold fronts.

There are four types of fronts that commonly occur over North America: *cold fronts, warm fronts, stationary fronts,* and *occluded fronts.* A front is usually about 1 km wide, but it may be hundreds of kilometers long. Within this area the opposing air masses mix rapidly. The temperature difference between the air masses may be as much as 10°C.

Cold and Warm Fronts Where a cold air mass moves in on a warm air mass, a *cold front* occurs. The cold air mass wedges itself under the warm air mass and raises it. The rising warm air expands and cools. Water vapor condenses and precipitation occurs. Figure 10–2 shows a typical cold front.

When a warm air mass moves over cold air, a *warm front* occurs. As the warm front continues to move, the warm air rises and cools. Water vapor condenses and precipitation occurs.

A cold front and a warm front move in different ways. A cold front moves toward a warm area, while a warm front moves toward a cold area.

Figure 10–2. These diagrams show four different fronts. A cold front (top left), a warm front (top right), a stationary front (bottom left), and an occluded front (bottom right). Notice the position of the different air masses in each type of front, and the direction that each air mass is moving.

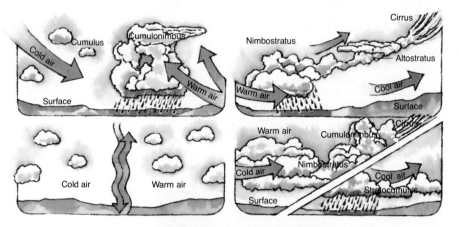

Stationary and Occluded Fronts A front that does not move is called a *stationary front.* This type of front usually does not last more than a day. Stationary fronts separate the warm air of one mass from the cool air of another.

An *occluded front* develops when cold air moves faster than a warm front and overtakes it from behind. The warm air is forced upward and precipitation occurs. A stationary front and an occluded front are shown in Figure 10–2.

○ *How are air masses usually classified?*
○ *Where do fronts form?*

208   Chapter 10   Weather

DEMONSTRATION

For a demonstration of atmospheric pressure see page 205c preceding this chapter.

Answers–10.1

○ Air masses are usually classified by temperature (tropical or polar) and by place of origin (maritime or continental).

○ Fronts form between two different air masses.

10.2 Cyclones

Have you ever seen the movie *The Wizard of Oz*? Dorothy and her dog, Toto, are carried away from Kansas by a "cyclone." The term *cyclone* is used by some people to describe especially severe storms, such as tornadoes. In meteorology the term **cyclone** is used to describe all weather systems that circulate around low-pressure centers.

If two air masses moving in opposite directions collide, they are deflected away from each other by the Coriolis effect. A low-pressure center forms between the two air masses. This low-pressure area attracts the air masses, and, in the Northern Hemisphere, a counterclockwise circulation develops. In the Southern Hemisphere, the circulation around a cyclone is clockwise.

In temperate regions of North America, frontal weather is usually associated with cyclones. Cyclones have a warm front ahead of them and a cold front behind them. Widespread precipitation occurs in advance of the warm front, and a narrow band of more intense precipitation occurs along the cold front. Eventually, the cold air overtakes the warm front, forming an occluded front.

DISCOVER

Making a Water Cyclone

Fill a sink in your classroom or at home. Pull the plug and note the water swirling down the drain. In which direction does the water spin? Explain the reason why it spins in that direction.

Figure 10–3. This diagram shows the development of a cyclone along a cold front. In North America, cyclones move from west to east because of the prevailing westerlies.

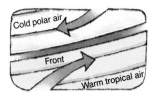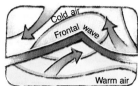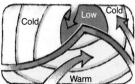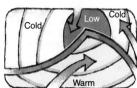

Cyclones and their frontal weather are frequent during the winter, moving eastward in both the Northern and Southern hemispheres. Cyclones follow the planetary wind patterns at a forward speed of about 40 km/h. A cyclone gets its energy from the heat released by condensation. Cyclones may last several days and drop several centimeters of water on areas they pass over.

Areas of low pressure are balanced in the atmosphere by areas of high pressure, called *anticyclones*. Circulation of air in anticyclones is opposite to that in cyclones—clockwise in the Northern Hemisphere and counterclockwise in the Southern Hemisphere. Also, instead of ascending, air in an anticyclone descends. As it descends the air warms and the relative humidity decreases. For this reason, the weather is usually fair within high-pressure areas.

Figure 10–4. The circulation around anticyclones, or high pressure areas, is opposite to that of cyclones.

○ *What are cyclones?*
○ *What causes the circulation of cyclones and anticyclones?*

Section 1   Frontal Weather   **209**

Answers—10.2

○ Cyclones are areas of low (lower) pressure.

○ The circulation is caused by the rotation of the earth and the Coriolis effect.

DISCOVER:
Making a Water Cyclone

Thinking Skill (*Observing*)

Students should observe that the water usually goes down the drain in a counter-clockwise direction. Answers may vary, but it is probably not due to the Coriolis effect.

10.2 Cyclones

DISCUSSION Direct the students' attention to Figure 10–3. ***Thinking Critically:*** Ask the students to explain why they think a warm front precedes a cold front in a cyclone. (The air masses move west to east because of the planetary winds.) Ask the students what the term *low* means as it pertains to weather. (an air mass whose pressure [weight] is lower than the air around it)

EXTENSION In the temperature zone where warm, moist air masses often collide with cold, dry air masses, middle latitude lows or wave cyclones will form. These cyclones develop when different air masses flow side by side at different speeds and directions. The wave cyclone will form at the front and is associated with clouds, precipitation, and stormy weather. These lows are migrating systems and will move rapidly with the air masses.

THINKING SKILL (*Inferring*) Ask the students to explain why they think fair weather is associated with anticyclones, and stormy or bad weather is associated with cyclones. (The air in a cyclone is rising, cooling, and releasing energy. The water vapor condenses, causing precipitation. In an anticyclone, the air is descending, warming, and absorbing energy. Water evaporates, and creates fair conditions.)

EXTENSION Temperature usually drops as altitude increases; it increases as altitude decreases. The change in temperature is about 1°C per 300 m. The students should try to picture a "package of air" rising or settling rather than a person moving through the atmosphere.

SKILL ACTIVITY:
Interpreting Weather Maps

Objectives

- Interpret weather data.
- Draw conclusions about future weather conditions.

Discussion Review with the students the symbols used to represent cold fronts, warm fronts, stationary fronts, and occluded fronts. ***Thinking Critically:*** Ask the students to explain why they think New York's weather is similar to that of Boston. (Both cities are under the influence of the same air mass.) Ask why the weather in Pittsburgh is so different from that of New York. (A cold front has passed, and Pittsburgh is under the influence of a cold mass.)

Answer to Application

Maps will vary slightly, but make sure the correct symbols are used for fronts and pressure areas.

Answer to Using What You Have Learned

Denver will have fair and cold weather, while the weather in Chicago will be rainy, with possible clearing late in the day as the cold front passes. The front should reach Philadelphia in about 48 hours.

SKILL ACTIVITY: Interpreting Weather Maps

BACKGROUND

A meteorologist uses maps and symbols to describe weather patterns on the earth's surface. The symbols are used to represent cold fronts, warm fronts, stationary fronts, and occluded fronts. Air pressure and wind velocity as well as air temperature and local precipitation are also shown. The symbols are international symbols, and they can be interpreted by meteorologists all over the world.

PROCEDURE

1. Study the weather map and symbols in the Reference Section on page 584.
2. Try to determine the sky and weather conditions at each city.

APPLICATION

On a map of the United States, indicate the weather conditions for the nation as described in the accompanying paragraph. Use the correct symbols for each weather condition.

WEATHER REPORT

A cold front moving southeast is over California, Nevada, Idaho, and Montana. To the west of the front, there is a high-pressure area with a barometer reading of 760 mm. Associated with the front is a low-pressure area with a barometer reading of 749 mm. The winds over Colorado are from the south, accompanied by clouds and rain.

A warm front is moving northeast ahead of the cold front. The warm front reaches from the low over Montana through Wyoming, Colorado, and New Mexico. A high-pressure area, with a barometer reading of 765 mm, is over Kentucky and Illinois. Partly cloudy skies are found over Tennessee. The Midwest can expect clear skies today. There is fog along the New England coast. These are the conditions in the following cities:

Atlanta—clear, calm
Boise—snow, wind NW
Boston—fog, calm
Chicago—clear, wind SE
Denver—rain, wind S
El Paso—cloudy, wind SE
Knoxville—partly cloudy, wind E
Los Angeles—cloudy, calm
New York—fog, calm
Pittsburgh—clear, wind NE
St. Louis—clear, wind NE
Tampa—partly cloudy, wind SW

USING WHAT YOU HAVE LEARNED

Assuming that the fronts are moving east at 1000 km/day, what do you predict for tomorrow for Denver and Chicago? When will the cold front reach Philadelphia?

10.3 Climate

If you live along the Gulf Coast, in the Southwest, in Hawaii, or in Puerto Rico, you know the weather will be warm throughout most of the year. If you live in the Northeast or the Midwest, you know the summers will be warm, but the winters will be cold and snowy. These weather conditions describe the climate of your region.

Climate is the combination of all the weather conditions that characterize a region. Temperature and humidity are the two most important conditions that affect climate. Other conditions are the number of hours of sunshine, the amount of cloudiness, the amount of precipitation, the latitude, the altitude, and the wind speed and direction. Weather conditions may change from day to day, but climate is the average weather over a long period. Classification of the major climates of the world is shown in the Reference Section on page 578.

The areas near the equator have tropical, humid climates. Rain forests dominate these areas. North of the tropical belt are areas of high pressure and low relative humidity. The great deserts of the world, such as the Sahara, are found in these areas.

North of the desert belt lies the temperate area. This area includes the most fertile land in the world. In this area hot summers, cold winters, and moderate precipitation favor agricultural development. North of the temperate belt lies the polar region. Climate there is cold and dry.

South of the equator are the same climatic belts as are found in the Northern Hemisphere. However, in the Southern Hemisphere you move south from the equator toward the pole. Figure 10–5 shows examples of the major climatic zones of the world.

A MATTER OF FACT

About 100 million years ago, the climate of the earth was considerably warmer than it is today. Fossil evidence shows that tropical creatures could live near the Arctic Circle. Temperatures in the Arctic probably remained above freezing even during the winter.

Figure 10–5. Anchorage, Alaska (left), has a cold climate, Savannah, Georgia (center), has a temperate climate, and Rio de Janeiro (right) has a warm climate.

Climate changes from tropical to polar with not only increased latitude but also increased altitude. Coastal New Guinea, for example, has a tropical humid climate. A little over 100 km inland lies Mount Wilhelm, which is 4700 m high. As you climb its slopes, the climate changes from tropical to polar.

Section 1   Frontal Weather   **211**

10.3 Climate

DISCUSSION Review the definition of climate with the students. Ask them to describe the local climate. Then ask students who have lived in other areas to describe the climate where they previously lived. Ask the students to describe their idea of an "ideal" climate and to explain their thinking.

THINKING SKILLS *(Communicating/Inferring)* Before the students read the text, ask them to list some things that can cause climate to differ from one location to another. (Answers will vary but the factors that determine climate include latitude, altitude above sea level, direction of the prevailing winds, proximity to ocean currents, and mountains and mountain ranges.)

THINKING SKILL *(Inferring)* Latitude is a key factor in determining the climate of an area or region. Ask the students to discuss how latitude determines climate. (Latitude determines the amount of sunlight an area receives both daily and annually, as well as the direction of the prevailing winds. Increasing latitude usually means a colder climate.)

THINKING SKILL *(Inferring Conclusions)* Many deserts are located in areas of high pressure and low humidity. Ask the students to explain why these factors would cause a desert to exist. (Air moves out of a high pressure area to areas of lower pressure. As the air descends it warms, causing evaporation. Air must rise and cool in order for condensation and precipitation to occur.)

EXTENSION You may wish to have interested students use library materials to research the concepts of plate tectonics and continental drift. Have these students report to the class on how the "drifting continents" can be used to explain changes in climate over geologic time.

SECTION REVIEW

Summary Review the differences between fronts and air masses. Have the students discuss the information about cyclones and the weather associated with both cyclones and anticyclones. Have them list the climatic zones from equator to pole and discuss the factors that affect climate.

Reinforcement Have the students complete the Section Review. Allow any students who need assistance to look up the answers or to work with another student.

Figure 10–6. Because of the influence of a warm ocean current, Ireland (top) has a much warmer climate than Labrador (bottom), even though they are at nearly the same latitude. Much of central Asia (right) has a warm dry climate, because the warm, moist winds of the Indian Ocean and Arabian Sea are deflected by the Himalaya Mountains.

Climate can also be affected by ocean currents. The Gulf Stream keeps the Gulf Coast, Florida, and the East Coast of the United States humid. Western Africa, on the opposite side of the Atlantic at the same latitude, is a desert. Further north the Gulf Stream passes near the northwestern coast of Europe. Ireland, England, and Norway have a temperate, humid climate due to the effects of this major current. Greenland and Labrador, directly across the Atlantic, are polar deserts.

Climate may also be affected by topography. All of central Asia is shielded from the ocean by high mountain ranges. Look at Figure 10–6. What do you think the effect is on the climate of central Asia as the warm, moist winds of the Indian Ocean are forced up over the Himalaya Mountains? **1**

○ *List the different climatic zones of the world.*
○ *How might an ocean current affect a coastal climate?*

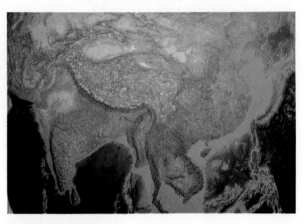

Section Review

READING CRITICALLY

1. How do cyclones develop? Why do they develop?
2. Describe a tropical climate.

THINKING CRITICALLY

3. Why would a cold frontal surface be steeper than a warm frontal surface?
4. Why is the climate of southern California moderate during most of the year while the climate of South Carolina shows great seasonal variations?

1 Central Asia is warm and dry.

212   Chapter 10   Weather

Answers—10.3

○ The major climatic zones of the world are tropical, temperate, arctic, and polar.
○ Oceans tend to cause moderate climates, keeping them warmer in the winter and cooler in the summer.

Answers to Section Review

Reading Critically

1. A cyclone is produced by diverging air masses of different temperatures. Cyclones develop when a cold front overtakes a warm front, lifting the warm air and forming an occluded front.
2. A tropical climate is one in which the temperature stays high all year. The humidity is also high most of the year.

continues

TECHNOLOGY: Using Computers to Model Climate

Computers producing climatic models for analysis

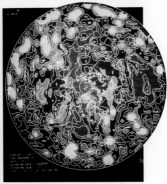

A computer-generated model of weather patterns

Scientists at the National Center for Atmospheric Research, in Boulder, Colorado, are using computers to model the world's climate. Drs. Stephan H. Schneider and Stanley L. Thompson have programmed the computer with vast amounts of weather data. Using mathematics and known principles of physics, they can simulate the present world climate. In addition, they are able to simulate the effects of volcanic eruptions, industrial pollution, and even nuclear war on worldwide climatic patterns.

Schneider and Thompson also simulate the climatic patterns that may have been present ages ago. They consider how the slow movement of the earth's crustal plates might have caused the climate to evolve to its present state. On their computer monitors, they watch ice ages come and go.

They examine how accumulating atmospheric carbon dioxide may change Earth's future climate. If industrial pollution slowly creates a greenhouse effect on the earth, will Earth's climate be like that of the planet Venus? Could nuclear war plunge the world into a nuclear winter, with a climate so cold that no living thing could survive? Computer models of the world's climate simulating these conditions are helping climatologists answer these and other questions.

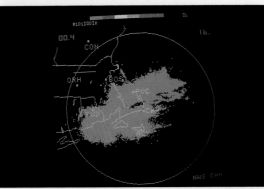

A computer-enhanced radar picture of New England

Section 1   Frontal Weather   **213**

TECHNOLOGY:
Using Computers to Model Climate

Discussion   After the students have read the article, hold a class discussion on the benefits of using computers to simulate climatic conditions. Also, discuss some of the drawbacks, such as unforeseen environmental factors, that could alter the results of the scientists' careful modeling conditions.

Extension   Some students may be interested in doing additional research into this topic. Related topics for research may include environmental pollution, computer modeling, and the history of Earth.

Thinking Critically

3. The cold front will wedge its way under the warm air mass. More cold air moving in will pile up and the frontal surface will become steep. A warm front rides over the colder air so the frontal surface will not be as steep.

4. The climate of southern California is modified by the Pacific Ocean because the prevailing winds are from the west. South Carolina is not influenced as much by the Atlantic Ocean because the prevailing winds are still from the west.

SECTION FOCUS

Section Overview This section discusses the formation of thunderstorms, tornadoes, and hurricanes and the weather associated with each type of storm.

Section Objectives For a list of section objectives, see pupil's edition page.

New Science Terms For a list of new science terms in this section, see pupil's edition page. Many of these terms may be familiar to the students in their everyday speech. Explain to the students that in this section they will learn and use the scientific definitions of these terms.

SECTION DEVELOPMENT

10.4 Thunderstorms

DISCUSSION Direct the students' attention to Figure 10–7. ***Thinking Critically:*** Ask the students to explain why it is possible for a cloud to "grow" to such a great altitude. (The warm, moist air is lifted rapidly, often by a cold front.)

EXTENSION Thunderstorms often form along a squall line. A squall line is an area of intense instability ahead of a fast-moving cold front.

THINKING SKILL *(Inferring)* Before the students read the text, ask them to explain where they think the energy of a thunderstorm comes from. (Energy is released as air rises and water vapor condenses.)

NEW SCIENCE TERMS	SECTION OBJECTIVES
thunderstorm lightning tornado waterspout hurricane	After completing this section, you should be able to: ■ **Explain** how thunderstorms get their energy. ■ **Describe** the formation of tornadoes. ■ **State** the reasons why hurricanes are dangerous.

10.4 Thunderstorms

Have you ever watched cumulonimbus clouds form on a hot, humid afternoon? They may reach several kilometers into the air in only a few minutes. You know that soon you can expect a thunderstorm. A **thunderstorm** is a violent storm that develops from cumulonimbus clouds. Thunderstorms get their energy from the condensation in the clouds. Figure 10–7 shows the development of a typical thunderstorm.

In fair weather, the ground is negatively charged to balance the positively charged ionosphere. During a thunderstorm, however, the negative charges that form in the cloud repel the negative charges in the ground. The negative charges are pushed deep into the ground, which becomes positively charged as a result.

If the difference in charge between the cloud and the ground is great enough, lightning occurs. **Lightning** is a stroke of electrons traveling at speeds of about 1000 km/s. The charge forms a

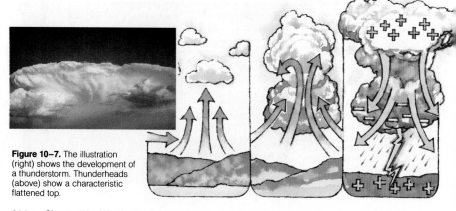

Figure 10–7. The illustration (right) shows the development of a thunderstorm. Thunderheads (above) show a characteristic flattened top.

214 Chapter 10 Weather

tunnel between the cloud and the ground. Lightning can also occur within a cloud or between clouds. The temperature inside the tunnel through which the lightning travels rises to 30 000°C, and air pressure increases dramatically. This sudden increase in pressure causes a shock wave that travels through the atmosphere. You hear this shock wave as *thunder.*

Lightning emits not only visible light but also X rays, ultraviolet light, and radio waves. These radio waves can cause radios to crackle during a thunderstorm.

You see the lightning as soon as it shoots out of the cloud because light travels extremely fast: about 300 million m/s. Sound, however, travels much more slowly: only about 340 m/s. Count the number of seconds between the time you see the lightning and the time you hear the thunder. Divide that number by three and the result will be your distance from the lightning in kilometers.

A MATTER OF FACT

About 45 000 thunderstorms occur each day throughout the world.

Figure 10–8. Lightning can often be spectacular and dangerous. The Empire State Building in New York City (below) is protected by a huge lightning rod that safely channels the lightning into the ground.

You may have been told that lightning strikes the highest object in an area. This is the reason you should not stand under a tall tree during a thunderstorm. Lightning is likely to strike high structures, such as church steeples, factory chimneys, or ship masts. Because of this, it is unsafe to be in the open, to be on top of tall structures, or even to be near tall structures or trees during thunderstorms.

You may have seen structures that have metal rods on the top. These metal rods are called *lightning rods.* They are connected to the ground by means of a thick wire, a rod, or a pipe. If lightning strikes the structure, it will hit the rod first because that is the highest point. The rod will channel the electrical charge into the ground, where it will discharge harmlessly.

○ *Where do thunderstorms get their energy?*
○ *What causes lightning?*

Section 2 Violent Weather **215**

DISCUSSION Ask the students to explain why they see lightning before they hear thunder. Ask them why they often see lightning and never hear the sound. (Light travels faster than sound. The lightning may be many kilometers away.)

EXTENSION The tunnel formed by the lightning is a few centimeters across and is surrounded by ionized air. The stroke that shoots down from the cloud is met about 50 m above the ground by a stroke of positive changes that rises through the same channel at the speed of 10 000 km/s.

THINKING SKILLS *(Comparing/ Drawing Conclusions)* Ask the students if they have ever experienced an effect similar to that of lightning. (The spark that may be visible on a dry day when someone rubs his or her feet on a carpet and then touches something metal is similar to lightning. Both are examples of static electricity.)

EXTENSION The return stroke also will be led through the lightning rod out of the ground to discharge harmlessly in the cloud.

Answers—10.4

○ Thunderstorms get energy from condensation in the clouds.

○ Lightning is caused by a stroke of electrons traveling at speeds of about 1000 km/s.

10.5 Tornadoes

DISCUSSION Direct the students' attention to Figure 10–9. ***Thinking Critically:*** Ask the students to list the conditions necessary for a tornado to form. (Cool, dry air from Canada meets warm, moist air from the Gulf of Mexico. Strong thunderstorms are produced.)

THINKING SKILL *(Sequencing)* Ask the students to use the information provided in Figure 10–9 to infer the sequence of events that lead to the development of a tornado. (Answers from photographs will vary: warm, moist air is lifted rapidly; towering clouds develop quickly; thunder, lightning, and heavy rains with winds occur; tornadoes develop within cumulonimbus clouds.)

THINKING SKILL *(Determining Cause-and-Effect Relationships)* Closed homes and buildings have been known to be blown apart as a tornado passes by. Ask the students to explain why they think this happens. (The winds of a tornado are so strong that few structures, especially frame houses, can survive.)

10.5 **Tornadoes**

The "cyclone" in *The Wizard of Oz* was a tornado. A **tornado** is a small, extremely violent storm. Tornadoes are a kind of cyclone because the air is rapidly circulating around a small area of very low pressure.

Tornadoes occur where two very different air masses collide. One air mass must be warm and humid, the other cool and dry. The advancing cool air pushes the warm air up and away. Warm air rising rapidly along the front condenses into a line of thunderclouds called a *squall line*. The energy given off by condensation causes violent motions along the squall line, and many small, but very intense, low-pressure areas develop. Some of these areas may form tornadoes.

[plural, *vortices*]

A tornado starts as a center, or *vortex*, of low pressure. The tornado's vortex develops from the bottom of the thundercloud toward the ground. This downward extension is caused not by the air moving toward the ground, but by the air below the cloud being drawn into the vortex.

Air pressure at the center of the vortex is estimated to be 10 to 15 percent lower than the pressure outside the storm. This sudden pressure drop causes the air drawn into the vortex to become saturated with water vapor. When saturation occurs, water vapor in the cloud begins to condense, making the funnel visible. Figure 10–9 shows the development of a tornado.

In the Northern Hemisphere, the vortex spins counterclockwise as does that of a cyclone. The width of the vortex may be less than 20 m or more than 1 km. Tornadoes move parallel to the cold front at speeds of 40 to 60 km/h. The path of a tornado along the ground averages 5 km in length. However, tornado paths often are not continuous. The funnel may touch down and destroy one house on a street and not even disturb the house next door. A tornado lasts, on the average, only five to ten minutes.

Figure 10–9. This series of photographs shows the development of a tornado, from a cloud called a *hook* to a fully developed funnel.

BACKGROUND INFORMATION

Within the main vortex of the cloud are smaller temporary vortices that can rotate either clockwise or counterclockwise. These vortices are created by the violent movement within the main vortex and are responsible for the roaring noise of tornadoes.

Figure 10–10. A tornado, with winds as high as 500 km/h, can cause almost total destruction to anything in its path.

Tornadoes are so destructive because of their wind speeds. Horizontal wind speed within a tornado can be as high as 500 km/h. Upward wind speed may be as high as 300 km/h. There are few precise measurements of wind speed within a tornado because the measuring instruments are usually destroyed by the wind. Wind speeds are estimated from the damage caused by the tornado. Updraft speed is based on the fact that objects such as roofs, farm machinery, cattle, and even people have been lifted and carried for hundreds of meters.

Few structures can survive the fury of a tornado. When a tornado occurs, you should go to a room that has no outside openings, or stand in an inside doorway away from any windows. Why do you think you should stand in a doorway?

Although tornadoes occur in every state, they are most frequent in an area of the United States called *Tornado Alley*. As many as 300 tornadoes form there each year. Figure 10–11 shows the location of this area, which extends from northern Texas to southern Illinois.

A MATTER OF FACT

Tornadoes can occur in every state of the United States, but are unlikely in Alaska and Hawaii.

Figure 10–11. The area of the United States from north Texas to southern Illinois is often referred to as *tornado alley* because of the large number of tornadoes that occur there each year.

Section 2   Violent Weather   **217**

DISCUSSION Tell the students that Texas, Oklahoma, and Kansas average more tornadoes per year than any other state. *Thinking Critically:* Ask the students how the topography of this area may explain these high averages. (The flat interior of the United States allows cold air from Canada to flow southward and be deflected southeasterly by the Rocky Mountains. Warm, moist air from the Gulf of Mexico has no barriers as it moves northeasterly.)

EXTENSION Doorways are usually stronger than other parts of a house so they offer protection from falling debris. If you don't have a basement or storm cellar, you should try to stand in an inside doorway away from windows to protect against flying glass.

EXTENSION Many states, such as Florida, Michigan, and Ohio, have a high occurrence of tornadoes but are not part of Tornado Alley.

DISCUSSION Tell the students that Rhode Island, Washington, Vermont, Nevada, and Oregon experience the lowest average number of tornadoes per year. Ask the students to explain why these states have so few tornadoes. (Possible answers include that these states do not experience the contrast in air mass temperatures necessary to produce the thunderstorms required; some have mountain barriers or temperatures moderated by oceans.)

ACTIVITY:
Making Weather Observations

Preparation of Materials These observations should be made by each student or group rather than copied from some media source. Advise the students to make their observations at the same time each day. At the end of one week, see if the students can determine any patterns that would allow them to make forecasts for the following few days.

Answers to Conclusions/Applications

1. Wind from the north or northwest should bring cooler temperatures, while winds from the south or southwest should bring warmer temperatures. There will be local variations due to mountains, large bodies of water, and frontal activity.
2. High or rising air pressure should bring fewer clouds. Low or falling air pressure should bring increased cloudiness and the possibility of precipitation.

Figure 10–12. Waterspouts are funnel clouds that develop over water. Waterspouts cause relatively little damage because the wind speed of a waterspout is usually not as great as that of a tornado.

A **waterspout** is a vortex that occurs over open water. Waterspouts are much less powerful than tornadoes because the temperature difference between air masses at sea is usually less than it is over land. Wind speed within waterspouts rarely exceeds 80 km/h. Waterspouts are common in the coastal waters of tropical and semitropical areas.

○ *Under what conditions do tornadoes develop?*
○ *Why are tornadoes dangerous?*

ACTIVITY: Making Weather Observations

How can you make weather observations with limited materials?

MATERIALS (per group of 3 or 4)
magnetic compass, thermometer, aneroid barometer, anemometer

PROCEDURE
1. Copy the chart below into your notebook. You will record simple weather observations for five days.

TABLE 1: WEATHER OBSERVATIONS

	Day 1	Day 2	Day 3	Day 4	Day 5
Temperature					
Pressure					
Sky					
Clouds					
Wind speed					
Precipitation					

2. Each day, take the outside temperature in a shady location and record it in the chart.
3. With the barometer, find the air pressure each day and record it in the chart.
4. Using the terms *clear*, *partly cloudy*, *partly sunny*, and *cloudy*, indicate on the chart the sky condition each day.
5. If there are clouds, determine the type from the photographs on pages 196 and 197 and record the cloud names on the chart.
6. Indicate on the chart any precipitation that is falling, or that has fallen since the previous observations.
7. Using the compass, and the anemometer, indicate on the chart the wind conditions each day.

CONCLUSIONS/APPLICATIONS
1. Is there any relationship between wind direction and air temperature?
2. Is there any relationship between air pressure and cloud cover or precipitation?

Answers—10.5

○ Tornadoes develop along squall lines between different air masses.
○ Tornadoes are dangerous because of the strong horizontal and vertical winds.

10.6 Hurricanes

What do you think is the most powerful storm in nature? Perhaps a tornado comes to mind. Although certainly destructive, tornadoes are confined to a relatively small area. The most powerful storm is a hurricane.

A **hurricane** is a large storm that develops over a tropical ocean. Hurricanes are also cyclones because of the way in which the air in the storms circulates—counterclockwise around areas of low pressure.

A hurricane develops as part of the planetary wind system. If you examine Figure 8–17 on page 183, you will see that there are several regions of the world where the planetary winds move in opposite directions. One of these regions is off the coast of western Africa. During summer, the southern trade winds cross the equator, where they collide with northern trade winds that blow in the opposite direction. The two are forced apart by the Coriolis effect. This creates low-pressure areas called *tropical depressions*. Although any of these depressions can become a hurricane, only about six, on the average, develop into hurricanes each year. Figure 10–13 shows areas of the world where tropical depressions form.

A MATTER OF FACT

Christopher Columbus was the first European to encounter western-Atlantic hurricanes—once in 1494 and again in 1495.

Figure 10–13. This map shows the areas of the world where tropical depressions are likely to develop.

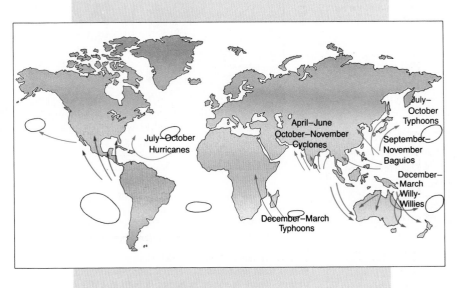

10.6 Hurricanes

DISCUSSION   Direct the students' attention to Figure 10–13. ***Thinking Critically:*** Ask the students to determine two factors that are common in the development of hurricanes. (They form over water in tropical areas and they have winds of at least 123 km/h.)

EXTENSION   Hurricanes in the Northern Hemisphere usually move in a northwesterly direction. Hurricanes that approach the Atlantic coast will often turn to the east when they move into the prevailing westerlies.

EXTENSION   The word hurricane is derived from the Spanish *huracan* and from Portuguese *furacano*. Hurricanes are also called typhoons in the Pacific, cyclones in India, and willy-willies in Australia.

EXTENSION   For names and dates of some of the major hurricanes that have occurred in the United States, have the students look at the maps of Texas, Louisiana, Alabama, and Florida in the Science Sites booklet.

THINKING SKILL *(Inferring)* The eye of a hurricane is a core of calm air. Ask the students in which direction air moves in the eye of a hurricane. (gently downward)

EXTENSION The southwestern edge of the eye of a hurricane usually has the highest wind speeds. The storm's overall speed to the northwest is added to the wind's southwest speed. On the northeastern edge, the speed of the storm subtracts from the speed of the wind.

DISCUSSION The winds of a hurricane usually increase in speed as long as the storm remains over warm water. Ask the students to explain why hurricanes are often described as being "water-powered." (As air rises, water vapor condenses, releasing heat. The air becomes even less dense. As the air in the center becomes lighter, surrounding air moves in faster. The faster this air moves, the more water vapor it picks up. More vapor releases more heat, increasing the storm's energy.)

EXTENSION The eye of a hurricane is usually 10 to 20 km wide. Inside the eye, the weather may be sunny and warm, giving the false impression that the storm is over.

DISCOVER:
Reading About Storms

Thinking Skill *(Collecting and Interpreting Data)*

The students should list the pertinent data from each article. This might include wind direction and speed; the kind, amount, and duration of precipitation; and the effects of these elements.

A fully developed hurricane is almost perfectly circular and is about 600 km in diameter. The pressure difference between the center and the outside of the vortex is about 10%. At the center of the storm is the *eye*—a core of warm, calm air about 10 to 20 km across. The highest winds circle the eye in a band 10 to 50 km wide that extends from sea level to the tropopause. Winds in this band may attain speeds of more than 350 km/h. Wind speed decreases rapidly about 50 km from the eye.

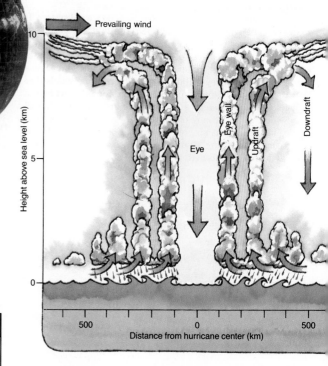

Figure 10–14. This diagram (right) shows a cross-section of a fully developed hurricane. Notice in the satellite photograph (above) that the center of the hurricane, called the *eye*, has no clouds.

DISCOVER

Reading About Storms

From a recent newspaper, find an article about a severe storm. Cut out the article, bring it to class, and discuss your article with your classmates. Determine the kind of damage each of the storms causes.

Outside the vortex are spiral bands of clouds. These clouds extend to the edge of the hurricane. The strength of the wind decreases from the center to the edge. The violent winds within a hurricane often cause tornadoes. These tornadoes increase the danger of hurricanes.

Like thunderstorms and tornadoes, hurricanes get their energy from condensation. Hurricanes develop only within the tropics, where absolute humidity is high and the sun provides heat for evaporation during most of the year.

BACKGROUND INFORMATION

Warm, moist air is very light. Denser air from surrounding areas pushes in and lifts the light air upward. As the denser air moves in across the water, the Coriolis effect causes it to begin to spin counterclockwise in the Northern Hemisphere. A hurricane begins when warm, moist air is lifted by denser air that is spinning couterclockwise into the storm's center.

Figure 10–15. These photographs show the damage caused by hurricane Camille, which struck the Gulf coasts of Mississippi, Alabama, and Florida in 1969. Camille had winds estimated to have been in excess of 400 km/h.

The destructive potential of a hurricane is tremendous. The sustained wind speed of most hurricanes is 100 to 120 km/h but may reach over 150 km/h. Hurricanes produce waves 3 to 5 m high in the open seas. These waves are piled on top of the *storm surge*. The storm surge is a dome of water 1 to 5 m high resulting from the hurricane's low pressure. When a hurricane crosses a coastline and moves onshore, the sea water pushes inland, causing extensive flooding. Strong winds, in addition to the flooding, can cause almost total destruction.

Movement of a hurricane, as it follows the general wind currents, is usually 5 to 25 km/h. Fortunately, the arrival of a hurricane can usually be predicted well in advance. People should follow the advice of authorities and promptly evacuate all low-lying coastal areas that are threatened.

○ *Where do hurricanes develop?*
○ *Why are hurricanes destructive?*

Section Review

READING CRITICALLY

1. What is the source of energy for thunderstorms, tornadoes, and hurricanes?
2. What causes lightning?

THINKING CRITICALLY

3. How do hurricanes create tornadoes? How are these tornadoes different from those created on land?
4. How do hurricanes distribute heat from the tropics to temperate regions?

Section 2   Violent Weather   **221**

THINKING SKILL *(Comparing)* Ask the students to compare a hurricane to a frontal cyclone. List their responses on the chalkboard. (Both are low-pressure systems; both receive energy from condensation of water; hurricanes do not form along fronts but in areas where air is rising; hurricanes are smaller than cyclones, but more powerful.)

SECTION REVIEW

Summary   Have the students summarize in their own words how thunderstorms, tornadoes, and hurricanes develop.

Reinforcement   Have the students complete the Section Review. Allow any students who need assistance to look up the answers or to work together in pairs.

Answers–10.6

○ Hurricanes develop over tropical oceans.
○ Hurricanes are destructive because of the strongs winds and storm surge, and because they may produce tornadoes.

Answers to Section Review
Reading Critically

1. All of these storms get their energy from the heat released as water vapor condenses.
2. When the bottom of a cloud accumulates too many negative charges, the negative charges repel negative charges in the ground, leaving positive charges in the ground. If a large potential is created between the cloud and ground, a lightning bolt will be created.

Thinking Critically

3. Violent wind movement within a hurricane can create tornadoes. Tornadoes within hurricanes are created by a single air mass rather then at a junction of two air masses.
4. They carry moisture that alters the humidity of the atmosphere in a given region. The temperature of the water may also influence the temperature of the surrounding atmosphere.

Thinking Skill *(Experimenting)*

Preparation of Materials

Many of these supplies can be obtained from a local supply store. You may want to ask the school cafeteria to supply the ice required for this investigation shortly before the experiment begins. Before the students trim the cardboard for the fish-tank lid, they may want to place the cardboard on top of the opening of the fish tank and, using a pencil, outline the opening to be sure the cardboard will cover the tank after it is cut.

CAUTION: Be sure the students wear both safety goggles and laboratory aprons to perform steps 7–13 of this investigation.

Answers to Analyses and Conclusions

1. The smoke rose because it was warmer than the surrounding air.
2. The smoke rose, moved around the tank, then sank on the other side of the tank.
3. Answers may vary, but one thermometer should record a temperature increase, and the other a temperature decrease.
4. The smoke moved as part of a convection cell within the tank.

Answer to Application

As the cool, moist air of the sea breeze reaches the dry, warm air over the land, it is forced upward. Clouds form as condensation occurs.

INVESTIGATION 10: Air Masses and Temperature

PURPOSE

To demonstrate that cold air sinks and warm air rises

MATERIALS (per group of 3 or 4)

Scissors
Tape
Fish tank, small
Cardboard, larger than fish tank
Pencil
Thermometers (2)
Ring stands (2)
Test-tube clamps (2)
Paper towels
Beaker, 100 mL
Ice
Petri dish
Safety goggles
Laboratory apron
Candle
Crucible
Matches
Wire gauze
Beaker, 1000 mL

PROCEDURE

1. Using the scissors and tape, trim the piece of cardboard to make a lid for the top of the fish tank.
2. With a pencil, make two holes, one on each side of the lid, so that the thermometers can be inserted into the tank.
3. Secure the thermometers to the ring stands with clamps as shown in the diagram. Cut paper towels into small strips and pack them loosely into the 100 mL beaker.
4. Place the ice in the Petri dish and put it in one side of the tank.
5. **CAUTION: Be sure to wear safety goggles and laboratory apron during the rest of this investigation.**
6. Melt the bottom of the candle just enough so that it can be stuck inside the crucible.
7. Light the candle and put it in the other side of the tank.
8. Cut two holes, each about five to ten centimeters in diameter, in the cardboard. Position the lid so that one hole is over the ice and the other hole is over the candle.
9. Place the wire gauze on the hole over the ice.
10. Insert the thermometers, one on each side of the tank.
11. Light the paper towels in the small beaker and allow them to burn for several seconds. Place the small beaker on top of the wire gauze.
12. Invert the large beaker and use it to cover the small beaker.

ANALYSES AND CONCLUSIONS

1. What happened to the smoke that accumulated in the large beaker?
2. Describe how the smoke moved inside the tank.
3. What temperatures were recorded by the thermometers?
4. Explain why the clouds of smoke moved as they did.

APPLICATION

Explain how this investigation relates to the formation of the clouds that form along with a sea breeze.

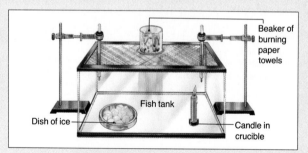

Beaker of burning paper towels

Dish of ice — Fish tank — Candle in crucible

SUMMARY

- An air mass is a large body of air with uniform properties. (10.1)

- Cold fronts wedge under warm fronts. As the warm air rises, it cools, and condensation occurs. (10.1)

- An occluded front occurs as a cold front moves faster than and overtakes a warm front. (10.1)

- Cyclones develop along differing air masses, mostly in winter. (10.2)

- Areas of low pressure are balanced by areas of high pressure called anticyclones. (10.2)

- The three major climatic zones are tropical, temperate, and polar. (10.3)

- The climate of northwestern Europe is temperate humid because of the Gulf Stream. Climate may also be affected by topography. (10.3)

- Thunderstorms are violent storms that develop on hot, humid days. (10.4)

- Lightning rods are used to conduct electric charges into the earth without damage to buildings or people. (10.4)

- Thunderstorms, tornadoes, and hurricanes derive their energy from the condensation of water vapor. (10.5)

- Tornadoes are small, violent cyclones. (10.5)

- Hurricanes are large cyclones that develop in the planetary wind system. (10.6)

- The highest wind speeds in a hurricane are around the eye of the hurricane. (10.6)

Write all answers on a separate sheet of paper.

SCIENCE TERMS

Correctly use each of the following terms in a sentence.

air mass **(207)**
climate **(211)**
cyclone **(209)**
front **(207)**
hurricane **(219)**
lightning **(214)**
thunderstorm **(214)**
tornado **(216)**
waterspout **(218)**

SCIENCE QUIZ

Modified True-False

Mark each statement *true* or *false*. If a statement is false, change the underlined word to make the statement true.

1. A <u>front</u> is the boundary between two air masses.

2. The most intense precipitation occurs along <u>warm fronts</u>.

3. The climate of an area is determined mainly by <u>temperature</u> and humidity.

4. An <u>air mass</u> is a large body of air with uniform conditions of temperature and humidity.

5. A <u>hurricane</u> is the most powerful of all storms.

Multiple Choice

Write the letter of the choice that best answers the question or completes the statement.

6. A cyclone gets its energy from the heat released as water in the atmosphere
   a) evaporates.   b) condenses.
   c) precipitates.   d) diverges.

continues

Chapter 10 Review   **223**

CHAPTER REVIEW

SUMMARY

The students may review the major concepts in the chapter by reading the summary statements. The statements are cross-referenced to the chapter to facilitate reinforcement of any concepts of which the students feel unsure. Encourage the students to work in groups to quiz each other.

SCIENCE TERMS

The sentence in which the science term is used should reflect an understanding of the definition of the term. You may wish to hold a science "bee" during which one team will give the term and the other team must supply the definition.

SCIENCE QUIZ

Answers to Modified True-False

1. true
2. false, cold fronts
3. true
4. true
5. true

Answers to Multiple Choice

6. b

continues

Answers to Multiple Choice
(continued)

7. b
8. d
9. a
10. c

Answers to Completion

11. tropical
12. stationary
13. hot, humid
14. hurricanes
15. condensation

Answers to Short Answer

16. An occluded front occurs when a cold front moves faster and overtakes a warm front. The warm air is pushed upward, and the cold air takes over.
17. Lightning tends to be attracted to high places because of the differences in charges that occur between the cloud and that place. This also becomes the path of least resistance for the discharge of the lightning.
18. Tropical climates are characterized by warm, humid conditions. There is little seasonal temperature change between winter and summer, although there may be wet and dry seasons. An example would be the climate of the Bahamas.

Answers to Writing Critically

19. In the winter, the winds in the air masses on either side of the front are stronger, and a circulation develops very quickly.
20. In summer, the air is heated more by the sun, creating strong convection in the atmosphere. The cumulonimbus clouds can develop higher in the atmosphere with a resulting increase in strength.

ANSWERS TO EXTENSION

1. Make sure records include temperature, wind speed and duration, air pressure, and humidity.
2. Procedures will vary depending on the area.
3. Answers will vary, but most blizzards occur in winter, most tornadoes in spring, and most hurricanes in late summer and fall.

224

SCIENCE QUIZ continued

7. The climate of Scandinavia is
   a) tropical.   b) temperate.
   c) continental.   d) polar.

8. Lightning emits
   a) visible light.
   b) X rays.
   c) radio waves.
   d) all of the above.

9. As a hurricane approaches, people are usually asked to evacuate all
   a) low-lying coastal areas.
   b) inland areas.
   c) urban areas.
   d) rural areas.

10. The two most important factors in determining the climate of a region are
    a) temperature and air pressure.
    b) topography and humidity.
    c) temperature and humidity.
    d) prevailing winds and temperature.

Completion

Complete each statement by supplying the correct term.

11. Hurricanes usually form in the _____ zone.

12. Fronts that do not move are called _____ fronts.

13. Tropical forests of South America have a _____, _____ climate.

14. The most destructive natural force is the _____.

15. A thunderstorm gets its energy from _____.

Short Answer

16. Describe the formation of an occluded front.

17. Why are high places so dangerous during thunderstorms?

18. Describe the conditions of a tropical climate.

224   Chapter 10 Review

Writing Critically

19. Why do cyclones typically form along the polar front more in winter than in summer?

20. Why do thunderstorms occur more often in summer than in winter?

EXTENSION

1. Listen to a sequence of NOAA weather radio broadcasts to see what kinds of information are presented. Keep a record of forecasts for one week and check each the following day for accuracy. If you cannot receive the weather station, use a newspaper or television forecast.

2. Find out if there are any special procedures that must be followed during a weather emergency in your area. Some areas have special snow emergency routes, hurricane evacuation routes, or tornado drills. Make a list of the procedures for your area and discuss them with your family.

3. Go to your school library or public library and find several weather-related stories that were important enough to make the front page of your local newspaper. Try to determine the time of year when most of the storms occur.

APPLICATION/CRITICAL THINKING

1. Look at the weather map on page 207 and describe the current weather conditions in Georgia and Colorado. Predict the conditions for Denver, Colorado, and Atlanta, Georgia, over the next 24 hours.

2. If the overall climate in your area were to suddenly change, what immediate effects might it have on your life? What long-range effects might it have on your community?

3. What are the risks involved with building a house on a beach, next to a river, or on a steep mountain slope?

ANSWERS TO APPLICATION/CRITICAL THINKING

1. In Georgia it is cloudy and mild; in Colorado it is clear and cool.
2. There might be less fresh food available or problems with roads being impassable. In the long term, there may be trouble surviving because of food shortages.
3. These areas are all prone to sudden and severe flooding, which could destroy the building.

FOR FURTHER READING

Gibrilisco, S. *Violent Weather: Hurricanes, Tornadoes, and Storms.* Blue Ridge Summit, Pennsylvania: TAB Books, 1984. This book explains the formation of each of these weather conditions, their similarities and differences, and the destructive power of each.

Ludlum, D. *The Weather Factor.* Boston: Houghton Mifflin, 1984. A meteorologist gives a fascinating account of how the weather has influenced American history and life from colonial times to the Space Age.

Miller, P. "Tornado." *National Geographic* 171 (June 1987):690. This article presents a fascinating review of the team of scientists that chases tornadoes to study them.

"Tornado." *Current Science* (May 1987). This article gives a brief account of one family's encounter with a "twister."

Weisbird, S. "Stalking the Weather Bombs," *Science News,* Vol. 129, 5/17/86, pp. 314–317. This article explains how huge cyclones develop and discusses the extensive variety of instruments scientists use to track and study them.

Challenge Your Thinking

This photograph shows a frozen landscape near Barrow, Alaska. Some scientists have predicted that the climate of all of North America might be like this after a nuclear war. They call this cold climate that might result from a nuclear war *nuclear winter.* If you were to survive the war itself, what problems would you face because of the severe new climate?

The problems would be the bitter cold and a scarcity of food. Most plants could not grow, and animals would soon die off.

THE SCIENCE CONNECTION: TOTO

Discussion This *Science Connection* is about finding ways to make tornado forecasts more reliable. Although the weather computer system that will be available to local weather stations in the 1990s will help make tornado forecasting more reliable, making forecasts of the occurrence of tornados is now very difficult. Ask the students to explain why this is so. (Possible suggestions might include the fact that scientists do not know the precise state of the atmosphere at a particular time or that scientists have difficulty observing the upper atmosphere.)

Discussion Tornadoes can occur at any time during the year, but they occur more frequently in the spring. The students should realize that this is true because the greatest differences in the characteristics of two air masses—polar and tropical—occur during this season of the year.

Discussion The students should be able to discuss the differences among the various types of violent weather, including thunderstorms, tornadoes, and hurricanes. Have the students name a characteristic of a tornado that makes it unique. (a black funnel cloud) The students should understand how this funnel cloud is formed and why it can be so dangerous.

Extension After completing the Unit and reading and discussing the *Science Connection,* have the students turn back to the Unit Opener on page 163 and answer the questions. When the class discussion has ended, you may wish to have the students work in pairs to prepare a report on forecasting tornadoes. Be sure they include information on the formation of tornadoes, the use of TOTO and Doppler radar, and weather-satellite photography.

UNIT 3   THE SCIENCE CONNECTION: TOTO

People can do nothing to stop the awesome fury of a tornado. But with enough warning, they can escape the storm. A team of tornado chasers is helping to make tornado forecasts more reliable. This team from the National Severe Storms Laboratory (NSSL) travels the highways of "Tornado Alley" in western Oklahoma, looking for bad weather. With them they carry TOTO, or Total Tornado Observatory. TOTO is a metal barrel strengthened to withstand the intense winds of a tornado. Inside the barrel are instruments that measure temperature, wind speed, and atmospheric pressure.

The research team carries TOTO into the middle of violent Midwest thunderstorms. If they are lucky, they spot a tornado nearby. Then they drop TOTO in the path of the funnel cloud. As a tornado whips past, TOTO tells them more about the tremendous forces inside a tornado cloud.

The TOTO team begins by looking for weather conditions likely to produce tornadoes. Meteorologists from the National Severe Storms Center help the team locate unstable air masses and gathering thunderstorms. They look for storms with a *mesocyclone,* a rotating column of air inside the storm cloud. Out of this mesocyclone, a tornado funnel can drop to the ground without warning. The team leader yells out, "Tornado on the ground," and the team's vehicles race toward the funnel cloud. In two minutes the team can roll TOTO out of the truck and activate its instruments. However, a tornado can move quickly, too. The NSSL team must be ready to flee if a tornado threatens them. They want to get close to a tornado, but not too close.

Doppler radar screen

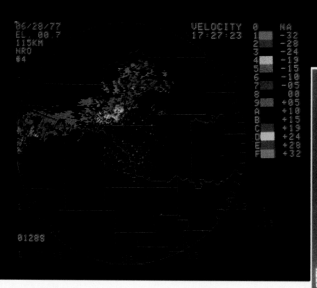

Computer-enhanced radar

Studying a tornado

The TOTO team is backed up by scientists using even more sophisticated weather instruments. Meteorologists at the National Severe Storms Center in Kansas City constantly watch thunderstorms across the U.S. Their aim is to give citizens a few minutes' warning before a tornado strikes. Those few minutes could mean the difference between life and death for the people in a tornado's path.

A special kind of radar, called *Doppler radar,* helps scientists locate tornado activity. Like all radar, a Doppler system sends out an electronic beep. The beep makes an "echo" when it bounces off raindrops and insects caught in a storm. By detecting differences in wave frequency, Doppler radar tells scientists in which direction the winds in the storm are moving and how fast. Rotating winds mean a tornado is brewing. On their radar screens, scientists look for bright red changing into bright green. These colors indicate twisting winds and a tornado on the ground.

Severe storm forecasters also rely on weather satellite photographs to spot bad weather. Infrared photos measure a storm's intensity by showing warm and cold air. Computers can flash several such photographs on a screen and combine them with other weather data. This shows scientists how rapidly a storm is developing. When they see a storm coming, forecasters issue a tornado watch, the first step in warning the public. By the 1990s, local weather stations will be linked up with a weather computer system. This system will give tornado

warnings more quickly and reliably.

During the tornado season, this computer information is also passed to the TOTO weather chasers. The researchers use the information to decide in which direction their team should head to look for tornado weather. Often they find themselves driving through violent rain- or hailstorms. When everyone else is taking shelter, this well-trained and dedicated team is on the lookout — trying to learn more about one of nature's most powerful storms.

A Kansas "twister"

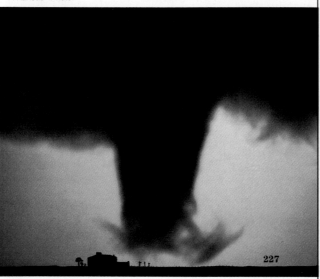

227

Unit 4: WEATHERING AND EROSION

UNIT OVERVIEW

This unit describes the processes of physical and chemical weathering that break rocks apart and contribute to the formation of soil. Also discussed are mass movements and the main agents of erosion and deposition: running water, wind, and glacial ice. Landforms resulting from weathering and erosion are described.

Chapter 11: Weathering and Soil Formation, page 230

In this chapter, the processes of physical and chemical weathering are described. Also discussed are the effects of weathering on rock material, including mass movements, the shaping of the land, and the formation of soil.

Chapter 12: Running Water, page 252

This chapter discusses running water and explains the effects it has on the earth's surface. Also presented are the factors that affect running water. Next, drainage basins, headward erosion, valley formation, and sediment deposition are explained. The chapter then concludes with a discussion of ground water. The water table is defined, and various geologic features associated with ground water are discussed.

Chapter 13: Wind and Ice, page 274

In this chapter, the forces of wind and ice are described. Causes and effects of these two major agents of erosion on the earth are discussed. The land formations resulting from erosion are then described in detail.

Chapter 14: Geomorphology of North America, page 298

This chapter presents an overview of the geomorphology of the United States. Geomorphic provinces are defined. The various geomorphic provinces of the United States are described in detail.

The Science Connection: Glacier National Park, page 318

The Science Connection presents the majestic beauty of Glacier National Park. Special landforms, such as the Purcell Sill, created by the advance and retreat of alpine glaciers, are described. The formation of isolated Chief Mountain is also explained.

ADVANCE PREPARATION

Chapter 11

For the Discover on page 233, you will need to place jars of water into a freezer overnight. For more information, see page 233.

For the Discover on page 244, you will need peat moss, sand, plastic cups, and some seeds (beans) for sprouting. The planted seeds will need about a week to germinate and to grow large enough to make observations. For more information, see page 244.

For Investigation 11 on page 248, you will need many small food jars (baby food) with lids. For more information, see page 248.

Chapter 12

For the Discover on page 254, you will need to place a metal baking dish outside during a significant rain. For more information, see page 254.

For the Discover on page 262, the students will need to find an area that shows erosion caused by runoff. For more information, see page 262.

For the Discover on page 266, you should have several sponges with varying pore sizes. For more information, see page 266.

Chapter 13

For the Discover on page 278, the students will need to find an area that shows erosion caused by the wind. For more information, see page 278.

For the Activity on page 281, you need to bring a hair-dryer with an "air only" setting. For more information, see page 281.

For the Discover on page 283, the students will need to make snowballs (from crushed ice) and place them in a freezer. The process of freezing and partial thawing may need to be repeated several times. For more information, see page 283.

Chapter 14

For the Activity on page 313, you may want to have several national, state, or local park informational brochures available as examples. For more information, see page 313.

BULLETIN BOARD SUGGESTIONS

Chapter 11

Using a rectangle of cardboard approximately 15 cm × 60 cm, construct a soil profile. Use unusual materials, such as pasta, yarn, beads, plastic foam packing forms, and coffee, to capture the students' attention. Glue chunks of varying sizes to represent the different-size rock fragments, humus, and minerals. You may also wish to add something to represent some of the creatures that occupy the soil. (Gummy worms or rubber fishing worms may be used.) Label each horizon and pose a question asking what can be found in that horizon.

Chapter 12

Post a FACT SHEET for the students to record as many facts about cave formations, environments, and inhabitants as they can. As an option, you might maintain individual FACT SHEETS for each class and hold a competition for the most facts gathered.

Chapter 13

Have the students write stories about how the earth, as they know it, would change if another Ice Age occurred. Post the best stories on the bulletin board.

Chapter 14

Create a display of travel brochures from Carlsbad Caverns and Mammoth Cave. Also, display a topographic map of Karst topography. For information and brochures, write to:

Carlsbad Caverns National Park, 3325 National Parks Highway, Carlsbad, New Mexico 88220 and Mammoth Cave National Park, Mammoth Cave, Kentucky 42259.

ISSUES IN EARTH SCIENCE

Several monuments in New York City have been nearly destroyed by air pollution, acid rain has affected many limestone and marble buildings, priceless art works in Europe are in ruins, and even the Sphinx is endangered by pollution. Have some students research the new techniques being developed to save these treasures. Other students could debate the general problems of pollution, and how it can be curtailed to prevent the destruction of more works of art.

In some areas of the western United States, federally owned land is leased to companies that remove the valuable minerals or timber, or graze animals on the open range. Have the students discuss the pros and cons of using publically owned land for making private profit.

Officials in California would like to divert most of the water of the Colorado River to California. This would eliminate use of the water in parts of Arizona and in much of the agricultural areas of Baja California, Mexico. Have the students discuss the rights of Californians to divert this water.

SUGGESTED PROJECTS

Have some students interview local farmers or land developers to find out what methods are used to prevent erosion by running water.

TEACHER RESOURCES

Readings

Chapter 11

Meyers, N. *GAIA: An Atlas of Planet Management*. Garden City, NY: Anchor Books, 1984. This book contains a wealth of information on environmental data, statistical predictions, and critical issues about life and resources of the earth.

Chapter 12

Thornbury, W.D. *Principles of Geomorphology*. New York: John Wiley & Sons, Inc., 1969. This classic text discusses all aspects of weathering and erosion. Information is thoroughly presented, and the text is easy to read.

Chapter 13

Bloom, A. *Geomorphology: A Systematic Analysis of Late Cenozoic Landforms*. Englewood Cliffs, N.J.: Prentice Hall, Inc., 1978. This book provides in-depth information on recent regional erosional history.

Chapter 14

Chronic, H. *Landscapes of North America*. Seattle, WA: The Mountaineers, 1984.

Harris, D.V. *The Geologic Story of the National Parks and Monuments*. Ft. Collins, CO: Colorado State University Foundation Press, 1976. This book contains information on the geology of our national parks and monuments.

Audiovisual/Software

Chapter 11

''Weathering, Mass Wasting, Soil and Ground Water, and Runoff'' Overhead Transparency Program (set of 15 transparencies with spirit masters), Wards.

''Rock Weathering and Mass Wasting'' 35mm color slides with Teaching Guide (set of 20 slides), Wards.

Chapter 12

A computer program, *Erosion,* is available for Apple II and TRS-80 computers from Geoscience Resources.

Chapter 13

The film, *Rise and Fall of the Great Lakes* is available from the Film Board of Canada and *Geological Work of Ice* (2nd Ed.) is available from Encyclopedia Britannica Films.

A computer program, *Continental Glaciation* is available for Apple II computers from Geoscience Resources.

Chapter 14

A world atlas of geomorphic features is available from Geoscience Resources.

Other Materials

Chapter 14

Topographic relief maps of national parks, and geologic road maps of the United States, are available from Geoscience Resources. A stereo atlas of landforms is also available from Geoscience Resources.

Chapter 11: WEATHERING AND SOIL FORMATION

PLANNING THE CHAPTER

Chapter Sections	Page	Chapter Features	Page	Program Resources	Page
Section 1: Weathering	231			Cross-Discipline: Science and Social Studies, *Examining Weathering as a Danger to Historical Structures* **(H)**	TRB 20
11.1 The Process of Weathering **(B)**	231	A Matter of Fact	232	Investigation 11.1: *Physical Weathering of Rocks* **(B)**	TRB 43 LM 45
11.2 Physical Weathering **(B)**	233	**Discover:** Demonstrating Frost Action **(B)**	233		
11.3 Chemical Weathering **(B)**	235	Section Review	236	Student Record Book: Textbook Investigations **(A)**	TRB 21
		Investigation 11: Weathering by Carbonation **(A)**	248		
Section 2: Mass Movement and Landforms	237			Investigation 11.2: *Mechanical Weathering* **(A)**	TRB 45 LM 47
11.4 Mass Movement **(B)**	237	A Matter of Fact	238		
		Discover: Showing Mass Movement **(B)**	239		
11.5 Landforms and Weathering **(B)**	239	Section Review	241		
		Activity: Analyzing a Landform **(B)**	241		
		Skill Activity: Interpreting a Pie Graph **(B)**	242		
Section 3: Soils	243			Critical Thinking **(H)** Reading for Content: *Writing a Simple Outline* **(A)** Concept Extension: *Soil pH*	TRB 20 TRB 20 LM 193 TRB 23 LM 245
11.6 Soil Formation **(A)**	243	A Matter of Fact	243		
		Discover: Making Soil **(A)**	244		
		Technology: Modern Farming	245		
11.7 Soil Composition **(A)**	246	Section Review	247		
Chapter 11 Review	249			Vocabulary **(A)**	TRB 11 LM 143
				Tests **(A)** Computer Test Bank	TRB 46

(LM) Laboratory Manual/Study Guide, **(TRB)** Teacher's ResourceBank™

B = Basic   **A** = Average   **H** = Honors

The coding Basic, Average, and Honors indicates sections or subsections that might be appropriate for different levels of learners. For additional suggestions regarding choice of topic and depth of coverage, see the Pacing Chart on pages T16–T20.

CHAPTER CONCEPTS, OBJECTIVES, AND TERMS

Section	Concepts	Objectives	Science Terms
Section 1: Weathering	■ Weathering is the process that breaks down rocks. **(11.1)** ■ The two major types of weathering are physical and chemical. Factors that affect the rate of weathering are rock structure, climate, topography, and vegetation. **(11.1)** ■ Physical weathering breaks rocks apart mechanically. **(11.2)** ■ Exfoliation, frost action, root-pry, and abrasion are four types of physical weathering. **(11.2)** ■ In chemical weathering, minerals within a rock are chemically changed. **(11.3)** ■ Hydration, oxidation, and carbonation are types of chemical weathering **(11.3)**	■ **Explain** the process of weathering. ■ **List** the factors that affect the rate of weathering. ■ **Compare** the processes of physical weathering and chemical weathering.	soil weathering exfoliation frost action abrasion hydration oxidation carbonation
Section 2: Mass Movement and Landforms	■ Mass movement is the movement of rocks due to gravity. Mass movement is classified by the slope of the land, the size of materials, and the amount of water involved. **(11.4)** ■ The four types of mass movement are creep, flow, landslide, and subsidence. **(11.4)** ■ Differential weathering of rock layers results in the formation of distinctive landforms such as buttes and mesas. **(11.5)** ■ Landforms resulting from mass movement are usually not as spectacular as those produced by differential weathering. **(11.5)**	■ **Explain** why gravity is necessary for mass movement. ■ **Describe** the role of a cap rock in the formation of mesas and buttes. ■ **List** the rock characteristics that would cause differential weathering.	mass movement creep flow landslide subsidence differential weathering
Section 3: Soils	■ Soils are the product of the weathering process. **(11.6)** ■ Soil formation is affected by the type of parent material, the climate, topography, the soil organisms, and time. **(11.6)** ■ The various layers, or horizons, of a soil are shown in a soil profile. **(11.7)** ■ Most of the world's fertile soils are found in temperate regions, such as the United States. **(11.7)**	■ **Identify** the factors that are involved in soil formation. ■ **Summarize** the characteristics of different types of soils.	humus horizons leaching

Title	Page	Materials
Discover: Demonstrating Frost Action	233	*(per student)* small plastic jar or bottle, water, freezer
Discover: Showing Mass Movement	239	*(per student)* large tray, wet sand
Activity: Analyzing a Landform	241	*(per student)* paper, pencil
Skill Activity: Interpreting a Pie Graph	242	*(per student)* paper, pencil
Discover: Making Soil	244	*(per student)* 3 paper cups, sand, peat moss, 6–9 bean or corn seeds
Investigation 11: Weathering by Carbonation	248	*(per group of 3 or 4)* safety goggles, laboratory apron, laboratory balance, small food jars with lids (4), tap water, limestone, carbonated water, granite

TEACHING SUGGESTIONS

Section 1: **Weathering**

Demonstration: Chemical Weathering

Purpose
To observe chemical weathering

Background
How many examples of weathering are the students aware of? Perhaps they have seen a rusty bicycle or lawn mower. The face of a building may be crumbling. A sidewalk may be cracked. A common agent of chemical weathering is ordinary water. Rainwater, for example, usually contains carbon dioxide. It can also be polluted with gases from industrial smokestacks and other sources. In this demonstration you will explore the effect of water and its dissolved substances on different materials.

Materials
Rainwater
Sodium chloride
Graduated cylinder
Carbonated water
Nitric acid
Distilled water
Test tubes (20)
Test tube racks
Four pieces each of marble, limestone, concrete, iron (an iron nail), and copper
Paper towels
Double-pan balance
Forceps
Safety goggles
Laboratory apron

Procedure
1. The solutions needed are: (a) Rainwater. If rainwater is not available, tap water can be used in its place. (b) Salt water. Dissolve a few grams of sodium chloride in 25 mL of distilled water. (c) A clear, carbonated drink can be used for water containing dissolved carbon dioxide. (d) Dilute nitric acid. Add several drops of nitric acid to 25 mL of distilled water.
2. Label test tubes from 1 to 20. Place a set of five test tubes in each test tube rack.
3. Soak all the test materials in distilled water. Wipe each sample dry and find its mass using a double-pan balance. Record the mass of each sample on a table on the chalkboard.
4. Place each test material in a test tube as outlined in the table on page 227f. Add about 5 mL of liquid to each test tube. Again, follow the order in the table.
5. After two days, use forceps to remove the marble from the first test tube. Dry the marble carefully with a paper towel. Find the mass and record it in the second column of the table. Then put the piece of marble back into the first test tube. Repeat this procedure for every substance in the test tubes. *CAUTION: Dilute nitric acid will burn your fingers and damage your clothing.* Lift the sample from the nitric acid with forceps and rinse it with water before you dry it.
6. On the fifth day repeat step 5 for each sample. Record the masses in the third column of the table.
7. At the end of two weeks repeat the procedure, recording the masses in column 4. Now determine the total change in mass of each rock. Subtract the mass on the 14th day from the mass at the beginning of the investigation. Then determine the percentage of mass lost. Divide the amount of mass lost by the total amount of mass at the beginning of the investigation and multiply the answer by 100. Record these results in the last two columns.

TABLE 1: Test Tube Set-up for Demonstration																				
Test Tube	1	2	3	4	5	6	7	8	9	10	11	12	13	14	15	16	17	18	19	20
marble	X					X					X					X				
limestone		X					X					X					X			
concrete			X					X					X					X		
iron				X					X					X					X	
copper					X					X					X					X
rainwater	X	X	X	X	X															
salt water						X	X	X	X	X										
carbonated drink											X	X	X	X	X					
nitric acid																X	X	X	X	X

Questions to Ask the Students

1. Which agent caused the most change in marble?
   (Nitric acid.)
2. Which agent caused the most change in limestone?
   (Nitric acid.)
3. Which agent caused the most change in concrete?
   (All about the same.)
4. Which agent caused the most change in iron?
   (Water.)
5. Which agent caused the most change in copper?
   (Answers will vary.)
6. On which material did rainwater or tapwater have the greatest effect?
   (Iron.)

Field Trip

Take a field trip to a cemetery. This is an excellent site to study the way that different rock materials (mostly granite and limestone) weather. The effects of time become very clear by comparing the older tombstones to the more recent ones. Invite the social studies teacher to use this field trip to develop an interdisciplinary unit.

Section 2: Mass Movement and Landforms

Class Activity

Have the students build landforms out of clay or other common materials. They can show their landform models to the class, explain the processes involved in the formation and weathering of the landform, and identify one site in the United States where such a landform can be found. Possible landforms might include buttes, mesas, caves, or cuestas.

Section 3: Soils

Outside Speaker

Have a speaker from the county agricultural extension station talk to the class about the importance of soil and about farming methods that conserve the soil.

WEATHERING AND EROSION

This is the top of Triple Divide Peak in Glacier National Park. This peak marks the junction of three different drainage basins. Water falling on the northern slope of this mountain flows into the Arctic Ocean by way of Hudson Bay. Water falling on the western slope flows to the Pacific Ocean. Water falling on the eastern slope flows to the Gulf of Mexico and then to the Atlantic Ocean.

- **Where do the glaciers come from that give the park its name?**
- **What are the divides that separate the major drainage basins of North America?**
- **How do wind, water, and ice help shape the land?**
- **Why are certain areas of North America set aside as national parks?**

By reading the chapters in this unit, you will learn the answers to these questions. You will also develop an understanding of concepts that will allow you to answer many of your own questions about the study of the processes that shape the surface of the earth.

Triple Divide Peak

229

INTRODUCING THE UNIT

Have the students look at the Unit Opener photograph on these pages. Point out that Triple Divide Peak is located on the Continental Divide. Read the text and use the questions to guide a class discussion. Do not expect the students to be able to answer the questions now. The questions are motivational and the students should be aware that they will be able to answer them after studying the Unit.

Discussion Only 100 years ago George Grinnell discovered the large glacier, named Grinnell Glacier, in what was to become Glacier National Park. It was Grinnell who realized that rain falling on Triple Divide Peak flowed from that one spot to the Pacific Ocean, to the Gulf of Mexico, and through Hudson Bay to the Arctic Ocean. Display a map of the United States and ask the students to trace the paths this rainwater follows.

Extension You can extend the discussion by bringing in other Unit themes. Some of the forces of nature that have shaped and are still shaping the earth are wind, water, and ice. What kinds of places on the earth are being shaped by water? What kinds of places are being shaped by wind? Is ice still shaping the earth? When the class has completed the Unit, ask them these questions again. The follow-up discussion may help the students realize how much they have learned.

CHAPTER 11

CHAPTER OVERVIEW

In this chapter, the processes of physical and chemical weathering are described. Also discussed are the effects of weathering on rock material including mass movements, the shaping of the land, and the formation of soil.

Section 1: Weathering This section defines the term *weathering* as the process by which rocks are broken down. The rate at which rocks weather is related to rock composition, climate, topography, and vegetation. Physical and chemical weathering are defined and discussed.

Section 2: Mass Movement and Landforms The four types of mass movements are named, and their major characteristics are described. In this section, the processes that shape the earth's surface are identified. The numerous types of landforms created by these processes are discussed. Identification of landforms is reinforced with a map-interpretation activity.

Section 3: Soils In this section, soil is defined as a product of the weathering process. The factors determining the type of soil formed in an area are discussed. The layers of soil are identified and characterized. The relationship of climate and the fertility of soil is emphasized.

CHAPTER

11

Weathering and Soil Formation

The rock formation in this picture looks as if it could have been created by some great sculptor; it was, in fact, created by the climate. As rain, wind, heat, and cold work together to make soil, they often produce beautiful works of art as well.

Weathered rocks

230

CHAPTER MOTIVATING ACTIVITY

Use the schoolyard or an outdoor play area to show the students examples of weathering. Be sure to locate these examples on your own beforehand. Examples will vary depending on your surroundings. Some examples you might try to find are weathered rocks, fences, and bricks; broken, pitted sidewalks or roads; and discolored walls and buildings. Have the students view the examples you have chosen. Also have them try to find examples of their own. Ask the students to predict what processes might have caused the changes in each example of weathering. (Answers will vary but probably will include rain, cold, human contact, and decay.)

SECTION
1 Weathering

SECTION OBJECTIVES
After completing this section, you should be able to:
- **Explain** the process of weathering.
- **List** the factors that affect the rate of weathering.
- **Compare** the processes of physical weathering and chemical weathering.

Section **1:** **Weathering**

NEW SCIENCE TERMS
soil
weathering
exfoliation
frost action
abrasion
hydration
oxidation
carbonation

11.1 The Process of Weathering

If you have traveled much, you may have noticed that, except in mountainous areas, very little rock is visible on the earth's surface. In fact, in places like the Great Plains, you can travel for hundreds of kilometers and seldom see a rock, since most of the rock is covered by soil. **Soil** is a combination of small rock fragments and organic material in which plants can grow. Soil is formed when the bedrock of an area is broken into small fragments and organic matter is added to it.

Have you ever seen an unpainted house or barn that was "weathered" by the rain and sun, or read about an old sailor with a "weatherbeaten" face? The expression means that the house or the person has been changed by exposure to the weather.

Where rocks are exposed to the environment, such as on a mountain slope, they also weather. The processes that break rocks into smaller fragments, eventually producing soil, are called **weathering.** When rocks are weathered by rain and wind, the process is called *physical weathering*. If the rocks are chemically changed by acids in the air, rain, or soil, the process is called *chemical weathering*. Both types of weathering help to shape the surface of the earth.

Figure 11–1. This old barn (right) was weathered by exposure to sun, rain, and wind. These weathering agents are also responsible for changing rock into soil (left).

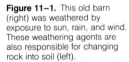

Section 1   Weathering   **231**

SECTION FOCUS

Section Overview This section discusses the effects of weathering on rocks. The differences between physical and chemical weathering are pointed out. The factors that control the rate of weathering as well as their effects are described.

Section Objectives For a list of section objectives, see pupil's edition page.

New Science Terms For a list of new science terms in this section, see pupil's edition page. Because the new terms may be unfamiliar to the students, you may want to go over the pronunciation of the words before beginning the section.

SECTION DEVELOPMENT

11.1 The Process of Weathering

THINKING SKILL *(Inferring)* Emphasize to the students that materials other than rocks can be weathered. Explain that their homes, as well as the cars, buses, and trains in which they travel also undergo weathering. Ask the students what effects weathering has on their homes and transportation vehicles. (roofs and outside brick walls deteriorate; wooden doors, shingles and shutters decay; metal car, bus, and train parts rust)

DISCUSSION Explain that organic matter includes all material from living organisms such as plant material, chemicals produced in the process of plant growth, and animal wastes. *Thinking Critically:* Ask the students to explain how organic material becomes mixed with the rock material in the soil. (Answers might include that animals and plants die and fall to the ground; animals bury plant and animal materials for storage; plant roots grow in the rock material.)

BACKGROUND INFORMATION
Igneous and metamorphic rocks are relatively resistant to physical weathering. However, their included minerals break down quite rapidly by chemical decay in the presence of water and other minerals. These rocks will not form prominent terrain features in warm humid climates but may do so in dry or cold areas. Point out to students the hard crystalline rocks of the Adirondack and White Mountains found in the cold northeastern United States.

DEMONSTRATION
For a demonstration of chemical weathering, see page 227e preceding this chapter.

DISCUSSION You may wish to outline the lesson on the chalkboard, listing the types of weathering and the factors that cause weathering. ***Thinking Critically:*** Have the students discuss the kind of house they might build and the location they might choose in order to keep weathering to a minimum. (Answers should include that igneous or metamorphic rock would be better than wood or sedimentary rock; building in a valley away from wind would be better than building on a hilltop; and building in an area with mild temperatures and little rainfall would be better than building in an area with extremes of temperature and precipitation.)

EXTENSION *(Interpreting Data)* Obtain permission to take the students to an historic cemetery in your area to study the relationship between the rate of weathering and the composition of rock. Most tombstones are made of either granite, slate, limestone, or sandstone. Have the students find old tombstones of each rock type and examine those that were erected at approximately the same time. Then have them write their observations in a chart that lists rock type, date erected, and observations. Be sure the students note discoloration and clarity of carvings. Have the students make a second chart of observations made of tombstones that are either 50 to 100 years older or younger than those already observed. Ask them how they can tell if the tombstones have weathered. (Answers will include: the words have worn off; the tombstones are broken or cracked; or the colors have faded or changed.) Ask the students to compare the extent to which each type of rock was affected by weathering. (Granite and slate are affected the least; limestone is affected the most.) ***Thinking Critically:*** Ask the students if they believe the amount of time that a rock is exposed has an effect on the extent of weathering. (Yes.) Have them use the data in their two charts to explain their answers. (Students will note, for example, that an older limestone tombstone is much more worn than a newer one.)

Figure 11–2. The cracks in these rocks may have been produced by cycles of freezing and thawing. If the seeds of some plant begin to grow in the cracks, the rocks may split even more.

Although weathering is continuous, there are several factors that influence the rate at which weathering occurs. These factors are the composition of the rock, the climate, the topography, and the presence of vegetation.

The mineral content of rocks and special features of rock formations affect weathering. Soft minerals, such as calcite, are easily weathered; hard minerals, such as quartz, resist weathering. Certain features of rock formations, such as joints, faults, and layers, allow water to enter the formations. This water causes an increase in weathering.

The climate also affects the rate of weathering. In places that receive little rainfall, such as the southwestern United States, the weathering process is usually slow. The weathering is slow because water is the fastest weathering agent. Some areas have large temperature changes between seasons or even between day and night. These changes speed up the rate of weathering. Water entering the rock during the day may freeze at night, causing rocks to break apart.

Some features of topography may speed up weathering. For example, mountain slopes tend to have more rock exposed at the surface than do flat lands. Also, mountains are likely to have high winds and freezing temperatures. Together these conditions speed up the weathering process.

Even the vegetation of an area may be a factor in the rate of weathering. Plants protect rocks from wind and rain. However, decaying vegetation may speed up the rate of weathering, because the roots of many plants make acids that can dissolve rocks.

○ ***What is weathering?***
○ ***What factors determine the rate at which rock will weather?***

A MATTER OF FACT

The town of Calama in the Atacama Desert of Chile had received no rain for 400 years. When rain finally fell in 1971, the entire town was washed out in a mud flow.

232   Chapter 11   Weathering and Soil Formation

Answers—11.1

○ Weathering is the process of breaking down rock material.

○ The structure of the rock, the climate, the topography, and the presence of vegetation determine the rate of weathering.

11.2 Physical Weathering

Physical weathering breaks rocks apart mechanically. Bedrock is broken into fragments in four ways: by pressure, cycles of freezing and thawing, plant growth, and abrasion.

The first type of physical weathering begins shortly after rocks form. Igneous rocks, forming at great depths, are under tremendous pressure because of overlying layers of rock. Once this pressure is removed, the rocks expand. This sudden expansion causes the surface of the rocks to crack and form joints. Expansion causes some rocks, especially granite, to flake off in sheets in a process called **exfoliation** (ehks foh lee AY shuhn). Temperature changes also cause exfoliation by loosening the mineral crystals in a rock. Figure 11–3 shows the typical rounded surface caused by exfoliation.

Have you ever placed a can or a bottle of liquid in the freezer to chill it and then forgotten about it for several hours? Besides the fact that the contents were frozen, you may have noticed that the can appeared swollen or that the bottle was broken. As water freezes, it expands and exerts a tremendous pressure on the container in which it is placed. Temperature changes also affect the water that collects in cracks in rocks. For instance, if a crack fills with water and the water freezes, the crack in the rock enlarges. This process of freezing and cracking is the second type of physical weathering, called **frost action**. Frost action mostly affects sedimentary rocks because they have many pore spaces, bedding planes, and fractures. Frost action also does severe damage to road surfaces during the winter. Why do you think frost action is so damaging to roads?

Figure 11–3. Exfoliation produces thin sheets of rock that flake off in layers.

Ice

DISCOVER

Demonstrating Frost Action

Fill a small plastic jar or bottle to the top with water. **CAUTION: Do not use a glass jar or bottle.** Replace the lid tightly, and place it in the freezer overnight. What happened to the container? What could have caused this to happen?

Figure 11–4. If you fill a bottle with water and place it in a freezer overnight, the bottle will break. In a similar manner, ice in a rock crack will break the rock.

Section 1   Weathering   **233**

11.2 Physical Weathering

DISCUSSION   You may wish to ask the students to discuss examples of physical weathering they may have seen on monuments, buildings, and roads. Then discuss exfoliation and frost action, calling the students' attention to Figures 11–3 and 11–4. *Thinking Critically:* Ask the students what change, other than exfoliation, could be caused by pressure. (The rocks could crack and split into smaller rock pieces.)

EXTENSION   Long ago, stonemasons had neither power drills nor saws to cut the blocks of rock used for building. Instead, they would fill existing cracks or joints with water and wait for frost action to split the blocks. Obviously, this method was useful only in cold climates.

DISCUSSION   Explain to the students that as miners dig through rock, the pressure in the rock is relieved. *Thinking Critically:* Ask the students to determine what danger this relieved pressure may pose to the miners. (Once pressure is removed, the rock expands. The miners could be crushed or injured by expanding and exfoliating rock.)

DISCOVER:
Demonstrating Frost Action

Thinking Skills   *(Observing, Drawing Conclusions)*

CAUTION: Do not use a glass jar or bottle. This Discover will work best with a push-on (rather than screw-on) lid. If the container is filled all the way to the top, the lid will be pushed off by the ice. The students might also find that straight sides bulge. Ice is one of the few solids that is less dense than its corresponding liquid; water expands when frozen.

DISCUSSION Reinforce the concept that gravity, streams, and wind all cause rock fragments to bump against one another. Explain that ocean water also physically weathers rocks through abrasion. **Thinking Critically:** Ask the students to describe the way this occurs. (Ocean waves hitting the shore constantly bounce loose rock fragments off each other and against the solid rock of some shores.)

THINKING SKILL *(Interpreting a Photograph)* Have the students observe the rocks shown in Figure 11–6. Ask them to describe the process involved in producing stones of this shape. (Rocks become rounded when they are physically weathered by abrasion. The constant bumping of one rock fragment against another breaks off sharp edges and smoothes the rocks.)

THINKING SKILL *(Drawing Conclusions)* Reinforce the concept of root-pry and its effect on breaking apart not only rocks but building materials such as concrete. Then ask the students to describe why planting a tree very near to a building may not be a good idea. (If any cracks were to occur in the building's foundation, the roots of the tree could grow into the cracks and split apart the foundation.)

Figure 11–5. Roots growing in a rock crack exert tremendous force on the surrounding rocks. This force can expand the cracks or even cause pieces of the rocks to break off.

The third type of physical weathering occurs when the cracks started by frost action are expanded by plants. As seeds sprout and begin to develop in the cracks, the growing roots enlarge the cracks. This type of physical weathering is called *root-pry.* Another example of root-pry can be seen where the roots of a large tree close to a sidewalk or driveway have pushed up and cracked the concrete.

The final type of physical weathering occurs when rocks weather by physical contact with other rocks. As rock fragments bounce down a hill, roll in a stream, or are carried by the wind, they bump against each other. This weathering by physical contact is called **abrasion**. Abrasion knocks off sharp edges, producing rounded rock fragments. The longer that rock fragments undergo abrasion, the smaller and rounder they become.

○ *What are the agents of physical weathering?*
○ *Which two agents cause exfoliation?*

Figure 11–6. As the rocks in a flowing stream tumble and bump into each other, sharp edges are rounded off.

234   Chapter 11   Weathering and Soil Formation

Answers—11.2

○ Exfoliation, frost action, and abrasion are agents of physical weathering.

○ Release of pressure and temperature changes cause exfoliation.

BACKGROUND INFORMATION

The surface of the moon is covered with broken rock material ranging in size from large boulders to fine dust. All the rock fragments are the result of physical weathering. No chemical weathering occurs on the moon because there is no atmosphere and therefore no water, oxygen, or carbon dioxide to react with the minerals in the rocks. The physical weathering of moon rocks results from meteorite bombardment, which crushes the surface rocks into many fragments.

11.3 Chemical Weathering

During the process of chemical weathering, the minerals within a rock are chemically changed. Minerals may be added, removed, or changed into other minerals within a rock. Most chemical weathering involves water, because water has the ability to dissolve many minerals. Halite, for example, separates into sodium and chloride ions in water. Most minerals do not dissolve as rapidly as halite, and the process of chemical weathering is usually quite slow.

In addition to dissolving minerals, water can also combine with certain minerals to form new compounds in a process called **hydration**. Micas and feldspars, for instance, change to clay in the presence of water. This process is illustrated in Figure 11–7.

Not all chemical reactions require water, however. Another type of chemical weathering occurs when oxygen combines directly with an element in a process called **oxidation**. Rocks that have undergone oxidation can usually be identified because oxidation produces a color change. For example, oxygen combines with magnetite to form a reddish-brown rust, similar in appearance to the rust on an old nail. Even in this case, however, the presence of water speeds up the process. An unpainted bicycle will rust more rapidly in a humid climate than in the desert, because the oxygen dissolved in water combines more readily with iron than the oxygen in the air. In the same way, rocks weather more quickly in humid climates.

Figure 11–7. The red-clay soil of Georgia and certain other states results from hydration of the minerals in the soil.

Figure 11–8. Chemical weathering occurs to things other than rocks. Oxidation of the iron in this old car creates rust. Even a shiny new bicycle will weather if it is not protected.

11.3 Chemical Weathering

DISCUSSION Remind the students that, as discussed in Chapter 4, feldspars, micas, and quartz are the major components of granite. Emphasize that feldspars and micas weather by hydration and that quartz is very resistant to any type of chemical weathering. *Thinking Critically:* Ask the students what the end products of chemically weathered granite would be. (The end products would be clay from the hydration of feldspars and micas, and quartz sand from the unweathered quartz crystals in the granite.)

THINKING SKILL *(Formulating Hypotheses)* Have the students study the photographs of the rusted objects in Figure 11–8. Reinforce the fact that oxidized iron and magnetite have a reddish-brown color. Explain that photographs taken of the surface of Mars show that it is covered by reddish-brown rocks and dust. Ask the students to write down some hypotheses about the rocks and atmosphere of Mars. (The rocks are probably made of iron that later rusted. There may have been oxygen or water in Mars' atmosphere at one time because oxidation of the metals could not have taken place without the presence of oxygen.)

EXTENSION Two atoms of oxygen are removed from the atmosphere for every atom of iron that is oxidized. Over the millions of years of the earth's geologic history, weathering of iron-containing rocks could extract all of the oxygen from the earth's atmosphere. However, this did not occur. Ask the students why there is still a great deal of oxygen in the atmosphere. (Plants are constantly releasing oxygen during photosynthesis.)

DEMONSTRATION

Have the students watch as you break an antacid tablet in half. Show the pieces to the students. Then drop the two halves into a glass of water and have the students observe the reaction. *Thinking Critically:* Ask the students to identify which procedure represented physical weathering and which represented chemical weathering. Have the students explain their answers. (In physical weathering, the rocks are only broken apart; there is no chemical change. Thus, breaking the tablet in pieces is similar to physical weathering. In chemical weathering, a chemical reaction changes the minerals in a rock. Thus, dissolving the tablet in water is similar to chemical weathering.)

Have the students discuss the roles of water and decaying plants in chemical weathering. Acid rain is causing a severe kind of chemical weathering. You may wish to have the students find newspaper and magazine articles on acid rain.

EXTENSION (Tie-in/Chemistry) Many chemical weathering reactions can be described by chemical equations. Listed below are equations describing the chemical weathering of various minerals.

Carbonation:
Calcite + carbonic acid= calcium ion + bicarbonate ion
$$CaCO_3 + H_2CO_3 \rightarrow Ca^{+2} + 2HCO_3^-$$

Oxidation:
iron pyroxene + oxygen + water = limonite (iron oxide) + dissolved silica
$$4FeSiO_3 + O_2 + 2H_2O \rightarrow$$
$$2Fe_2O_3 \cdot H_2O + 4SiO_2$$

Write these equations on the board; however, leave several subscripts blank on one side of the equation. Explain that the number and type of atoms must be equal on both sides of the equation. Then let the students balance the equations. You may also want to introduce the terms *reactants* (starting materials) and *products* (end materials).

SECTION REVIEW

Summary Have each student write one or two paragraphs summarizing the main concepts of this section, including the many types of weathering. (Weathering is the process by which rocks are broken down. Rock structure, climate, topography, and vegetation affect the rate of weathering. Exfoliation, root-pry, abrasion, frost action, hydration, oxidation, and carbonation are the major types of physical and chemical weathering.)

Reinforcement Have the students make large charts listing each type of weathering with a photo example of each type cut out from a magazine and glued onto the chart. Have the students display their charts in class.

236

Figure 11–9. Sulfur pollution in the atmosphere produces sulfuric acid when it combines with rain. The marble in the statue is being dissolved by acid rain.

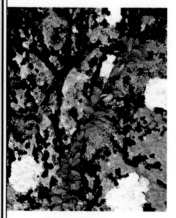

Figure 11–10. When lichens grow on rocks, they produce a type of acid that dissolves the minerals in the rocks.

Water is also involved in another chemical reaction. As rain falls, small amounts of carbon dioxide from the atmosphere are dissolved in the water. This forms a weak acid called *carbonic acid* (H_2CO_3). Carbonic acid reacts with the minerals in rocks such as limestone, dissolving them in a process called **carbonation.** The product of this reaction is washed away by the rain, leaving only the impurities from the original rock.

The decaying of plants also produces acids that chemically weather rock. Even bare rock surfaces may be weathered by plant acids. As more plants grow in an area, the effects of chemical weathering reach deeper, breaking down more rock and building up more soil. However, good soil takes thousands of years to form. Why do you think this is so? **1**

○ *What happens to minerals in chemical weathering?*
○ *What are the agents of chemical weathering?*

Section Review

READING CRITICALLY

1. Describe the differences between chemical and physical weathering.
2. Why is water the most important agent of chemical weathering?

THINKING CRITICALLY

3. Why is there very little soil on mountaintops? Does that mean that little weathering occurs there? Explain your answer.
4. Explain the weathering process that would occur in a place with a lot of rain and large temperature changes.

236   Chapter 11   Weathering and Soil Formation

1 Wind, rain, and mass movement may cause soil to be transported.

Answers–11.3

○ Chemical weathering either alters the original minerals to other types of minerals or completely dissolves them.

○ Hydration, oxidation, and carbonation are the agents of chemical weathering.

Answers to Section Review

Reading Critically

1. Chemical weathering alters the minerals in rocks to other types of minerals.

Physical weathering breaks the rocks into smaller pieces, changing the minerals.

2. Water is involved in both physical and chemical weathering.

Thinking Critically

3. Although much weathering occurs on mountaintops, the weathered rock material tumbles down the mountain-sides. Most mountaintops experience too severe a combination of weather conditions for much plant growth.

SECTION OBJECTIVES

After completing this section, you should be able to:

■ **Explain** why gravity is necessary for mass movement.

■ **Describe** the role of a cap rock in the formation of mesas and buttes.

■ **List** the rock characteristics that would cause differential weathering.

NEW SCIENCE TERMS

mass movement
creep
flow
landslide
subsidence
differential weathering

11.4 Mass Movement

If you live in an area with steep hills and snow, you have probably rolled snowballs down a slope and watched them gain speed and snow. Rocks and soil are affected by gravity in much the same way in a process called *mass movement*. **Mass movement** is the downhill movement of rocks and soil due to the force of gravity. Mass movement is classified by the size of the rocks moved, by the amount of water involved, and by the steepness of the slope. The four types of mass movement are *creep, flow, landslide,* and *subsidence.*

Creep occurs when rock materials saturated with water move slowly downhill. The movement is often so slow that changes are hard to see. An example of creep is shown in Figure 11–11. This particular change took place over many years.

Figure 11–11. This fence was once straight and now it bends in the middle. The soil in which the fence posts sit is slowly moving by a process called *creep.*

237

4. Considerable chemical weathering would occur in an area receiving rain because water is the main agent in chemical weathering. Also if the range of temperature changes caused freezing and thawing, there would be much frost action and exfoliation (physical weathering).

Section 2: **Mass Movement and Landforms**

SECTION FOCUS

Section Overview This section discusses the different types of mass movements. Also characterized are the landforms that result from differential weathering and mass movements.

Section Objectives For a list of section objectives, see pupil's edition page.

New Science Terms For a list of new science terms in this section, see pupil's edition page. To familiarize the students with the terms, you may wish to discuss definitions before beginning the section.

SECTION DEVELOPMENT

11.4 Mass Movement

DISCUSSION Have the students study Figure 11–11. Ask them to name examples other than the one shown in the photograph that indicate soil creep has occurred. (tilted fences and telephone poles, tree trunks that curve downhill at their bases, and hillside embankments that slant downslope)

THINKING SKILL *(Predicting)* Have the students predict which types of mass movement they might be able to observe in the area near their school, taking into account recent weather conditions, topography, and local rock types. After the students have recorded their predictions, allow motivated students to seek out the predicted mass movement areas. Have them either photograph or draw pictures of the evidence of mass movement and share the information with the rest of the class.

DISCUSSION Have the students compare the appearance of the rock fragments in the two photographs in Figure 11–12. Ask the students to identify the visual clues that emphasize a major difference between the two types of mass movements shown. (It should be obvious to the students that there was a great deal of water flowing on the slope where the mud flow occurred, while the slope where the landslide occurred shows no sign of water. This indicates a major difference between a mud flow and landslide: mud flows occur when the rock materials are saturated with water; landslides occur when dry rock materials overcome friction and slip.)

SKILL *(Classifying)* Review each type of mass movement. Have the students classify mass movements by the slope of the land involved, the material size, and the amount of water involved. (Creep: any slope, all sizes of rock fragments, saturated; Earth flows: gentle hillsides, clay or silt, saturated; Mud flow: steep hillside channels, rock debris, saturated; Landslide: steep slope, soil or rock, dry; Subsidence: below ground with no slope necessary, easily dissolved rock layers, running water)

THINKING SKILL *(Analyzing)* Have the students carefully study the area surrounding the sinkhole shown in Figure 11–13. Ask the students to determine how constructing homes and roads in sinkhole-prone areas may accelerate the formation of these features. (The added weight of buildings, roads, and the vehicles traveling on the roads may be just enough added stress to cause the rock that they are on to collapse. The vibrations of heavy traffic may also trigger the collapse.)

Figure 11–12. The house on the left was caught in an earth flow. Soil that is saturated with water often flows down hill. The talus at the base of the mountain on the right resulted from a landslide—the movement of dry soil and rocks.

A MATTER OF FACT

In 1985 an eruption of a Colombian volcano, *Nevado del Ruiz*, created a mud flow that took the lives of 23 000 people. A wall of ash and mud 40 m high traveled down a narrow river canyon and covered a 40-km² area.

Figure 11–13. This sinkhole in Winter Park, Florida, is the result of subsidence. The limestone bedrock was dissolved by underground water, and the overlying soil caved in.

238

When water-saturated rock material moves at a noticeable rate, the process is called **flow.** There are two types of flows: earth flows and mud flows. Earth flows are the movement of water-saturated clay or silt down gentle hillsides. Mud flows are the movement of water-saturated rock debris down channels on steep hillsides. Mud flows, often called *mud slides*, are common along the California coast because of the clay soils.

A **landslide** occurs when dry soil or rock moves down a steep slope. The movement is usually rapid and unexpected, because, unlike mud flows, which are usually caused by heavy rain, most landslides have no direct cause. The force of gravity simply overcomes the friction between rocks, and the material moves. The fallen material collects at the base of the slope and forms a mass called *talus* (TAL uhs).

When rock material sinks to a lower level without sliding down a hillside, the process is called **subsidence** (suhb SY duhns). Subsidence can occur when minerals, such

DEMONSTRATION

Demonstrate a landslide and a mudflow by using a tray of rocks and soil, a sprinkling can, and water. Pack the rocks and soil into a low-sided tray approximately 24 cm x 36 cm. Tilt the tray at about a 30-degree angle. The materials should remain stable within the tray. Then shake the tray slightly. Compare this movement to an earthquake. Have the students record the reaction of the rock materials to the "earthquake." Next, replace the material. Using the sprinkling can, saturate the materials with water. Explain that the soil cannot hold any more water. ***Thinking Critically:*** Ask the students to predict the effect of adding more water. (A simulated mud flow will occur.) Add the water to demonstrate to the students the accuracy of their predictions.

as halite, are dissolved by water, and the overlying rock then collapses into the hollow. Sinkholes, such as the one shown in Figure 11–13, often result from subsidence. Subsidence can also occur as young rocks compact, especially if wells are pumping out water or oil. Many areas along the Gulf Coast are experiencing subsidence due to the compaction of sediments.

○ *What are the four types of mass movement?*
○ *What is the difference between creep and earth flow?*

11.5 Landforms and Weathering

Some interesting features of the land, or landforms, are created by weathering and mass movement. For example, landforms called *mesas* (MAY suhz) and *buttes* (BYOOTS) are the result of physical weathering. Mesas are lands with flat tops and nearly vertical sides. A butte is a lone, flat-top hill that rises abruptly above the surrounding flat land.

In mesas and buttes, the top rock, called the *cap rock*, is more resistant to weathering than the rocks below it. The less-resistant rock weathers away, leaving the formations shown in Figure 11–14. The cities of Mesa, Arizona, and Butte, Montana, are named after these prominent features. What places in your area are named after a particular landform?

Sometimes layers of rock that are resistant to weathering alternate with those that are less resistant. The easily weathered rock will form a gentle slope as it undercuts the resistant rock above it. The resistant rock will weather into vertical cliffs. The

DISCOVER

Showing Mass Movement

In a large tray of wet sand, construct some sort of landform, such as a hill. Now begin to weaken the hill slowly by removing sand from the bottom. Continue this until part of the hill collapses. Which type of mass movement does this collapse best illustrate? Explain your answer. What other kinds of mass movement could be demonstrated by varying the water content and topography of the sand?

Figure 11–14. Mesas, such as the one on the left, and buttes, shown on the right, result from the weathering of nonresistant rock layers. Rock that is resistant to weathering forms the caps of these formations.

Section 2   Mass Movement and Landforms   **239**

Answers—11.4

○ Creep, flow, landslide, and subsidence are four types of mass movement.

○ Creep is very slow mass movement; earth flows are much more rapid.

DISCUSSION Explain to the students that subsidence often increases when great amounts of ground water are pumped from certain rock layers. The ground water had filled spaces within the rock layers. *Thinking Critically:* Ask the students to formulate a hypothesis as to why subsidence increases in these areas. (With the spaces in the rock layer emptied of water, the rocks become more compact. Thus the ground above the rock layer sinks.)

DISCOVER:
Showing Mass Movement

Skill *(Observing)*

The students should note that this best illustrates subsidence because overlying sand collapses as the base disintegrates.

11.5 Landforms and Weathering

DISCUSSION Explain to the students that in arid climates, limestones are resistant to weathering and often form cap rocks. *Thinking Critically:* Ask the students why limestone almost never forms cap rocks in humid climates. (Limestone is quickly weathered by carbonation, a common chemical weathering process in humid climates.)

EXTENSION Have the students build landforms out of clay, plaster of paris, or other materials. Each student can present his or her model to the class and lead a discussion about the processes involved in producing the landform he or she chose to model. The students should be able to identify at least one site in the United States where such a landform is located. You may want to suggest that the students model such landforms as caves, mesas, and cuestas.

240

DISCUSSION You may wish to have other photographs or slides showing differential weathering. ***Thinking Critically:*** Have the students point out in the pictures the results of differential weathering (cliffs, gentle slopes) and give the reason for them. (Softer rock weathers into a gentle slope. Resistant rock weathers into vertical cliffs.)

SKILL *(Organizing)* Emphasize that different weathering processes result in the formation of different landforms. Have the students identify landforms by the processes that form them. (mesas, buttes, caves, ledges, rock columns, ridges: differential weathering; karst topography: chemical weathering and subsidence)

THINKING SKILL *(Drawing Conclusions)* Ask the students to describe what the earth's landforms would look like if there were no differential weathering and mass movements. (There would be no mesas and buttes, ridges and valleys, or karst topography. The earth would appear much flatter since all landforms would weather uniformly.)

SKILL *(Writing)* Using magazines and books containing scenic photographs, have the students identify talus, buttes, mesas, sinkholes, and other features formed by weathering. Have the students write a descriptive paragraph on each feature and the process that formed it. They should classify the major weathering process as physical or chemical.

Figure 11–15. Some of the unusual formations of the Grand Canyon are due to the differential weathering of rock layers. Different layers weather at different rates, even though they are exposed to the same environmental factors.

Figure 11–16. Resistant rocks sometimes form the tops of ridges, such as those shown in the photograph, while the less resistant rocks form the valleys between the ridges. Notice the location of mountains and ridges on the map.

difference in weathering of different rocks in the same formation, exposed to the same environment, is called **differential weathering**. Differential weathering is responsible for creating many of the features seen in the Grand Canyon.

Differential weathering can sometimes be seen in a single formation. If weathering agents enter bedding plane cracks or fractures, differential weathering can form caves, ledges, or long columns of rocks. If the rock layers are not horizontal, the resistant rock will form ridges, as shown in Figure 11–16, while the more easily weathered rock will form valleys between the ridges.

Landforms resulting from mass movement are usually not as spectacular as those produced by differential weathering.

240 Chapter 11 Weathering and Soil Formation

DEMONSTRATION

Explain to the students that different-sized materials have different angles of repose. (The repose is the steepest slope upon which the material can remain without slipping downslope.) Pour sand, coarse sand, and angular pebbles in separate piles on a flat surface. Have the students observe each pile. Ask them to determine the relationship between angle of repose and rock-particle size and shape. (Larger, flatter, and more angular rock fragments have a steeper angle of repose.) Also ask the students to describe the influence of the type of rock fragment on the probability of a mass movement. (A cliff or hillside of fine sand is more likely to undergo mass movement than a similarly steep cliff or hillside of large, angular rock fragments.) Inform the students that highway departments often spread gravel along steep road embankments to hold the finer rock materials that are beneath in place.

Figure 11–17. Small, round lakes, such as these in northern Florida, are examples of karst topography. Karst topography results from a combination of chemical weathering and subsidence.

However, in Florida, a combination of chemical weathering of the limestone bedrock and subsidence has created a landscape dotted with small, round lakes. This type of landform is called *karst topography.*

○ **What are landforms?**
○ **What is differential weathering?**

Section Review

READING CRITICALLY

1. Explain why landslides and mud slides are not the same.
2. What is the difference between a mesa and a butte?

THINKING CRITICALLY

3. What would be the main agents of weathering in the formation of a mesa or butte? Explain your answer.
4. Imagine that you are considering buying a home in a subdivision where subsidence might be occurring. What should you look for in the area to decide if the house is safe to move into? Why?

ACTIVITY: Analyzing a Landform

How can you identify the effects of weathering on a landform?

PROCEDURE

Examine the topographic map on page 581. Study the contour interval and the map scale; then answer the following questions.

CONCLUSIONS/APPLICATIONS

1. What type of landform is shown in this map? Explain the features that led you to this conclusion.
2. How many resistant rock masses can you identify on this landform? Identify the resis-

tant rocks by referring to the contour elevation marking the top of the layer.
3. Would any of these rocks be appropriately referred to as cap rocks? Explain your answer.
4. Use the term *differential weathering* to explain how the landform has weathered into this shape.
5. How does the type of landform shown help you determine the climate in this area?

GOING FURTHER

Use clay to build a model of the landform in this activity. Use one color of clay for the resistant rock layers and another color for the easily weathered layers.

Section 2   Mass Movement and Landforms   **241**

SECTION REVIEW

Summary Have the students present a summary of the section listing the section objectives. (Gravity works to pull weathered rock fragments downslope either slowly, as in soil creep, or rapidly, as in mudslides. The rock that forms cap rocks is less resistant to weathering than the rocks below it. As the less resistant rocks weather away, the cap rock protects the rocks beneath it, forming elevated landforms. Differential weathering occurs when less resistant rock layers alternate with rock layers more resistant to weathering.)

Reinforcement Have students work in groups. Have students quiz one another on the new terms covered in this section, using index cards with definitions on one side and new science terms on the other.

ACTIVITY:
Analyzing a Landform

Preparation of Materials Displaying photographs of the Grand Canyon will help the students visualize the manner in which alternating layers of soft and resistant rock will weather. Resistant rock forms steep cliffs; softer rock forms gentle slopes.

Answers to Conclusions/Applications

1. This is a butte. It is a rather small landform with a flat top and steep sides.
2. There are two resistant rock layers: 4560′ and 4240′.
3. The 4560′ layer serves as a cap rock, forming a flat top above all the other rock formations.
4. The soft rocks have weathered back to where they are protected under the resistant rocks. This differential weathering process forms alternating gentle and steep slopes in the side of the butte.
5. The shape of this landform indicates that the climate is quite arid. The effects of differential weathering would be reduced if there were enough rainfall to support vegetation. The growth of plants would smooth the slopes.

Answers—11.5

○ Landforms are shapes or features on the land.
○ Differential weathering is the difference in weathering of rock layers in the same formation or in adjacent formations.

Answers to Section Review

Reading Critically

1. A mudslide is actually a type of flow. A flow is rapid movement of oversaturated

rock material; a landslide is rapid movement of dry rock material.
2. A mesa is larger landform than a butte.

Thinking Critically

3. Frost action and abrasion are the main agents in the formation of buttes and mesas.
4. If the ground under a home is subsiding, there would be cracks in the foundation and in the interior walls.

241

SKILL ACTIVITY:
Interpreting a Pie Graph

Objectives

- Interpret a graph.
- Understand the use of percentages.

Discussion   To emphasize to the students that all the sections of a pie graph total 100 percent, have them construct their own pie graphs. Tell the students to cut out paper circles. Then survey the class to find out what they had for breakfast. Write on the board the percentages of the students who fall into each breakfast-type category. Have the students draw these percentages on their pie graphs, being sure that each section is the appropriate size. Emphasize that all sections of the pie must represent a category and the total of all the sections must equal 100 percent.

Answers to Application

1. South America
2. Australia
3. North and Central America
4. Africa

Answer to Using What you Have Learned

Answers will vary according to area and student, but they should show an understanding of a pie graph.

SKILL ACTIVITY: Interpreting a Pie Graph

BACKGROUND

There are many methods available to scientists for presenting data. Sometimes scientists use a circle, or pie, graph to present information about the way something is divided or distributed. Usually the percentage of the whole is written on each division of the graph.

PROCEDURE

Examine the pie graphs for North and South America. Each one represents the uses of fertile soil on the continent. Determine how much of each continent is available for cropland and grazing land.

APPLICATION

1. Which continent has the greatest percentage of fertile soil in forested land?
2. Which continent has the greatest percentage of fertile soil in grazing land?
3. Which continent has the greatest percentage of fertile soil in cropland?
4. Which continent has the greatest percentage of fertile soil not used for cropland?

USING WHAT YOU HAVE LEARNED

Do some library research to find out the uses of land in your state; then make a pie graph for land use in your state. Label the percentage for each type of land use.

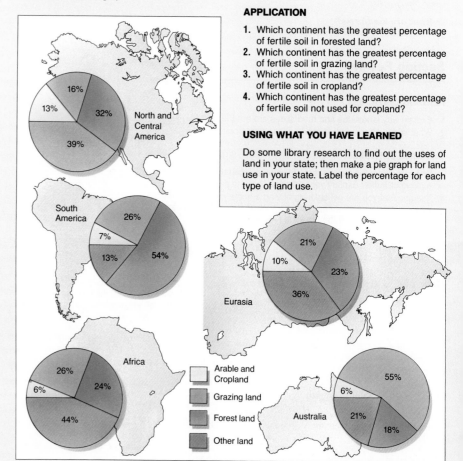

Legend:
- Arable and Cropland
- Grazing land
- Forest land
- Other land

SECTION OBJECTIVES

After completing this section, you should be able to:
- **Identify** the factors that are involved in soil formation.
- **Summarize** the characteristics of different types of soils.

NEW SCIENCE TERMS

humus
horizons
leaching

11.6 Soil Formation

The formation of soil is one of the most important aspects of the weathering process. Life as we know it would not be possible without soil for growing grains, fruits, vegetables, and pasture grasses. Care must be taken to preserve and protect the soil, because the process of soil formation is very slow. It takes from 100 to 1000 years for one centimeter of soil to form.

> **A MATTER OF FACT**
>
> A single rye plant can produce roots over 640 m long. When the plant dies, the root system decays and becomes part of the humus of the soil.

Figure 11—18. Deep, rich soil is needed to grow grains, fruits, and vegetables. The great variety of food available to North Americans is due to the abundance of good soil found on the continent.

Five factors are involved in the formation of soil: the parent material, the climate, the topography, the organisms in the soil, and time. The parent material is the original rock from which the soil is formed. Soils formed from shale, for instance, will be very different from soils formed from limestone.

The climatic factors that influence soil formation are temperature change and precipitation. Soils form more quickly in moist climates because weathering is faster in the presence of water. Soils also form more quickly where there are significant temperature changes. Related to climate is the topography, or shape of the land. Topography affects the drainage of an area. If you look at a steep slope, you will probably not find much soil, because a

Section Overview This section discusses the factors that influence various aspects of soil formation. The different layers of soil are described. Soil types are characterized and compared.

Section Objectives For a list of section objectives, see pupil's edition page.

New Science Terms For a list of new science terms in this section, see pupil's edition page. Read the terms aloud to the students and explain that they will learn the scientific definitions of the terms by reading this section.

SECTION DEVELOPMENT

11.6 Soil Formation

DISCUSSION Point out that shales contribute clays to soil; sandstones contribute quartz sand grains. ***Thinking Critically:*** Ask the students why the chemical weathering of limestone does not form a soil component. (Limestone that is weathered by carbonation is completely dissolved in water, leaving no solids behind.) Ask the students why only small amounts of feldspar and mica are found in soil. (The minerals are altered by hydration.) Explain to the students the effects of temperature on soil formation. In areas with consistently large changes in temperature, physical weathering is increased and thus soil formation occurs more quickly.

THINKING SKILL *(Drawing Conclusions)* Emphasize the slow rate at which soil is formed. Ask the students to explain why farmers should be very concerned by the washing away of the soil on their farms. (It will take many years for new soil to form and replace the soil that was washed away.)

SKILL *(Calculating)* Reinforce the fact that soil forms at the rate of 1 cm every 100 to 1000 years. Have the students suppose that a particular area has a soil formation rate of 1 cm/200 years. However, each 75 years, 0.25 cm is washed away. Ask the students to calculate the thickness of soil gained in the area after 100 years. (1.75 cm)

DISCUSSION Organisms are more abundant in warm climates so there are more animal- and plant-related weathering processes in this type of climate. Explain to the students that humans also are a major influence on soil formation. *Thinking Critically:* Ask the students to name some human activities that affect soil formation. (farming, land deforestation, reforestation, irrigation, diversion of waterways, mining, and construction)

THINKING SKILL *(Comparing)* Have the students make a visual comparison of the vegetation shown in the photographs in Figure 11–19. Ask the students to explain the differences in the plant growth on the two slopes. (The plant growth on the steeper slope is much sparser than on the gentler slope. The difference occurs because the soil is thinner on steep slopes, and fewer plants can grow in the soil.)

EXTENSION You may wish to introduce the terms *residual soil* and *transported soil*. Residual soil is developed from the local underlying bedrock. Transported soil is formed from rock material brought into an area from elsewhere. Transported soil usually differs greatly from the underlying bedrock. Soils on the northern plains and midwest are composed of material deposited by glaciers and are thus transported soils.

DISCOVER: Making Soil

Thinking Skill *(Drawing Conclusions)*

Be sure the students allow several days for a root system to develop. The roots should not be left in direct sunlight. Students' answers will vary depending on their observations, but most should find the best roots in the mixture because the peat moss is organic matter, supplying nutrients and holding moisture.

244

Figure 11–19. The skimpy vegetation on steep slopes (right) is the direct result of the limited soil found in these areas. Gentle slopes (left) have more soil and, therefore, more vegetation.

DISCOVER

Making Soil

Take three paper cups and punch small holes in the bottom of each. Fill the first cup with sand and the second with peat moss. In the third cup, place a mixture of sand and peat moss. Place two or three bean or corn seeds in each, add water, and allow them to sprout. In a few days, carefully remove each sprout from the cup and wash off any loose material. Which sprout has the best-looking root system? Explain why. In which cup was the material most like soil? What purpose did the peat moss serve?

steep slope allows water and mass movement to remove newly formed soil. A gradual slope allows soil to accumulate without being carried away.

The plant and animal life within the soil also influences soil formation. Tunnels made by ants and earthworms, for instance, create spaces for air and water. These spaces allow more weathering to occur; therefore, more soil will be formed. In addition, dead plants and animals decay, adding to the humus (HYOO muhs). **Humus** is the organic part of the soil.

The final factor in soil development is also the most obvious—time. The longer the time for soil formation, the more soil there will be. This will be true except in areas where environmental factors act to transport soil somewhere else. What environmental factors might act to transport soil? 1

○ *What are the five factors involved in soil formation?*
○ *What is humus?*

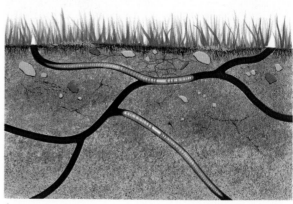

Figure 11–20. Worms, part of the soil biota, often make tunnels in the soil. These tunnels allow water and nutrients to reach the roots of plants.

1 Wind, rain, and mass movement may cause soil to be transported.

244   Chapter 11   Weathering and Soil Formation

Answers—11.6

○ Parent materials, climate, topography, organisms in the soil, and time are all factors in soil formation.

○ Humus is the organic portion of the soil.

DEMONSTRATION

Show the students two soil samples, one high in organic content and one with little organic material. Have the students record observations such as color, size of rock particles, feel of the soil, and type of organic material visible.

TECHNOLOGY: Modern Farming

Most people in the United States live in urban areas. They get their food, wrapped in plastic and paper containers, from a supermarket. With each new day the demands on the land increase; more people must be fed, more trees must be cut for paper and building materials, and more land must be turned into areas where people can live.

Total land area is being carefully evaluated to make the best use of every available hectare.

Plants called *legumes* are being planted on marginally fertile soil because they produce nitrogen compounds that other plants can use. The contour of the land is studied, and crops are planted on terraces, or flat steps, rather than on slopes, to prevent loss of soil due to excess runoff.

Many farmers are now using a practice called *minimum till farming.* In this program, the soil is protected in the winter by the stubble of the previous crop. New crops are planted in narrow

Modern harvesting equipment

strips between the stubble to reduce the exposure of the soil to the drying wind.

New techniques in irrigation are turning semiarid parts of the world into productive cropland. In some places, water is dripped onto individual plants instead of being sprayed over entire fields, where much of the water evaporates before it reaches the roots of the plants.

Much work is being done with the practice of *hydroponics;* that is, growing food plants without soil. In hydroponics, water with essential plant nutrients is sprayed directly onto the roots of plants suspended in air. The roots absorb the nutrients, and the plants carry out photosynthesis as if they were growing in rich soil.

Efforts are also being made to "engineer" superior plants and animals in the laboratory. By cross-breeding organisms with many desirable characteristics, it is possible to develop plants and animals that produce more food in less time and at a lower cost than ever before.

If the world is to continue feeding its increasing population, further advances will be needed to produce even more food from the shrinking supply of land.

Contour plowing and drip irrigation (inset)

Section 3   Soils   **245**

TECHNOLOGY:
Modern Farming

Discussion   After the students have read this article, hold a class discussion on the need for and benefits of modern farming. You may also wish to discuss the economics of modern farming, the decline of the small farmer, and government supports.

Extension   Some students may be interested in researching the effect of poor management on farming and the development of the Dust Bowl in the central United States in the 1930s.

11.7 Soil Composition

DISCUSSION Have the students go into the schoolyard and collect samples of soil. Direct their attention to Table 11–1. Have the students compare their soil samples to those listed in the chart. ***Thinking Critically:*** Ask the students which soil in the chart is most similar to their samples. Answers will vary.

THINKING SKILL *(Analyzing)* Point out that some of the most fertile soils in the United States are along rivers. Tell the students that in the past century large cities and towns have been constructed along many major waterways. Ask the students to predict the consequences of the continuing expansion of construction in these areas. (Of the total land area of the earth, very little is covered with fertile soil. As more of the fertile land that is available is built upon, farmland is lost forever.)

11.7 Soil Composition

Soil consists of inorganic (nonliving) and organic (once-living) material. However, at least 80 percent of all soil is inorganic—it is weathered rock.

Quartz and clay minerals are the most common minerals in soils, although other minerals may also be present. Soils also commonly contain compounds of potassium, phosphorus, and nitrogen, as well as air and water. Table 11–1 shows the composition of various soils.

TABLE 11–1: SOIL CHARACTERISTICS

Color	Drainage	Composition
Brown	Well drained	Silt, some humus
Red-brown	Well drained	Clay, iron minerals
Tan	Well drained	Sand, little humus
Gray	Poorly drained	Clay, few minerals
Black	Poorly drained	Muck, much humus

If you dug into the ground and then drew a picture of the soil layers you found, you would be drawing a *soil profile.* Soils develop in horizontal layers called **horizons.** Each horizon in a soil profile is labeled with a capital letter. Examine the soil profile in Figure 11–21.

The top layer is the A horizon. This layer contains humus and small amounts of clay and sand. The roots of many plants are restricted to this horizon. The next layer, the B horizon, contains coarse clay, sand, and a small amount of humus. The B horizon also contains dissolved elements that were washed from the

Figure 11–21. Good topsoil has a thick A horizon with plenty of humus and just the right proportion of sand and clay. Most of the roots of cover crops, such as grains and grasses, grow in this horizon.

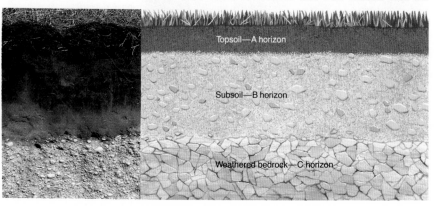

Topsoil—A horizon

Subsoil—B horizon

Weathered bedrock—C horizon

upper horizon by water. The movement of minerals from an upper horizon to a lower horizon is called **leaching.**

Below the B horizon is a thicker layer of partially weathered rock—the C horizon. Chemical weathering occurs in the C horizon, although here the process is very slow. Most plant roots do not reach this layer, and the exposure to air and water is also limited at this depth. Beneath the C horizon is bedrock.

Although most soils have the horizons just described, soils do vary from region to region. The climate of a region controls the rate of chemical weathering. Areas that are cold throughout most of the year have poorly developed soils because the frozen surface does not allow much weathering to occur. These areas also lack protective plant growth, so bare rock is found more often.

Humid, tropical areas develop thick soils because of the high temperature and abundant rain they receive. In such areas, however, many of the minerals are removed from the top layer of soil by water; thus the soil is not rich enough in nutrients to grow abundant crops.

In temperate climates, there is usually less rainfall and therefore less leaching of the soil than in the tropics. However, many temperate areas receive enough rainfall to develop thick soils. Ample plant growth in these regions produces soil that is rich and black with humus. Due to their richness, temperate soils are among the world's most fertile soils. Most of the world's food comes from soils of temperate areas, like those in the midwestern United States and Canada.

○ *What are soil horizons?*
○ *Why are tropical soils so poor in minerals?*

Section Review

READING CRITICALLY

1. What are the differences among soil horizons?
2. What is mineral leaching?

THINKING CRITICALLY

3. If temperate climate is important in the development of good soils, where else in the world might you expect to find soils as good as those in North America?
4. Why are both inorganic and organic materials necessary for the development of good soils?

Figure 11–22. Poor crops, such as those shown in the photograph at left, often result if the topsoil has been leached of its nutrients. Notice the difference in color between the leached soil (top) and the rich soil (bottom).

EXTENSION The students may be interested to know that the soils of the mainland United States can be divided into two major types—pedacal and pedalfer. Pedacal is a calcium-rich soil that covers most of the western United States. Pedacal gets its name from the Latin word *ped,* meaning "soil," joined with *cal,* representing the calcium in the soil. Pedalfer is an iron- and aluminum-rich soil that covers most of the eastern half of the country. The *al* in *pedalfer* stands for aluminum and the *fer* stands for *ferrous* (iron). *Thinking Critically:* Ask the students to predict the type of soil they could find in their town or city. Have them bring in samples to determine if their predictions were correct. (Answers will vary; predictions will be based on the area in which the students live.)

DISCUSSION Discuss Figure 11–21 with the students, pointing out the various soil horizons. Have the students study the leached and nutrient-rich soils shown in Figure 11–22. *Thinking Critically:* Ask the students to describe the visual clues that indicate which soil is nutrient-rich and which is not. (The leached soil is lighter-colored with little visible organic matter; the nutrient-rich soil is dark, moist, and much organic matter can be seen.)

SECTION REVIEW

Summary Have the students summarize the main points of soil formation and soil composition. You may wish to have them write a brief outline of the section.

Reinforcement The students may share their outlines with each other, filling in the outlines where they could be more complete. They could then quiz each other on the main points.

Answers–11.7

○ Layers in soil are called horizons.
○ The great amount of rainfall in the tropics leaches vast amounts of minerals from the top horizons.

Answers to Section Review

Reading Critically

1. Horizon A contains humus, fine clays, and sand. Horizon B contains coarse clay and sand, as well as minerals leached from Horizon A. Horizon C is made of partially weathered rock.
2. As rainwater soaks into the soil, it dissolves some of the minerals. As gravity pulls the water downward through the soil, the dissolved minerals are washed out of the top layers of soil. This washing-out process is called *leaching.*

Thinking Critically

3. Other areas in temperate latitudes such as Europe and parts of South America and Africa are likely to have good soils. However, some of these areas do not receive adequate rainfall and, thus, would not have richly developed soils.
4. A good soil is one rich in a great variety of elements. Both organic and inorganic materials are necessary to supply the variety of elements needed to produce arable land.

INVESTIGATION 11:
Weathering by Carbonation

Preparation of Materials

CAUTION: Be sure that students have properly put on their safety goggles and laboratory aprons. Warn them that they must not shake the jar too vigorously or it may shatter. In fact, you may want to substitute plastic containers for the glass jars. You may also want to include a third type of mineral or rock, such as quartz or sandstone, to extend the scope of the investigation.

Hint

Limestone is a rock composed solely of the mineral, calcite. Granite, however, is composed of feldspar, which weathers easily by carbonation. Quartz is not affected at all. Thus, the students may notice that the weathering of the granite appears in patches.

Answers to Analyses and Conclusions

1. The limestone in the carbonated water should change the most.
2. Carbonation, a kind of chemical weathering, occurs in this investigation. The calcite in the limestone dissolves in the carbonated water.
3. The natural reaction of carbon dioxide and water forms carbonic acid, which reacts with calcite. The process observed in nature is similar to the process occurring in this investigation. The natural process takes much longer than the process in the investigation.

Answer to Application

The students may list a variety of rock types. Any would be acceptable except those with a high calcite content such as limestone, marble, or even dolomite. These calcite- or dolomite-rich rocks should be predicted to react because they are easily weathered by carbonation.

INVESTIGATION 11: Weathering by Carbonation

PURPOSE

To examine the effects of carbonic acid on the weathering of different rocks

MATERIALS (per group of 3 or 4)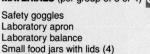

Safety goggles
Laboratory apron
Laboratory balance
Small food jars with lids (4)
Tap water
Limestone
Carbonated water
Granite

PROCEDURE

1. Copy the chart shown.
2. **CAUTION: Wear safety goggles and a laboratory apron during this investigation.** Using the balance, determine the mass of the limestone specimen.
3. Fill a small jar half full of tap water, and add the limestone to the jar. Put the cap on the jar and gently shake the jar 200 times.
4. Remove the limestone from the jar, dry it, and determine its mass.
5. Repeat steps 3 and 4. After shaking the jar 400 times, record the mass.
6. Repeat steps 3 and 4 for a third time, shaking the jar 600 times. Record the results.
7. Compute the percentage mass change, using the following formula:
$$\frac{\text{Initial Mass} - \text{New Mass}}{\text{Initial Mass}}$$
8. Repeat steps 2 through 7, using limestone in carbonated water. Record your results.
9. Repeat steps 2 through 7, using granite in both tap water and carbonated water. Record your results.

ANALYSES AND CONCLUSIONS

1. Which combination of liquid and rock produced the greatest change in mass?
2. What kind(s) of weathering occurred in this investigation? List and describe the weathering in as much detail as possible.
3. In what way is the weathering process in this investigation similar to weathering in nature? In what way is it different?

APPLICATION

Using what you have learned in this investigation, predict two types of rock that would probably not weather very much. Predict one type of rock that would probably weather a lot. State the reasons for your predictions.

TABLE 1: EFFECTS OF WEATHERING				
	Limestone		Granite	
	Water	Carbonated Water	Water	Carbonated Water
Initial mass				
Mass after 200 shakes				
Mass after 400 shakes				
Mass after 600 shakes				

SUMMARY

- Weathering is the process that breaks down rocks. (11.1)

- The two major types of weathering are physical and chemical. Factors that affect the rate of weathering are rock structure, climate, topography, and vegetation. (11.1)

- Physical weathering breaks rocks apart mechanically. (11.2)

- Exfoliation, frost action, root-pry, and abrasion are four types of physical weathering. (11.2)

- In chemical weathering, minerals within a rock are chemically changed. (11.3)

- Hydration, oxidation, and carbonation are types of chemical weathering. (11.3)

- Mass movement is the movement of rocks due to gravity. Mass movement is classified by the slope of the land, the size of materials, and the amount of water involved. (11.4)

- The four types of mass movement are creep, flow, landslide, and subsidence. (11.4)

- Differential weathering of rock layers results in the formation of distinctive landforms such as buttes and mesas. (11.5)

- Landforms resulting from mass movement are usually not as spectacular as those produced by differential weathering. (11.5)

- Soils are the product of the weathering process. (11.6)

- Soil formation is affected by the type of parent material, the climate, the topography, the soil organisms, and time. (11.6)

- The various layers, or horizons, of a soil are shown in a soil profile. (11.7)

- Most of the world's fertile soils are found in temperate regions, such as the United States. (11.7)

Write all answers on a separate sheet of paper.

SCIENCE TERMS

Correctly use each of the following terms in a sentence.

abrasion **(234)**
carbonation **(236)**
creep **(237)**
differential weathering **(240)**
exfoliation **(233)**
flow **(238)**
frost action **(233)**
horizons **(246)**
humus **(244)**
hydration **(235)**
landslide **(238)**
leaching **(247)**
mass movement **(237)**
oxidation **(235)**
soil **(231)**
subsidence **(238)**
weathering **(231)**

SCIENCE QUIZ

Modified True-False

Mark each statement *true* or *false*. If a statement is false, change the underlined term to make the statement true.

1. The most effective weathering process in a desert would be <u>chemical</u> weathering.

2. The soil layer richest in humus is the <u>A horizon</u>.

3. Mud slides are classified as <u>landslides</u>.

4. Subsidence is an example of <u>chemical weathering</u>.

5. Carbonic acid is formed when carbon dioxide combines with <u>calcite</u>.

continues

CHAPTER REVIEW

SUMMARY

To review the major concepts in the chapter, the students should read the summary statements. The statements are cross-referenced to the chapter to facilitate reinforcement of any concepts of which the students feel unsure. Have the students work in groups to quiz one another on the information.

SCIENCE TERMS

The sentence in which the science term is used should reflect an understanding of the definition of the term. You may want to make up a crossword puzzle of the science terms to hand out to the class.

SCIENCE QUIZ

Answers to Modified True-False

1. false, physical
2. true
3. false, flows
4. false, mass movements
5. false, water

continues

Answers to Multiple Choice

6. d
7. a
8. b
9. c
10. d

Answers to Completion

11. chemical
12. water
13. physical
14. humus
15. tropics

Answers to Short Answer

16. Mass movement involves the down-slope movement of rock material that has broken free of the solid bedrock. The material is usually freed from the bedrock by physical weathering, although chemical weathering may play a part.
17. Quartz is a common mineral that is very resistant to the weathering processes. It does, however, slowly wear down by abrasion. Many of the more common igneous and metamorphic rock-forming minerals weather into clays. Thus, quartz and clay minerals are common as weathering products that form soils.
18. Vegetation helps to protect rocks from the sandblasting effects of wind. Vegetation decay forms acids that speed up the chemical weathering of rocks. Roots further open cracks in the soil and rock material, thus increasing the surface area exposed to the weathering processes.

Answers to Writing Critically

19. The surface of the mineral may be weathered and therefore dull. By breaking it open, you would find an unweathered surface.
20. Students' answers may vary. If they do not classify soils as nonrenewable, they should at least explain that soils should be carefully protected because their renewing rate is extremely slow.

ANSWERS TO EXTENSION

1. Pothole development is a major problem in regions where there is frequent freezing and thawing. The water enters cracks in the pavement, expands as it freezes, and widens the hole. This se-

250

Multiple Choice

Write the letter of the choice that best answers the question or completes the statement.

6. Hydration is what type of weathering?
a) physical   b) fast
c) rock   d) chemical

7. The process of carbonation forms karst topography with what type of rock?
a) limestone   b) quartzite
c) granite   d) shale

8. Frost action is most effective when there
a) are extended periods of subzero weather.
b) is repeated freezing and thawing.
c) is a rainy season.
d) are large breaks in the rock.

9. Rounded quartz pebbles found along a stream bed would probably have been weathered by
a) exfoliation.   b) hydration.
c) abrasion.   d) mass movement.

10. A factor that is not important in the formation of soils is the
a) topography.   b) amount of rainfall.
c) type of bedrock.   d) age of bedrock.

Completion

Complete each statement by supplying the correct term or phrase.

11. Carbonation is a form of _____ weathering.

12. The main agent of chemical weathering is _____.

13. Frost action is an example of _____ weathering.

14. Decayed plant and animal matter within the soil is called _____.

15. Leaching of almost all soil minerals occurs in the _____.

Short Answer

16. Explain why mass movement is more closely related to physical weathering than to chemical weathering.

17. Why are soils made almost entirely of quartz and clay minerals?

18. Explain how vegetation can slow down some forms of weathering but speed up others.

Writing Critically

19. If you found a mineral that did not have a recognizable luster, why would you break it to look at a fresh surface?

20. Considering the length of time that it takes to form soils, do you think soils should be classified as renewable or nonrenewable resources?

EXTENSION

1. Write a report on the formation of potholes in the roads of your area. Take photographs of some potholes, analyze their locations, and identify the steps in their development.

2. Examine the exterior of several older stone buildings in your area. Determine if any chemical or physical weathering has occurred. Make an oral report to your class on your findings.

3. Describe several of the landforms in your area. Draw or photograph each. Identify the role that weathering has played in their formation.

4. Read John Steinbeck's book *The Grapes of Wrath,* and write a report on the effects of the "dust bowl" on farming in the 1930s.

5. If you live in a rural area, talk to one of the local farmers about the ways in which he or she prevents the loss of soil. Report your findings to the class.

---

quence is repeated over and over until the pothole is large enough to do major damage to an automobile tire.
2. Older buildings frequently show discolored building material due to chemical weathering. The edges and surfaces of the material are pitted or rounded. Areas where excess water may run down the side from the roof may show even more signs of weathering.
3. The answers to this question will vary depending upon the local landforms. In many regions, you may find it useful to discuss the present location of streams and have the students hypothesize how the landforms might have looked before the streams were well-developed, cutting through ridges or widening valleys. There may be talus slopes, mesas, or areas of subsidence to identify.
4. The report should include information about the effects of the Dust Bowl on farming.
5. Reports will vary.

APPLICATION/CRITICAL THINKING

1. In some parts of the country the soil has a definite reddish-brown color. What mineral might be the source of this color? Explain your answer.

2. Give an example of subsidence that might be caused by human activity. What steps might be taken to prevent subsidence in your example?

3. The Sphinx, a huge statue in the desert in Egypt, has withstood the processes of weathering for thousands of years. Yet in the last 50 years the decay of the statue's features has increased greatly. What might have caused this rapid increase in weathering? Explain your answer.

FOR FURTHER READING

Gerster, G. "Patterns of Plenty: The Art in Farming." *National Geographic* 166 (September 1984): 391. This photo essay on the subject of different crop patterns beautifully demonstrates the diversity of agricultural activities.

Gibbons, B. "Do We Treat Our Soil Like Dirt?" *National Geographic* 166 (September 1984): 350. This article analyzes the ways humans alter the soil with industrial and agricultural processes.

Harpstead, M., and F. Hole. *Soil Science Simplified.* Ames, Iowa: Iowa State University Press, 1980. This classic book presents a basic introduction to soil sciences without a strong emphasis on the technical aspects of the field.

Steinhart, P. "We Can't Grow When It's Gone." *National Wildlife* (March 1985): 74. This article impresses upon the reader the importance of good soil for growing abundant crops.

Modern farming practices, such as those described in the Technology, would probably help prevent a recurrence.

Challenge Your Thinking

This is a picture of the "dust bowl" of the 1930s. Poor farming practices and unusually dry weather created these conditions.

Although many steps have been taken to preserve our soil, scientists estimate that some areas of the country lost 10 cm of soil last year. What do you think could be done to prevent another dust bowl?

Chapter 11 Review   **251**

ANSWERS TO APPLICATION/CRITICAL THINKING

1. Any of the minerals containing iron can weather by the process of oxidation. The result is a reddish-brown clay.

2. Ground water fills the open spaces in many rocks. If considerable ground water is removed for industrial or home use, then subsidence will increase.

3. Pollution from the large metropolitan areas to the north is probably speeding up the weathering of the Sphinx.

Chapter 12: RUNNING WATER

PLANNING THE CHAPTER

Chapter Sections	Page	Chapter Features	Page	Program Resources	Page
Section 1: Surface Water	253			Critical Thinking **(H)**	**TRB 22**
12.1 Runoff **(B)**	253	**Discover:** Measuring Runoff **(B)**	254		
12.2 Drainage Basins **(B)**	255	A Matter of Fact	255		
		A Matter of Fact	256		
		Section Review	256		
		Biographies: Then and Now Eugene Hilgard, Helen M. Turley	257		
Section 2: Erosion and Deposition	258			Reading for Content: *Using Diagrams in Sequence* **(A)** Concept Extension: *Stream Flow* **(H)**	**TRB 22** **LM 195** **TRB 25** **LM 247**
12.3 Headward Erosion **(B)**	258	A Matter of Fact	259		
12.4 Valley Formation **(B)**	260				
12.5 Water Deposits **(B)**	262	**Discover:** Studying Erosion **(B)**	262		
		Section Review	263		
		Skill Activity: Analyzing Diagrams **(B)**	264		
Section 3: Ground Water	265			Cross-Discipline: Science and Health, *Planning for Cave Living* **(A)**	**TRB 22**
12.6 The Water Table **(B)**	265	**Discover:** Measuring Porosity **(B)**	266	Investigation 12.1: *Permeability and Capillary Water* **(H)** Investigation 12.2: *Soft Water and Hard Water* **(B)**	**TRB 47** **LM 49** **TRB 51** **LM 53**
12.7 Aquifers **(B)**	267				
12.8 Karst Topography **(B)**	268	**Activity:** Analyzing Karst Landforms **(B)**	268		
		A Matter of Fact	269		
		Section Review	269		
		Investigation 12: Porosity and Permeability **(A)**	270	Student Record Book: Textbook Investigations **(A)**	**TRB 23**
Chapter 12 Review	271			Vocabulary **(A)**	**TRB 12** **LM 145**
				Tests **(A)** Computer Test Bank	**TRB 50**

(LM) Laboratory Manual/Study Guide, **(TRB)** Teacher's ResourceBank™

B = Basic **A** = Average **H** = Honors

The coding Basic, Average, and Honors indicates sections or subsections that might be appropriate for different levels of learners. For additional suggestions regarding choice of topic and depth of coverage, see the Pacing Chart on pages T16–T20.

CHAPTER CONCEPTS, OBJECTIVES, AND TERMS

Section	Concepts	Objectives	Science Terms
Section 1: Surface Water	■ Surface runoff is responsible for considerable erosion of the land. **(12.1)** ■ The amount of eroded material carried by a stream is called its load. **(12.1)** ■ A drainage basin is an area drained by a single stream system. This system includes a trunk and tributaries. **(12.2)** ■ Drainage systems form different shapes, depending upon the steepness of the slope and the underlying rock structure. **(12.2)**	■ **List** the factors that affect runoff. ■ **Describe** the various drainage patterns.	runoff erosion load drainage basin
Section 2: Erosion and Deposition	■ Streams increase their length by a process known as headward erosion. **(12.3)** ■ The low point of a stream or river is its base level. A stream cannot erode below its base level. **(12.3)** ■ Stream valley development is divided into three stages: youthful, mature, and old-age. **(12.4)** ■ Youthful streams have V-shaped valleys. Mature streams have wide floodplains. Old-age streams have many meanders. **(12.4)** ■ If a stream carries a large load of sediment, a delta forms at the mouth of the stream. **(12.5)** ■ Playas and alluvial fans are features of dry, mountainous areas. **(12.5)**	■ **Relate** a stream's slope, volume, base level, and velocity to its ability to erode the land. ■ **Compare** the stages of valley development. ■ **Explain** the processes involved in the formation of deltas. ■ **Describe** the conditions necessary for the formation of alluvial fans and playas.	headward erosion base level flood plain meanders delta alluvial fans playas
Section 3: Ground Water	■ The amount of water that can enter the ground is determined by the porosity and permeability of the bedrock. **(12.6)** ■ Porosity is the ratio of the volume of air space in a rock to the total volume of the rock. Permeability is the ability of a rock to allow the movement of water through the rock. **(12.6)** ■ A permeable rock layer may act as an aquifer. **(12.7)** ■ Stalactites and stalagmites are examples of karst deposits. **(12.8)**	■ **Define** the term water table. ■ **Describe** the rock characteristics of porosity and permeability. ■ **Explain** how karst deposits develop.	water table porosity permeability aquifer

CHAPTER MATERIALS

Title	Page	Materials
Discover: Measuring Runoff	254	*(per student)* metal baking dish
Discover: Studying Erosion	262	*(per student)* paper, pencil
Skill Activity: Analyzing Diagrams	264	*(per student)* paper, pencil
Discover: Measuring Porosity	266	*(per student)* sponges, bowl, water, graduate
Activity: Analyzing Karst Landforms	268	*(per group of 3 or 4)* topographic map of Mammoth, Kentucky
Investigation 12: Porosity and Permeability	270	*(per group of 3 or 4)* clear plastic tube; hose and clamps (2), fitted to tube; beaker, 500 mL; water; gravel; sand; clay; graduate; stopwatch

TEACHING SUGGESTIONS

Section 1: Surface Water

Class Activity

Streams can be studied using most topographic maps. The students should be able to identify associated features such as flood plains, waterfalls, or sandbars. They should also be able to classify types of valleys and drainage patterns from topographic maps.

You may wish to have the students analyze the gradient of various streams by using topographic maps. The formula for computing gradient is:

gradient = vertical drop / horizontal distance
(Vertical drop = change in elevation using contour lines; Horizontal distance = use map scale)

Section 2: Erosion and Deposition

Field Trip

Take the students on a field trip to study streams and valley development. Visit a quarry to find the water table level. (Water is often pumped out of active quarries because they dip below the water table.) Visit a flood control dam or a local sewage treatment plant.

Outside Speaker

Have someone from the Army Corps of Engineers talk to the class about flood control projects in your area.

Section 3: Ground Water

Demonstration: Karst Topography

Purpose

To show how karst topography forms

Background

Caves can be produced when ground water dissolves underground limestone. Dripping from the roofs of caves, ground water can also produce the interesting formations called *stalactites* and *stalagmites*. The carbon dioxide dissolved in the ground water allows these structures to form. In this demonstration the students will make some of the substances involved in the formation of caves.

Materials

Test tubes (2)
Limewater made from $Ca(OH)_2$
Test-tube rack
Test-tube clamp
Ring stand
Thistle tube
Glass tubing
Rubber stopper to fit test tubes (2)
Marble chips
Dilute hydrochloric acid
Rubber tubing
Bunsen burner
Safety goggles
Laboratory apron

Procedure

1. Half-fill one large test tube with limewater. Put this test tube in the rack.
2. Set up a carbon dioxide generator. First clamp a test tube to the ring stand. Fit the thistle tube and glass tubing into the holes of the rubber stopper. *(CAUTION: First*

moisten the tubing with water or glycerin. Then gently push the tubing into the stopper. Never use force. Never push the tubing toward the palm of your hand. Wrap the glass tubing in a paper towel to prevent injury to your hands in case the glass breaks.) Add marble chips to the test tube to a depth of about 3 cm. Then add about 3 mL of dilute hydrochloric acid. Put the stopper in the test tube. Connect two pieces of glass tubing with a short length of rubber tubing. Place the end of the second piece of glass tubing into the test tube of limewater. The bubbles you see are carbon dioxide. Some of the carbon dioxide will be absorbed by the limewater and a substance called carbonic acid (H_2CO_3) will form. In nature, rain absorbs CO_2 from the air. When the rain soaks into the ground, the resulting ground water contains weak solutions of carbonic acid.

3. Bubble the CO_2 until the limewater turns milky. The carbonic acid and limewater have formed calcium carbonate.

4. Continue bubbling CO_2 into the limewater until the solution becomes clear. Soluble calcium bicarbonate $Ca(HCO_3)_2$ has now formed. The carbonic acid has dissolved the calcium carbonate. In nature, carbonic acid dissolves underground calcium carbonate (limestone) forming holes, or caves, in the earth.

5. Now try to reverse the process. Take the CO_2 generator apart. Hold the limewater test tube with a clamp. Heat the test tube gently. The reappearance of the limestone corresponds to the formation of stalagmites and stalactites in nature.

Questions to Ask the Students

1. How are limestone caves formed in nature?
   (Rain absorbs CO_2 from the air. Rainwater contains carbonic acid which dissolves limestone, forming holes, or caves.)
2. What process produces the stalagmite and stalactite formations in caves?
   (As calcium carbonate drips into a cave, the water evaporates, leaving a solid calcium carbonate deposit.)
3. Name the products that were formed in this investigation. (carbon dioxide, carbonic acid, calcium carbonate, and calcium bicarbonate)

CHAPTER 12

CHAPTER OVERVIEW

This chapter discusses running water and explains the effects it has on the earth's surface. Also presented are the factors that affect running water. Next, drainage basins, headward erosion, valley formation, and sediment deposition are explained. The chapter then concludes with a discussion of ground water. The water table is defined, and various geologic features associated with ground water are covered.

Section 1: Surface Water In this section, runoff and drainage basins are described. Runoff is defined as the movement of water toward a lower point on the surface of the earth. Factors that affect runoff, such as the shape of the land, the amount of rainfall, and the extent of plant growth are discussed. The process of erosion is defined, and the formation of a river from a gully is explained. Next, drainage basin is defined as the area drained by a river and the river's tributaries. The section ends with an explanation of how drainage patterns develop.

Section 2: Erosion and Deposition Erosion and deposition by running water are covered in this section. First, the relationship between a stream's slope, volume, base level, velocity, and the stream's ability to erode land is discussed. Then each of the three stages in the development of a stream valley is described. The features of a youthful, mature, and old-age valley are compared. Finally, the processes involved in the formation of deltas and desert deposits are covered.

Section 3: Ground Water This section presents the topic of ground water and its associated concepts. The location of the water table is explained. Porosity and permeability are defined as they relate to the amount of water a rock can hold and how much water can move through the rock. Finally, aquifers and karst topography are discussed.

Running Water

Have you ever looked at a stream and tried to imagine the number of places the water had been? Some of the water may have spent thousands of years as ice in an Alaskan glacier. Some may have traveled the length of the Amazon River or fallen as sleet in the mountains of Montana. Some of the water may even have been part of the stream that carved this natural bridge.

A natural bridge, carved by water

252

CHAPTER MOTIVATING ACTIVITY

Stream tables are highly effective in getting students interested in the study of running water. If you don't have a commercial stream table, one can be easily made using a wooden box, plastic liners, and sand. Allow the students to experiment with different strengths of running water and different-sized soil or sand grains.

1 Surface Water

SECTION OBJECTIVES

After completing this section, you should be able to:
- **List** the factors that affect runoff.
- **Describe** the various drainage patterns.

NEW SCIENCE TERMS

runoff
erosion
load
drainage basin

12.1 Runoff

You may recall from Chapter 9 that water molecules evaporate into the atmosphere from the surface of the ocean. Once in the atmosphere, water vapor may be carried over a landmass by the wind. After condensing into cloud droplets, the water molecules fall to the earth's surface as drops of rain. The movement of water is called the *water cycle*.

The moment a raindrop falls to the earth it begins its return to the sea. Some of the water soaks into the ground, but most moves across the surface toward a lower level. This movement of water on the surface of the earth is called **runoff**.

The shape of the land and the amount of rainfall affect runoff. The steeper the slope, the greater the speed of the runoff. The volume of runoff increases during heavy rain because there is not enough time for much water to soak into the ground. Runoff also increases if rain falls on ground that is already saturated with water, or if the water cannot get into the ground. Large cemented areas, such as those that are present in cities, provide little opportunity for water to get into the ground.

Figure 12–1. If a lot of rain falls in a short period, or if there is a sudden melting of a large accumulation of snow, a flood may result.

Section 1 Surface Water **253**

BACKGROUND INFORMATION

Running water is the most important agent of erosion. Running water erodes in the following ways: direct lifting, abrasion and impact, solution, and undercutting. Direct lifting occurs when turbulent streams dislodge materials and carry them downstream. Running water erodes through abrasion and impact when solid particles in the stream wear down rocks and loosen other particles. Large amounts of rock are carried by streams in solution. A stream that erodes laterally by one of the methods described above may undercut its bank to the point where the force of gravity causes it to collapse into the river.

Section **1**: **Surface Water**

SECTION FOCUS

Section Overview This section concentrates on runoff, which is the most important agent of erosion. The amount of eroded material carried by a stream or river is defined as the stream's load. Factors that affect runoff and the amount of load a stream or river carries are described. Drainage basins and different drainage patterns are also discussed.

Section Objectives For a list of section objectives, see pupil's edition page.

New Science Terms For a list of new science terms in this section, see pupil's edition page. You may wish to familiarize the students with these terms before they read the section. Students are probably familiar with some of these terms. Tell them that in this section they will learn the scientific meanings of the terms.

SECTION DEVELOPMENT

12.1 Runoff

DISCUSSION Begin this section by telling the students that they are already aware of some of the concepts introduced in the section, such as runoff, drainage, and erosion. Convince them of this by discussing observations they have probably made during a rainfall. Ask the students to describe how rainwater flows. (Rainwater travels downhill.) Finally, have your students compare how a bare field looks during a light drizzle and during a heavy rainfall. (During a heavy rainfall, a bare field is muddy because there is little time for the ground to absorb the water.) *Thinking Critically:* Ask the students to explain why both runoff and ground water move downslope toward sea level. (Both move downslope due to gravity.)

EXTENSION Accelerated erosion occurs when humans disturb the surface of the earth or cause the rate of erosion to increase in any way. Accelerated erosion poses a serious problem for agricultural growth because it removes large quantities of soil and reduces the fertility of the soil. Certain conservation measures can be taken to prevent the loss of soil. Farming practices that decrease the amount of runoff include contour plowing, terracing, and the planting of cover crops. Cover crops reduce erosion by physically impeding the flow of water. Contour plowing creates ridges across rather than up and down a hillside in order to direct the flow of water. Terracing, which is practiced on hilly lands, directs the flow of water with gentle slopes constructed in the hill and grasses planted between the terraces. All of these farming practices reduce erosion.

EXTENSION As noted in the pupil's edition, water is not the only agent of erosion. ***Thinking Critically:*** Ask the students to name other agents. (ice, wind, organisms, and gravity)

DISCUSSION Direct the students' attention to Figures 12–3 and 12–4. Stress the idea that small streams feed into larger streams and rivers. Many students have the misconception that rivers divide and flow into smaller water systems.

DISCOVER: Measuring Runoff

Skill *(Measuring)*

Suggest that the students trap rainwater in a container large enough to prevent overflowing. Some students may need help with the mathematical calculation required in this activity. Remind them that the formula for volume is Volume = length x width x height, and the formula for area is Area = length x width. Students can determine the volume of water in the pan by pouring it into a graduate, or by measuring its depth and applying the volume formula.

Figure 12–2. A muddy stream (left) is an indication of excess erosion. If soil is unprotected by a cover of plants, deep gullies can be eroded (right).

DISCOVER

Measuring Runoff

After a significant rain, measure the volume of water that collects in a metal baking dish left outside, and convert your measurement to L/m². Next, find the area of your basketball court or driveway in square meters, and determine the volume of runoff from a typical rainfall.

Runoff decreases in areas with extensive plant growth. Plants slow down the water's movement, so more water soaks into the ground. Some types of soil can also reduce the amount of runoff. If soil is loosely packed or has a large humus content, water can more easily soak into the ground.

As water rushes downhill, it carries sediment, which is then deposited at the bottom of the hill. If enough sediment washes away, a small channel, or *gully*, forms. The process that moves sediment and reshapes landscapes is called **erosion.** Water is not the only agent of erosion, but it is the most important.

Gullies usually form in places where the surface is not protected by a cover of plants, such as in a plowed field. With continued erosion, a gully develops into a *stream.* A stream carries water throughout most of the year, while a gully carries water only when it rains. As time passes and more land is eroded, a stream expands and develops into a *river.* The volume of water in most rivers is large because many streams empty into them.

Water has been called "the great leveler" because of its ability to erode highlands and fill in lowlands. Even a small stream can carry a tremendous quantity of eroded material.

The amount of eroded material carried by a stream or a river is called the **load.** The load a stream carries depends on the stream's water volume and the steepness of the slope down which it flows. A large stream or a fast-moving stream carries a larger load than a small or slow-moving stream carries. Scientists estimate that the rivers of the world carry 1000 metric tons of sediment to the sea each second.

○ ***What is runoff?***
○ ***What is erosion?***

Answers—12.1

○ Movement of water on the surface of the earth is called runoff.

○ Erosion is the process that moves soil and rock materials.

DEMONSTRATION

Demonstrate how contour plowing reduces erosion. Fill two rectangular pans with soil and prop up one end of each pan. Make lengthwise furrows in the soil of one pan and horizontal furrows in the other pan. Pour water into the pans at the propped-up end. Have the students tell which pan shows the most erosion. (the pan with lengthwise furrows)

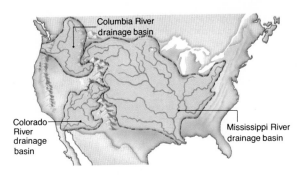
Columbia River
drainage basin

Colorado
River
drainage
basin

Mississippi River
drainage basin

12.2 Drainage Basins

A large tree with its many leafy branches is part of a system for producing food. Similarly, a single river, called a *trunk*, and the streams that flow into it, called *tributaries*, act as a system for draining water from the land. The area drained by a trunk and its tributaries is called a **drainage basin.**

An imaginary line called a *divide* separates one drainage basin from another. For example, the continental divide, which runs the length of the Rocky Mountains, divides the basin that drains into the Pacific Ocean from the basin that drains into the Gulf of Mexico. Many regions of the country have divides that separate major drainage basins. Some of these drainage basins are shown in Figure 12–3.

The streams in a basin often develop unique drainage patterns due to the dominant landform of the area. For instance, streams flow in all directions away from a volcano, forming a radial drainage pattern. *Radial* means "coming from a central point," and as you can see in Figure 12–4, the volcano acts as the central point for this drainage pattern.

Drainage basins containing long, parallel ridges and valleys form *trellis* patterns. Unlike a flower trellis, which has branches originating from a single point, a trellis drainage pattern has tributary streams that flow parallel to each other and join the trunk at right angles.

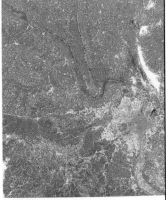

Figure 12–3. This map (left) shows the major drainage basins of the United States. A drainage basin includes all the main streams and tributaries that drain an area (right).

> **A MATTER OF FACT**
>
> Of the total amount of water on the earth, slightly more than 97 percent is in the oceans, about 2 percent is frozen in icecaps and glaciers, and about 0.01 percent is held in lakes. Rivers contain only about 0.0001 percent.

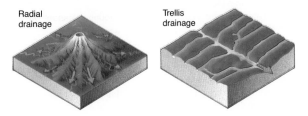

Radial
drainage

Trellis
drainage

Figure 12–4. The streams in a radial drainage pattern all come from the same place, such as from a volcano (left). The tributaries of a trellis drainage pattern flow nearly parallel to one another, and join the trunk stream at right angles (right).

Section 1   Surface Water   **255**

12.2 Drainage Basins

DISCUSSION Refer the students to Figure 12–3 and have them locate the Columbia River drainage basin, the Colorado River drainage basin, and the Mississippi River drainage basin. Then have the students locate the three rivers or trunks that drain the respective areas. While they are studying Figure 12–3, discuss the concept of drainage basins. It is important that the students have a clear understanding of drainage basins before you proceed to an explanation of drainage patterns.

THINKING SKILL *(Interpreting a Map)* After the students have located the three drainage basins in Figure 12–3, ask them to name the bodies of water into which the trunks flow. (The Columbia River flows into the Pacific Ocean; the Colorado River flows into the Gulf of California; the Mississippi River flows into the Gulf of Mexico.)

DISCUSSION Refer the students to Figure 12–4 and discuss a radial drainage pattern. ***Thinking Critically:*** Ask the students to think of a sea organism that has radial symmetry. (a starfish) Then compare the volcano to the center of a starfish, and the streams to the rays of the starfish. Finally, ask the students to observe the trellis drainage pattern. Have them locate the trunk and the various tributary streams in Figure 12–4.

BACKGROUND INFORMATION

The term *drainage basin* is applied to the entire area drained by a given stream and its tributaries. The Mississippi River drainage basin drains most of the United States between the Rocky Mountains and the Appalachian Mountains. In some desert areas, depressions drain entirely inward and are called centripetal drainage patterns. In some dry regions rivers flow for very long distances.

DISCUSSION Refer the students to Figure 12–5 and discuss how topography affects a drainage pattern. ***Thinking Critically:*** You may wish to have the students compare a dendritic drainage pattern with a trellis drainage pattern. (In a trellis pattern, tributary streams join the trunk at right angles. In a dendritic drainage pattern, tributary streams form fairly large angles where they join the trunk.)

EXTENSION (Tie-in/Biology) A nerve cell, or neuron, is the functional unit of the human nervous system. The neuron is composed of a cell body, an axon, and dendrites. Dendrites are extensions that receive information in the form of electrical impulses. Dendrites receive the information from other neurons, other types of cells, or external stimuli. An impulse travels from the dendrites to the cell body and then down the axon. From the axon the impulse travels to another neuron or other type of cell. The dendrites of a human neuron average approximately 1000 to 10 000 connections. Obtain a photograph or illustration of a neuron with obvious dendrites. Have the students describe the appearance of the dendrites. (branching) ***Thinking Critically:*** Ask why the dendritic drainage pattern is so named. (because of its similarity in appearance to nerve cell dendrites)

EXTENSION Deranged drainage patterns commonly develop in areas where the pre-existing drainage system has been disturbed. Excellent examples of this type of pattern can be found in recently glaciated areas such as Minnesota and Wisconsin where rocks carried by glaciers were deposited all over the region. The rocks blocked stream beds, resulting in a very unusual drainage pattern.

A MATTER OF FACT

The world's largest drainage basin is the Amazon River system in South America. It covers an area of 6 150 000 km² and includes nearly 15 000 tributaries.

Sometimes topography is the dominant factor in the development of a drainage pattern. Branching, or *dendritic*, patterns develop in basins where slopes are not very steep. Here the tributaries form fairly large angles where they join the trunk. The entire drainage system forms a pattern similar to the dendrites, or branched endings, of a nerve cell. Where slopes are steeper, *parallel* drainage may develop.

Finally, drainage patterns may be determined by special characteristics of the basin. In areas in which the bedrock is jointed or faulted, *rectangular* patterns develop. Some of the stream channels make sharp turns as they follow lines weakened by fractures in the rock. Patterns found in poorly drained areas tend to be irregular, or *deranged*. Deranged drainage basins have areas in which surface water collects, such as swamps, marshes, and lakes.

○ **What is a drainage basin?**
○ **Name four types of drainage patterns.**

Figure 12–5. Dendritic drainage patterns (bottom left) develop where slopes are gentle, whereas parallel patterns (top left) develop where slopes are steeper. Rectangular patterns (bottom right) occur where the bedrock is jointed or fractured. Deranged drainage patterns (top right) are found in areas where there is almost no slope at all to the land.

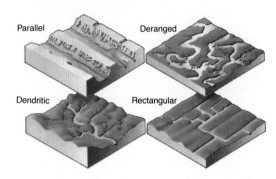

Parallel   Deranged

Dendritic   Rectangular

Section Review

READING CRITICALLY

1. What are three factors that affect runoff?
2. What types of drainage patterns develop around volcanoes and severely faulted areas?

THINKING CRITICALLY

3. From the map on page 255, determine how many major drainage basins there are in the United States.
4. Explain how the drainage pattern might change if a volcano were to develop in an area with trellis drainage.

SECTION REVIEW

Summary Review the factors that affect runoff. Ask the students to define erosion and load. Finally, have the students name and describe five drainage patterns.

Reinforcement Have the students complete the Section Review. Then go over the questions together.

Answers–12.2

○ A drainage basin is the entire area drained by a single trunk and its tributaries.

○ Dendritic, radial, trellis, and deranged are four types of drainage patterns.

Answers to Section Review

Reading Critically

1. The shape of the land, the amount of rainfall, the type of soil, and the amount of plant growth all affect runoff.

2. Radial patterns develop around volcanoes; rectangular patterns develop around faulted areas.

Thinking Critically

3. Three are shown, but there are many more smaller basins.

4. The trellis pattern would be changed to a radial pattern.

EUGENE HILGARD (1833–1916)

Eugene Hilgard was born in Bavaria in 1833. He immigrated to America with his family in 1836, when he was three years old. Hilgard attended schools and universities in both the United States and Europe, where he studied chemistry and geology. In 1853 he received a doctorate from the University of Heidelberg in Germany.

After returning to the United States, Hilgard spent much of his career studying the structure and composition of soils. Combining his knowledge of chemistry and geology, he was one of the first scientists to recognize the importance of soil analysis to agriculture. He understood that different types of plants require different soil conditions. For example, some plants grow only in well drained sandy soil, while others require wet soil with a lot of humus. At the time, most farmers did not understand the complex nature of soils. They only knew that a crop would produce smaller yields each year that it was planted. After several years of planting the same crop in the same location, they would abandon the field and begin planting in a new location.

Hilgard argued that soil should be considered a special geologic formation, deserving as much attention as the underlying rock formations. His writings brought him international recognition as a geologist and a soil scientist. His ideas lead to many changes in farming practices and to the recognition of soil as a complex material vital for human survival.

HELEN M. TURLEY (1943–)

Helen Turley was born in Augusta, Georgia, in 1943. She received her undergraduate training from St. John's College in Annapolis, Maryland. She then attended the College of Agriculture and Life Sciences at Cornell University in Ithaca, New York. There she expanded her knowledge of soil science and biology.

In New York she was employed at the New York State Agriculture Experiment Station, where she studied the problems associated with growing grapes. Grapevines are commonly grown on hillsides, so the growers are especially concerned with soil erosion. In addition to concerns about soil erosion, the chemical and physical makeup of the soil is also of concern. If the soil chemistry is not right, the grapevines will not grow. Turley began concentrating on the needs of the grape-growing industry and soon became a specialist in the characteristics of soils and soil erosion.

With this expertise she relocated to serve the grape-growing industry in California. There she began a consulting business, applying her knowledge of soils and erosion to the problems found in the various grape-growing regions of California. She is presently conducting research on cover crops planted between the grapevines to protect the soil from erosion. Her research may lead to the selection of the best cover crops for preventing soil erosion in other agricultural areas as well.

Section 1 Surface Water **257**

Discussion Have the students read the two biographies and then discuss the contributions of each individual. *Thinking Critically:* Ask the students to describe any farming techniques that are used to protect the soil. (contour plowing, crop rotation) Have the students explain why cover crops are important. (Cover crops slow down runoff, decreasing erosion.)

Extension The four major components of soil are rock particles, organic matter, water, and air. Soils have horizons, or layers, which run parallel to the surface. Soil horizons differ from one another in characteristics such as color, texture, and porosity. The succession of horizons is called the soil profile. Soil horizons are designated by the letters A, B, C, and their subdivisions. The "A" horizon has the most organic material and is the zone from which dissolved materials are removed. Obviously, soil is very important to agriculture. Soil surveys are done to predict a soil's adaptability to various crops, grasses, and trees. Soil surveys were first used in the United States in 1898.

Extension Have interested students find out more about Eugene Hilgard, Helen Turley, or William Morris Davis.

Section 2: **Erosion and Deposition**

SECTION FOCUS

Section Overview In this section, headward erosion, valley formation, and water deposition are discussed. Headward erosion is the process by which streams increase their length. Concepts related to headward erosion are explained. The three stages in the development of a stream valley are described. The section ends by explaining how deltas, alluvial fans, and playas are formed.

Section Objectives For a list of section objectives, see pupil's edition page.

New Science Terms For a list of new science terms in this section, see pupil's edition page. You may wish to pronounce each of the new science terms before beginning this section. Most of the students are probably not familiar with any of these terms.

SECTION DEVELOPMENT

12.3 Headward Erosion

DISCUSSION Ask the students to describe different types of streams and rivers that they have seen in your local area or while they were traveling or in the movies or television. Discuss the following characteristics about streams and rivers: overall appearance, length and width, the amount of water carried, and velocity of the flow.

THINKING SKILL *(Comparing)* Have the students compare the two streams shown in Figure 12–6 in relation to some of the concepts discussed above.

NEW SCIENCE TERMS

headward erosion
base level
flood plain
meanders
delta
alluvial fans
playas

SECTION OBJECTIVES

After completing this section, you should be able to:
- **Relate** a stream's slope, volume, base level, and velocity to its ability to erode the land.
- **Compare** the stages of valley development.
- **Explain** the processes involved in the formation of deltas.
- **Describe** the conditions necessary for the formation of alluvial fans and playas.

12.3 Headward Erosion

Think about some of the streams and rivers that you have seen in movies, on television, or while traveling. Some of these streams probably carry very little water and may flow only part of the year, while others carry large quantities of water year-round. You may have seen streams flowing down narrow, steep-sided valleys or across wide plains. Some streams are clear, with very little sediment, but others are always muddy because of the large amount of eroded material they transport.

In spite of these differences, all streams and rivers erode material from one place and deposit it somewhere else. As these processes of erosion and deposition continue, the shape of the stream itself changes.

Streams increase their length by a process called **headward erosion.** As more and more material is eroded from the origin, or *source*, of a stream, large gullies become part of the

Figure 12–6. Some streams (left) may be almost dry during part or even most of the year. Other streams run almost bank full (right), especially during seasons of heavy rain.

258   Chapter 12   Running Water

BACKGROUND INFORMATION

Few, if any, streams have completely stable conditions of flow. The flow of most streams varies from season to season or even from day to day. In arid regions, intermittent streams are common. Under more humid climatic conditions, the channels of large volume streams are only occasionally dry. In middle latitudes, low flow usually occurs during the summer, when evaporation and transpiration are highest. Large flows may occur in spring, fall, and winter, depending upon the time of heaviest precipitation.

In high-latitude areas in the Northern Hemisphere, high flow usually occurs in the spring on northward flowing rivers because melting of ice occurs from headwater to mouth and the flow of water released upstream is blocked by ice downstream. In low-latitude areas high flow is related to the season of highest precipitation.

streambed. The other end of a stream, called the *mouth*, is where the stream flows into some other body of water. Deposits often occur at the mouth of a stream.

A stream or river cannot erode below the elevation of its mouth. This low point of the stream or river is called its **base level.** Sea level is the ultimate base level of all rivers and streams. A lake or a layer of resistant rock may act as a temporary base level.

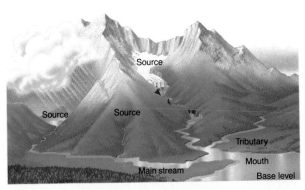

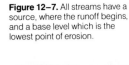

Figure 12–7. All streams have a source, where the runoff begins, and a base level which is the lowest point of erosion.

A MATTER OF FACT

The highest waterfall in the world is the Angel Falls in Venezuela, which is over 800 m high. The widest waterfall is the Khon Falls in Laos, Southeast Asia, which is over 10 km wide.

Where a stream encounters a layer of rock that resists erosion, waterfalls or rapids often result. The resistant rock serves as a temporary base level for the water upstream. As the water slowly erodes the resistant rock, the waterfall or rapids move upstream. For instance, Niagara Falls, formed on resistant rock, has moved upstream more than 11 km in the last 12 000 years.

○ *What is headward erosion?*
○ *What is a stream's base level?*

Figure 12–8. Resistant rock may form a temporary base level and a waterfall (left). The temporary base level of the Niagara River is beginning to erode away as this photograph of Niagara Falls shows (below).

DISCUSSION Direct the students' attention to Figure 12–7 and discuss the two ends of a river—the source and the mouth. Have the students locate the different sources of the main stream and its tributaries. Then have them locate the mouth of the river and discuss the river's base level. *Thinking Critically:* Ask the students to explain why sea level is the ultimate base level. (because it is the lowest point relative to land)

Have the students write reports on rivers of their choice. Tell them to include as much information as they can about the particular river, including its drainage basin, dimensions, and the location of its mouth. Suggest that the students include diagrams and/or magazine pictures in their reports.

EXTENSION Niagara Falls and the Niagara River form one of the most scenic places in the world. The Niagara River connects Lake Erie with Lake Ontario and separates New York state from the province of Ontario. Niagara Falls, formed during the last Ice Age, consists of two falls that are separated by an island called Goat Island. The Niagara River plunges over a 50-m cliff, and water flows at the rate of about one million metric tons per minute. Have interested students find out more about Niagara Falls and report their findings to the class.

DISCUSSION Some of the students may have difficulty understanding that a river lengthens at its source or that a waterfall moves upstream as the flow of water continually erodes away the rocks. Make sure the students understand that the direction of lengthening and the movement of falls is opposite the direction of the flow of the river.

Answers—12.3

○ Headward erosion is the removal of material from the source of a stream or river. Headward erosion increases the length of the stream or river.

○ The low point of a stream or river is its base level.

DEMONSTRATION

Demonstrate the concept of base level, using a stream table (or even a clear shoe box). Stack the sand so that there is a slope. Pour in enough water so that there is a "lake" at one end. Carve out a small stream and demonstrate that the bed of the stream cannot dip below the lake level (base level).

12.4 Valley Formation

DISCUSSION Ask the students to describe the life stages of a human being. (While the life stages of a human being are not always clearly delineated, many students will separate a person's life into three stages: childhood, adult—working person, and old-age—retirement.) Compare these stages to the stages that characterize a stream valley's "youth," "maturity," and "old-age."

EXTENSION Rivers have been extremely important to human society and have been closely tied to the progress of civilization. Ancient civilizations arose along the flood plains of rivers such as the Tigris-Euphrates in Mesopotamia, the Indus in India, and the Nile in Egypt. These people benefited from the fertile soils of the flood plains but were also subjected to the heavy flooding that occurred. Today dams and drainage canals prevent or restrict natural flooding in order to protect cities and their inhabitants. Some of the most productive farmlands in the world are found along the banks of large rivers and in river valleys. Today the flood plains of the Nile support practically all of Egypt's dense population. *Thinking Critically:* Ask the students to consider the long-term effects of programs to control flooding by the use of dams. (Although dams prevent damage from flooding, they also can prevent flood plains from receiving the beneficial effects of overflow, namely rich nutrients that make the soil fertile. In the long run, dams may adversely affect agricultural production.)

Figure 12–9. A youthful stream (above) is characterized by straight, steep-sided, V-shaped walls (bottom left). A mature stream (right) has many bends, and a wide, gently sloping valley (bottom right).

12.4 Valley Formation

The development of a stream valley can be divided into three stages: youthful, mature, and old-age. Youthful streams have narrow, V-shaped valleys with steep walls, since most of the erosional energy is used to deepen the streambed. Areas between valleys are poorly drained because they are not really part of the drainage basin. Waterfalls or rapids may occur where the stream encounters layers of resistant rock.

As the youthful stream approaches its base level, it begins to erode the sides of its banks, producing a wider valley. Weathering of the valley walls and erosion by tributaries also help to widen the valley and to form a mature stream.

Youthful

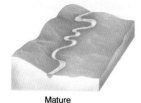
Mature

Mature stream valleys are characterized by a wide **flood plain**, the area over which a stream spreads in times of high-volume runoff. When a stream overflows, it deposits sediment on its flood plain, enriching the soil and making it a fine place to grow crops. Before the development of the Aswan Dam, the Nile River in Egypt flooded every spring, providing nutrients for the summer crops.

BACKGROUND INFORMATION

From the time water begins its flow until it reaches base level, erosion continues downward through stages of gradually diminishing slope. Each stage is characterized by different features and landforms. William Morris Davis, a physiographer from Harvard University was the first to describe the three stages of the process as youthful, mature, and old-age.

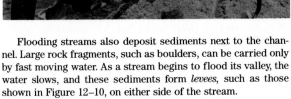

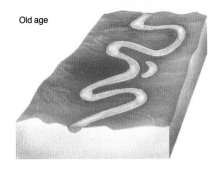

Figure 12–10. Natural levees (above) form as mature streams overflow their banks. To prevent flooding, people sometimes construct levees of sand and rocks (left).

Flooding streams also deposit sediments next to the channel. Large rock fragments, such as boulders, can be carried only by fast moving water. As a stream begins to flood its valley, the water slows, and these sediments form *levees*, such as those shown in Figure 12–10, on either side of the stream.

The water in a mature stream erodes the banks, forming broad curves called **meanders.** A cross-sectional view of a meander shows that the outside of the curve is deeply eroded by fast-moving water, while the inside is shallow due to deposition in the slower-moving water.

An old-age valley is characterized by slow-moving water and a wide flood plain with many lakes, swamps, and meanders. The meanders often form such broad loops that one meander cuts into the next. Most of the water follows this shorter route. The abandoned meander forms an *oxbow*, a bow-shaped bend with a temporary lake. Some river valleys, such as the Mississippi, may be youthful at the source, while the section near the mouth may be in the old-age stage.

○ *What are the three stages of valley development?*

Figure 12–11. Old-age streams have many meanders and cutoffs, called *oxbows* (below). The meanders form because old-age streams are near their base level, and most of the erosion occurs in the sides of the banks.

Old age

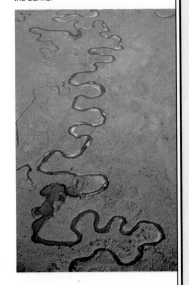

Section 2   Erosion and Deposition   **261**

DISCUSSION   Ask the students if they have ever seen, either first-hand or on a television news report, the efforts of people trying to prevent a rising river from flooding its banks. The students will probably mention having seen people dumping sand, soil, or sandbags along the banks of a rising river. Compare these actions to the formation of natural levees, such as the ones shown in Figure 12–10.

THINKING SKILL (Applying Concepts) Ask the students to describe the features they would look for to find the three stages of valley development along the Mississippi River. (Youthful stage: V-shaped valleys with steep walls; mature: a wide flood plain and broad curves, as well as meanders and natural levees; old-age: many lakes, swamps, oxbow lakes, meanders, and a wide flood plain.)

Answer–12.4

○ Youthful, mature, and old-age are the three stages of valley development.

DISCOVER: Studying Erosion

Thinking Skills *(Observing, Relating)*

Suggest that the students apply what they have learned in the first section of this chapter to find an area that has been subjected to erosion. One possibility would be a hilly area that does not have much vegetation. When the students draw their diagrams, they should also note the kind of erosion and state why it is occurring. To extend this activity, you may wish to suggest that interested students take photographs of the area before and after a rainstorm.

12.5 Water Deposits

EXTENSION   (Tie-in/Physical Science) Rivers carry their sediment loads in different ways. Dissolved ions are carried in the water, along with clay, silt, and fine sand. Water carrying a large amount of sediment appears cloudy. Heavy sediments such as gravel and boulders are not carried in the water but are rolled, bounced, and dragged along the bottom of the river. The velocity of moving water determines the size of the particles it can move. Rapidly moving water has a greater kinetic energy so it is capable of moving large particles. As rivers slow down, their sediments settle out according to size. Larger particles settle first and the smallest particles settle last. *Thinking Critically:* Ask the students why larger particles settle first. (They are heavier.)

THINKING SKILL *(Applying Concepts)* Ask the students to explain how deltas can affect the shipping industry. (As deltas build up, shipping channels must be dredged in order to remain open.)

EXTENSION   Point out that Louisiana is shaped like a delta because much of the area was formed by materials carried by the Mississippi River. You may want to have the students look at the maps of Louisiana and Alabama in the Science Sites booklet. Ask them to point out the deltas on both maps.

DISCOVER

Studying Erosion

Find an area near your home or school where there seems to be some erosion due to runoff. Diagram the present characteristics of the area. After the next significant rain, return to the same area and diagram any additional erosion that has occurred because of the rain.

12.5 Water Deposits

Erosion is only part of the leveling process of water. Once eroded by running water, sediment is transported only as long as a stream maintains its volume or speed. If the water volume decreases, some of the load will be deposited in the streambed, forming *sandbars*. Before a channel was dug to improve river transportation, there were many of these deposits in the lower Mississippi River valley.

Deltas   When a large, sediment-laden stream reaches its mouth, the flow spreads out and the water speed drops. As a stream slows down, it is unable to carry a large load of sediments, so the load is deposited near the mouth, forming a **delta.** Many streams form deltas, but the deltas located at the mouths of the Mississippi and Nile rivers are especially well known.

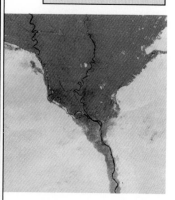 

Figure 12–12. Deltas form at the mouth of the Mississippi River (right) and the Nile River (left) because of the large loads carried by these rivers.

As a delta builds around a river's mouth, deposits build up, and the mouth shifts slightly to one side. As this shifting back and forth of the mouth continues, the delta becomes fan-shaped. Sometimes the slowly moving stream splits into several channels as it moves through the delta. For example, the delta of the Mississippi River is 60 km across and has many channels flowing into the Gulf of Mexico. The many parts of the Mississippi River delta have been built during the past 5000 years—the present lobe in the past 500 years.

A small stream also deposits material at its mouth; however, it may not carry enough sediment to form a delta. Most small streams empty into rivers or lakes. When a stream flows into a lake, pebbles and sand carried in the water are deposited near the mouth. Silt and clay settle to the bottom farther out in the lake.

262   Chapter 12   Running Water

BACKGROUND INFORMATION

A delta is a deposit of sediment at the mouth of a river or tidal inlet. The name delta, from the Greek letter, was first used in the fifth century B.C. to describe the delta of the Nile because it resembled the triangular letter.

Most of the sediment a stream carries is deposited when the inflowing stream decelerates. As the sediments are deposited, a delta forms. The shape and structure of a delta depend on the interaction of two forces, the sediment-carrying stream, and the current and wave action of the body of water into which the stream flows.

Answers—12.5 (p. 263)

○ Deltas are deposits at the mouths of streams or rivers.

○ Alluvial fans are fan-shaped deposits at the base of mountains. Playas are flat deposits of sediment on the desert floor.

Alluvial Fans and Playas   Erosion by water occurs nearly everywhere. Even deserts receive some rain, and the runoff causes erosion and deposition. In dry, mountainous areas the runoff caused by a sudden storm carries large amounts of sediment down the slopes. At the base of the mountains, the water slows, leaving fan-shaped deposits called **alluvial fans.** Because there are no permanent streams to carry away these deposits, they form a thick accumulation of sediments, such as the one shown in Figure 12–13.

Although storms are rare in desert areas, they sometimes produce large amounts of runoff. The runoff collects temporarily on the desert floor in muddy pools on plains called **playas.** The water on these playas rapidly evaporates, leaving behind minerals such as gypsum and halite. The great salt flats east of the mountains in southern California are playas.

○ *What are deltas?*
○ *Describe two erosional features of deserts.*

Figure 12–13. Alluvial fans (left) and playas (right) form in dry, mountainous areas. The occasional heavy rains carry large amounts of sediment, which is deposited at the base of steep slopes.

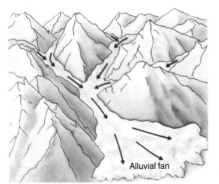
Alluvial fan

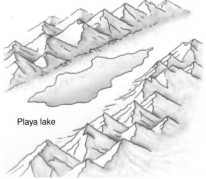
Playa lake

Section Review

READING CRITICALLY

1. Why are playas also called salt flats?
2. How is the formation of levees related to flood plains?

THINKING CRITICALLY

3. How could the amount of lake deposits show the volume of runoff of a stream?
4. What is the importance of slope and water volume to the amount of material that can be eroded by a stream?

DISCUSSION Discuss alluvial fans with the students. Then ask them to explain the difference between deltas and alluvial fans. (Deltas are formations that occur near a river's mouth where a river slows down and deposits its sediments. A delta becomes fan-shaped as the mouth of the river shifts back and forth. An alluvial fan is a fan-shaped deposit formed at the base of a mountain in dry areas where runoff, caused by sudden storms, carries large amounts of sediment down the mountain.) By having the students compare deltas and alluvial fans you will perhaps offset some confusion about these two formations resulting from the fact that both are fan-shaped.

EXTENSION   Along with playas there are a number of other physiographic features characteristic of desert erosion and deposition. Although rainfall in the desert is low, it is the major factor in the formation of desert features. Rainfall often occurs as fierce downpours that, because they occur in a short amount of time, cause a great deal of runoff. Have interested students find out about one of the following features that are the result of runoff erosion: badlands, bolson, or dry wash. The students should report their findings to the class.

SECTION REVIEW

Summary Review the definition of *headward erosion,* the three stages of a valley's development, and the features formed by deposition.

Reinforcement Have each student formulate two questions based on the information in this section. Then have each student take a turn quizzing the class. Expect a duplication of questions. Duplicate questions will assist students who are having difficulty with the material presented in the section.

Answers to Section Review

Reading Critically

1. The water evaporates from a playa lake leaving behind the mineral salts eroded from the mountains.
2. When a river floods its bank onto its floodplain, the velocity of the water immediately decreases and the materials are dropped at the edge of the channel. This build-up of materials is called a levee.

Thinking Critically

3. The greater the amount of deposits, the higher the volume of runoff.
4. The greater the slope, the greater the amount of material carried; the greater the volume of water, the greater the amount of material carried.

Objectives

- Analyze diagrams.
- Identify the relationships between the features illustrated in the diagram.

Answers to Application

1. Only trunks in trellis patterns flow parallel.
2. Trunks in radial patterns flow outward from a central point.
3. No; trellis and rectangular join at right angles, the others do not.
4. Deranged
5. Radial

SKILL ACTIVITY: Analyzing Diagrams

BACKGROUND

Diagrams are commonly used in textbooks to show relationships between objects, sequences of events, or other information best presented in a visual format. Many times the diagrams are simple and easily interpreted. Some diagrams, however, are more complex and require careful study in order to be fully understood.

PROCEDURE

Copy the stream pattern analysis table. Use the three steps described in the Procedure to analyze the diagram of stream patterns in Figures 12–4 and 12–5. As you follow each step, place your information on the Stream Pattern Analysis table. In analyzing diagrams about water-created landforms, the following steps should be used.

1. Study each pattern. List the types of landforms and water systems presented in each diagram. List any other features in the diagrams.
2. Identify relationships between the details. What is the relationship among the various tributaries in each diagram? In what direction do they flow?
3. Analyze similarities and differences. Examine the characteristics you have identified for each stream pattern. Select one or two characteristics that would allow you to distinguish each pattern from the others.

APPLICATION

1. If there is more than one trunk stream in a diagram, do they flow parallel to each other?
2. Do the trunk streams show any other pattern?
3. In each diagram, do the tributary streams enter the trunk streams at the same angle?
4. Which stream pattern has poor drainage, with considerable standing water?
5. Which stream pattern has tributaries flowing nearly parallel to the trunk stream?

USING WHAT YOU HAVE LEARNED

Analyze the drainage pattern shown, and decide which type of pattern it most closely represents.

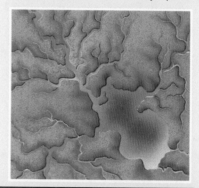

TABLE 1: STREAM PATTERN ANALYSIS

Stream Pattern	Relationship Between Trunk Streams	Relationship Between Tributaries	Relationship Between Tributaries and Trunk	Thoroughness of Drainage	Associated Landform
Dendritic					
Trellis					
Deranged					
Radial					
Rectangular					

Stream Pattern	Relationship Between Trunk Streams	Relationship Between Tributaries	Relationship Between Tributaries and Trunk	Thoroughness of Drainage	Associated Landform
Dendritic	widespread	branching	any angle	complete	irregular
Trellis	nearly parallel	nearly parallel	right angles	complete	rolling hills
Deranged	none	none	any angle	incomplete	low, swampy
Radial	diverging	none	acute angles	complete	rounded hill
Rectangular	nearly parallel	many sharp turns	right angles	complete	fractured rocks

SECTION OBJECTIVES

After completing this section, you should be able to:

- **Define** the term *water table.*
- **Describe** the rock characteristics of porosity and permeability.
- **Explain** how karst deposits develop.

NEW SCIENCE TERMS

water table
porosity
permeability
aquifer

12.6 The Water Table

Although the cycle of water is usually completed by runoff, water is sometimes removed from the cycle temporarily. Not all of the water that falls on the earth runs off into streams. Some water soaks into the ground, where it enters tiny air spaces in the soil and rocks. As water moves through these air spaces in the soil, some of it sticks to the soil fragments. This creates an area in the soil with both air and water in the spaces, called the *zone of aeration.*

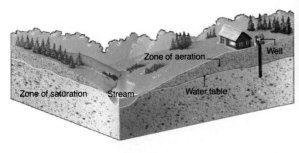

Figure 12–14. In swampy areas (above), the water table is very close to the surface. The water table is at the top of the zone of saturation (left).

Much of the water, however, continues moving down into the ground. There it reaches an area called the *zone of saturation,* where all the spaces are filled with water. This zone is the top of the **water table.**

The depth of the water table varies in different locations. Have you ever noticed that a hole dug on a beach quickly fills with water? In areas near the sea, the water table is at sea level, which may be just below the surface of the sand. In high desert regions, the water table may be hundreds of meters below the earth's surface. Every place where the earth's surface is below the water table, there is standing water such as a lake, a pond, or a swamp.

Section 3 Ground Water **265**

DEMONSTRATION

For a demonstration of karst topography see page 251c preceding this chapter.

Section **3**: Ground Water

SECTION FOCUS

Section Overview This section discusses ground water and the features associated with it. The term *water table* is defined, and the relationship between porosity, permeability, and water flow is discussed. Aquifers are then discussed. The section concludes with a description of karst deposits. An activity based on analyzing karst landforms reinforces the concept.

Section Objectives For a list of section objectives, see pupil's edition page.

New Science Terms For a list of the new science terms in this section, see pupil's edition page. You may wish to familiarize your students with these terms before they read the section. You may wish to read the terms aloud so the students are familiar with the pronunciations.

SECTION DEVELOPMENT

12.6 The Water Table

DISCUSSION Remind the students that not all rain water becomes runoff. Some of the water seeps into the ground where it drains into small spaces between soil particles. Refer your students to Figure 12–14. Discuss the water table and the different zones above the water table. ***Thinking Critically:*** In Figure 12–14, the water table is labeled on the right side of the pond. Ask the students to point out the water table on the left side of the pond. (The students should point to any place on the line marking the boundary between the zone of aeration and the zone of saturation.)

DISCOVER: Measuring Porosity

Skills *(Measuring, Communicating)*

In order for the students to keep track of their results, you may wish to suggest that they record their data in a table. Remind the students that after they divide the volume of the pores by total volume, they must multiply the result by 100 to express porosity as a percentage.

THINKING SKILL *(Applying Concepts)* As stated in the text, the depth of the water table varies in different locations. Ask the students if they think the depth of the water table in an area remains constant. (While the depth of the water table may remain relatively constant in some areas, other water tables fluctuate with seasonal variations. If there is a drought in an area, the water table is lowered. If there is heavy rainfall, the water table rises.)

EXTENSION The permeability of limestone results from the fact that it easily weathers along fractures and this increases its permeability.

EXTENSION Have the students contact the local water authority for information about the local or state water budget and other water statistics. Have the students find out the elevation of the local water table.

DISCOVER

Measuring Porosity

Using the following formula, determine the porosity of several different sponges.

$$\text{porosity} = \frac{\text{volume of pores}}{\text{total volume}}$$

To find the volume of the pores, soak each sponge in a bowl of water, and then wring as much water as possible out of each sponge into a graduate. To find the total volume of each sponge, measure and then multiply length × width × height.

Figure 12–15. Porous rock, (left) such as sandstone, has many air spaces between mineral grains. Permeable rock (right), such as fractured limestone, has many interconnected pores or passages.

Porosity In most places the soil holds only a small fraction of the water that soaks into the ground; much of the water is held in the pores of the bedrock. It may seem strange that a solid object, such as a rock, can be porous. The structure of some rocks is similar to that of a sponge. If you try pouring a small amount of water into a sponge, you will find that most of the water will be trapped in the pores of the sponge. The amount of water that a rock can hold depends on its porosity. **Porosity** is the ratio of the volume of air space in the rock to the total volume of the rock.

The porosity of a rock depends both on the shape of the mineral grains and on the degree of cementing in the rock. For example, sand has more porosity than sandstone because some of the air space in sandstone is filled with cement. Porosity is always expressed as a percentage. An imaginary rock made of mineral spheres of uniform size that are packed as closely as possible will have 26 percent porosity. Porosities greater than 26 percent indicate poor packing of the mineral grains or irregular mineral shapes.

Permeability A rock that is porous can hold a lot of water, but it will not necessarily allow the passage of water. If you look at a cork, you can see that it is very porous; however, if you put a cork into a bottle and turn the bottle upside down, none of the contents of the bottle will flow out through the cork.

Like cork, shale is porous, but water cannot move through shale because it is not very permeable. **Permeability** is the ability of rock to allow water to move through the rock. The permeability of a rock depends on how well the pores are connected to each other. Some rocks are permeable even though they are not porous at all. Limestone, for example, has almost no pores, but it is very permeable because of fractures that occur in the rock.

○ *What is the water table?*
○ *What is the difference between porosity and permeability?*

Porous permeable   Nonporous, permeable

266   Chapter 12   Running Water

Answers–12.6

○ The water table is the top of the zone of saturation. Below this level all pore spaces are filled with water.

○ Porosity is the ratio of the volume of air space to the volume of the rock. Permeability is the ability of water to move through a rock.

BACKGROUND INFORMATION

The study of ground water is a part of hydrology. However, a thorough understanding of geology is required in order to predict where water may be found. Of primary concern is porosity, which is determined by the number and size of openings in rocks, and permeability, which is determined by how well the spaces are connected. There are two types of openings in rocks: those that were there when the rock formed and those that occur as a result of weathering.

12.7 Aquifers

When a permeable rock layer is between two nonpermeable layers, it forms an **aquifer** (AHK wuh fuhr). An aquifer transports water in much the same way as an underground pipeline does. Water enters an aquifer where the permeable rock comes to the surface. Some aquifers transport large amounts of water over great distances. For instance, much of the midwestern part of the United States lies on top of an aquifer that starts in the Rocky Mountains.

Water can be obtained from aquifers by drilling wells into them. Sometimes the water pressure in the aquifer is so great that the water emerges without being pumped. This type of well is called an *artesian well*. Sometimes a small aquifer comes to the surface naturally, where it forms a *spring*.

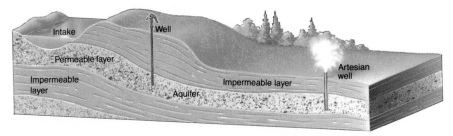

Ground water that has traveled through an aquifer is usually safe to drink because the rocks act as a filter. However, ground water may contain minerals that give it a distinctive taste, depending on the type of bedrock in the area.

Besides having an unusual taste, water with a lot of dissolved minerals may have undesirable characteristics. Water containing calcium, magnesium, or iron, called *hard water*, reacts chemically with soap, forming sticky lumps and film instead of suds. If the bedrock is limestone, dissolved calcium carbonate is redeposited as mineral layers in water pipes and drains.

○ *What is an aquifer?*
○ *What is hard water?*

Figure 12–16. An aquifer is a permeable layer of rock through which water passes. Some wells drilled into aquifers flow without being pumped. These wells are called *artesian wells*.

Figure 12–17. The minerals in hard water can accumulate on the inside of pipes and may even clog the pipe completely. Pipes carrying soft water do not show these obstructions.

267

Answers–12.7

○ An aquifer is a permeable rock layer between two impermeable layers through which water travels.

○ Water containing a lot of dissolved minerals is called hard water.

12.7 Aquifers

THINKING SKILL *(Interpreting Diagrams)* Direct the students' attention to Figure 12–16 and discuss the aquifer that is illustrated in the diagram. Have the students compare the direction of flow of ground water with that of runoff. (Both ground water and runoff flow downslope as a result of gravity.)

EXTENSION In some areas the rocks in the earth's surface are very hot. When ground water comes in contact with these hot rocks it is heated and turned into steam. The steam causes pressure, which forces the water above it upward. This eruption of water and steam is called a geyser. "Old Faithful" in Yellowstone National Park, Wyoming, is the most famous geyser. "Old Faithful" erupts approximately every hour and a half for about five minutes and surges to heights of 30 to 45 meters. For the exact location of this geyser, have the students look at the map of Wyoming in the Science Sites booklet.

EXTENSION (Tie-in/Ecology) The students should be aware of the extreme importance of ground water as a human water supply. They should also be aware of the ever-increasing pollution of ground-water supplies. You may want them to read Rachel Carson's classic work *Silent Spring* to increase their understanding of why ground water is essential to the survival of all organisms. Some students might be required to write a book report on *Silent Spring* and then present it to the class.

DISCUSSION The mineral deposits in some hard water have caused problems such as clogging up household plumbing, reducing the flow of water, and discoloring laundry. *Thinking Critically:* Have the students think of a standard household item that is used in many areas of the country where there is hard water. (A water softener is used to remove minerals from the water before it enters household plumbing.)

12.8 Karst Topography

DISCUSSION   Remind the students that limestone is a sedimentary rock composed mainly of calcium carbonate. As water travels through the air, the water molecules combine with carbon dioxide to produce carbonic acid. This acid in ground water chemically weathers limestone.

EXTENSION   You may wish to caution the students that although fresh spring water is usually potable, water from springs in karst topography is unsafe. The water travels through caves and underground streams and is not filtered by percolating through sediment or rock material.

ACTIVITY:
Analyzing Karst Landforms

Skills   *(Classifying, Observing)*

Discussion   Some students may need to review the other sections of this chapter to answer the Conclusions/Applications questions.

Answers to Conclusions/ Applications

1. The topography is hilly.
2. The Green River is youthful. It does not have a flood plain.
3. Answers may include valleys without streams, springs, and sinkholes.
4. These valleys would make dendritic drainage patterns.
5. Water travels downward through fractures in the rock and dissolves calcite in the rock. As the mineral is removed, caves and underground drainage systems develop. Cave ceilings collapse, forming sinkholes on the surface above the cave.

12.8 Karst Topography

You may recall from Chapter 11 that karst topography occurs in areas with limestone bedrock. These areas have many ravines, sinkholes, underground streams, and caves. These features are the result of chemical weathering, mass movement, and water deposition.

Erosion   As ground water travels down through fractures in the limestone, the rock is rapidly dissolved, leaving large underground passageways and caves. These passageways sometimes carry underground streams. These streams often come to the surface in the form of springs. After much water erosion, the ceilings of some caves collapse, forming sinkholes. The area around Mammoth Cave, Kentucky, shown in Figure 12–18, has karst topography.

Figure 12–18. This is a topographic map of the area around Mammoth Cave, Kentucky.

ACTIVITY: Analyzing Karst Landforms

What special features characterize karst landforms?

MATERIALS (per group of 3 or 4)

Topographic map of Mammoth, Kentucky

PROCEDURE

1. Review the features associated with karst deposits.
2. Examine the topographic map of the Mammoth Cave, Kentucky, area and answer the questions that follow.

268   Chapter 12   Running Water

CONCLUSIONS/APPLICATIONS

1. In general, is the topography in this area flat or hilly?
2. Locate the Green River and classify it as youthful, mature, or old-age. Explain your choice.
3. Locate two features on this map that support the hypothesis that this is a karst landform.
4. Assume that during heavy rains the various valleys probably contain temporary streams. What type of drainage pattern do these various valleys make?
5. Describe the sequence of events in the formation of karst landforms.

DEMONSTRATION

Demonstrate that chalk is composed of limestone. Limestone reacts with an acid to form bubbles of carbon dioxide. Put some acid, such as vinegar or diluted hydrochloric acid, on a piece of chalk. The chalk will bubble as CO_2 is produced.

Deposits Underground deposits, called *karst deposits*, occur mainly on the inside of caves. Spectacular deposits form in some caves because ground water redeposits some of the calcium carbonate it had dissolved earlier from the limestone. As this water drips from cave ceilings, the water evaporates and limestone "icicles," called *stalactites*, form. Limestone also builds upward from the floor of a cave, forming *stalagmites*. There are many of these caverns, with their beautiful formations, in the limestone regions of the Appalachian Mountains and in New Mexico.

○ *How do karst deposits form?*

A MATTER OF FACT

The longest stalactite, measuring 59 m, is in the Cueva de Nerja in Spain. The tallest stalagmite, which measures 29 m, is La Grande Stalagmite in the Aven Armand Cave located in Lozère, France.

Figure 12–19. Karst deposits are common in limestone caverns, such as Mammoth Cave, Kentucky, (above) and Carlsbad Caverns, New Mexico (left).

Section Review

READING CRITICALLY

1. Explain how a rock can be permeable without being porous.
2. Why do karst deposits occur mainly in limestone formations?

THINKING CRITICALLY

3. Explain why water from a sandstone aquifer is not hard, while water from a limestone aquifer is hard.
4. Is sandstone porous or permeable or both? Explain your answer.

Section 3   Ground Water   **269**

EXTENSION Carlsbad Caverns National Park was established in 1930, in the foothills of the Guadalupe Mountains in southeastern New Mexico. Discovered by a cowboy in 1901, these limestone caves are among the most magnificent in the world. The caves are the site of many stalactite and stalagmite formations. One of the chambers, called the Big Room, is nearly 1609 m long, 183 m wide at one point, and 107 m high at the highest point. The caverns have not yet been fully explored and probably never will be. However, several miles of the caverns are electrically lighted in order to accommodate the many visitors to the park. For the exact location of Carlsbad Caverns National Park, have the students look at the map of New Mexico in the Science Sites booklet.

EXTENSION If possible, have the students create a display of travel brochures from Carlsbad Caverns, New Mexico, and Mammoth Cave, Kentucky. For information and brochures students can write to:
Carlsbad Caverns National Park
3325 National Parks Highway
Carlsbad, New Mexico 88220, and
Mammoth Cave National Park
Mammoth Cave, Kentucky 42259.

EXTENSION Suggest that the students research other cavern sites in or near their home state including, Luray Caverns, Virginia; Howe Caverns, New York; Dixie Caverns, Virginia; and Cave of the Mounds, Wisconsin.

SECTION REVIEW

Summary To summarize this section, have the students define the following terms: water table, zone of saturation, zone of aeration, porosity, permeability, aquifer, and karst topography.

Reinforcement Reinforce the concepts of this section by having the students draw and label diagrams of an aquifer.

Answer—12.8

○ Karst deposits form as calcite is redeposited from the ground water dripping from cave ceilings.

Answers to Section Review

Reading Critically

1. Rocks can develop fractures through which water will pass even if there are few pores between the minerals in the rock.
2. The calcite easily dissolves, forming large passageways and caves. Few other types of rock dissolve so easily.

Thinking Critically

3. Quartz, the major mineral in sandstone, does not dissolve in water; calcite, the major mineral in limestone, dissolves easily into the ground water.
4. Sandstone is both permeable and porous. It is made of rounded grains of sand with many spaces between, making it porous. The spaces between the sand grains are well-connected, making it permeable.

269

INVESTIGATION 12: Porosity and Permeability

Skills (*Comparing, Observing*)

Preparation of Materials

Sand and gravel can usually be obtained from garden centers or building-supply stores.

Discussion This activity may require more than one trial for each type of sediment because students frequently fail to accurately measure the time required for water to travel through the column. Use this as an opportunity to emphasize that scientists often conduct many trials of the same experiment in order to verify the accuracy of their results.

Before the students begin this investigation, you may wish to have them examine the rock fragments, using a hand lens or microscope. They should describe the roundness of the rocks, estimate the variation in shape and size of the particles, and compute the average volume of the particles. They could then hypothesize which rocks will pack more and which will be more porous.

Answers to Analyses and Conclusions

1. Answers will vary depending upon the roundness and packing of the various particles.The gravel will likely be the most porous.
2. 1=gravel, 2=sand, 3=clay
3. Highly porous sediments are often very permeable because the pores are connected.
4. Sandstone should be less porous because it is better packed and some of the pores are filled with cementing material.

Answer to Application

A block of rock could be massed, soaked in water overnight, and then remassed to determine porosity. By leaving the rock supported on a grill, the time it takes for the water to drip out of the rock could be determined. (The time should be measured until the mass returns to the original value.)

INVESTIGATION 12: Porosity and Permeability

PURPOSE

To compare the porosity and permeability of gravel, sand, and clay

MATERIALS (per group of 3 or 4)

Clear plastic tube
Hose and clamps, fitted to tube (2)
Beaker, 500 mL
Water
Gravel
Sand
Clay
Graduate
Stopwatch

PROCEDURE

1. Copy the table shown. Fill in the table as you complete this investigation.
2. With a clamp, attach the hose to the plastic tube.
3. Pour gravel into the beaker to the 250 mL level (slide the gravel down the side of the beaker so that you do not break the beaker); then pour the gravel into the plastic tube.
4. Secure a clamp on the hose so that water will not run out of the plastic tube; then slowly pour water from a graduate over the gravel in the plastic tube until the water reaches the top of the gravel.
5. Record the amount of water needed to fill all the pores in the gravel. Compute the porosity of the gravel using the following formula:

$$\text{porosity} = \frac{\text{amount of water} \times 100}{250 \text{ mL of gravel}}$$

6. Place the beaker under the plastic tube; then remove the clamp from the hose. Record the time it takes for the water to run out of the tube.
7. Repeat steps 2 through 6 for the sand and clay materials. .

ANALYSES AND CONCLUSIONS

1. Which of the materials was the most porous? Which was the least porous?
2. Using your permeability records, rank the materials in order of permeability. Use number 1 to indicate the most permeable material.
3. According to your data, what is the relationship between permeability and porosity? Explain why this relationship exists.
4. Would sandstone have a porosity different from the sand in this experiment? Explain your answer.

APPLICATION

How could this procedure be used to determine the porosity and permeability of a rock layer?

TABLE 1: POROSITY AND PERMEABILITY

Characteristics	Gravel	Clay	Sand
Amount of Water in Pore Space			
Porosity			
Time for Water to Permeate			
Order of Permeability			

270   Chapter 12   Running Water

SUMMARY

- Surface runoff is responsible for considerable erosion of the land. (12.1)

- The amount of eroded material carried by a stream is called its load. (12.1)

- A drainage basin is an area drained by a single stream system. This system includes a trunk and tributaries. (12.2)

- Drainage systems form different shapes, depending upon the steepness of the slope and the underlying rock structure. (12.2)

- Streams increase their length by a process known as headward erosion. (12.3)

- The low point of a stream or river is its base level. A stream cannot erode below its base level. (12.3)

- Stream valley development is divided into three stages: youthful, mature, and old-age. (12.4)

- Youthful streams have V-shaped valleys. Mature streams have wide flood plains. Old-age streams have many meanders. (12.4)

- If a stream carries a large load of sediment, a delta forms at the mouth of the stream. (12.5)

- Playas and alluvial fans are features of dry, mountainous areas. (12.5)

- The amount of water that can enter the ground is determined by the porosity and permeability of the bedrock. (12.6)

- Porosity is the ratio of the volume of air space in a rock to the total volume of the rock. Permeability is the ability of a rock to allow the movement of water through the rock. (12.6)

- A permeable rock layer may act as an aquifer. (12.7)

- Stalactites and stalagmites are examples of karst deposits. (12.8)

Write all answers on a separate sheet of paper.

SCIENCE TERMS

Correctly use each of the following terms in a sentence.

alluvial fans **(263)**
aquifer **(267)**
base level **(259)**
delta **(262)**
drainage basin **(255)**
erosion **(254)**
flood plain **(260)**
headward erosion **(258)**
load **(254)**
meanders **(261)**
permeability **(266)**
playas **(263)**
porosity **(266)**
runoff **(253)**
water table **(265)**

SCIENCE QUIZ

Modified True-False

Mark each statement *true* or *false.* If a statement is false, change the underlined term to make the statement true.

1. Streams are able to erode both the <u>head</u> and the sides of their channels.

2. Mature streams mainly erode their <u>beds.</u>

3. Runoff is <u>increased</u> in areas of thick vegetation.

4. When a river overflows its banks, it forms a <u>delta.</u>

5. Aquifers are formed in layers of <u>permeable</u> rock.

continues

CHAPTER REVIEW

SUMMARY

The students may review the major concepts in the chapter by reading the summary statements. The statements are cross-referenced to the chapter in order to facilitate reinforcement of any concepts of which the students feel unsure. Encourage the students to work in groups to quiz one another.

SCIENCE TERMS

The sentence in which the science term is used should reflect an understanding of the definition of the term. You may wish to hold a science "bee" during which one team will give the term and the other team must supply the definition. As an alternate activity, stage a game of "Jeopardy" during which the definition is provided and the "contestants" must provide the correct term.

SCIENCE QUIZ

Answers to Modified True-False

1. false, base
2. false, banks
3. false, decreased
4. false, natural levees
5. true

continues

Answers to Multiple Choice

6. c	9. d
7. d	10. a
8. d	

Answers to Completion

11. base level	14. alluvial fan
12. dendritic	15. runoff
13. porous	

Answers to Short Answer

16. Both deltas and alluvial fans are fan-shaped because the water flow shifts back and forth as deposits block the channels. Deltas occur in water, alluvial fans occur on dry land.

17. Resistant rock formations can serve as temporary base levels until the stream erodes down through the formation.

18. As soon as a stream overflows its banks, the water slows and deposits materials. This occurs at the edge of the stream channel so, there is a buildup of material along the banks.

Answers to Writing Critically

19. In the youthful stage, the energy of stream erosion is focused on cutting down and shaping a smooth drop in the stream bed. Once this is done, erosion begins on the banks, thus shifting the channel sideways. This shifting widens the valley, creating a flood plain.

20. Rectangular (left) and dendritic (right) drainage patterns are shown. Rectangular drainage patterns develop in areas in which the bedrock is jointed or faulted. Dendritic drainage patterns develop where slopes exist but are not steep.

ANSWERS TO EXTENSION

1. Answers will vary. Students may focus on either metallic or non-metallic pollutants. Some dangerous pollutants are arsenic, beryllium, cadmium, cobalt, chromium, iron, lead, nickel, selenium, titanium, and zinc. Other pollutants to be studied are PCBs or pesticides like DDT.

2. Answers will vary. Three major aquifers are the Ogallala formation (Pliocene age), Santa Ana ground-water basin (Pleistocene age), and the Dakota Sandstone (Cretaceous age). The Ogallala was formed on the High Plains as rock materials were carried

---

Multiple Choice

Write the letter of the choice that best answers the question or completes the statement.

6. Which of the following would not be porous?
   - a) sandstone
   - b) pumice
   - c) obsidian
   - d) scoria

7. What type of drainage pattern forms in an area of long parallel ridges?
   - a) dendritic
   - b) parallel
   - c) deranged
   - d) trellis

8. Deposits built up at the mouths of rivers are
   - a) meanders.
   - b) distributaries.
   - c) natural levees.
   - d) deltas.

9. The primary cause of hard water is
   - a) nearness to rocks.
   - b) depth of ground water.
   - c) relationship to deltas.
   - d) mineral content.

10. The Great Lakes could be considered to be
    - a) a temporary base level.
    - b) a playa.
    - c) an old-age valley.
    - d) formed by headward erosion.

Completion

Complete each statement by supplying the correct term.

11. A river is unable to erode below its _____.

12. A(n) _____ drainage pattern would form where slopes are not steep.

13. Water passes easily through most sandstones because they are very _____.

14. A(n) _____ is formed as water slows and drops sediments in fan-shaped deposits at the base of a steep desert slope.

15. Erosion by _____ is affected by both volume of water and steepness of slope.

Short Answer

16. How are deltas and alluvial fans similar? How are they different?

272   Chapter 12 Review

---

17. How is the base level of a stream affected by resistant rock formations?

18. Why do streams form natural levees as they overflow their banks?

Writing Critically

19. Why do flood plains occur only in the mature stage of stream development?

20. Identify each type of drainage pattern shown, and describe the situation that might produce each one.

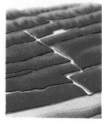

EXTENSION

1. Research and report on some impurities or pollutants that may enter ground water supplies and not be filtered out as the water passes through bedrock.

2. Do library research on major aquifers in the United States. Find out where their source is and how they flow. Report to your class on the aquifer in your region.

3. On a map of South America, locate the main drainage basins, and determine the divide that separates each basin.

---

eastward from the formation of the Rocky Mountains. The Santa Ana basin is located in the semi-arid area of southwestern California. This large enclosed basin contains extensive alluvial fan deposits. These fans reached a depth of approximately 300 m as materials were eroded from the San Gabriel Mountains. Because faults have broken the formations in the area, the groundwater in these sediments is trapped, making this an

excellent aquifer. The Dakota artesian aquifer is the largest and most important source of artesian water in the United States. Water enters it along the eastern edge of the northern and central Rocky Mountains. It serves as a water source for much of North Dakota, South Dakota, and Nebraska.

3. The Amazon and Orinoco river basins are separated by the Guiana Highlands. The Amazon and the Paraná are separated by the Brazilian Highlands.

APPLICATION/CRITICAL THINKING

1. Explain why meander cutoffs form a single river channel rather than several small channels.

2. A special type of drainage pattern develops over an area underlain by limestone. This type of pattern is called *centripetal,* or circular. Using your knowledge of karst topography, describe what you think the characteristics of centripetal drainage would be.

3. Many large farms are found in the hilly regions of the northeastern United States and in the foothills of the eastern mountains. Explain how modern contour plowing could decrease the amount of runoff into streams of these areas.

FOR FURTHER READING

Bain, I. *Water on the Land.* New York: The Bookwright Press, 1984. Major concepts about surface and ground water are clearly presented in this book.

Gunston, B. *Water.* Morristown, N.J.: Silver Burdett, 1980. This classic book presents a thorough treatment of water characteristics, water activity on land and in the atmosphere, and water uses by living things and industry.

Maltby, E. *Waterlogged Wealth: Why Waste the Earth's Wet Places?* Washington, D.C.: International Institute for Environment and Development, 1986. An excellent review of wetlands of the world.

Challenge Your Thinking

Flowing through the Grand Canyon, the Colorado River provides a puzzle as to its stage of development. The river shows the wide, looping meanders characteristic of an old-age river; however, the valley walls are steep and V-shaped, characteristics of a youthful river. How might this apparent conflict of characteristics be explained?

ANSWERS TO APPLICATION/ CRITICAL THINKING

1. After even a small channel is cut through a meander, the water will flow through the cutoff because it provides a straighter course. The greater volume of water rapidly cuts a deeper channel. The force of gravity always pulls the water into the lowest area, thus all the water concentrates in the lowest channel.

2. This type of drainage would show very little surface drainage because of the underground streams. This would result in valleys or ravines without water, as well as many sinkholes and springs.

3. Contours would trap the runoff and keep it from forming gulleys on the hillsides.

ANSWER TO CHALLENGE YOUR THINKING

The meanders in the Colorado River show that it was once a mature river, but now it is eroding its base like a youthful river. Perhaps the area has been uplifted, changing the base level of the river.

Chapter 13: WIND AND ICE

PLANNING THE CHAPTER

Chapter Sections	Page	Chapter Features	Page	Program Resources	Page
Section 1: Wind	275			Cross-Discipline: Science and Social Studies, *Examining World Deserts* **(A)**	**TRB 24**
13.1 Arid Environments **(B)**	275			Reading for Content: *Writing a Complex Outline* **(A)**	**TRB 23** **LM 197**
13.2 Wind Erosion **(B)**	277	A Matter of Fact	277	Investigation 13.2: *Wind Erosion* **(A)**	**TRB 55** **LM 57**
		Discover: Locating Wind Erosion **(B)**	278		
13.3 Sand Dunes **(B)**	280	Section Review	280		
		Activity: Analyzing Dune Migration **(B)**	281		
Section 2: Ice	282			Critical Thinking **(H)**	**TRB 24**
				Investigation 13.1: *Glacial Erosion* **(B)**	**TRB 53** **LM 55**
13.4 Glacier Formation **(B)**	282	**Discover:** Making Firn **(B)**	283	Concept Extension: *Erosion Control* **(B)**	**TRB 27** **LM 249**
		A Matter of Fact	284		
13.5 Glacial Deposits **(B)**	284	Section Review	287		
		Careers: Science Teacher, Snow Hydrologist, Computer Operator	288		
Section 3: Glacial Landforms	289				
13.6 Alpine Glaciers **(A)**	289				
13.7 Continental Glaciers **(A)**	291	**Discover:** Finding Glacial Deposits **(A)**	291		
		A Matter of Fact	292		
		Section Review	292		
		Skill Activity: Constructing Models **(A)**	293		
		Investigation 13: Examining Glacial Topography **(A)**	294	Student Record Book: Textbook Investigations **(A)**	**TRB 25**
Chapter 13 Review	295			Vocabulary **(A)**	**TRB 13** **LM 147**
				Tests **(A)** Computer Test Bank	**TRB 53**

(LM) Laboratory Manual/Study Guide, **(TRB)** Teacher's ResourceBank™

B = Basic **A** = Average **H** = Honors

The coding Basic, Average, and Honors indicates sections or subsections that might be appropriate for different levels of learners. For additional suggestions regarding choice of topic and depth of coverage, see the Pacing Chart on pages T16–T20.

CHAPTER CONCEPTS, OBJECTIVES, AND TERMS

Section	Concepts	Objectives	Science Terms
Section 1: Wind	■ Over one-third of the land area of the earth is covered by desert. **(13.1)** ■ Arid conditions are the result of planetary winds, ocean currents, topographic features, and distance from a water source. **(13.1)** ■ Wind erosion occurs where there is little vegetation to protect the soil. **(13.2)** ■ Wind erosion may be temporary or seasonal, caused by factors such as dry monsoon winds. **(13.2)** ■ The removal of all soil in an area is called *deflation*. **(13.2)** ■ The most common dune is the barchan. **(13.3)**	■ **List** the factors that create arid environments. ■ **Identify** the conditions that cause wind erosion. ■ **Describe** the major deposit created by wind.	arid rain shadow saltation deflation barchan dune
Section 2: Ice	■ Presently, there are two large continental glaciers and many smaller glaciers on Earth. **(13.4)** ■ Glaciers are formed when layers of snow and ice accumulate over the years. **(13.4)** ■ Alpine glaciers form in the mountains; continental glaciers form on flat topography. **(13.4)** ■ Glaciers are powerful erosional agents. They erode by plucking and snowplowing large amounts of rock materials. **(13.5)** ■ Glacial deposits may be left when a glacier melts, or they may be transported by water and wind. **(13.5)** ■ Glacial deposits usually take the form of moraines and outwash plains. **(13.5)**	■ **Describe** the formation of glaciers. ■ **Explain** the differences between direct and water-carried deposits.	glacier iceberg firn plucking till moraine outwash plain loess
Section 3: Glacial Landforms	■ Alpine glaciers form U-shaped valleys, horns, and hanging valleys. **(13.6)** ■ Continental glaciers form kettle lakes, drumlins, kames, and eskers. **(13.7)**	■ **Describe** the deposits formed by alpine glaciers. ■ **Explain** how the deposits of continental glaciers are different from those of alpine glaciers.	There are no new science terms in this section.

Title	Page	Materials
Discover: Locating Wind Erosion	278	*(per student)* paper, pencil
Activity: Analyzing Dune Migration	281	*(per group of 3 or 4)* fine, dry sand; hair dryer; tray with low sides; goggles
Discover: Making Firn	283	*(per student)* snowball, or ball of crushed ice, freezer
Discover: Finding Glacial Deposits	291	*(per student)* United States/Canadian Map
Skill Activity: Constructing Models	293	*(per student)* modeling clay or wet sand
Investigation 13: Examining Glacial Topography	294	*(per group of 3 or 4)* topographic maps of Chief Mountain, Montana, and Ayer, Massachusetts, in the Reference Section.

TEACHING SUGGESTIONS

Section 1: Wind

Class Activity

Study the global predictions of plate tectonics movement on page 138. Have the students predict where future arid locations will develop.

Outside Speaker

Have a geologist come to talk to the class about the formation of landforms caused by wind erosion, and about what can be done to prevent wind erosion of valuable croplands.

Section 2: Ice

Demonstration: Erosion and Deposition

Purpose

To model the erosion and deposition of glaciers

Materials

Ice cubes
Sand
Dish
Hand towel

Procedure

1. Show how the ice of a glacier might pick up rocks from the earth by placing an ice cube in a dish of sand. Let it stand in a warm room for about 5 minutes. Then place it in a freezer. Leave it there overnight.

2. The next day, remove the ice cube with a hand towel. Shake the loose sand particles away from the ice cube. Now rub the surface that has been in contact with the sand across an ordinary rock such as a piece of limestone. You should observe scratches resembling glacial striae.

Questions to Ask the Students

1. Explain what striae are and how they are formed.
   (The sand scratched the rock in much the same way as glacial material scratches rocks.)
2. If you allow the ice cube to stand in a saucer until it melts, what remains?
   (sand)
3. How is this related to the action of a glacier?
   (The sand is similar to the deposits left by a retreating glacier.)

Class Activity

Ask the students to bring in slides or pictures taken on family trips. Make it a class project to classify landforms in the slides.

Field Trip

If your area has any landforms shaped by wind or ice, take a field trip to show the students what these look like.

Section 3: Glacial Landforms

Class Activity

Use the stream table to demonstrate the formation of various glacial features. Use a large sheet of cotton batting to represent a continental glacier. Bury an ice cube in the sand to represent the way kettle lakes are formed. Build up mounds for moraines, kames, and eskers. Give a lecture/demonstration showing how the various features are formed as the glacier retreats.

You can also build alpine features in the stream table.

Field Trip

In glaciated areas, till deposits are often used as sources for gravel. Contact a local gravel yard to schedule a field trip.

CHAPTER 13

CHAPTER OVERVIEW

In this chapter, the forces of wind and ice are described. Causes and effects of these two major agents of erosion on the earth are discussed. The land formations resulting from erosion are then described in detail.

Section 1: Wind This section explains how arid climates are created on Earth. Four factors causing arid climates are described. Wind erosion is then described, with an emphasis on its permanent and seasonal effects. The section concludes with a discussion of how sand dunes are formed, and a table of the most common types of dunes is included.

Section 2: Ice This section describes glacier formation and distinguishes between two types of glaciers: continental glaciers and alpine glaciers. Iceberg and firn formations are also explained. The movement of glaciers is then discussed, with an emphasis on how warmer months in certain regions affect this movement. The section concludes with a description of the two kinds of glacial deposits—direct deposits and water deposits.

Section 3: Glacial Landforms This section explains how alpine and continental glaciers produce landforms. Locations and distinguishing features of both types of glacial landforms are described.

Wind and Ice

Deserts, with their rolling hills of sand, have always been fascinating to writers and moviemakers searching for exotic settings. They are fascinating to geologists as well, because of the unusual landforms created by wind erosion. Since dry land covers so much of the earth, wind is second only to water as an agent of erosion and deposition.

The wind creates many desert landforms.

274

CHAPTER MOTIVATING ACTIVITY

Begin this chapter by showing the students photographs of sand dunes, deserts, and sandstone and glacial formations. Have the students speculate on the causes of these formations. To help the students visualize the power and size of glaciers, you may also wish to show a film. *Fire and Ice,* a National Park Service film, depicts the formation and glacial erosion of Mount Rainier. Have the students discuss how both wind and ice affect the earth's surface.

SECTION OBJECTIVES

After completing this section, you should be able to:
- **List** the factors that create arid environments.
- **Identify** the conditions that cause wind erosion.
- **Describe** the major deposit created by wind.

NEW SCIENCE TERMS

arid
rain shadow
saltation
deflation
barchan dune

13.1 Arid Environments

Viewed from space, our planet seems to be covered mostly by water. However, it may surprise you to learn that over one third of the land is desert. With so much water on Earth, special environmental conditions are needed to create dry, or **arid,** climates. Arid climates are usually caused by one of four factors: planetary winds, ocean currents, topography, or great distance from a water source.

You may recall from Chapter 9 that two dry planetary wind belts converge at the edge of the tropics. These are the areas where most of the world's major deserts can be found.

The circulation of ocean waters can also create arid climates. Along the western coast of some continents, the prevailing winds push the warm surface waters away from the land. This allows cold water to rise to the surface, chilling the air and condensing the water vapor. Rain and fog occur offshore, and the land remains dry. Ocean currents are discussed in detail in Chapter 19.

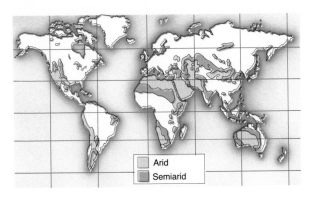

Arid
Semiarid

Figure 13–1. Many of the major deserts of the world are located just north and south of the tropics. Other deserts are created by topography, geography, or ocean currents.

BACKGROUND INFORMATION

In the past, scientists defined a desert as a place receiving 30 cm or less of annual rainfall. This is not quite accurate, however, because much more than 30 cm of rain falls annually in many desert regions. Today scientists use a system that compares the solar energy an area receives to its annual precipitation. For example, an extremely dry area, such as the Peruvian Desert, has an aridity index of 200. This means that the sun's energy can evaporate 200 times the amount of rainfall received in a typical year.

Section 1 : Wind

SECTION FOCUS

Section Overview This section describes the causes and effects of arid environments on the earth. The forces of wind erosion are then described, including explanations of the processes of saltation and deflation. The section concludes with a discussion of sand dunes, including their causes, and a description of the most common types of dunes.

Section Objectives For a list of section objectives, see pupil's edition page.

New Science Terms For a list of new science terms, see pupil's edition page. You may wish to familiarize the students with these terms before they read the section by reading the terms aloud.

SECTION DEVELOPMENT

13.1 Arid Environments

DISCUSSION Remind the students that in Chapter 8 they studied planetary wind patterns. You may wish to review these concepts before beginning this section. Have the students study Figure 13–1, which shows where the major deserts of the world are located. Point out that most arid regions lie in narrow belts that straddle the Tropic of Cancer and the Tropic of Capricorn. ***Thinking Critically:*** Ask the students to think of one or more reasons why many desert regions are located in the western half of continents. (One explanation is that many mountain ranges, formed by the forces of plate collisions, are located in coastline areas. High mountains keep moisture from reaching areas east of their peaks.)

EXTENSION (Tie-in/Meteorology) Desert regions are usually characterized by high temperatures as well as by little rainfall. The sun's rays are able to warm the ground in a desert region to a much greater extent than is possible where moister climates deflect much of the solar heat. Summer temperatures of 50°C are quite common in deserts, and the ground surface can be as much as 28°C warmer than the air temperature. At night, the lack of moisture in the desert causes the day-time heat to dissipate very quickly and the temperatures may drop 30°C or more. *Thinking Critically:* Ask the students to explain what advantage this drastic change in temperature at night has on desert life. (Without the coolness of the night, many of the plants and animals that now live in the desert could not survive.)

THINKING SKILL (Drawing Conclusions) Tell the students that altitude and latitude control differences in desert climates. The greater the elevation above sea level, or the greater the distance from the equator, the colder a desert will be. Have the students use this information to point out on a globe or world map the desert they think would be the hottest (the Sahara) and the desert that would be the coldest. (the Gobi in Mongolia) Antarctica is actually a desert, even though there is much ice, because of the small amounts of precipitation.

EXTENSION (Tie-in/Geology) The Great Basin Desert, which lies just east of the Cascade Range and the Sierra Nevadas in the western United States, is the result of a rain shadow. Remind the students that plate tectonics was studied in Chapter 7. Have them refer to Chapter 7 and name the plates that created these mountain ranges. (Pacific Plate, which subducts beneath the western edge of the North American Plate) Point out that these mountains, rising to heights of 4200 m, make up a north-south barrier between the Pacific Ocean and the interior.

Figure 13–2. A rain shadow (right) often creates a desert such as the Sonoran Desert (left) of Arizona and northern Mexico.

Figure 13–3. The Gobi Desert of central Asia (right) is hundreds of kilometers from the Pacific and Indian oceans. The dry conditions of the desert create a nomadic way of life for the people who live there.

Arid climates can also form because of special topography. For example, as moist air rises over a mountain range, it cools and the water vapor condenses. Rain falls on the side of the mountain facing into the wind, forming a rain shadow on the opposite side. A **rain shadow** is an area shielded from precipitation by a mountain or other topographic feature. Air descending on the other side is warm and dry, and the moisture in the land evaporates.

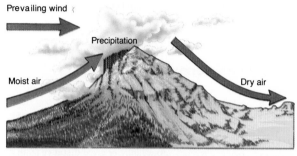

Another factor in creating arid environments is the large size of some continents. As air masses travel great distances from the oceans, most of the water condenses and falls as precipitation, leaving little moisture in the air. For example, the prevailing westerlies drop most of their water on Europe, creating arid conditions farther to the east, in parts of Asia.

276

BACKGROUND INFORMATION

Marco Polo, the Italian explorer, traveled in the very cold Asian deserts in the thirteenth century. His adventures have inspired many tales. However, even he and his companions were intimidated by the Gobi Desert. They were told that it would take them at least one year to travel from one end of the Gobi to the other. They chose to travel the narrowest end of the desert, and their journey took 30 days. Marco Polo recorded 28 places where travelers could find water holes. Six hundred years later, an explorer named Sir Aurel Stein crossed the Gobi following the same route and found the same number of water holes.

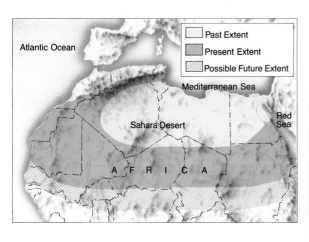

Atlantic Ocean

Mediterranean Sea

Sahara Desert

Red Sea

A F R I C A

Past Extent

Present Extent

Possible Future Extent

Figure 13–4. Desert areas sometimes increase in size. The Sahara desert (left) has been steadily increasing in size, creating food shortages and climatic problems for the people of Africa.

Deserts and other arid lands have not always been in their present locations. Deserts change location because the environmental factors that form them change, sometimes with human help. The deserts today are increasing in size. For example, the Sahara desert in northern Africa is constantly extending southward. Poor farming practices are one of the leading causes of this movement.

○ *What conditions create arid climates?*
○ *What is a rain shadow?*

13.2 Wind Erosion

Have you ever been to a beach on a very windy day and felt the sting of sand grains hitting your legs? Besides the pain, the blowing sand irritates your skin, and if you were to stay long enough, it could literally tear the skin from your legs. This same constantly blowing wind produces erosion of rocks and soil.

Wet soil tends to resist wind erosion; therefore, as the amount of water in the soil decreases, wind erosion increases. Part of this erosion occurs because a decrease in surface water means fewer plants grow. Loss of ground cover allows soil and rocks to be exposed to the wind, which then picks up and moves the unprotected soil.

Wind erosion can be temporary or seasonal. The creation of the "dust bowl" in the western United States in the 1930s was the result of several years of unusually dry weather. When the land dried out, the plants died, and the exposed soil was blown

A MATTER OF FACT

The sun is ultimately responsible for the erosion caused by wind. The sun heats the earth unevenly, causing pressure differences, which —in turn—cause the air to move.

Section 1   Wind   **277**

DISCUSSION   Deserts are formed by a process called desertification. Soil erosion is the most obvious cause of desertification. In dry periods, soil cracks and decays into particles that are blown about, primarily by wind storms. Fertile topsoils are also carried away. The results are clogged water holes and mounting sand dunes. Vegetation will grow again when it finally rains, but a population explosion among herbivore animals also occurs as a result of the rains. These animals strip away all vegetation near water holes. In addition, heavy downpours can often carry away the soil. *Thinking Critically:* Ask the students why overplanting the land would not be an effective strategy for preventing erosion under these conditions. (It begins a process that is merely a cycle, but does not provide a solution.)

EXTENSION   (Tie-In/Geography) Tell the students that desertification is a worldwide problem; parts of more than 100 countries are affected. Each year about 7 million hectares of useful land are lost. On a world map, point out to the students the earth's largest desert regions.

13.2 Wind Erosion

THINKING   SKILL   *(Comparing)* Point out that dust storms and sandstorms are not the same thing. Sand rarely gets more than one meter or so off the ground. Ask the students to explain the difference between the particles that form dust storms and particles that form sandstorms. (Particles that form dust storms are very fine and lightweight; they can be carried easily by wind. The particles that form sandstorms are heavier and coarser; the wind cannot lift these particles as high into the air.)

Answers–13.1

○ Dry planetary wind belts, circulation of ocean waters along coasts, mountains causing rain shadows, and large continents providing little water as a source all help create arid climates.

○ When moisture-laden air is blown over mountain ranges, it cools and drops its moisture. As the air descends on the other side, it warms and dries. This dry side of a mountain is said to be in a rain shadow.

DEMONSTRATION

For a demonstration of the erosional effects of ice, see page 273c preceding this chapter.

THINKING SKILL *(Communicating)*
Have the students research landforms that result from wind erosion in arid environments. Tell the students to write papers in which they explain how these formations occur. Tell them to use a variety of adjectives to describe the wind-formed structures. You may also wish to have the students illustrate the structures if the visual information is available to them.

EXTENSION The dust from arid regions also affects semiarid regions. Thick, silty deposits in semiarid regions, known as loess, originate as dust is blown from a neighboring desert area. Loess can cover huge areas. For example, loess in northern China, originally dust from the central Gobi Desert, covers more than 321 800 km² with soil thicknesses up to 100 km² near the Yellow River.

DISCOVER:
Locating Wind Erosion

Skill *(Observing)*

The students may work in small groups by comparing their observations and then presenting them to the class.

Figure 13–5. A dry, plowed field loses much valuable topsoil to wind erosion.

DISCOVER

Locating Wind Erosion

In most areas of the United States, the prevailing winds blow from the west, so any evidence of wind erosion should be on the western side of a landform, and any deposition should be on the eastern side. On your way home from school, see if you can locate any wind-eroded areas and confirm that most of the erosion has been on the western side.

Figure 13–6. The monsoon winds, which bring torrential rains to India, also bring a season of drought (right). Wind blown soil often obscures the sun, producing a red haze.

278

away. However, as the area began receiving normal amounts of rain again, the plant cover returned, and the wind erosion stopped.

Areas influenced by monsoon winds may experience seasonal periods of wind erosion. When the monsoon wind blows from the dry continent, soil may be lost due to wind erosion. When the wind reverses and blows from the sea, torrential rains may cause erosion by runoff. Without water, wind becomes the dominant erosional force. As the wind blows over dry, barren land, it transports and deposits soil and rock fragments. Particles of clay and silt are carried to great heights, forming dust clouds. The dust clouds created in the dust bowl traveled all the way across the United States and deposited soil particles on the east coast. How could the western dust have been identified on the east coast? 1

1 It could have been identified by the types of minerals found in the soil particles.

Sand-sized particles are usually blown along near the surface of the ground. If the wind speed is great enough, the sand may bounce as high as 2 m. This bouncing is called **saltation.** Saltation acts like a sandblasting machine, forming bizarre sculptures as different layers of rock are eroded at different rates. Saltation is the process that irritates your legs at the beach.

Tremendous amounts of soil, sometimes even all the soil in an area, can be eroded by the wind. The removal of all the soil of an area is called **deflation.** The resulting depressions in the earth's surface are called *deflation basins.*

○ *Why are the effects of saltation like sandblasting?*
○ *How are deflation basins created?*

Figure 13–7. The eroding power of wind blown particles produced this formation (left). This power is similar to that of commercial sandblasting (right).

Figure 13–8. A deflation basin, such as the one shown here, results when all of the loose particles of an area are removed by the wind.

Section 1   Wind   **279**

Answers—13.2

○ Both saltation and sandblasting blow sand into objects to erode their surfaces.

○ Deflation basins are created as the fine particles are blown away, leaving a depression lined with larger rocks.

DISCUSSION   Point out to the students that in Figure 13–7, sand is being used to abrade, or wear away, the dirt on the surface of a building. *Thinking Critically:* Ask the students how wind affects rocks and what kinds of rocks probably erode the fastest. (Strong winds can erode rocks by picking up rock fragments and blowing them against the surface of other rocks. Sedimentary rocks usually erode the fastest from wind-blown fragments.)

EXTENSION   Deflation is a process that occurs over a period of many years. It primarily occurs in dry areas where there is very little plant life to hold the soil in place. Lightweight particles, such as silt, are carried away by wind, while heavier objects, such as small rocks, remain on the surface.

13.3 Sand Dunes

DISCUSSION   Point out that the gentle slope of a dune that faces the wind is called the windward side. The side of the dune away from the wind is called the lee-ward side. The slope is much steeper on the leeward side. Explain that the wind-ward side is constantly eroding. If the wind continues to blow hard, grains of sand constantly move up the windward slope, over the crest, and down. The movement of the sand grains causes the dune to rearrange itself, but the dune keeps its form while it is moving slowly downwind. Dunes can move as much as 0.3 m a day in areas where there are strong winds.

Figure 13-9. When the supply of sand is plentiful and the wind blows constantly from the same direction, giant ridges of sand (right) can be produced. The area from which the sand is removed is often left barren, resulting in a hard surface called *desert pavement* (left).

SECTION REVIEW

Summary   Review the four factors that cause arid climates. Have the students explain the rain-shadow effect. Then discuss the causes and effects of wind erosion and ask the students to define saltation and deflation. Finally, review formations of sand dunes.

Reinforcement   Have the students write the main ideas in each paragraph for each numbered subsection. Then have them compare their ideas with those of others in the class.

13.3 Sand Dunes

Sand dunes are the best known form of wind deposit. Dunes are characterized by gently sloping sides, called the *windward side,* facing into the wind. The steep side, facing away from the wind, is called the *leeward side.* Dunes move, or migrate, as sand is rolled up the gentle slope and deposited over the crest onto the other side.

The most common type of dune is the **barchan** (BAHR kahn) **dune.** A barchan is a crescent-shaped dune with the bulging side facing into the wind. This type of dune forms in areas that have a limited supply of sand. As sand is separated from the larger pebbles and rocks in an area, the rocks that are left are called *desert pavement.* In areas with a large supply of sand or with variable winds, different kinds of dunes form. Some of these different dunes are shown in Table 13-1.

○ *What is a barchan dune?*

Section Review

READING CRITICALLY

1. List four factors that form arid lands.
2. Explain how a dune migrates.

THINKING CRITICALLY

3. How can a coastal environment be very much like a desert?
4. Explain why the term *shadow* is especially useful in picturing the concept of a rain shadow.

Answer–13.3

○ A barchan is crescent-shaped dune.

Answers to Section Review

Reading Critically

1. Global wind patterns, ocean currents along coasts, mountain chains, and large continents help in the formation of arid lands.
2. Sand grains are rolled up the windward side and then deposited on the leeward side. As more and more material is rolled to the leeward side, the dune moves or migrates in a downwind direction.

Thinking Critically

3. Both have exposed areas of soil or sand, which are blown and shaped by winds.
4. Just as light does not fall in a person's shadow, rain does not fall in a rain shadow.

TABLE 13-1: TYPES OF DUNES		
Shape	**Name**	**Description**
	Barchan	Forms where sand is limited and wind is strong and constant
	Star	Forms where sand is plentiful and wind is strong and shifting
	Linear	Forms primarily along seacoasts where the sea breeze and land breeze push the sand into long lines
	Transverse	Forms where sand is plentiful and wind blows from one direction
	Parabolic	Forms along seacoasts where vegetation holds the sand

ACTIVITY: Analyzing Dune Migration

How do changes in wind direction affect the shape of a barchan dune?

MATERIALS (per group of 3 or 4)

fine, dry sand; hair dryer; tray with low sides; goggles

PROCEDURE

1. Pour the sand into the tray and construct a barchan-shaped dune.
2. **CAUTION: Wear goggles to prevent sand from blowing into your eyes.** Turn the hair dryer on a low, cool setting and direct the air at a low angle toward the dune. Experiment with the dryer to determine how close you should be to the dune to make sand grains roll up the side of the dune.
3. Hold the hair dryer in a fixed position and study the way in which the dune slowly travels or migrates across the tray.
4. Shift the dryer about 10 cm away from your original position to change the wind direction. Study the way in which a wind shift causes the dune to change shape.

CONCLUSIONS/APPLICATIONS

1. Explain the differences in appearance between the two sides of a sand dune.
2. Why is the side away from the wind not vertical?
3. Describe the sequence of events that occurs as the dune migrates.
4. What happens to the dune when you shift the wind direction?

Section 1 Wind **281**

BACKGROUND INFORMATION

The most extensive study of desert sand dunes occurred in the 1930s. Ralph Bagnold, a British Army officer and scientist, was sent to Egypt's Western Desert. Bagnold became fascinated by the dunes. He studied sand deposits of every kind, from small ripples to 160.9- km long "whalebacks." Upon returning to England, he convinced the Imperial College of Science and Technology in London to let him use their laboratories and wind tunnels to study the effects of the interaction of sand and wind.

DISCUSSION Sand dunes differ in shape according to environmental conditions. Ask the students to study Table 13-1. *Thinking Critically:* Ask the students to explain how the description of the dunes relates to the illustrations in the left column.

EXTENSION The students may be interested to know that many desert travelers, from Marco Polo's time until today, have recorded hearing mysterious, eerie "singing dunes." The sudden, low-pitched sounds have on occasion been so loud that normal speech could barely be heard. Scientists have not yet been able to explain what causes these sounds.

ACTIVITY:
Analyzing Dune Migration

Skill (*Constructing Models*)

Preparation of Materials Caution the students to use to blow dryers carefully and to keep the dryers at the low setting throughout the activity. You may wish to demonstrate the use of the dryer and how to form the sand dunes before the students begin the activity. Be sure they understand how a barchan-shaped dune is actually formed before they begin. You may wish to obtain other photographs of these types of dunes for the students to look at and use as references. Tell them to dispose of all sand, if any, in a wastebasket. Be sure they understand never to dispose of sand in a sink.

Answers to Conclusions/Applications

1. The windward side is a gentler slope than the leeward side.
2. The round sand grains will not hold a vertical slope.
3. Sand rolls up the windward side and then falls over the crest onto the leeward side. As more grains stack up on the leeward side, the entire dune migrates in the leeward direction.
4. The direction of migration changes.

Section 2: Ice

SECTION FOCUS

Section Overview This section focuses on the immense eroding power of ice. Glacier formations are first discussed, including descriptions of icebergs and the process of producing glacial ice. Movement of glaciers is then described. The section concludes with a discussion of glacial deposits, which are classified as either direct deposits or water deposits.

Section Objectives For a list of section objectives, see pupil's edition page.

New Science Terms For a list of new science terms in this section, see pupil's edition page. You may wish to familiarize students with these terms before they read the section. You may also wish to read the terms aloud.

SECTION DEVELOPMENT

13.4 Glacier Formation

DISCUSSION Tell the students that the movements of huge glaciers during the last ice age eroded much of North America. Display a large map of North America and have the students point out features that they think are evidence of this erosion. (for example, the Great Lakes and the Finger Lakes of New York state) Tell the students that the last of the great ice sheets disappeared from North America about ten thousand years ago. ***Thinking Critically:*** Ask the students to name other erosional processes that have changed the face of North America. (for example, volcanic activity in such places as Mt. St. Helens and Hawaii; earthquake activity along California's coastline; and beach erosion along Florida's coast.)

NEW SCIENCE TERMS
glacier
iceberg
firn
plucking
till
moraine
outwash plain
loess

SECTION OBJECTIVES
After completing this section, you should be able to:
- **Describe** the formation of glaciers.
- **Explain** the differences between direct and water-carried deposits.

13.4 Glacier Formation

Can you imagine a time when much of Canada and the northern United States was covered with ice 3 km thick? Can you imagine a warm, summer day with temperatures up to almost 0°C and only half a meter of new snow? For much of the recent history of the earth, glaciers covered large parts of North America and Europe. A **glacier** is a sheet of ice that covers a large continental area or a mountain region.

Today, 98 percent of all ice on Earth is located in the polar regions; the rest is found on high mountains. There are only two large *continental glaciers* remaining: one in Greenland and the other in Antarctica. However, many mountains around the world have smaller glaciers called *alpine glaciers*.

A continental glacier may be as much as 4 km thick and so heavy that it causes the rocks below it to sink into the mantle. If the ice sheet covering Greenland were suddenly to melt, the island would rise by nearly 1 km. The South Pole is under approximately 3 km of ice. However, there is no glacier at the North Pole. Glaciers form on land, but the North Pole is in the Arctic Ocean, where there is a thin layer of frozen sea water.

Figure 13–10. Continental glaciers, such as the one shown here, may be hundreds of meters thick and may cover areas as large as Antarctica.

282   Chapter 13   Wind and Ice

BACKGROUND INFORMATION

About two and a half million years ago, the geological period known as the Pleistocene Epoch began. Since the beginning of this epoch, the ice sheets of both North America and Europe have moved into the middle latitudes at least four times, and geologists believe this could have happened as many as ten times. Each of these ice sheet advancements can be considered an ice age.

Even now the epoch continues. Glaciers still cover about 10 percent of the land on Earth. Another 14 percent contains ice held in the ground. This is called permafrost.

There are pieces of glaciers, called *icebergs*, in the polar seas. An **iceberg** is a great mass of freshwater ice freed from a glacier by a process called *calving*. The process of calving is illustrated in Figure 13–11.

Why do glaciers form in some cold areas but not in others? Many areas of the world receive snow and ice during the winter months. However, in most areas the warm summer causes the snow and ice to melt. Glaciers form only in areas where the temperature remains too low for all of the snow to melt.

As the snow piles up, the pressure increases on the bottom of the snow pile, and the snow compacts. Snow that melts quickly refreezes, forming ice. Fresh snow is about 90 percent air. Once half of the air spaces in the snow pile have been filled with ice, the pile is called **firn.** Firn takes about a year to form.

Eventually firn becomes compressed and changes into a solid body of ice. This transition from snow to firn to ice may take three to five years in places where warm temperatures cause snow to melt, forming melt water. In colder parts of the world, this transition may take as long as 100 years. Why would this change take longer in places where it is colder?

Figure 13–11. The calving of an iceberg from a glacier is shown in this series of photographs.

DISCOVER

Making Firn

Make a snowball or a ball of crushed ice and place it in a freezer until the snowball is solidly frozen. Then take it out of the freezer and apply heat and pressure to it by squeezing it. When the snowball begins to melt, place it back into the freezer. Repeat this procedure several times, until some of the air-space in the snowball is filled with ice. How is this similar to the way firn is produced in a glacier?

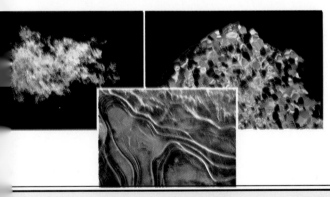

Figure 13–12. This series of photographs shows the transformation of snow (left) to firn (right), and then to glacial ice (bottom).

EXTENSION   Tell the students that icebergs are a hazard to ships and ocean liners. Have the students research the sinking of the *Titanic* in 1912. Have them write a report on the history of the *Titanic*, including the most recent expeditions to study the sunken ship off the coast of Newfoundland. Tell the students to include information on radar and other equipment that ships use to help spot hazardous icebergs in the sea.

DISCUSSION   Tell the students that a glacier will advance as long as it accumulates more snow and ice than melts. If the ice retreats, a new moraine may develop. Later the glacier may renew its advance. This is happening to the Columbia Glacier in Alaska. You may wish to have the students research information on the rapidly changing glaciers in Alaska.

DISCOVER: Making Firn

Thinking Skill *(Inferring)*

If necessary the students can conduct this experiment at home and report to the class, or the students may discuss the possible answers in class.

BACKGROUND INFORMATION

Oceanographic research ships bring up core samples that help scientists uncover some of the mysteries of the earth's past ice ages. Shells of long-dead sea creatures help scientists to determine the temperature of the waters in which the creatures lived and died. Seabed cores reveal the movements of the earth's crustal plates, the earth's past climate, and where the ice sheets were located.

DISCUSSION Tell the students that the ice in a glacier flows somewhat like a conveyor belt. However, instead of being driven by a motor, it is driven by gravity and the mounting ice and snow behind it. The melting and advancing is a delicate balance based on whether enough new ice is added to the glacier to replace melted ice. *Thinking Critically:* Ask the students how these processes affect the movement of the glacier. (If the balance is even, the glacier will be stable. The glacier will advance or retreat, depending on this balance.)

EXTENSION Scientists are concerned about certain glacial activity occurring in Antarctica. A number of observations have been made, showing that the East Antarctica ice cap is growing thicker. This is due to increased snowfall, which is the result of a warmer climate. Scientists estimate that its growth may be matched by the calving of icebergs from the ice shelves, if the ice cap is truly thickening. Should these shelves melt, a collapse of the West Antarctic ice sheet could occur, causing an estimated rise in worldwide sea levels.

13.5 Glacial Deposits

DISCUSSION Rock materials are carried by glaciers in three ways. These materials are carried in the ice, pushed in front of the glacier, or carried in melt water flowing from the glacier. *Thinking Critically:* Ask the students to explain how glacial deposits reveal the secrets of the earth's earlier history. (Glacial deposits show the characteristics of Earth's terrain before it was reshaped by the glaciers.)

Figure 13-13. Crevasses, shown here, can be very dangerous to anyone exploring a glacier.

A MATTER OF FACT

A rapidly moving glacier may travel as fast as 4 km per year.

Figure 13-14. The movement of glaciers can be measured using this apparatus (right). Stakes placed along ABC will move to positions A'B'C' as the ice moves. Grooves, or striations, in rocks also show the movement of glaciers (above). Notice that the striations are all parallel. This shows the direction of movement of the glacier.

During the warmer months, the surface of a glacier may melt. Puddles of water or small channels often form on a glacier's surface. The water eventually flows through cracks, or crevasses, in the ice and into water channels within the glacier. In the summer, you can hear the sound of running water coming from within a glacier.

Even though glaciers are solid ice, they can still move and expand. As pressure builds up from the glacier's weight, some of the ice becomes deformed and flows like hot plastic. Individual ice grains respond to the pull of gravity by sliding past one another, allowing the ice to flow slowly downhill. Movement also occurs as melt water filters through the glacier and along the ground under the glacier. There it provides lubrication between the ice and the ground, allowing the glacier to slide smoothly on a film of water. Glacial movement increases during the warmer months because there is an increase in the amount of melt water.

○ *What are icebergs?*
○ *How do glaciers move?*

Valley side   Valley side

A
A'   B   C
B'   C'

Glacier flow

13.5 Glacial Deposits

Because of their mass, glaciers are especially effective at carrying rock materials. Some material simply falls from mountain slopes onto the surfaces of glaciers. However, glaciers also obtain rock material through a process called *plucking*. During **plucking**, glacial ice freezes around rocks and plucks them out of the ground as the glacier moves. Glaciers also move material much as a snowplow does. Rocks at the base of a glacier are picked up and pushed along in front of, and within, the glacier.

Answers—13.4

○ Icebergs are masses of freshwater ice that have broken off from glaciers and are floating in the ocean.

○ After glacial ice becomes deformed due to the great pressure of the overlying ice, it flows. Melt waters under the ice also provide lubrication between the ice and the ground that helps the ice slide smoothly.

Figure 13–15. The deposits left by a melting glacier range in size from the large boulders in the middle of this photograph to the fine soil in the foreground.

Glaciers leave two different types of deposits. The first type is directly deposited by the ice. The other is eroded by ice and then reworked, or transported, by water as the ice melts. In this reworking, the materials are sorted by size, as are most water-carried materials. In contrast, the materials directly deposited by the ice are rough, irregular, and completely unsorted.

Direct Deposits   Because of their massiveness, glaciers are able to transport materials of all sizes. Glacial deposits are commonly found as heaps of boulders, sand, and fine silt mixed together. The shape of the piles of debris varies from ridges and hills to thick, lumpy blankets covering the land. This unsorted rock debris deposited directly by glaciers is called **till**.

Figure 13–16. Unsorted glacial deposits, such as those shown here, are called *till*.

Section 2   Ice   **285**

EXTENSION   In North America, the most recent ice advance, called the Wisconsin glaciation, occurred about 18 000 years ago. At the height of the Wisconsin glaciation, the ice was at least 1500 m deep in New England. It covered the White and Green mountains. When the climate warmed, the ice sheet left tills and moraines along the farthest advance lines. Cape Cod and Long Island are two examples of deposits left by this movement. The melting ice from the Wisconsin glaciation left many types of landforms in North America and also contributed to rising sea levels. Measurements have put the rise close to 100 m. ***Thinking Critically:*** Ask the students to infer what effect the rising seas must have had on our coasts. (The rising seas must have completely covered and reshaped the coasts.)

DEMONSTRATION

For a simple demonstration of how glaciers move, place ice cubes into two Petri dishes. Cover the ice with clear plastic plates. Place a 1-kg mass on top of one of the plates. Have the students observe and record how long it takes each ice cube to melt. Then ask them to describe how the demonstration explains the movement of glaciers. (Pressure caused the bottom of the ice cube to melt faster. In a glacier, this top ice layer exerts tremendous pressure on the ice below it. Movement occurs as the ice below melts and the water acts as a lubricant.)

THINKING SKILL *(Relating)* Remind the students that they studied chemical and physical weathering in Chapter 11. Have them explain how chemical weathering differs from physical weathering. Then have them explain why rock flour must be the result of physical weathering. (In chemical weathering, new substances are formed; in physical weathering, large rocks are broken into smaller rocks. Rock flour has been broken down from till and is not a result of a chemical change.)

DISCUSSION (Tie-in/Language Arts) Have the students look up the meaning of the word erratic in the dictionary. Have them explain why the large boulders may have been named erratic. (Erratic means "characterized by lack of regularity." An erratic boulder is one that is not where it might be expected to be.)

EXTENSION Another type of moraine is a recessional moraine. It is deposited during a temporary glacial standstill. These deposits show the history of glacial retreats along the valley. There have been ten or more recessional moraines found in some valleys.

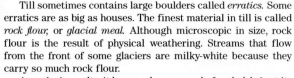

Figure 13–17. The color of this water (left) is due to the presence of fine, white material called *rock flour*. The large boulders, seemingly placed on the landscape (right) are called *erratics*, because of their erratic occurrence.

Figure 13–18. Terminal moraines (right) are often found near streams and rivers. The streams may have carried the water of the melting glacier. Lateral and medial moraines can sometimes be seen near the end of an active valley glacier (left).

Till sometimes contains large boulders called *erratics*. Some erratics are as big as houses. The finest material in till is called *rock flour*, or *glacial meal*. Although microscopic in size, rock flour is the result of physical weathering. Streams that flow from the front of some glaciers are milky-white because they carry so much rock flour.

As a glacier melts, it leaves a long mound of rock debris at its edge. A mound of unsorted rock materials that builds up along the edge of a glacier is called a **moraine.** Moraines appear as long ridges or hills on an otherwise flat landscape. In northwestern Pennsylvania and southeastern Wisconsin, state parks have been built on and between moraines.

The moraine marking the farthest advance of a glacier is called a *terminal moraine*. Those formed along the sides of a glacier are called *lateral moraines*. When two alpine glaciers join together, the lateral moraines combine to form a moraine down the middle of the glacial valley. This type of moraine is called a *medial moraine*.

Water Deposits Large amounts of gravel, sand, and rock flour are carried away from glaciers by melt water. Much of this material is deposited on outwash plains. An **outwash plain** is a thick blanket of glacial sediments. The outwash plain is formed as water rushing from the melting glacier deposits material in front of a stationary glacier.

Figure 13–19. The flood of water from a melting glacier (right) often deposits a thick blanket of eroded material on an outwash plain.

Wind Deposits Rock flour in dry outwash is easily transported by wind. Thick layers of this fine glacial material form deposits called **loess** (LEHS). Much of the midwestern United States and Canada is covered by loess. When mixed with humus, loess forms a thick, rich soil, perfect for growing grains such as wheat and corn.

○ *What is glacial till?*
○ *What is an outwash plain?*

Figure 13–20. Much of the thick, rich soil of the Great Plains is loess. Loess is easily blown by the wind.

Section Review

READING CRITICALLY

1. Would the rocks in an outwash plain be sorted by size? Explain your answer.
2. How are moraines formed?

THINKING CRITICALLY

3. Explain why temperature is a controlling factor in the amount of time it takes firn to turn into ice.
4. Using the process that transforms snow into ice, explain why ice might be called a metamorphic rock.

DISCUSSION Explain to the students that loess, transported by winds, does not show the rounding characteristic of water-carried sediment. *Thinking Critically:* Ask the students how this might affect the way in which loess erodes. (The lack of rounding results in the formation of steep cliffs when eroded. Examples can be found along the upper Mississippi River Valley.)

EXTENSION In searching for clues of the climatic conditions of the last Ice Age, scientists discovered that the deep beds of loess, deposited during times of glaciation, are similar to recent deposits transported from the Gobi Desert. This was one of the first signs that the glacial environment had been arid.

SECTION REVIEW

Summary Review the way that glaciers form. Have the students explain calving and describe icebergs and firn. Then have the students describe how glaciers move across the land. They should be able to describe the difference between direct deposits and water deposits from a glacier.

Reinforcement Have the students look through magazines such as National Geographic or the Smithsonian for photographs of ice and glaciers. The students can then make a poster of the cut-out photographs and identify as many parts of a glacier as they can.

Answers–13.5

○ Glacial till is unsorted rock debris deposited by glaciers.

○ An outwash plain is a layer of rock material carried away from a glacier by melt waters. The material is more rounded and sorted than till.

Answers to Section Review

Reading Critically

1. Rock material in an outwash plain is sorted by size because the size of the material that water can carry is determined by its velocity. When water slows, it drops larger rocks but continues to carry small ones.
2. Moraines are mounds of unsorted rock material built up at the edges of a glacier. They are formed as material is pushed in front of the moving glacier. Melt water within the glacier also carries rock material to the moraine.

Thinking Critically

3. If the temperature remains below the melting point, little snow melts and then refreezes. Thus it takes a long time for air spaces between the snow flakes to fill, forming firn. With higher temperatures and more melting, this process occurs more rapidly.
4. Heat and pressure are the two processes that form metamorphic rocks. Heat is required to melt the snow to form firn. Firn is converted into ice by the great pressures of the overlying snow. Thus both heat and pressure are involved in the formation of glacial ice.

287

Discussion Have interested students write to one or all of the addresses provided for additional information. Have the students discuss the opportunities offered by the various careers.

Extension You may wish to hold a "Career Day" in the class during which the students report on the information they have obtained. If possible, invite a person in each career to speak to the class about why he or she chose the career, what specific training he or she received, and what a typical day at work involves. As an alternative to outside speakers have students assume the roles of a representative of each career and make a presentation to the class covering the same topics mentioned above. The students may be interested in researching information in the following careers: Geologist, Soil Conservationist, and Meteorologist.

CAREERS

SCIENCE TEACHER

If you enjoy studying science and working with people and ideas, you might consider a career as a *science teacher*. Usually, people studying to be science teachers take many different science courses, including biology, chemistry, physics, and geology, as well as general courses in mathematics and English. Eventually, science teachers specialize in one or more science areas to prepare themselves for teaching in their favorite field.

A bachelor's degree is required to begin teaching and, in most regions of the country, teachers are expected to continue taking college courses to remain current in their subject area.

For Additional Information
National Science Teachers
 Association
1742 Connecticut Avenue, NW
Washington, DC 20009

SNOW HYDROLOGIST

A *snow hydrologist* studies the movements and characteristics of ice and snow on all parts of the earth. This career blends a scientific background with the challenges of outdoor survival techniques. Snow hydrologists venture onto dangerous ice fields and spend long months enduring cold temperatures. However, they also get to enjoy the breathtaking beauty of glaciers.

This career requires a college degree in geology or hydrology. A strong background in foreign languages is helpful, too, since snow hydrologists have the opportunity to work with international teams of scientists. Training in mountain climbing and outdoor survival skills is also important.

For Additional Information
The American Geological
 Institute
5205 Leesburg Pike
Falls Church, VA 22041

COMPUTER OPERATOR

A *computer operator* provides an important link between a computer and the scientists who need data from the computers. Computer operators originate and experiment with new programs, make certain that programs are run in the proper order and on schedule, and monitor a program's progress.

Most computer operators receive on-the-job training after completing their high-school pro-

gram. Many community colleges and trade schools offer courses in computer operation, programing, and repair.

There is much room for advancement, since many corporations hire computer programers to solve unique computer problems.

For Additional Information
Contact your local
 community college.

288   Chapter 13   Wind and Ice

3 Glacial Landforms

SECTION OBJECTIVES

After completing this section, you should be able to:
- **Describe** the deposits formed by alpine glaciers.
- **Explain** how the deposits of continental glaciers are different from those of alpine glaciers.

NEW SCIENCE TERMS

There are no new science terms in this section.

13.6 Alpine Glaciers

Even after a glacier melts, it leaves permanent marks on the land. These marks create many landforms that can easily be identified as having been created by glaciers.

As an alpine glacier cuts into a narrow mountain valley, it carves out a wide, U-shaped trough. At the end of the valley, a terminal moraine is deposited, and lateral moraines line the sides of the valley. Also, the floor of the valley is layered with a type of till called *ground moraine*.

Sections along the valley floor that were deeply gouged by the glacier fill with melt water to form lakes. Such lakes have a characteristic bluish-white color due to the presence of rock flour.

An eroded mountain peak at the source of several glaciers forms a dramatic pointed spire called a *horn*. Horns are named for the famous Matterhorn on the border between Switzerland and Italy. In addition to horns, bowl-shaped depressions called *cirques* (SUHRKS) often form near the beginning of alpine glaciers.

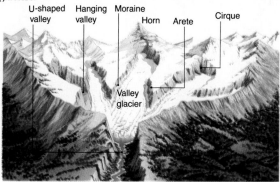

Figure 13–21. The action of alpine glaciers produces many distinct features (left). The famous Matterhorn in the Swiss Alps (right) is one of those features.

SECTION FOCUS

Section Overview This section describes the two major types of glaciers found on earth—alpine glaciers and continental glaciers. The main types of deposits formed by both glaciers are described and compared. Specific geographical areas in the world where these glacial deposits can be seen are also cited.

Section Objectives For a list of section objectives, see pupil's edition page.

New Science Terms There are no new science terms in this section.

SECTION DEVELOPMENT

13.6 Alpine Glaciers

DISCUSSION Tell the students that glaciers are important resources in many parts of the world. For example, many glaciers regulate the flow of large rivers. Glacial erosion produces spectacular U-shaped valleys and beautiful mountain scenery. The area surrounding the Matterhorn in Switzerland is a result of glacial erosion. In addition, the moraines from a glacier form rich, though rocky, soil that is good for farming. From this soil, records of past climates can be studied. ***Thinking Critically:*** Ask the students to infer why glaciers near populated areas are important resources. (The freshwater run-off from glaciers is important during dry periods.)

EXTENSION There are also other specific types of glaciers, including piedmont glaciers, polar, subpolar, and temperate glaciers. You may wish to have the students research these types of glaciers and report their findings.

BACKGROUND INFORMATION

In 1837, a young Swiss scientist named Louis Agassiz stunned colleagues by announcing his proposed Ice Age theory. Agassiz pointed out that the Jura Mountains in Switzerland were strewn with huge erratics that were completely unlike the limestone on which they rested. He argued that the erratics were carried to these locations by glaciers. To prove this, he pointed out the scratches on the exposed bedrock in the Jura. He claimed that the glaciers still present in the area had once extended far beyond their present locations. The scientific community was shocked by his assertions. At the time, it was generally believed that the boulders were transported by a great flood.

DISCUSSION The high position of hanging valleys is sometimes confusing. *Thinking Critically:* Ask the students to explain the location of these valleys and why they never erode to the bottom of the valley. (Tributary glaciers need to feed only into the top of the main valley glacier.)

EXTENSION (Tie-in/Geography) The largest of the Finger Lakes is Cayuga Lake. It is almost 65 km long and more than 120 m deep. Some of the fiords in Norway created by the moving glaciers are as much as 1200 m.

Figure 13–22. Many spectacular waterfalls, such as Yosemite Falls in California, are found in hanging valleys. Hanging valleys once contained tributary glaciers along the main valley glacier.

Figure 13–23. The Finger Lakes of New York (right) and the fiords of Alaska (top left), and Norway (bottom left) are valleys that once contained glaciers.

Along a glacial valley, small hanging valleys may be seen. These valleys form from tributary glaciers that joined the main valley glacier. They are much higher than the main valley and often contain small lakes drained by spectacular waterfalls. Yosemite Falls in California plunges 740 m from a hanging valley to the main valley below.

Some old glacial valleys are entirely filled with water, forming long, narrow "finger lakes." The Finger Lakes of upstate New York are old valleys, deepened and reshaped by glaciers. The fiords along the coasts of Norway and Greenland are also glacial valleys that were flooded by the rise in sea level after the glaciers melted.

○ *What is a horn?*
○ *What are hanging valleys?*

Answers–13.6

○ A horn is a mountain peak formed at the source of several glaciers.

○ Hanging valleys are formed by tributary glaciers that join the main valley glacier. They sit high along the main valley wall.

13.7 Continental Glaciers

Continental glaciers form long lines of terminal moraines at the glaciers' edges. These deposits are constantly moved and reworked by streams that meander back and forth across the outwash plain. As the glacier melts back, huge blocks of ice are sometimes isolated in the outwash plain. As deposits build around these blocks, the ice becomes buried. When the ice melts, a depression remains in the outwash. These gently sloping depressions, called *kettles*, fill with water, forming ponds and lakes. Thousands of kettle lakes can be found in Minnesota and Wisconsin.

The various piles of till left by continental glaciers are named for the shape of the deposit. Rounded mounds or hills of till are called *kames* (KAYMS). Kames are formed as melt water, traveling down through a crevasse in the glacier, piles up rock material at the bottom.

As melt water travels through crevasses and cracks in the glacier, it often develops a water channel within the glacier. Because melt water carries rocks, these channels become long, winding, rock-filled tubes. When the glacier melts, the rocks in these channels sink to the ground, leaving a winding, narrow ridge called an *esker*. Eskers can be found in the glaciated states of the Mississippi Valley, the north central states, New York,

DISCOVER

Finding Glacial Deposits

On a map of one of the states along the United States-Canada border, look for the names of places or landmarks that are related to glacier erosion and deposition. For example, in Wisconsin there is an area called Kettle Moraine, and in Pennsylvania there is Moraine State Park. Make a list of these names, and find out which state seems to have had the most glacial erosion and deposition.

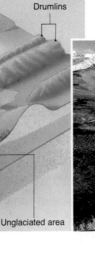

Figure 13–24. Kettle lakes, such as those shown below, are one of the distinct features of continental glaciers, diagrammed at left.

Kettle lakes — Kames — Drumlins — Eskers — Meltwater lake — Outwash plain — Terminal morraine — Unglaciated area

13.7 Continental Glaciers

DISCUSSION   The movement of a continental glacier differs greatly from that of an alpine glacier. ***Thinking Critically:*** Ask the students to explain the results of these different glacial movements. (Instead of constantly advancing as alpine glaciers do, continental glaciers follow a pattern of expanding, retreating, and then expanding again. Thus they leave many deposits with each retreat.)

EXTENSION   The records of continental glaciation are very complex, with several periods of glacial advances. In order from oldest to youngest, the periods are named Nebraskan Ice Sheet, Kansan, Illinoian, and Wisconsin.

EXTENSION   (Tie-in/Geography) Tell the students that the greatest amount of ice in the past one million years was located in North America. Ice stretched from the Arctic Islands south to the Hudson Bay and the Great Lakes, through the midwestern plains. It stretched west across the Yukon and western British Columbia to Puget Sound. You may wish to show this to students on a large map.

DISCOVER:
Finding Glacial Deposits

Skill (*Organizing*)

The students may work in small groups, each group studying a different region of the northern states. Each group may present its findings to the class.

BACKGROUND INFORMATION

Most glaciers move only a few centimeters per day and follow a normal pattern of advance and retreat in response to a slowly varying climate. There are glaciers in certain parts of the world, however, that show very different patterns of behavior. These include glaciers in areas of the Soviet Union, Iceland, and the Alaskan Range. These glaciers may, after a period of quiet lasting from ten to one hundred years, begin suddenly to move very fast—as much as six meters per hour. These types of glaciers are commonly called surging or galloping glaciers.

EXTENSION Although the advancing ice sheets during the last Ice Age usually followed river valleys, they sometimes broke away, plowing through ridges and damming rivers. The rivers then found new outlets that cut new valleys and drastically changed the drainage system. The effect was different, however, when ice sheets dammed river valley outlets. For example, a great torrent of water broke through the glacial dam in the Clark Fork River valley in the northwestern United States. The result was a greatly deepened river valley. Huge boulders were also formed that eroded and carved the Columbia Plateau as they were dragged through it.

SECTION REVIEW

Summary Review the section by having the students compare and contrast an alpine glacier with a continental glacier. Be sure they point out the different types of deposits from each glacier.

Reinforcement Have the students quiz each other on the important concepts in the sections.

A MATTER OF FACT

The famous Bunker Hill in Charlestown, Massachusetts, is a drumlin. Next to Bunker Hill is another drumlin, Breed's Hill. The Revolutionary War battle of Bunker Hill was actually fought on Breed's Hill.

and Maine. *Drumlins*, another type of continental glacier deposit, are long, narrow hills of unsorted material. One end of a drumlin, the side facing the glacier, is always steeper than the opposite end.

A continental glacier has a great effect on the drainage of the occupied region. Old river systems are blocked or buried, while depressions carved by the ice fill with water and form lakes. The most notable of these are the Great Lakes on the border between the United States and Canada.

○ *What are kettles?*
○ *What is the difference between a kame and an esker?*

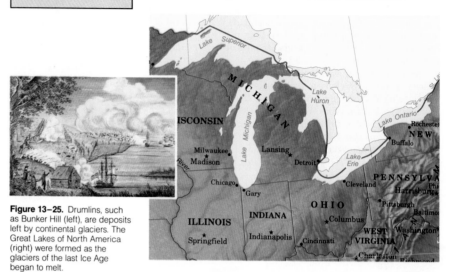

Figure 13–25. Drumlins, such as Bunker Hill (left), are deposits left by continental glaciers. The Great Lakes of North America (right) were formed as the glaciers of the last Ice Age began to melt.

Section Review

READING CRITICALLY

1. How are the landforms produced by alpine and continental glaciers different from each other?
2. Why are hanging valleys high above the main glacier valley?

THINKING CRITICALLY

3. What causes continental glaciers to leave so many different kinds of deposits?
4. Explain why Minnesota has kettle lakes, while New York has elongated "finger lakes."

Answers—13.7

○ Kettles are depressions formed when isolated blocks of glacial ice were buried in the outwash plain.

○ A kames is a rounded mound of till; an esker is a thin, winding ridge.

Answers to Section Review

Reading Critically

1. Alpine landforms such as horns and hanging valleys that are carved out of the rocks are not found in areas of continental glaciers. Continental glaciers show more evidence of advances and retreats. Thus, continental deposits often bury previous deposits.

2. Tributary glaciers join the main glacier at the top of the glacial ice. They do not need to erode to the bottom of the valley. Thus when the ice melts, the hanging valley sits high along the main valley wall.

continues

2. Research the sinking of the *Titanic*. Pretend you are a survivor of the event and write a story about the tragedy.

3. Research and write a report about how the modern windmill could become an economical source of electricity in some areas of the world.

4. After reading about the earth's last Ice Age, make a poster showing the earth as it might have looked with much of the high latitudes covered with ice. Be sure to show how the vast expanses of ice would have affected the level of the oceans.

APPLICATION/CRITICAL THINKING

1. How has the erosion and deposition by glaciers influenced the drainage pattern of a glaciated region, such as upstate New York?

2. Explain how the lack of soil in the western United States shows the effect of erosion by wind.

3. Geologists claim that Long Island, which is off the coasts of New York and Connecticut, is actually a terminal moraine left from the last Ice Age. What evidence would you look for on Long Island if you were trying to prove this theory? What type of evidence might you look for on the mainland to help prove this theory?

FOR FURTHER READING

Bailey, R. *Glaciers*. Alexandria, Virginia: Time-Life Books, 1982. This book describes in detail the formation of glaciers and their effects on landforms.

Page, J. *Arid Lands*. Alexandria, Virginia: Time-Life Books, 1982. This book describes the processes that create and maintain arid environments.

Radlauer, R. and L. S. Gitkin. *The Power of Ice*. Chicago: Childrens Press, 1985. The book describes the experiences of a teen-age research assistant who spent two summers on the glaciers in Alaska.

Watson, L. *Heaven's Breath: A Natural History of the Wind*. New York: Morrow, 1985. This book is a comprehensive explanation of the impact of wind on the earth.

Challenge Your Thinking

The following is R. A. Bagnold's description of a dune: "In places, vast accumulations of sand weighing millions of tons move inexorably, in regular formation over the surface of the country, growing, retaining their shape, even breeding, in a manner which, by its grotesque imitation of life, is vaguely disturbing to an imaginative mind." Study the photograph, and then explain what you think Bagnold means by this.

Chapter 13 Review   **297**

ANSWERS TO APPLICATION/ CRITICAL THINKING

1. Students should select a fairly small region to study, such as upstate New York, northern Minnesota, or one of the Great Lakes.
2. The widespread loess deposits extend far beyond the areas covered by ice sheets.
3. On Long Island you could look for unsorted material that has been transported a long distance. On the mainland you could look for striations on rocks.

ANSWER TO CHALLENGE YOUR THINKING

Answers will vary, but basically the shifting sand seems to have a mind of its own as it moves, grows, and changes form.

2. Stories will vary, but all should include information about the nature of iceberg movement in the North Atlantic.
3. Research will vary.
4. Compare the posters with a reliable source to check for accuracy.

Chapter 14: GEOMORPHOLOGY OF NORTH AMERICA

PLANNING THE CHAPTER

Chapter Sections	Page	Chapter Features	Page	Program Resources	Page
Section 1: Geomorphic Regions	299			Investigation 14.1: *Interpreting Rock Structure* **(A)** Investigation 14.2: *Separation of Soil Particles* **(A)**	**TRB 59** **LM 61** **TRB 61** **LM 63**
14.1 Geomorphology **(A)**	299	**Skill Activity:** Drawing Profiles **(A)**	300		
14.2 Geomorphic Provinces **(A)**	301	**Biographies:** Then and Now Theodore Roosevelt, Marjorie Stoneman Douglass	302		
		Discover: Using Words **(A)** Section Review	303 303		
Section 2: Provinces of North America	304			Critical Thinking **(H)** Cross-Discipline: Science and Geography, *Unraveling History* **(H)**	**TRB 26** **TRB 26**
14.3 The Western Provinces **(B)**	304	A Matter of Fact A Matter of Fact	305 306	Reading for Content: *Mapping Concepts* **(B)** Concept Extension: *Identifying Landforms* **(A)**	**TRB 25** **LM 199** **TRB 29** **LM 251**
14.4 The Alaskan and Hawaiian Provinces **(B)**	307	**Discover:** Identifying Geomorphic Provinces **(B)**	307		
14.5 The Plains and Lowland Provinces **(B)**	309				
14.6 The Appalachian Provinces **(B)**	310	Section Review **Activity:** Planning an Informational Brochure **(B)** **Investigation 14:** Correlating Features on Maps **(A)**	313 313 314	Student Record Book: Textbook Investigations **(A)**	**TRB 27**
Chapter 14 Review	315			Vocabulary **(A)** Tests **(A)** Computer Test Bank	**TRB 14** **LM 149** **TRB 56**

(LM) Laboratory Manual/Study Guide, **(TRB)** Teacher's ResourceBank™

B = Basic **A** = Average **H** = Honors

The coding Basic, Average, and Honors indicates sections or subsections that might be appropriate for different levels of learners. For additional suggestions regarding choice of topic and depth of coverage, see the Pacing Chart on pages T16–T20.

CHAPTER CONCEPTS, OBJECTIVES, AND TERMS

Section	Concepts	Objectives	Science Terms
Section 1: Geomorphic Regions	■ Rock structure, erosional agents, and time influence the geomorphology of an area. **(14.1, 14.2)** ■ North America is divided into regions called geomorphic provinces. **(14.2)**	■ **Identify** the factors that create the landforms of a geomorphic province. ■ **Locate** the major geomorphic provinces in North America.	geomorphology geomorphic province
Section 2: Provinces of North America	■ The Western Provinces of the United States are characterized by relatively young mountains with many folds, and faults, intrusive magma bodies, and extrusive volcanic formation. **(14.3)** ■ The Western Provinces include the Rocky Mountain Province, the Colorado Plateau, the Basin and Range Province, the Columbia-Snake River Plateau, the Pacific Mountain System, and the Pacific Border Province. **(14.3)** ■ Alaska is composed of many different provinces. **(14.4)** ■ The Hawaiian Islands are part of an arc of volcanic islands, many of which remain submerged. **(14.4)** ■ The Central Provinces are composed primarily of thick layers of undisturbed sedimentary rocks. **(14.5)** ■ The Plains and Lowland provinces include the Great Plains and Central Lowlands and the Coastal Plains. **(14.5)** ■ The eastern mountains of the United States are very old and highly eroded. **(14.6)** ■ The Appalachian Provinces include the Blue Ridge Mountains, the Piedmont Plateau, the Valley and Ridge Province, the Appalachian Plateau, the Ouachita and Ozark mountains, the New England Province, and the Canadian Shield. **(14.6)**	■ **Describe** the existing landforms of each province. ■ **Identify** national parks that show unique landforms.	There are no new science terms in this section.

Title	Page	Materials
Skill Activity: Drawing Profiles	300	*(per student)* graph paper, pencil
Discover: Using Words	303	*(per student)* paper, pencil
Discover: Identifying Geomorphic Provinces	307	*(per student)* paper, pencil
Activity: Planning an Informational Brochure	313	*(per group of 3 or 4)* paper, colored pencils, rulers, local topographic and geologic maps
Investigation 14: Correlating Features on Maps	314	*(per student)* geologic/topographic map

TEACHING SUGGESTIONS

Section 1: Geomorphic Regions

Demonstration: Landforms

Purpose

To characterize several landforms

Materials

Modeling clay
Small rock pieces

Procedure

Draw four different kinds of mountainous landforms on the chalkboard. Have the students complete a table by filling in the information required under the following two columns: *Distinguishing Characteristics* and *Method of Formation*. Then use modeling clay or some other medium and small pieces of rock to make models of the four landforms drawn on the chalkboard. If you want to make permanent models of the landforms, you can substitute self-hardening clay for the modeling clay.

Questions to Ask the Students

1. How are all of these landforms similar?
   (All have mountainous features.)
2. How are these landforms different?
   (They formed in different ways.)

Class Activity

Divide the class into three groups and assign each group to study one of the three major rock structures: mountains, plains, or plateaus. Have them examine the Geomorphic Province map on page 301 and report to the class about variations observed in their assigned structure. They should notice that there are directional trends to the mountains. They should also notice that there are mountains in the plains provinces as well as plains and plateaus nestled in the mountain provinces.

Outside Speaker

Have a naturalist from a local park visit the class and discuss the topographic and geologic features that make the park unique.

Section 2: Provinces of North America

Class Activity

Have the students build a model of one of the national parks. Show how the erosional features are a reflection of the type of structure underlying the area.

Field Trip

If you are near a county, state, or national park, take a field trip to the interpretive center of the park; then observe the landforms that make the park unique.

CHAPTER

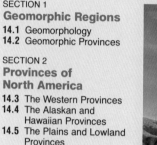

14

Geomorphology of North America

CHAPTER OVERVIEW

This chapter presents an overview of the geomorphology of the United States. Geomorphic provinces are defined. The various geomorphic provinces of the United States are then described in detail.

Section 1: Geomorphic Regions   In this section, the factors that influence the geomorphology of an area are discussed. Geomorphic province is defined, and the three major types of rock structures are identified.

Section 2: Provinces of North America This section characterizes many of the geomorphic regions that comprise the North American continent. The six regions are subdivided into geomorphic provinces, all of which are described in some detail. National parks that can be found in each province are also included.

This shoreline, like all continental features, is the result of a struggle among geologic processes. Folding, faulting, uplifting, and orogeny build landforms, while the agents of weathering and erosion tear them down. The primary agents of erosion for this feature are the ocean waves.

The northern California coast

298

CHAPTER MOTIVATING ACTIVITY

This chapter is ideal for using a "show and tell" format. You may want to begin showing segments of travel videos of North America in class. If you have traveled within the United States, showing the students slides or photographs from your trip would be helpful. The first-hand descriptions of the areas visited could prove quite interesting. You may request that the students bring in their own slides and photographs to share with the class. If you have no slides of your own, many libraries have slides that can be borrowed. Making a display of United States topographic maps may prove beneficial in understanding geomorphology.

1 Geomorphic Regions

SECTION OBJECTIVES

After completing this section, you should be able to:

- **Identify** the factors that create the landforms of a geomorphic province.
- **Locate** the major geomorphic provinces in North America.

NEW SCIENCE TERMS

geomorphology
geomorphic province

14.1 Geomorphology

The word *geomorphology* sounds like a tongue twister, but it is made of three simple parts. *Geo* means "the earth." You have already been introduced to the word *geology*, which is "the study of the earth." The next part of *geomorphology* is *morph*, which means "form." Another word using this stem is *metamorphic*, which means "a change in form." You may recall that metamorphic rocks change their form. The final part of *geomorphology* is the ending *logy*. This ending is attached to many words and always means "the study of." For example, biology is "the study of living things." **Geomorphology** is the study of the landforms of the earth.

When you look at a photograph of the area in which you live, you may recognize a specific topographic pattern. This pattern is the result of a rock structure that has been eroded to a recognizable landform. Each region of the world has certain recognizable characteristics that set it apart from every other region.

○ *What is geomorphology?*

Figure 14–1. Gently rolling hills (left) and rugged mountains (right) are just two of the varied landforms that may be found in North America.

Answer–14.1

○ Geomorphology is the study of the landforms on the earth's surface.

DEMONSTRATION

For a demonstration of landforms see page 297c preceding this chapter.

Section **1**: Geomorphic Regions

SECTION FOCUS

Section Overview   This section first defines geomorphology. The three categories of geomorphic provinces are described. The differences that distinguish regions within categories are then explained.

Section Objectives   For a list of section objectives, see pupil's edition page.

New Science Terms   For a list of new science terms in this section, see pupil's edition page. You may wish to familiarize students with these terms before they read the section. You may also wish to read the terms aloud so the students are familiar with the pronunciation of each.

SECTION DEVELOPMENT

14.1 Geomorphology

DISCUSSION   Emphasize to the students that all features of the earth's surface are the result of geologic processes, such as erosion and deposition. Point out that regions, both small and large can be classified by their geologic features. Have the students take a walk around the schoolyard or a local park. Ask them to classify different areas of the yard or park by the geologic features they can identify. For example, they may classify sections as hills, depressions, flat ground, or high rocks. Accept all reasonable classifications.

THINKING SKILL   *(Applying Information)*   Ask the students to use the description of the term geomorphology to speculate about the type of details that would be studied by a geomorphologist. (A geomorphologist studies faults, folds, and other structures and features that shape the land.)

Objectives

- Draw a profile of a geologic area.
- Interpret a topographic map.
- Measure map distances.
- Draw conclusions about how profiles are used to represent landscapes.

Discussion Explain to the students that map profiles serve many purposes. The students will know from their skill activity that profiles are used to show shape of the land accurately. Ask the students to suppose that geologists knew the surface rock-type pattern as well as the dip of each layer. *Thinking Critically:* Ask the students what the geologists could draw, using a profile and this information. (a profile map of the underlying rock structure; knowing where the rock layer lies at the surface and the tilt of the rock layer, the continuation of that layer underground could be drawn)

Answer to Application

Drawings will vary but should accurately represent the profile from A to B.

Answers to Using What You Have Learned

1. Answers will vary.
2. The slopes are steepest where the lines are closest together. Steep slopes rapidly cross elevations. Therefore, the lines are closer together.

300

BACKGROUND

When analyzing the geomorphology of an area, you need to be able to interpret both the rock structure underlying the area and the resulting shape of the land surface. Sometimes geologists draw a diagram called a *profile* to show accurately the shape of the land surface. A profile is a representation of an object as seen from the side. A profile of an area of land is like looking at hills outlined against the horizon. The contour lines on a topographic map are used to draw a profile.

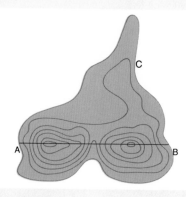

PROCEDURE

1. Identify the area to be profiled. The first step in drawing a profile is to draw a line across the map to show where the profile is to be taken. On the map, the line marked AB is drawn so that it crosses the peaks of two hills. Near the center of a piece of graph paper, draw a horizontal line. Fold your paper along this line, lay the fold along line AB on the map, and mark the ends of the line on your graph paper with dots labeled A and B.
2. Determine the vertical and horizontal scales. The horizontal scale is determined by the scale of the map. The scale of the map in this exercise is 1:20 000. The vertical scale is your choice; however, if you use a vertical scale that is the same as the horizontal scale, the slopes in your profile will be proportional to

---

the actual slopes of the land. On your graph paper, draw a vertical line upward from dot A. Make each line along the vertical a 20-foot contour interval starting with 0.

3. To mark the positions of the contour lines, place your line AB along line AB on the map. Mark on the horizontal where each contour line crosses line AB. Carefully label each mark with the elevation listed below the line.
4. Draw the profile. Find the mark that represents the position where the 20-foot contour line crossed your line AB. Directly above this mark, make a dot on the 20-foot line of your vertical scale. Mark the elevations of other contour lines in the same fashion. Connect the points on your graph paper with a smooth, curving line. You have now drawn a profile along line AB.

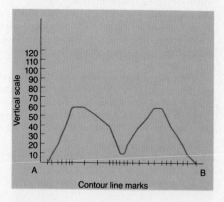

Contour line marks

APPLICATION

Follow the same procedure and make a profile of the area between points C and A.

USING WHAT YOU HAVE LEARNED

Compare your profile to the contour-line pattern on the map.

1. Does it look as you expected? Explain.
2. Are the slopes steepest where the lines are closest together? Explain.

14.2 Geomorphic Provinces

North America is divided into regions called *geomorphic provinces*, which are given specific names. A **geomorphic province** is an area with a distinctive pattern of landforms. Sometimes a province covers an area that is hundreds of kilometers wide.

Geomorphic provinces can be divided into one of three categories: mountains, plateaus, or plains. Several examples of each of these provinces are found in North America. However, each example has some distinctive features that make it different from the others. The distinctive features that modify the provinces are rock type, folds, fractures, and faults. These features are then sculptured by climate, weathering agents, and erosion to produce the present landform.

Figure 14–2. This physical map of North America shows several large geomorphic regions. Each region is divided into geomorphic provinces, each with its own distinct features.

Figure 14–3. All rock structures can be divided into plateaus (left), plains (center), or mountains (right).

14.2 Geomorphic Provinces

SKILL *(Observing)* Have the students observe the three photographs in Figure 14–3. Request that they describe the features of the three province types that distinguish one from the other. (Answers will vary, but the students should note that plains are completely flat; plateaus, though elevated, are completely flat at the top; mountains are much higher than the surrounding landscape and have peaks.)

EXTENSION You may want to give the students the specific definition of each structure discussed in the text. Tell them that plains are relatively flat or gently sloping areas that are almost at sea level. Plateaus are relatively flat areas that are usually higher than 600 m above sea level. Mountains are landforms that usually reach at least 600 m above the surrounding land. They have relatively steep slopes and narrow summits.

DISCUSSION Give the students some time to study the geomorphic provinces. Ask them to estimate the percentage of land that can be classified as mountains, plains, and plateaus. Then have the students classify the province in which they live.

BIOGRAPHIES: Then and Now

Discussion Have the students read the two biographies and discuss how the contributions of each person helped to further our understanding and appreciation of our natural resources and the environment.

Thinking Skill *(Relating)* Emphasize that, as the lives of President Roosevelt and Ms. Douglass illustrate, individuals do not have to be scientists in order to do something that can help preserve the environment. Ask the students what they can do to keep our national parks (and non-park areas) from being destroyed.

Extension Interested students might be encouraged to research further the life and work of Roosevelt or Douglass or one of the people listed below.

Frederick Law Olmsted

John Muir

Stephen Mather

John James Audubon

BIOGRAPHIES: Then and Now

THEODORE ROOSEVELT (1858–1919)

Theodore Roosevelt was the twenty-sixth President of the United States. One of his many triumphs as president was in the area of conservation of the country's natural resources.

When Roosevelt was elected president, private interests were in control of most of the natural resources in the eastern part of the United States. Mineral rights, timber, waterways, and many other resources had been sold at prices below market value. There were few safeguards established to protect recreational sites and to guarantee the replenishment of renewable resources.

During Roosevelt's administration, a conservation program was developed to protect the vast western riches. Government-supervised programs for water and land usage were established. Public land was leased rather than sold for grazing and timber. Leasing contracts required the timberland to be replanted after cutting. Safeguards also were provided against overgrazing.

Roosevelt's focus on conservation and natural resources paved the way for the establishment of many national and state parks. The Theodore Roosevelt National Park was established in North Dakota in 1978. This section of sculptured badlands, petrified logs, and low-grade coal seams was described by Roosevelt as "barren, fantastic, and grimly picturesque."

MARJORIE STONEMAN DOUGLASS (1890–)

Marjorie Stoneman Douglass was born on April 7, 1890, in Minneapolis, Minnesota. She attended Wellesley College in Wellesley, Massachusetts, where she graduated in 1912 with a bachelor's degree. One of her first jobs was with the *Miami Herald* in Miami, Florida. She was a reporter and editor, and it was during this time that she learned first-hand about the beautiful and mysterious Everglades that lies just to the west of Miami.

After leaving the *Herald*, Ms. Douglass became an instructor at the University of Miami. While at the university she wrote and published many short stories for *The Saturday Evening Post*. Although she was never trained as a scientist, many of her stories were about the land she loved—the Everglades.

The Everglades is not a forbidding swamp as many people think; instead, it is a shallow river filled with tall grass. This river is over 100 km wide, but only about 10 to 15 cm deep, flowing southeast from Lake Okeechobee to the Atlantic Ocean.

With the publication of her novel, *The Everglades: River of Grass*, Ms. Douglass became a proponent of preserving the Everglades and was instrumental in the establishment and preservation of Everglades National Park. She has won many wildlife awards, and because of her work, the National Park Service established a special award in conservation and named it in her honor.

Answers—11.2 (p. 303)

○ Geomorphic provinces are areas with distinctive rock structures.

○ The three basic categories of rock structure are mountains, plains, and plateaus.

The age of the rocks is also important in shaping a province. In general, the older the rocks, the greater the effects of weathering and erosion. For instance, because of its recent eruption, Mount St. Helens has very few erosional features. On the other hand, all that is left of ancient mountain ranges in central and eastern Canada is their nearly level granite cores.

The overall pattern of rock structure in North America consists of relatively flat plains along the eastern and southern coasts, several mountain ranges running north-south, and a wide plain in the central part of the continent. The uniqueness of a region is quite often highlighted in areas set aside as national parks. As you study the different geomorphic provinces in this chapter, you will be introduced to some of these parks.

○ *What are geomorphic provinces?*
○ *What are the three basic categories of rock structure?*

DISCOVER

Using Words

See how many words you can write that have the prefix *geo*. Now see how many words you can write that have the stem *morph*. Finally, see how many words you can write that have the suffix *logy*. Compare your lists with those of your classmates.

Figure 14–4. The erosion of Mount St. Helens (left) is minimal, because the mountain is young. The erosion of the Canadian Shield (right) is extensive, because the mountains are millions of years old.

Section Review

READING CRITICALLY

1. What factors determine the shape of a region?
2. How does the age of a rock structure affect its shape?

THINKING CRITICALLY

3. Why are national parks often good places to study the landforms of a region?
4. Using your knowledge of plate tectonics, explain why the mountain ranges in North America run in a north-south direction.

Section 1   Geomorphic Regions   **303**

Answers to Section Review

Reading Critically

1. The age of the rock structure, weathering, and forces within the earth determine the shape of a region.
2. The older a rock structure is, the more weathered and eroded it is.

Thinking Critically

3. Examples of special and unique landforms that characterize a province are often highlighted at national parks.
4. The North American plate moved eastward and crashed into the Eurasian plate. This caused the eastern edge of the continent to fold and form the Appalachian Mountains. The mountains run north-south because the stress was compression from the east and the west. Similar forces formed the Rocky Mountains as the North American plate moved westward and collided with the Pacific plate.

DISCOVER: Using Words

Thinking Skill (*Comprehending Ideas*)

Students should be able to list words such as the following with a prefix, stem, or suffix: geography, geomorphic, metamorphic, geology, metamorphosed, geometry, and biology.

SKILL (*Communicating*) Divide the class into three groups and assign each group one of the three major provinces to study. Have the students examine the map of geomorphic provinces on p. 301. They should then make oral reports to the class about variations that are observable in their particular province. The students will most probably notice that the mountains have a directional trend. (north-south) They also will probably observe that mountains are found within plains regions and that plains and plateaus are nestled in the mountain regions.

THINKING SKILL (*Analyzing Data*) Have the students view the two photographs in Figure 14–4. Ask them to describe the differences between these two mountains. (The mountain on the left is high above the surrounding area, and it has steep, angular slopes and sharp peaks. The mountain on the right is low and smooth with gentle slopes and flattened at the top.) Ask the students to explain why these differences exist. (The mountain on the right is older than the one on the left and thus has been subjected to the effects of weathering and erosion for a longer time.)

SECTION REVIEW

Summary Have the students identify the factors that shape landforms and describe the effect of each factor. Then have them describe the type of geomorphic province found in specific areas of North America.

Reinforcement Encourage the students to describe the geomorphic province or provinces that their state lies in.

SECTION FOCUS

Section Overview This section concentrates on the variations in landforms throughout the North American continent. Region by region, each distinctive feature of the landforms that define the geomorphic provinces is described. The forces and agents that were responsible for shaping each province are identified. Also described are national parks in which examples of the provinces' characteristic landforms can be seen.

Section Objectives For a list of section objectives, see pupil's edition page.

New Science Terms There are no new science terms in this section.

SECTION DEVELOPMENT

14.3 The Western Provinces

DISCUSSION The students may be confused about the distinction between mountain ranges, mountain chains, and mountain belts. Explain that mountain ranges are series of parallel mountains that are similar in structure and shape. A mountain chain is a group of adjacent mountain ranges. A mountain belt is a larger group of many interconnected mountain chains. In fact, most mountains are part of two world-wide mountain belts. One, the circum-Pacific belt circles the edge of the Pacific Ocean. The other, the Eurasian-Melanesian belt runs through southern Asia and Europe and into northwestern Africa.

NEW SCIENCE TERMS
There are no new science terms in this section.

SECTION OBJECTIVES
After completing this section, you should be able to:
■ **Describe** the existing landforms of each province.
■ **Identify** national parks that show unique landforms.

Figure 14–5. This map shows the region of the western geomorphic provinces.

14.3 The Western Provinces

The western provinces of the United States are primarily mountainous. In general, the mountain-building events started along what is now the eastern edge of the Rocky Mountain Province and moved westward. The western mountains have had a great deal of volcanic activity. On the eastern side of the mountains, there was much folding and faulting, and many intrusions formed.

The Rocky Mountain Province The Rocky Mountains are about 4500 km long and are part of a mountain chain extending from Alaska to the tip of South America. The summits of most of the mountains are over 1800 m above their bases. The rugged topography throughout the Rocky Mountains indicates that it is a young mountain range. Between the mountain peaks are U-shaped valleys carved by alpine glaciers.

The structure of the Rocky Mountains is both varied and complex. The southern part of the mountain chain has many

Figure 14–6. The Rocky Mountains contain landforms created by glacial erosion.

304   Chapter 14   Geomorphology of North America

anticlines, batholiths, and volcanic formations. The middle part of the chain has lower mountain ranges that were formed by movement along faults. The northern part of the chain is made of thick sedimentary rocks and huge batholiths moved by thrust faults.

One of the parks in the southern part of the province is Rocky Mountain National Park in Colorado. The mountain peaks in this park are formed mostly of gneiss and granite.

The Colorado Plateau   The vast Colorado Plateau is an arid area characterized by deep canyons, buttes, and mesas. The upheavals that created the mountainous areas surrounding the Colorado Plateau had little effect on this region. The only change was that the region was raised just over one kilometer higher than it was before the mountains formed.

The Colorado Plateau has more national parks than any other province. Among the best known are Zion National Park and Bryce Canyon National Park. Most show interesting erosional features where wind and water have carved the brightly colored sedimentary rocks.

The Basin and Range Province   The Basin and Range Province includes a large area extending from southern Oregon southward into Mexico. The predominant structure in the Basin and Range Province is block faulting. The area broke into large blocks that tilted westward like a stack of fallen dominoes.

Several very well-known parks can be found in this province, including Death Valley National Monument of California and Nevada. Death Valley, located at the bottom of a fault block, is the lowest, hottest, and driest of all the national parks.

Figure 14–7. The western provinces include the Colorado Plateau (left), which was created by uplifting, and the Basin and Range Province. Death Valley National Monument (right) is located in a basin between the mountain ranges of this province.

> **A MATTER OF FACT**
>
> Eleven hundred species of insects have been identified in the fossils found in the volcanic ash formations in Florissant Fossil Beds National Monument in Colorado.

Section 2   Provinces of North America   **305**

DISCUSSION   The height of a mountain is measured as its elevation above sea level rather than its height from base to summit. Thus a mountain with its base below sea level might not be measured as being very high, yet it would look higher than its measurement would indicate. On the other hand, a mountain with its base at a high elevation would measure very high, yet it would look lower than its measurement would indicate. For example, the summit of Pikes Peak in Colorado is said to be 4000 m high, or 4000 m above sea level. Actually, from base to summit, it is only 2700 m.

EXTENSION   Grand Canyon National Park is located on the Colorado Plateau. Here rocks that formed nearly 2 billion years ago can be found at the bottom of the canyon. If it were possible to stack the rocks of the Grand Canyon, Zion Canyon, and Bryce Canyon together, a nearly continuous record of the geologic history of the western United States could be observed. There are rocks in the Grand Canyon that date from the Precambrian Era through the Paleozoic Era. There are rocks in Zion Canyon that date from the Mesozoic Era, and in Bryce Canyon the rocks date from the Cenozoic Era.

BACKGROUND INFORMATION

You may want to give the students a brief geologic history of the Rocky Mountains. They began as a sediment-filled trough 150 million years ago. These sediments that hardened to rock were laid down by an inland sea that once covered the area where the Rockies are now located. Eventually the sea retreated as the area was uplifted, slowly folding the sedimentary rocks, and raising it as high as 6000 m. These folded mountains were worn down to low hills over 35 million years. Then 25 million years ago, the area was once more uplifted and the cycle of erosion started again, leaving the granite core of the original mountains that we see today.

EXTENSION You may want to tell the students about Thousand Springs, Idaho, on the Columbia Plateau. Natural springs gush from between rock layers in canyon walls and pour down the cliff walls to the Snake River below. It is thought that the source of these springs is the Lost River, which flows beneath the lava beds northeast of Craters of the Moon National Park, and never resurfaces.

DISCUSSION Yosemite National Park has six of the ten highest waterfalls in the United States. You may want to reinforce the lesson on glacial erosion from Chapter 13 by asking the students how these waterfalls probably formed. (Tributary glaciers that joined the main glacier carve small valleys called *hanging valleys* above the main glacial valley. Often glacial lakes drain through the hanging valley forming the waterfalls.) The highest of the six falls is Yosemite Falls at 739.14 m.

A MATTER OF FACT

During the hot summers in Idaho, children sometimes go sledding on the ice in one of the caves at Craters of the Moon National Monument. They must wear hard hats since the ceiling of the cave is only one meter above the ice.

Figure 14–8. Craters of the Moon National Monument (below) is located on the Columbia-Snake River Plateau. Yosemite National Park (inset) is located in the Pacific Mountain System.

The Columbia-Snake River Plateau   This province, in Oregon and Washington, contains a variety of plains, hills, mountains, and plateaus. The predominant feature, however, is the extensive lava flow that blankets the province. The lava must have had a low viscosity, because it covered wide areas before solidifying. In some places, the lava deposits are more than 3000 m thick.

Although the region is mostly covered by lava flows, a few small volcanic cones do occur. The Craters of the Moon National Monument in Idaho is one such site. In this location, small cinder cones stand starkly on top of rugged blankets of lavas.

The Pacific Mountain System   The Pacific Mountain System forms a continuous line of young mountains paralleling the western coast of the United States. The volcanic origin of these mountains is evident, as some of the volcanoes are still active. This line of volcanic cones, known as the *Cascades*, is built on an eroded volcanic plateau. The Sierra Nevada, also in this province, formed from a single fault block.

The beautiful landscapes of the fault block and volcanic mountains may be admired in several national parks in this province, including Yosemite National Park in California, Crater Lake National Park in Oregon, and Mount Rainier National Park in Washington.

Figure 14–9. These giant redwood trees are located in Redwood National Park, part of the Pacific Border Province.

The Pacific Border Province The Pacific Border Province extends along the western edge of North America. The southern part of the province consists of the coast ranges and lowlands east of the mountains. The lowlands of this area, such as the Great Valley of California, are formed by faults. Earthquakes are frequent throughout this province as the western edge of southern California continues to move along the San Andreas fault.

In this province is Redwood National Park, which not only shows off the majestic redwoods but also provides beaches, tide pools, sea cliffs, and beach sand dunes.

○ *How did the Rocky Mountains form?*
○ *What evidence is there that the Pacific coastal mountains are volcanic?*

14.4 The Alaskan and Hawaiian Provinces

Alaska is so large that it actually contains many provinces within its borders. The southeastern sliver of Alaska is a mountainous coastal region with many islands and fiords. The Alaska Peninsula and the Aleutian Islands form a volcanic arc, and the Yukon Basin is a plateau and lowland area drained by the Yukon River.

> **DISCOVER**
>
> *Identifying Geomorphic Provinces*
>
> Make a list of ten states, provinces, or territories in North America. Then, using the maps in this chapter, indicate the geomorphic province or provinces in which each lies.

Answers–14.3

○ The Rocky Mountains were formed by folding, faulting, intrusions, and volcanic activity.

○ The mountains are volcanic cones. Some of the volcanic mountains are still active.

THINKING SKILL *(Analyzing)* Emphasize that the Aleutians are an island arc. Have the students study the map of Alaska on p. 308 and explain why the Aleutians are called an arc. Ask the students to explain the formation of island arcs. (When two oceanic plates collide, one of the plates is subducted beneath the other. As the subducting crust melts, some melted material rises and erupts through the other plate, forming volcanic mountains.)

EXTENSION Have the students build three-dimensional models of the landforms found in the national parks. These models should show how the erosional features reflect the type of underlying rock structure. The students should label surface features and make accompanying topographic maps, using the standard symbols. Suggest that they begin by tracing road maps of the area to get the proper shape of the land and the distance scale.

> **DISCOVER:**
> Identifying Geomorphic Provinces

Thinking Skill *(Observing)*

Students should be able to determine and describe the physical features of the province or provinces in which their state is located.

14.4 The Alaskan and Hawaiian Provinces

SKILL *(Classifying)* Distribute copies of the map in Figure 14–10 and ask the students to study them carefully. Ask them to use information in Section 14.4, as well as their interpretation of the map, to draw the boundaries of the various provinces of Alaska.

THINKING SKILL *(Inferring)* Ask the students to make a visual comparison of the coast lines shown in the two photographs in Figure 14–11. Have them record their observations of the two landforms. Then ask them to identify the agents that are mainly responsible for the shaping of these landforms. (The coast shown on the left appears low and smoothed and was probably shaped by glaciers. The coast on the right is made of dark jagged rocks, probably of volcanic origin.)

Figure 14–10. This map shows the Alaskan provinces. Alaska is so large that it contains several provinces, including the Alaska Range (below)

South central Alaska has a coastal strip leading to a band of mountains known as the Alaska Range. The Brooks Range in northern Alaska is roughly parallel to the Alaska Range. The most northern area is a gently sloping plain called the *Arctic Slope*, which borders the Arctic Ocean. Mount McKinley, in the Alaska Range, is the highest mountain in North America. The peak is nearly 6200 m high and has many glaciers.

Figure 14–11. The Arctic Slope (left) and the Alaskan coast near Sitka National Park (right) are very different from the mountainous provinces in the center of the state.

The Hawaiian Islands have developed over a hot spot below the Pacific plate. The most northwestern volcanoes are the oldest and have been weathered down to *atolls*, or round coral islands, and guyots.

The southernmost volcanoes are actively building up the present islands, and southeast of the island of Hawaii are submerged volcanoes that will eventually reach the ocean surface. These islands are all shield volcanoes with characteristic gentle slopes and relatively quiet eruptions. A great deal of the scientific knowledge about volcanoes has come from research done in Hawaii Volcanoes National Park, which is on the island of Hawaii.

○ *Describe the provinces of Alaska.*
○ *What are atolls?*

Figure 14–12. The Hawaiian Islands (above) comprise a province of their own. The structure of mid-plate volcanoes can be studied at Hawaii Volcanoes National Park (left).

14.5 The Plains and Lowland Provinces

In the heart of the continent are the great plains and prairies that stretch from the Arctic Circle nearly to the Tropic of Cancer. The lowland provinces of North America are primarily located along the Atlantic and Gulf coasts. Although these regions all seem the same, there are many subtle differences that make each one unique.

The Great Plains and Central Lowlands   The Great Plains and the Central Lowlands extend from the Rocky Mountains to the eastern mountains, and from the arctic to the Gulf of Mexico. They form a vast plain composed mostly of sedimentary rocks that lie relatively undisturbed by structural events.

Three notable interruptions to the nearly horizontal orientation of the rocks are the Black Hills of South Dakota, the Central Mineral Region of Texas, and the Baraboo Dome of Wisconsin. Each of these locations is a dome formed as granite rocks were forced upward by intrusions into the crust.

Many of the parks in the plains states are small state-owned areas showing off lovely streams or glacial features, such as the Kettle Moraine area in Wisconsin.

Figure 14–13. The famous Wisconsin Dells are located in a glaciated area of the Great Plains.

Section 2   Provinces of North America   **309**

THINKING SKILL *(Inferring)* Remind the students that the youngest Hawaiian island is to the southeast of the oldest island. By studying Figure 14–12, the students will be able to see that the islands are in a relatively straight line. Ask the students to infer the direction in which the Pacific Plate, on which Hawaii rides, is moving over the hotspot. (It is moving in a northwesterly direction.)

14.5 The Plains and Lowland Provinces

DISCUSSION Draw the students' attention to distinguishing characteristics of the Great Plains and Central Lowlands. The Great Plains consist of broad river valleys and very low plateaus; the plains consist of weak sedimentary rocks. The Central Lowlands are made of sedimentary rocks overlain by their glacial deposits which form low hills in some areas. Although the plains-lowland region appears nearly horizontal, the strata rises gradually toward the Rocky Mountains, reaching an altitude of 3000 m above sea level.

Answers–14.4

○ There are mountain chains, a volcanic arc, a plateau, and a plain.

○ Atolls are volcanic islands that have weathered and eroded down below sea level. They are often topped with coral reefs.

DISCUSSION The students may be interested in the subtle differences between interior and coastal plains. Interior plains, such as the Great Plains, are slightly more elevated than coastal plains. Also sediments of the coastal plains are deposited by the shallow ocean water. Sediments on interior plains come from the erosion of the surrounding mountains and plateaus.

EXTENSION (Tie-in/Language Arts) Assign each student an imaginary cross-country trip, giving each an exact route. After doing some research, have the students write essays describing the landforms and other geologic features they have observed on their "trip." You may want to recommend that they write to the visitors' information center of the particular states to which they will be "traveling."

14.6 The Appalachian Provinces

EXTENSION Explain to the students the origin of the name "Great Smoky Mountains." The mountains are covered with a thick vegetation. This vegetation releases water that rises up the mountain and condenses as it cools. The condensation forms a thick bluish fog that sometimes completely obscures the mountains' slopes. Explain to the students that the Smokies are the oldest surviving mountain range in the United States. Some of the sedimentary rocks found there were formed over 500 million years ago. Show the students photographs of the mountains and ask them to identify features that indicate the age of the mountains. (The mountains appear to be smooth, rolling hills.)

Figure 14–14. Many karst features, such as these caverns, occur on the Coastal Plains.

The Coastal Plains The Coastal Plains, formed of gently sloping layers of sedimentary rock, are located along the Atlantic Ocean and the Gulf of Mexico. These young layers of rock blanket the older rock structures of the interior.

In this region there are many excellent examples of karst topography, with sink holes, underground rivers, and many cave systems. For instance, Florida has karst topography because of its limestone bedrock, but much of the state was underwater about 100 000 years ago. For this reason, Florida is often referred to as "the land from the sea."

○ *Name one of the dome structures of the Central Lowlands.*
○ *What type of rock is found on the Coastal Plains?*

Figure 14–15. This map of the eastern half of North America shows the location of the plains and central lowlands and the Appalachian provinces.

14.6 The Appalachian Provinces

Scientists believe that about 225 million years ago compressional forces folded and raised the area west of the Atlantic coast, forming the Appalachian Mountains. Since that time, these mountains have been deeply eroded, supplying vast amounts of sediments to the Coastal Plains and regions between the mountains.

Answers–14.5

○ Students may choose one of the following: Black Hills, Baraboo Dome, or Central Mineral Region of Texas.

○ Sedimentary rocks are found on the Coastal Plains.

Figure 14–16. Many small farms are located in the valleys among the ridges of the Blue Ridge Mountains.

The Blue Ridge Mountains   The Blue Ridge Mountains extend from Virginia to Tennessee. Although this area must once have been of very high elevation, it has been extensively eroded. All that remain of the original rocks in this province are very old granites, with a few slightly metamorphosed sandstones and conglomerates.

There are two beautiful national parks in the Blue Ridge Mountains. Shenandoah National Park is located in northern Virginia, and Great Smoky Mountains National Park is in North Carolina and Tennessee. Both provide beautiful landscapes as well as important historical details about the westward growth of the United States.

The Piedmont Plateau   The rocks of the Piedmont Plateau, east of the Blue Ridge Mountains, are a complex mixture of metamorphic rocks, such as slates and marbles. The structure shows both faulting and folding, but the area has been eroded to such an extent that the underlying rock structure is not reflected in the plateau.

The Valley and Ridge Province   The Valley and Ridge Province of northern Virginia, Maryland, and eastern Pennsylvania provides a classic example of folded mountains that have been extensively eroded. The accordion-like ridges and valleys run in long, narrow lines along a northeast-to-southwest axis. The ridge lines are composed of the most resistant rocks; the valleys the most easily eroded rocks.

Figure 14–17. Hardwood forests (right) are dominant in the Valley and Ridge Province, which stretches from northern Virginia to eastern Pennsylvania (left).

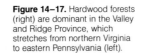

DISCUSSION   To point out the importance of setting aside national parks, tell the students about the denudation of the Blue Ridge Mountains in what is now Shenandoah National Park. The earliest settlers in the eighteenth century commonly burned the forests to clear the land so that they could plant cash crops such as blueberries. Thus, when the park was set aside for preservation in 1936, many of the slopes were bare. Trees were planted, and after 40 years the young forest covered the land. Ask the students to discuss the benefits of setting aside land for national parks. (The natural environment is saved from destruction.)

Figure 14–18. Many rivers erode the landscape of the Appalachian Plateau (right). Hot Springs, Arkansas (below), is located in the Ouachita Mountains.

THINKING SKILL *(Communicating)* Describe again the major types of mountains: folded, fault-block, volcanic, and dome. Make a list of each of the mountain ranges discussed in this chapter and assign one to each student or small group of students. Ask them to research their assigned mountains to learn which type of mountains they are. Emphasize that some may be complex systems of ranges with several types of montains combined. Have the students give oral reports on the geologic history of the mountains. Encourage them to use illustrations to clarify the information.

EXTENSION Tell students about the White Mountain National Forest in New Hampshire, where the effects of glacial erosion such as deep glacial valleys and cirques are apparent. There are also deep gorges, such as Dixville Notch and Franconia Notch, carved by fast-flowing streams.

The Appalachian Plateau   The Appalachian Plateau is a large province to the west of the Valley and Ridge Province. The Appalachian Plateau includes an area equal to the Piedmont, Blue Ridge Mountains, and Valley and Ridge provinces combined. Its rock layers are nearly horizontal or only slightly folded. The plateau's surface dips westward, forming a cliff along its eastern edge that overlooks the neighboring Valley and Ridge Province.

The Ouachita and Ozark Mountains   An extension of the Appalachian Provinces occurs in Oklahoma and Arkansas. The rock structures of the Ouachita (WAHSH uh taw) and Ozark mountains of these states are predominantly tightly folded sediments. The entire area has been thrust northward for a distance of at least 30 km, which probably occurred at the same time as the folding and faulting of the Appalachian Mountains.

Continental glaciers did not reach as far south as the Ouachitas or the Ozarks. Instead, their topography is due to differential weathering, mass wasting, and stream erosion.

Hot Springs National Park is located in the Ouachita Mountains. The spring water in this area is unique since it does not have an offensive odor or taste like most other hot springs.

The New England Province   The rock structure of the New England Province is similar to that of the Piedmont Plateau; however, its topography is very different. There are high mountains, such as the Green Mountains of Vermont, made of very old gneiss. Other mountains in this province are large granite intrusions. The glaciers of the Ice Ages carved and recarved the New England Province, rounding the highest ridges and gouging out river valleys. Glacial erratics and till can be found everywhere.

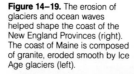

Figure 14–19. The erosion of glaciers and ocean waves helped shape the coast of the New England Provinces (right). The coast of Maine is composed of granite, eroded smooth by Ice Age glaciers (left).

Answers—14.6 (p. 313)

○ They are very old and greatly eroded.

○ The valleys are underlain by easily eroded formations; the ridges are formed of more resistant rocks.

○ Acadia National Park contains evidence of both sea and glacial erosion.

Acadia National Park of Maine is located in this province. The park offers an excellent opportunity to view landforms shaped by both the sea and glaciers.

The Canadian Shield Extending from the Canadian Great Plains all the way to the east coast of Canada lies the old, stable, granite base on which much of the continent is built. The great ice sheets that originated on the shield have eroded it to a flat plain. Due to its low elevation and poor rock porosity, standing water forms many lakes and marshes.

○ *Why are the Appalachian Mountains no longer very high?*
○ *What type of erosion created the Valley and Ridge Province?*
○ *What is unusual about the erosion in Acadia National Park?*

Section Review

READING CRITICALLY

1. Compare and contrast the Rocky Mountain Province and the Pacific Mountain System.
2. Compare the Columbia Plateau to the Appalachian Plateau.

THINKING CRITICALLY

3. If you were to draw a cross section of the continent from San Francisco, California, to Norfolk, Virginia, how would the elevation of the different provinces change?
4. The valleys in the Blue Ridge Mountains are not accumulating many sediments. Explain what happens to the sediments.

Figure 14–20. The granite base of the Canadian Shield is all that remains of the once-great mountain range upon which much of North America rests.

Everglades

ACTIVITY: Planning an Informational Brochure

How can you design a brochure to provide tourists with useful details about the geology of a park?

MATERIALS (per group of 3 or 4)

paper, colored pencils, rulers, local topographic and geologic maps

PROCEDURE

1. Pretend that you have been hired to develop a brochure for a park in your region. Review a topographic-geologic map of the area you wish to write about.

2. Design a booklet or brochure to inform the public about the rock types, rock structures, topography, and erosional history of the park. Design your booklet to have at least six pages, including pictures to assist you in your presentation.

CONCLUSIONS/APPLICATIONS

1. How would this type of brochure be useful to tourists visiting the park?
2. In addition to the required information, what else might you include in the brochure to make it more informative?

Section 2 Provinces of North America 313

Answers to Section Review

Reading Critically

1. The Pacific Mountain System is composed of volcanic cones. The Rocky Mountain Province has varied and complex structures composed of folds, faults, intrusions, and volcanoes.
2. The Columbia Plateau is covered by very young lava flows. The Appalachian Plateau is much older and consists mostly of sedimentary rocks.

Thinking Critically

3. The elevation would start at sea level but then rise rapidly to show the high and varied elevations of the coastal mountains and of the Rocky Mountains. It would be low and relatively uniform as it crossed plains and prairies to the Appalachian Plateau, where it would rise until crossing the Appalachian Mountains. Then it would be lower as it approached sea level once more.
4. They are rapidly being removed by the highly developed drainage system.

DISCUSSION Most of the Canadian Shield, as indicated by its name, is located in Canada where there are many provincial and national parks. The students should be encouraged to write to the Canadian Tourist Bureau or the individual parks to learn about the geologic structures that are prominent in these parks. Remind the students that the rocks throughout the eastern half of Canada are the oldest rocks on the entire continent.

SECTION REVIEW

Summary Ask the students to list, in the form of a chart, each geologic province, giving a description of each and identifying some characteristic national parks.

Reinforcement Have the students use flash cards to identify the characteristic landforms found in the geologic province written on the front of the cards.

ACTIVITY:
Planning an Informational Brochure

Thinking Skill (*Communicating*)

Preparation of Materials You may wish to assign the students different regions rather than limit all students to the same area. To obtain the necessary maps, you should write to the geologic surveys in the various states and ask for brochures that present relevant science information as well as information of general interest.

Answers to Conclusions/Applications

1. It would help tourists to locate features and to understand why the features developed as they did.
2. Answers will vary, though the students may include maps, hiking trails, study questions, or other regional features.

Preparation of Materials

You may want to review with the students the basic map-reading procedures from Chapter 2. Use the maps from previous chapters to practice map-reading skills.

Answers to Analyses and Conclusions

1. This is an anticline. The rocks in the center are older and dip away from the center (or axis).
2. The Clinton formation and the Clinch Sandstone both form highlands and mountains.
3. The drainage pattern is trellis. The tributaries enter the trunk stream at a right angle. The trunk streams run parallel to each other.
4. The parallel ridges and valleys produce parallel drainage patterns.

Answer to Application

The students' answers may vary. Some of the following points should be considered: an engineer would choose the most resistant rocks to support the bridge foundation; the maps allow an engineer to select the locations where the stronger rocks provide a narrow passage for the bridge; using the maps, calculations concerning the exact length of the bridge could be made.

INVESTIGATION 14: Correlating Features on Maps

PURPOSE

To use geologic and topographic maps to interpret the geomorphology of an area

PROCEDURE

1. Locate streams, rivers, mountains, and other features represented on the topographic map.
2. Study the rock structure for the area, and compare it with the position of the landforms on the topographic map.
3. Determine which rock structures make valleys and which rock units make ridges and mountains by comparing the geologic map to the topographic map.

ANALYSES AND CONCLUSIONS

1. Is the rock structure between Black Creek Mountain and Jack Mountain an anticline or a syncline? Explain your conclusion.
2. Which two units seem to be the most resistant to weathering?
3. What type of drainage pattern is found in the Jackson River Valley (left of Jack Mountain)? Describe its features.
4. Relate the drainage pattern to the structural pattern in the area.

APPLICATION

How could an engineer use this map and rock column to design a new bridge across the Jackson River?

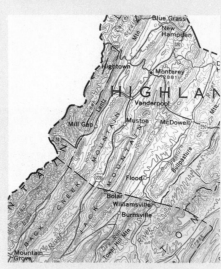 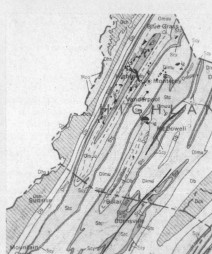

SUMMARY

- Rock structure, erosional agents, and time influence the geomorphology of an area. (14.1, 14.2)

- North America is divided into regions called geomorphic provinces. (14.2)

- The western provinces of the United States are characterized by relatively young mountains with many folds, and faults, intrusive magma bodies, and extrusive volcanic formation. (14.3)

- The western provinces include the Rocky Mountain Province, the Colorado Plateau, the Basin and Range Province, the Columbia-Snake River Plateau, the Pacific Mountain System, and the Pacific Border Province. (14.3)

- Alaska is composed of many different provinces. (14.4)

- The Hawaiian Islands are part of an arc of volcanic islands, many of which remain submerged. (14.4)

- The central provinces are composed primarily of thick layers of undisturbed sedimentary rocks. (14.5)

- The Plains and Lowland provinces include the Great Plains and Central Lowlands and the Coastal Plains. (14.5)

- The eastern mountains of the United States are very old and highly eroded. (14.6)

- The Appalachian Provinces include the Blue Ridge Mountains, the Piedmont Plateau, the Valley and Ridge Province, the Appalachian Plateau, the Ouachita and Ozark mountains, the New England Province, and the Canadian Shield. (14.6)

Write all answers on a separate sheet of paper.

SCIENCE TERMS

Correctly use each of the following terms in a sentence.

geomorphic province **(301)**
geomorphology **(299)**

SCIENCE QUIZ

Modified True-False

Mark each statement *true* or *false*. If a statement is false, change the underlined term to make the statement true.

1. <u>Geology</u> is the study of the landforms of the earth.

2. Geomorphic provinces can be divided into <u>ten</u> main categories.

3. Rocky Mountain National Park is located in the <u>western provinces</u>.

4. <u>Plateaus</u> can be formed by folding and faulting.

5. Florida has cave systems because of its <u>karst topography</u>.

Multiple Choice

Write the letter of the term that best answers the question or completes the statement.

6. Which one of the following has little in common with the others?
   a) Hawaiian Islands
   b) Rocky Mountain National Park
   c) Cascade Mountains
   d) Colorado Plateau

continues

CHAPTER REVIEW

SUMMARY

The students may review the major concepts in the chapter by reading the summary statements. The statements are cross-referenced to the chapter to facilitate reinforcement of any concepts of which the students feel unsure. Encourage the students to work in groups to quiz one another.

SCIENCE TERMS

The sentence in which the science term is used should reflect an understanding of the definition of the term. You may wish to stage a game of "Jeopardy" during which the definition is provided and the "contestants" must provide the correct term.

SCIENCE QUIZ

Answers to Modified True-False

1. false, geomorphology
2. false, three
3. false, Rocky Mountain Province
4. false, mountains
5. true

Answers to Multiple Choice

6. d

continues

Answers to Multiple Choice (continued)

7. d
8. c
9. a
10. b

Answers to Completion

11. Valley and Ridge
12. volcanic cones
13. dome
14. volcanic arc
15. New England

Answers to Short Answer

16. Both lie along the western edge of the continent where the North American plate is pressing against the Pacific plate.
17. They do not include a large enough area.
18. Both were formed as structural lowlands by faulting.

Answers to Writing Critically

19. The eastern mountains are older; therefore, there has been more weathering and erosion.
20. The bedrock in Florida is limestone, so when it weathers it forms karst topography.

ANSWERS TO EXTENSION

1. Answers will vary. If there isn't a park close to your area, the students could write to parks of their choice to obtain information.
2. Answers will vary. As a variation of this, you might have the students build models out of clay to represent the landforms of the region.
3. Answers will vary.

SCIENCE QUIZ continued

7. What do the Black Hills and the Central Mineral Region have in common?
  a) dome formation
  b) old granite cores
  c) formed at the same time as the Rocky Mountains
  d) All of the choices are correct.

8. How are the Rocky Mountains and the Appalachian Mountains different?
  a) One is folded; the other is faulted.
  b) One contains ridges: the other does not.
  c) One is older than the other.
  d) One has been eroded; the other has not.

9. Redwood National Park is located in the
  a) Pacific Border Province.
  b) Colorado Plateau.
  c) Basin and Range Province.
  d) Pacific Mountain System.

10. Which one of the following has little in common with the others?
  a) Rocky Mountains
  b) Canadian Shield
  c) Colorado Plateau
  d) Valley and Ridge Province

Completion

Complete each statement by supplying the correct term or phrase.

11. The province in the Appalachian Mountains characterized by repeated folds is the _____ Province.

12. The Cascades are formed from _____ .

13. One type of structural interruption in the Great Plains and the Central Lowlands is a _____ .

14. The Alaskan Peninsula and Aleutian Islands are formed from _____ .

15. Maine's Acadia National Park is located in the _____ Province.

Short Answer

16. Why might it be argued that the source of lava for the Cascades is the same as the source for Mount St. Helens?

17. Why are the Black Hills not a separate province?

18. What do the Great Valley and Death Valley have in common?

Writing Critically

19. Explain why the eastern mountains are more eroded than the western mountains.

20. Why is there karst topography in the state of Florida?

EXTENSION

1. Visit a local, state, or national park in your area and ask for an informational brochure. Read the brochure and write a report on the geological characteristics of the park.

2. Go to the library and research the geomorphology of your region. Determine the dominant rock type and structure and note any unusual landforms in your area. Make careful notes and report your findings to the class.

3. This photograph shows a Landsat image of south Florida, including Everglades National Park. Read about the Everglades and report to the class on its special features and environmental problems.

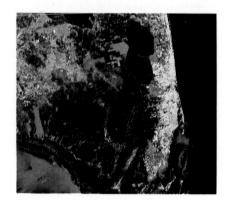

316   Chapter 14 Review

APPLICATION/CRITICAL THINKING

1. Look at the Landsat photograph of south Florida on page 316. The large lake in the picture is Lake Okeechobee, which supplies water to most of the people of southeast Florida, as well as to part of Everglades National Park. The survival of wildlife in the park depends on a constant supply of water from the lake. The cities of southeast Florida are increasing in population faster than any other section of the country, which means greater demands for water in the future. How can the water needs of the people and the park best be satisfied in the immediate and distant future?

2. What might happen to the Everglades wildlife and the cities of south Florida if the water that flows into Lake Okeechobee is cut off?

FOR FURTHER READING

Chronic, H. *Pages of Stone: Geology of Western National Parks and Monuments.* Seattle, Washington: Mountaineers, 1984. This well-illustrated book presents the geology of the western United States through a study of the national parks and monuments.

Harris, B. *Landscapes of America,* 2d ed. New York: Crescent Books, 1986. This book contains a spectacular collection of photographs that show a wide range of American landscapes.

Maltby, E. *Waterlogged Wealth: Why Waste the Earth's Wet Places?* Washington, D.C.: International Institute for Environment and Development, 1986. An excellent review of wetlands of the world.

Challenge Your Thinking

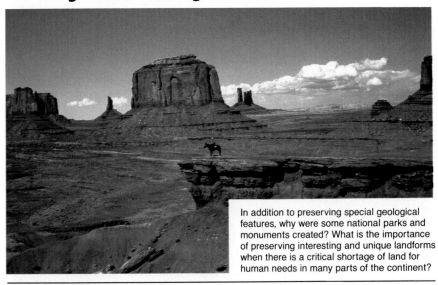

In addition to preserving special geological features, why were some national parks and monuments created? What is the importance of preserving interesting and unique landforms when there is a critical shortage of land for human needs in many parts of the continent?

Chapter 14 Review   **317**

ANSWERS TO APPLICATION/ CRITICAL THINKING

1. By allowing the lake and the Everglades to maintain a natural condition, more clean water would be available to South Florida.
2. Neither would have enough water for their needs.

ANSWER TO CHALLENGE YOUR THINKING

Some parks have been established for their beauty; others were selected for historical importance. The parks provide excellent locations to learn about geologic processes. They also provide sites for individuals to appreciate the power and beauty of the natural environment.

Discussion This *Science Connection* is about Glacier National Park. The fifty glaciers in the park are small in comparison to the huge glaciers that, over a million years ago, carved the natural beauty of the park, and for which the park is named. Mountains like Chief Mountain were formed by the warping, folding, and buckling of rocks that resulted from the hardening of ocean sediments millions of years ago. Ask the students to name some of the forces, caused by both natural processes and human activity, whose effects on the earth are hardly noticeable today, but whose long-term impact will be great.

Discussion Scientists have been studying Grinnell Glacier to learn more about how glaciers move and how these movements weather rock. Each summer, instruments that measure pressure and temperature are placed in a cave beneath the glacier and left until the following summer, when the cave is again accessible. Scientists estimate that this glacier, which is now approximately 112 hectares, was twice as large when it was discovered 100 years ago. Ask the students why scientists who are concerned about protecting the natural environment from weathering and erosion are interested in the movements of glaciers.

Discussion The United States Congress established Glacier National Park in 1910. The park is a refuge for 250 kinds of birds, 60 kinds of mammals (including black bears, wolves, moose, and elk, among others), and 1000 kinds of plants. Some people are concerned that human activities nearby, such as logging, drilling for gas, and mining for coal, might harm the park's environment. Ask the students why it is important to preserve such places as Glacier National Park.

Extension After completing the Unit and reading and discussing the *Science Connection,* have the students turn back to the Unit Opener on page 229 and answer the questions. When the class discussion has ended, you may want to have the students research information about other national parks.

The majestic beauty of Glacier National Park provides many contrasts. Glacier-carved mountain peaks rise to heights of over 3000 m, forming horns and long, pointed ridges. Many hanging valleys, where tributary glaciers once joined the main valley glaciers, may be seen along upper valley walls. Beautiful waterfalls plunge hundreds of meters to the bottoms of U-shaped glacier valleys.

The valley floors are sprinkled with cold, blue lakes fed by the many waterfalls spilling from higher lakes in the hanging valleys. These sapphire-blue lakes, clouded with rock flour, stand in sharp contrast to the glistening white of the 50 glaciers found throughout the park. Although these glaciers still erode the valleys, their effects do not compare to those of the massive bodies of ice that sculptured the original valleys during the last Ice Age.

The largest glacier in Glacier National Park is Grinnell Glacier. The ice of this glacier moves down slope about 10 m each winter and then melts back about the same distance in the summer. The massive rock walls that form towering peaks and ridges above the glacier contrast with the piles of rubble and loose rock at its base. Long, rolling mounds of this moraine line many of its valleys.

Glacier National Park straddles the Continental Divide, which separates the drainage basins of the Pacific Ocean and Gulf of Mexico. The high mountains of the divide also create distinctly contrasting climates on the east and west sides of the park. The western side receives warm, moist winds. Vegetation thrives and the temperatures are moderate. As the air rises to cross the mountains, it cools and loses much of its water content. Thus, the eastern side of the park, in a rain shadow, has a drier, colder climate. Vegetation struggles to survive the cold winds of winter and the hot, drying winds of summer.

A hanging valley waterfall

Purcell Sill

Glacier National Park

Chief Mountain

The rocks in Glacier National Park show a history of calm periods of sedimentation alternating with times of violent upheaval. The rocks show evidence of extremely old deposits as well as relatively recent changes.

Most of the rocks found in Glacier National Park are about a billion years old. These ancient red and gray-green mudstones and siltstones reveal amazing details about the environment in which they formed. Some have ripple marks made by wave motion, while others show mudcracks and impressions made by raindrops. Still other rocks contain stromatolites. These details indicate that during the time the sediments were deposited, this part of the continent was a shallow mudflat at the edge of a sea. Occasionally, the sea must have dried out, causing the mud to crack. Thick limestones were also deposited in this area, indicating that during some periods it was a deep-water marine environment.

The imprint of flowing lava can also be found in the park. Iron-rich magma forced its way into cracks in the limestone, leaving dark vertical dikes and horizontal layers of igneous rocks. One of the most visible of the lava formations is the Purcell sill, a thin, black layer of igneous rock sandwiched between layers of white limestone. This sill is a dramatic feature in the valley wall high above Grinnell Glacier.

About 70 million years ago, as the Rocky Mountains were forming, a large block of land slid eastward as a low-angle thrust fault. This block moved eastward about 55 km and came to rest on top of layers of much younger rocks. This overlying block confused scientists for years. They could not figure out why the mass of billion-year-old rock that forms Chief Mountain rests on rocks only 100 000 000 years old. The roots of Chief Mountain lie over 50 km to the west. Chief Mountain is actually an isolated mountain; the rest of the thrust block has eroded away.

The contrasting features of Glacier National Park make it a beautiful and interesting place to visit. As conservationist John Muir said, "Give a month at least to this precious reserve. The time will not be taken from the sum of your life. Instead of shortening it, it will indefinitely lengthen it, and make you truly immortal."

Grinnell Glacier

319

Unit 5: EARTH HISTORY

UNIT OVERVIEW

In this unit the basic principles of geology are described, and the procedures for measuring the age of geologic materials are explained. Also described are the divisions of Earth history and the outlook for the future of the mineral and energy resources of the planet.

Chapter 15: Historical Geology, page 322

This chapter discusses how information about the earth's geologic past has become recorded in rock. Also described are the many methods geologists use to determine the age of rocks and minerals.

Chapter 16: Earth's Past, page 340

This chapter describes the geologic history of Earth from its formation to the evolution of humans. Discussed are the two main geologic divisions—the Cryptozoic and Phanerozoic eons. The eras that make up each eon and the important developments that occurred during each era are also discussed. To aid in the discussion, the chapter compares the eons and eras to a calendar year.

Chapter 17: Earth's Future, page 362

This chapter explains the differences between renewable and nonrenewable resources. Emphasis is placed on the importance of protecting renewable resources such as water, air, and land. The importance of effective management of nonrenewable resources, such as fuels, is emphasized, and the development of alternative energy sources for the future is discussed.

The Science Connection: How the Dinosaurs Died, page 386

The Science Connection describes one theory that explains the rather sudden extinction of the dinosaurs about 65 million years ago. Perhaps an asteroid slammed into the earth, raising huge clouds of dust that blocked the sun and greatly cooled the planet.

ADVANCE PREPARATION

Chapter 15

For the Activity on page 330, the students will need to bring 100 pennies. For more information, see page 330.

For Investigation 15 on page 336, you will need several large chicken bones or small lamb bones. For more information, see page 336.

Chapter 16

For the Discover on page 347, you will need some fresh hay. The hay infusion that is made must sit for a day or two before making any observations. For more information, see page 347.

Chapter 17

For the Activity on page 366, you will need some pond or aquarium water with microorganisms, and some liquid laundry bleach. For more information, see page 366.

For the Discover on page 371, you need to find a local recycling center, and have the students organize a weekend collection of aluminum cans and newspapers. For more information, see page 371.

For the Discover on page 379, the students will need to make a list of the energy resources that they use during one day. For more information, see page 379.

BULLETIN BOARD SUGGESTIONS

Chapter 15

Chart the radioactive disintegration of Uranium-238 to Lead-206. The vertical axis should be labeled "Atomic Number," while the horizontal axis should be labeled "Atomic Mass." Show all alpha and beta disintegrations and label each new element created by successive disintegrations.

Illustrate various elements and their isotopes.

Illustrate the processes of fission, fusion, and radioactive dating.

Chapter 16

Create a time line listing the characteristics of the eras of the Phanerozoic Eon, including pictures and drawings of any fossils of the time. See page 574 in the Reference Section.

Chapter 17

On a globe or map of the world use flagged insect pins to indicate where the major mineral and energy resources of the world are found. Try to include relative amounts of each resource as compared with global reserves. Also indicate major industrial cities where air and fresh water resources are in danger of becoming polluted. Indicate nuclear reactors.

ISSUES IN EARTH SCIENCE

Have the students research the facts about breeder reactors and then stage a debate on the pros and cons of developing breeder reactors in the United States.

Have the students research the events of the oil crisis of the 1970s and then report on the possibilities of a similar event occurring today and the consequences it would have on our way of life.

Have some students research other theories about the reasons for the extinction of the dinosaurs and then present their ideas to the class, which will act as a learned body and decide on the merits of each theory.

SUGGESTED PROJECTS

Some students may be interested in starting a fossil collection. The state geologist's office could provide information about the types of fossils available in your area.

Have some students contact the Environmental Protection Agency to learn how they can participate as members of a citizen's committee to monitor air and water as a deterrent against pollution.

TEACHER RESOURCES

Readings

Chapter 15

"Evolution," *Scientific American*. (September, 1978) p. 46. Theme issue devoted to the mechanisms and patterns of evolution. Prominent authors in the field of evolutionary theory discuss topics ranging from the origin of life on Earth to the evolution of animal behavior.

Leakey, L.B.S. *By the Evidence: Memoirs, 1932–1951*. New York: Harcourt Brace Jovanovich, Publishers, 1976. A stimulating personal and professional account of the author's discoveries of fossils and artifacts in the heart of Africa that pushed back the time of human genesis by more than a million years.

Chapter 16

Life: Origin and Evolution. Scientific American. San Francisco: W.H. Freeman and Company, 1979. The book is a compilation of informative articles elucidating current theories about the creation of the solar system and the evolution of life on earth.

Margulis, Lynn. *Origin of Eukaryotic Cells*. Yale University Press, 1970. The book is a supporting argument for the author's theories regarding the evolution of nucleated cells.

Chapter 17

Isaar, Arie. "Fossil Water Under the Sinai-Negev Peninsula." *Scientific American* 253 (July, 1985) p. 104. An analysis of the vast underground reserves of water that were formed in the last ice age and are being prepared for use in the Sinai desert.

Hamakawa, Yoshihiro. "Photovoltaic Power." *Scientific American* 256 (April, 1987) p. 86. Within fifteen years, solar-cell conversion plants could be supplying power in the megawatt range.

Audiovisual/Software

Chapter 15

A computer program, *Earth History*, is available for Apple II and TRS-80 computers from Geoscience Resources.

Chapter 16

A computer program, *T. Rex The Dinosaur Survival Adventure*, is available for Apple II and IBM-PC computers from Geoscience Resources.

A package of 5 slide series on *Life in the Past* is also available from Geoscience Resources.

Chapter 17

A computer program, *Oil Model Game*, is available for Apple II computers from Geoscience Resources.

Other Materials

Chapter 15

All commissions and administrations of the United States government submit annual reports to congress. Obtain a report of the Nuclear Regulatory Commission from your public library. Students should appreciate the complexities of managing radioactive materials and nuclear energy.

Fossil study packs are available from Geoscience Resources.

Chapter 16

A geologic time chart game and models of geologic change are available, along with wooden dinosaur models, from Geoscience Resources.

Chapter 17

Write the Environmental Protection Agency for information on monitoring and clean-up measures that insure the healthful quality of our air and water resources.

Environmental Protection Agency
401 M Street S.W.
Washington, D.C. 20460

Chapter 15: HISTORICAL GEOLOGY

PLANNING THE CHAPTER

Chapter Sections	Page	Chapter Features	Page	Program Resources	Page
Section 1: The Record in Rocks	323			Critical Thinking **(H)** Concept Extension: *Identifying Tracks* **(B)**	**TRB 28** **TRB 31** **LM 253**
15.1 Principles of Geology **(B)** 15.2 Records of Environmental Change **(B)**	323 324	**Discover:** Studying Clues **(B)** Section Review **Skill Activity:** Using Classification Systems **(B)**	323 326 327		
Section 2: The Geologic Clock	328			Cross-Discipline: Science and Mathematics, *Relating Half-Lives and Powers of Two* **(H)**	**TRB 28**
15.3 Radioactivity and Absolute Dating **(H)**	328	**Activity:** Demonstrating the Half-life of a Radioactive Isotope **(H)**	330	Reading for Content: *Using Resources Properly* **(A)** Investigation 15.1:	**TRB 26** **LM 201** **TRB 63**
15.4 Dating Minerals and Rocks **(H)**	330	A Matter of Fact A Matter of Fact	330 331	*Reconstructing Fossils* **(A)** Investigation 15.2: *Making Your Own Fossil* **(B)**	**LM 65** **TRB 65** **LM 67**
15.5 Fossils **(B)**	332	**Discover:** Making Fossils **(B)** A Matter of Fact Section Review **Technology:** New Techniques in Radiocarbon Dating **Investigation 15:** Fossilizing Bones **(A)**	333 334 334 335 336	Student Record Book: Textbook Investigations **(A)**	**TRB 29**
Chapter 15 Review	337			Vocabulary **(A)** Tests **(A)** Computer Test Bank	**TRB 15** **LM 151** **TRB 64**

(LM) Laboratory Manual/Study Guide, **(TRB)** Teacher's ResourceBank™

B = Basic **A** = Average **H** = Honors

The coding Basic, Average, and Honors indicates sections or subsections that might be appropriate for different levels of learners. For additional suggestions regarding choice of topic and depth of coverage, see the Pacing Chart on pages T16–T20.

CHAPTER CONCEPTS, OBJECTIVES, AND TERMS

Section	Concepts	Objectives	Science Terms
Section 1: The Record in Rock	■ The principle of uniformitarianism states that today's geological processes are similar to those of the past. **(15.1)** ■ The principle of superposition states that younger rocks are deposited on top of older rocks. **(15.1)** ■ Fossils and other sedimentary features provide evidence about past environmental conditions. **(15.2)** ■ A series of formations can provide a continuous record of environmental changes over a long period of time. **(15.2)**	■ **State** the principle of uniformitarianism. ■ **Use** the principle of superposition to find the relative ages of layers of sedimentary rocks. ■ **Relate** some characteristics of sedimentary rocks to the environment in which they formed.	principle of uniformitarianism principle of superposition crossbedding
Section 2: The Geologic Clock	■ The time it takes for half of the atoms of a radioactive isotope to decay is called half-life. **(15.3)** ■ Some unstable atomic nuclei undergo radioactive decay. **(15.3)** ■ Radioactive isotopes can be used to absolutely date geologic material. **(15.3)** ■ Igneous rocks can be dated absolutely, but sedimentary rocks must be dated relative to nearby igneous formations. **(15.4)** ■ As a radioactive isotope decays, it may leave fission tracks in the surrounding material. **(15.4)** ■ Fossils are the remains, or traces, of organisms that lived in the past. **(15.5)** ■ Fossils can be used to date sedimentary rock. **(15.5)**	■ **Discuss** how some elements decay. ■ **Explain** various methods of dating rocks.	radioactive decay half-life absolute dating relative dating

Title	Page	Materials
Discover: Studying Clues	323	*(per pair of students)* paper, pencil
Skill Activity: Using Classification Systems	327	*(per student)* paper, pencil
Activity: Demonstrating the Half-life of a Radioactive Isotope	330	*(per group of 3 or 4)* 100 pennies, shoe box, pencil and paper, clock with second hand
Discover: Making Fossils	333	*(per student)* modeling clay
Investigation 15: Fossilizing Bones	336	*(per student)* Plaster of Paris, tray, water, petroleum jelly, vise, small saw, needle-nose pliers, small hammer, chicken or lamb bones, snail shells or seashells

TEACHING SUGGESTIONS

Section 1: The Record in Rock

Class Activity

Using two different colors of modeling clay, build atomic nuclei of several simple atoms (hydrogen, carbon, oxygen) showing the protons and neutrons within each nucleus. Distinguish atomic number from atomic mass, and then display isotopes of the atoms already built. Have students identify the different nuclei and their isotopes.

Class Activity

Prepare a dish of water and a dropper of oil. One at a time, let several oil droplets fall and float on the surface of the liquid. The droplets eventually adhere to one another. The adhesion is due to the phenomenon of surface tension. Cut the large droplets in half. Note again that the resulting droplets take on a spherical shape. The behavior of these ''droplets'' is analogous to the behavior of subatomic protons and neutrons in the atomic nucleus.

Field Trip

Arrange for a field trip to a local university physics or hospital radiology department. Ask the professors and doctors to discuss and demonstrate how radioactive materials are handled and stored. Or, obtain a film or filmstrip about these procedures. If a local university has a paleontology department, ask someone there to arrange a demonstration of radioactive dating methods.

Section 2: The Geologic Clock

Demonstration: Fossil Imprints

Purpose

To compare ''fossil'' imprints

Background

Preserved footprints are a type of fossil. A footprint well preserved in mud can tell you the shape of the foot of the animal that made it. It may show you details of the skin on the bottom of the foot. It may also tell you something about the weight of the animal. Further, the clearness of the footprint may depend on the material in which it was made.

In this Demonstration, you will make footprints in different types of materials and observe how the weight of an object affects the imprint it makes.

Materials

Beakers (4)
Gravel
Clay
Sand
Plaster of Paris
Metric ruler
Foil pans (4)
Balance

Procedure

1. Label four small beakers from 1 to 4. Leave beaker 1 empty. Add 10 g of sand to beaker 2, 25 g of sand to beaker 3, and 50 g of sand to beaker 4.
2. Mix the plaster of Paris with water. (Follow the directions on the box of plaster.) Pour the plaster into the foil pan. Working quickly, smooth out the surface of the plaster.
3. Coat the bottom of beaker 1 with petroleum jelly and place it gently on the plaster. Quickly repeat this procedure for the other three beakers. Do not press the beakers into the plaster.

4. Let the plaster dry. This may take about 30 minutes or longer.
5. When the plaster is dry, remove the beakers. Measure the depth of each imprint. Record the information in a table, which you should copy on the chalkboard.
6. Make mixtures of sand and water; clay and water; clay, water, and gravel. Pour the mixtures into three different foil pans. Label the pans so you can identify the mixture each contains.
7. Choose an object that will make an imprint in the mixtures. The objects could be a rock, a shell, a broken piece of pottery. Push the object into the surface of each mixture so that it makes an imprint.

Questions to Ask the Students

1. How did the mass of the beaker and sand change the depth of the imprint?
(The greater the mass, the deeper the imprint.)

2. Would a light animal or a heavy animal make a deep imprint?
(heavy) Why?
(The heavier animal would sink deeper into soft mud.)

Field Trip

If there is a good area for collecting fossils near your school, arrange for a fossil hunting dig at the site.

Outside Speaker

Have a paleontologist speak to the class about collecting, classifying, and interpreting fossils. Have him or her bring some of their more unusual fossils to show the students.

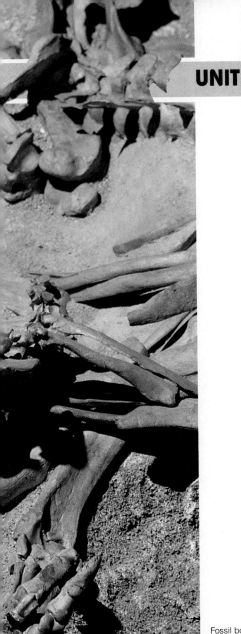

EARTH HISTORY

In 1977 geologist Luis Alvarez
set out to study a thin layer of clay
rock about 65 million years old. He
concluded that the clay had been
deposited when a giant meteor
struck the earth. Alvarez knew that
at about that same time nearly all
the dinosaurs had vanished in a
mysterious mass extinction. Alvarez
and his scientist father put these
facts together and stated a new
theory to explain the extinction of
the dinosaurs. This theory started a
scientific controversy. The Alvarezes
claimed that the dinosaurs died
when a cloud of meteor dust covered
the earth.

- How can fossils be dated to show
  the evolution of animal and plant
  life?
- What may have caused the evolution
  and extinction of animal species?
- How will studying Earth's past enable
  humans to prepare for the future?

By reading the chapters in this unit,
you will begin to learn the answers
to these questions. You will also
begin to develop an understanding
of concepts that will help you to
answer many of your own questions
about the history of the earth.

Fossil bones

321

INTRODUCING THE UNIT

Have the students look at the Unit Opener
photograph on these pages. Point out that
these are fossil bones. Read the text and
use the questions to guide a class discussion. Do not expect the students to be able
to answer the questions now. The questions are motivational and the students
should be aware that they will be able to
answer them after they have finished the
Unit.

Discussion Over the years, scientists
have formulated many hypotheses for the
periodic mass extinctions of plants and
animals, including dinosaurs. Ask the students to name any of these theories with
which they are familiar. By dating fossils,
scientists can also determine the evolution
of plants and animals. Ask the students to
suggest how scientists use their knowledge of Earth history to study evolution
and mass extinctions.

Extension You can extend the discussion about Earth history by bringing in
other Unit themes. What are the ''record
in rock'' and the ''geologic clock''? How
are fossils dated and classified? What are
some energy and natural resources, how
are they used today, and how can they be
better managed in the future? You may
wish to record student responses and review them when the questions are asked
again after the Unit is finished.

CHAPTER 15

CHAPTER OVERVIEW

This chapter discusses how information about the earth's geologic past has become recorded in rock. Also described are the many methods geologists use to determine the age of rocks and minerals.

Section 1: The Record in Rocks   In this section, two major principles of geology, uniformitarianism and superposition, are described and their significance in dating rock formations is explained. Next is a discussion of rock records that help scientists understand the conditions under which rock layers form.

Section 2: The Geologic Clock   In this section the half-life of radioactive elements is defined. Radioactive isotopes that are commonly used in absolute dating of rocks are listed along with substances that can be dated using this method.

CHAPTER

15

Historical Geology

You may not think of the Grand Canyon as a history book, but to a geologist it is. The pages of the book are the layers of rock piled one on top of the other. Fossils from the early history of the earth are found in layers near the bottom of the canyon; more recent history can be read in the layers nearer the top. The Grand Canyon contains one of the most complete fossil records on Earth.

The Grand Canyon, in Arizona

322

CHAPTER MOTIVATING ACTIVITY

Obtain fossils from a local college or from a scientific supply house. Display the fossils in class. Discuss the environment in which each fossil was formed. If possible, take a field trip to a local university that has a spectrophotometer, and ask someone there to demonstrate radioactive dating methods.

1 The Record in Rocks

Section 1: The Record in Rocks

SECTION OBJECTIVES

After completing this section, you should be able to:
- **State** the principle of uniformitarianism.
- **Use** the principle of superposition to find the relative ages of layers of sedimentary rocks.
- **Relate** some characteristics of sedimentary rocks to the environment in which they formed.

NEW SCIENCE TERMS

principle of uniformitarianism
principle of superposition
crossbedding

15.1 Principles of Geology

Have you ever noticed how uniform or consistent some things are? For example, if you take a slice of bread from anywhere in a loaf, you notice its uniform texture. From top to bottom and from side to side, the texture in that loaf of bread is the same.

A slice of bread from a different loaf of the same brand still shows the same uniformity of texture. The bread-making process insures the uniformity of the product. The same processes, even at different times, produce the same results.

Uniformity is an underlying principle of many geological processes. In the eighteenth century, James Hutton, a Scottish physician and geologist, proposed that the geological processes that form rocks today formed similar rocks in the past. He called this process *uniformitarianism*. The **principle of uniformitarianism** states that similar processes, in different times, produce similar results.

DISCOVER

Studying Clues

Play a game of "twenty questions" with a classmate. You need to think of a common rock or mineral. Be especially careful that you know for sure where this rock or mineral is found. Your partner will ask questions about your rock or mineral to try to guess where it is found. The questions asked must be able to be answered with a "yes" or "no" and must not include the name of the rock or mineral. The object is to find out where the rock or mineral is found, using the fewest number of questions possible.

Figure 15–1. In formulating his principle of uniformitarianism, Hutton may have studied the White Cliffs of Dover, in England.

Section 1   The Record in Rocks   **323**

SECTION FOCUS

Section Overview   This section concentrates on two basic principles of geology that help geologists understand the events of the earth's past. The clues in rock layers that scientists use to discover the conditions of the earth's past environments are emphasized.

Section Objectives   For a list of section objectives, see pupil's edition page.

New Science Terms   For a list of new science terms in this section, see pupil's edition page.

SECTION DEVELOPMENT

15.1 Principles of Geology

DISCUSSION   Before beginning the paragraphs on uniformitarianism, you may want to review the processes that shape the earth's surface. Ask the students to name these processes. (volcanic eruptions, faulting, earthquakes, weathering, erosion) Stress the idea that the processes that we observe today are the same processes that have occurred in the past.

> **DISCOVER:**
> Studying Clues

Thinking Skill (*Formulating Questions*)

Emphasize to the students the importance of asking the right questions and being able to piece together the information. Compare this activity with the skills exhibited by historical geologists.

BACKGROUND INFORMATION

Explain to the students that geologists of the eighteenth century first tried to calculate the age of the earth by observing the world around them. (This deviated from the biblical view and returned to the Greek way of thinking.) The great Scottish naturalist, James Hutton, was among the first to differ with the church view that the earth was less than 6000 years old. His views were stated in a book called *Theory of the Earth*. After observing the rates of erosion and deposition of sediments, and after intensely studying stratigraphy, Hutton estimated the age of the earth to be several million years old. In the early 1800s the Compte de Buffou of France determined that the age of the earth was 75 000 years. He determined this by analyzing the melting and cooling rates of iron. In 1854 Herman von Helmholtz used what he thought was the rate of the sun's contraction to arrive at an age of 20 to 40 million years. In the twentieth century, when radioactive decay was fully understood, the earth was discovered to be billions of years old.

15.2 Records of Environmental Change

EXTENSION Ripple marks are another sedimentary rock feature that help scientists determine the environment in which rocks were formed. Ripple marks are small waves on the surface of rock beds that were formed when wind or water moved over sediments. Scientists use other clues, such as fossils, texture, and rock fragment type and size to determine whether wind, land, or water along a seashore or stream formed the ripples.

DISCUSSION Have the students observe the two photographs in Figure 15–3. Ask them to compare the processes that formed both types of sedimentary rocks.

Figure 15–2. These alternating layers of sandstone and limestone illustrate the principle of superposition. The younger layers are on top of the older layers.

Figure 15–3. Fossils of ancient sea life (left) and mud cracks (right) provide evidence that these areas were once covered by water.

Years after Hutton developed the principle of uniformitarianism, scientists continue studying various geological processes. Observations confirm Hutton's theory. For example, sediments may be carried by water and wind and deposited in new places, but they are always deposited in similar ways. Over long periods the sediments are compressed and cemented, forming sedimentary rocks.

Sediments are deposited on top of each other. Look at the formation in Figure 15–2. The layers of sandstone and limestone alternate, but the younger rocks are always on top; the rocks on the bottom must have been deposited first. The fact that younger rocks are deposited on top of older rocks is called the **principle of superposition.** Geologists use this principle to help them determine the ages of the rock formations.

○ *What is uniformitarianism?*
○ *How can the principle of superposition be used to date the relative ages of formations?*

15.2 Records of Environmental Change

Fossil seashells are sometimes found on mountains, and ice scratches are often found on desert rocks. For this to be true, either the rocks have moved or the environment in which the rocks formed has changed.

Geologists can "read" environmental changes as easily as you read a mystery story. In some ways, rocks are like a mystery story; they contain clues about the environment in which they were formed. These clues help scientists reconstruct the past. For example, fossil seashells in mountain rocks indicate that the rocks formed under water and were later uplifted.

Fossils give information about past conditions. For instance, limestone formed in shallow water often contains shells of animals such as clams or snails. Deep-water limestone contains fossils of microscopic organisms that lived in the open ocean.

Answers—15.1

○ Uniformitarianism is the principle that similar processes occur today as occurred in the past.

○ Since younger rocks are deposited on top of older ones, the relative ages of several layers in a formation can be determined.

Figure 15—4. The Dakota sandstone (left) formed on a beach, while the Cody shale (right) probably formed in deep, still water.

Mineral texture may also help geologists determine what sort of environment existed when a rock formed. For instance, in sandstone, the rounded quartz grains probably formed on a beach. Geologists know this because the action of waves rounds the sand grains, and sand-sized sediments are usually deposited near a shore.

The size of the fragments in sedimentary rocks is another clue to the conditions under which the rocks formed. Consider certain rocks that are found only where large bodies of water once existed. You may recall from Chapter 4 that large fragments are deposited first as a stream slows. In a similar way, the movement of waves also separates fragments by size. Large fragments are deposited close to shore, while smaller ones settle farther offshore, in deeper water. Ocean currents may carry clay-sized fragments far from land. In quiet waters, away from the beach, the clay settles to the bottom, eventually forming shale.

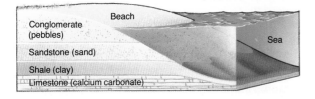

Figure 15—5. The shoreline profile (left) shows the relationship between sediment size and location of deposition.

Figure 15—6. The pattern of crossbedding (below) shows that the sand in this formation was blown by winds that constantly changed direction.

Still other environmental clues are found in rocks formed on land. Sandstone formed on land might show crossbedding patterns. **Crossbedding** is a design in rocks that shows directional changes in wind patterns. Notice the crossbedding in the sandstone shown in Figure 15–6. Rocks formed on land contain fossils of land organisms. For instance, the bones of dinosaurs, horses, or even humans would be found only in rocks formed on land. Fossils of seashells would not be found in rock formed on land.

Section 1   The Record in Rocks   **325**

DISCUSSION A feature that can be used to determine the depositional environment of a rock layer is graded bedding. Graded bedding is variation in particle size within a rock layer, grading from largest at the bottom to smallest at the top. Graded bedding in clay or sandstone layers often indicates a stream or delta environment. The most commonly formed graded bedding, however, is seen in layers deposited on the ocean's abyssal plains by turbidity currents. These graded beds vary from coarse sandstones at the bottom to fine-grained shales at the top. ***Thinking Critically:*** Ask the students how graded bedding can help scientists determine whether rock layers have been overturned by faulting and folding. (The coarse materials should be at the bottom of the layer and the fine materials at the top. If they are inverted, then overturning has occurred.)

EXTENSION Because crossbedding is formed by wind and by running water, it can indicate either a desert or seashore, or the bed of a stream environment. Nongraded bedding often indicates land deposition while graded beds usually indicate deposition in water. Crossbeds can also be used to determine whether sedimentary rock layers have been overturned through faulting and folding. If the crossbedding within a sedimentary layer is eroded, the top of the s-curve is truncated and thus the flattened end of the crossbedded layer indicates the top of the layer. Have students view Figure 15–6 and ask them to decide if the layers have been overturned. (No.)

EXTENSION Reinforce the idea that unconformities represent periods of geologic time not recorded in a particular series of rock layers. Unconformities may occur during a period of nondeposition or as previously deposited rocks have been eroded. Some unconformities are visible within rock layers; others are not quite as obvious. An angular unconformity is formed when tilted rock layers are eroded and then overlain by horizontal rock layers. The easily recognizable erosional surface represents the unconformity. A disconformity is difficult to see because it separates an older series of parallel layers that have been overlain by a younger series of parallel rock layers, with a significant lack of rocks in between of an intermediate age. An unconformity represents an erosional surface between igneous or metamorphic rocks and younger, overlying sedimentary rock layers. Have students study Figure 15–8 to determine the type of unconformity shown. (Figure is of an angular unconformity.)

SECTION REVIEW

Summary Have the students discuss the two major principles of geology and how sedimentary rock characteristics indicate the environment in which they formed. (The principle of superposition states that younger rocks are deposited on top of older rocks. The principle of uniformitarianism states that similar processes, in different times, produce similar results. Such factors as texture, mineral content, size of rock fragments, and combination of rock types indicate environment.)

Reinforcement Have the students construct charts on posterboards showing sedimentary rock characteristics in one column and the environments in which they formed in another column.

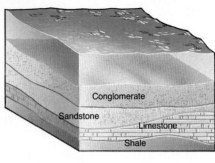

Figure 15–7. This diagram shows the types of rocks that form in areas where the sea level is dropping.

Figure 15–8. An unconformity (above) is a break that occurs where layers of sedimentary rocks are missing.

A series of several different formations can show environmental changes that occurred over a long period. In Figure 15–7 the bottom deposit is shale. On top of the shale is coral limestone. This suggests that between the formation of the shale and the limestone, the sea level probably dropped. Geologists would know this because shale often forms in deep water and coral limestone in shallow water. If sandstone were found on top of limestone, geologists might conclude that the sea level fell even further.

If there were a column of sediments deposited continuously since the formation of the earth, the entire history of the planet could be reconstructed. Unfortunately no such column exists. Where sediments are missing, a break in the sedimentary record occurs. Breaks result in gaps in the record that may range from a few years to hundreds of millions of years. Breaks in the sedimentary record are called *unconformities*.

○ **How do geologists read environmental changes in rock formations?**
○ **What is an unconformity?**

Section Review

READING CRITICALLY

1. What is uniformitarianism?
2. Suppose you find a sedimentary rock that has fossil corals. Was this rock formed in shallow water or deep water? Explain.

THINKING CRITICALLY

3. Imagine that you are doing field work and are studying the Cody shale and the Dakota sandstone shown in Figure 15–4. What was the probable environment in which these two rocks formed?
4. Describe a rock formation that would result from the gradual rising of sea level. Explain why this formation develops.

326   Chapter 15   Historical Geology

Answers—15.2

○ Geologists look for environmental clues, such as crossbedding.

○ An unconformity is a break in the sedimentary record, resulting in a time gap ranging from one hundred thousand years to hundreds of millions of years.

Answers to Section Review

Reading Critically

1. The principle of uniformitarianism states that the geologic processes occurring today are the same processes that have occurred in the past.
2. The rock was formed in shallow water because coral grows only in shallow water.

Thinking Critically

3. Cody shale may have formed in quiet, deep water, while Dakota sandstone may have formed on a beach.

continues

SKILL ACTIVITY: Using Classification Systems

BACKGROUND

Classification systems are used by scientists to organize information. Being organized makes it easier to find information when it is wanted. Also, organization makes it easier to understand the objects being classified.

Imagine what it would be like to list all of the plants and animals in the world. You could list them in alphabetical order, or by color, size, or shape. There are many ways to classify things. In this activity you will create a classification system for some objects in your classroom.

PROCEDURE

1. List six objects in your classroom.
2. Divide the objects into two groups, so that each object in a group has the same characteristics as every other object in that group.
3. Divide each large group into smaller groups. Once again, all objects within a group should have some of the same distinctive characteristics.
4. Continue this procedure until each object is separated from all of the others.
5. Name each object according to the groups to which it belongs, for example, a pencil is also named a hard, small, skinny, pointed, yellow tool.

APPLICATION

Look at the pictures of the seven fossils. You are to classify these in the same manner. If you do not know the names of the fossils, name them as you might have named a pencil in procedure 5.

USING WHAT YOU HAVE LEARNED

1. What are some of the difficulties scientists encounter in trying to classify objects?
2. What characteristics did you choose to classify the fossils?
3. Which characteristics were common to the most fossils?

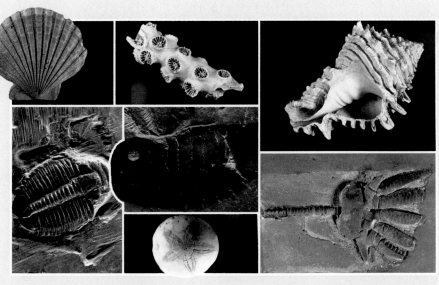

Objectives

- Organize objects into distinctive groups with common characteristics.
- Draw conclusions about the difficulties in trying to classify objects.

Discussion Students should be aware that classification systems are arbitrary. The systems are designed by humans in an effort to create order out of what might appear to be disorder. Any good system of classification should (1) make it possible to categorize new data; (2) allow the data to be easily retrieved; and (3) make it possible to compare and contrast groups of data in the system.

Answer to Application

Answers will vary; accept all reasonable answers.

The fossils are: top row—scallop (mollusk), coral (coelenterate), whelk (mollusk), bottom row—trilobite (arthropod), eurypterid (arthropod), sand dollar (echinoderm), crinoid (echinoderm).

Answers to Using What You Have Learned

1. Answers will vary but should include the observation that classification systems are arbitrary and created by scientists merely as convenient ways of organizing and studying the world.
2. Answers will vary, but be sure classification is based on logical observable differences.
3. Answers will vary but should include such things as size, shape, and texture.

---

4. Sandstone might be on the bottom, with shale on top.

Section 2: The Geologic Clock

SECTION FOCUS

Section Overview In this section radio-activity and the use of radioactive isotopes to determine the age of fossils, minerals, and rocks are discussed. The processes by which fossils are formed are described. The section ends by discussing the use of fossils in dating rocks.

Section Objectives For a list of section objectives, see pupil's edition page.

New Science Terms For a list of new science terms in this section, see pupil's edition page. Remind the students to read the list of terms in this section before they begin the section.

SECTION DEVELOPMENT

15.3 Radioactivity and Absolute Dating

DISCUSSION You may want to review with the students the definition of *isotope*. (atoms of the same element with the same number of protons and different numbers of neutrons) Ask the students whether all isotopes of an element have the same atomic numbers and atomic masses. Have them explain their answers. (All isotopes of an element have the same number of protons, which is equal to the atomic number. The isotopes do, however, have different numbers of neutrons. Thus different isotopes have different atomic masses that equal the number of protons plus the number of neutrons.)

DISCUSSION In class, review the periodic table of the elements. Point out those elements known to have several isotopes. Ask the students to name some uses of these isotopes. (Answers may include medicine [as radioactive tracers], archeology and geology [radioactive dating], energy production [nuclear reactors], and weapons [nuclear submarines and missiles].)

NEW SCIENCE TERMS
radioactive decay
half-life
absolute dating
relative dating

SECTION OBJECTIVES

After completing this section, you should be able to:
- **Discuss** how some elements decay.
- **Explain** various methods of dating rocks.

15.3 Radioactivity and Absolute Dating

You may recall from Chapter 3 that atoms are made of protons, neutrons, and electrons. The lighter atoms tend to have similar numbers of protons and neutrons in their nuclei. For instance, calcium has 20 protons and 20 neutrons. The heavier atoms tend to have many more neutrons than protons. The heaviest naturally occurring atom is an isotope of uranium. Uranium-238 has 92 protons and 146 neutrons in its nucleus.

If a nucleus has a large number of neutrons and protons, it may be unstable. Unstable nuclei may give off particles or energy, producing stable atoms. When this happens the atoms are said to undergo **radioactive decay.** For example, lead, a stable element, forms from the decay of unstable Uranium-238.

Figure 15-9. Uranium is mined from large deposits of uranium ore.

Some radioactive nuclei are very unstable and decay rapidly; others decay more slowly. The time it takes for half of the atoms of a radioactive isotope to decay is called **half-life.** Each successive half-life reduces the remaining number of atoms by one-half. Some isotopes have a half-life shorter than one second; others have a half-life of billions of years.

328 Chapter 15 Historical Geology

DEMONSTRATION

Using two different colors of modeling clay, construct for class display the nuclei of several simple atoms (hydrogen, carbon, or oxygen) showing both the protons and neutrons. Distinguish atomic number from atomic mass. Then construct and display isotopes of the atoms already on display. Have the students identify the different nuclei for their isotopes. next, construct a helium nucleus to demonstrate an alpha particle given off during radioactive decay.

The fact that radioactive isotopes decay at known rates can be used to find the age of geologic material. **Absolute dating** is the process of determining the age of a sample using the radioactive isotopes in the sample. Rocks, minerals, fossils, water, and ice are dated using this method. The isotopes most commonly used in absolute dating are listed in Table 15–1.

TABLE 15–1: ISOTOPES COMMONLY USED IN ABSOLUTE DATING

Isotope	Symbol	Half-life (years)	Material to which applied
Tritium	H-3	12.3	Ground water, sea water, ice
Carbon-14	C-14	5730	Wood, bones, shells
Potassium-40	K-40	1.3×10^9	Rocks, minerals
Rubidium-87	Rb-87	48×10^9	Rocks, minerals
Thorium-232	Th-232	14×10^9	Rocks, minerals
Uranium-235	U-235	704×10^6	Rocks, minerals
Uranium-238	U-238	4.5×10^9	Rocks, minerals

Geologists can determine the age of water by using an isotope of hydrogen called *tritium* (TRIHT ee uhm). Tritium forms in the upper atmosphere and combines with hydrogen and oxygen atoms to form water. This radioactive water falls as rain. Scientists can calculate the age of surface or ground water by measuring the amount of tritium it contains. For example, if a sample of the ground water of the Mississippi River Valley has only half as much tritium as rain does, this water must be 12 years old. Explain why this is true.

Carbon-14, or radiocarbon, also forms in the upper atmosphere, where it combines with oxygen to form carbon dioxide. Carbon dioxide becomes part of plant tissue during photosynthesis. Animals eat plants, so radiocarbon becomes part of animal tissue as well. This process is shown in Figure 15–10.

During the life of an organism, the level of radiocarbon stays about the same. When an organism dies, however, the amount of radiocarbon decreases by one half every 5730 years. A piece of wood with half the radiocarbon of a tree would be 5730 years old.

○ *What is radioactive decay?*
○ *What is meant by the half-life of a radioactive isotope?*

Figure 15–10. This diagram shows the decay of carbon-14, which has a half-life of 5730 years.

Carbon-14 remaining	Time
1/2	5730 years
1/4	11 460 years
1/8	17 190 years
1/16	22 920 years

EXTENSION (Tie-in/Chemistry) Explain that some radioactive elements decay by emitting alpha particles, or a helium nucleus of 2 protons and 2 neutrons. Other elements decay by emitting beta particles, or electrons. Both alpha and beta decay may be accompanied by the emission of gamma radiation, or high-energy X rays. Carbon-14 and rubidium decay through beta decay, as shown in the equation below.
$$C^{14} - \beta \rightarrow N^{14}$$
$$Ru^{87} - \beta \rightarrow Sr^{87}$$
Thorium as well as both radioactive isotopes of uranium decay through alpha decay as shown in the equations following.
$$U^{235} \rightarrow Pb^{207} + 7He^4$$
$$U^{238} \rightarrow Pb^{206} + 8He^4$$
$$Th^{232} \rightarrow Pb^{208} + 6He^4$$

You may want to write these three equations on the board, leaving the number of helium atoms blank, and have the students calculate the number of helium atoms necessary to balance the equation.

THINKING SKILL (*Interpreting Data*) Have the students examine the data in Table 15–1. Ask them to choose the proper test for discovering the absolute ages of the following materials if their approximate ages are known. Point out that enough of the material must have decayed and enough of the original radioactive element must remain. Wood, about 20 000 years old; minerals, about 210×10^7 years old; rock, about 4 billion years old. (wood, C-14; minerals, K-40, U-235, U-238; rock, Rb-87, Th-232)

Answers–15.3

○ Radioactive decay is the breakdown of atomic nuclei with the release of particles, such as protons and neutrons, and energy.

○ Half-life is the time it takes for half of the atoms of a radioactive isotope to decay.

ACTIVITY:
Demonstrating the Half-life of a Radioactive Isotope

Thinking Skill (*Calculating*)

Preparation of Materials Have the students wear safety goggles to avoid injury. Every shake will reduce the number of pennies by approximately half. In nature half-life is fixed by the laws of physics; the model half-life of the pennies is arbitrary. However, instruct the students that the amount of time between shakes should be approximately equal.

Answers to Conclusions/Applications

1. One-half of the pennies were removed.
2. Answers will vary but should be around 7 times.
3. Yes, if they knew the ''half-life.''
4. The bone is 11 460 years old.

15.4 Dating Minerals and Rocks

Skill (*Calculating*) Have students imagine a radioactive element with a half-life of 200 000 years. Ask them how much of the original radioactive element would be left after 800 000 years if the original amount was 7.5 grams. (about 0.47 g) Ask them how much of the decay element would be present. (7 g) Ask the students how many half-lives have passed. (4) Next have them imagine an element with a half-life of 91 days. After a year and a half, there are 3 g of the radioactive element left. Ask the students to calculate the number of half-lives that have passed, as well as the original mass of the radioactive element. (4 half-lives; 48 g) If the students are having difficulties understanding the concept of half-life, have them work on more examples such as these.

EXTENSION Give the students time to work on library-book reports about the early scientists who pioneered the study of radioactivity. You may want to suggest such scientists as Antoine Henri Becquerel, Marie Sklodowska-Curie, Ernest Rutherford, B.B. Boltwood, and Lisa Meitner.

330

How can radioactive decay be used to determine the age of a sample?

MATERIALS (per group of 3 or 4)

100 pennies, shoe box, pencil and paper, clock with second hand

PROCEDURE

1. Copy the data table shown below.

TABLE 1: HALF-LIFE		
Shake	**Time**	**Pennies in box**
0	_____	100
1	_____	_____
2	_____	_____
3	_____	_____
4	_____	_____
5	_____	_____

2. Place 100 pennies in a shoe box so that all pennies are heads up.
3. Record the time.
4. Cover the shoe box and shake it vigorously so that the pennies are well mixed.
5. Open the shoe box and remove all pennies that are tails up.
6. Record on your data table the number of pennies remaining in the box.
7. Record the time, cover the box, and shake it again.
8. Repeat Steps 5–7 until only one penny remains. Record the time for each trial.

CONCLUSIONS/APPLICATIONS

1. Approximately what fraction of the remaining pennies was removed from the box after each shaking?
2. How many times did you shake the box before only one penny remained?
3. If someone stopped you and counted only 12 pennies remaining, could he or she have calculated the time that you started the experiment? Explain.
4. Imagine that the shoe box is a fossilized bone that contains 24 pµg (picomicrogram) of radiocarbon. When it was buried, it contained 100 pµg of radiocarbon. If the half-life of radiocarbon is 5730 years, how old is the bone?

15.4 Dating Minerals and Rocks

Tritium has a short half-life; therefore, it cannot be used to date material much older than 100 years. After 100 years, there would not be enough tritium left in the material to measure. Radiocarbon has a longer half-life, so it can be used to date materials as old as 50 000 years.

There are radioactive isotopes, such as uranium-238, that have a much longer half-life—long enough to date rocks and minerals as old as the earth itself. Most igneous rocks contain traces of these isotopes. One problem with dating rocks is that scientists do not know the amounts of the radioactive isotopes that were in the rocks when they formed. Therefore, the age of the rocks cannot be found by determining how much of the isotopes remain.

What scientists can do, however, is measure the amounts of both the parent isotope (such as uranium) and the decay product (lead). The ratio of decay product to parent isotope

A MATTER OF FACT

Scientists believe that about 65 million years ago a worldwide disaster destroyed more than two-thirds of all plant and animal life on Earth, including dinosaurs. Geologists are examining rocks to find out what might have caused this extinction.

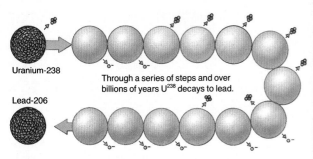

Uranium-238

Lead-206

Through a series of steps and over billions of years U²³⁸ decays to lead.

gives the age of the sample. The older the mineral or the rock, the more decay product, in this case lead, is present. Figure 15–11 shows how this process can be used to date a rock sample.

When uranium-238 decays, for example, particles shoot out in different directions, producing tracks in the surrounding material. Since this type of decay is similar to fission, these tracks are called *fission tracks*. Each fission track represents a nucleus that has decayed. The tracks can be counted. Then the remaining amount of parent isotope is determined in the laboratory. The ratio of these two quantities is used to determine the age of the sample. This method of dating is called *fission-track dating*.

Absolute dating cannot be used to date sedimentary rocks. Radioactive isotopes in a sedimentary rock can be used to date the minerals in the rock, but not the rock itself. You may recall from Chapter 4 that sediments can be transported by water and wind. Therefore, sedimentary rocks may be composed of pieces of many different parent rocks.

Metamorphic rocks cannot be dated by radioactive processes either. Temperature and pressure changes affect the amount of both parent isotope and decay product, and fission tracks are sealed over by the heat of metamorphism.

Section 2   The Geologic Clock   **331**

Figure 15–11. The diagram (left) shows the decay of uranium into lead. The scientist (right) is using a Geiger counter to locate radioactive rocks. Some rocks can be dated by determining the ratio of uranium to lead in a rock sample.

Figure 15–12. Fission tracks in rocks can be used to date some types of rocks.

A MATTER OF FACT

Over a thousand metric tons of meteors enter the earth's atmosphere every year, but most of them never hit the ground because they burn up in the atmosphere. Those that do reach the earth are used to determine the age of the earth and the solar system.

EXTENSION (Tie-in/Language Arts) Have the students research the origin of the names of radioactive elements studied in this chapter. (*Th* for Thor, of Nordic mythology; *St* for Strontian, a town in Scotland; *U* for Uranus, the planet; *Rb* from the Latin *rubidue* or red, which refers to the red lines in its spectrum) The etymology of other elements can usually be found in a dictionary.

DISCUSSION   Besides not knowing the original amount of a radioactive isotope in a mineral, there are several other problems that scientists encounter when attempting to determine a rock's age. For example, groundwater solutions can remove some of the decay product, resulting in an age determination that is too young. Heating can cause some of the decay product to escape from the rock. ***Thinking Critically:*** Ask the students whether heating will result in an age determination that is too young or too old. (Too young because it will appear as though less of the radioactive element has decayed than actually has.)

EXTENSION   The age of meteorites has been determined through radioactive dating methods. Surprisingly, it was discovered that all meteorites are about the same age, 4.5 billion years old. This data suggests that the source of these meteorites and the earth formed at about the same time.

THINKING SKILL (*Analyzing*) Remind the students that the heating and pressure involved in metamorphism change the relative amounts of parent isotope and decay product. Ask the students what can be dated using the relative amounts of these atoms. (the time at which the rocks were metamorphosed)

DISCUSSION

Explain to the students that sedimentary rocks are always older than the igneous rocks that intrude into them. Have the students explain why this is true. (The sedimentary rocks must already be in place in order for the igneous rocks to intrude into them.) Point out that although the sedimentary rocks cannot be dated, fossils within the rocks can be. Carbon-14 tests can determine the absolute age of the fossil and thus the age of the rock. Sedimentary rock layers between layers of extensive igneous rocks can be dated with relative accuracy if the absolute ages of the igneous rocks above and beneath the sedimentary layers are known.

15.5 Fossils

DISCUSSION Inform the students that even the hair of the woolly mammoths is well preserved in the frozen soil of the Arctic. *Thinking Critically:* Ask the students why these mammoth remains are so well preserved. (Animals and plants decay because bacteria cause the chemicals that compose these organisms to break down. Because bacteria cannot grow in freezing temperatures, there are few present to cause decay.) Point out that the same mechanism preserves food in freezers for indefinite periods.

EXTENSION For the location of one of the largest mammoth fossils ever discovered, have the students look at the map of Nebraska in the Science Sites booklet.

EXTENSION Explain to the students that entire animal skeletons are often preserved in large pools of tar called tar pits. With the exception of tar pits and amber, almost all fossils are found in sedimentary rock. Igneous rock is too hot while it is forming and burns up any organic matter within it. Fossils also are rarely found in metamorphic rock. *Thinking Critically:* Ask the students to explain why fossils are not present in metamorphic rocks. (The intense heat and pressure of metamorphism either totally deforms or destroys any fossils in the original rock.)

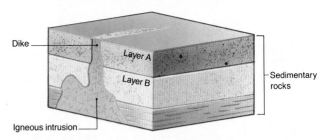

Figure 15–13. This igneous rock intrusion may be helpful in determining the relative ages of the surrounding sedimentary rocks.

Figure 15–14. Fossils, such as brachiopods, enable scientists to study organisms that lived millions of years ago.

Figure 15–15. The mammoth (right) was preserved in the frozen soil of Siberia for thousands of years, whereas the insect (left) was preserved in amber, or hardened tree resin.

Sedimentary and metamorphic rocks can be dated by relative dating. In **relative dating,** a nearby igneous formation is dated absolutely, and then the age of the neighboring formation is estimated by its position relative to the igneous rock.

○ *How can relative dating be used to date a sedimentary formation?*
○ *What is fission-track dating?*

15.5 Fossils

Fossils are the remains or traces of organisms that lived in the past. Hard parts of organisms, such as bones and shells, may be preserved intact for millions of years.

Buried in sediments, fossils often undergo chemical changes. Underground water may deposit silica in the pores of bone or wood, transforming it into stone. Often the silica replaces soft tissue. Replacement takes place slowly so that much of the original tissue structure is preserved. Replacement may be so complete as to allow scientists to study the structure of ancient cells.

Sometimes whole organisms are preserved nearly intact. Figure 15–15 shows an insect fossilized in hardened tree resin, called *amber.* The animal is preserved by the chemicals in the resin. Figure 15–15 also shows one of many Ice Age mammoths preserved for thousands of years in the frozen soil of Siberia. They were so well preserved that when thawed, their tusks were sold for the ivory and their meat was used to feed sled dogs.

Answers—15.4

○ The age of surrounding rocks is determined absolutely, then the relative age of the sedimentary rock may be determined.

○ Radioactive decay produces particles that leave tracks in igneous rocks. These tracks may then be counted to determine the rock's age.

BACKGROUND INFORMATION

The oldest fossils discovered so far are single-celled organisms estimated to be over 3 billion years old. The rocks containing these fossils were discovered in the outback of Australia. The oldest fossils in North America are found in the Canadian Shield rocks.

Figure 15–16. The remains of some animals have been found preserved in pits of tar such as the famous La Brea Tar Pits in Los Angeles, California.

Plant and animal tissue buried under thick sediments undergoes a process called *carbonization*. In the absence of oxygen, part of the tissue becomes methane and other gases, and the rest remains as pure carbon. Most of the fossils found in coal formed by carbonization.

Shells buried in sediments may be dissolved by water and replaced with debris or mineral crystals that preserve the shape of the original shell. This type of fossil, shown in Figure 15–17, is called an *external cast*. There are also internal casts. An *internal cast* is made when the internal cavity of a shell becomes filled with material, and then the shell dissolves. Internal casts often show the internal structure of the organism in great detail.

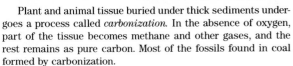

DISCOVER

Making Fossils

Take a small piece of modeling clay and form it into a disk the size of a silver dollar. Carefully press your thumb into the clay. Make sure you have a good imprint. Stack your thumbprint "fossil" with several others from members of your class and store them overnight. During the next class period, take one of the thumbprint "fossils" and try to find the person who made the print.

Figure 15–17. This series of drawings shows the formation of two types of fossils—internal casts and external casts.

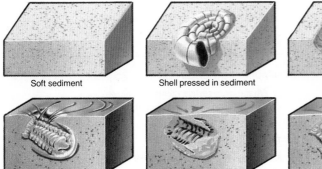

Soft sediment | Shell pressed in sediment | External cast

Animal buried | Sediment replaces animal body | Internal cast

---

DISCOVER:
Making Fossils

Thinking Skill *(Formulating Hypotheses)*

Before looking at the thumbprints the second day, ask the students to develop a set of criteria to follow as clues to each print's identity. (height, width, depth of grooves, and distinguishing scars)

DISCUSSION   You may want to mention to your class a number of other processes that form fossils. For example, sometimes organisms are buried in sediments but dissolve after the sediments have hardened. The dissolved organism leaves a hollow space in the shape of its exterior surface. This hollow is called a mold. A similar process forms imprints or thin impressions of an organism. Some fossils are not of the organism itself but are marks or other evidence left by the organism. These fossils, called trace fossils, include trails, burrows, footprints, and fossilized feces.

DISCUSSION   Have the students study Figure 15–16. Ask them to identify the process by which fossils of the material shown are formed. (carbonization)

SKILL   Have the students make observations of the internal and external casts shown in Figure 15–17. Have them describe the characteristics that distinguish one from the other.

DEMONSTRATION

For a demonstration of fossil imprints see page 319e preceding this chapter.

EXTENSION Once the absolute age of a fossil has been definitely determined by radioactive dating techniques, the fossil's age can be used to estimate the age of the sedimentary rock layer in which the fossil was formed. Fossils or organisms that existed for only one very short period of geologic time are called index fossils. Index fossils of the same organisms are all nearly the same age no matter where they are found. Thus rock layers scattered around the world can be determined to be of the same age if a particular index fossil is found in all of the layers.

DISCUSSION Remind the students that fossils can be used to determine past environments. Fossils can also indicate changes in climate. For example, fossil alligator-like animals, which were warm subtropical dwellers, have been discovered in Canada where it is very cold. Trace fossils can be used to discover what the organisms that made them looked like. *Thinking Critically:* Ask the students to describe what a scientist might learn from studying a fossil burrow. (the size of the animal, whether the organism lived alone or in groups, whether it stayed with its young, and the type of claws or feet it had)

SECTION REVIEW

Summary Using the Section Objectives, have the students summarize the major concepts of this section. (Unstable radioactive isotopes decay as they emit protons, neutrons, electrons, and energy. Rocks and minerals can be dated using K-40, Rb-87, Th-232, U-235, and U-238. Fossils of an absolute age can be used to absolutely date the rocks in which they are found.)

Reinforcement Have the students make up a chart showing the decay of a radioactive element with a half-life of 2 billion years and a starting mass of 1 kg for a period of 8 billion years. One column should show time; a second column should show the amount of radioactive element; and a third column should show the amount of decay product.

334

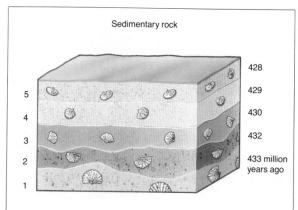

Figure 15–18. The diagram (right) shows a method for dating rocks using fossils. The scientist (left) is looking for fossils in layers of sedimentary rocks.

A MATTER OF FACT

Fossilized bones of prehistoric hominids show that early human ancestors began to walk upright about 4 million years ago.

Plant and animal species exist for a limited time. Because life evolves, fossils can be used as time indicators. Different fossils appear in sedimentary rock of different ages. By dating the fossils absolutely, or by dating similar fossils that occur elsewhere, the age of sedimentary rock can be determined.

The major geologic events in Earth's history can be dated and placed on a time line. The time line begins with the formation of the earth and continues to the present. The line is divided into sections representing specific events in geologic time. In the next chapter, you will study the history of Earth and geologic time.

○ *What are fossils?*
○ *How can fossils be used to date sedimentary rock formations?*

Section Review

READING CRITICALLY

1. What makes certain atoms radioactive?
2. State two reasons fossils are important to the study of the geologic record.

THINKING CRITICALLY

3. If a piece of wood contains 1/32 of its original radiocarbon, how old is that piece of wood?
4. How would you explain the occurrence of a formation where the younger fossils were underneath the older ones?

Answers—15.5

○ Fossils are the remains or traces of organisms that lived in the past.

○ The fossil is dated absolutely and then the relative age of the rock is determined.

Answers to Section Review

Reading Critically

1. If there are excess protons or neutrons in the nucleus of an atom, it may be unstable. The nucleus may then undergo radioactive decay.

2. Fossils are a record of the plant and animal life of the past. Some plant and animal species lived for limited amounts of time, so their fossilized remains can be used to date sedimentary strata in a given area.

continues

TECHNOLOGY: New Techniques in Radiocarbon Dating

Radiocarbon dating is the most common way to date fossils. Scientists use this method to find the age of fossils less than 50 000 years old. The use of this method depends on two things: the age of the fossil and how well it is preserved. In some cases, the fossil is too old for this dating method—there is not enough radiocarbon for instruments to detect. Background radiation, which occurs naturally, often interferes with the detection of very small amounts of radiocarbon.

To avoid this problem, a larger sample must be used, but this usually destroys the fossil. Some fossils are simply not preserved well enough for radiocarbon dating. Still other fossils are contaminated with recent carbon sources. This makes the radiocarbon method inaccurate.

Recently scientists have discovered a dating method that requires a very small sample of fossil. The new method is called *accelerator mass spectrometry*. In this process, the atoms in the sample are stripped of one electron each, and then the electrically charged ions are accelerated to high speeds. The ions are forced to move in a curved path. The heavier ions in the sample, such as radiocarbon, have greater energy than the lighter ions.

Because of their mass, the radiocarbon ions cannot be made to change direction as easily as the lighter ions. Therefore, they will separate from the lighter ions. After separation, the radiocarbon ions can be counted directly, and the age of the sample determined in the usual manner.

Some scientists think that this new method will at least double the range of radiocarbon dating, because it will allow detection of very small amounts of radiocarbon.

Searching for fossils

Examining fossils in the laboratory

Section 2   The Geologic Clock   **335**

TECHNOLOGY:
New Techniques in Radiocarbon Dating

Discussion   Tell the students that geochemists date fossils by accelerator mass spectrometry. Also point out that this method is used to determine the elements and their reactive amounts in rock samples. This method identifies the type of rock by identifying the minerals present in it.

Extension   The techniques and instruments used in mass spectrometry were first developed during World War II by scientists working on the atom bomb. An off-shoot of the bomb project was a series of studies on geochronology that led to the development of the Potassium-Argon method of mass spectrometry.

Thinking Critically

3. A piece of wood containing 1/32 of its original radiocarbon has undergone 5 half-lives. Because one half-life of C-14 is 5730 years, 5 times 5730 years equals 28 650 years. Thus the wood is 28 650 years old.

4. The sequence of the rock layers must have changed due to faulting. This would cause older, deeper-lying strata to be uplifted and pushed over more recently formed strata.

Skill (Modeling)

Preparation of Materials

Leaching is the process of subjecting a material to the washing action of a filtering liquid. During leaching, the soft, soluble substances in the material will be extracted or washed out. During petrifaction, entire cells of organic material can be leached and replaced by minerals. Soluble materials such as the calcite of mollusk shells are completely washed away, leaving just a mold in the sediment. Although neither leaching nor petrifaction actually occurs during the investigation, this will serve as an analogy. After this investigation, the students should be able to describe how molds and casts of organisms are formed both in a laboratory and in nature.

Answers to Analyses and Conclusions

1. Answers will vary, but the students should mention that fossils, are fragile and will break if not handled carefully. Paleontologists must be prepared to assemble broken samples after the samples are removed from their embedded sediments.
2. Answers will vary depending on the samples used.
3. Answers will vary depending on the samples used.

Answer to Application

The mold could be rubbed with a thin layer of petroleum jelly and filled with plaster. The hardened plaster will form a cast.

INVESTIGATION 15: Fossilizing Bones

PURPOSE

To prepare internal and external casts of bones and shells

MATERIALS (per group of 3 or 4)

Plaster of Paris
Tray
Water
Petroleum jelly
Vise
Small saw
Needle-nosed pliers
Small hammer
Chicken or lamb bones
Snail shells or seashells

PROCEDURE

1. Secure one of the bone or shell samples in the vise. Use the saw to expose the interior of the sample. Be careful not to splinter the sample.
2. Remove the sample from the vise and wash and dry it thoroughly to remove any soft organic tissue.
3. Coat the surface of each sample with a thin layer of petroleum jelly.
4. Prepare a mixture of plaster of Paris according to the directions on the package and pour the mixture into the tray.

5. Submerge all of the samples completely in the plaster. Be sure that all cavities in the sample are filled.
6. Allow the plaster to dry for at least 24 hours.
7. Use the small hammer to crack the dried plaster of Paris and expose the sample. Do not break up the plaster.
8. Use the pliers to gently remove the sample from the plaster.
9. Repeat steps 1 through 8 for as many samples as you have.

ANALYSES AND CONCLUSIONS

1. What are some of the problems geologists might have in removing fossils from rocks?
2. Distinguish between the internal and external casts made with your samples.
3. Ask other students to try to identify what samples you used to make your plaster casts.

APPLICATION

How could these procedures be modified to make casts of fossil imprints, such as dinosaur footprints?

SUMMARY

- The principle of uniformitarianism states that today's geological processes are similar to those of the past. (15.1)

- The principle of superposition states that younger rocks are deposited on top of older rocks. (15.1)

- Fossils and sedimentary features provide evidence about past environmental conditions. (15.2)

- A series of formations can provide a continuous record of environmental changes over a long period of time. (15.2)

- The time it takes for half of the atoms of a radioactive isotope to decay is called half-life. (15.3)

- Some unstable atomic nuclei undergo radioactive decay. (15.3)

- Radioactive isotopes can be used to absolutely date geologic materials. (15.3)

- Igneous rocks can be dated absolutely, but sedimentary and metamorphic rocks must be dated relative to nearby igneous formations. (15.4)

- As a radioactive isotope decays, it may leave fission tracks in the surrounding material. (15.4)

- Fossils are the remains, or traces, of organisms that lived in the past. (15.5)

- Fossils can be used to date sedimentary rock. (15.5)

Write all answers on a separate sheet of paper.

SCIENCE TERMS

Correctly use each of the following terms in a sentence.

absolute dating (329)
crossbedding (325)
half-life (328)
principle of superposition (324)
principle of uniformitarianism (323)
radioactive decay (328)
relative dating (332)

SCIENCE QUIZ

Modified True-False

Mark each statement *true* or *false*. If a statement is false, change the underlined term to make the statement true.

1. The nucleus of an atom contains protons and <u>electrons</u>.

2. Isotopes of an element have different numbers of <u>protons</u> in the nucleus.

3. The oldest rocks can be dated using <u>uranium-238</u>.

4. The age of <u>fossils</u> can be found by absolute dating.

5. Radiocarbon is commonly used to date <u>rock</u> material less than 50 000 years old.

6. The age of <u>sedimentary</u> rocks can be estimated from the age of fossils found in the rocks.

Multiple Choice

Write the letter of the choice that best answers the question or completes the statement.

7. The isotope of hydrogen known as *tritium* has an atomic number of three, so it probably has
   a) 2 protons and 1 neutron.
   b) 1 proton and 1 neutron.
   c) 1 proton and 2 neutrons.
   d) 2 protons and 2 neutrons.

continues

CHAPTER REVIEW

SUMMARY

Students may review the major concepts in the chapter by reading the summary statements. The statements are cross-referenced to the chapter to facilitate reinforcement of any concepts of which the students feel unsure. Students should be encouraged to work in groups to quiz one another.

SCIENCE TERMS

The sentence in which the science term is used should reflect an understanding of the definition of the term. You may want to make up a "word scramble" for the students in which they must unscramble the science term and then provide its definition.

SCIENCE QUIZ

Answers to Modified True-False

1. false, neutrons
2. false, neutrons
3. true
4. true
5. false, organic
6. true

Answers to Multiple Choice

7. c

continues

Answers to Multiple Choice
(continued)

8. c
9. b
10. b
11. a
12. c

Answers to Completion

13. radioactive decay
14. radioactive decay
15. radiocarbon
16. 12 years
17. relative
18. superposition

Answers to Short Answer

19. Half-life is the time it takes for a given amount of an element to decay to one-half of its original amount.
20. An unconformity is the erosional surface representing a gap in the geological record.
21. Radiocarbon is useful for dating fossils that are less than 50 000 years old. After that time too much of the original radioactive element has decayed to be measurable.
22. The principle of unformitarianism states that the earth processes that occur today are the same as those that occurred in the past.

Answers to Writing Critically

23. Answers should include a discussion of the instability of heavy nuclei.
24. Answers should include discussion of the heat extremes of magma.
25. Folding can change horizontal layers to a vertical position.

ANSWERS TO EXTENSION

1. Students will find that the fossils on these two coasts are similar if not exactly the same. This is important evidence in support of the theory of continental drift.
2. Rocks brought from the moon may be among the oldest rocks in the solar system.
3. The size of the horse has increased as the horse has evolved. The number of toes have decreased from four to one.
4. The report should include information on dating fossils.

338

SCIENCE QUIZ continued

8. What type of particles make radioactive isotopes unstable?
   a) protons and electrons
   b) neutrons and electrons
   c) protons and neutrons
   d) electrons

9. As the number of particles inside a nucleus increases, the stability of the nucleus often
   a) increases.
   b) decreases.
   c) remains the same.
   d) cancels the forces of gravity.

10. The best isotope to use in dating wood, bone, and shell is
    a) tritium.
    b) radiocarbon.
    c) rubidium-87.
    d) uranium-235.

11. Absolute dating can be used to find the age of
    a) igneous rocks.
    b) sedimentary rocks.
    c) metamorphic rocks.
    d) all rocks.

12. Unconformities in the rock record are due to
    a) fission.
    b) radiocarbon.
    c) missing sediments.
    d) extra neutrons

Completion

Complete each statement by supplying the correct term.

13. An element whose nucleus loses a particle containing 2 protons and 2 neutrons and energy has undergone _____.

14. The process of decay of the nucleus of an atom is called _____.

15. Another name for carbon-14 is _____.

16. Water containing tritium will lose half of its radioactivity in about _____ years.

17. Sedimentary rocks can be dated by _____ dating.

18. The geologic principle that states that younger sediments are deposited on top of older sediments is the principle of _____.

Short Answer

19. Explain the meaning of the term *half-life.*

20. What is an unconformity?

21. Explain why radiocarbon would not be useful in dating very old samples.

22. Explain the principle of uniformitarianism.

Writing Critically

23. How might excess neutrons and protons make an atomic nucleus unstable?

24. Explain why fossils are not found in igneous rocks.

25. According to the principle of superposition, sediments are deposited in horizontal layers. However, in some places rock layers are nearly vertical. Explain this apparent contradiction between the principle and facts.

EXTENSION

1. At your local or school library, research the types of land fossils that have been found on the east coast of South America and the west coast of Africa. Compare and contrast these fossils, and then use tracings and drawings to support the theory that these two continents were once joined.

2. Use the library to find information about the Apollo moon missions of the 1970s. Explain why these space missions were helpful in determining the age of the earth and the solar system.

3. Trace the evolution of the horse over the past 50 million years. Display this information in class in the form of a time line.

4. From the list For Further Reading, read the *National Geographic* article "Fossils: Annals of Life Written in Rock" and report to the class on the article.

APPLICATION/CRITICAL THINKING

1. A geologist finds a fossilized bone in a canyon wall. Carefully digging out the fossil, the geologist takes a small sample of the bone and discovers that the sample contains about 2 pμg of radiocarbon. If the geologist assumes that a similar amount of living tissue would contain about 16 pμg of radiocarbon, what would most likely be the age of the bone?

2. Scientists do not have a complete record of the living forms that have inhabited our planet. There are many gaps in the fossil record. Discuss the processes that may have caused gaps in the record.

3. The sedimentary rock walls of the Grand Canyon have often been referred to by geologists as a "history book of the earth's geological past." Explain what is meant by this phrase, and tell why the walls of the Grand Canyon, unlike any other place on Earth, provide so much fossil evidence of past geological processes.

FOR FURTHER READING

Cairne-Smith, A. G. *Seven Clues to the Origin of Life: A Scientific Detective Story.* New York: Cambridge University Press, 1985. The author uses a Sherlock Holmes approach to develop a possible scenario for the origin of life during the early history of Earth.

Jeffery, D. "Fossils: Annals of Life Written in Rock." *National Geographic* 168 (August 1985): 182. This article gives a colorful explanation of how Earth's ancient life has been preserved in rocks.

Weaver, K. "The Search for Our Ancestors." *National Geographic* 168 (November 1985): 560. This article contains brilliant photographs and a review of the most recent evidence of human evolution.

Challenge Your Thinking

This photograph shows a break in a sedimentary rock formation. At many points in the sedimentary record, there are places where fossils just seem to disappear. What do you think could cause these gaps in the rock record?

ANSWERS TO APPLICATION/ CRITICAL THINKING

1. If the scientist assumes that a similar sample of living tissue contains 16 pμg of carbon-14 (which may not be strictly accurate), then the fossilized sample containing one-eighth that amount has gone through three half-lives. Since the half-life of radiocarbon is 5730 years, then three times 5730 years is equal to 17 190 years. The scientist would conclude that the fossil is 17 190 years old.

2. Natural forces of erosion and biochemical decomposition have limited the number and quality of plant and animal life that has been preserved in the fossil record.

3. The Grand Canyon cuts through hundreds of layers of sedimentary rocks, allowing scientists to study millions of years of Earth's history.

ANSWERS TO CHALLENGE YOUR THINKING

Faulting, folding, and erosion all cause gaps in the rock record.

PLANNING THE CHAPTER

Chapter Sections	Page	Chapter Features	Page	Program Resources	Page
Section 1: The Cryptozoic Eon	341			Reading for Content: *Using Charts to Organize Information* **(A)**	TRB 28 LM 203
16.1 Divisions of Earth History **(A)**	341			Investigation 16.1: *Microorganisms in the Environment* **(A)**	TRB 67 LM 69
16.2 The Hadean Era **(A)**	342	A Matter of Fact	342		
16.3 The Archean Era **(A)**	344	A Matter of Fact	344		
		A Matter of Fact	345		
16.4 The Proterozoic Era **(A)**	346	**Discover:** Creating Life? **(A)**	347		
		Section Review	347		
		Skill Activity: Organizing Background Data **(A)**	348		
Section 2 The Phanerozoic Eon	349			Critical Thinking **(H)** Cross-Discipline: Science and Mathematics, *Examining Earth's History* **(A)**	TRB 30 TRB 30
16.5 The Paleozoic Era **(A)**	349	**Discover:** Comparing Relatives **(A)**	349	Concept Extension: *Life in the Paleozoic* **(H)**	TRB 33 LM 255
		A Matter of Fact	350	Investigation 16.2: *Representing Geologic Time* **(A)**	TRB 71 LM 73
16.6 The Mesozoic Era **(A)**	352				
16.7 The Cenozoic Era **(A)**	353	Section Review	356		
		Activity: Making a Geologic Time Line **(A)**	356		
		Biographies: Then and Now Harold Urey, Lynn Margulis	357		
		Investigation 16: Classifying Fossils **(A)**	358	Student Record Book: Textbook Investigations **(A)**	TRB 31
Chapter 16 Review	359			Vocabulary **(A)**	TRB 16 LM 153
				Tests **(A)** Computer Test Bank	TRB 68

(LM) Laboratory Manual/Study Guide, **(TRB)** Teacher's ResourceBank™

B = Basic **A** = Average **H** = Honors

The coding Basic, Average, and Honors indicates sections or subsections that might be appropriate for different levels of learners. For additional suggestions regarding choice of topic and depth of coverage, see the Pacing Chart on pages T16–T20.

CHAPTER CONCEPTS, OBJECTIVES, AND TERMS

Section	Concepts	Objectives	Science Terms
Section 1: The Cryptozoic Eon	■ Earth's history has two main divisions, or eons: the Cryptozoic Eon and the Phanerozoic Eon. **(16.1)** ■ The Cryptozoic Eon is the time interval from the formation of Earth to about 590 million years ago. **(16.1)** ■ The Cryptozoic Eon is divided into three eras: the Hadean, the Archean, and the Proterozoic. **(16.1)** ■ Most of the earth's crust was formed during the Hadean Era. Dense basalt formed the oceanic crust while the lighter granite formed the continental crust. **(16.2)** ■ The fundamental chemical reactions necessary to living cells were developed by bacteria during the Archean Era. **(16.3)** ■ Multicellular animals evolved during a glaciation that took place in the Proterozoic Era. **(16.4)**	■ **Describe** how the earth's crust formed and grew through time. ■ **Discuss** the role of bacteria in the evolution of life on Earth. ■ **Explain** when and how multicellular animals evolved.	planetesimals fermentation cellular respiration
Section 2: The Phanerozoic Eon	■ The Phanerozoic Eon is divided into the Paleozoic, Mesozoic, and Cenozoic eras. **(16.5)** ■ Trilobites were marine animals that lived during the Paleozoic Era. **(16.5)** ■ Great forests developed during the Carboniferous Period. **(16.5)** ■ Reptiles were the first animals to deposit their eggs on dry land. **(16.5)** ■ Dinosaurs became the dominant animals during the Mesozoic Era. **(16.6)** ■ A major extinction of plants and animals occurred at the end of the Mesozoic Era. **(16.6)** ■ *Homo sapiens* evolved about 0.125 million years ago. **(16.7)**	■ **List** the characteristics of trilobites. ■ **Compare** the mass extinctions that occurred at the end of the Ordovician and Devonian Periods. ■ **Describe** the ice ages of the Cenozoic Era.	glaciation

Title	Page	Materials
Discover: Creating Life	347	*(per student)* fresh hay, water, laboratory beaker or glass pan, microscope, burner or hot plate, microscope slide, medicine dropper
Skill Activity: Organizing Background Data	348	*(per student)* tracing paper, pencil, life science or biology textbook
Discover: Comparing Relatives	349	*(per student)* paper, pencil
Activity: Making a Geologic Time Line	356	*(per group of 3 or 4)* construction paper, tape, markers, metric ruler, scissors
Investigation 16: Classifying Fossils	358	*(per student)* set of ten numbered fossils, hand lens or stereomicroscope

TEACHING SUGGESTIONS

Section 1: The Cryptozoic Eon

Class Activity

Place marbles around the perimeter of a Petri dish (or any pan with a shallow rim). Add additional marbles to the center. The marbles around the perimeter are analogous to the earth's fragile crust. The marbles in the center are materials from the earth's mantle. Due to convection, mantle rises and new crust is created at mid-ocean ridges. However, the upwelling pressure and plate separation causes the remaining crust to buckle and crack. What happens to the "crustal" marbles along the rim of the dish as you try to fit some of the "mantle" marbles into the "crust?" Of course, they are forced back toward the center, or "subducted into the mantle." Had this crust contained evidence of living organism, such as fossils, what would have become of them? *(They would have been destroyed in the mantle.)*

Class Activity

Collect specimens of single-celled organisms from a nearby, safe, unpolluted stream, pond, or lake. Examine the creatures under a microscope and compare and contrast their shape and behavior.

Section 2: The Phanerozoic Eon

Demonstration: Preserving Plants

Purpose

To demonstrate how to preserve plants

Background

If you cut a flower from a plant and place it in water, the flower lives for a while but then it withers and dies. But the beauty of a flower may be preserved in various ways. Some flowers can be dried. Others can be preserved in stone, as fossils, or in sand or ice. In these ways information about flowers and plants of long ago has been recorded for us. In this demonstration, you will preserve flowers in ice and in sand.

Materials

Small beaker
Small flowers
Fine-grained sand
Shallow box
Small paint brush

Procedure

1. Add just enough water to the beaker to float a flower. Remove the flower and place the beaker in the freezer. When the water begins to freeze, pour off the cold water, add more water, and return the beaker to the freezer.
2. When the added water begins to freeze, put the flower in the cold water.
3. Leave the beaker in the freezer. In about 20 to 25 minutes the flower will be caught in the ice. Discard the unfrozen water and add more water to cover the flower. (You will find it impossible to cover the flower with water to begin with because the flower will float. Discarding the cold water and adding other water will result in clearer ice.)
4. After the ice is frozen with the flower in it, remove the ice from the beaker and have the students examine the flower from all sides. Keep the ice and flower in the freezer and have the students examine the flower from day to day.
5. Preserve some flowers in sand. Cover the bottom of a shallow box with several centimeters of clean, fine-grained sand. Put several flowers on the sand. Sprinkle sand carefully over the flowers so that the sand reaches all parts of the flower. Cover the flowers with sand to a depth of several more centimeters.

6. Leave the box undisturbed for at least two weeks.
7. Carefully remove the sand. If necessary, brush away the sand with a small paint brush.

Questions to Ask the Students

1. Which method preserves a plant for a longer time?
   (freezing)
   Which method do you think would give scientists the most information about plants of long ago?
   (Freezing, because it preserves organisms with a minimum of change.)
2. Describe the flower before and after freezing.
   (The flowers look pretty much the same.)
3. Describe the flowers before and after drying in the sand.

How are they the same?
(fresh, colorful)
How are they different?
(dry or decayed, no color)

Field Trip

If there is an archeological site in your area, arrange for a field trip to observe the work being done.

Outside Speaker

Have an archeologist talk to the class about the importance of preserving historical sites for archeological study. The speaker might also discuss the scientific knowledge that has been gained from these sites.

339d

CHAPTER 16

CHAPTER OVERVIEW

This chapter describes the geologic history of Earth from its formation to the evolution of humans. Discussed are the two main geologic divisions—the Cryptozoic and Phanerozoic eons. The eras that make up each eon, and the important developments that occurred during each are also discussed. To aid in the discussion, the chapter compares the eons and eras to a calendar year.

Section 1: The Cryptozoic Eon This section discusses the formation of the earth and the other planets during this eon, which covers the longest time in Earth's history. It also describes the development of the oceanic and continental crusts and the evolution of living cells, simple organisms, and multicellular animals.

Section 2: The Phanerozoic Eon The Paleozoic, Mesozoic, and Cenozoic eras are discussed in this section. Topics include the evolution of trilobites and other forms of marine life, the development of large forests and other land plants, the evolution of reptiles, and the major extinctions of some species of plants and animals. Finally, this section describes present-day plants and animals including the evolution of *Homo sapiens* during the Cenozoic Era.

Earth's Past

These scientists are studying history—Earth history. The study of Earth history is a bit different from the history you may be familiar with. Studying Earth history is a physically demanding job, but it is one that offers great rewards. Imagine discovering, as these men have done, the skeleton of an animal that no human has ever seen before.

Examining a new fossil find

340

CHAPTER MOTIVATING ACTIVITY

Show the students pictures taken by *Viking* landers, which visited the planet Mars in the late 1970s. Discuss the methods and equipment used by NASA scientists to test the soil of Mars in an effort to find some evidence of life. Ask the students what they would consider to be conclusive evidence of the existence of life on *any* planet: respiration using oxygen? presence of water? physical evidence such as fossil remains?

Show pictures of the other planets and discuss known facts about their atmospheres and surface features. Compare and contrast these features with those of Earth. Ask the students what forces might have caused these similarities and differences.

Discuss some of the difficulties scientists have encountered in interpreting data from the latter eras of the Cryptozoic Eon. (Soft, fleshy organisms are not preserved easily.)

1 The Cryptozoic Eon

SECTION OBJECTIVES

After completing this section, you should be able to:
- **Describe** how the earth's crust formed and grew through time.
- **Discuss** the role of bacteria in the evolution of life on Earth.
- **Explain** when and how multicellular animals evolved.

NEW SCIENCE TERMS

planetesimals
fermentation
cellular respiration

16.1 Divisions of Earth History

The history of the earth covers such a long period that most people have trouble imagining when some important event occurred. If you try to imagine the whole of Earth's past compressed into one calendar year, it becomes easier to understand. Like a calendar year, which is divided into months of different lengths, the calendar of Earth's history can be divided into unequal segments.

Earth's history has two main divisions, or *eons* (EE ahnz): the Cryptozoic (krihp toh ZOH ihk) Eon and the Phanerozoic (fan IR uhz OH ihk) Eon. Just as a calendar year is divided into months, eons are divided into *eras* (EHR ahz). Like months, eras are not all the same length.

The calendar of Earth history begins with the *Cryptozoic Eon.* This eon includes the time between the formation of the solar system, probably about 4.6 billion years ago, and the appearance on Earth of abundant fossils, about 590 million years ago. The Cryptozoic Eon is followed by the *Phanerozoic Eon,* which covers the period from about 590 million years ago to the present.

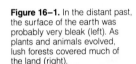

Figure 16–1. In the distant past, the surface of the earth was probably very bleak (left). As plants and animals evolved, lush forests covered much of the land (right).

Section 1   The Cryptozoic Eon   **341**

BACKGROUND INFORMATION

The prefix *crypto-* comes from the Greek word *kryptos,* meaning "secret or hidden." The suffix *-zoic* comes from the Greek *zoikos,* pertaining to animals. Thus the Cryptozoic Eon refers to the fact that scientists have been unable to discover any major record of living organisms from this time.

Section 1: The Cryptozoic Eon

SECTION FOCUS

Section Overview   This section concentrates on the longer of the two eons of Earth's history, the Cryptozoic Eon. Discussed are the important developments that occurred during this time, including the formation of the earth and its crust, and the evolution of simple life forms.

Section Objectives   For a list of section objectives, see pupil's edition page.

New Science Terms   For a list of new science terms in this section, see pupil's edition page. You may wish to call the students' attention to the new terms they will meet in this section and then pronounce the terms.

SECTION DEVELOPMENT

16.1 Divisions of Earth History

DISCUSSION   Direct the students' attention to Figure 16–1. Discuss the characteristics of the earth in the left-hand photograph. Ask the students to explain how the mountains in the picture may have formed.

THINKING SKILL *(Interpreting Word Parts)* Have the students look up the prefixes *crypto-* and *phanero-* and the suffix *-zoic* in a dictionary. Ask the students to hypothesize why this first eon is named the Cryptozoic Eon. (Refer to Background Information on this page.) Ask them to hypothesize how two eons that seem so opposite in meaning both describe Earth's past. (Earth's history includes periods in which evidence of life was both hidden and visible.)

EXTENSION   There are several other words that begin with the prefix *crypt-.* ***Thinking Critically:*** Ask the students to list some of these words and discuss their relationship to the Cryptozoic Eon. (Answers will vary but could include crypt, cryptography, and cryptogram.)

Figure 16–2. Some of the oldest rocks on Earth have been found in Greenland. These rocks are nearly 3 billion years old.

The Cryptozoic Eon covers 87 percent of the earth's history. On a scale of a one-year period, this would approximate the time from New Year's Day to Thanksgiving on the calendar.

The Cryptozoic Eon is divided into three eras: the Hadean (HAY dee uhn), the Archean (AHR kee uhn), and the Proterozoic (proht uhr oh ZOH ihk). The *Hadean Era* is the time between the formation of the solar system and the age of the oldest known rocks on Earth. In a calendar year, this era would last about two months. The *Archean Era* ranges from about 3.8 billion years ago to about 2.7 billion years ago, or approximately three months. The *Proterozoic Era* was the period from about 2.7 billion years ago to about 590 million years ago; this would equal about 5.5 months on a year's calendar.

○ **What are the two main divisions of Earth history?**

16.2 The Hadean Era

During the first 100 million years of the Hadean Era, the earth and the other planets probably formed. Scientists believe that the planets formed from small bodies of gases and solids called **planetesimals** (plahn uh TEHS uh muhlz). The planetesimals circled the sun in rings, much like the rings that now circle Saturn and other planets. The largest planetesimal within each ring probably captured the smaller ones, and an early planet formed. Why would the largest planetesimal capture the smaller ones?

As the earth formed, its surface probably resembled that of today's moon. Craters and large plains of basalt marked the surface. The earth was hot, and there were large pools of molten

A MATTER OF FACT

The idea that the solar system formed from a rotating disk of dust and gas is not new. This was first proposed by the great French philosopher René Descartes in the year 1644.

Figure 16–3. In the Hadean Era, the surface of the earth was probably much like the present surface of the moon.

342

16.2 The Hadean Era

DISCUSSION Direct the students' attention to the three pictures in this section. Have the students describe each one. *Thinking Critically:* Ask the students to tell how lava flow from erupting volcanoes and torrential rains can alter the character of the landscape. (Lava flow can damage or destroy trees and plants; offshore lava flows can create new land masses. Rains can cause mudslides, erosion, and flood damage.)

THINKING SKILL (*Comparing*) Point out that each planet is unique. Ask the students to compare and contrast Earth with what they may know about other planets. (Answers will vary but students may mention that Earth has an atmosphere, Mercury does not, and Venus has a dense cloud layer of sulfuric acid; Earth has land surfaces and Jupiter is primarily gaseous; Earth has water and Mars has only frozen water on the poles; Earth has one satellite (moon) and Jupiter has 16; and Earth and other planets have volcanoes.)

Answer–16.1

○ The two main divisions of Earth history are the Cryptozoic Eon and the Phanerozoic Eon.

BACKGROUND INFORMATION

The solar system is believed to have formed from cosmic dust and gas. Gravity caused the accumulations of matter to give rise to planetesimals and finally, planets.

EXTENSION Stories about acid rain constantly appear in the news. Acid rain is produced when moisture in the atmosphere (in the form of rain, sleet, hail, or snow) mixes with sulfur dioxide and some nitrogen oxides. One source of sulfur dioxide is the smoke emitted from some coal, which is burned to produce electricity. Another source is the exhaust from automobiles. Acid rain kills trees and plants, as well as fish and other forms of life that live in freshwater lakes and streams. It also causes buildings and monuments to deteriorate and contaminates drinking water.

lava. There was a thick atmosphere of water vapor, carbon dioxide, and sulfur dioxide. As the earth cooled, the water vapor probably condensed, and torrential rains began. Water collected in craters and other surface depressions. The carbon dioxide and sulfur dioxide remained in the atmosphere.

The earth's original surface, or crust, consisted of silicate rock. Basalt, a dense rock, formed the crust of low-lying areas, such as the ocean basins. Less-dense granite formed the elevated areas—the continents. Scientists believe the crust greatly increased in size during the Hadean Era and is, in fact, still growing. Almost all of the continental crust, however, probably formed during the Cryptozoic Eon.

The interior of the earth remained hot due to heat trapped during its formation. In fact, the interior was so hot that convection currents developed. Hot rocks from the interior rose to the surface, and cooler surface rocks sank. The convection currents brought together blocks of granite to form crustal plates. These crustal plates formed a lid and trapped the heat below.

By the end of the Hadean Era, scientists believe, the surface of the earth had large oceans. There were barren continents with mountain ranges, plains, rivers, and volcanoes. There were many clouds, strong winds, and torrential rains, which were acidic due to the sulfur dioxide in the atmosphere.

○ *Name three events that occurred during the Hadean Era.*

Figure 16–4. The earth's primitive atmosphere probably contained carbon dioxide, sulfur dioxide, and water vapor. These gases may have escaped from the interior of the earth through volcanic vents (left), creating torrential rains that filled the earth's oceans.

Section 1   The Cryptozoic Eon   **343**

DEMONSTRATION

For a demonstration of preserving plants, see page 339c preceding this chapter.

Answer–16.2

○ During the Hadean Era the planets formed, the crust of the earth greatly increased in size, and large oceans formed on the earth.

16.3 The Archean Era

DISCUSSION Remind the students that as far as scientists know, Earth is the only planet that is able to support life. ***Thinking Critically:*** Ask the students to describe the conditions on Earth that combined to help life forms evolve from simple chemical elements. (Conditions on Earth that contributed to the evolution of life forms include the presence of complex organic molecules, carbohydrates, and amino acids; an atmosphere that permitted ultraviolet rays to enter; high-energy lightning from many violent storms; and the presence of the oceans.)

EXTENSION (Tie-in/Nutrition) Have the students review the definition of *carbohydrates*. Explain that carbohydrates are important food used to supply energy. ***Thinking Critically:*** Ask the students to name some foods in their diet that are important sources of carbohydrates. (vegetables, especially legumes—peas and beans; fruits such as apples; whole grains and cereals such as wheat and flour used in bread)

THINKING SKILL (*Applying Concepts*) Remind the students that although simple organisms such as yeast evolved during the Archean Era, they are still very much a part of life today. Ask the students to give examples of current applications of fermentation using yeast. (Yeast is used to make bread rise and to produce alcoholic beverages.)

EXTENSION (Tie-in/Nutrition) Have the students review the definitions of *amino acids* and *proteins*. Explain that proteins are essential to all living cells because they help build new tissue during growth, help replace tissue destroyed during metabolism, and also provide a source of energy. ***Thinking Critically:*** Ask the students to name some important sources of protein in their diet. (meat and poultry, fish, eggs, milk, and cheese)

16.3 The Archean Era

In early March on the one-year calendar of Earth history, the Archean Era began. Many important events occurred during the Archean Era, the most important of which, according to scientists, was the evolution of life.

Among the many different molecules on the surface of the primitive earth, complex organic molecules containing carbon were probably common. However, to progress from complex molecules to even the simplest living organism was a very long process.

Figure 16–5. During the Archean Era, the waters of the earth probably contained a fermenting organic broth (left). Dr. Stanley Miller (right) tested the hypothesis that organic chemicals could be produced from simple chemicals in the earth's atmosphere. By simulating conditions of the primitive atmosphere, Miller proved that organic molecules could be produced from simple chemicals.

Among the first, and most important, chemicals of life were the *carbohydrates*. Carbohydrate molecules are made of carbon, hydrogen, and oxygen atoms. Carbohydrates are easily broken down by simple organisms. Primitive bacteria, probably some of the first living organisms to evolve, used carbohydrates in a process called *fermentation*. **Fermentation** is the partial breakdown of sugar, a simple carbohydrate, to release energy. This process takes place without oxygen, and it is not very efficient. However, it is likely that it was the basis for most early forms of life. Today simple organisms such as bacteria and yeasts get their energy through fermentation.

Another important group of compounds probably present on the early earth were the *amino acids*. Amino acids are molecules made of carbon, hydrogen, oxygen, and nitrogen atoms. These amino acids probably joined together by chance to make long complex molecules known as *proteins*. Much of the tissue of living organisms is made of proteins. In addition, many chemical processes require the presence of special proteins.

Perhaps the most important organic molecule produced was *chlorophyll.* Chlorophyll allowed early organisms to make their

A MATTER OF FACT

The sequence of amino acids in similar proteins can be used to trace the evolution of organisms from one species to another.

344 Chapter 16 Earth's Past

BACKGROUND INFORMATION

The formation of amino acids from NH_3, H_2O, CO_2, and CH_4 requires a great deal of energy. In the laboratory, electricity and ultraviolet light can be used in the synthesis. The primitive atmosphere probably had ample supplies of both. An oxygen- and ozone-free atmosphere allowed the sun's ultraviolet rays to pass through. Violent lightning storms probably characterized the Cryptozoic Eon.

Special proteins called enzymes control the rate and sometimes the products of a chemical reaction by slightly changing the shape or orientation of the reacting molecules. The reacting molecules are brought into close proximity where chemical bonds can be broken or rearranged.

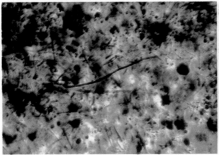
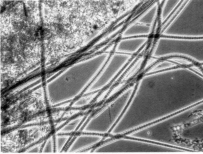

own food. Molecules of chlorophyll consist of a magnesium atom surrounded by nitrogen, carbon, hydrogen, and oxygen atoms. Solar energy was used by chlorophyll-containing organisms to produce carbohydrates from carbon dioxide and water.

Photosynthesis first appeared in organisms known as *cyanobacteria*. An example of an early fossil cyanobacterium is shown in Figure 16–6. These ancient bacteria were similar to modern cyanobacteria, such as *Nostoc*, also shown in Figure 16–6.

Scientists believe these cyanobacteria gave off oxygen as a byproduct of photosynthesis. For about 2 billion years the oxygen combined with iron dissolved in the water of the primitive earth. Iron oxide layers accumulated on the bottom of lakes and oceans, where they formed structures called *stromatolites*. As most of the available iron was oxidized, what do you think happened to the oxygen produced by photosynthesis? Why?

○ *What organisms appeared during the Archean Era?*
○ *Why was chlorophyll important in the evolution of new organisms?*

Figure 16–6. Ancient cyanobacteria (left) were probably the first organisms to use chlorophyll to make their own food. *Nostoc* (right), a modern cyanobacterium, produces its own food in much the same way.

A MATTER OF FACT

Oxygen present in the atmosphere when cyanobacteria began to photosynthesize was only one one-thousandth of its present level.

Figure 16–7. Stromatolites, such as these found in the Bahamas, were formed as oxygen from photosynthesis combined with iron in the rocks on the bottom of the ocean. These layers alternate with layers of sand, algae, and calcium carbonate.

Section 1   The Cryptozoic Eon   **345**

EXTENSION Bacteria, like yeasts, also evolved during this era. They too are still very important elements of life today. Bacteria are found in the bodies of all living things and on all parts of Earth. Most are beneficial for purposes such as enriching the soil, pickling foods, and decomposing organic wastes in the soil and in sewage disposal plants. Harmful bacteria, called pathogens, can cause plant diseases such as wilt and animal diseases such as cholera, typhoid fever, tetanus, and tuberculosis.

EXTENSION In addition to enabling plants to conduct photosynthesis, chlorophyll is the material that gives green plants their color. Chlorophyll looks green because the green part of the light spectrum is reflected and the blue-violet parts are absorbed.

SKILLS *(Observing/Comparing)* Have the students list the chemical properties of carbohydrates, amino acids (proteins), and chlorophyll. *Thinking Critically:* Ask which elements are common to all three and are therefore the building blocks of life. (carbon, hydrogen, and oxygen)

THINKING SKILL *(Drawing Conclusions)* Remind the students that some bacteria will grow in the presence of oxygen. Ask the students to speculate whether *all* bacteria can do so. (There are different kinds of bacteria. Aerobes can grow only in the presence of oxygen, anaerobes cannot grow when oxygen is present, and facultative bacteria can grow with or without oxygen.)

BACKGROUND INFORMATION

Adenosine triphosphate (ATP) is the ultimate product of glycolysis and the tricarboxylic acid (or Kreb's) cycle. The energy in glucose is released during this complex series of intracellular reactions and is stored in the chemical bonds of ATP. Mitochondria are the organelles that act as the "energy factories" responsible for this product. Photosynthesis involves the reactants, carbon dioxide and water, which are energized by the sun's rays in the presence of the catalytic enzyme chlorophyll. The resulting products are glucose and molecular oxygen.

Answers–16.3

○ Bacteria appeared during the Archean Era.

○ Chlorophyll was important because it allowed early organisms to make their own food.

16.4 The Proterozoic Era

DISCUSSION Remind the students that this era is important in evolutionary history. Direct their attention to Figure 16–10 and point out that sponges are one of the earliest forms of multi-cellular organisms. Ask the students if they have ever seen a real sponge. *Thinking Critically:* Ask the students to describe what they saw and whether they recognized different types of sponges. Ask them if the sponge looked alive.

THINKING SKILL *(Comparing)* Review the definition of cellular respiration with the students. Ask them to explain the advantages of cellular respiration over fermentation. (Cellular respiration releases energy much more efficiently than does fermentation.)

THINKING SKILL *(Inferring)* Ask the students to describe the advantages of possessing a more efficient means of releasing energy. (requires fewer nutrients for the organism's survival or the organism can accomplish more using the same amount of nutrients, for example, can use "excess" energy to reproduce or to develop other means of survival)

EXTENSION The evolution of cellular respiration was a very important development during this era. Point out that some primitive organisms and many modern protozoa use cellular respiration. *Thinking Critically:* Ask the students if they can name any other modern organisms that do not use this method of releasing energy. (Answers will vary but the students should realize that many modern organisms, including anaerobic bacteria do not use cellular respiration.)

DISCUSSION Direct the students' attention to Figure 16–9. *Thinking Critically:* Ask the students to explain why scientists think green algae and protozoa may have a common evolutionary link. (They are similar in structure, both have a cell wall, a nucleus and cytoplasm, and both use cellular respiration.)

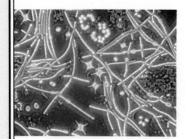

Figure 16–8. Diatoms are single-celled organisms that produce much of the oxygen needed for cellular respiration.

Figure 16–9. The life style of *Euglena* (bottom left) may provide a clue to the evolution of animals. In sunlight, *Euglena* produces food by photosynthesis like the green alga (top left). Without sunlight, Euglena can gather food like *Paramecium* (bottom right).

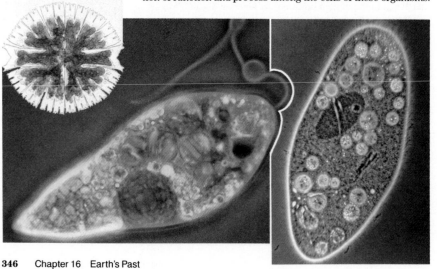

16.4 The Proterozoic Era

The longest "month" on the earth-history calendar is the Proterozoic Era. Many important events occurred during this era.

Scientists believe that during the Proterozoic Era life forms continued to evolve and occupy new habitats. Bacteria began to form simple groups. Some of these groups became enclosed in protein membranes, forming single complex cells. These cells were much more complex than the early bacteria. Most of these organisms continued the process of photosynthesis. In appearance, they were probably similar to modern green algae.

During this era some algae lost their ability to photosynthesize; they evolved a process for releasing energy from carbohydrates. This process, known as **cellular respiration,** is the breakdown of carbohydrates in the presence of oxygen. Respiration releases almost ten times more energy than fermentation does.

Having lost the ability to photosynthesize, these organisms, which were probably similar to present-day protozoa, had to depend on food made by cyanobacteria and green algae for their survival. Figure 16–9 shows a possible evolutionary link between algae and protozoa.

In a period of about 150 million years—or less than two weeks on the earth-history calendar—protozoa evolved into multicellular organisms. Multicellular organisms were more than just collections of simple cells. There was much specialization of function and process among the cells of these organisms.

BACKGROUND INFORMATION

Bacteria and yeast carry out anaerobic breakdown of glucose in the absence of oxygen. This type of energy production is called fermentation. Unlike aerobic respiration, which produces carbon dioxide and water, the products of fermentation are lactic acid, carbon dioxide, and alcohol (instead of water). Fermentation yields less ATP than does respiration.

Answers—16.4 (p. 347)

○ Cellular respiration is the breakdown of carbohydrates in the presence of oxygen.

○ Multicellular organisms evolved during the Proterozoic Era.

 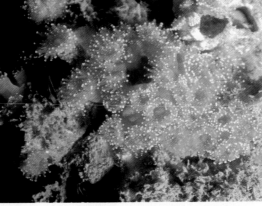

Most of these organisms were similar to the group of animals to which modern corals belong. Others were like present-day marine worms. There was nothing simple about these early animals; they were as complex as their modern counterparts.

These animals probably fed by eating algae or each other. As this food-hunting practice, known as *predation*, developed, a number of organisms evolved protective body coverings. These coverings were hard and left many fossils. The appearance of abundant fossils in the geologic record marks the beginning of the next eon, the Phanerozoic Eon.

○ **What is cellular respiration?**
○ **What type of complex organisms evolved during the Proterozoic Era?**

Section Review

READING CRITICALLY

1. List several kinds of organic compounds and their functions in living organisms.
2. What is photosynthesis? What effect did it have on the development of life?

THINKING CRITICALLY

3. Why were the rains during the Hadean Era acid rains?
4. Even though multicellular organisms preyed on each other, what provided the original energy for all life on Earth? Explain your answer.

Section 1 The Cryptozoic Eon **347**

Figure 16–10. The earliest animals may have been similar to simple sponges (left) and corals (right).

DISCOVER

Creating Life?

Put some fresh hay and water in a laboratory beaker or a glass pan. Gently boil the mixture until the water appears amber in color. After the water sits for a day or two, observe a drop of it under a microscope. Draw several of the organisms that you see. Where might these organisms have originated?

THINKING SKILL (Inferring) Review the text material on the evolution of multicellular organisms. Ask the students to identify the advantages that multicellular organisms enjoy because they have different, specialized structures. These organisms can sense things in their environment such as presence of nutrients; can move toward nutrients and away from threats; and can protect themselves.

DISCUSSION Direct the students' attention to Figure 16–10 and remind them of the opening discussion on sponges. Ask them to describe the structure of corals by looking at the picture and thinking about the nature of multicellular organisms. Explain that the structure they see is actually the organism's protective body covering. Within its protective covering, coral also has tentacles to capture food, a soft digestive tract to process the nutrients, and reproductive organs.

THINKING SKILL (Inferring) Remind the students that the emergence of organisms with protective body coverings marked the end of the Cryptozoic Eon and the start of the Phanerozoic Eon. Ask them why this is significant given the prefixes in the names of the two eons. (It made it much easier for scientists to collect evidence of the existence of these early plants and animals because their remains were fossilized. Therefore, their existence was no longer "hidden.")

DISCOVER: Creating Life?

Thinking Skill (Inferring)

Try to get fresh hay from a local feed store or farm. You may have to call several riding academies or stables to locate some.

The water may not have been boiled long enough to kill the organisms.

SECTION REVIEW

Summary Have the students write short paragraphs about the history of Earth and the evolution of life during the Cryptozoic Eon.

Reinforcement Have the students quiz each other on the vocabulary terms.

Answers to Section Review

Reading Critically

1. Enzymes control the rate of chemical reactions. Carbohydrates give plants and animals energy to do work. Lipids form protective layers, like membranes. Nucleic acids carry genetic information.
2. Photosynthesis is the production of simple sugar using carbon dioxide, water, and energy from the sun. Photosynthesis provided oxygen; life could not have evolved without oxygen.

Thinking Critically

3. The rains during the Hadean Era were acidic because of the sulfur dioxide in the atmosphere.
4. The sun, because it provides the energy for photosynthesis.

347

SKILL ACTIVITY: Organizing Background Data

Objectives

- Collect and analyze data on the physical characteristics of plants and animals.
- Establish evolutionary relationships among plants and animals.

Discussion Review the concept of evolution by asking questions such as: When did the first living organisms evolve? (Archean Era, 3.8 billion years ago) What were probably the first living organisms? (primitive bacteria) How were advanced organisms different from more primitive organisms? (Answers will vary but should include discussions of methods of converting energy to support life, such as cellular respiration vs. fermentation—efficient vs. inefficient methods—and adapting to changing environmental conditions.) ***Thinking Critically:*** Ask the students to explain what data support their answers to these questions. (Answers will vary but should include discussions of the fossil record and evidence of changing environmental conditions.)

Answers to Application

1. bacteria
2. Groups A through E are plants. Groups H through Q are animals. Groups F and G have characteristics of both plants and animals.
3. Answers will vary because many student placements of the plants and animals on the evolutionary tree will not match the instructor's key. This should convince students that physical characteristics alone are many times insufficient to accurately classify all living things.

Answers to Using What You Have Learned

Answers will vary depending on fossils used.

BACKGROUND

Scientists often have to analyze data collected from different sources before they draw any conclusions about a problem. Sometimes the data must be organized before it becomes meaningful. In this activity you will be gathering data about plant and animal characteristics to establish evolutionary relationships.

Multicellular organisms have existed for only about the last fifth of the history of life on Earth, having evolved from single-celled organisms prior to the Phanerozoic Eon. Yet they have adapted to a variety of environments in a relatively short time. Plants and animals are the principal organisms considered in this activity.

PROCEDURE

1. Use tracing paper to make a copy of the evolutionary tree shown below.
2. Using a life-science or biology textbook, examine the physical features of the plants and animals labeled in this activity. Note their similarities and differences.

A) Cyanobacteria or blue-green bacteria
B) Red algae
C) Brown algae
D) Seed plants
E) Green algae
F) Fungi
G) Slime molds
H) Annelids
I) Arthropods
J) Vertebrates
K) Primitive chordates
L) Echinoderms
M) Mollusks
N) Coelenterates
O) Sponges
P) Ciliates
Q) Amoeboids
R) Bacteria

348

3. Check the position for each group of organisms on the evolutionary tree. Decide which are most likely to have had a common ancestor.
4. Write the name of each plant or animal group above the lettered circle on the tree.

APPLICATION

1. Which group of organisms has evolved most independently of the others?
2. List the organisms that could be considered plants. List the animals. Are there any organisms that could be on both lists?
3. Compare your list with your teacher's list. Do you think that physical characteristics alone are sufficient to accurately classify all of the organisms pictured here? Explain.

USING WHAT YOU HAVE LEARNED

Organize the fossil data from Investigation 16, on page 358, in a similar fashion, then try to answer the Application questions again for this new data.

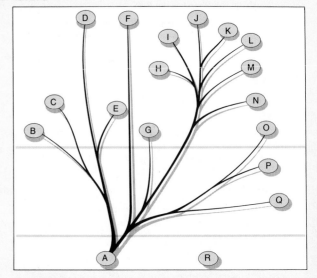

SECTION
2 The Phanerozoic Eon

Section 2: The Phanerozoic Eon

SECTION OBJECTIVES

After completing this section, you should be able to:
- **List** the characteristics of trilobites.
- **Compare** the mass extinctions that occurred at the end of the Ordovician and Devonian Periods.
- **Describe** the ice ages of the Cenozoic Era.

NEW SCIENCE TERM

glaciation

16.5 The Paleozoic Era

Many of the familiar events of Earth history occurred during the *Phanerozoic Eon.* This eon begins the last month or so of the year-long geologic calendar. As with the Cryptozoic Eon, the Phanerozoic Eon has been divided into three eras: the *Paleozoic* (pay lee oh ZOH ihk), the *Mesozoic* (mehs uh ZOH ihk), and the *Cenozoic* (see nuh ZOH ihk). These eras are the final weeks of the geologic calendar year. Each era is further divided into periods, making it easier to study the many events that occurred. To see the relationship between eras and periods, time boundaries, and the major events of each period, look at the geologic time scale on page 574 of the Reference Section.

The Cambrian, Ordovician, and Silurian Periods

The earliest shelled organisms evolved during the Paleozoic Era. Most impressive of these organisms were the *trilobites* (TRY luh byts). Trilobites were crablike animals with highly complex, segmented bodies. These shelled animals were like the horseshoe crabs of today. Most trilobites were a few centimeters long, but some reached a length of several decimeters.

Trilobites were an important part of the shallow-water ocean communities of the first Paleozoic period, the *Cambrian* (KAM bree uhn) *Period.* Trilobites lived in shallow ocean water from about 570 million years ago to about 250 million years ago.

DISCOVER

Comparing Relatives

Compare the photograph of a horseshoe crab shell with the illustration of the trilobites. Note the similarities and differences between these organisms. The horseshoe crab has remained nearly unchanged for millions of years. What differences between the two organisms might have allowed horseshoe crabs to survive, and caused the trilobites to become extinct?

Figure 16–11. A modern horseshoe crab (right) is similar in appearance to the trilobites of the Paleozoic Era (left).

The Phanerozoic Eon   **349**

BACKGROUND INFORMATION

The earliest preserved specimens of living organisms are remnants from the beginning of the Phanerozoic Eon. Organisms living prior to this eon might have been too soft to be preserved or were destroyed by geologic processes. The only exception are microscopic bacteria that have been uncovered by advanced microscopic techniques.

BACKGROUND INFORMATION

A trilobite is so named for the "three lobes" that segmented its body lengthwise. It was an arachnid, and like modern arachnids such as horseshoe crabs and scorpions, it had a cephalothorax and tail.

SECTION FOCUS

Section Overview This section discusses the evolution of marine life during the Phanerozoic Eon, including trilobites and reptiles. Also covered are the development of great forests and the occasional mass extinctions of plants and animals.

Section Objectives For a list of section objectives, see pupil's edition page.

New Science Terms For a list of new science term in this section, see pupil's edition page. Call the students' attention to the new term they will meet in this section.

SECTION DEVELOPMENT

16.5 The Paleozoic Era

DISCUSSION Direct the students' attention to the pictures in this section and ask them to discuss what they see. Review the meanings of the prefix *phanero-* and the suffix *-zoic.* ***Thinking Critically:*** Ask the students to name some differences between this eon and the Cryptozoic Eon. (Species of plants and animals that were more numerous, more diverse, and more highly developed evolved during this eon.)

THINKING SKILL *(Comparing)* Direct the students' attention to Figure 16–11. Ask them to recall the discussion of animals of the Cryptozoic Eon and to distinguish trilobites from the earlier animals. (Trilobites have segmented bodies and legs.)

DISCOVER: Comparing Relatives

Thinking Skill *(Drawing Conclusions)*

Horseshoe crabs may be found naturally in coastal regions or can possibly be obtained from a biology supply house or a local university. If a shell is unavailable, refer to the photograph.

DISCUSSION The Paleozoic Era has been referred to as the Age of Fish. Sharks evolved during the Ordovician Period of this era. They are one of the earliest forms of fish that still exist today. ***Thinking Critically:*** Ask the students to describe a physical feature that suggests that sharks are a more primitive species of fish than trout or tuna. (Sharks have a skeleton of cartilage rather than bones.)

EXTENSION Spores are unicellular reproductive structures that contain a mass of protoplasm and a nucleus surrounded by a cell wall. Ferns and mushrooms are two examples of modern organisms that reproduce using this method. During the resting stage of reproduction, the cell wall becomes tough and waterproof and allows the cell to survive unfavorable conditions such as extremes of temperature and moisture. ***Thinking Critically:*** Ask the students why organisms that reproduce by spores were well adapted to the environment during this period. (Organisms that can adapt easily to extreme changes in temperature and moisture have a better chance of survival than those that cannot. During this period, Earth experienced extreme changes in environment.)

Figure 16–12. Ordovician fishes (right) may have been similar to modern fishes (left).

A MATTER OF FACT

Life could not evolve on our planet today as it did over 3 billion years ago. The concentration of oxygen in the present atmosphere would destroy the early forms of bacteria.

Figure 16–13. Modern liverworts (left) have many of the same characteristics as the first land plants (right).

Only a few species of trilobites survived into the next period, the *Ordovician* (ohr duh VIHSH uhn). During the Ordovician Period, an ice age began, and sea level was greatly lowered. The shallow marine environment was reduced in area, and many habitats were destroyed. The extinction of many organisms probably occurred due to these changed conditions. How would an ice age lower the sea level?

The earliest fishes appeared late in the Ordovician Period. Because their bodies were covered with protective bony plates, they were probably poor swimmers. As they evolved, the bony plates disappeared. During the next period, the *Silurian* (sih LUHR ee uhn), all major groups of fishes—sharks, rays, and bony fishes—appeared.

During the Cambrian and Ordovician periods, there were no land plants. Land plants first appeared in the Silurian Period. These earliest plants had no true roots, stems, or leaves. Instead, they had flexible structures that served some of the same purposes. These plants reproduced by spores.

BACKGROUND INFORMATION

The first jawless fishes, called *agnathas* (from the Greek word *gnathos* meaning "jaw," and the prefix *a-* meaning "without"), evolved during the Ordovician Period. Similar jawless fishes, such as the lamprey, exist to this day.

BACKGROUND INFORMATION

Gondwana, a continent that was comprised of Africa and South America, and Laurasia, composed of Eurasia and North America, collided and separated at least twice during the Phanerozoic Eon. The first separation occurred during the Ordovician Period of the Paleozoic Era; the second separation took place during the Cretaceous Period of the Mesozoic Era.

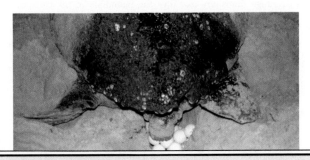

The Devonian, Carboniferous, and Permian Periods

During the *Devonian* (deh VOH nee uhn) *Period* amphibians and seed-bearing trees first appeared. Amphibians were the first vertebrates that could live on dry land. A temperature increase occurred toward the end of the Devonian Period that may have caused the extinction of many animal species.

The last two periods of the Paleozoic Era were the *Carboniferous* (kar bah NIHF ur uhs) and the *Permian* (PUR mee uhn). Many species of land plants evolved during these periods, and vast forests of seed-bearing trees covered the land.

Reptiles, which evolved during the Carboniferous Period, differ significantly from amphibians and fish in their method of reproduction. Amphibians and fish deposit eggs that can survive only in water. Reptile eggs can survive on land. Due to their ability to reproduce away from water, the reptile populations spread rapidly during the Permian Period.

○ *What may have caused the extinction of the trilobites?*
○ *What important groups of plants and animals evolved during the last half of the Paleozoic Era?*

Figure 16–14. During the Carboniferous Period, great forests covered much of the land (left). The dominant plants of these forests were probably the tree ferns, similar to tree ferns growing today in Australia (right).

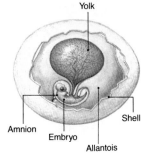

Figure 16–15. Reptile eggs (above) are protected by several internal membranes. These eggs, such as those of the sea turtle (left), can incubate and hatch on dry land.

351

THINKING SKILL *(Relating)* Remind the students that amphibians evolved after fishes. Ask the students to provide evidence that this is true. (Although both have gills and can live in the water, amphibians also developed the characteristics necessary to live on land. For instance, a tadpole's gills disappear when it changes from a water-breathing creature with no limbs into an air-breathing, four-legged animal.)

THINKING SKILL *(Relating)* Remind the students that both amphibians and reptiles can live on land. Ask the students to describe some characteristics of reptiles, in addition to their method of reproduction, that differentiate them from amphibians. Ask the students to name some examples of both amphibians and reptiles. (Reptiles have dry skin and scales, and they do not have gills during any stage of their development. amphibians: toads, frogs, and salamanders; reptiles: snakes, lizards, turtles, alligators, and crocodiles)

DEMONSTRATION

Heat several wooden splints in a test tube clamped over a Bunsen burner. Hold a lighted splint at the mouth of the tube. Point out that the splint glows more brightly while the splints in the test tube turn from brown to black. The splints in the tube are, in fact, releasing a gas similar to methane (natural gas) because the wood is organic in origin. Remind the students that under considerable heat and pressure, organic material is transformed into fossil fuel such as coal.

Large deposits of natural gas, coal, and oil were formed during the Phanerozoic Eon.

Answers–16.5

○ Sea level was reduced, which in turn may have wiped out the trilobite's habitat.

○ Fish appeared during the Cambrian and Ordovician periods. Land plants appeared during the Silurian.

16.6 The Mesozoic Era

DISCUSSION Direct the students' attention to Figure 16–16. Ask them if they can name some of the reptiles pictured. These reptiles are, clockwise from the upper left: Apatosaurus, Trachodon, pterodactyl, Icthyosaurus, Triceratops, and Struthiomius. *Thinking Critically:* Ask the students to name some features that may indicate that birds have evolved from reptiles. (Both birds and reptiles lay eggs; reptiles' forelimbs evolved into wings; reptiles' scales evolved into feathers; reptiles' heavy jaws and teeth were replaced by a lighter-weight bill.)

THINKING SKILL *(Communicating)* Review the meaning of the term *cold-blooded* with the students. Ask them to name the disadvantages of being cold-blooded. (These animals must rely on the sun's warmth to maintain their body temperature because they cannot regulate it internally. Therefore, cold-blooded animals become sluggish in cold climates and have greater difficulty moving about to find food or to avoid danger.)

Figure 16–16. The Mesozoic Era was the age of reptiles (top). Some of the dinosaurs of this era developed elaborate structures to keep themselves cool in the warm climate of the time.

16.6 The Mesozoic Era

The geologic calendar is now into the middle of December. This is the *Mesozoic Era*, the age of reptiles.

By far, the most impressive group of animals of the Mesozoic Era were the reptiles. They evolved into a large number of species that occupied all major habitats. Most reptiles lived on land, but there were also marine and flying reptiles. Some reptiles evolved into dinosaurs, birds, and mammals in the *Triassic* (try AS ihk) *Period*, the first period of the era.

Mammals and dinosaurs evolved at about the same time. Both were small animals at first, but dinosaurs spread much faster and forced the mammals out of many habitats. The major advantage dinosaurs had over the mammals of the time was their agility.

Dinosaurs faced the problem of overheating during the warm *Jurassic* (joo RAS ihk) *Period*. To combat this problem, dinosaurs evolved elaborate structures to cool their blood. These structures included back plates, collar plates, and ducts in the skull. These structures made some of the dinosaurs look quite strange, as shown in Figure 16–16.

352 Chapter 16 Earth's Past

BACKGROUND INFORMATION

The Palisade orogeny spanned the end of the Triassic and the early Jurassic periods. During this time the Appalachian mountain chain, which covered an area from the Carolinas to southeast Canada, was uplifted and split by faulting and folding of the earth's crust.

BACKGROUND INFORMATION

Larger surface areas have a higher rate of evaporation than do smaller surface areas, and because evaporation is a cooling process, surface temperature is lowered. Internal heat migrates toward the surface and body temperature is lowered. Because of this, an elephant's large ears (which it frequently sprays with water to aid the evaporation process) serve as thermoregulating devices. A stegosaur's dorsal plates served the same purpose.

The dinosaurs became extinct about 65 million years ago, at the end of the *Cretaceous* (krih TAY shuhs) *Period*, the final period of the era. Many other organisms also became extinct at this time. Many shallow-water marine animals, reef-building corals, protozoa, and algae died. High-latitude animals and animals living on the deep sea floor suffered less. Some scientists believe that a period of high temperature was the cause of this mass extinction. Other scientists believe just the opposite: that a short period of unusually cold weather killed most of the tropical plants of the world, and the animals soon died from starvation.

○ *Why is the Mesozoic Era called the age of reptiles?*
○ *What might have caused the mass extinction at the end of the Mesozoic Era?*

Figure 16–17. There are many locations where fossilized bones may be found.

16.7 The Cenozoic Era

We are now at the end of the geologic calendar year. The present era is the *Cenozoic Era.* Although the Cenozoic Era started about 65 million years ago, it would be represented on the geologic calendar by the last few days of the year. During the Cenozoic Era, the mammal populations expanded, filling the habitats of the extinct dinosaurs.

The Cenozoic Era is divided into two periods: the Tertiary Period, which lasted from about 65 million years ago to about 3 million years ago; and the Quaternary Period, which is the present period. Each period is subdivided into epochs.

About 3 million years ago, ice began to push southward across the northern continents. Since then great sheets of ice have advanced and retreated every 100 000 years. Each time the ice formed, it would take about 75 000 years to reach its southernmost point. The ice would remain fewer than 10 000 years

DISCUSSION   Remind the students that many scientists think that a period of high temperature caused the mass extinction of plants and animals during this period. *Thinking Critically:* Ask the students to describe why animals living at high latitudes or those living deep within the ocean had a better chance of survival than did those living elsewhere on the earth. (Animals living at those latitudes and ocean depths were not as affected by the extremely high temperatures.)

16.7 The Cenozoic Era

EXTENSION   Continental glaciers are huge ice sheets that form when accumulations of snow are compacted. The ice moves continuously because its mass constantly accumulates. Glaciers alter the character of the topography by eroding the landscape, transporting debris from one place to another, and depositing drift. Drift consists of deposits of clay, gravel, sand, and boulders. Much of North America is covered by drift.

Answers—16.6

○ Reptiles were the dominant animals during the Mesozoic Era.

○ Scientists are unsure as to the exact cause of the mass extinction because there is no way as of yet to prove exactly what happened.

BACKGROUND INFORMATION

Luis and Walter Alvarez's theory regarding the Mesozoic extinction can be compared to the belief held by scientists that a "nuclear winter" would follow a nuclear war. Nuclear blasts could start fires that might rage for weeks, sending vast amounts of dust and debris into the upper layers of the atmosphere. This debris would block out the light of the sun, thereby plunging the world into a cold darkness that could result in another mass extinction.

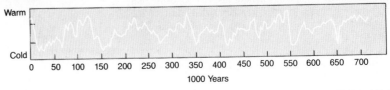

DISCUSSION Reinforce the concept of advancing and retreating ice during this era. Direct the students' attention to the graph in Figure 16–18 which shows fluctuations in temperature. ***Thinking Critically:*** Ask the students what happened to the ice when it retreated and disappeared. (As the temperatures warmed, the ice melted from the southernmost latitudinal edge of the ice sheet northward. The level of the oceans rose which submerged the land and changed the habitats of plants and animals.)

EXTENSION (Tie-in/Geography) At one time ice covered large parts of North America. The ice blanketed an area as far south as a line drawn (from east to west) through New Jersey and Pennsylvania, continuing along the Ohio and Missouri rivers to North Dakota, northern Montana, Idaho, and Washington state into the Pacific Ocean. Give each student a photocopy of a map of North America and ask them to draw a line showing how far the ice advanced during this era. ***Thinking Critically:*** Ask the students to describe the North American continent if a new Ice Age were to occur within the next thousand years. (Answers will vary but the students should understand that large parts of North America would become uninhabitable and many people and animals would be displaced from their present habitats.)

EXTENSION As the ice retreated, it left behind lakes and forced rivers to cut new channels. The Grand Canyon was cut during this era and many of the lakes in Minnesota and Maine were created.

Figure 16–18. The graph (above) is a record of Earth's temperature changes over the last 700 000 years. During this time, periods of glaciation (top left) alternated with periods of moderate climate (top right).

Figure 16–19. This time line shows the divisions of Earth's history.

and would retreat over another 10 000 years. The ice would then disappear from all northern continental areas except Greenland. The continents would remain free of ice for about 5000 years, and then the ice cycle would repeat.

Figure 16–18 shows the global changes in temperature that have taken place during the past 700 000 years. Each peak represents conditions similar to those of today. Each valley represents a glaciation. **Glaciation** is a period in which great continental ice sheets advance and retreat repeatedly.

During glaciations, summer temperatures were probably similar to the winter temperatures of today. Summers were stormy; winters were even stormier. Periods between glaciations were short. Warm temperatures, like those of today, lasted about 5000 years or less. If the past is an indication of the future, the world should be entering a new glaciation within the next thousand years.

Cryptozoic Eon											Phanerozoic	
					Paleozoic Era							
Precambrian or Hadean, Archean, and Proterozoic eras	Cambrian Period			Ordovician Period		Silurian Period	Devonian Period			Carboniferous Period		
	Early	Middle	Late	Early	Late		Early	Middle	Late	Mississippian	Pennsylvanian	

Figure 16–20. Modern tarsiers (left) and lemurs (right) may resemble the earliest primates.

The evolution of present-day plants and animals, including humans, occurred during these extreme weather cycles. Humans evolved with a group of mammals known as *primates*. Primates appeared soon after the dinosaurs became extinct, about 65 million years ago, or about December 24 of the geologic year. Modern lemurs and tarsiers, shown in Figure 16–20, are very similar to primitive primates.

Monkeys and apes appeared later, about 30 million years ago. The first *hominids*, or humanlike primates, appeared about 3.5 million years ago, or December 29. *Homo sapiens*, the present human species, evolved about 0.125 million years ago, at approximately 11:30 P.M., the last half-hour of the geologic calendar year. The geologic year shows how short a time humans have existed, compared to the age of the earth.

At the end of the last glaciation, humans were hunters, mainly of large mammals. More than 200 species became extinct

EXTENSION The Cenozoic Era marks the evolution of warm-blooded animals. Review the discussion on cold-blooded animals and then ask the students to name the advantages of being warm-blooded. (Warm-blooded animals can regulate their own body temperature, are less dependent on the environment, and are therefore less susceptible to cooler temperatures.)

EXTENSION The warm-blooded animals of the Cenozoic Era flourished because of their ability to adapt to changes in the environment. *Thinking Critically:* Ask the students how warm-blooded animals adapted to the changes in the environment that occurred during this era. (Answers will vary, but the students might mention that warm-blooded animals built nests, sought other forms of shelter, migrated to warmer climates, and evolved a layer of fat and a heavy coat of fur or feathers to protect them from cold temperatures.

THINKING *(Inferring)* Remind the students that *Homo sapiens* evolved during the last "half hour" of the geologic claendar year. Ask the students to hypothesize why this is true. (Answers will vary, but the students should mention that the complexity and diversity of *Homo sapiens* is the result of several billion years of evolution.)

Eon														
			Mesozoic Era						Cenozoic Era					
Permian Period		Triassic Period			Jurassic Period			Cretaceous Period		Tertiary Period				Quaternary Period
Early	Late	Early	Middle	Late	Early	Middle	Late	Early	Late	Paleocene	Eocene	Oligocene	Miocene	Pliocene	Pleistocene	Holocene

Section 2   The Phanerozoic Eon   **355**

EXTENSION By hunting mammals to the point of extinction, humans started to shape their environment for the first time. *Thinking Critically:* Ask the students to name several other ways in which people currently affect the environment. (Answers will vary but the students may mention that people now seed clouds to produce rain, cut down timber for fuel and commercial uses, and blast through parts of mountains to create highways.)

SECTION REVIEW

Summary Have the students list the major points covered in this section. Make sure they include the new science term in their summaries.

Reinforcement Have the students quiz each other using the questions at the end of each numbered subsection.

ACTIVITY: Making a Geologic Time Line

Thinking Skill *(Interpreting Data)*

Preparation of Materials A large, flat surface will be required for this activity. Student's desks may have to be rearranged in rows to create a space to work.

Answers to Conclusions and Applications

1. 342 million years, 183 million years, 65 million years
2. Nearly twice as long (342 x 10^6 divided by 183 x 10^6 equals 1.87)
3. The epochs lasted for less than 1 million years. Since our timeline equates 1 mm with 1 million years, it is difficult to separate them with the markers.
4. 0.33% or one-third of one percent (*Homo sapiens* appeared 2.0 x 10^6 years age. The Phanerozoic Eon began 590×10^6 years ago. 2.0×10^6 divided by 590×10^6 equals 0.0033 and 0.0033 multiplied by 100% equals 0.33%)

Figure 16–21. All people, regardless of individual characteristics, evolved from the same ancestral hominids over 3.5 million years ago.

as a result of this hunting. About 12 000 years ago, humans discovered that there was little left to hunt. Therefore, for about 4000 years life was probably hard, since food was scarce. Only when agriculture was developed, about 8000 years ago, did food again become plentiful.

Humans are an important part of the earth environment. Together with every other living organism, humans probably evolved from bacteria that lived more than 4 billion years ago. What kinds of organisms do you think will evolve over the next 4 billion years?

○ *What is glaciation?*
○ *What is a hominid?*

Section Review

READING CRITICALLY

1. How did the climate of the Cenozoic glaciations differ from today's climate?
2. When did dinosaurs, mammals, and birds come into existence?

THINKING CRITICALLY

3. Do you think humans are still evolving? Explain.
4. How might the dinosaurs' body-heat problems have led to their extinction?

ACTIVITY: Making a Geologic Time Line

How can you compare the relative length of eras and periods of the Phanerozoic Eon?

MATERIALS (per group of 3 or 4)

construction paper, tape, markers, metric ruler, scissors

PROCEDURE

1. Cut the construction paper along its length into strips measuring 5 cm in width.
2. Tape the strips end to end so that you have a ribbon approximately 4 m in length.
3. Roll up the ribbon so that it is easy to handle, and unroll it as required.
4. Using Figure 16–19 on pages 354–355 and using a scale that 5 mm equals 1 million years, measure the length of each era and

period of the Phanerozoic Eon. Draw a single line to mark the beginning and end of each period. Use a double line to separate eras.
5. Label the eras and periods.
6. Add important events in Earth history or make drawings of fossils in the appropriate places.

CONCLUSIONS AND APPLICATIONS

1. How long was the Paleozoic Era? The Mesozoic? The Cenozoic?
2. How many times longer than the Mesozoic was the Paleozoic?
3. On your measurement scale, what problems arose when you were trying to measure the lengths of the periods of the Cenozoic Era?
4. According to Figure 16–19, for what percent of the Phanerozoic Eon has *Homo sapiens* inhabited this planet?

356 Chapter 16 Earth's Past

Answers–16.7

○ Glaciation is a period when great sheets of ice extend and retreat several times in a regular cycle.

○ A hominid is a humanlike primate.

Answers to Section Review

Reading Critically

1. Summer temperatures were similar to today's winter temperatures, and winter temperatures were even colder than that.
2. Dinosaurs, mammals, and birds came into existence during the Cenozoic Era.

continues

BIOGRAPHIES: Then and Now

HAROLD UREY (1893–1981)

Most scientists become interested in only one or two fields of science, such as biology or chemistry. Other scientists, however, have undying curiosity about all sciences. They try to learn as much as they can about many scientific fields. Harold Clayton Urey was the second kind of scientist.

He received his bachelor's degree in 1917 from Montana State University. At that time his major interests were zoology and chemistry. After he was awarded a doctorate from the University of California at Berkeley, he traveled to Denmark. There he worked from 1923 until 1924 with Niels Bohr, the famous atomic physicist.

Urey's work in atomic research over the next decade led to his discovery of deuterium (doo TEER ee uhm), a heavy isotope of hydrogen. In 1934 he received the Nobel prize for his work.

In the early 1950s, Urey turned his attention to the studies of geochemistry, astrophysics, and the origin of life. He wanted to know how the earth and solar system had come to be. His vast knowledge of biology, physics, and chemistry helped him conduct research and write many articles on geochemistry. He reviewed many theories on how the sun and planets were formed. He studied the chemical reactions of gases that existed in Earth's primitive atmosphere, and he was the first to show that amino acids could have formed in the atmosphere. Although he never proved how life originated, he did add evidence to the theory that life could have started by itself on the primitive earth.

LYNN MARGULIS (1938–)

Imagine that you are standing knee-deep in a pond of pink, green, and purple ooze. You turn to the person next to you, a scientist from Boston University. She tells you that you are soaking in some of the most interesting bacteria in the world. You might scream and run for dry land. Professor Lynn Margulis would stay right where she is, studying the microscopic creatures of the murky pool.

Dr. Margulis considers herself a "microbiological evolutionist," and her work has shed a great deal of light on the origin of life. Because the subject is so complex and difficult, it requires interest and knowledge in many scientific fields.

She has tried to explain how primitive organisms might have evolved into more complex life forms. She has declared that the evolution of complex animals might have relied more on biological cooperation, or *symbiosis,* than was previously thought. Symbiosis occurs when organisms live together for their mutual benefit.

Margulis received a doctorate in genetics from the University of California at Berkeley in 1965. She is currently teaching at Boston University. In 1985 she was named "The Wizard of Ooze" by author Richard Wolkomir in an article in *Omni* magazine.

Dr. Margulis has suggested that the millions of kinds of microorganisms may be adding gases other than oxygen to our atmosphere. "Kill off animals and plants and the planet will recover," she has said, "but kill off the microbes and in weeks the earth will be just as sterile as the moon."

Section 2   The Phanerozoic Eon   **357**

BIOGRAPHIES: Then and Now

Discussion   After the students have read the two biographies, hold a class discussion on the contributions made by the two scientists.

Extension   You may wish to take a field trip to a local planetarium or a nearby safe, unpolluted stream, pond, or lake to collect specimens for examination. You might also want to invite several scientists to visit the class and discuss a ''typical'' day at work. Interested students might be encouraged to research further the life and work of Urey or Margulis or one of the scientists listed below.

Heinrich Schliemann

Luis Alvarez

Thinking Critically

3. Yes. Environmental pressures continue to select from any species the most successful variations.

4. Chemical reactions taking place in the body can only occur within a specific range of temperatures. If the animals were not able to get rid of excess body heat, then they died.

INVESTIGATION 16:
Classifying Fossils

Skill *(Classifying)*

Preparation of Materials

You will need a set of 10 numbered fossils for each student group. Sets of fossils may be purchased from supply houses. If sets are not available, they can be created by giving each group as many different fossils as you can.

Answers to Analyses and Conclusions

1. There is no one way to group the fossils, but they should be grouped by some logical method, such as by plant and animal, or by shape.
2. Answers will vary, but should include some of the same reasons as for question one.
3. Yes, there are different ways to group these fossils. Other ways should include some of the same ideas as listed for question one.
4. Answers will vary, but should include things such as shape, size, and similarity to known organisms.
5. Scientists use some of the same ways as the students, particularly the similarity to known fossils.

Answers to Applications

1. New fossils can be placed on the table according to the characteristics they most closely match.
2. The different characteristics can be used to make a key to classify fossils.

INVESTIGATION 16: Classifying Fossils

PURPOSE

To classify fossils based on similar characteristics

MATERIALS

Set of ten numbered fossils
Hand lens or stereomicroscope

PROCEDURE

1. Copy the table for recording your observations.
2. Make four or five detailed observations of each fossil specimen and record your description in the table. You may use a hand lens or a stereomicroscope to help see details.
3. Separate the fossils into groups with similar characteristics.
4. List the fossils by number on the table. Give several reasons for your grouping.
5. Compare the fossils in your groups to those in your classmates' groups.

ANALYSES AND CONCLUSIONS

1. Is there any one correct way to group these fossils? Explain.
2. On what similarities were your groupings based?
3. Are there different ways to group these fossils? What are some other ways?
4. What information about the fossils helped you in grouping them?
5. How do you think scientists might group them?

APPLICATIONS

1. How could you use your data table to help you classify any fossils you might find in the field?
2. How would you use your data table to develop a classification key of fossil types?

Number	Description	Group	Reasons
1.			
2.			
3.			
4.			
5.			
6.			
7.			
8.			
9.			
10.			

TABLE 1: FOSSIL GROUPS

CHAPTER
16 REVIEW

SUMMARY

- Earth's history has two main divisions, or eons: the Cryptozoic Eon and the Phanerozoic Eon. (16.1)

- The Cryptozoic Eon is the time interval from the formation of Earth to about 590 million years ago. (16.1)

- The Cryptozoic Eon is divided into three eras: the Hadean, the Archean, and the Proterozoic. (16.1)

- Most of the earth's crust was formed during the Hadean Era. Dense basalt formed the oceanic crust while the lighter granite formed the continental crust. (16.2)

- The fundamental chemical reactions necessary to living cells were developed by bacteria during the Archean Era. (16.3)

- Multicellular animals evolved during a glaciation that took place in the Proterozoic Era. (16.4)

- The Phanerozoic Eon is divided into the Paleozoic, Mesozoic, and Cenozoic eras. (16.5)

- Trilobites were marine animals that lived during the Paleozoic Era. (16.5)

- Great forests developed during the Carboniferous Period. (16.5)

- Reptiles were the first animals to deposit their eggs on dry land. (16.5)

- Dinosaurs became the dominant animals during the Mesozoic Era. (16.6)

- A major extinction of plants and animals occurred at the end of the Mesozoic Era. (16.6)

- *Homo sapiens* evolved about 0.125 million years ago. (16.7)

Write all answers on a separate sheet of paper.

SCIENCE TERMS

Correctly use each of the following terms in a sentence.

cellular respiration **(346)**
fermentation **(344)**
glaciation **(354)**
planetesimals **(342)**

SCIENCE QUIZ

Modified True-False

Mark each statement *true* or *false*. If a statement is false, change the underlined term to make the statement true.

1. The Cryptozoic Eon <u>ended</u> 4.6 billion years ago.

2. The earth's original crust was made of <u>silicate</u> rock.

3. Trilobites looked a lot like <u>mammals</u>.

4. The last <u>era</u> of the Paleozoic is the Permian.

5. <u>Stromatolites</u> were formed from iron oxides and silica.

Multiple Choice

Write the letter of the choice that best answers the question or completes the statement.

6. The earth's primitive atmosphere contained
   a) water vapor.
   b) carbon dioxide.
   c) sulfur dioxide.
   d) All of the above are correct.

7. The time between the formation of the solar system and the age of Earth's oldest known rocks is called the
   a) Hadean Era.   b) Proterozoic Era.
   c) Archean Era.   d) Cryptozoic Eon.

continues

Chapter 16 Review   **359**

CHAPTER REVIEW

SUMMARY

The students may review the major concepts in the chapter by reading the summary statements. The statements are cross-referenced to the chapter to facilitate reinforcement of any concepts of which the students feel unsure. Encourage the students to work together in groups to quiz one another.

SCIENCE TERMS

The sentence in which the science term is used should reflect an understanding of the definition of the term. You may wish to hold a science "bee" during which one team will give the term and the other team must supply the definition. As an alternate activity, stage a game of "Jeopardy" during which the definition is provided and the "contestants" must provide the correct term.

SCIENCE QUIZ

Answers to Modified True-False

1. false, began
2. true
3. false, horseshoe crabs
4. false, period
5. true

Answers to Multiple Choice

6. d
7. a

continues

Answers to Multiple Choice (continued)

8. c
9. b
10. a

Answers to Completion

11. Proterozoic
12. chlorophyll
13. proteins
14. Paleozoic Era
15. reptiles

Answers to Short Answer

16. The Hadean Era was characterized by the formation of the earth's crust, mantle, and core; and the formation of nucleic and amino acids. During the Archean Era, the earliest cells and bacteria evolved. The Proterozoic Era was the beginning of eucaryotes and multicellular organisms.
17. Stromatolites are formed by the layering of carbon and silicates produced by cyanobacteria.
18. Paleozoic Era—fish; Mesozoic—reptiles; Cenozoic—mammals

Answers to Writing Critically

19. This brief essay should include a discussion of habitat reformation due to the migration of the continents (for example, plate tectonics) and a discussion of Alvarez's theory of an asteroid's collision with Earth.
20. The primitive atmosphere did not have enough oxygen to support life as we know it today. In order for animal species to survive, new sources of this important gas would have to be developed. (for example, breathing tanks) Modern species would also have to learn how to cope with increased temperatures. (Protective, pressurized clothing would have been worn.) Some of the poisonous gases, like ammonia and methane, that existed then would also have to be dealt with. (Gas masks would be needed.)

ANSWERS TO EXTENSION

1. Students should discover that fermentation is a less efficient way of producing energy than respiration. It also results in the manufacture of different products, such as alcohol.

---

8. Photosynthesis produces
   a) carbon dioxide. b) water vapor.
   c) oxygen. d) lipids.

9. When did the earliest protozoa appear on Earth?
   a) about 3.8 billion years ago
   b) about 900 million years ago
   c) about 1.5 billion years ago
   d) about 590 million years ago

10. A major extinction occurred at the end of the
    a) Paleozoic Era. b) Cenozoic Era.
    c) Hadean Era. d) Archean Era.

Completion

Complete each statement by supplying the correct term.

11. Multicellular animals appeared during the _____ Era.

12. Photosynthesis requires a magnesium-containing molecule called _____.

13. Amino acids link together to form _____.

14. Trilobites lived during the _____.

15. The Mesozoic Era is sometimes called the age of _____.

Short Answer

16. Give one characteristic feature of each era of the Cryptozoic Eon.

17. Describe the formation of stromatolites.

18. Name one dominant form of animal life for each of the eras of the Phanerozoic Eon.

Writing Critically

19. Compare and contrast two theories of geologic history that try to explain the extinctions at the end of the Mesozoic Era.

20. Explain why modern animals could not have evolved before green algae and plants.

EXTENSION

1. Research the process of fermentation and the types of single-celled organisms that

---

use it to make energy. What are the chemical products of fermentation? Do these products have any uses?

2. Obtain information on the work conducted in the early 1950s by Harold C. Urey at the University of Chicago. In what ways did his research help explain how life might have started on our planet?

APPLICATION/CRITICAL THINKING

1. Radioactive dating methods reveal that the oldest known rocks on Earth are about 3.8 billion years old. The same methods have shown that some rocks brought back from the moon are 4.6 billion years old. What might account for this difference?

2. Describe how glaciation might cause living organisms to evolve.

3. The following statements summarize the geologic history of the earth during the Phanerozoic Eon. However, some of the events are not in the correct sequence. Copy the statements on another sheet of paper, and put them in the correct sequence.

   a. Shallow seas cover the land.
   b. Fish appear, and simple animals inhabit the seas.
   c. Dinosaurs are dominant, and mammals begin to evolve.
   d. Ice covers large parts of the continents, and hominids evolve.
   e. Great forests cover the land.
   f. Trilobites inhabit the seas.
   g. Reptiles evolve protected eggs.
   h. Reptiles and amphibians dominate.
   i. Many sea creatures, including trilobites, become extinct.
   j. Dinosaurs become extinct, and mammals dominate.
   k. Corals and sponges dominate the seas, and simple land plants evolve.
   l. Birds, mammals, and dinosaurs evolve.

---

2. Much of this information can be obtained in the Biography accompanying this chapter.

ANSWERS TO APPLICATION/ CRITICAL THINKING

1. The creation of new crust at mid-ocean ridges is accompanied by the destruction of old crust at trenches (subduction). If any living forms started to evolve earlier than 3.8 billion years ago, they were probably consumed and destroyed during the subduction of land masses.

2. The more variety that exists among individuals of a species, the greater their chance of survival under stressful conditions. Answers here will vary, but should contain this common theme.

continues

FOR FURTHER READING

Gore, R. "The Planets: Between Fire and Ice." *National Geographic* 167 (January 1985): 4. This article contains colorful illustrations of how the solar system might have looked during its creation.

Halliday, T. R., and K. Adler. *The Encyclopedia of Reptiles and Amphibians.* New York: Facts on File, 1986. This book tells almost everything you could ever want to know about snakes, lizards, frogs, turtles, and related animals.

Leakey, R., and A. Walker. "Homo Erectus Unearthed." *National Geographic* 168 (November 1985): 624. Famous archaeologist Richard Leakey gives evidence that humans' most recent ancestor was around over 1 600 000 years ago.

Sattler, H. R. *Pterosaurs, The Flying Reptiles.* New York: Lothrop, Lee & Shepard, 1985. The author describes pterosaurs, which were flying reptiles that ruled the skies during the age of the dinosaurs.

Wolkomir, R. "The Wizard of Ooze." *Omni* 7 (January 1985): 48. The work of micro-biological evolutionist Lynn Margulis is explored in this picturesque display of creatures whose ancestors may have lived in Earth's primordial ooze.

Challenge Your Thinking

This is a picture of a manatee, sometimes called a *sea cow*. Manatees and elephants are believed to have had a common ancestor. What kinds of environmental changes might have occurred that resulted in the development of two such different-looking animals?

Chapter 16 Review   **361**

ANSWER TO CHALLENGE YOUR THINKING

Answers may vary but should include the idea that the oceans may have invaded the land and only mammals that had evolved in the water would have survived.

3. Shallow seas cover the land; corals and sponges dominate the seas, and simple land plants evolve; trilobites inhabit the seas; many sea creatures, including trilobites, become extinct; fish appear, and simple animals inhabit the seas; great forests cover the land; reptiles and amphibians dominate; reptiles evolve protected eggs; birds, mammals, and dinosaurs evolve; dinosaurs are dominant, and mammals begin to evolve; dinosaurs become extinct, and mammals dominate; ice covers large parts of the continents, and hominids evolve.

PLANNING THE CHAPTER

Chapter Sections	Page	Chapter Features	Page	Program Resources	Page
Section 1: Managing Natural Resources	363			Concept Extension: *Who Should Pay?* **(H)**	**TRB 35** **LM 257**
17.1 Natural Resources **(B)**	363				
17.2 Water **(B)**	364	A Matter of Fact	364		
		Activity: Chlorinating Water **(B)**	366		
17.3 Air **(B)**	367	A Matter of Fact	368		
17.4 Land and Soil **(B)**	369				
17.5 Mineral Resources **(B)**	371	**Discover:** Recycling Resources **(B)**	371		
		Section Review	372		
		A Matter of Fact	372		
		Skill Activity: Analyzing Bar Graphs **(B)**	373		
Section 2: Managing Energy Resources	374			Critical Thinking **(H)** Cross-Discipline: Science and Language, *Expressing Your Opinion* **(H)**	**TRB 32** **TRB 31**
17.6 Nonrenewable Energy Sources **(B)**	374			Reading for Content: *Grouping Items to Help Memorization* **(B)**	**TRB 29** **LM 205**
17.7 Renewable Energy Sources **(B)**	375	A Matter of Fact	376	Investigation 17.1: *Combustion of Coal* **(A)**	**TRB 73** **LM 75**
17.8 Nuclear Energy **(H)**	379	**Discover:** Using Energy Resources **(B)**	379	Investigation 17.2: *Alternatives to Fossil Fuels* **(A)**	**TRB 75** **LM 77**
		Section Review	380		
		Technology: Fusion Power	381		
		Investigation 17: Generating Electricity **(A)**	382	Student Record Book: Textbook Investigations **(A)**	**TRB 33**
Chapter 17 Review	383			Vocabulary **(A)**	**TRB 17** **LM 155**
				Tests **(A)** 💻 Computer Test Bank	**TRB 71**

(LM) Laboratory Manual/Study Guide, **(TRB)** Teacher's ResourceBank™

B = Basic **A** = Average **H** = Honors

The coding Basic, Average, and Honors indicates sections or subsections that might be appropriate for different levels of learners. For additional suggestions regarding choice of topic and depth of coverage, see the Pacing Chart on pages T16–T20.

CHAPTER CONCEPTS, OBJECTIVES, AND TERMS

Section	Concepts	Objectives	Science Terms
Section 1: Managing Natural Resources	■ Partly renewable natural resources include forests, water, air, land, and soil. **(17.1)** ■ Metals and fossil fuels are nonrenewable resources. **(17.1)** ■ In the United States, each person uses about 300 L of water daily. **(17.2)** ■ Much water is polluted by organic waste, sediments, chemicals, and heat. **(17.2)** ■ Polluted water can be treated, and the quality of water has improved since the mid-1970s. **(17.2)** ■ Smog is a chemical fog produced by the reaction of sunlight and air pollutants. **(17.3)** ■ Acid rain affects buildings, plants, and even fish. **(17.3)** ■ Only a small percentage of the earth's land surface is suitable for farming. The rest is either too rocky or has a climate too severe for growing crops. **(17.4)** ■ Chemical fertilizers improve the soil in countries with advanced technologies, but some chemical pesticides may be harmful to more than pests. **(17.4)** ■ The use of nonrenewable resources may be extended by recycling. **(17.5)** ■ Where it is not practical to recycle, conservation of resources may extend their usefulness. **(17.5)**	■ **Distinguish between** renewable and nonrenewable resources. ■ **Explain** the importance of water, air, and land, and **describe** how pollution can endanger their usefulness. ■ **Discuss** the recycling of natural resources.	pollution smog pesticides recycle conservation
Section 2: Managing Energy Resources	■ Nonrenewable energy sources, such as fossil fuels and uranium, are quickly being exhausted. **(17.6)** ■ Inexhaustible energy alternatives for the future may include hydroelectric power, wind power, tidal power, geothermal power, solar power, and biomass. **(17.7)** ■ Nuclear energy from fission may be limited by the small reserve of uranium and by the danger of radioactivity. **(17.8)**	■ **Explain** the difference between renewable and nonrenewable energy sources. ■ **List** five energy sources that may be used in the future. ■ **Discuss** the advantages of fusion energy over fission energy.	hydroelectric power tidal power geothermal power solar power solar cell biomass fission fusion energy

CHAPTER MATERIALS

Title	Page	Materials
Activity: Chlorination of Water	366	*(per group of 3 or 4)* binocular microscope, Petri dish, pond water (containing microscopic organisms), laundry bleach (containing 5 percent sodium hypochlorite), small beaker, medicine dropper
Discover: Recycling Resources	371	*(per class)* no materials needed
Skill Activity: Analyzing Bar Graphs	373	*(per student)* paper, pencil, graph paper
Discover: Using Energy Resources	379	*(per student)* paper, pencil
Investigation 17: Generating Electricity	382	*(per group of 3 or 4)* bar magnet, insulated bell wire (10 m), electrical tape, plastic lid from a margarine container, compass, wooden block (5 cm × 10 cm × 10 cm), thumbtacks (4), pencil, hole punch

TEACHING SUGGESTIONS

Section 1: **Managing Natural Resources**

Demonstration: Natural Resources

Purpose

To show how nations must trade equitably for the world's available natural resources before those resources are spoiled or depleted.

Background

The students will quickly realize that no single nation is independent when it comes to managing natural resources. Our ability to function as industrialized societies makes us dependent upon one another. (i.e., The United States might trade its agricultural—land and soil—products for petroleum). The students should also realize that resources are depleted through natural causes or misuse. This makes it necessary for nations to change their trading partners and priorities. They may also discover that trade "negotiations" sometimes tend to break down and friction among nations can result.

Materials

Construction paper
Scissors
Markers
Timer

Procedure

1. Design a stack of 100 playing cards with the construction paper, scissors, and markers, so that the deck consists of seven suits. The suits represent the following categories of natural resources: [1] Water (20 cards), [2] Air (20 cards), [3] Land and Soil (20 cards), [4] Minerals (15 cards), [5] Nonrenewable Energy (10 cards), [6] Renewable Energy (10 cards), and [7] Nuclear Energy (5 cards).
2. Divide the class into five "nations." Shuffle the deck thoroughly and deal *ten cards* to each nation-group.
3. Explain that points are earned by any nation whose "hand" consists of a *complete set* of *six* resource cards consisting of *one* Water, *one* Air, *one* Land and Soil, *one* Mineral, and any *two* Energy cards. Additional cards represent "reserves."
4. Award an arbitrary ten points to any group having this complete set after the first deal.
5. Using the timer, allow the students *two* minutes to send one representative to an "International Resource Conference," where representatives may trade any number of resources in an attempt to complete their set. The students may trade two or more cards for one.
6. At the end of two minutes, allow representatives to return to their "nation" and collect *ten points* if they compiled a complete set.
7. Each group must then *discard five cards* and give it to the instructor before the next deal.
8. Deal each group a fresh set of *five cards* from the original pack and recycle the "used" set for the next turn.
9. Repeat **Steps #5** through **7** until one of the nations accumulates one hundred points (or less at the teacher's discretion).

Questions to Ask the Students

1. What strategy works well in this game? *(Answers will vary.)*
2. What kinds of compromises had to be made? *(Answers will vary.)*

Class Activity

Have the students make a list of the ways that natural resources are being wasted in the school, and some suggestions of what could be done to improve the situation.

Outside Speakers

Have a speaker from the Environmental Protection Agency or a state department of natural resources speak to the class about the wise use of resources and energy.

Section 2: Managing Energy Resources

Field Trip

Plan a field trip to the Department of Water and Power. Collect information on how the installation gathers and processes water from the environment for use by consumers.

CHAPTER 17

CHAPTER OVERVIEW

This chapter explains the differences between renewable and nonrenewable resources. Emphasis is placed on the importance of protecting renewable resources such as water, air, and land. The importance of effective management of nonrenewable resources, such as fuels, is emphasized, and the development of alternative energy sources for the future is discussed.

Section 1: Managing Natural Resources

In this section renewable resources such as water, air, and land are distinguished from nonrenewable resources, such as metals, minerals, and fossil fuels. Emphasis is placed on the dangers of pollution and on the ways in which nonrenewable resources can be protected and preserved. Specific sources of pollution and methods of conservation are detailed in this section.

Section 2: Managing Energy Resources

Nonrenewable and renewable energy resources are compared in this section. The problem of dependence upon nonrenewable fossil fuels is discussed. Renewable energy resources such as water, wind, solar energy, and nuclear energy are offered as alternative energy resources for future use.

CHAPTER

17

Earth's Future

What will the future really be like? No one can be certain, but it is possible to imagine because geological processes, such as mountain building and erosion, remain constant. What is less certain, however, is the future of human life-styles. If humans use all the mineral resources of the earth, the life-styles of the future will be very different from those of today.

Spaceship Earth, EPCOT Center, Florida

362

CHAPTER MOTIVATING ACTIVITY

The purpose of this activity is to elicit an awareness of our dependence upon nonrenewable natural resources and what might happen if these resources were depleted. Have the students work in small groups to write a science fiction story. Suggest having the main character in the story live in a world in which one of five major resources is scarce or polluted. Assign each group one of the following resources: air, water, soil, minerals, and renewable or nonrenewable fuels. To prepare for writing, have groups outline two components of their stories: an introduction describing the world without their assigned resource, and a description of the daily life of their main character and how the character is affected by the lack of the resource. Encourage the students to have their characters express opinions about how the lack of the resource affects them, and how the situation might be remedied.

1 Managing Natural Resources

SECTION OBJECTIVES

After completing this section, you should be able to:

- **Distinguish between** renewable and nonrenewable resources.
- **Explain** the importance of water, air, and land, and **describe** how pollution can endanger their usefulness.
- **Discuss** the recycling of natural resources.

NEW SCIENCE TERMS

pollution
smog
pesticides
recycle
conservation

17.1 Natural Resources

When you think of natural resources, you probably think of coal, oil, iron, and various minerals. While it is true that these are important resources, the most important natural resources are usually taken for granted. These resources include forests, water, air, land, and soil. These resources are vast, and at least partly *renewable*. With careful management they may be used over and over. However, waste and misuse of these resources seriously threaten their futures. Forests can be replanted, but water, air, and land abuse must be controlled to ensure that these resources will serve the future needs of humans.

Other natural resources, including metals, minerals, and fossil fuels, are *nonrenewable*. Nonrenewable resources are limited in quantity. When they are used up, they cannot be replaced in the foreseeable future. Even though these resources cannot be replaced, there are ways to extend their use. However, as with the renewable resources, waste and misuse will shorten the time of their usefulness.

○ *What are renewable and nonrenewable resources?*

Figure 17–1. This rusting scrap metal (left) represents a waste of precious resources as does air, water, and land pollution, which wastes the natural beauty of the earth (right).

Section 1   Managing Natural Resources   **363**

Answer–17.1

○ Renewable resources are materials that are not easily exhausted. Nonrenewable resources are materials that are limited in quantity.

DEMONSTRATION

For a demonstration of managing natural resources, see page 361c preceding this chapter.

Section **1**: **Managing Natural Resources**

SECTION FOCUS

Section Overview This section distinguishes between natural resources that are renewable, such as water, air, and soil, and those that are nonrenewable, such as metals, minerals, and fossil fuels. The section emphasizes the fact that all natural resources are vulnerable to misuse and pollution.

Section Objectives For a list of section objectives, see pupil's edition page.

New Science Terms For a list of new science terms in this section, see pupil's edition page. To help the students understand these terms, you may wish to have them categorize lists of natural resources.

SECTION DEVELOPMENT

17.1 Natural Resources

DISCUSSION Elicit from the students a list of natural resources and write them under the headings renewable and nonrenewable on the chalkboard. As the students provide examples, have them justify their classification into these categories. (renewable: trees, fish, and so on) ***Thinking Critically:*** Ask the students how a renewable resource, such as a lake, forest, animal species, or soil, might become nonrenewable. (A lake might be dredged and filled; a forest may be replaced by buildings; animals can become extinct; soil can be polluted with certain wastes that can contaminate it for thousands of years.)

EXTENSION The Environmental Protection Agency and state environmental agencies are mandated to protect our natural resources. These agencies spend millions of dollars annually to clean up water and land resources that have been severely damaged by misuse and pollution. The National Parks System, as well as hundreds of local wildlife organizations, has set aside parcels of land where wildlife and special geologic features are protected from development and overuse.

363

17.2 Water

DISCUSSION You may wish to introduce this section by reviewing the three parts of the water cycle—evaporation, condensation, precipitation. Draw a diagram of the water cycle on the chalkboard. Include a body of water and a forest to show evaporation, clouds, and rain. Also include underground water sources. Explain to the students that many communities are dependent upon underground sources of water. Chemicals that seep through the soil can pollute a community's water supply. ***Thinking Critically:*** Ask the students to suggest ways in which underground water sources can be protected. (Local governments can restrict the use of harmful chemicals that are used in agriculture and gardening; underground storage tanks can be monitored.)

EXTENSION Major cities and many large communities in the United States must operate water treatment plants in order to protect their water supplies. At these sites, water that has been used before is chemically treated to kill harmful bacteria and to neutralize or remove toxic chemicals. One method of purifying water is to filter it through beds of fine sand or charcoal. Harmful organisms and other particles cling to these materials during the filtering process.

THINKING SKILL *(Solving Problems)* Tell the students to imagine that they live in a community with limited water resources. A company wants to open a plant in the community. The plant will produce toxic wastes. Divide the class into four groups—a town council, representatives of the new plant, a citizens' group, and representatives of an environmental agency. Then have the latter three groups present a proposal to the town council that will allow the plant to operate while maintaining water quality. Finally, tell the students in the town council group that they will devise a compromise plan based on the proposals presented by the other groups. (The students may suggest that the plant build a water treatment facility partially financed by taxes. Plant representatives may suggest that the jobs they provide to the community will offset expenditures on the water treatment plant.)

Figure 17–2. Many places on Earth have too little water. In Holland, however, there is too little land. The solution is to reclaim land from the North Sea. Dikes are built (left) and the seawater is pumped out using the energy of the wind (right).

A MATTER OF FACT

If all the surface water in the world were divided equally, there would be enough for each person to fill 10 olympic-sized swimming pools.

17.2 Water

Nearly three-fourths of the earth's surface is covered by water. Through the water cycle, this resource is purified and returned to the land. Where precipitation is limited, humans have brought in water through elaborate irrigation systems. In some places, such as Holland, there is not enough land, so a series of dikes and levees is used to reclaim land from the sea.

Plants and animals, including humans, require fresh water for survival. Unfortunately, 97 percent of the world's water is salty. Much of the fresh water is unavailable for use because it is frozen in glaciers or trapped deep underground. Surface water supplies most of the water needed for industry, homes, and irrigation.

In the United States, about 300 L of water are used every day by every person for drinking, cooking, and sanitation. If industrial uses are included, about 5000 L of water per day per person are used.

With so many people using the relatively limited supply of surface water, much of it must be reused. For example, it is estimated that by the time the water in the Mahoning River flows from its origin in central Ohio to the city of Youngstown—a distance of less than 200 km—it has been used 10 times.

Unfortunately, in many areas, surface water is polluted. **Pollution** is the unwanted dirt and waste that fouls water, air, and land. Surface water is polluted by untreated organic wastes, sediments, chemicals, and heat.

Organic waste, or sewage, comes from untreated human and animal wastes. The wastes of food-processing plants, including blood, hair, feathers, and bones, also pollute. These wastes require oxygen as they decay, and the loss of oxygen may kill

fish and other organisms. This type of pollution can be decreased if organic wastes are removed by sewage treatment plants. These plants break down the waste before it is released into the water.

Sediments, although natural in surface water, may be increased by the construction of roads and buildings, or by runoff from bare soil or hillsides stripped of their protective cover of plants. Sediments cover water plants, preventing photosynthesis, and clog the gills of fish and other animals. The amount of sediments can be reduced by allowing the water to stand in settling ponds. As the sediments fall to the bottom, the water becomes clear again.

Chemical pollutants are usually of two different types. Plant nutrients come from fertilizers and detergents. Toxic, or poisonous, chemicals come from industrial processes. Excess nutrients cause an increase in algal growth. This sudden population explosion, called a *bloom*, creates problems for fish living in the water. When algae die, much of the oxygen in the water is used in the process of decay. The resulting lack of oxygen causes large fish-kills. This is especially true in summer because warm water cannot hold much dissolved oxygen.

Toxic chemicals include acids, lye, metals, cyanide, and even radioactive elements. These are dangerous to plants and animals that live in or use the water. Chemicals can be removed by settling or by special types of treatment plants. Some water may be reused within a factory, thus retaining the chemical pollution and reducing the need for water resources.

Figure 17–3. A modern sewage treatment plant (left) cleans waste water by aeration (center) and by filtration (right).

Figure 17–4. Even seemingly clean water (bottom) can be deadly to fish and other organisms if it contains dangerous chemicals. Industrial wastes can be removed by filtration and sedimentation (below).

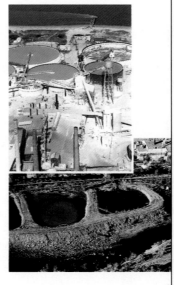

365

EXTENSION Point out to the students that most nonindustrialized countries in the world today have agricultural economies. Human and animal wastes pass from the soil into the water supply. In this manner the eggs or larvae of many disease-causing parasites pass from livestock or people to infect individuals who drink or bathe in local waters. In Africa, Asia, and Latin America, millions of people are debilitated or die from diseases that are spread by polluted water. (Example: Schistomiasis)

DISCUSSION Review with the students how erosion by water and wind causes sediments to wash into bodies of water. *Thinking Critically:* Ask the students to suggest ways to limit runoff that causes sediments to accumulate. (One way is to plant ground-cover plants that hold soil in place.)

EXTENSION Tell the students that in some areas of the United States, such as the New England states, many lakes have been killed by the algal bloom phenomenon. When fish die as a result of oxygen depletion, birds and other wildlife that depend upon water organisms for food cease to inhabit the areas around these lakes. In this way, whole communities of wildlife, which are made up of interdependent plants and animals, have been displaced.

THINKING SKILL (*Using Concepts*) Ask the students to draw a food web to illustrate how toxic chemicals pass from water through the food chain. Remind the students that animals that eat other animals ingest more concentrated doses of toxic chemicals. (A food web might show algae being eaten by small water organisms, which are eaten by small fish, which are eaten by larger fish. Fish may be eaten by birds, people, or other animals.)

365

DISCUSSION Explain to the students that millions of gallons of water are used daily in the manufacture of steel and in other industries. These enormous quantities of water are used to cool equipment such as iron-making furnaces. The heated water is then added to a nearby water source, such as a stream or river. The rise in temperature kills many organisms and causes others to seek new environments in which to live.

EXTENSION Marine estuaries are valuable resources that are endangered by misuse. These coastal waters are breeding grounds for many marine animals and birds. Landfill and development, as well as pollution, are causing many of these areas to disappear, endangering the future of America's wildlife. Point out to the students that coastal states are now trying to regulate development along their seashores in order to protect the natural beauty of the land and the animals who inhabit it.

ACTIVITY:
Chlorinating Water

Thinking Skill (*Experimenting*)

Hint You may wish to take the students to a local pond to collect water. The students can use plastic jugs for this purpose.

CAUTION: Use a medicine dropper to put drops of sodium hypochlorite into the students' beakers. Caution the students not to allow bleach to touch their skin or clothing.

Answers to Conclusions/Applications

1. They died.
2. The water can be filtered or aerated.

Figure 17–5. Thermal pollution can be eliminated through the use of cooling towers (right). In a cooling tower, excess heat is removed by the evaporation of water (left).

Excess heat, called *thermal pollution*, is created by industrial processes such as the manufacture of steel. Power plants, which use large quantities of water to produce steam to drive electrical generators, also produce thermal pollution. Heated water may kill some organisms directly. Heated water also decreases the amount of oxygen in the water, causing more fish to die. Water can be cooled in evaporation towers or retained in ponds before being discharged into streams or lakes.

The quality of water has been gradually improving since the mid-1970s. However, more must be done to ensure a clean, sufficient supply for the future.

○ *What is pollution?*
○ *Name four sources of water pollution.*

ACTIVITY: Chlorinating Water

How can laundry bleach be used to purify water?

MATERIALS (per group of 3 or 4)

binocular microscope, Petri dish, pond water (containing microscopic organisms), laundry bleach (containing 5 percent sodium hypochlorite), small beaker, medicine dropper

PROCEDURE

1. Place the Petri dish on the microscope stage; then pour the pond water into the dish and observe the microorganisms. Draw what you observe.
2. Your teacher will put a few milliliters of laundry bleach, containing a 5 percent solution of sodium hypochlorite, into the beaker.

3. Use the medicine dropper to add the bleach to the Petri dish, one drop at a time, while observing the solution's effect on the movement of the organisms.

CONCLUSIONS/APPLICATIONS

1. What happened to the microorganisms as the bleach was added?
2. What other steps might be taken to make sure a water sample is safe to drink?

366 Chapter 17 Earth's Future

BACKGROUND INFORMATION

Chlorination is the most common and effective method of destroying harmful microorganisms in water. Chlorine disinfectant is usually used in the form of sodium hypochlorite in low concentrations.

Answers–17.2

○ Pollution is the dirt and waste that spoils water, air, land, and soil.
○ Organic wastes, sediments, chemicals, and heat are sources of water pollution.

17.3 Air

Unlike water, air is evenly distributed over the earth. Except for high elevations where the atmosphere is thin, there is plenty of air for all organisms. However, there are no barriers to hold pollution in one place. Local and planetary winds constantly mix the air. Moreover, any pollution in the atmosphere tends to remain there until it is washed out by precipitation.

There have always been natural sources of air pollution, such as volcanic eruptions, forest and grass fires, and wind-blown soil or dust. In fact, some natural pollution is necessary for the formation of condensation nuclei for rain and snow. However, as you may recall from Chapter 8, too much smoke and carbon dioxide can lead to a greenhouse effect. If the worldwide temperatures were to rise only a few degrees, the polar ice-caps would melt, and many coastal areas would be flooded by rising seas.

Pollution caused by human activity has a much more immediate effect on the atmosphere. The burning of fossil fuels has increased the amount of carbon dioxide in the atmosphere. The burning of fossil fuels can also produce a type of air pollution known as smog. **Smog** is a chemical fog produced by the reaction between sunlight and pollutants. Smog is particularly serious in places like Los Angeles and Denver, where it is sunny nearly every day and where thousands of automobiles produce tons of pollution every day. Smog is also potentially dangerous to human health. Since the 1940s, smog is thought to have caused the deaths of thousands of people, both in big cities and small towns.

Figure 17–6. Natural air pollution includes sea salt, dust, volcanic ash, and smoke from forest fires (above).

Figure 17–7. In Los Angeles (left) pollution from automobiles is converted to smog by a chemical reaction started by the sun (right).

DISCUSSION   Help the students to list sources of air pollution. (emissions from cars, factories, burning of fuel to heat homes and offices) *Thinking Critically:* Ask the students how the weather can influence the degree of air pollution in a location. (Wind may blow pollutants toward a populated region; moisture in the air may cause the air to hold more pollution than dry air could hold.)

THINKING SKILL   *(Making Comparisons)* Ask the students to explain the difference between fog and smog. (Fog is low-lying clouds; smog is chemical fog produced by a chemical reaction between sunlight and pollution.)

367

DISCUSSION Ask the students if they are familiar with the chemical, sulfuric acid. The students may have some understanding that acids are strong chemicals that actually eat away at solid materials. Tell the students that acid precipitation actually eats away at stone structures such as buildings and monuments. Acid precipitation is also killing off millions of trees in forests of North America and Europe.

EXTENSION Have the students work in small groups to list ways in which communities can help control air pollution. (using car pools or public transportation; using solar energy)

Figure 17–8. Acid rain can cause damage to statues (left) and to trees (right).

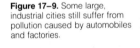

A MATTER OF FACT

An open-pit copper mine produces 12 million metric tons of waste material for every 10 thousand metric tons of copper metal that is produced.

There is also evidence that the burning of fossil fuels is responsible for producing acid rain. The sulfur dioxide released by the burning of some coal and oil becomes sulfuric acid (H_2SO_4) when it combines with moisture in the air. The acid precipitation that results damages buildings, plants, and even fish. Even the ancient Sphinx in Egypt, which has withstood natural weathering for over 5000 years, is now in danger of being destroyed by acid rain.

As has happened with water, the quality of air has been slowly improving since the mid-1970s. Enforcement of laws by state, federal, and international environmental agencies have reduced some sources of pollution. Modern equipment, such as air scrubbers for industry and catalytic converters for automobiles, has also helped reduce pollution. However, continual improvement is needed to keep up with the pollution resulting from an increase in numbers of cars and factories.

○ *What is smog?*
○ *What causes acid rain?*

Figure 17–9. Some large, industrial cities still suffer from pollution caused by automobiles and factories.

368

Answers—17.3

○ Smog is a chemical fog produced by chemical reactions in the atmosphere.

○ Acid rain is precipitation of sulfuric acid formed by water and sulfur products released into the atmosphere by industry and automobiles.

BACKGROUND INFORMATION

In the 1980s efforts were begun to control acid rain in North America. Canada and the United States argued over the source of the pollution that causes acid precipitation in both countries. Leaders of Canada and the United States began an effort to join forces to stem the increasing damage caused by acid precipitation.

17.4 Land and Soil

Dry land covers less than one-third of the earth's surface, or about 15 billion hectares. Of that, only about 2 billion hectares can be farmed. The rest is either too rocky, or the climate is too wet, too dry, or too cold for farming. In the United States and Canada, the same land is used for crops each year; the nutrients used by the plants are replaced by chemical fertilizers. So efficient are farmers in the United States and Canada that only about one person in ten is involved in farming. These farmers produce enough food to more than meet the needs of all their countries' people.

The United States, Canada, Australia, and Argentina are the great food exporting nations of the world. In most other countries, and especially in Asia, Africa, and Central America, not enough food can be produced to meet the people's needs. Many underdeveloped countries cannot afford the technology to produce large crops on the same land each year. When good crops can no longer be produced, people abandon the land, so new land must be found each year to produce food. The abandoned land is left without a protective layer of plants, and more soil is lost to wind and water erosion.

Figure 17–10. In many industrialized countries, modern equipment is important to food production (left). In underdeveloped countries, farming depends more heavily on the labor of humans and animals (right).

Figure 17–11. In countries such as the United States, Canada, and Australia, surplus food is sometimes given away (left), or stored until it is needed (right).

369

17.4 Land and Soil

DISCUSSION Point out to the students that it is the thin layer of topsoil that makes soil fertile. In fact, if desert sands were covered with topsoil, their mineral content would make them extremely fertile. ***Thinking Critically:*** Ask the students what the effect would be if a highly productive agricultural region such as the Midwest were to lose its topsoil due to erosion. (Farmers would not be able to grow their crops.)

THINKING SKILL (*Understanding Concepts*) Ask the students to name some farming techniques that might help underdeveloped countries produce enough food to feed their populations. (crop rotation; fertilizing the soil)

EXTENSION Tell the students that certain plants are able to return nitrogen to the soil, helping to maintain the soil's fertility. Legumes, such as lentils or soy beans, can be planted in place of other crops every few years in order to help keep soil fertile. Some farmers in the United States practice this farming method, although most use chemical fertilizers.

DISCUSSION Tell the students that the natural resources and wildlife of an area exist in a delicate balance that can easily be disturbed. *Thinking Critically:* Ask the students why birds in particular were found to have large amounts of DDT and other pesticides in their bodies. (Insects and small mammals feed on plants that take in pesticides from soil. As birds eat large numbers of these animals, pesticides accumulate in their bodies.)

EXTENSION The bald eagle has become an endangered species from eating prey contaminated with pesticides and from being hunted excessively. In recent years, however, the bald eagle has been making a comeback due to the banning of certain pesticides and the implementation of federal regulations that outlaw the hunting and capturing of this magnificent bird.

THINKING SKILL *(Using Conservation Concepts)* Ask the students to suggest ways in which forests can be protected from excessive harvesting of timber. (recycle paper; use of building materials other than wood)

EXTENSION The Amazon Rain Forest, the largest rain forest in the world, has been increasingly destroyed to create grazing land for cattle that are raised for beef. Conservationists are alarmed by the misuse of this precious natural resource. The use of plant proteins as a dietary supplement has been suggested to lessen the need for grazing land.

EXTENSION You may wish to have the students look at the map of South America in the Science Sites booklet for the location of the Amazon Rain Forest.

Figure 17–12. The survival of species such as the bald eagle (above) was once threatened by the use of a pesticide called *DDT.*

In countries with advanced technologies, improper use of chemicals can poison the soil and make it unfit for growing crops. **Pesticides,** chemicals used to kill crop-destroying insects, can accumulate in the soil. There, the pesticides can become a hazard to animals other than the insects for which they were intended. In the late 1960s, a pesticide called *DDT* had to be banned from general use because it threatened some species of birds, such as the brown pelican and the bald eagle, with extinction.

Another important use of land and soil is for the grazing of animals. Land used for range or pasture may not be fertile enough for crops, or it may be too hilly for farming. Range land may either be planted with special grasses for feeding cattle, sheep, or goats, or the animals may graze on native plants. Sometimes, especially during dry periods, the land may be overgrazed; then the plants die, and the soil may be eroded by wind and water.

Figure 17–13. In some areas, overgrazing by cattle has destroyed the grasses covering the soil (right). Much erosion can be prevented if trees are planted to hold the soil. Millions of tree seedlings, ready for transplanting on barren ground, are grown at tree farms (below).

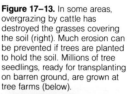

Forests also occupy large areas of marginally fertile land. This renewable resource provides lumber for buildings, fuel for stoves or for making charcoal, and pulpwood for paper products. However, many forests are threatened as more land is used for urban development and the expansion of industry.

○ *State three uses for land and soil.*
○ *Name three of the world's major food-producing countries.*

Answers—17.4

○ Farming, development, and grazing are three uses for land and soil.

○ The United States, Canada, and Australia are three major food-producing countries.

BACKGROUND INFORMATION

In the mid-1800s cattle ranchers and sheep herders feuded bitterly over the rights to grazing land for their herds. Today, grazing lands are cultivated and controlled to assure all livestock an adequate food supply.

17.5 Mineral Resources

How many nonrenewable resources have you used this week? Have you had a drink from a metal can and thrown the can away? Have you thrown away a plastic cup? If you have, you have wasted mineral resources. You may recall from earlier in this chapter that part of the definition of nonrenewable resources is that once they are used up, they are gone—forever. Metal deposits and fossil fuels take millions of years to develop but only a few seconds to be destroyed.

Although nothing can be done to replace the resources that have been used up, it is possible to extend the time of usefulness for those that remain through recycling. To **recycle** means to use something again. There are many examples of recycling in nature. You may recall the discussion of the water cycle in Chapter 9. All the water on Earth goes through a series of processes that eventually return it to the sea. The water cycle is an example of natural recycling.

Figure 17–14. The carbon-oxygen cycle (left) is an example of natural recycling. Many metals and other mineral resources can be recycled as well (below).

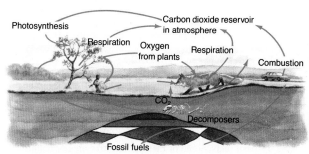

Many elements and compounds are also recycled in nature. For example, water and carbon dioxide are taken in by green plants and, through the process of photosynthesis, become carbohydrates. At the same time, oxygen is released into the atmosphere. Animals use the oxygen and the carbohydrates in cellular respiration. As these are used to provide energy for animals, water and carbon dioxide are released into the environment. This completes the cycle—the elements have been recycled and are ready to be used again.

Some nonrenewable resources may also be recycled. For example, the aluminum in beverage cans can easily be recycled to produce new beverage cans. Many aluminum companies will pay a few cents per kilogram for recycled aluminum. Making cans from recycled aluminum is much cheaper than producing cans from aluminum ore. In addition to saving money, recycling cans saves aluminum ore and energy resources. Melting aluminum cans uses far less energy than refining ore.

DISCOVER

Recycling Resources

Find out if there is a recycling center for aluminum cans, newspapers, or glass in your community. If there is, make it a class project to collect these materials on a weekend or holiday, and to take them to the recycling center. If the recycling center pays for the material, you may wish to contribute the profits to an organization involved in protecting or beautifying the environment.

17.5 Mineral Resources

DISCUSSION   Ask the students to suggest items they use daily that can be recycled. *Thinking Critically:* Ask the students to name some advantages of recycling. (Recycling eliminates the need for exploration and extraction of valuable ores and minerals from soil. It also helps lessen the need for strip mining, a method of mining ores and coal in which the top layers of soil and rock are removed and the land is left scarred and barren.)

THINKING SKILL   *(Making Analogies)* Ask the students to think of forms of natural recycling that are similar to the natural recycling of water. (Organic materials in forests are recycled as materials from the remains of dead plants and animals are broken down and become part of the soil, then reenter plants through their roots. These materials are also returned to animals when they eat plants.)

EXTENSION   You may wish to start a recycling center in your classroom. The students may wish to save paper to be used as scrap or to bring to a local recycling center.

DISCOVER:
Recycling Resources

Thinking Skill   *(Communicating)*

Have the class work together to compose a letter to send to environmental organizations. Allow the students to work together in groups, coordinating the areas of the community to be covered and the means of transportation.

DISCUSSION Lead the students in a discussion about how they can use conservation in their everyday lives. (not letting tap water run; turning off lights and appliances when not in use)

EXTENSION Some countries have been experimenting with alcohol as an alternative to fossil fuels. Cars in Brazil can have their tanks filled at alcohol stations. This fuel is made from organic materials and can be made relatively inexpensively where there is a lot of plant material available to produce alcohol.

SECTION REVIEW

Summary Have the students work in small groups to write reports about how school resources can be conserved. They may interview kitchen and office staff to find out what materials are discarded. Have the students suggest which of these materials might be recycled and how it could be done.

Reinforcement Have the students complete the Section Review. Remind them to use specific examples to support their answers. Allow the students to use library resources to find relevant data.

Figure 17–15. Abandoned cars (left) and old newspapers (right) can be recycled.

Figure 17–16. Riding a bicycle to work or to school is a good way to conserve energy resources.

A MATTER OF FACT
Three metric tons of copper ore are required to provide enough copper (2.5 kg) for the wiring of a single automobile.

You would probably never think of just throwing away gold jewelry or silver coins. Like gold and silver, most metals should be recycled. Old cars can be melted and their iron used to produce new steel, thus saving much iron ore and energy.

Many nonmetals can also be recycled. Glass can be ground up, melted, and formed into new products. Although trees are renewable resources, many can be saved for other uses by recycling newspapers and cardboard.

Recycling is not the only way to save nonrenewable resources. With some resources, such as fossil fuels, recycling is impractical if not impossible. Sometimes, however, conservation of resources can extend their use. **Conservation** means using materials wisely. Energy resources can be conserved by careful use. Energy resources can also be extended by using alternatives to fossil fuels. Some of these alternative energy resources will be discussed in the next section.

○ *What is recycling?*
○ *Name three advantages of recycling aluminum cans.*

Section Review

READING CRITICALLY

1. Explain why wood is a renewable resource.
2. How can conservation increase the availability of a nonrenewable resource?

THINKING CRITICALLY

How does pollution threaten the quality of air, water, and soil?
If certain pesticides are dangerous to birds and other animals, why are they not completely banned from use?

Answers to Section Review

Reading Critically

1. Unlike fossil fuels, trees can be replaced by planting new trees.
2. By using nonrenewable resources wisely and recycling resources when possible, the use of nonrenewable resources can be extended for longer periods.

Thinking Critically

3. Dirt, dust, and hazardous chemicals interfere with the natural recycling of air, water, and soil. In addition, natural cycles do not remove some pollutants from the environment.
4. Some have been banned. However, others are considered safe, when used according to the directions distributed by the companies that produce them. Each pesticide must be examined carefully to understand the dangers it presents to the environment.

SKILL ACTIVITY: Analyzing Bar Graphs

BACKGROUND

Bar graphs are very useful for presenting information that must be quickly and accurately compared. Seeing information presented in a bar graph often makes the information easier to remember than reading the information in print. When using a bar graph, the first task is to determine what information is being presented, and then to determine what each bar on the graph represents.

PROCEDURE

1. Look for the title of the graph. The title tells you what information is being presented in the graph.
2. Look at the bottom of the graph. There you will find the names of the items that are being compared.
3. Look at the side of the graph. There you will find the quantities that are being compared.
4. The colored bars on the graph help to visually present the comparisons.

APPLICATION

1. What is being compared in this bar graph?
2. What units are used for this comparison?
3. Which area consumes the most energy? How much energy is used there?
4. Which area consumes the least energy? What is the difference between the most energy consumed and the least energy consumed?

USING WHAT YOU HAVE LEARNED

Make a bar graph using Table 17–1 on page 374. Be sure to include a key showing the items that are being compared.

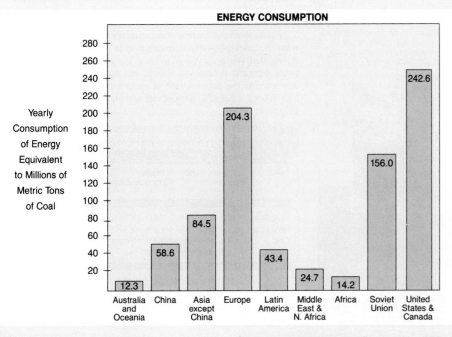

ENERGY CONSUMPTION

Yearly Consumption of Energy Equivalent to Millions of Metric Tons of Coal

- Australia and Oceania: 12.3
- China: 58.6
- Asia except China: 84.5
- Europe: 204.3
- Latin America: 43.4
- Middle East & N. Africa: 24.7
- Africa: 14.2
- Soviet Union: 156.0
- United States & Canada: 242.6

Section 1   Managing Natural Resources   **373**

SKILL ACTIVITY: Analyzing Bar Graphs

Objectives

- Interpret bar graphs.
- Compare data.
- Make a bar graph.

Discussion Help the students locate information on the bottom and sides of the bar graph. Have them take notes of increments in metric tonnage on the side of the graph. You may wish to explain to the students that a metric ton is equivalent to 1000 kilograms. ***Thinking Critically:*** In what ways are bar graphs useful? (Relationships can be seen quickly.)

Answers to Application

1. The annual consumption of coal in various regions around the world.
2. Metric tons of coal.
3. The United States; 242 million metric tons.
4. Australia and Oceania; 231.3 million metric tons.

Answer to Using What You Have Learned

The students may compare either the amount of known reserves or yearly consumption of each fuel listed in Table 17-1. You may wish to have motivated students research the breakdown of consumption by region and record this data on a bar graph.

Answers–17.5 (p. 372)

○ Recycling means to use something again.

○ Recycling aluminum saves money, aluminum ore deposits, and energy resources.

Section 2: Managing Energy Resources

SECTION FOCUS

Section Overview This section discusses the limited reserves of nonrenewable energy sources. In contrast, renewable energy sources, such as solar, wind, and water energy, are inexhaustible. The advantages and disadvantages of nuclear energy are also discussed.

Section Objectives For a list of section objectives, see pupil's edition page.

New Science Terms For a list of new science terms in this section, see pupil's edition page. Students may already have some knowledge of these terms from prior reading and media sources.

SECTION DEVELOPMENT

17.6 Nonrenewable Energy Sources

DISCUSSION Have the students study Table 17-1 to compare availability and consumption of world fuel reserves. *Thinking Critically:* Ask the students what factors might affect the problem of diminishing fuel reserves and increasing fuel consumption. (New reserves may be found; other forms of energy may replace fossil fuels.)

THINKING SKILL *(Interpreting Data)* Draw the students' attention to the fact that uranium has the least available known world reserves of all the fuels listed in Table 17-1, yet the percent of annual consumption is relatively high. *Thinking Critically:* Have the students suggest reasons for (1) the relatively high consumption of uranium and (2) the relatively low consumption of coal. (Nuclear energy has been more widely used in the last two decades; the burning of coal has been restricted in many areas and replaced by other forms of energy.)

NEW SCIENCE TERMS
hydroelectric power
tidal power
geothermal power
solar power
solar cell
biomass
fission
fusion energy

SECTION OBJECTIVES
After completing this section, you should be able to:
- **Explain** the difference between renewable and nonrenewable energy sources.
- **List** five energy sources that may be used in the future.
- **Discuss** the advantages of fusion energy over fission energy.

17.6 Nonrenewable Energy Sources

Most scientists agree that in the future, industry, transportation, and residential needs will be supplied almost exclusively by electrical power. At present, however, most electricity is produced from nonrenewable energy sources, such as fossil fuels and uranium. These nonrenewable energy resources are quickly being exhausted. The resources that took 500 million years to accumulate are being exhausted by humans in just a few decades.

The amounts of known world reserves of fuels are compared with the present yearly consumption in Table 17–1. The table shows that the reserves of natural gas and coal are far from being exhausted; however, this is not true for petroleum and uranium. At the present rate of consumption, there is enough petroleum for only 28 years of use and enough uranium for only 40 years.

Figure 17–17. Oil shale (below), the rock that burns, could provide an alternative source of petroleum (inset).

TABLE 17–1: WORLD RESERVES OF FUELS VERSUS YEARLY CONSUMPTION			
Fuel	Known World Reserves	Yearly Consumption	Yearly Consumption as a Percentage of Reserves
Petroleum	72.0×10^9 tons	2.5×10^9 tons	3.5 percent
Natural gas	5.0×10^{13} m³	1.5×10^9 m³	0.003 percent
Coal	7.5×10^{12} tons	3.7×10^6 tons	0.00005 percent
Uranium	1.5×10^6 tons	38 000 tons	2.5 percent

The problem is even worse for the United States. At the present rate of consumption, the United States' reserves of petroleum probably will be exhausted in 15 years and those of uranium in 18 years.

BACKGROUND INFORMATION

According to the Stanford Research Institute, over 60 percent of the domestic and imported petroleum in the United States is used for automobile, truck, airplane, railroad, and ship transportation.

When these sources are used up, they will be gone forever. In the past, geologists have been able to discover new fuel reserves to balance those consumed. If humans are to continue present life styles, new, inexhaustible sources of energy must become available before these reserves run out.

○ *Which fuel has the largest reserves?*
○ *How soon will the world run out of petroleum?*

Figure 17–18. Coal is the main fuel used to produce electricity (above). Many coal mines are abandoned (left) as they are mined out.

17.7 Renewable Energy Sources

Renewable energy sources are those that are produced from inexhaustible supplies, such as the sun. Solar, wind, hydroelectric, and ocean tidal energy are all examples of renewable resources. At the present time, hydroelectric energy is the only renewable resource contributing significantly to the world's power production.

Hydroelectric Power   Power produced from the energy of falling water is called **hydroelectric power.** Before falling water was used to produce electricity, water power was used for such functions as turning millstones to grind grain. In the production of electricity, the falling water turns turbines. Hydroelectric power is clean and nonpolluting. However, it requires a natural waterfall or a large river where a dam can be built. Unfortunately, the places where dams can be easily and inexpensively built are few. One of the largest hydroelectric power plants in North America is the Grand Coulee Dam in Washington state.

Wind Power   In some areas of the world, it is possible to use wind as an energy source to produce power. Wind power has been used for thousands of years to lift water and to grind grains. In areas where the wind is strong and constant, wind-driven electrical generators are practical. A windmill is connected to the generator's shaft, and the electricity produced is

Figure 17–19. Hydroelectric power is clean and inexpensive, but its use is limited to places with large water supplies and steep topography.

17.7 Renewable Energy Sources

DISCUSSION   Point out to the students that as of 1962 hydroelectric power generated half as much electricity as nuclear power in the United States. ***Thinking Critically:*** Ask the students what the advantages of using hydroelectric power are. (It is not polluting and relatively inexpensive.) Then ask them why they think its use is not more widespread. (Many regions do not have large rivers where dams can be built.)

THINKING SKILL *(Making Judgments)* Ask the students where they think the best locations to build wind-driven generators might be. (mountain and hill tops; large, flat areas; near oceans)

THINKING SKILL *(Formulating Hypotheses)* Ask the students to speculate on what types of obstacles there might be to using wind-generated power in large urban areas. (Not enough energy is generated to meet the needs of heavily populated areas; density of buildings obstructs the wind; wind turbines are extremely noisy.)

Answers—17.6

○ Coal has the largest reserves.
○ At the present rate of consumption, there is enough petroleum to last 28 years.

EXTENSION In California, individuals may purchase windmills and sell the electricity they generate to local power companies.

DISCUSSION Help the students understand that a bay must be narrow in order to be a good site for a tidal generator. Also, there must be a significant daily tidal change. ***Thinking Critically:*** Where is the best place along a coastline to place a tidal power station? (On a narrow bay where the tidal change will be exaggerated.)

THINKING SKILL *(Making Judgments)* Ask the students why tidal power is not more widely used world-wide. (Many countries are land-locked and do not have access to shore lines; most bays do not have high enough tides to generate power. Sealing off a bay eliminates its use for transportation.)

EXTENSION Temperature gradients in ocean water could theoretically generate energy in a similar way to the recycling of Freon in a refrigerator. A low-boiling fluid can be vaporized and used to drive a turbine. This vapor can then be funneled to cold, deep ocean water along the shoreline and condensed. The condensed vapor can then be recycled. This is known as ocean-thermal power.

Figure 17–20. Wind turbines (left) can provide power for rural areas, while large wind-turbine "farms" (right) can provide power for entire cities.

A MATTER OF FACT

Volcanic eruptions and lightning flashes are not useful as sources of power because they are too uncertain.

---

used directly or is used to charge storage batteries. The electricity produced is limited and is most practical in rural areas where it is not profitable to distribute commercially generated electricity. However, in California there are many wind farms, where wind turbines produce significant amounts of electricity for commercial use.

Tidal Power Along some coastlines, in France for instance, the ocean tides rush up bays that extend many kilometers inland. At the mouth of the bay, the tide can be directed through barriers with openings leading to turbines, producing electric energy. At high tide, the water returns to the ocean through the same turbines. This electric energy is produced by tidal power. **Tidal power** is produced by the energy of changing tides.

A French power station at the mouth of the Rance River produces 150 megawatts (MW) of electricity. A huge tidal power project under construction in northeastern Canada is discussed in the Science Connection on pages 386–387.

Figure 17–21. Tidal power may produce much electricity for coastal cities in the future.

BACKGROUND INFORMATION

Two areas in the United States where tidal power is used are the Passamaquoddy Bay between Maine and New Brunswick, Canada, and Cook Inlet in Alaska.

Geothermal Power

Geothermal Power   In some regions, such as in Italy, New Zealand, and Iceland, hot igneous rocks produce usable energy within the earth. Rainwater penetrates porous rocks near the heat source and is converted to steam. The steam may surface through natural vents, or it can be extracted by drilling. The most famous steam field is at Larderello in Tuscany, Italy. Steam from Larderello is used to drive turbines that produce electricity. This steam power from the earth is called **geothermal power.** Geothermal power accounts for only 0.4 percent of the power used in Italy, but in New Zealand geothermal power accounts for 5 percent of the power produced.

Hot water and steam from geothermal fields are also used as a source of direct heat for buildings. The entire city of Reykjavik, Iceland, is heated in this way.

Figure 17–22. Geothermal energy is widely used in Iceland (right). In California some electricity is produced from steam fields such as the one shown here (left).

Solar Power

Solar Power   On a cold day, you have probably tried to stay in the sun to keep warm. The power produced by energy from the sun, called **solar power,** can be used directly as a source of heat or to produce electricity. The most common use of direct solar energy is for heating water. An array of dark-colored pipes placed on the roof of a building will provide hot water for personal use, and it may provide some heat for the building as well. Why would the pipes need to be dark in color?

Figure 17–23. Solar energy can be collected passively to heat water (right) or entire homes (left).

DISCUSSION   Have the students look at the maps in the Science Sites booklet to locate countries where geothermal power is used. (Italy, New Zealand, and Iceland) Remind the students of what they learned about activity along crustal plates in Chapter 7. **Thinking Critically:** Ask the students why they think scientists are exploring the possibility of using geothermal power on the Hawaiian Islands. (The Hawaiian Islands are located in an area of volcanic activity.) Have the students suggest other regions where geothermal power might be used. (along continental plate boundaries where there is volcanic activity)

THINKING SKILL (*Making Judgments*) Ask the students to think of drawbacks that might be associated with the use of geothermal power. (Communities using this energy would have to be located close to sources of hot water. Some of these areas may be near active volcanoes and, therefore, may be dangerous places in which to live.)

THINKING SKILL (*Using Prior Knowledge*) Ask the students why they think dark-colored panels are used in solar water heaters. (Dark colors absorb more heat than light colors.)

DISCUSSION Tell the students that only a small fraction of the energy collected in solar cells is actually available to produce electricity. ***Thinking Critically:*** Ask the students what the advantages and disadvantages of solar energy are. (Advantages: limitless supply, nonpolluting; Disadvantages: not practical in regions that have short daylight hours for part of the year; don't work well on cloudy days; expensive storage systems)

EXTENSION Parabolic mirrors have also been used to collect solar energy. Heat absorbed by the darkened glass or aluminized sheets is funneled into tubes where circulating water is heated. Steam that is generated can be used to power turbines or provide hot water to dwellings.

THINKING SKILL (*Understanding Concepts*) Ask the students which would be the best direction to face rooftop solar panels in the Northern Hemisphere. Have the students explain their answers. (South would receive direct sunlight for the greatest number of hours.)

THINKING SKILL (*Making Judgments*) Ask the students why solar batteries are the most practical source of energy for spacecraft equipment. (The need to carry large quantities of fuel is eliminated.)

Figure 17–24. Solar energy can provide electricity directly for a variety of uses.

A **solar cell** can produce electricity from sunlight. A solar cell, such as those shown in Figure 17–24, includes a junction between two semiconductors. Photons falling on the junction produce electricity.

One disadvantage of solar power is the size of the solar cells needed to produce large amounts of electricity. A solar power plant with a capacity of 100 MW would cover a surface of 4 km^2. A solar power plant might better be constructed in space. There, the panels of solar cells would be free from clouds and air pollution, with no limit to the number of panels that could be displayed. In addition, the assembly could be made to face the sun at all times. The power generated could be transmitted to Earth by means of microwave beams.

Solar energy already has been used extensively to power spacecrafts. Panels of solar cells keep batteries charged for radio and data transmission from space.

BACKGROUND INFORMATION

Engineers are currently experimenting with amorphous-silicon solar cells that can replace roofing tiles on most conventional buildings. Recent research shows that amorphous-silicon absorbs more sunlight than crystalline silicon, which has been widely used in solar batteries.

Biomass Another renewable energy source is **biomass,** or once-living matter. Wood and plant waste can be burned as fuel. In the United States much of the plant material is first converted to methane or alcohol, but in many countries the waste is burned directly.

About 10 percent of the energy consumed in the world comes from biomass. The use of biomass as a source of energy may not be apparent in industrialized nations, but in underdeveloped countries biomass is a major source of energy.

○ *Name five sources of renewable energy.*
○ *What are solar cells?*
○ *What is biomass?*

17.8 Nuclear Energy

You may recall from Chapter 15 that some atoms, such as those of uranium, tend to break apart due to natural radioactive decay. Some radioactive isotopes, however, can be forced to decay. When the nuclei of these atoms are bombarded with neutrons, the atoms undergo fission. **Fission** is the splitting of the nucleus of one atom into smaller nuclei. Fission releases a tremendous amount of energy, which can be used to produce steam to drive turbines. There are, however, some serious problems with fission power. Besides the possibility of accidental releases of radioactivity, such as occurred at Chernobyl, USSR, in 1986, the world supply of uranium for fission is very limited.

There is a type of nuclear energy that is potentially so abundant as to be considered truly inexhaustible. This energy, called **fusion energy,** is produced when small atomic nuclei fuse to form new atoms. In a fusion reaction, two isotopes of hydrogen, deuterium (doo TIHR ee uhm) and tritium (TRIHT ee uhm), fuse to form helium. This is the same process that produces the energy of the sun.

The main advantage of using fusion energy as compared to fission energy is that no dangerous radioactive isotopes are produced. The reaction produces only harmless helium that diffuses into the atmosphere and escapes into space. Also, the deuterium and tritium for the reaction can be obtained easily from sea water.

Figure 17–25. In some countries peat is cut, stacked, and dried so it can be used as fuel.

DISCOVER

Using Energy Resources

Make a list of all the energy resources that you use in one day. For example, if you ride a bus to school, be sure to include that on your list.

When you have completed your list for one day, indicate which resources are in short supply and which are still plentiful. Then list ways to conserve these resources.

Figure 17–26. Although inexpensive and plentiful, there are some dangers to using nuclear energy to generate electricity, as the accident at Chernobyl (left) showed.

379

DISCUSSION Lead the class in a discussion about fuel sources of biomass that are used for heating. Most students will be familiar with wood. Some students may not know that in some countries in Asia and Africa, animal dung is dried and then used as fuel. *Thinking Critically:* Ask the students to suggest the disadvantages of using wood as fuel. (Wood must be cut and dried before it can be used; it would take a large amount of wood to heat a large building; forest denudation.)

17.8 Nuclear Energy

DISCUSSION Explain the terms *fission* and *fusion* to the students. Remind them that in fission, large nuclei are broken apart to form small nuclei. In fusion, nuclei are joined to form a new atom. In both processes, enormous amounts of energy are released.

THINKING SKILL (*Understanding Cause-and-Effect Relationships*) Point out to the students that the effects of the 1986 disaster at Chernobyl are still being experienced around the world. Ask the students to suggest how wildlife, land resources, and people are being affected by the radioactivity that was released into the atmosphere. (Radioactivity was carried across Europe and beyond, where it settled on soil and water. The radioactive elements were taken in by plants and, in turn, by people and other animals. In many communities in the USSR and in Europe, milk was contaminated and possibly unfit for human consumption.)

EXTENSION Birds migrating from northern Europe to Africa have been found to contain high levels of radioactivity as a result of the Chernobyl disaster.

DISCOVER:
Using Energy Resources

Skill (*Organizing*)

Hint Make sure the students are careful to list all the sources of energy used during the day.

BACKGROUND INFORMATION

On July 16, 1945, the first atomic explosion took place in New Mexico. The first atomic bomb had the force of 20 000 tons of TNT. For the location of this explosion, have the students look at the map of New Mexico in the Science Sites booklet.

Answers—17.7

○ Hydroelectric power, wind power, tidal power, geothermal power, and solar power are all renewable energy sources.
○ Solar cells produce energy from sunlight.
○ Biomass is once-living matter that can be burned as fuel.

380

DISCUSSION Nuclear reactors are very controversial because of the dangers of accidental release of radioactivity. Discuss with the students the difficulties in developing fusion plants to replace current reactors.

EXTENSION Fusion engineers are experimenting with a variety of fusion techniques. These include laser fusion, muon fusion, and magnetic field containment. In most fusion projects, hydrogen ions are circulated through doughnut-shaped vacuum tubes at enormous speeds. As ions collide, they fuse and release tremendous amounts of energy.

EXTENSION For the approximate location of nuclear reactors in your state and neighboring states, refer the students to the Science Sites booklet. Continent maps also show other countries that have nuclear reactors. The number of reactors per country is shown in parentheses.

SECTION REVIEW

Summary Review with the class the advantages and disadvantages of hydroelectric, wind, tidal, geothermal, solar, biomass, and nuclear energy sources.

Reinforcement Have the students complete the Section Review. Allow students to use the information they compiled in their group summaries to answer the questions.

Figure 17–27. When construction of this fusion reactor is completed, it may help make sea water a source of abundant electric power.

The main disadvantage of fusion energy is the very high temperature needed for the fusion reaction to take place. While fissioning of uranium can be started at room temperature, a temperature of 100 million °C is needed to initiate a fusion reaction. No known matter can withstand temperatures this high, so an energy field must be developed to contain the reaction. Scientists in the United States are working hard on these problems because fusion holds the hope of providing an infinite amount of energy for the future.

○ **What is the difference between fission and fusion?**
○ **What is the main problem in producing fusion energy?**

Section Review

READING CRITICALLY

1. What are the major sources of energy used to produce electrical power?
2. What limits the expansion of hydroelectric power?
3. How is fusion energy produced?

THINKING CRITICALLY

4. Why is it important to develop energy alternatives to fossil fuels and uranium?
5. How could coal be used to extend the reserves of petroleum and natural gas?

Answers–17.8

○ In fission, large atomic nuclei are split into smaller nuclei; in fusion, hydrogen isotopes fuse to form helium.

○ The main disadvantage of fusion is the high temperatures required for containment.

Answers to Section Review

Reading Critically

1. Fossil fuels, hydroelectric, and nuclear power are major sources of electricity.

2. The proximity to large rivers on which dams can be built limits the expansion of hydroelectric power.

3. Fusion energy is produced when two hydrogen isotopes fuse to form helium.

Thinking Critically

4. Fossil fuels and uranium are nonrenewable resources. Only a limited supply of each is still available.

5. Coal can be burned in place of petroleum and natural gas.

TECHNOLOGY: Fusion Power

In the past few decades humans have begun to tap the energy of the atom. Vast amounts of energy are released when the nuclei of atoms are split (fission) or fused (fusion).

The reality of this energy source was demonstrated with the explosion of the first atomic bomb in 1945. The uncontrolled, chain reaction of splitting atoms produced a violent explosion. Since then nuclear scientists have learned how to control the rate at which atoms split. Today, fission reactors are used in many parts of the world to generate electricity.

Scientists are now trying to control nuclear fusion reactions in the same way they can control fission reactions. Controlled fusion can produce much more energy with less pollution than fission reactions. The problem with fusion is that a great amount of heat is necessary to start the reactions. For example, a tem-

Reactor control room

Producing a fusion reaction

perature in excess of 100 million°C is needed to start a fusion reaction.

However, some scientists are currently working on a method of producing fusion reactions at very low temperatures. Experiments have shown that hydrogen fusion can take place at temperatures as low as 13 K (−260°C) or at room temperature.

The key to the success of these low-temperature reactions lies in the development of a process called *muon-catalyzed fusion*, or cold fusion. In order for the positively charged nuclei of deuterium to fuse, they must be brought close together. To overcome their natural electrical repulsion, most experimental techniques use high temperatures. However, muon particles make that unnecessary.

A *muon particle* is a subatomic particle that is about one-fifth the size of a proton, but with a negative electrical charge. Be-

Testing the equipment

cause it has the same charge as an electron, it can replace an electron in a molecule of deuterium. The added mass of the muon causes the nuclei of the deuterium atoms to be drawn close together. When the nuclei of the atoms are close to one another, they can be more easily fused. If this process of cold fusion can be perfected, a relatively inexpensive alternative to nuclear fission could be producing electricity in the future.

Section 2   Managing Energy Resources   **381**

TECHNOLOGY: Fusion Power

Discussion   After they read the article, you may wish to have the students debate the question of whether or not to expand the use of nuclear energy. Divide the class into two groups. Allow groups time to discuss their ideas. Then call on members of each group alternately to respond to each other's points.

Extension   You may wish to show the students photographs of different types of nuclear plants, including the newest research facilities where research on fusion is being conducted.

Skill *(Constructing Models)*

Preparation of Materials

This investigation may be done as a demonstration for the entire class by using one workable model of an electric generator. Groups of students can manipulate the parts of the generator to answer the questions to the Analyses and Conclusions section.

Hint

Make sure the pencil rotates smoothly through the holes in the coil. If the pencil does not rotate smoothly, the generator will not work.

Answers to Analyses and Conclusions

1. hydropower, wind power, steam power
2. The compass tends to orient itself perpendicular to the flow of current in the coil near the end of the wire.
3. Diagrams will vary.

Answer to Application

The pointer on the galvanometer will move; the bulb will light up. The generator produces a measurable electric current.

INVESTIGATION 17: Generating Electricity

PURPOSE

To construct a working model of an electric generator

MATERIALS (per group of 3 or 4)

Bar magnet
Insulated bell wire (10 m)
Electrical tape
Plastic straw
Compass
Wooden block (5 cm × 10 cm × 10 cm)
Thumbtacks (4)
Pencil
String
Hole punch

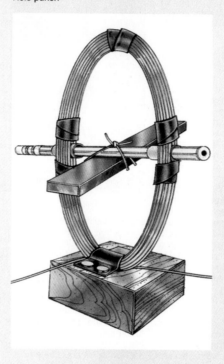

PROCEDURE

1. Place the bar magnet on the table, and prepare an oval coil of wire (about 30 turns) so that the bar magnet will fit into it lengthwise.
2. Use electrical tape wrapped at several places as shown to prevent the coil from unwinding.
3. Cut two small pieces (5 cm) from a plastic straw. Insert a piece of straw into each long side of the coil and secure it with electrical tape as shown.
4. Cut a piece of electrical tape and use it to mount the coil, with thumbtacks, to the wooden block as shown.
5. Insert a pencil through the holes in the coil, and use electrical tape or string to secure the bar magnet as shown.
6. Twist together the exposed ends of the wire coil.
7. Twirl the pencil to rotate the magnet while observing the behavior of a compass placed at varying positions around the closed loop.

ANALYSES AND CONCLUSIONS

1. Twirling the pencil is an example of mechanical energy. What other forms of mechanical energy could be used to rotate the magnet?
2. How did the compass behave when placed near the coil? Near the attached ends of the loop?
3. Draw a diagram of a generator in which the coil, and not the magnet, is rotated.

APPLICATION

Connect the ends of the wire coil to a galvanometer or a small flashlight bulb. Explain what happens.

SUMMARY

- Partly renewable natural resources include forests, water, air, land, and soil. (17.1)

- Metals and fossil fuels are nonrenewable resources. (17.1)

- In the United States, each person uses about 300 L of water daily. (17.2)

- Much water is polluted by organic waste, sediments, chemicals, and heat. (17.2)

- Polluted water can be treated, and the quality of water has improved since the mid-1970s. (17.2)

- Smog is a chemical fog produced by the reaction of sunlight and air pollutants. (17.3)

- Acid rain affects buildings, plants, and even fish. (17.3)

- Only a small percentage of the earth's land surface is suitable for farming. The rest is either too rocky or has a climate too severe for growing crops. (17.4)

- Chemical fertilizers improve the soil in countries with advanced technologies, but some chemical pesticides may be harmful to more than pests. (17.4)

- The use of nonrenewable resources may be extended by recycling. (17.5)

- Where it is not practical to recycle, conservation of resources may extend their usefulness. (17.5)

- Nonrenewable energy sources, such as fossil fuels and uranium, are quickly being exhausted. (17.6)

- Inexhaustible energy alternatives for the future may include hydroelectric power, wind power, tidal power, geothermal power, solar power, and biomass. (17.7)

- Nuclear energy from fission may be limited by the small reserve of uranium and by the danger of radioactivity. (17.8)

Write all answers on a separate sheet of paper.

SCIENCE TERMS

Correctly use each of the following terms in a sentence.

biomass **(379)**
conservation **(372)**
fission **(379)**
fusion energy **(379)**
geothermal power **(377)**
hydroelectric power **(375)**
pesticides **(370)**
pollution **(364)**
recycle **(371)**
smog **(367)**
solar cell **(378)**
solar power **(377)**
tidal power **(376)**

SCIENCE QUIZ

Modified True-False

Mark each statement *true* or *false.* If a statement is false, change the underlined word to make the statement true.

1. Metals and fossil fuels are <u>nonrenewable</u> resources.

2. Water is recycled for repeated use through the <u>purification</u> cycle.

3. In the United States, a single person uses about <u>20 liters</u> of water every day.

4. Another name for chemical fog is <u>acid rain</u>.

5. Without conservation, petroleum reserves in the United States will be exhausted in <u>15 years</u>.

6. Hydroelectric power comes from <u>falling water</u>.

continues

CHAPTER REVIEW

SUMMARY

The students may review the major concepts in the chapter by reading the summary statements. The statements are cross-referenced to the chapter to facilitate reinforcement of any concepts of which the students feel unsure. Encourage the students to work in groups to quiz one another.

SCIENCE TERMS

The sentence in which the science term is used should reflect an understanding of the definition of the term. You may wish to have the students write short paragraphs using the science terms to explain a major concept in the chapter.

SCIENCE QUIZ

Answers to Modified True-False

1. true
2. false, water
3. false, 300 L
4. false, smog
5. true
6. true

continues

Answers to Multiple Choice

7. d
8. a
9. a
10. b
11. c
12. d

Answers to Completion

13. fusion
14. biomass
15. geothermal energy
16. 1 million
17. pollution
18. hydroelectric

Answers to Short Answer

19. Renewable resources are those having an inexhaustible supply. Nonrenewable resources are those that cannot be replenished.
20. After planting, forests must be cultivated for long periods of time before they can be harvested. In addition, much of the land previously used as forest land is being lost due to urban and industrial development.
21. hydroelectric power, wind power, tidal power, and geothermal power
22. Solar energy may be used to heat water, or to produce small amounts of electricity.

Answers to Writing Critically

23. Forests need fresh water and a particular balance of chemicals and minerals for healthy growth. Pollution poisons water and leaves undesirable chemical wastes that ruin the soil's ability to support plants.
24. Answers will vary, but students should mention the following: (1) crops, (2) animal grazing, and (3) development of living areas.
25. Answers will vary. Benefits include using less energy and using wasted heat. Drawbacks include using limited fossil fuels.

ANSWERS TO EXTENSION

1. The students' reports should mention the fact that biological organisms inhabit delicately balanced ecosystems. The students should understand that biological control of pests that are injurious to crops is an attempt to manipulate those systems to the farmer's advan-

---

Multiple Choice

Write the letter of the choice that best answers the question or completes the sentence.

7. Which of the following is not a renewable resource?
   a) air   b) water
   c) land   d) coal

8. Which country exports the least food to other nations of the world?
   a) Soviet Union   b) United States
   c) Canada   d) Australia

9. The resource consumed most per year by percentage of its remaining reserves is
   a) uranium.   b) natural gas.
   c) petroleum.   d) coal.

10. Using the tides to generate electricity would be an example of ____ power.
   a) geothermal   b) tidal
   c) biomass   d) lunar

11. The resource that could be used to provide fuel to generate electricity using nuclear fusion is
   a) uranium.
   b) coal.
   c) water.
   d) All of the choices are correct.

12. Which of the following energy sources could be considered truly renewable?
   a) nuclear fusion
   b) hydroelectric power
   c) geothermal power
   d) biomass

Completion

Complete each statement by supplying the correct term or phrase.

13. Nuclear ____ produces less radiation than nuclear fission.
14. In underdeveloped nations, ____ is a major source of energy.
15. Much of the city of Reykjavik, Iceland, is heated by ____.
16. Fossil fuels required more than ____ years to accumulate.

17. Unwanted and unnatural chemicals found in the environment are called ____.
18. Natural waterfalls can sometimes be used as sources of ____ power.

Short Answer

19. Explain the difference between renewable and nonrenewable resources.
20. Explain why forests are considered less renewable than air or water.
21. In addition to fossil fuels, list four alternative sources of power.
22. Describe the two ways that solar energy may be used by humans.

Writing Critically

23. How could air pollution affect the renewability of forests?
24. Discuss the multiple uses of land and soil.
25. When electricity is generated by burning fossil fuels, much waste heat is produced. This waste heat could be used to heat buildings in a process called *cogeneration.* Describe some benefits and some possible drawbacks to using cogeneration as an alternative energy source.

EXTENSION

1. Go to the library and read about alternatives to chemical pesticides. Write a brief report describing the effort of scientists to use biological forms of pest control.
2. From the electric meter for your home, find out how much electricity is used each day by your family. Your electric power company can send you a booklet on how to read the meter. Then look around your home for ways by which electricity could be saved, such as turning off lights. After locating several areas of waste and correcting the problems, check again to see how much electricity has been saved. Report to the class on your success.

---

tage. The students should discuss the advantages and dangers of using these techniques.
2. Reports will vary.

APPLICATION/CRITICAL THINKING

1. Compare and contrast the processes of nuclear fission and fusion. Because of the nuclear reactions involved, fusion would be a considerably cleaner form of energy production. Referring to the different reactions, explain why this is so.

2. Coal can be converted to a gaseous fuel by a chemical process, and the coal gas can be used instead of natural gas. What environmental problems might develop if coal gas is used as a replacement for natural gas?

3. What types of common items could be recycled that are not presently being recycled? Why do you think they are not now being recycled?

FOR FURTHER READING

Caulfield, C. *In the Rainforest.* Chicago: University of Chicago Press, 1986. This book describes the rain forests of Africa, South and Central America, and Asia. The reasons for the destruction of these fragile ecosystems are discussed.

League of Women Voters Education Fund. *The Nuclear Waste Primer.* New York: Nick Lyons Books, 1985. This book offers the nonexpert a brief introduction to nuclear wastes.

McKay, A. *The Making of the Atomic Age.* London: University of Oxford Press, 1984. This comprehensive text follows the development of the nuclear age from the discovery of the electron to the industrial development of nuclear power.

Challenge Your Thinking

This is a photo of a nuclear breeder reactor in France. A breeder reactor makes electricity and at the same time makes plutonium fuel, which can also be used to produce more electricity. This eliminates somewhat the dependence on nonrenewable uranium as a source of fuel. The problem is that the plutonium is very dangerous to living organisms and produces wastes that take hundreds of thousands of years to decay. Discuss the advantages and disadvantages of using breeder reactors as an alternative way of generating electricity.

Chapter 17 Review   **385**

ANSWERS TO APPLICATION/ CRITICAL THINKING

1. The heavy isotopes produced in a nuclear fission reaction decay, leaving radioactive residue that needs to be carefully stored to prevent it from filtering into the environment. The products of nuclear fusion reactions are more stable. Radioactive by-products of fusion reactions are limited compared to those of fission reactions.

2. Answers will vary but should include a discussion of the methods employed in acquiring the coal. (strip and subterranean mining)

3. We could recycle more cars, more newspaper, glass, some plastics, and water.

ANSWERS TO CHALLENGE YOUR THINKING

Advantages—they produce plutonium, which can be refined into more fuel. Disadvantages—plutonium is extremely deadly; an accident could kill millions of people.

THE SCIENCE CONNECTION:
How the Dinosaurs Died

Discussion This *Science Connection* is about atmospheric and climatic changes that may have caused the mass extinction of the dinosaurs. Dinosaurs lived during the Mesozoic Era. Most students are probably familiar with several types of dinosaurs through books, television programs, or movies. You may wish to show the class photographs or filmstrips that depict as many types of dinosaurs as possible. Ask the students to hypothesize why so many species of dinosaurs became extinct during this era. (Possible suggestions include that very few, if any, plants and animals could have survived if Alvarez' theory is correct.)

Discussion The students should realize that not all species of plants and animals died during the mass extinction of the dinosaurs. Have the students suggest ways that some species of plants and animals adapted to the lower temperatures and relative scarcity of food during this period in order to survive and eventually evolve into present life forms on Earth.

Extension After completing the Unit and reading and discussing the *Science Connection*, have the students turn back to the Unit Opener on page 321 and answer the questions. When the class discussion has ended, you may wish to have the students write a short report describing Alvarez' theory of how atmospheric and climatic changes may have caused the dinosaurs to become extinct. Encourage the students to share their reports with the class.

UNIT 5 THE SCIENCE CONNECTION:
How the Dinosaurs Died

No one knows for certain why the dinosaurs suddenly disappeared 65 million years ago. Fossils provide few clues about the cause of the extinction.

Scientists are currently debating a startling new theory. Some scientists now believe that a giant meteor or comet struck the earth near the end of the Mesozoic Era. The meteorite would probably have exploded as it struck the earth, causing global climatic changes. The sky would have darkened as ash clogged the air, blocking the sun's rays and lowering the temperature. Millions of animals would have died from the cold or lack of food, and many species would have become extinct. The animals that survived became the evolutionary ancestors of present life forms on Earth. Some scientists believe that catastrophes like this occur every 26 million years. They say that the catastrophes may be caused by an unknown planet or star passing through the solar system and carrying along many comets. Each time Earth is visited by this alien body, there is a rain of comets, and millions of creatures die.

Luis Alvarez developed a new theory after observing some sedimentary rock formations in Italy. Sediments in central Italy contain one of the best records of the boundary between the Cretaceous Period of the Mesozoic Era and the Tertiary Period of the Cenozoic Era. Alvarez discovered a curious layer of clay between the rocks of these two periods. This clay had a high concentration of the element iridium, which is very rare in the earth's crust. "Why," asked Alvarez, "would the iridium level in this layer of clay be 25 times higher than normal?" Alvarez hypothesized that a meteorite slammed into the earth about 65 million years ago. Alvarez based his hypothesis on the fact that extraterrestrial bodies sometimes contain 1000 times more iridium than does the earth's crust.

Alvarez reasoned that iridium-rich dust from the collision spread through the atmosphere. The dust was deposited as the layer of sediment that separates the rocks of the Cretaceous Period from those of the Tertiary Period.

Iridium-rich layer (marked by coin)

Volcanic ash darkens the sky

386

Looking for fossil clues

Jurassic Period dinosaur

Alvarez believed that this gigantic collision and the extinction of the dinosaurs happened at about the same time. First, he calculated how much dust a meteorite collision would have thrown into the air, and then he determined how much sunlight would have been blocked by the dust. Under these conditions, he said, plants and marine organisms would have died almost immediately. Soon large animals such as the dinosaurs would have had nothing to eat, and would have died too. Eventually the meteorite dust would have fallen to Earth like black snow, burying more than half of Earth's animal species.

The scientific community debates these theories as it searches for new evidence. At the same time, scientists in other fields have been inspired by the Alvarez research. Some scientists have theorized that a nuclear war might cause the same kind of global winter as the Cretaceous meteorite. Evolutionary biologists, who once thought that dinosaurs had died out because of their small brains, now realize that there may have been other reasons. Astronomers are searching for the cause of the catastrophe. Was the meteor jolted into Earth's path by some unknown star or mysterious planet? Does a catastrophe like this happen every 26 million years, as some fossil records suggest?

The answers to these questions may affect the way humans view themselves. Evolutionary biology encourages humans to believe that they survive because they are the most intelligent species on Earth. Could it be that without the intervention of a meteorite from space, dinosaurs might still be ruling the earth? The answers to this and other questions lie somewhere in Earth's history, waiting perhaps for you to find them.

Alvarez in his lab

Unit 6: OCEANOGRAPHY

UNIT OVERVIEW

In this unit the earth's oceans, major seas, and large lakes are located; the composition of the ocean's waters is described; and the formation of waves, tides, and currents is discussed. The living resources of the oceans are described, and the possible effects of pollution are explained.

Chapter 18: The Water Planet, page 390

This chapter concentrates on characteristics of oceans, seas, and lakes. Formation and characteristics of features of the ocean floor are presented. Special emphasis is placed on coastal features and the importance of coastline preservations.

Chapter 19: Ocean Waters, page 410

This chapter presents the characteristics of ocean waters, including temperature, density, and a description of the salts dissolved in ocean waters. Also discussed are currents, waves, tides, and the effects of these motions on the people who live nearby.

Chapter 20: Ocean Resources, page 432

The sediments of the ocean and the resources contained in these sediments are discussed. This chapter also contains information about the geologic history of the earth that scientists have uncovered by studying these sediments. Minerals and other materials found in or under the oceans are described, as is how fresh water may be obtained from ocean water. The plankton, nekton, and benthos that live in the ocean are described. The causes of and difficulties created by ocean pollution are explained.

The Science Connection: Power From the Tides, page 450

The Science Connection describes the project, currently under construction, to create electric power from the huge tidal changes on the Bay of Fundy. Also discussed is the currently operating tidal power plant on the Rance River in France.

ADVANCE PREPARATION

Chapter 18

For the Discovers on pages 396 and 403, and for Investigation 18 on page 406, you will need quantities of sand, silt, and gravel. For more information, see pages 396, 403, and 406.

For Investigation 18 on page 406, you will need a large, rectangular aquarium. For more information, see page 406.

Chapter 19

For the Discover on page 414, you will need to place several cups in a freezer until their contents are frozen solid. For more information, see page 414.

Chapter 20

For the Discover on page 443, the students will need to find newspaper and magazine articles about polluted water and beaches. For more information, see page 443.

For Investigation 20 on page 446, the students will need to bring in empty soup cans. For more information, see page 446.

ISSUES IN EARTH SCIENCE

In light of the discovery of the Titanic in the north Atlantic, and the recovery of some of its artifacts, discuss whether these treasures should belong to the finders, to the families of the victims, or to the government.

Manganese nodules may soon be mined from the ocean floor. Discuss possible ownership of minerals and the use of ocean resources generally.

Debate whether the ocean should be used for dumping garbage, sewage, and other wastes, including nuclear wastes.

SUGGESTED PROJECTS

Have groups of students work on constructing fairly detailed models of the ocean floors. When each group is finished, join the models all together for a large display for the school.

Have one group of students make large charts of surface ocean currents on clear plastic sheets. Have another group of students make large charts of geostrophic and density currents. When all the charts are completed, put them together to show the whole spectrum of currents.

BULLETIN BOARD SUGGESTIONS

Chapter 18

Have the students collect copies, tracings, drawings, and photographs of the "Tools of Oceanology." These should include submersibles, sedimentary corers, water pressure gauges, and aqualungs. Each photo should accompany a brief description (on a 3″ × 5″ index card) of how the tool is used in ocean exploration. Have the students decide how and in which categories the items should be displayed.

Chapter 19

Display a map of the world that shows the major ocean currents. Flag the routes of early explorers to the New World, as well as trade and travel routes used today by modern cargo and cruise lines.

Chapter 20

Assign bulletin boards categorized according to the section divisions of the chapter. The displays should be photographic essays illustrating the ways that we utilize ocean resources.

TEACHER RESOURCES

Readings

Chapter 18

Menard, H.W. *Islands*. New York: W.H. Freeman and Company, 1986. A beautifully illustrated study of the geography, geomorphology, and biogenesis of the oceanic islands of the world.

Chapter 19

Ballard, R. "A Long Last Look at Titanic," *National Geographic Magazine* 170 (December, 1986) p. 698. Picturesque memorial to the lost luxury liner's last resting place.

Chapter 20

"Oceans." *Scientific American* 221 (September, 1969) p. 54. Theme issue containing articles covering every aspect of oceanology from the origin and structure of the oceans to the physical and food resources tapped by humans.

Audiovisual/Software

Chapter 18

A computer program, *Shore Features*, is available for Apple II computers from Geoscience Resources.

A slide series, *Coastal Morphology*, is also available from Geoscience Resources.

Chapter 19

A computer program, *The Oceans*, is available for Apple II and TRS-80 computers from Geoscience Resources.

A computer program, *Ocean Tides and Currents*, is available for Apple II computers from Educational Images.

Chapter 20

A computer program, *Coral Reefs and their Residents*, is available for Apple II computers from Educational Images.

Other Materials

Chapter 18

Write the National Geographic Society and request a catalog of their posters and maps. Choose and purchase maps of ocean currents, features of the ocean floor, tidal charts, and more:
National Geographic Society
17th and M Streets, N.W.
Washington, DC 20036

Chapter 19

Write the United States Coast Guard and request pamphlets documenting information about their history and current responsibilities.:
Ocean Operations Division
U.S. Coast Guard Headquarters
Washington, DC 20590

Chapter 20

Have students choose a food company that cans ocean products, such as tuna or salmon. Write the public relations department of the company and request pamphlet information on where and how those resources are taken from the ocean. Most companies will be happy to comply because distributing this type of information is an additional method of advertising their products.

PLANNING THE CHAPTER

Chapter Sections	Page	Chapter Features	Page	Program Resources	Page
Section 1: Bodies of Water	391			Cross-Discipline: Science and Geography, *Studying Geoboundaries* **(A)**	TRB 33
18.1 Oceans and Seas **(B)**	391	A Matter of Fact	393	Investigation 18.1: *The Ocean's Surface* **(H)**	TRB 79
18.2 The Ocean Floor **(B)**	394	A Matter of Fact	395		LM 81
		Discover: Imitating Deep Ocean Sediments **(B)**	396	Concept Extension: *Conserving Water* **(B)**	TRB 37
		Skill Activity: Charting the Ocean Floor **(B)**	397		LM 259
18.3 Lakes and Rivers **(B)**	398	Section Review	399		
		Careers: Oceanographer, Marine Conservationist, Underwater Photographer	400		
Section 2: Coasts	401			Critical Thinking **(H)**	TRB 34
				Reading for Content: *Predicting Outcomes* **(A)**	TRB 31
18.4 Coast Development **(B)**	401	**Discover:** Settling Sediments **(B)**	403		LM 207
				Investigation 18.2: *Wave Action* **(A)**	TRB 83
18.5 Preserving Coasts **(B)**	404	A Matter of Fact	405		LM 85
		Section Review	405		
		Activity: Taking Core Samples **(B)**	405		
		Investigation 18: Wave Action **(B)**	406	Student Record Book: Textbook Investigations **(B)**	TRB 35
Chapter 18 Review	407			Vocabulary **(A)**	TRB 18
				Tests **(A)**	LM 157
				Computer Test Bank	TRB 79

(LM) Laboratory Manual/Study Guide, **(TRB)** Teacher's ResourceBank™

B = Basic **A** = Average **H** = Honors

The coding Basic, Average, and Honors indicates sections or subsections that might be appropriate for different levels of learners. For additional suggestions regarding choice of topic and depth of coverage, see the Pacing Chart on pages T16–T20.

CHAPTER CONCEPTS, OBJECTIVES, AND TERMS

Section	Concepts	Objectives	Science Terms
Section 1: Bodies of Water	■ The earth has one large ocean that is divided into four oceans and numerous seas. **(18.1)** ■ Seas may be less salty than the oceans if they receive a large amount of fresh water, or they may be saltier than the oceans if there is a high rate of evaporation. **(18.1)** ■ The detailed study of the ocean floor began in the 1940s after the invention of sonar. **(18.2)** ■ Between the continents and the Atlantic Ocean floor are the gently sloping continental shelf and the steeper continental slope. **(18.2)** ■ In the middle of the Atlantic Ocean is an underwater mountain range called the Mid-Atlantic Ridge. **(18.2)** ■ The floor of the Pacific Ocean has many deep trenches and volcanic mountains. **(18.2)** ■ Lakes are large bodies of water completely surrounded by land. **(18.3)** ■ Rivers and lakes are important as sources of fresh water and as means of transportation. **(18.3)**	■ **Locate** the oceans of the earth, and **identify** the differences between oceans and seas. ■ **Describe** some characteristics of the ocean floor, and **compare** the floors of the Pacific and the Atlantic. ■ **Discuss** the importance of large bodies of fresh water on the earth.	ocean seas sonar continental shelf continental slope lakes
Section 2: Coasts	■ Coasts are areas where the ocean meets the land. Coasts may be classified as primary or secondary. **(18.4)** ■ Primary coasts show erosion by land agents, such as glaciers and running water. Secondary coasts show much erosion by waves and wind. **(18.4)** ■ Drowned valleys are common along primary coasts, while secondary coasts have sand bars and barrier islands. **(18.4)** ■ Groins, seawalls, jetties, and breakwaters are all used to prevent erosion of coasts. **(18.5)**	■ **Describe** the two main processes that shape coastlines. ■ **Identify** some of the common coastline features and **explain** how they are formed. ■ **Analyze** the problem of coastline erosion and **evaluate** efforts being made to preserve coastlines.	primary coast fiord bay secondary coast

Title	Page	Materials
Discover: Imitating Deep Ocean Sediments	396	*(per student)* glass jar, water, soil, sand, powdered clay, spoon
Skill Activity: Charting the Ocean Floor	397	*(per student)* paper, pencil, graph paper
Discover: Settling Sediments	403	*(per student)* jar, soil, sand, gravel, shallow pan, water, stirrer
Activity: Taking Core Samples	405	*(per group of 3 or 4)* pan of sediments, soda straw, marking pen, razor blade
Investigation 18: Wave Action	406	*(per group of 3 or 4)* coarse soil or sand; large aquarium; bucket; water; large beaker, 1500 mL; metric ruler

TEACHING SUGGESTIONS

Section 1: **Bodies of Water**

Demonstration: How Water Gauges Work

Purpose

To demonstrate how water gauges work

Background

Water pressure increases with depth. This water gauge can be used to demonstrate this fact. As the thistle tube is submerged in the tank, pressure on the balloon at varying depths will compress the air inside the rubber and glass tubing and force the colored liquid up the open end of the U-tube. You may challenge students to correlate increasing water depth in the tank (measured in cm) with the ruler level of colored water in the U-tube (also in cm).

Materials

Glass tubing (20 cm)
Fine sand
Bunsen burner
Thistle tube or small funnel
Balloon
Rubber band
Fish tank
Food coloring
Ruler
Tape
Small beaker
Rubber tubing (50 cm)
Scissors
Safety goggles
Laboratory apron

Procedure

1. Fill the glass tubing with fine, dry sand.
2. Holding the tube at both ends while twirling slowly, heat the center over a small blue flame until the glass is malleable. Slowly bend the tube into the shape of a "U." Gently set the tube down and allow several minutes for it to cool. When the tube is cool, tap out the sand.
3. Fill the fish tank with water.
4. Prepare a small beaker of colored water. Fill the U-tube halfway with the colored solution.
5. Using tape, secure the U-tube to the ruler and tape the assembly to the side of the fish tank, leaving the top of the tube at a higher level than the top of the tank.
6. Cut the balloon and spread it over the end of the thistle tube (or small funnel). Secure the balloon with a rubber band.
7. Attach one end of the rubber tubing to the end of the thistle tube and the other end to one arm of the U-tube.
8. Place the thistle tube into the fish tank. Change the depth of the thistle tube several times, measuring the water in the U-tube at each depth. Record the measurements on the chalkboard.

Question to Ask the Students

Why does pressure increase with depth? (*The weight of the water produces the pressure.*)

Outside Speaker

Invite a parent or other member of the community who has had experience at sea or on any of the nation's waterways. Any boating enthusiast or member of the armed services (request that sailors or marines bring their uniforms) can be asked to lecture to the class about their experiences.

Class Activity

Have the students read books about the ocean and present short reports to the class.

Section 2: Coasts

Field Trip

If you live near the ocean or a large lake, plan a class trip to study the effects of wave erosion on the coast.

UNIT 6

OCEANOGRAPHY

The sun, the moon, and the ocean combine to create one of Earth's most reliable forces—tides. Pulled by the gravity of the sun and the moon, tides flow in and out of bays and inlets. Engineers dream of harnessing tidal power to generate inexpensive electricity. On the Rance River in France the dream is a reality. In Canada the dream is rapidly becoming a reality. Canadian engineers are planning a giant tidal power project on Canada's Bay of Fundy, which has the highest tides in the world.

- How can tides be used to produce electricity?
- How are tides, waves, and currents produced?
- What causes the ocean to be salty?
- What important resources does the ocean hold?

By reading the chapters in this unit, you will learn the answers to these questions. You will also develop an understanding of concepts that will allow you to answer many of your own questions about the oceans of the earth.

Moonrise over a bay

389

INTRODUCING THE UNIT

Have the students look at the Unit Opener photograph on these pages. Point out that because ocean tides are caused primarily by the gravitational pull of the moon, they are higher on the side of the earth that faces the moon. Read the text and use the questions to guide a class discussion. Do not expect the students to be able to answer the questions now. The questions are motivational and the students should be aware that they will be able to answer them after completing the Unit.

Discussion Some of the students may have vacationed at a beach. Ask them if they have ever felt the push and pull of waves and surf. Waves can have enough force to push even adults off balance. This can be dangerous because, if the undertow, or seaward return of water, is strong it can drag a person under the waves and toward the deep water. Ask the students if they have ever bodysurfed or it they have ever ridden a surfboard and felt the tremendous kinetic energy of waves.

Extension You can extend the discussion by bringing in other Unit themes. What causes ocean waves and currents? What elements are contained in ocean water? What should be done to protect the ocean? How does the ocean affect weather? You may wish to record the answers and review them at the end of the unit.

CHAPTER 18

CHAPTER OVERVIEW

This chapter concentrates on characteristics of oceans, seas, and lakes. Formation and characteristics of features of the ocean floor are presented. Special emphasis is placed on coastal features and the importance of coastline preservations.

Section 1: Bodies of Water In this section oceans and seas are compared. Features of the ocean floor are discussed. The large lakes of the world are discussed.

Section 2: Coasts Processes that form and shape coasts are discussed in this section. Special emphasis is placed on the preservation of coasts from damage caused by natural and human factors.

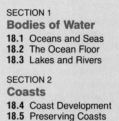

390

The Water Planet

Water covers nearly three fourths of the earth's surface. Yet humans have explored more of the surface of the moon than they have of the floor of the ocean. Unique in the solar system, Earth is truly the water planet. The ocean links the continents and provides many of the raw materials needed for human survival. The final frontier may not be space; it may be the oceans.

Sunset over the Pacific Ocean

CHAPTER MOTIVATING ACTIVITY

Tell the students that the oceans of the world comprise 362.1 million km^2 of the earth's surface, compared to a land area of only about 87.3 million km^2. Much about this vast portion of the earth's surface is still a mystery. Have the students work with the school librarian to gather together 20–30 issues of *National Geographic* that contain articles about oceans, seas, and lakes. Distribute the magazines to the class. Have individuals or small groups scan the relevant articles and take notes on the important points about the earth's large bodies of water. Have the students volunteer suggestions for topics about which they would like to learn more. Tell them that this chapter will address some of the questions that they have. Lead a discussion about the coastal and river cities whose inhabitants depend on the waterways not only for recreation and drinking water, but also for their livelihood.

1 Bodies of Water

SECTION OBJECTIVES

After completing this section, you should be able to:

- **Locate** the oceans of the earth, and **identify** the differences between oceans and seas.
- **Describe** some characteristics of the ocean floor, and **compare** the floors of the Pacific and the Atlantic.
- **Discuss** the importance of large bodies of fresh water on the earth.

NEW SCIENCE TERMS

oceans
seas
sonar
continental shelf
continental slope
lakes

18.1 Oceans and Seas

The surface of the Earth is actually one huge ocean with large islands, or continents, scattered here and there. The ocean is so large that you could sail on it for weeks and not see any land. The ocean is also very deep, reaching depths of more than 10 km in some places. From space the ocean looks calm, but if you walk along a beach, you can see that the ocean is constantly in motion.

If the earth were completely smooth, its surface would be covered evenly with water to a depth of over 2.5 km. Only about 60 percent of the Northern Hemisphere is covered with water since most of the continents are found there. About 81 percent of the Southern Hemisphere is covered with water.

Oceans are major expanses of salt water with surface areas larger than 10 million km². Even though the earth has only one large ocean, most geographers usually divide it into four oceans.

Figure 18–1. Before the twentieth century, the oceans influenced nearly all aspects of human society. Cities were built near the shore and the oceans provided a transportation link between countries. Humans also depended on the oceans for salt and for much of their food.

391

SECTION FOCUS

Section Overview This section deals with the locations and sizes of the oceans and seas of the earth. The formation and characteristics of features of the ocean floor are described. The importance of lakes and rivers is discussed. Special emphasis is given to some of the earth's largest lakes.

Section Objectives For a list of section objectives, see pupil's edition page.

New Science Terms For a list of new science terms in this section, see pupil's edition page. You may wish to locate specific examples of these terms, such as the Pacific Ocean or Lake Superior, on a world map or on the figures presented in the section as the students encounter the new terms.

SECTION DEVELOPMENT

18.1 Oceans and Seas

DISCUSSION You may wish to use a globe to show students that the oceans are continuous and cover most of the earth's surface. Have the students locate and name the four oceans. *Thinking Critically:* Ask the students why they think less is known about the oceans than about the earth's landmasses. (The ocean depths are nearly inaccessible; the oceans are so vast.)

SKILL

BACKGROUND INFORMATION

Remind the students that they read about plate tectonics in Chapter 7. Tell them that over a period of millions of years, the size and shape of the oceans have undergone continuous changes as a result of crustal movement. Remind the students that today's oceans formed as the supercontinent Pangaea broke apart about 250 million years ago. The migration of the continental landmasses that resulted has reshaped the earth's oceans.

DEMONSTRATION

For a demonstration of how water gauges work, see page 387e preceding this chapter.

DISCUSSION Help the students to locate the seas in Table 18–2 on a world map. Remind the students that seas contain salt water because ocean water is able to pass freely into seas.

SKILL *(Using Maps)* Have the students identify oceans that are near each of the seas listed in Table 18–2.

DISCUSSION Challenge the students to suggest the geopolitical advantages a country might enjoy by having a sea within or near its borders. (The sea could provide access for shipping via the ocean, as well as a natural border.). You may wish to have the students research the recent events in the Persian Gulf.

SKILL *(Interpreting Data)* Have the students look at Table 18–1. Ask them to compare the oceans listed in terms of area and depth. ***Thinking Critically:*** Based on the range of latitudes of each ocean, ask the students to suggest differences they would expect to find in the temperatures of the oceans. (Oceans near the poles have colder waters.)

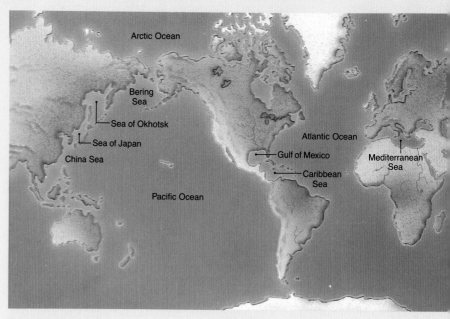

Figure 18–2. This map shows the location of the earth's oceans and major seas.

The area around the continent of Antarctica is sometimes called the *Antarctic* or *Southern Ocean*, but as the map in Figure 18–2 shows, the Southern Ocean is really just the southernmost portions of the Atlantic, Pacific, and Indian oceans. Figure 18–2 also shows many of the earth's seas. **Seas** are expanses of salt water less than 3 million km^2 in area. Seas are partially enclosed by land and are therefore somewhat separated from the oceans.

TABLE 18–1: THE WORLD'S OCEANS		
Ocean	**Area ($\times 10^6$ km²)**	**Average depth (m)**
Pacific	166.2	4188
Atlantic	86.6	3736
Indian	73.4	3872
Arctic	12.2	1117

TABLE 18–2: THE LARGEST SEAS		
Sea	**Area ($\times 10^6$ km²)**	**Average depth (m)**
China Sea	2.8	1437
Caribbean Sea	2.5	2575
Mediterranean Sea	2.5	1502
Bering Sea	2.3	1492
Gulf of Mexico	1.5	1614
Sea of Okhotsk	1.4	973
Sea of Japan	1.0	1667

The Caribbean Sea and the seas off eastern Asia are separated from the open ocean by island arcs. Sea water passes freely between the islands, making the saltiness of these seas close to that of the oceans. Other seas, such as the Black Sea and the Baltic Sea, receive a great deal of fresh water from rivers that flow into them. Therefore, their saltiness is lower than that of the oceans. The Mediterranean Sea, on the other hand, joins the ocean only through the narrow, shallow Strait of Gibraltar. The saltiness of the Mediterranean is higher than that of the open ocean because of a high evaporation rate.

Six million years ago, the passage between the Atlantic and the Mediterranean Sea closed due to the movement of Africa and the lowering of the water level in the oceans. Over the next thousand years, the Mediterranean nearly dried up; only in the eastern Mediterranean did some small salt lakes remain. These lakes were fed by water spilling down from the Black Sea and from the Nile River. The rest of the Mediterranean was a dry sea bed, covered with a layer of salt 50 m thick. The Mediterranean remained dry for several thousand years, and then the ocean level rose again. This cycle was repeated many times over the next million years. Each time the ocean level rose, water poured over the Strait of Gibraltar, forming a great system of rapids and waterfalls that quickly refilled the Mediterranean.

○ *Name the four oceans.*
○ *How is a sea different from an ocean?*

A MATTER OF FACT

Sea level does not remain constant and has varied considerably over the ages. Borings by marine mollusks into the columns of the Greek Temple at Serapis indicate that the structure was submerged more than once since its construction 2000 years ago.

Figure 18–3. The famous Strait of Gibraltar (below) separates the Atlantic Ocean from the Mediterranean Sea (left).

393

DISCUSSION Help the students to understand that inland seas remain salty because of their connection to open seas. Tell them that all runoff eventually empties into seas or oceans. ***Thinking Critically:*** Ask the students why the fresh water from rivers has little effect on the saltiness of the oceans. (Oceans are so vast that the relatively small amount of fresh water that empties into them from rivers does not affect their saltiness.) Because seas are much smaller, those that receive fresh water from large rivers have less salt than do the oceans. Their proximity and access to oceans also greatly affects the saltiness of seas.

SKILL *(Using Maps)* Ask the students to identify the major rivers that provide fresh water to the seas listed in Table 18–2.

EXTENSION (Tie-In/Social Studies) The Mediterranean Sea lies between the continents of Africa and Eurasia. In the ancient world, those countries that had access to the Mediterranean Sea controlled the commerce of the region. The Phoenicians, Egyptians, Romans, and Greeks all established trade routes over these waters. Their naval fleets helped them to control huge empires of the ancient world.

EXTENSION For more detailed information on the location of the major oceans and seas, have the students look at the maps of the continents in the Science Sites booklet.

Answers–18.1

○ The four oceans are the Pacific, Atlantic, Indian, and Arctic.

○ An ocean is larger than a sea, having a surface area greater than 10 million km². The surface area of a sea is less than 3 million km².

18.2 The Ocean Floor

DISCUSSION   Point out to the students that the oceans are so vast and deep that it is difficult to produce a precise map of the ocean floor. Data from soundings are used in conjunction with data about movements of the earth's crust to create maps that approximate details of the ocean floor. Only small portions of this vast area have been accurately charted.

THINKING SKILL *(Using Prior Knowledge)* Remind the students that the continental crust is less dense than the oceanic crust. Ask them why the deep ocean floor is so far below the continents. (Oceanic crust is composed of basalt, overlaid with sediments, while the continental crust is granite and sedimentary rock. These are less dense than basalt.)

18.2 **The Ocean Floor**

While the surface of continents is easy to study, studying the ocean floor presents unique problems. Scientists began to map the ocean floor in the 1940s using sonar, which stands for *SO*und *NA*vigation *R*anging. **Sonar** is the process of bouncing sound waves off a solid surface. A sonar instrument sends a series of sound waves through the water. These sound waves strike the ocean floor and bounce back to the device. This is similar to what happens when you hear your voice echo in an empty room or in a mountain canyon.

Just as you can guess the size of a dark room or unlit cave by how fast your echo returns, the depth of the ocean floor can be determined by the time it takes for the sonar echo to return. From this information, a picture of the ocean floor, such as the one in Figure 18–5, can be produced.

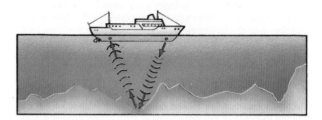

Figure 18–4. Sonar operators receive echoes from the ocean floor. Oceanographers use sonar to chart the depth of the ocean.

Sonar readings show that the North American continent extends for several kilometers into the Atlantic Ocean. This extension forms a gradually sloping bottom called the **continental shelf.** Farther offshore, the continental shelf is replaced by a steeper slope called the **continental slope.** The North American continental slope forms one side of the Atlantic Ocean basin. What do you think forms the other side of the Atlantic Ocean basin?

Figure 18–5. This sonar screen (below) shows a profile of a small section of the ocean floor diagrammed at right.

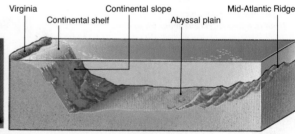

Virginia   Continental slope   Mid-Atlantic Ridge
Continental shelf   Abyssal plain

BACKGROUND INFORMATION

Marine mammals and bats use a type of echolocation to navigate and also to locate food. Porpoises emit high-frequency sounds that help them locate objects and food in the oceans by perceiving the echos of these sounds as they bounce off objects. Research with these animals has helped scientists to develop sonar devices that are used for navigation as well as for exploration of the oceans.

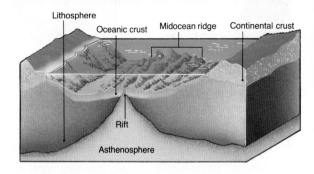

Lithosphere
Oceanic crust   Midocean ridge   Continental crust
Rift
Asthenosphere

Nearly in the middle of the Atlantic Ocean is a long underwater mountain range called the Mid-Atlantic Ridge. The mountains of this range are parallel to the edges of the continents on either side of the ocean. This mountain range is 14 400 km long and is part of the midocean ridge system. Parts of this midocean ridge are found in the other oceans as well.

Recent studies of the Mid-Atlantic Ridge show that a deep valley occurs along the ridge's center. A cross section of this ridge and valley is shown in Figure 18–6. These mountains do not form a continuous line; they are separated by cracks along the chain, where many small earthquakes occur. Lava escapes from the mantle, forming new crust on the ocean floor. Dating has shown that the rocks nearest the center of the ridge are the youngest; older rocks are found farther away from the center, on either side of the ridge.

The rocks on either side of the ridge show another interesting pattern. The basalt of the ocean crust contains magnetic minerals that act like compass needles. These minerals usually line up with the earth's magnetic field before the rock solidifies. Geologists discovered, however, that not all the minerals point toward the magnetic north pole. In some places, the minerals are magnetized so that they point toward the magnetic south pole. Earth's magnetic field reverses periodically. When these minerals formed, the earth's magnetic field must have been reversed.

Magnetic mapping of the ocean floor along the ridges shows a pattern like the one shown in Figure 18–7. One strip of rock is magnetized toward the present North Pole, and the next strip is magnetized toward the South Pole. There are matching patterns on opposite sides of the ridge. The ocean floor is spreading from the center of the ridge. As the ocean floor spreads, it forces the continents on either side of the Atlantic farther apart. This evidence was used to support the theory of continental drift discussed in Chapter 6.

Figure 18–6. Along the Mid-Atlantic Ridge (left) the crust is spreading with the addition of magma from below the crust. Venting gases, rich in chemicals, provide nutrients for the growth of many special organisms (right).

A MATTER OF FACT

The cooling and shrinking of the upper layer of the earth can pull down an island by more than a kilometer in less than a million years.

Figure 18–7. A record of the earth's magnetic reversals can be found in the rocks on both sides of the Mid-Atlantic Ridge.

North

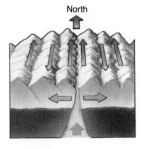

DISCUSSION   Tell the students that the features of the ocean floor are similar to those of the land. Both have mountains, valleys, and plains. *Thinking Critically:* Ask the students to use the theory of plate tectonics to explain how features form on the ocean floor. (Folding and faulting along plate boundaries form ocean ridges and trenches.)

DISCUSSION   Point out to the students that the earth's magnetic field is constantly shifting. Because of this, the magnetic poles have reversed many times in geologic history. *Thinking Critically:* Ask the students how scientists use data about the magnetic orientation of different layers of rock to determine the relative ages of rocks. (Scientists compare data about rocks in one area with data about rocks in other areas.)

EXTENSION   The Mid-Atlantic Ridge marks the boundary between the African and South American plates. Similar ridges

DISCUSSION The floor of the Pacific Ocean differs greatly from that of the Atlantic. Seamounts along the Pacific floor were formed as a result of volcanic activity. ***Thinking Critically:*** Have the students look at Figure 7–2 on page 138 of the text. Ask them where they would expect to find Pacific Ocean ridges. (along plate boundaries).

EXTENSION In the southeastern Pacific Ocean, ridges rise to heights of 3048 m above the ocean floor and extend for almost 1600 km. Trenches in the Pacific can be very deep. The Mariana Trench, located south of Guam, has a depth of 10 912 m. The extremes between the depths and heights of the Pacific Ocean floor are unique.

THINKING SKILL *(Making Judgments)* Ask the students how data about the ocean floor can help scientists understand the geologic processes of the earth. (The ocean floor is still changing as new features form due to plate movement. Data about these changes helps scientists understand the earth's geologic processes.)

DISCOVER: Imitating Deep Ocean Sediments

Thinking Skill *(Comparing)*

Be sure the students are cautious and gentle when they add the rocks and gravel to the jars so they do not break the glass. The stirring should be thorough but gentle. The students should find that the larger stones settle first, then the gravel. The sand will settle next, filling in spaces between the gravel and stones. The finer soil will settle out of suspension last, just as fine sediments do in the ocean.

Figure 18–8. The ocean floor has recently been explored with the use of small submersible vessles (left). These vessels are transported to the research site by a support ship (right).

DISCOVER

Imitating Deep Ocean Sediments

Half fill a glass jar with water, and into it place several spoonfuls of soil, sand, and powdered clay. Shake the jar. Set the jar down and observe the settling. Make a sketch of the order of the layering, starting on the bottom. How is this similar to the settling that occurs on the ocean floor?

Figure 18–9. The floor of the Pacific Ocean is different from the floor of the Atlantic Ocean. The continental edges are much steeper, there is no midocean ridge, and there are many deep trenches.

396 Chapter 18 The Water Planet

The Pacific Ocean is the largest and deepest of all the oceans, covering nearly one third of the earth's surface. The floor of the Pacific is quite different from that of the Atlantic. Instead of gently sloping continental shelves, deep trenches occur all along the edges of the Pacific Ocean where the oceanic lithosphere is subducted under the continents. Trenches such as these are rare in the Atlantic and other oceans. These trenches trap much of the sediment that flows in rivers from the continents. Perhaps the Pacific Ocean is so deep because it is not flooded with these sediments.

Although there is no mountain ridge in the middle of the Pacific, an extension of the midocean ridge forms the Eastern Pacific Rise. Other chains are made up mostly of volcanoes, and in some places they rise above the surface, forming islands such as the Hawaiian and Aleutian islands.

○ **What is the midocean ridge?**
○ **Where are most ocean trenches located?**

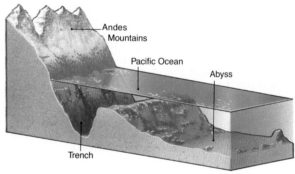

Andes Mountains

Pacific Ocean

Abyss

Trench

Answers—18.2

○ A mid-ocean ridge is a mountain ridge in an area of the ocean where plates are moving apart.

○ Most ocean trenches are located along the edges of the Pacific Ocean.

SKILL ACTIVITY: Charting the Ocean Floor

BACKGROUND

In the past few decades, oceanographers have learned more about the floor of the ocean than at any other time in history. Long ago, seafarers measured the depth of the shallow seas by throwing overboard a mass of lead tied to a rope. The rope had knots tied in it at regular intervals. They lowered the lead until it touched bottom. Then, they pulled it back on board and counted the knots of the rope that had gone into the water. If they ran out of rope, the water's depth in that area remained unknown.

PROCEDURE

Today, scientists use *echo sounding,* a form of sonar, to measure the depth of the ocean. Since scientists know that sound travels through water at about 1500 m/s, depth can be calculated by using the following formula.

$$\frac{s \times t}{2} = d$$

Here *s* is the speed of sound, 1500 m/s; *t* is the time it takes for the sound produced by the scientist to leave the ship and echo from the bottom; and *d* is depth of the water. Calculate the depth of the water if the time is 2 seconds.

APPLICATION

1. Prepare a table like the one shown.

2. Calculate and record the depth to the ocean bottom for each recorded sounding, using the given formula.

3. Use a piece of graph paper to draw the ocean bottom. Plot ocean depth on the vertical axis and distance from origin on the horizontal axis. Connect the plotted dots with smooth lines to diagram the features (i.e., mountains, valleys, and plains) of the ocean bottom.

TABLE 1: CHARTING THE DEPTH OF THE OCEAN

Sounding	Distance from Origin (km)	Echo-Sounding Time (seconds)	Calculated Ocean Depth (m)
1	2	1.2	
2	4	2.0	
3	6	3.6	
4	8	4.8	
5	10	5.3	
6	12	2.3	
7	14	3.1	
8	16	4.5	
9	18	5.0	
10	20	5.1	
11	22	4.9	
12	24	4.8	
13	26	6.7	
14	28	3.6	
15	30	3.0	

USING WHAT YOU HAVE LEARNED

How could you use soundings to help locate a deep channel where you could safely sail your boat?

Objectives

- Use mathematical formulas.
- Make a graph.

Discussion Explain to the students that in order to calculate the depth of the ocean, they must use data about the speed and time it takes for a sounding to reach the bottom and return. ***Thinking Critically:*** Ask the students why the distance, or speed x time, is divided by 2 to calculate the depth of the ocean. (This takes into account two trips the sound is making—one to reach the ocean floor and one to return.)

Answer to Application

You may wish to have the students label their diagrams to show geologic features such as ridges and trenches.

Sounding	Distance from Origin (km)	Echo Sounding Time (seconds)	Calculated Ocean Depth (m)
1	2	1.2	900
2	4	2.0	1500
3	6	3.6	2700
4	8	4.8	3600
5	10	5.3	3975
6	12	2.3	1725
7	14	3.1	2325
8	16	4.5	3375
9	18	5.0	3750
10	20	5.1	3825
11	22	4.9	3675
12	24	4.8	3600
13	26	6.7	5025
14	28	3.6	2700
15	30	3.0	2250

Answer to Using What You Have Learned

Compare distances of soundings to find areas of greatest depth.

397

18.3 Lakes and Rivers

DISCUSSION The composition of lake water depends upon the sediments that erode from surrounding rock and sediments that are carried in by rivers and streams. Tell the students that sediments containing chlorides, sulfates, and carbonates cause bitter alkaline lake water. Water flowing from volcanic rifts contains hydrous aluminum silicates called *zeolites*. Mineral deposits in these lakes are extracted for use by chemical industries.

EXTENSION Many of the world's lakes were formed as glaciers cut depressions in their paths. As glaciers melted, these depressions filled with water. Other lakes formed over limestone, which was dissolved by groundwater over time, forming sinkholes. ***Thinking Critically:*** Ask the students to suggest a geologic event that might result in a crater that fills with water to form a lake. (the formation of a caldera by a volcanic explosion)

EXTENSION Point out to the students that many lakes dry up and disappear due to natural changes. As climates change or rivers change their course, lake water may evaporate and plants begin to fill the lake. The natural succession of different types of plant communities, from water plants to grasses to shrubs and small trees, may eventually result in a mature forest.

18.3 Lakes and Rivers

Large bodies of fresh or salt water completely surrounded by land are called **lakes.** Lakes receive water from rivers and streams. Some bodies of water that are called seas are actually large lakes. For example, the Caspian and Aral seas in Asia are really lakes, even though they have salty water.

The Caspian Sea is the world's largest lake. The Caspian has an area of over 370 000 km^2 and an average depth of more than 1000 m. The largest freshwater lake is Lake Superior, on the border between the United States and Canada. Lake Superior, with an area of nearly 85 000 km^2 and an average depth of about 400 m, is much smaller than the Caspian Sea.

Figure 18–10. The Great Lakes (right) form a northern coast for the United States that stretches nearly halfway across the continent. The shores of the lakes are similar in many ways to an ocean shore (above).

The North American chain of the five Great Lakes, which includes Lake Superior, contains 20 percent of the world's available fresh water. These lakes, shown in Figure 18–10, were formed about 12 000 years ago at the end of the last glaciation. All the lakes are connected by natural and artificial channels, and they empty into the North Atlantic through the Saint Lawrence River. These channels form a 3058-km-long waterway called the *Saint Lawrence Seaway* which reaches nearly halfway across North America. The seaway allows for many ocean ports in inland cities, such as Toronto, Cleveland, Detroit, and Chicago.

Figure 18–11. From Lake Erie to the Atlantic Ocean, the elevation of the St. Lawrence Seaway drops by about 300 m.

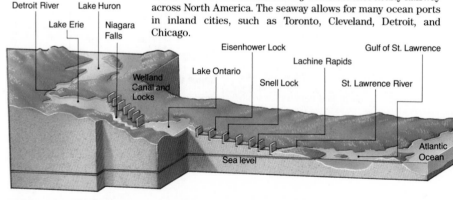

BACKGROUND INFORMATION

The St. Lawrence Seaway is the largest inland waterway to the oceans, extending for 291 km. The United States and Canada jointly built the seaway in 1954. Ocean-going ships can sail 3709 km from Lake Superior to the Atlantic Ocean. A series of canals, locks, and dams helps control the water level in the seaway.

Answers—18.3 (p. 399)

○ A lake is a body of water completely surrounded by land.

○ The Caspian Sea is the largest saltwater lake; Lake Superior is the largest freshwater lake.

Rivers and lakes are very important to civilization. Besides being a source of fresh water, they provide a means of transportation within and between countries. Many early human settlements developed along the banks of rivers. The names, locations, and lengths of some of the world's major rivers are shown in Table 18–3.

Figure 18–12. Large ocean-going vessels (left) use the locks (right) of the St. Lawrence Seaway to reach inland ports.

TABLE 18–3: SOME MAJOR RIVERS OF THE WORLD		
River	**Location**	**Length (km)**
Nile	Africa	6695
Amazon	South America	6276
Chang Jiang	Asia	4989
Congo	Africa	4394
Mississippi	North America	3779
Missouri	North America	3726
Yukon	North America	3185

○ *What is a lake?*
○ *What is the largest saltwater lake and the largest freshwater lake on Earth?*

Section Review

READING CRITICALLY

1. Explain how sonar is used to map the ocean floor.
2. What happened when the water level in the ocean rose and the passage between the Atlantic and the Mediterranean reopened?

THINKING CRITICALLY

3. What process created the trenches in the Pacific Ocean?
4. Why could the oceans and the ocean floors be called the last frontier?

Section 1   Bodies of Water   **399**

DISCUSSION Many centers of civilization have sprung up along the banks of the world's great rivers. The Nile River was vital to the people of ancient Egypt. London, Paris, Rome, and New York City are all located on the banks of major rivers. ***Thinking Critically:*** Ask the students how the discovery of the Mississippi and Missouri rivers affected the history of the United States. (These rivers were used as trade routes. As trade developed, communities sprang up along the rivers' banks. Some of these communities, such as St. Louis and New Orleans, grew into large cities.)

EXTENSION You may wish to have the students look at a world map to identify the countries through which the rivers listed in Table 18–3 flow. Have the students locate large cities on the banks of these rivers.

Section Review

Summary Have the students write summaries of the major ways that oceans, seas, and lakes formed. (Oceans and seas formed as a result of shifting of crustal plates. Many lakes formed due to the erosion of glaciers.)

Reinforcement Have the students complete the Section Review. Allow the students to look up the answers and discuss them with partners.

Answers to Section Review

Reading Critically

1. A sonar device is used to send sound waves through the water. The time it takes for the sound wave to echo back from the ocean bottom is noted and the distance to the bottom calculated using the known velocity of sound traveling through water.
2. The Mediterranean Sea was created.

Thinking Critically

3. The subduction of oceanic plates created the trenches in the Pacific Ocean.
4. The oceans can be called the last frontier because they are the only places left on Earth that have not been fully explored.

Discussion Have the students compare the careers described on this page. Remind them that although different skills may be required for each, much of the knowledge for all three is similar. Encourage the students to write to the addresses listed for information about opportunities in these fields.

Extension You may wish to have the students choose a problem that a person in one of these fields might investigate. Have the students write questions to which they would like to find answers. Then encourage the students to use the library or visit an aquarium or marine wildlife sanctuary to research answers to their questions. The students can report their findings to the class. Encourage them to include photographs of specimens in their reports. The students may be interested in researching information on the following additional careers:

Marine Biologist

Deep Sea Diver

CAREERS

OCEANOGRAPHER

Oceanography is the study of the oceans. An *oceanographer* is a scientist who studies the oceans. Oceanographers are concerned with the interaction between the atmosphere and the waters of the world, the formation and features of the ocean floor and its sediments, and the influence of tidal action and weather on coasts and estuaries.

An oceanographer must study many branches of science, including geology, meteorology, biology, chemistry, mathematics, and physics. Usually oceanographers have a master's degree or a doctorate.

Oceanographers conduct many experiments at surface research stations and at submerged oceanographic stations. They gather data that help them to better understand the oceans and the seas.

For Additional Information
American Oceanographic
   Organization
P.O. Box 2249
Springfield, VA 22152

MARINE CONSERVATIONIST

Bird watchers, marine biologists, recreational boaters, waterfront property owners, and many other citizens are concerned about the health and vitality of the oceans.

A *marine conservationist* helps all of us stay aware of the condition of our marine environment. Because the oceans and waterways of the world are so dynamic, marine conservation is a challenging and exciting pursuit. A bachelor's degree or higher is usually necessary, because a marine conservationist must keep abreast of the current research reported by oceanographers in order to understand how the oceans are affected by both natural and human pressures.

For Additional Information
National Coalition of Marine
   Conservationists
P.O. Box 23298
Savannah, GA 31403

UNDERWATER PHOTOGRAPHER

Underwater photography is both an enjoyable and difficult task. A professional *underwater photographer* must be well trained in two very important skills: scuba diving and photography.

Underwater photographers not only have to contend with the sometimes hazardous conditions of the water but also must know how the water affects the quality of their work. Sophisticated equipment is used to account for the different wavelengths of light that are absorbed at different ocean depths. Many types of filters and film must be used to produce quality images. Only a high-school diploma is needed to be an underwater photographer.

For Additional Information
Underwater Photographic
   Society
P.O. Box 2401
Culver City, CA 90231

SECTION
2 Coasts

SECTION OBJECTIVES

After completing this section, you should be able to:
- **Describe** the two main processes that shape coastlines.
- **Identify** some of the common coastline features and **explain** how they are formed.
- **Analyze** the problem of coastline erosion and **evaluate** efforts being made to preserve coastlines.

NEW SCIENCE TERMS

primary coast
fiord
bay
secondary coast

18.4 Coast Development

Have you ever placed a seashell or a cup against one ear and heard what sounded like the ocean? If you have been to the seashore, you know that the sound you hear in the shell is similar to the sound of the ocean meeting the land. Areas where the ocean meets the land are very special places called *coasts*. Coastal areas are in a state of constant change as they are shaped by erosional agents such as wind, glaciers, running water, and waves. Some coasts have wide, flat, sandy beaches, while others have rocky shores or jagged cliffs.

Coastal areas can be classified into two types. The first type, called a *primary coast*, is formed by erosional processes of the land. The second type, called a *secondary coast*, is formed by erosional and depositional processes of the sea.

Figure 18–13. Erosional agents of the land cause the features found along primary coasts (top left and top right). Secondary coasts (bottom left and bottom right) are eroded by the sea.

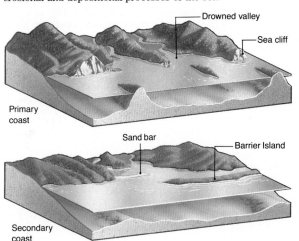

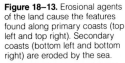

Section **2**: Coasts

SECTION FOCUS

Section Overview This section describes the processes that form coasts. The importance of coasts to marine life is discussed. Special emphasis is placed on the erosion of coasts and how coasts can be protected and preserved.

Section Objectives For a list of section objectives, see pupil's edition page.

New Science Terms For a list of new science terms in this section, see pupil's edition page. You may wish to provide specific examples and photographs to illustrate the terms as they are being discussed.

SECTION DEVELOPMENT

18.4 Coast Development

DISCUSSION Have the students describe coastal areas they have visited or of which they have seen photographs or films. ***Thinking Critically:*** Ask the students why all coasts do not look the same. (Different types of rocks and different weathering agents shape coasts in different ways.)

THINKING SKILL (*Understanding Cause and Effect*) Ask the students how rivers, glaciers, and wind affect coasts. (They deposit sediment and affect the shape of coasts through erosion.)

ations, the shapes of coasts underwent drastic changes. Tell the students that as glaciers moved into the seas they cut deep valleys, which formed fiords and shaped many bays. ***Thinking Critically:*** Ask the students how the melting of glaciers affected coasts. (As the glaciers melted, the water level of the oceans rose, covering portions of the coasts in some places and drowning river valleys in other places.)

THINKING SKILL *(Making Predictions)* Tell the students that the "greenhouse effect" described in Chapter 8 is raising the temperature of the air. Scientists believe that as a result of this process polar ice caps are beginning to melt. Ask the students how this series of events might affect the world's coasts. (Sea levels may rise, covering coastal land areas and creating new coasts farther inland.)

EXTENSION The greatest volume of sediments deposited in coastal regions is found in deltas. Deltas are nearly flat alluvial deposits near the mouths of rivers. The shape of a delta depends upon the volume of river flow and the action of tidal forces. Where river flow is strong, long fingerlike extensions of sediment form. Rounder outlines form where river flow is abutted by strong tides and wave action. Deltas contain some of the most fertile soil in the world. The Nile Delta has been an agricultural center for thousands of years.

Figure 18–14. Chesapeake Bay (left) formed as a drowned river valley. Where glacial valleys are flooded by the ocean, deep channels called *fiords* form (right).

A **primary coast** often results from erosional processes of the land and changes in sea level. For instance, the sea level was lower during the last glaciation because so much water was trapped in glaciers. When the glaciers melted, the sea level rose, covering, or drowning, many old river valleys and glacier valleys. The coasts of these drowned valleys show characteristics of stream and glacial erosion, rather than erosion by waves. The coasts of Norway, Greenland, New England, and Alaska formed this way. When the sea level rose, water entered many glacier valleys, forming fiords (fee OHRDZ). A **fiord** is a long, deep, narrow, U-shaped sea channel that was originally formed by ice.

The sea has also traveled up the mouths of many old river valleys, forming bays and estuaries. A **bay** is an indentation of the sea, partially enclosed by land. If the bay is a submerged river valley, it is called an *estuary.* The Chesapeake and Delaware bays and New York Harbor are all examples of estuaries.

The drowned valley of the Hudson River slopes into the sea near New York City, forming part of the harbor. This is a very deep valley, so the waters several kilometers upstream are a mixture of fresh and salt water. This mixture of waters, called *brackish waters,* are important nursery areas for many kinds of ocean fish and invertebrates. The living resources of the sea are discussed in Chapter 20.

Figure 18–15. The estuaries that occur along much of the east coast (right) are probably the most productive waters in North America. Many estuaries, such as New York Harbor (below), are less productive because of pollution.

Figure 18–16. A strong sea breeze and plentiful sand have created many large dunes along the Atlantic coast, such as Jockey Ridge on North Carolina's Outer Banks.

Winds also help shape primary coasts. A dune line forms as wind-blown sediments accumulate. This dune line may migrate inland if winds blow constantly offshore. For example, the dunes along the Outer Banks of North Carolina are moving westward because of the prevailing sea breezes from the Atlantic Ocean.

Forces from within the earth may also shape some coasts. Faults may raise or lower parts of the land or create fractures, which are easily eroded. Much of the California coast is on the San Andreas fault or branch faults. The sea erodes inland along the faults, forming bays such as Tomales Bay near San Francisco.

A coast that is shaped by erosional or depositional processes of the sea is a **secondary coast.** Secondary coasts are usually so modified by the sea that the primary-coast features, such as river and glacier valleys, are completely gone. Even the hardest rock is undercut and broken into fragments by the pounding of waves. If the rocks are uniformly resistant to wave erosion, a straight coast will form. If the rock formations along a coast differ in resistance, the waves form an irregular coast. In these cases, headlands of hard rock jut into the sea, and areas of softer rock erode rapidly, forming sandy coves and inlets.

Erosion by the sea creates some very distinctive coastal features. As waves continue to erode headlands, some of the rocks become isolated from the shore. These form islands called *sea stacks.*

Figure 18–17. Along the west coast of the United States, there are many spectacular arches and seastacks.

Section 2   Coasts   403

DISCOVER

Settling Sediments

To a jar half filled with water, add a handful of sediments of various sizes (soil, sand, and gravel). Stir the mixture and slowly pour it into one end of a shallow pan of water. Observe the settling of the sediments in the pan. Which sediments settled closest to the point where they entered the pan? Which sediments settled farthest away? How is this similar to the settling of sediments at the point where a stream enters a lake or the ocean?

DISCUSSION Direct the students' attention to Figure 18–16. Tell them that they can see the effects of wind erosion by noting the shapes of dunes. Point out to the students that the grasses anchor dunes and help to preserve coasts. Remind the students that the composition of rocks along coasts affects the amount of water and wind erosion and the shape of the coast. *Thinking Critically:* Ask the students to compare features they would expect to see along coasts made of sedimentary rocks with those made of igneous or metamorphic rocks. (Sedimentary rock is less resistant to weathering and would be more readily eroded, resulting in a more irregular coast than where there are igneous and metamorphic rocks.)

DISCOVER: Settling Sediments

Skill *(Observing)*

Remind the students that the settling rate of sediments depends upon the size, shape, and mass of the fragments. Have the students prepare a table with columns for these characteristics and record data on each of the materials in their sediment mixtures. In a fourth column, have the students record which material settled first, second, and last. (Largest particles settled first; the soil or silt settled last. This is very similar to the settling of stream or river sediment.) **PLEASE NOTE:** These sediments will be used in the Activity on page 405. Have a place ready for the students to keep them.

DISCUSSION Tell the students that sandbars and barrier islands are usually parallel to the coast, conforming to the wave patterns that form them.

Have you ever been to a beach and seen people many meters offshore, standing in shallow water? Often sand washed away from a beach is deposited a short distance offshore. These offshore deposits, which parallel the coast, are called *sand bars*. Sometimes sand bars curve away from a beach, forming attached deposits called *sandspits*. If the sand bars build up enough to break the surface of the water, they form *barrier islands*. There are many large barrier islands along the Atlantic and Gulf coasts from Fire Island, New York, to Miami Beach, Florida, to Padre Island, Texas. Why do you think that there are no barrier islands on the Pacific coast?

○ *How are primary coasts formed?*
○ *How are secondary coasts formed?*

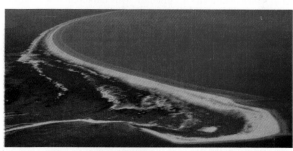

Figure 18–18. The Atlantic and Gulf coasts of the United States, from New York (above), around Florida, and to the Texas border with Mexico (right), are protected by barrier islands.

Figure 18–19. Groins, jetties (top), seawalls, and breakwaters (bottom) are constructed to provide protection against the eroding action of waves.

18.5 Preserving Coasts

DISCUSSION Have the students note the differences in orientation of groins and seawalls in relation to the coast. *Thinking Critically:* Have the students suggest advantages and disadvantages of artificially constructed barriers that are placed parallel and perpendicular to coasts. (Perpendicular groins can trap sand; parallel seawalls are more susceptible to erosion, but can span a wide length of coastline.)

EXTENSION The Atlantic and Gulf coasts pose many problems for environmental engineers. The coasts are particularly fragile due to the violent storms to which they are subjected and the burgeoning residential and recreational development over the last few decades. As a result of these factors, much of the coasts' natural habitats are endangered. Efforts are being made by citizens' groups and environmentalists to stem the growing destruction of these fragile areas.

18.5 Preserving Coasts

Coastal areas are considered prime recreational and development areas. As a result, their erosion is of major concern. Various structures have been designed to protect beach and coastal areas. *Groins*, vertical walls of rock placed perpendicular to the shore, are designed to trap sand carried by currents. Groins do succeed in trapping sand on one side, but the erosion on the opposite side is usually increased.

Seawalls are another type of structure designed to provide protection from erosion. Seawalls are usually made of rocks or concrete and are placed parallel to the shore. Seawalls are placed between the water and the shore to protect the shore from erosion. However, waves sometimes wash over and destroy seawalls by eroding away their support.

Rocky projections built to extend out into the sea are called *breakwaters* and *jetties*. They are usually built to protect harbors and beaches from the effects of storms. The water behind breakwaters and jetties usually remains calm. However, jetties collect soil and sand deposition. As a result, the next beach downshore from a jetty receives no sand and soon erodes.

Answers–18.4

○ Primary coasts are formed by the erosional processes of the land.

○ Secondary coasts are formed by the erosional processes of the sea.

Answers–18.5 (page 405)

○ Erosion is the major concern to coasts.

○ Groins, seawalls, and jetties are structures built to help stop coastal erosion.

Figure 18–20. Natural vegetation is the best protection against erosion of coasts. The addition of fences, however, sometimes helps hold the sand until the plants become established.

Designing coastal structures that correct one problem without creating other problems is very difficult. Each coast is different and presents its own unique set of problems. The best protection is to plant vegetation along the shore and to restrict building close to the sea.

○ *What is the major concern to coasts?*
○ *What are groins, seawalls, and jetties?*

Section Review

READING CRITICALLY

1. What is the difference between a primary coast and a secondary coast?
2. What new problems do the building of groins and jetties produce?

THINKING CRITICALLY

3. How would planting vegetation along a coast prevent erosion?
4. When do coastlines stop changing? Explain your answer.

ACTIVITY: Taking Core Samples

How can you take core samples of near-shore sediments?

MATERIALS (per group of 3 or 4)

sediments, soda straw, marking pen, razor blade

PROCEDURE

1. Use the sediments prepared for the Discover on page 396, or follow the directions there for making sediments. Gently push a soda straw into the sediments until it reaches the bottom.
2. Mark the top of the layer of sediments on the straw with the marking pen.
3. **CAUTION: Razor blades are very sharp; use extreme caution.** Carefully cut the straw and core lengthwise with the razor blade.
4. Make a drawing of your core sample. Be sure to label the top and bottom of the sample, and indicate each layer of sediments.

CONCLUSIONS/APPLICATIONS

1. How many layers are there in your core sample?
2. Which sediments were deposited first? Which sediments were deposited last? Explain your answers.
3. If your core sample was from the deep ocean instead of near-shore, how would the layering of the sediments be different?

Section 2   Coasts   **405**

A MATTER OF FACT

As a result of wave action, the shelves surrounding the Hawaiian Islands have been widening at a rate of 1.5 cm/year for the past 6 million years.

DISCUSSION
Divide the class into small groups. Have the students debate the pros and cons of the artificial methods of preserving coasts discussed in this section.

SECTION REVIEW

Summary   Have the students summarize the ways in which coastal features are formed and how these factors affect physical characteristics of the coast.

Reinforcement   Have the students complete the Section Review. Allow the students to refer to the text as they complete the review questions.

ACTIVITY: Taking Core Samples

Skill   *(Collecting Data)*

Preparation of Materials   The sediments from the Discover on page 403 may be used again. If the Discover was not done, follow the directions on page 403 for making the needed sediments. Use single-edge razor blades, or cover one edge of a double-edged blade with several layers of masking tape. **CAUTION: Razor blades are very sharp; use extreme caution.**

Hint   Be sure students take the core sample slowly and carefully so as not to disturb the sediments.

Answers to Conclusions/Applications

1. three
2. Larger, more dense sediments settled first; finest, least dense sediments settled last.
3. In the deep oceans, there would be little layering by size.

Answers to Section Review

Reading Critically

1. A primary coast is formed by erosional processes of the land, while a secondary coast is formed by erosional processes of the sea.
2. Groins and jetties prevent sand from washing away from a beach, thereby preventing the accumulation of sand at other sites. The other sites, therefore, continue to wash away without replenishing sand deposits.

Thinking Critically

3. The roots of plants anchor soil and slow down erosion by wind and water.
4. Coasts never stop changing because erosional processes are continuous.

INVESTIGATION 18: Wave Action

Skill (Observing)

Preparation of Materials

Remind the students that the ocean bottom they are simulating in this investigation should be level. They may wish to wet the soil with water to facilitate the shaping of the slope. Also remind the students to pour the water gently into the tank so as not to change the shapes of the features they have formed.

Hint

You may wish to have the students record their observations on a table or chart as they perform the investigation.

Answers to Analyses and Conclusions

1. Shorter wavelength patterns should speed up the erosion.
2. Higher waves will erode the coast faster and farther inland.
3. Different height waves showed the greatest difference in erosion because the higher waves eroded more.

Answer to Application

The engineers could test several types of protective structures before construction to see what worked best.

INVESTIGATION 18: Wave Action

PURPOSE

To show how the action of waves erodes coasts

MATERIALS (per group of 3 or 4)

Coarse soil or sand
Large aquarium
Bucket
Water
Large beaker, 1500 mL
Metric ruler

PROCEDURE

1. Pour coarse soil or sand into the aquarium to a depth of 1 or 2 cm. This will serve as the floor of the "ocean."
2. Add more sand to one side of the tank. This pile will represent a coast, and should be 6 or 7 cm high. Make the coast slope gently to the "ocean floor."
3. Fill a bucket with water; then use a large beaker to fill the ocean side of the tank with about 4 or 5 cm of water.
4. Place a ruler on edge, at the bottom of the tank, on the side away from the coast. Gently move the ruler back and forth to create waves. Try to keep the wavelengths constant, and note the effect of the waves on the coast.
5. Repeat the motions several times, creating waves of varying wavelength. Note the effect of the different wavelengths on the erosion of the coast. Compare your results with those of other groups in the class.
6. Use the beaker to empty the aquarium, returning the water to the bucket. Reshape the coast, duplicating as much as possible the original features. Then refill the ocean.
7. Place the ruler flat on the bottom this time and move it up and down, creating waves of different heights. Note the effect of waves of different heights on the coast, and compare your results with the results of other groups.

ANALYSES AND CONCLUSIONS

1. How did waves of varying wavelengths affect the coast?
2. How did waves of different heights affect the coast?
3. What had the most effect on the coast, waves of varying wavelengths or of different heights? Explain your answer.

APPLICATION

How could engineers use a coastal model, such as the one you constructed, to show the effects of waves on various coastal construction projects?

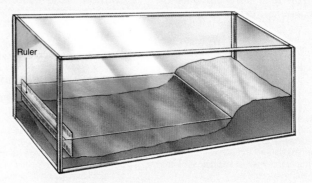

Ruler

SUMMARY

- The earth has one large ocean that is divided into four oceans and numerous seas. (18.1)

- Seas may be less salty than the oceans if they receive a large amount of fresh water, or they may be saltier than the oceans if there is a high rate of evaporation. (18.1)

- The detailed study of the ocean floor began in the 1940s after the invention of sonar. (18.2)

- Between the continents and the Atlantic Ocean floor are the gently sloping continental shelf and the steeper continental slope. (18.2)

- In the middle of the Atlantic Ocean is an underwater mountain range called the Mid-Atlantic Ridge. (18.2)

- The floor of the Pacific Ocean has many deep trenches and volcanic mountains. (18.2)

- Lakes are large bodies of water completely surrounded by land. (18.3)

- Rivers and lakes are important as sources of fresh water and as means of transportation. (18.3)

- Coasts are areas where the ocean meets the land. Coasts may be classified as primary or secondary. (18.4)

- Primary coasts show erosion by land agents, such as glaciers and running water. Secondary coasts show much erosion by waves and wind. (18.4)

- Drowned valleys are common along primary coasts, while secondary coasts have sand bars and barrier islands. (18.4)

- Groins, seawalls, jetties, and breakwaters are all used to prevent erosion of coasts. (18.5)

Write all answers on a separate sheet of paper.

SCIENCE TERMS

Correctly use each of the following terms in a sentence.

bay **(402)**
continental shelf **(394)**
continental slope **(394)**
fiord **(402)**
lakes **(398)**
oceans **(391)**
primary coast **(402)**
seas **(392)**
secondary coast **(403)**
sonar **(394)**

SCIENCE QUIZ

Modified True-False

Mark each statement *true* or *false*. If a statement is false, change the underlined word to make the statement true.

1. The saltiness of the Mediterranean is the result of a high rate of <u>evaporation</u>.

2. The Pacific Ocean has nearly <u>twice</u> the area of the Atlantic Ocean.

3. The steep incline off the continental shelf is called the <u>ocean basin</u>.

4. The largest lake in the world is <u>Lake Superior</u>.

5. Barrier islands are built up from <u>sand bars</u>.

6. The system of rivers, canals, and lakes that allow ocean ships to reach cities such as Chicago is the <u>St. Lawrence Seaway</u>.

continues

CHAPTER REVIEW

SUMMARY

The students may review the major concepts in the chapter by reading the summary statements. The statements are cross-referenced to the chapter to facilitate reinforcement of any concepts of which the students feel unsure. Encourage the students to work in groups to quiz one another.

SCIENCE TERMS

The sentence in which the science term is used should reflect an understanding of the definition of the term. You may wish to have groups of students review each other's sentences and discuss points of confusion about precise definitions with one another.

SCIENCE QUIZ

Answers to Modified True-False

1. true
2. true
3. false, continental slope
4. false, the Caspian Sea
5. true
6. true

continues

Answers to Multiple Choice

7. d
8. a
9. d
10. a
11. d
12. d

Answers to Completion

13. the deep Pacific Ocean trenches
14. Mid-Atlantic Ridge
15. secondary coasts
16. brackish water
17. bay
18. Southern

Answers to Short Answer

19. A lake is a body of fresh or salt water completely surrounded by land. A bay is a body of sea water partially surrounded by land.
20. continental shelf, continental slope, ridge, and trench
21. wind, glaciers, running water from rivers and streams, and waves
22. Waves can erode under breakwaters and seawalls. Groins and jetties may cause the accumulation of sand in one place while depriving other places of needed sand.

Answers to Writing Critically

23. Industries discharge much of their wastes into rivers so that the wastes will be deposited and diluted in the oceans. Many wastes, however, are not easily diluted or degraded, and therefore they accumulate, leading to ocean pollution.
24. Groins, seawalls, breakwaters, and jetties are structures built to protect coastlines. Groins are walls that are built perpendicular to the shoreline to prevent currents from washing sand away from the beach. Seawalls are built parallel to the shore to prevent waves from eroding the beach. Breakwaters and jetties are projections of rock that are built to keep waters near the beach calm during storms.
25. Estuaries are nursery waters for many fish and shellfish. If the waters are polluted or the estuaries drained, these resources will be lost.

SCIENCE QUIZ continued

Multiple Choice

Write the letter of the choice that best answers the question or completes the sentence.

7. The percentage of the earth's surface that is covered by water is about
   a) 25 percent.　　b) 50 percent.
   b) 10 percent.　　d) 75 percent.

8. Which ocean is the deepest?
   a) Pacific　　b) Atlantic
   c) Indian　　d) Arctic

9. Which is the largest sea?
   a) Sea of Japan
   b) Bering Sea
   c) Mediterranean Sea
   d) China Sea

10. What is the longest river in the world?
    a) Nile　　b) Amazon
    c) Yangtze　　d) Congo

11. Coasts can be reshaped by
    a) wind.
    b) glaciers.
    c) running water.
    d) All of the choices are correct.

12. Which type of feature is not found along secondary coasts?
    a) sand bars　　b) barrier islands
    c) estuaries　　d) sea stacks

Completion

Complete each statement by supplying the correct term or phrase.

13. The deepest portion of the Pacific Ocean is the ____.
14. The idea that the earth's magnetic poles reverse can be supported by the mapping of ____.
15. Coasts eroded by the action of the ocean are called ____.
16. A mixture of fresh and salt water is ____.
17. A body of water partially enclosed by land is called a ____.

18. The water around the continent of Antarctica is sometimes referred to as the Antarctic or ____ Ocean, but this water is really just part of the Atlantic, Pacific, and Indian oceans.

Short Answer

19. Explain the difference between a bay and a lake.
20. List four features of the ocean floor.
21. Describe four factors that affect the shape of coasts.
22. Explain how groins, jetties, and breakwaters, which are meant to protect coasts, may actually do more harm than good.

Writing Critically

23. Explain how the industrial use of rivers can pollute the oceans.
24. Compare and contrast some of the methods used to preserve coasts.
25. Why is it important to protect estuaries and other areas of brackish water?

EXTENSION

1. Use a map to determine the elevations of several major coastal cities. If the level of the sea were to rise 6 m, which cities would be flooded?
2. Read about the differences in the coasts of Maine and Florida, and prepare a poster that shows the different features and the processes that created them.
3. The level of Lake Superior is over 300 m above sea level, yet ocean vessels dock at ports along the lake. Find out about the locks along the Saint Lawrence Seaway, and report your findings to the class.
4. Early fur-trading routes followed nearly the same path as the St. Lawrence Seaway. Read about these routes, and report on their importance in the development of the interior of North America.

ANSWERS TO EXTENSION

1. Answers will vary; however, a 6-meter rise in sea level would have devastating effects on nearly all coastal regions.
2. The coastline of Maine was cut largely by the recession of glaciers. Florida, over the ages, has been alternately submerged and elevated by changes in sea level.
3. Students should include a description of how locks are used.
4. Reports should include the fact that canoes had to be portaged around rapids and waterfalls; these are the places where locks have now been built.

APPLICATION/CRITICAL THINKING

1. If one sound from an oceanographic vessel takes 1.6 seconds to echo back to the sounding instrument, a second sound takes 2.4 seconds, and a third sound takes 4.8 seconds, what might that section of the ocean bottom look like? If sound travels 1.5 km/s in ocean water, what are the differences in depth of the three locations?

2. Explain why the melting of glaciers might be responsible for changes in sea level in the future.

FOR FURTHER READING

Gilbert, S. "America Washing Away." *Science Digest* (August 1986). This article cites many examples of how coastal erosion is affecting homes and recreation areas. It clearly explains some of the factors involved in coastal erosion.

Gilbreath, A. *The Continental Shelf: An Underwater Frontier.* Minneapolis: Dillon, 1986. This book provides an overview of the geology and ecology of the continental shelf. All the species living on the shelf are discussed.

Glaser, M. *The Nature of the Seashore.* Fiskdale, Mass.: Knickerbocker, 1986. This field guide covers sand, wind, waves, and animal and plant life of the seashore.

Jeffery, D. "Maine's Working Coast." *National Geographic* 167 (February 1985): 209. This article tells a lively story of the pleasures and perils of living and working along the coast of Maine.

Challenge Your Thinking

From your knowledge of the processes that shape coasts, what process or processes were probably involved in creating these features? Is this a primary or a secondary coast? How can you tell? Where in North America might you find a coast such as this?

ANSWERS TO APPLICATION/ CRITICAL THINKING

1. The speed of sound in sea water is approximately 1460 m/s. An explosion that takes 1.6 seconds to echo back to the sounding instrument traveled 1460 m/s for 1.6 seconds, or 2336 m. The bottom lies at half that depth, or 1168 m. The second sounding reached a depth of 1752 m, and the third, a depth of 3504 m.

2. Melting glaceries would add water to the oceans, thereby raising sea level.

ANSWER TO CHALLENGE YOUR THINKING

This coast shows evidence of both primary and secondary erosion. Much of the coast of California has features such as these.

PLANNING THE CHAPTER

Chapter Sections	Page	Chapter Features	Page	Program Resources	Page
Section 1: Characteristics of Ocean Waters	411			Reading for Content: *Making Bar Graphs* **(B)**	**TRB 32** **LM 209**
19.1 Composition of the Oceans **(A)**	411	A Matter of Fact	413		
19.2 Temperature of the Oceans **(A)**	413	**Discover:** Observing Expansion **(A)**	414		
19.3 Pressure and Density of the Oceans **(A)**	416	**Activity:** Relating Water Temperature and Density **(A)**	416		
		Section Review	417		
		Biographies: Then and Now Matthew Fontaine Maury, Robert Duane Ballard	418		
		Investigation 19: The Effect of Temperature on Water **(B)**	428	Student Record Book: Textbook Investigations **(B)**	**TRB 37**
Section 2: Motions of the Ocean	419			Critical Thinking **(H)** Cross-Discipline: Science and Social Studies, *Examining Ocean Currents and World Climate* **(A)**	**TRB 36** **TRB 35**
19.4 Ocean Currents **(B)**	419	A Matter of Fact	419	Investigation 19.1: *Convection Currents in Water* **(A)**	**TRB 85** **LM 87**
		A Matter of Fact	420	Investigation 19.2: *Tides and the Tide Curve* **(A)**	**TRB 87** **LM 89**
		A Matter of Fact	421	Concept Extension: *Density Currents* **(H)**	**TRB 39** **LM 261**
		Skill Activity: Mapping Ocean Currents **(B)**	422		
19.5 Ocean Waves **(B)**	423	**Discover:** Demonstrating Waves **(B)**	424		
19.6 Ocean Tides **(B)**	425	Section Review	427		
Chapter 19 Review	429			Vocabulary **(A)**	**TRB 19** **LM 159**
				Tests **(A)** Computer Test Bank	**TRB 82**

(LM) Laboratory Manual/Study Guide,   **(TRB)** Teacher's ResourceBank™

B = Basic   **A** = Average   **H** = Honors

The coding Basic, Average, and Honors indicates sections or subsections that might be appropriate for different levels of learners. For additional suggestions regarding choice of topic and depth of coverage, see the Pacing Chart on pages T16–T20.

CHAPTER CONCEPTS, OBJECTIVES, AND TERMS

Section	Concepts	Objectives	Science Terms
Section 1: Characteristics of Ocean Waters	■ Sodium chloride is the most common salt in sea water. **(19.1)** ■ The average salinity of the ocean is about 34.5 g/kg. **(19.1)** ■ Salt water freezes at a lower temperature than fresh water. **(19.2)** ■ Cold, salty water sets up a circulation in ocean water called *thermohaline.* **(19.2)** ■ The thermocline is a layer below the surface of the ocean through which temperature decreases rapidly. **(19.2)** ■ The pressure of the ocean water increases with depth. **(19.3)**	■ **Name** the major chemicals dissolved in ocean water. ■ **Explain** how heat is stored in sea water. ■ **Describe** the factors that affect the density of ocean water.	salinity thermohaline thermocline
Section 2: Motions of the Ocean	■ Many surface ocean currents are caused by planetary winds blowing on the surface of the ocean. **(19.4)** ■ The largest ocean current in the world is the Antarctic Circumpolar Current. **(19.4)** ■ Where currents converge, the water level is about 1 m above the average sea level; where currents diverge, the water level is about 1 m below the average sea level. **(19.4)** ■ Geostrophic currents are currents that move out of highs in the surface of the ocean and into lows. Like all currents, they are influenced by the Coriolis effect. **(19.4)** ■ Ocean currents help to distribute heat from low latitudes to high latitudes. **(19.4)** ■ Waves are produced by friction between local winds and the ocean surface. **(19.5)** ■ The water in a wave does not move forward; only the wave form moves. **(19.5)** ■ The tides are produced by the attraction of the moon and the sun on different parts of the earth's surface. **(19.6)** ■ Spring tides occur when the sun and moon align. Neap tides occur when the sun and moon are at right angles. **(19.6)**	■ **Diagram** the location of major ocean currents. ■ **Describe** the characteristics of a sea wave. ■ **Explain** how ocean tides are produced.	currents geostrophic currents waves rip current tides

CHAPTER MATERIALS

Title	Page	Materials
Discover: Observing Expansion	414	*(per student)* plastic cups, 50 ml (2); wax pencil; salt (5 g); water
Activity: Relating Water Temperature and Density	416	*(per group of 3 or 4)* safety goggles; laboratory apron; beaker, 500-mL; red and blue food coloring; crushed ice; test tubes (2); test-tube holder; Bunsen burner
Skill Activity: Mapping Ocean Currents	422	*(per student)* tracing paper, red pencil, blue pencil, purple pencil
Discover: Demonstrating Waves	424	*(per student)* rope (5-m)
Investigation 19: The Effect of Temperature on Water	428	*(per group of 3 or 4)* beaker, 250 mL; crushed ice; wax pencil; test tubes (2); tap water; beakers, 100 mL (2); salt; balance; stirring rod; graduate, 10 mL; thermometers (2); stopwatch

TEACHING SUGGESTIONS

Section 1: Characteristics of Ocean Waters

Demonstration: Underwater Habitats

Purpose

To illustrate how underwater habitats help aquanauts live and work under the sea.

Materials

Small beaker, 25 mL
Large beaker, 1500 mL
Rubber tubing, 40 cm
Small cork
Small piece of cardboard, 2 cm²

Procedure

1. Fill the large beaker to 1000 mL.
2. Insert the tubing into the small beaker, positioning the tubing at the spout.
3. Float the cork on the small piece of cardboard in the large beaker.
4. Invert the small beaker over the cork and submerge the beaker in the water.
5. Ask a student to hold the small beaker down, while another pupil blows air through the tube into the simulated "underwater habitat."

Questions to Ask the Students

1. How are real underwater habitats kept submerged on the bottom? (*They are held down with weights and anchored to the seafloor.*)
2. If the cork were an aquanaut, how would he/she breathe? How can he/she be resupplied with fresh air? (*Fresh air can be introduced through the tube from the surface. Stale air is pushed back into the sea.*)

Outside Speaker

Find out if any member of the school staff, or any parent or relative of students, has ever gone snorkling or scuba diving. Invite this person to describe these experiences, including in the lecture many of the safety precautions and training required (many of which are required for any swimming activity). Have the speaker bring any equipment, such as fins, diving suit, aqualung, and snorkles, for display and demonstration.

Section 2: Motions of the Ocean

Class Activity

Use a local tide table (even if you aren't near the ocean) and see if the students can determine the patterns of the tidal time changes. Have the class predict high and low tides based on the patterns observed, and then later check the tide table to see if the predictions were right.

Field Trip

If you live near the ocean, take the class on a trip to the shore to observe the changes that occur from low to high tide.

CHAPTER 19

CHAPTER OVERVIEW

This chapter presents the characteristics of ocean waters, including temperature, density, and a description of the salts dissolved in ocean waters. Also discussed are currents, waves and tides, and the effects of these motions on the people who live nearby.

Section 1: Characteristics of Ocean Waters This section discusses the makeup of ocean water, as well as some of its physical characteristics. The relative abundance of sodium chloride and the presence of other salts in ocean water is described, as is the comparison of fresh water with ocean water. The effects of the freezing and the warming of oceans is explained, as well as conditions that result from water pressure.

Section 2: Motions of the Ocean This section discusses the motions of the ocean. The currents—their size, what causes them, the interaction among them, and their effects—are discussed. The shape and characteristics of waves are explained. The movement of waves and the various phenomena that result when waves hit the shore are considered. What tides are, why they occur, and the characteristics of spring and neap tides are also emphasized.

Ocean Waters

Over 70 percent of the earth is covered with salt water. If the water were to suddenly evaporate, the remaining deposits would be 13 meters thick. These deposits, however, would not be just salt. They would contain many of the minerals that have eroded from the land since the first torrential rains began filling the oceans billions of years ago.

Pacific Ocean "tube"

410

CHAPTER MOTIVATING ACTIVITY

Begin this chapter by relating the story of the *Titanic,* the famous luxury liner that sank on its maiden voyage in 1912. Although the vessel was nearly three football fields long (268 m), it lay undiscovered at the bottom of the Atlantic Ocean for more than 70 years. Ask the students what characteristics of the ocean might have concealed the vessel for almost three-quarters of a century. (The unpredictable weather of the North Atlantic and the complex terrain of the ocean floor that includes submarine canyons, sand dunes, craters [probably made by boulders released by melting icebergs] and other geologic formations. An undersea landslide occurred in 1929.) Ask the students what advances in technology made the location of the luxury liner possible. (Sonar and video cameras were particularly important.)

1 Characteristics of Ocean Waters

SECTION OBJECTIVES

After completing this section, you should be able to:
- **Name** the major chemicals dissolved in ocean water.
- **Explain** how heat is stored in sea water.
- **Describe** the factors that affect the density of ocean water.

NEW SCIENCE TERMS

salinity
thermohaline
thermocline

19.1 Composition of the Oceans

If you have ever been swimming in the ocean, you know that the water is salty. When the oceans first began filling, rivers carried countless tons of dissolved compounds from the land to the oceans. As a result, the oceans became salty. Almost all of the elements of the earth's crust can be found in today's ocean water. For example, 1 km³ of ocean water contains over 10 kg of gold. Could this gold be extracted from the water? The answer is yes; however, the concentration of gold is only about 10 parts of gold per trillion parts of water. Extracting gold from the water would cost more than the gold is worth. Although it does not make good economic sense to "mine" gold from the ocean, other elements, such as bromine, magnesium, and salt, can be extracted economically from the ocean waters.

If 1 kg (about 1 L) of normal ocean water is allowed to evaporate, about 34.5 g of salts are left behind. The most abundant of these salts is sodium chloride (NaCl)—common table salt. Sodium chloride makes up 77.4 percent by weight of the total salts in ocean water. Table 19–1 lists some of the salts found in ocean water.

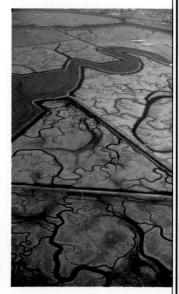

Figure 19–1. Probably the most important mineral resource humans get from the ocean is salt.

TABLE 19–1: SALTS DISSOLVED IN OCEAN WATER			
Name	Chemical Formula	Concentration (g/kg)	Percent
Sodium chloride	NaCl	26.70	77.4
Magnesium sulfate	MgSO₄	3.25	9.4
Magnesium chloride	MgCl₂	2.20	6.4
Calcium chloride	CaCl₂	1.10	3.2
Potassium chloride	KCl	0.70	2.0
Sodium bicarbonate	NaHCO₃	0.20	0.6
All others		0.35	1.0

Section 1   Characteristics of Ocean Waters   **411**

BACKGROUND INFORMATION

Nearly 80 of the 92 naturally occurring elements are found in ocean waters. Many are only detected in minute concentrations, but their total amounts are quite large due to the enormity of the oceans. Eleven elements comprise 99.9 percent of the dissolved substances (called solute) in sea water. The table shown at right lists these eleven elements and their total masses.

ELEMENT	TOTAL MASS
Chlorine	2.5×10^{16}
Sodium	1.4×10^{16}
Magnesium	11.7×10^{15}
Potassium	5.0×10^{14}
Calcium	5.4×10^{14}
Silicon	2.6×10^{12}
Zinc	6.4×10^{9}
Copper	2.6×10^{9}
Iron	2.6×10^{9}
Manganese	2.6×10^{8}
Cobalt	6.6×10^{7}

Section **1**: Characteristics of Ocean Waters

SECTION FOCUS

Section Overview This section describes the composition of ocean waters, emphasizing the large amount of sodium chloride found in the oceans. A consideration of the cooling and heating of the oceans leads to a discussion of thermohaline and thermocline. Also explored is the relationship between the depth of the water and the pressure exerted by the water.

Section Objectives For a list of section objectives, see pupil's edition page.

New Science Terms For a list of new science terms in this section, see pupil's edition page. You may wish to pronounce the terms for the students. In particular, assist them in differentiating between thermohaline and thermocline.

SECTION DEVELOPMENT

19.1 Composition of the Oceans

DISCUSSION Explain that ocean water differs from the fresh water found in rivers, lakes, and streams in both the concentration and types of chemicals in the water. ***Thinking Critically:*** Ask the students how the difference in the composition of ocean water and fresh water has affected plant and animal life. (Plants and animals suited to ocean water usually do not thrive in fresh water and vice versa. Therefore, different species have developed in each environment.)

SKILL *(Reading a Table)* Table 19–1 shows the concentration of selected salts in ocean water. Ask the students how the concentration is expressed. (grams of salt per kilogram of water) Draw attention to the column headed "Percent." Ask the students if this statement is true or false: "Seventy-seven and four-tenths percent of ocean water is sodium chloride." (False; 77.4 percent of the salts in ocean water is sodium chloride. However, all of the salts together total far less than 5 percent of the various substances that make up ocean waters.)

DISCUSSION You may wish to explain to the students what ocean sediments are. (the solid materials that settle to the bottom of the ocean) *Thinking Critically:* Ask the students why all the rain that has fallen in the oceans over the centuries has not altered the salinity. (The salinity does change slightly, but over such a long time that it is nearly undetectable.)

THINKING SKILL *(Predicting)* Ask the students if samples of ocean water collected just off the coast of a continent would probably contain exactly the same minerals as those collected a few kilometers offshore. (No. The mineral content of the nearby land and that of the seabed in surrounding areas will probably be reflected in the water.)

EXTENSION (Tie-in/Social Studies) There are many places on the earth where there is insufficient water for animals and plants. Some of those places are near the oceans. To make ocean water usable, the amount of salt it contains must be reduced. Desalination plants have been developed that can remove salt from ocean water. However, because the expense involved in running and maintainng these plants is prohibitive and because output is low, desalination is not usually a feasible method of obtaining fresh water.

Figure 19–2. The salinity of the North Atlantic Ocean is below average partly because of the addition of fresh water from melting icebergs.

Salinity is the measure of the amount of salts dissolved in ocean water. Salinity is usually expressed in grams per kilogram of sea water. The average salinity of the ocean is 34.5 g/kg. Areas where evaporation is high have higher-than-average salinity. Areas of high salinity include tropical oceans, the Mediterranean Sea, and the Red Sea. Areas where a lot of rain, melting ice, or river water enters the oceans have lower-than-average salinity. Examples of low-salinity areas are the Arctic Ocean, the Baltic Sea, and the Black Sea.

The water that flows from the land is called *fresh water.* Fresh water is not the same thing as pure water. Pure water contains water molecules and nothing else. Fresh water contains elements and compounds eroded from the rocks and soil by rainwater. Rainwater flowing across the land carries these chemicals to the oceans. Some of the chemicals stay dissolved in ocean water. Others precipitate out of the ocean water and accumulate on the ocean floor.

Of all the chemicals dissolved in Earth's waters, only silica and nitrate are more concentrated in fresh water than in ocean water. Silica and nitrate are not as concentrated in ocean water because they are used by ocean organisms. Silica, or silicon dioxide (SiO_2), is used by some organisms to make their shells. Nitrate (NO_3^-) is removed from sea water by ocean plants to make proteins.

There are two elements that occur in much greater quantities in ocean water than in fresh water—chlorine and sodium. These two elements stay dissolved in ocean water, making it "salty."

Although sodium and chlorine are continually added to ocean water, ocean salinity does not increase. Ocean water reached its present salinity more than a billion years ago. Excess sodium and chlorine precipitate out of ocean water and accumulate in ocean sediments. From there they are carried into the mantle along subduction zones. Both reappear at the surface as products of volcanoes or in minerals. There they are eroded by rain and returned to the sea, completing the cycle.

Figure 19–3. These organisms live near a volcanic vent along a midocean ridge. Their food chain is based on chemosynthesis rather than on photosynthesis, because no sunlight reaches this depth.

412

DEMONSTRATION

For a demonstration of underwater habitats, see page 409c preceding this chapter.

Ocean water also contains dissolved atmospheric gases. You may recall from Chapter 8 that most of the carbon dioxide on Earth is dissolved in the oceans; only a small fraction of Earth's carbon dioxide is in the atmosphere. The opposite is true of nitrogen and oxygen. Very little nitrogen and oxygen are dissolved in the ocean; most is in the atmosphere.

○ *What is the most abundant salt in ocean water?*
○ *What areas of the oceans have higher-than-average salinity?*
○ *What is fresh water?*

19.2 Temperature of the Oceans

Have you ever been to a desert where it is hot during the day and cold at night? Why do you think the temperature changes so much more in the desert than in other places? The reason is the lack of water in the desert. Large bodies of water help to moderate temperature changes.

If the earth had no oceans, the air temperature would rise above 100°C during the day and plunge below −100°C at night. In the polar regions, the temperature would drop to −200°C during the winter. Fortunately, the earth does have oceans that help to regulate the temperature. The oceans absorb a lot of solar energy when the sun is high and release that stored energy very slowly when the air is cold. Aided by the mixing effect of the planetary winds, this tends to keep the overall climate of the earth moderate.

A MATTER OF FACT

Gases from the atmosphere are important parts of ocean water. The transfer and mixing of these gases is produced by the ocean spray of breaking waves. There is a delicate balance between the concentrations in the atmosphere and in the water. This balance can be upset by pollution and excess decay of ocean organisms.

Figure 19–4. The oceans help to modify climate. Vancouver Island, British Columbia (left) and Calgary, Alberta (right) are at about the same latitude, yet the winters on Vancouver Island are much milder than those in Calgary.

Answers—19.1

○ Sodium chloride is the most abundant salt in ocean water.
○ The tropical oceans, the Mediterranean Sea, and the Red Sea have higher-than-average salinity.
○ Fresh water contains elements eroded from rocks and soil by rainwater.

19.2 Temperature of the Oceans

DISCUSSION Review how large bodies of water serve to moderate temperature changes on the surrounding land. ***Thinking Critically:*** Ask the students to think which parts of the United States are least affected by this beneficial situation. (Areas least affected are those farthest from the oceans, large lakes, or large rivers. The students might mention states such as Wyoming, Arizona, and Oklahoma.)

EXTENSION (Tie-in/Language Arts) Have the students look up the prefix *therm-* in the dictionary. (*therm* - ''heat'') Ask the students how this knowledge helps them to understand that thermohaline and thermocline are related but different phenomena. Although their dictionaries may not indicate this, *-haline* refers to salt content. The circulation of deep ocean current is called a thermohaline circulation because the temperature and salinity of sea water are two factors with a great effect upon the circulation of deep ocean currents. Their influence upon the density of deep ocean waters creates layers within the ocean that can move almost independently from surface layers, as well as from one another. The circulation of deep ocean currents is determined by factors quite different from those that determine surface currents.

DISCOVER: Observing Expansion

Skill: *Making Inferences*

Remind the students that since salt water freezes at a lower temperature than fresh water, the cup containing only water will freeze first. The water (ice) level in the cup of fresh water will be slightly higher than the cup of salt water.

SKILL *(Measuring)* The volume of water in a rectangular carton can be found by multiplying the length, the width, and the height of the water. (V = l × w × h) Have the students fill several such containers to different levels, compute the volume of water in each, and record this information on a chart. Then have the students freeze the containers, record the change in height of the water levels, and compute the volume again. Have the students calculate the percent of change of volume caused by freezing. (change in volume × 100 divided by volume before freezing = percent change in volume)

THINKING SKILL *(Predicting)* After completing the activity described above, show the students a container filled to a certain height with water and ask them to predict the amount of change in volume that will occur after freezing. Then freeze the contents to check the predictions.

DISCOVER

Observing Expansion

Fill two 50-mL plastic cups about halfway with water; then mark the waterline on the outside of each cup with a wax pencil. Place 5 g of salt in one of the cups and stir the water until the salt dissolves. Place both cups in a freezer for about 30 minutes. Is there any sign of freezing in either cup? Check the cups and record your observations every half-hour until one cup is frozen solid. Which cup of water froze first? Why? Is there any difference in water levels in the cups? Explain.

Figure 19–5. Spreading salt on an icy street causes the ice to melt. Melting occurs because salt water freezes at a lower temperature than does fresh water.

Figure 19–6. The Arctic Ocean does not freeze solid. Each summer, patches of open water can be found near the coasts.

The ability of sea water to absorb and release energy results from the unique structure of the water molecule and the dissolved salts in the water. The water molecule has a somewhat triangular shape. Because of covalent bonds, water molecules have a positive side and a negative side. This causes water molecules to bond together. A lot of heat is needed to break these bonds. In the oceans, weak bonds form between the positive ions in the water and the negative ends of the water molecules. Bonds also form between negative ions in the water and the positive ends of the water molecules. This bonding tends to keep the water molecules in ocean water from bonding to each other. As a result, ocean water freezes at a lower temperature than fresh water, that is, at −1.872°C instead of 0°C. Water that is saltier than ocean water freezes at an even lower temperature. The Great Salt Lake in Utah, for instance, freezes at −20°C. Salt is put on icy streets to lower the freezing point and melt the ice and snow. Explain another way that you could use this principle.

The freezing of ocean water forces the salt ions out of the water, so the ice that forms is almost freshwater ice. Ocean ice is generally no thicker than 3 m. The ice insulates the water under it from the colder air temperature above; therefore, the water below the ice never gets cold enough to freeze.

The water immediately below the ice is not only cold but it also contains a high concentration of salts. This high concentration of salts makes the water denser than surface water. The denser water sinks to the bottom of the ocean, causing a vertical circulation of the ocean water. This type of circulation is called **thermohaline** (thur moh HAY lyn), referring to both temperature (thermo) and salinity (haline). Because of this circulation, the deep water of the oceans has about the same temperature as the coldest surface water.

In tropical regions the ocean is warmed by the sun. Ocean waters near the equator may have a surface temperature as high as 30°C. Near the poles the surface water temperature is about 0°C. Swimmers know that the water temperature is warmer off the coast of Florida than off the coast of Maine.

The surface layer of water, which is penetrated by solar radiation, is about 100 m deep. This water is warmer and less dense than the colder water below. Because of this temperature difference, there is little mixing between the two. The zone separating the warm surface water and the cold water below is called the **thermocline** (THUR muh klyn). As you can see in Figure 19–8, temperature changes rapidly through the thermocline.

○ *How does salt affect the freezing point of water?*
○ *What is a thermocline?*

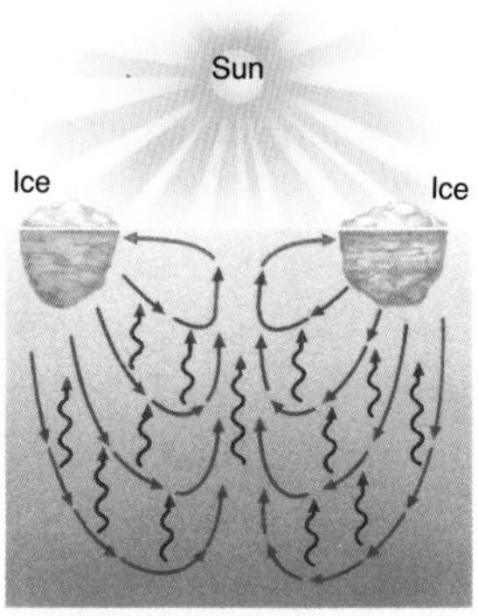

Figure 19–7. Cold, salty water is very dense, so it sinks to the ocean bottom, causing warmer, less-salty water to rise. This is a thermohaline circulation.

EXTENSION The graph in Figure 19–8 depicts the change in temperature with depth. You may wish to analyze with the class, the relationship of a thermohaline (Figure 19–7) and a thermocline.

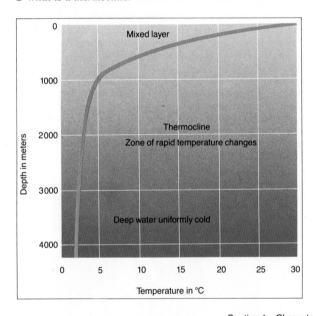

Figure 19–8. This graph shows the change in ocean temperature from the warm, sunlit surface to the cold, dark depths.

Section 1   Characteristics of Ocean Waters   **415**

Answers–19.2

○ Salt lowers the freezing point of water.

○ A thermocline is the regular change in temperature from the warm surface to the cold depths.

ACTIVITY:
Relating Water Temperature and Density

Thinking Skill (Experimenting)

Discussion   Temperature and salinity are both important in determining the density of seawater. The colder or more saline the water, the more dense it is. This activity simply demonstrates that cold water will sink below warmer water, and that even in this closed system, thermal currents can develop.

Answers to Conclusions/Applications

1. It ran down the side of the beaker, sinking to the bottom before spreading out over the bottom.
2. It remained close to the surface and spread there.
3. After several moments, the two solutions mixed at mid-depths, swirling around each other.
4. Cold ocean-water molecules have less kinetic energy than warm ocean-water molecules and are, therefore, more densely packed. Since there are more cold water molecules in a given volume of water than warm water molecules, cold water is more dense and will sink below warm water.

19.3 Pressure and Density of the Oceans

DISCUSSION   Review with the students the fact that water density and pressure increase with depth. Explain, however, that this is not the only type of pressure gradient that exists in ocean water. A pressure gradient is used to refer to a change in the value of the pressure with the change in depth from the ocean surface. Still other pressure gradients may develop where water piles up against land formations. As a result of these pressure gradients, geostrophic currents are created. **Thinking Critically:** Ask the students to think of conditions under which several pressure gradients are at work at once. In such conditions water pressure may be even greater than usual. (If the water that piles up against land formations is shallow, not as much pressure is generated as in the place where the water is deeper. Another example: various currents may converge in ways that increase the water pressure.)

ACTIVITY: Relating Water Temperature and Density

How can you show that cold water sinks while hot water remains near the surface?

MATERIALS (per group of 3 or 4)

safety goggles, laboratory apron, 500-mL beaker, red and blue food coloring, crushed ice, test tubes (2), test-tube holder, Bunsen burner

PROCEDURE

1. **CAUTION: Put on safety goggles and a laboratory apron and leave them on throughout this activity.**
2. Fill the 500-mL beaker nearly to the rim with tap water at room temperature.
3. Place four or five drops of red food coloring in one test tube. Place four or five drops of blue food coloring in the other test tube. Add about 2 cm³ of crushed ice to this tube.
4. Fill both tubes about two-thirds full with tap water.
5. Using the test-tube holder, heat the red water to a near boil over the Bunsen burner. Turn off the Bunsen burner when you are done.
6. Slowly pour the contents of both test tubes over the edge of the beaker at opposite sides.

CONCLUSIONS/APPLICATIONS

1. What happened to the cold water?
2. What happened to the hot water?
3. Did the colors mix or tend to remain separate? Explain.
4. Write a general statement about the movement of ocean water.

19.3 Pressure and Density of the Oceans

You may recall from Chapter 8 that air pressure is due to the weight of the atmosphere and that as you go up in the atmosphere, the air pressure decreases. Water also exerts pressure, but water pressure is much greater than air pressure, because water is denser than air.

Have you ever seen a movie about deep-sea diving? The heavy suit and metal helmet that deep-sea divers wear protect them from the tremendous pressure of the water. The pressure of the air pumped into the diving suit from the surface must equal the water pressure on the outside of the suit. If the air pressure in the suit were to suddenly drop, the diver could be crushed from the water pressure.

Figure 19–9. Deep-sea divers (left) were the first to explore the ocean bottom. Today scientists protect themselves in thick-walled submersible vessels (right).

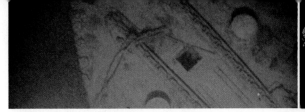

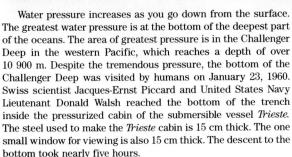

Water pressure increases as you go down from the surface. The greatest water pressure is at the bottom of the deepest part of the oceans. The area of greatest pressure is in the Challenger Deep in the western Pacific, which reaches a depth of over 10 900 m. Despite the tremendous pressure, the bottom of the Challenger Deep was visited by humans on January 23, 1960. Swiss scientist Jacques-Ernst Piccard and United States Navy Lieutenant Donald Walsh reached the bottom of the trench inside the pressurized cabin of the submersible vessel *Trieste*. The steel used to make the *Trieste* cabin is 15 cm thick. The one small window for viewing is also 15 cm thick. The descent to the bottom took nearly five hours.

How does the great pressure at the bottom of the ocean affect the density of the water there? Water, like all fluids, is not easily compressed because the molecules are in contact with each other. Density, therefore, does not increase much with pressure. Density at the bottom of the Challenger Deep is only 7 percent greater than the density of the surface water.

For many years scientists thought that pressure would have a greater effect on the density of ocean water than it actually does. In fact, it was once thought that sinking ships would descend only halfway to the bottom and float forever on a layer of denser water. Scientists now know that this is not the case.

○ *How does pressure affect density?*

Figure 19–10. The foredeck of the Titanic, on the bottom of the North Atlantic Ocean, is shown on the left. Parts of the ship are well preserved due to the lack of oxidation of the ship's steel (right).

Section Review

READING CRITICALLY

1. How does water hold heat?
2. Why is sea water more dense than fresh water?
3. What causes the differences in salinity at different depths in the oceans?

THINKING CRITICALLY

4. What is the relationship among temperature, pressure, salinity, and density?
5. If there were no oceans in the Northern Hemisphere, how might the climate be different?

Section 1   Characteristics of Ocean Waters   **417**

DISCUSSION Ask the students to describe their experiences of diving into a swimming pool. Explain that water pressure increases with depth and that divers in the ocean must be well-trained and physically fit to withstand the rigors of even shallow ocean dives.

SECTION REVIEW

Summary   Review the physical characteristics of ocean water. Have the students compare the characteristics of ocean water and fresh water.

Reinforcement   Encourage the students to make a chart listing some of the elements found in ocean water.

Answer–19.3

○ Water is not easily compressed, so although pressure increases with depth, the density of water does not change that much.

Answers to Section Review

Reading Critically

1. Water holds heat better than many other substances because of the unique structure of water molecules.
2. Sea water is denser than fresh water because of the salts in sea weater.
3. Salinity differs in different parts of the ocean because the amount of evaporation differs among regions.

Thinking Critically

4. Pressure and salinity increase with depth. Temperature tends to decrease with increased depth.
5. It would be drier, and the temperature extremes would be greater.

BIOGRAPHIES:
Then and Now

Discussion Have the students read the two biographies and discuss how the contributions of each person helped to further our understanding of the oceans. Also ask them to identify the contributions these people made to those who sail the oceans.

Thinking Skill *(Synthesizing Information)* After the students have read the two biographies, ask them to compare the education received by a scientist 100 years ago with the education a scientist receives today.

Extension Interested students might be encouraged to research further the life and work of Jacques Cousteau, the French marine explorer. These students might investigate not only Cousteau's own observations of the ocean but also the organization, planning, and coordination of the efforts of the many individuals who have been crucial to his success.

BIOGRAPHIES: Then and Now

MATTHEW FONTAINE MAURY (1806–1873)

Few people have contributed as much to the science of oceanography as Matthew Fontaine Maury. Born in Virginia on January 14, 1806, he was the fourth son in a family of four boys and four girls. At the age of 12, he entered Harpeth Academy in the frontier town of Franklin, Tennessee. In 1825 he followed in the footsteps of one of his elder brothers and became a United States naval officer.

Maury began his career as an oceanographer by making three long ocean voyages. The first voyage took him across the Atlantic to France. Along the way he made many measurements about the ocean waters. On board was the famous French general, the Marquis de Lafayette. Lafayette had helped the colonists win their independence from Great Britain. On his second voyage, Maury sailed around the world and began to draw charts of the winds and ocean currents of the world.

In 1836 Maury published his first major work on oceanography, *A New Theoretical and Practical Treatise on Navigation.* The book contained charts and maps of winds and ocean currents. It contained information about all of the oceans of the world. Maury claimed that by following his charts, navigators could save many wasted days of sailing. He claimed that a trip from New York to Rio de Janeiro could be decreased by 10 to 15 days. His challenge was accepted by several shipping companies who were pleased to discover that he was right.

ROBERT DUANE BALLARD (1942–)

Although he was born in Wichita, Kansas, Robert Ballard grew up in Montana, where he worked on a ranch. Some have called him a "high-tech cowboy." Ballard claims he's a cowboy who feels just as comfortable training porpoises as chasing wild horses.

He graduated from the University of California at Santa Barbara in 1965. With a degree in chemistry and geology, he joined the United States Army as a second lieutenant. On duty in Hawaii, he continued his studies at the Institute of Geophysics. He became interested in the use of small submarines called *submersibles* to explore the depths of the ocean.

In 1966 the United States Navy asked him to use a submersible to retrieve an unexploded hydrogen bomb from the floor of the Mediterranean Sea. Ballard has logged more hours in the deep than any other marine scientist. He has made over 200 dives, and he compares his journeys to the dark depths to those of astronauts who have visited the surface of the moon.

In September of 1985 Ballard led a team of American and French scientists from the Woods Hole Oceanographic Institution to the lost wreckage of the *Titanic.* The great luxury liner, which sank after striking an iceberg on April 15, 1912, was resting more than 4 km below the surface of the North Atlantic. For weeks Ballard's team made videotapes and photographs of the wreck. Since then, French explorers have recovered material from the *Titanic*.

418   Chapter 19   Ocean Waters

SECTION

② Motions of the Ocean

SECTION OBJECTIVES

After completing this section, you should be able to:

■ **Diagram** the location of major ocean currents.

■ **Describe** the characteristics of a sea wave.

■ **Explain** how ocean tides are produced.

NEW SCIENCE TERMS

currents
geostrophic currents
waves
rip current
tides

19.4 Ocean Currents

The next time you are in a swimming pool, slowly push the water with your hand. You will notice that you can produce a stream, or current, of water within the pool. **Currents** are streams of water that move like rivers through the oceans. Most currents are created by planetary winds blowing along the surface of the ocean. On land, the wind may move loose sediments, forming dunes. At sea, the wind drags along the surface of the ocean. The friction between water molecules causes whole columns of water to move at the same time, forming currents. As these currents move, they are influenced by the Coriolis effect described in Chapter 8.

Large ocean currents transport millions of cubic meters of sea water per second. By comparison, the Amazon River, the largest river by volume in the world, moves only 120 000 m³ of water per second.

A MATTER OF FACT

The major ocean currents are often hundreds of kilometers wide and hundreds of meters deep.

Figure 19–11. Notice the flow of red-colored water that identifies the Gulf Stream current in this infrared photograph.

Section 2   Motions of the Ocean   **419**

Section **2**: **Motions of the Ocean**

SECTION FOCUS

Section Overview This section discusses ocean currents, waves, and tides. The causes and characteristics of particular types, as well as phenomena associated with each, are explained. How the oceans affect human beings is also discussed.

Section Objectives For a list of section objectives, see pupil's edition page.

New Science Terms For a list of new science terms in this section, see pupil's edition page. You may wish to pronounce the terms for the students.

SECTION DEVELOPMENT

19.4 Ocean Currents

DISCUSSION Direct the students' attention to the map of ocean currents on page 420. Ask them to explain what is shown in the illustration. Then have them compare the map with Figure 19–11. This will help them visualize what actually occurs in the oceans.

EXTENSION Winds are named for the direction from which they originate. Ocean currents, on the other hand, are named for the direction toward which they flow. The trade winds (easterlies) drive equatorial currents westward. These currents are eventually driven poleward by the Coriolis effect. Near the poles, they turn eastward due to westerly winds. The absence of southern continents allows surface currents to circumscribe the globe around Antarctica. *Thinking Critically:* Many ocean winds and currents were named by ships' crews who traveled on the oceans. Ask the students to speculate on the origin of names such as the doldrums and the trade winds. (A doldrum is a listlessness or despondency: the current got its name because there is little wind in that area. The trade winds got their name because they blew trade ships toward the New World from Europe.) Interested students might research how the horse latitudes got their name.

THINKING SKILL (*Applying Concepts*) Use a globe to point out to the students some of the currents shown in Figure 19–12. Ask how the students might navigate a boat from Hawaii to California using ocean currents to aid their progress. Discuss whether the same route might be used for the return voyage. (No, because the same current that might speed the journey when the boat is going in one direction would slow the vessel down when it is going in the opposite direction. A different current would speed the return trip.)

EXTENSION Surface currents are sometimes referred to as Ekman flow, after the Swedish oceanographer, V. Walfrid Ekman. Ekman flow is the movement of water in response to winds. Nevertheless, because of Coriolis forces and friction between water layers at increasing depths, the direction of subsurface currents at 100 meters or more below the surface changes drastically, eventually moving in a direction opposite to the wind. The pattern of changing speed and direction of deep-ocean currents as depth increases is called the Ekman spiral.

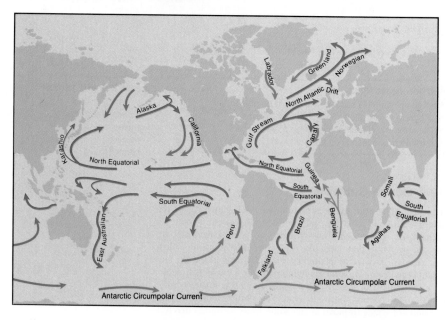

Figure 19–12. This map shows the major surface currents of the world's oceans.

A MATTER OF FACT

Ocean currents below the surface actually flow in the opposite direction from surface currents. This reverse, called the *Ekman spiral*, is caused by surface wind forces, the Coriolis effect, and friction between ocean layers at increasing depths.

The largest ocean current, the Antarctic Circumpolar Current, flows eastward around Antarctica and transports 125 million cubic meters of water per second. This great current is caused by the westerly winds that blow steadily from west to east around Antarctica. In the northern oceans, there is no current that circles the earth because the continents are in the way. However, the winds over the North Atlantic and North Pacific drive ocean currents in a similar manner, forming great circular systems called *gyres*. Water in each of the gyres tends to move toward the center of the gyre because of the Coriolis effect.

The currents shown in Figure 19–12 are surface currents. In the deep water below many surface currents are countercurrents. Countercurrents flow in the opposite direction to currents, keeping water from piling up in one place.

There are areas of the ocean where surface currents come together, or converge. There are also areas where a current may separate, or diverge. A major area of convergence occurs around Antarctica. There the Antarctic Circumpolar Current and parts of the great gyres of the south Pacific, Atlantic, and Indian oceans come together. Surface water at this convergence sinks to the bottom, carrying oxygen to the depths.

One major area where currents diverge is along the equator. There the Coriolis effect forces surface currents on either side of the equator to turn in opposite directions. A current may also diverge from a coast, again because of the Coriolis effect. The California Current, for instance, turns to the right and moves away from the California coast. Deep water rises to the surface to replace the water of this diverging current. Deep water is rich in nutrients, so areas of divergence have many marine organisms.

Often where currents converge, the water surface is 1 m higher than the average sea level. Where currents diverge, the water level is about 1 m lower. Water moving out of these high areas in the surface of the ocean and into low areas forms **geostrophic currents.** Geostrophic currents also flow under the influence of the Coriolis effect.

Ocean currents help move heat from low to high latitudes. Although winds account for 80 percent of the heat transport, ocean currents account for most of the remaining 20 percent. Heat transport across latitudes has the effect of warming the higher latitudes, making them suitable places for people to live. For instance, much of Europe is at higher latitudes than the northeastern United States, yet the climate is similar. This is due to the effects of the Gulf Stream. The Gulf Stream carries warm Atlantic water from near the equator northward along the American coast. The current then turns eastward, bringing warm water to the coast of Europe.

○ *What causes most surface currents?*
○ *What are convergences and divergences?*

Figure 19–13. The Gulf Stream diverges in the Atlantic Ocean between North America and Eurasia, and the equatorial currents converge between Africa and South America.

A MATTER OF FACT

The speed of most ocean currents is about 0.1 m/s to 0.5 m/s.

Figure 19–14. Because of one gyre of the Gulf Stream, in winter the beaches of the Mediterranean are crowded (left), while the New Jersey shore, at about the same latitude, is nearly deserted (below).

EXTENSION (Tie-in/History) Ask the students to point out on a map of the world the route(s) taken by early sea explorers like Christopher Columbus to reach the "New World." Ask them if the explorers could have gone in any direction they wished or were the destinations reached on these long, dangerous voyages predetermined by winds and currents. Explain that because Columbus' sailing vessels were simply pushed and pulled along by the trade winds and the north equatorial current they had, in fact, little choice of destination.

Answers–19.4

○ Surface currents are caused by planetary wind systems.

○ Areas where currents come together are called convergences. Areas where currents flow apart are called divergences.

Objectives

- Locate main surface currents on a map.
- Locate subsurface currents on a map.
- Draw conclusions about comparisons of currents.

Discussion Mapping is widely used in science to survey and explore a region of the earth and space. Review the ocean's surface currents by asking the students to locate the main surface currents of the world in their textbooks. ***Thinking Critically:*** Ask why it is useful for sea captains and other navigators to use maps for long oceanic voyages. (They can use maps to locate ocean currents that can help them speed their journeys as they navigate from one place to another.)

Answers to Application

1. They are in the same general area.
2. Subsurface currents move in the opposite direction from surface currents.
3. No, it is the direction of the flow, rather than the temperature that determines where deep currents will form.
4. Differences are caused by local gyres and differences in ocean depth.

Answer to Using What You Have Learned

The planetary wind pattern would be very similar to the pattern of surface currents.

SKILL ACTIVITY: Mapping Ocean Currents

BACKGROUND

The wind, the earth's rotation, and the water temperature are the main factors that determine the path of the oceans' surface currents. Subsurface, or counter currents, however, flow in the opposite direction, and their paths are determined by other factors. These deep ocean currents circulate mainly because of density differences and the friction between moving layers of water.

PROCEDURE

Use page 420 of your textbook to locate the main surface currents of the world. On a piece of tracing paper, sketch a rough outline of the oceans. Then trace the warm currents with a red pencil and the cold currents with a blue pencil.

APPLICATION

On another sheet of tracing paper, again sketch a rough outline of the oceans, and trace the counter currents from below with a purple pencil. Hold the two drawings together so the outlines of the oceans match.

1. How do the locations of the surface and subsurface currents compare?
2. How does the direction of flow between the surface and subsurface currents compare?
3. Does it seem to matter if the surface current is warm or cold? Explain.
4. How can you account for any differences in location or direction of flow between the two sets of currents?

USING WHAT YOU HAVE LEARNED

If you were to make a third drawing, this one of the planetary wind patterns, how do you think it would compare to the drawings of the surface and subsurface currents? Explain.

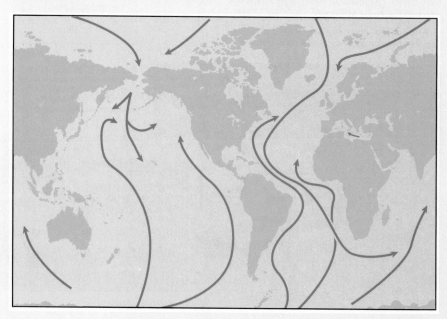

19.5 Ocean Waves

While surface currents are produced by planetary winds, winds also produce the surface motions called **waves**. Waves are produced by friction between the wind and the ocean surface. This friction causes ripples that are heightened by the impact of wind on one side of the wave and air suction on the other side. How is the energy of the wind transferred to the waves?

Figure 19–15. On a calm day, the surface of the ocean can be nearly as smooth as glass (right). A strong wind, however, can whip up large waves on the same beach (left).

Regardless of size, all waves are characterized by *wavelength, height,* and *period,* illustrated in Figure 19–16. *Wavelength* is the distance from the crest of one wave to the crest of the next wave. The average wavelength of ocean waves is 60 m to 120 m. The *height* of a wave is the vertical distance between the bottom, or trough, of a wave and its crest. The average height of ocean waves is 1 m to 2 m. The *period,* or frequency, of a wave is the time it takes two successive waves to pass the same point. The standard period for ocean waves is 6 seconds to 9 seconds.

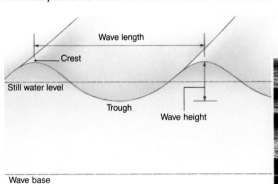

Figure 19–16. Deep-water waves do not break (left). Instead, successive crests and troughs roll past any given point (right).

19.5 **Ocean Waves**

DISCUSSION Have the students describe what a swell looks like and how it is formed. ***Thinking Critically:*** Why can swells sometimes cross an entire ocean without losing much energy? (Where there are no large bodies of land, there are few obstacles or obstructions to impede the movement of the waves or dissipate their energy.)

THINKING SKILL *(Communicating)* Draw and label a diagram of an ocean wave with a wavelength of 80 m, a height of 1.2 m, and a period of 8 seconds. Ask the students how they would indicate the period on the diagram. (by marking the wavelength line at 8 seconds)

DISCOVER:
Demonstrating Waves

Thinking Skill (*Communicating*)

Ask the students to draw a series of diagrams that show some of the stages through which the rope moves. The students might flip the rope in a slow motion to help them observe more clearly how it moves.

DISCUSSION Direct the students' attention to Figure 19–18. Ask the students to explain in their own words what is occurring in the illustration. ***Thinking Critically:*** Have the students compare ocean waves with electromagnetic waves. Have them list ways in which they are the same and ways that they are different. (Each of these may be described in terms of wavelength, height, and period. Electromagnetic waves have no transporting medium, they are propagated spherically, and travel much faster than ocean waves.)

THINKING SKILL (*Comparing*)
Compare the waves shown in Figures 19–16 and 19–17 with those breaking on shore in Figure 19–18.

Figure 19–17. Swells are open-water waves with long wavelengths.

DISCOVER

Demonstrating Waves

Tie one end of a 5-m-long rope to a fence or a tree. Hold the free end of the rope in one hand and flip it up and down about once every second, creating waves. How does each fiber in the rope move? How is this motion similar to the surface motion of water in a wave?

Waves develop best in the open ocean where there is no interference with the ocean bottom. Large storms are capable of producing waves with long wavelengths. These waves, called *swells*, have wavelengths of several hundred meters. Swells sometimes cross an entire ocean at speeds of tens of kilometers per hour without losing much energy. Waves generated by storms around Antarctica, for instance, often cross the entire Pacific and reach Alaska.

If you watch waves in the open ocean, they seem to move. However, the only thing that really moves is the wave form. The water that makes up the wave only moves up and down, as diagrammed in Figure 19–18. Water moves in the direction of the wave motion only if the bottom of the ocean interferes with the traveling wave. Where the bottom becomes shallow enough to affect the wave, friction slows the lower part of the wave. The wave then becomes unbalanced, breakers form, and the water crashes onto the beach. Beaches that face into the prevailing winds of the open ocean have the high, curling breakers that are perfect for surfing. Beaches in Hawaii, Australia, and the west coast of the United States and Mexico often have these kinds of waves.

Figure 19–18. In deep water, the water molecules in a wave travel in a circle. In shallow water, the water molecules at the surface move faster than those at the bottom (right). The top of the wave falls over the bottom, forming a breaker that spills onto the beach (left).

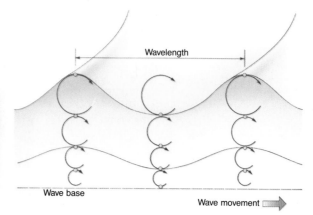

Wavelength

Wave base

Wave movement

424

424

Water crashing onto the beach quickly returns to the sea. Depending on the shape of the coastline, the returning water may form a **rip current** flowing out along a channel at right angles to the beach. The speed of a rip current, or undertow, may be as fast as 1 m/s—about the same as a fast-paced walk. Some rip currents are strong enough to drag even good swimmers out to sea. Sometimes people caught in a rip current panic and drown. If you are ever in this situation, all you have to do is swim parallel to the beach. Rip currents are usually very narrow, so you can easily swim out of them and return to shore. Never try to swim against a rip current.

Occasionally waves over 10 m high come ashore, causing tremendous damage and killing many people. These waves, called *tsunamis* (soo NAH meez), are caused by earthquakes in coastal areas or beneath the ocean floor. Tsunamis have very long wavelengths, often more than 100 km. In the open ocean, tsunamis are only about 0.5 m high, but they travel at the speed of several hundred kilometers per hour. A tsunami generated by an earthquake in Alaska can reach Hawaii in less than eight hours. As the ocean bottom becomes shallow, the tsunami becomes higher. Even though the height of a tsunami in the open ocean is small, the wave contains a lot of water because of its great wavelength. A tsunami hits land with tremendous force. Few structures can survive such a force.

○ *What causes waves?*
○ *What are the three characteristics of a wave?*
○ *What are tsunamis?*

Beach

Rip current

Figure 19–19. These arrows show the location of rip currents on this beach.

Figure 19–20. In 1946, a tsunami caused almost total destruction of this waterfront area of Hilo, Hawaii.

19.6 Ocean Tides

If you are a sailor or a fan of high-seas adventure films, you have probably heard someone say, "We sail with the tide." As the earth turns on its axis, the gravity of the moon pulls up a crest of ocean water, forming **tides.** Tides occur because the gravitational pull of the moon is strong on the side of the earth facing

DISCUSSION Have the students consider the damage waves may cause when they strike the land. This influences architects and planners who design buildings and determine whether they should be built with special structural reinforcements. *Thinking Critically:* The students may have seen pictures of beach houses on stilts. How does this design protect them from the force of the ocean waves? (The water flows around the stilts, and therefore makes much less of an impact than when it crashes into a solid wall.)

THINKING SKILL (*Relating Events*) Discuss with the students what they should do if they are ever caught in a rip current. Ask them why that technique would work.

19.6 Ocean Tides

DISCUSSION Ask the students what high and low tides actually look like when observed from a beach. (At high tide, the waterline is far advanced onto the shore. At low tide, it is recessed. One could place a marker at the waterline during high tide and observe how far the waterline at low tide is from the marker). *Thinking Critically:* Ask the students if high tides on all sides of an island, like Hawaii, would be of approximately equal heights at the same time. (As a rule, whether a shore lies on the north, south, east, or west side of an island does not determine the levels or the times of high tides. However, since factors like ocean currents and the formation of the land affect the tides, specific locations would have to be investigated separately.)

BACKGROUND INFORMATION
Nearly ten percent of the kinetic energy of the winds can be transformed into the kinetic energy of ocean waves. The height of the waves increases with wind strength.

Answers—19.5
○ The wind causes waves.
○ Wavelength, height, and period are characteristics of waves.
○ Tsunamis are waves caused by earthquakes.

EXTENSION (Tie-in/Social Studies)
Many cities have developed not only
where there are deep water harbors, but
also where the shape of the landmasses
give boats some protection from the potential dangers caused by the movements
of the ocean. Cities like San Francisco,
Baltimore, and New York City might be
mentioned.

DISCUSSION Review with the students the things they learned about the
ocean in Section 1. ***Thinking Critically:***
Why is it that large boats have frequently
sailed with the tide? (At low tide, the
water in a harbor is more shallow. The
bottom of a boat may be damaged.
Though the level of the water near shore
may not be deep during high tides, the
depth is maximized. Remember that even
in a good harbor there may be places
where large rocks protrude.)

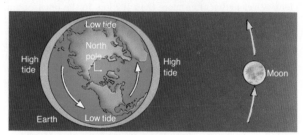

Figure 19–21. Because it is the closest body to the earth, the moon has the greatest influence on producing tides (right). The revolution of the moon around the earth causes the tides to be about 50 minutes later each night (above).

the moon, and weaker elsewhere on the earth. The height of the
tide varies with the position of the moon. Tides are very long
period waves, usually 24 hours long.

As the earth-moon system rotates about its center of mass,
the force of this moving system produces two tidal crests, one
on the side of the earth facing the moon and one on the opposite
side. These tidal crests circle the earth in 24 hours and 50 minutes. Why does it take the tide more than a day to circle the
earth? The extra 50 minutes are needed because the moon is
traveling around the earth in the same direction in which the
earth rotates. This means that the earth has to turn for 24 hours
and 50 minutes in order for the same place to be directly under
the moon. For example, if you look out your bedroom window
and see the moon straight ahead, it will not return to that exact
position until 50 minutes later the next night.

The extra 50 minutes in the lunar day is an ideal case. That
is, the earth would have two tides every 24 hours and 50 minutes
if there were no continents. The continents, however, slow down
the progress of the tides. Special charts show the times of the
day when high tides or low tides occur at different locations on
Earth. These charts are important to navigation, especially in
shallow water and in the entrance to harbors.

The average tide is 0.55 m high in the open ocean. However,
the height may reach 2 m to 3 m as the tide runs into a coast.
If the shape of the coast funnels the water into a small area, the
tide will be even higher. The world-record tide occurs at the
head of the Bay of Fundy, in Nova Scotia, Canada. There the tide
may reach a height of 14 m.

Figure 19–22. The greatest tidal differences occur in the Bay of Fundy, in Nova Scotia, Canada.

Spring tides

Sun

Neap tides

Earth

Moon

Earth

Moon

The sun also produces tides. The sun, of course, is much larger than the moon, but it is also much farther away. The gravitational attraction of the sun is much stronger than the moon's, but it does not change much from place to place. As a result, solar tides are only about half as high as lunar tides.

When the moon and sun are in a straight line with the earth, the heights of the lunar and solar tides add up. The tides that result from this alignment are called *spring tides*. The name "spring tide" does not refer to the spring season, but to the "springing up" of the water. Spring tides occur twice during a lunar month: at full moon and at new moon. When the moon and sun are at right angles to each other, as seen from the earth, very low tides, called *neap tides*, occur. Neap tides also occur twice a month—midway between the spring tides. Figure 19–23 shows the Spring and Neap tides.

Under special circumstances, strong, dangerous tidal currents can develop. These strong currents develop most often in narrow passages connecting bodies of water that have tides at different times of the day. The most famous of these is Charybdis, a tidal vortex that develops in the Strait of Messina, between Sicily and mainland Italy.

○ *What causes tides?*
○ *What are spring tides and neap tides?*

Figure 19–23. The greatest daily tidal differences occur at spring tide (left). The smallest daily tidal differences occur at neap tide (right).

Figure 19–24. Shown here is an artists view of the tidal vortex in the Strait of Messina, between mainland Italy and Sicily.

DISCUSSION Explain the alignment of the sun, the moon, and the earth during spring and neap tides, using Figure 19–23. *Thinking Critically:* Ask the students if they think that tides can cause beaches to erode. (Yes) Explain that retaining walls are sometimes constructed to prevent or to minimize such erosion. Buildings near beachfronts are sometimes constructed so that steel or reinforced concrete foundations are sunk deep into the ground. In this way the structures are made less vulnerable to the movements of the ocean.

Section Review

READING CRITICALLY

1. How does the depth of the ocean floor affect waves?
2. What are geostrophic currents?
3. What is a rip current?
4. Why are lunar tides more than twice as high as solar tides?

THINKING CRITICALLY

5. Clams live in the sand under shallow water. When would be the best time to collect clams? Explain your answer.
6. Why are storm waves more dangerous during spring tides?

Section 2   Motions of the Ocean   **427**

SECTION REVIEW

Summary Review the surface and subsurface ocean currents and what causes them. Have the students describe the differences between surface and subsurface currents.

Reinforcement Encourage students to draw and label a diagram of an ocean wave.

Answers to Section Review

Reading Critically

1. Waves develop more easily in the open ocean because there is no interference from the ocean bottom.
2. Currents caused by water moving from high areas of the ocean to lower areas are called geostrophic currents.
3. A rip current is water flowing rapidly out of a narrow channel along a beach.
4. The moon's gravitational attraction for the earth's oceans is stronger than the sun's, because the moon is so much closer to earth.

Thinking Critically

5. At low tide during a quarter moon, when the moon, earth, and sun are not directly aligned with one another.
6. At spring tide, high tides are at their highest, making storm waves more likely to cause damage on the land.

Answers—19.6

○ The gravitational attraction of the moon for the oceans of the world.

○ Spring tides are the very high, high tides and very low, low tides that occur once a month. During neap tides, the difference between high and low tides is much less.

INVESTIGATION 19:
The Effect of Temperature on Water

Skill *(Making a Table)*

Preparation of Materials

Fresh water freezes at 0°C. Seawater has a concentration of 34.5 grams of salt per kilogram of seawater and freezes at −1.8°C. Since the concentration of the salt used in this investigation exceeds the average concentration of normal seawater, the difference in freezing point between the two liquids will be more apparent.

At the molecular level, the sodium (Na+) and chloride (Cl−) ions interfere with the regular pattern of hydrogen bonding among water molecules.

It will be necessary to explain the concept of heat of fusion, so that the students will understand why there is a pause or plateau in the graph of temperature versus time. This plateau indicates the additional energy drawn from the system to sufficiently decrease the kinetic energy of the water molecules so they will form a crystal lattice. This energy is drawn from the water molecules themselves and not from the surface of the thermometer. Until all of the water is frozen, therefore, the temperature registered by the thermometer will remain constant. The temperature will begin dropping again after the water has solidified completely.

Answers to Analyses and Conclusions

1. The thermometer in the fresh water, because there are fewer salt ions.
2. Yes. During this pause, the water was freezing.
3. Between 1° and 0°C. (Answers will vary depending upon the calibration of the thermometers).
4. Answers will vary according to the concentration of salt (which is not strictly controlled in this investigation) and on the calibration of the thermometers. However, it should be obvious that saltwater freezes at a lower temperature than freshwater.

Answer to Application

Adding salt to the ice allows it to reach a lower temperature and causes the ice cream to freeze.

INVESTIGATION 19: The Effect of Temperature on Water

PURPOSE

To test the hypothesis that fresh water freezes at a higher temperature than salt water

MATERIALS (per group of 3 or 4)

Beaker, 250 mL
Crushed ice
Salt
Wax pencil
Test tubes (2)
Tap water
Beakers, 100 mL (2)
Balance
Stirring rod
Graduate, 10 mL
Thermometers (2)
Stopwatch

TABLE 1: EFFECTS OF TEMPERATURE ON WATER

Time	Temperature (°C)	
	Fresh Water	Salt Water

PROCEDURE

1. Prepare a data table similar to the one shown. Make the table large enough for 30 entries.
2. Fill the 250-mL beaker with crushed ice and salt.
3. Use the wax pencil to label the test tubes "F" (for fresh water) and "S" (for salt water).
4. Gently insert the test tubes into the ice about 3 cm apart. The test tubes should be touching the inside of the beaker, 3 cm from the bottom.
5. Put about 100 mL of tap water in one of the 100-mL beakers.
6. Measure 4 g of salt and add it to the other 100-mL beaker. Fill the beaker with water. Stir for 30 seconds.
7. Use the graduate to measure 2 mL of fresh water. Pour it into test tube "F."
8. Measure an equal amount of salt water and pour it into test tube "S."
9. Carefully insert a thermometer into each test tube. **CAUTION: Do not drop or hit the thermometers against the walls of the test tubes. They may crack the test tubes or break them.**
10. Record in your data table the original temperature of each solution.
11. Take temperature readings from each test tube once a minute for 30 minutes.
12. Graph your results.

ANALYSES AND CONCLUSIONS

1. Which thermometer records a faster drop in temperature? Explain why.
2. Do your results show a pause in the temperature drop at any time during the experiment? If so, what do you think was happening to the water during this pause?
3. At what temperature did the fresh water finally begin turning to ice?
4. At what temperature did the salt water finally begin freezing, if at all?

APPLICATION

How might this investigation explain why salt is added to the ice in a home ice cream freezer?

SUMMARY

- Sodium chloride is the most common salt in sea water. (19.1)

- The average salinity of the ocean is about 34.5 g/kg. (19.1)

- Salt water freezes at a lower temperature than fresh water. (19.2)

- Cold, salty water sets up a circulation in ocean water called thermohaline. (19.2)

- The thermocline is a layer below the surface of the ocean through which temperature decreases rapidly. (19.2)

- The pressure of the ocean water increases with depth. (19.3)

- Many surface ocean currents are caused by planetary winds blowing on the surface of the ocean. (19.4)

- The largest ocean current in the world is the Antarctic Circumpolar Current. (19.4)

- Where currents converge, the water level is about 1 m above the average sea level; where currents diverge, the water level is about 1 m below the average sea level. (19.4)

- Geostrophic currents are currents that move out of highs in the surface of the ocean and into lows. Like all currents, they are influenced by the Coriolis effect. (19.4)

- Ocean currents help to distribute heat from low latitudes to high latitudes. (19.4)

- Waves are produced by friction between local winds and the ocean surface. (19.5)

- The water in a wave does not move forward; only the wave form moves. (19.5)

- The tides are produced by the attraction of the moon and the sun on different parts of the earth's surface. (19.6)

- Spring tides occur when the sun and moon align. Neap tides occur when the sun and moon are at right angles. (19.6)

Write all answers on a separate sheet of paper.

SCIENCE TERMS

Correctly use each of the following terms in a sentence.

currents **(419)**
geostrophic currents **(421)**
rip current **(425)**
salinity **(412)**
thermocline **(415)**
thermohaline **(415)**
tides **(425)**
waves **(423)**

SCIENCE QUIZ

Modified True-False

Mark each statement *true* or *false.* If a statement is false, change the underlined term to make the statement true.

1. Sodium chloride accounts for <u>less than</u> three-quarters of the salt in the world's oceans.

2. Ocean temperature increases rapidly <u>above</u> a thermocline.

3. Ocean water freezes at a <u>higher</u> temperature than fresh water.

4. Most ocean currents are driven by the <u>wind</u>.

5. The high salinity of the Mediterranean Sea is a result of high <u>evaporation</u>.

Multiple Choice

Write the letter of the choice that best answers the question or completes the statement.

6. The sides of the earth away from the moon will be in
   a) low tide.   b) high tide.
   c) spring tide.   d) neap tide.

continues

CHAPTER REVIEW

SUMMARY

The students may review the major concepts in the chapter by reading the summary statements. The statements are cross-referenced to the chapter to facilitate reinforcement of any concepts of which the students feel unsure. Encourage the students to work in groups to quiz one another.

SCIENCE TERMS

The sentence in which the science term is used should reflect an understanding of the definition of the term. You may wish to hold a science "bee" during which one team will give the term and the other must supply the definition.

SCIENCE QUIZ

Answers to Modified True-False

1. false, more than
2. true
3. false, lower
4. true
5. true

Answers to Multiple Choice

6. d

continues

Answers to Multiple Choice (continued)

7. d
8. a
9. b
10. d
11. b
12. b
13. a
14. a
15. a

Answers to Completion

16. organisms
17. lowest
18. rip
19. density
20. Challenger Deep

Answers to Short Answer

21. Check students' drawings by referring to Figure 19–23 on page 427 of the pupil's edition.
22. Increased salinity lowers the freezing point of water.
23. The individual water molecules in a wave move up and down from the bottom, while the wave transfers its energy in a direction perpendicular to that motion.

Writing Critically

24. The difference between high tide and low tide in the Bay of Fundy is several meters. Boats that are afloat at high tide may become "beached" as the shoreline ebbs to low tide. A boat that goes aground twice a day would give its owner numerous repair and maintenance problems.
25. The waveform of an ocean wave moves perpendicularly to the wave's vibration. While water in the wave moves vertically with the respect to the bottom, the energy of the wave is transferred horizontally with respect to the bottom.

SCIENCE QUIZ continued

7. The percentage of the earth's surface that is covered by salt water is about
   a) 40 percent.
   b) 50 percent.
   c) 60 percent.
   d) 70 percent.

8. Surface currents are caused by
   a) the wind.
   b) waves.
   c) the tides.
   d) the ocean.

9. As salinity increases, the freezing point of sea water
   a) increases.
   b) decreases.
   c) varies.
   d) remains the same.

10. Which salt is not common in ocean water?
    a) NaCl   b) KCl
    c) $MgCl_2$ d) $FeCl_2$

11. If there were no oceans, the air temperature range over much of the earth would be
    a) $-50°C$ to $50°C$.
    b) $-100°C$ to $100°C$.
    c) $0°C$ to $50°C$.
    d) $-20°C$ to $20°C$.

12. The deepest part of the ocean is the
    a) Pacific Ocean.
    b) Challenger Deep.
    c) thermocline.
    d) thermohaline.

13. The ocean currents caused by winds are
    a) surface currents.
    b) geostrophic currents.
    c) divergences.
    d) convergences.

14. The time between the passage of two successive waves is the
    a) period.
    b) wavelength.
    c) height.
    d) crest.

15. High tides that occur monthly are called
    a) spring tides.   b) neap tides.
    c) tidal crests.   d) seasonal tides.

430   Chapter 19 Review

Completion

Complete each statement by supplying the correct term or phrase.

16. Silica and nitrate are taken out of sea water by _____.

17. During a first-quarter moon, high tides are at their _____.

18. Swimmers should always be aware of _____ currents near the shoreline.

19. The _____ of ocean water increases with a decrease in temperature.

20. The deepest portion of the oceans is in the _____.

Short Answer

21. Draw diagrams to show how the earth, moon, and sun would be aligned during spring tides and neap tides.

22. Tell how salinity in ocean water affects its freezing temperature.

23. Describe the movement of water in an ocean wave.

Writing Critically

24. Describe the problems of docking a boat in the Bay of Fundy.

25. If the water in a wave does not move, how do surfers ride the waves for a distance?

EXTENSION

1. Read about the Sargasso Sea and report to your class about where it is located, why it exists, and what effect it has had on sailing during the past 500 years.

2. In a shallow tub of water, try to create long- and short-period waves, and then diagram the differences on a sheet of paper.

3. If you live near the ocean, study the tide tables for a week. See if you can figure out the pattern for the daily difference in time for high and low tide. After you know the

---

ANSWERS TO EXTENSION

1. A discussion of the Sargasso Sea, lying between the Azores and West Indies, should include experiences of early explorers who found the area to have unusually calm winds and waves. Because the Gulf Stream and equatorial currents move in a clockwise rotation around the region, extensive matting of a brown seaweed called sargassum tends to accumulate in the area, earning it the name. The rudders of early sailing vessels were frequently entangled in the masses of seaweed, causing considerable problems.

2. Diagrams of long period waves should have long wavelengths. Diagrams of short period waves should show shorter wavelengths.

3. Answers will vary.

pattern, determine the times for high and low tide for the following week. Then check the newspaper to see how accurate you are.

APPLICATION/CRITICAL THINKING

1. Benjamin Franklin was one of the first persons to describe the flow of the Gulf Stream. Without the modern equipment that is used today, how do you think he was able to accurately diagram this ocean current?

2. What do you think the pattern of surface currents would be like if there were no continents to block the flow of ocean waters?

3. In some fast-moving streams, waves develop that do not move but remain stationary in the middle of the stream. Explain how these waves, called *standing waves*, might develop.

Challenge Your Thinking

FOR FURTHER READING

Ballard, R. "How We Found *Titanic.*" *National Geographic* 168 (December 1985): 696. This article gives an exciting, minute-by-minute account of the discovery of the lost ocean liner.

Ballard, R. "A Long Last Look at *Titanic.*" *National Geographic* 170 (December 1986): 697. This article describes the exploration of the *Titanic,* using deep-sea submersibles such as *Alvin.*

Severin, T. "Jason's Voyage: In Search of the Golden Fleece." *National Geographic* 168 (September 1985): 406. This article tells how the author and his shipmates sailed the Black and Aegean seas, mapping the course taken over 3000 years ago by Jason and the Argonauts.

Look at these photographs of the Arctic Ocean in winter and summer. Much of the winter ice melts in the summer because it averages only 3 m in thickness. Even at the North Pole, only a thin layer of the Arctic Ocean is frozen. This is because ice is less dense than water, so the ice floats on the surface, insulating the water below and preventing it from freezing. If, like most substances, solid water were denser than liquid water, what would happen to the Arctic Ocean? What effect might this have on the climate of the world and on the organisms that live on Earth?

Chapter 19 Review   **431**

ANSWERS TO APPLICATION/ CRITICAL THINKING

1. He made careful observations and questioned sea captains concerning their trips to Europe.
2. The currents would more accurately mimic the winds.
3. Currents are deflected back to the point where the wave forms, keeping the wave stationary.

ANSWERS TO CHALLENGE YOUR THINKING

If the oceans would freeze from the bottom, they would freeze solid and only a thin layer would melt each summer. There would probably be no life in polar seas, and the climate of northern latitudes would be much colder.

Chapter 20: OCEAN RESOURCES

PLANNING THE CHAPTER

Chapter Sections	Page	Chapter Features	Page	Program Resources	Page
Section 1: Ocean Sediments	433			Investigation 20.1: *Composition of "Sea" Water* **(H)** Investigation 20.2: *Desalination* **(A)**	**TRB 89** **LM 91** **TRB 91** **LM 93**
20.1 Iron, Coal, and Petroleum **(B)**	433	**Discover:** Researching Ancient Climates **(B)**	434	Concept Extension: *Survival at Sea* **(B)**	**TRB 41** **LM 263**
20.2 Fresh Water and Salt **(B)**	434	A Matter of Fact	435		
20.3 Deep-Sea Sediments **(B)**	436	A Matter of Fact	437		
		Section Review	437		
		Technology: Submersibles	438		
Section 2: Ocean Life	439			Critical Thinking **(H)** Cross-Discipline: Science and Geography, *Mapping the Ocean's Living Resources* **(A)**	**TRB 38** **TRB 37**
20.4 Living Resources **(B)**	439	**Skill Activity:** Organizing Information **(B)**	440		
		A Matter of Fact	441	Reading for Content: *Using Your Textbook* **(A)**	**TRB 34** **LM 211**
		A Matter of Fact	442		
20.5 Ocean Pollution **(B)**	442	**Discover:** Producing a Newsletter **(B)**	443		
		Section Review	445		
		Activity: Recycling Water **(B)**	445		
		Investigation 20: Water Filtration **(B)**	446	Student Record Book: Textbook Investigation **(B)**	**TRB 41**
Chapter 20 Review	447			Vocabulary **(A)**	**TRB 20** **LM 161**
				Tests **(A)** Computer Test Bank	**TRB 85**

(LM) Laboratory Manual/Study Guide, **(TRB)** Teacher's ResourceBank™

B = Basic **A** = Average **H** = Honors

The coding Basic, Average, and Honors indicates sections or subsections that might be appropriate for different levels of learners. For additional suggestions regarding choice of topic and depth of coverage, see the Pacing Chart on pages T16–T20.

CHAPTER CONCEPTS, OBJECTIVES, AND TERMS

Section	Concepts	Objectives	Science Terms
Section 1: Ocean Sediments	■ The oceans help provide many mineral resources, such as iron, coal, and petroleum. **(20.1)** ■ Coal was formed as the sea invaded swamps and buried organic matter under heavy sediments. **(20.1)** ■ Ocean sediments may form deposits of petroleum. **(20.1)** ■ Fresh water and salt may be obtained from sea water by evaporation. Salt may be mined from large deposits known as salt domes. **(20.2)** ■ Except for manganese nodules, the mining of deep-sea sediments is not cost-effective. **(20.3)** ■ Manganese nodules may provide an economical source of iron, manganese, nickel, and cobalt. **(20.3)** ■ Deep-sea sediments provide clues to past geologic and climatic changes. **(20.3)**	■ **Discuss** how the world's iron, coal, and oil resources formed. ■ **List** the minerals obtained from sea water and the processes used to obtain them. ■ **Describe** how salt domes form. ■ **Discuss** how deep-sea sediments reveal the climatic history of the earth.	brine salt domes
Section 2: Ocean Life	■ Plankton are microscopic organisms that are the basis for ocean food chains. **(20.4)** ■ Phytoplankton are plantlike organisms that photosynthesize, while zooplankton are animal-like organisms that feed on phytoplankton. **(20.4)** ■ Nekton are free-swimming animals. They provide much of the seafood consumed by humans. **(20.4)** ■ Benthos, or bottom-dwelling organisms, provide many products for human use. Seaweed and many shellfish are benthos. **(20.4)** ■ Organic waste, such as sewage, may be harmful to ocean life. **(20.5)** ■ Industrial chemicals and nuclear wastes are dangerous to ocean life and humans. **(20.5)** ■ Oil spills have killed many sea birds and fouled many beaches. **(20.5)**	■ **Identify** the categories of sea life and give some examples of each. ■ **Discuss** some uses of algae. ■ **List** the major sources of ocean pollution and **describe** the effects of ocean pollution on human and sea life.	plankton nekton benthos

CHAPTER MATERIALS

Title	Page	Materials
Discover: Researching Ancient Climates	434	*(per student)* encyclopedia, poster board
Skill Activity: Organizing Information	440	*(per student)* paper, pencil
Discover: Producing a Newsletter	443	*(per student)* newspapers, magazines, large sheet of paper
Activity: Recycling Water	445	*(per group of 3 or 4)* safety goggles, glass tubing, rubber tubing, plastic bag, string, ring stand and clamp, Bunsen burner, crushed ice
Investigation 20: Water Filtration	446	*(per group of 3 or 4)* hammer, finishing nail, empty soup can, ring stand and clamp, small beaker (150 mL), large beaker (500 mL), coarse gravel, fine gravel, muddy water, crushed charcoal, sand

TEACHING SUGGESTIONS

Section 1: Ocean Sediments

Demonstration: Offshore Oil Rig

Purpose

To illustrate how an offshore oil rig is erected

Background

As the cylinder fills with water (the air being displaced and allowed to escape from the second hole in the stopper), it will submerge until it settles on the bottom of the fish tank. Explain to the students that oil rigs are constructed to float out to sea to sites on the continental shelf that are suspected of containing deposits of oil. There, they are submerged in much the same manner as demonstrated here—by opening ballast tanks in the rig.

Materials

Fish tank
Graduated cylinder
Double-holed stopper
Small glass tube
Rubber tubing
Small funnel
Rubber band

Procedure

1. Place the fish tank next to or into the sink.
2. Fill the tank to a depth that is no more than two-thirds the height of the graduated cylinder.

3. Insert the small glass tube into the double-holed stopper and insert the stopper into the graduated cylinder.
4. Insert the funnel into one end of the rubber tubing and attach the other end of the tubing to the glass tube in the stopper.
5. Float the cylinder "to sea" in the fish tank.
6. Hold the funnel under the faucet and open the spigot to fill the cylinder.

Questions to Ask the Students

1. How is the ballast removed to refloat the rig? (*Usually by compressed air pushing out the water.*)
2. What keeps the ballast tanks from being crushed at great depths? (*The pressure of the water inside the tank is the same as the pressure outside the tank.*)

Class Activity

Have the students make a list of the resources they use daily that come from the oceans.

Section 2: Ocean Life

Class Activity

Make a food chain that includes marine organisms and humans.

Field Trip

Visit a marine life park, a public aquarium, or even a large fish retailer and observe the activities of some of the marine organisms.

Outside Speaker

Invite a biologist from a local college (preferably a specialist in marine biology) to talk to students about ocean life. Ask the scientist to bring slides (or other audiovisual materials) depicting the variety of ocean organisms and to emphasize their relationships as members of a common ecosystem. Make sure to prepare pupils for the visit by having them make a list of three or four questions they would like to put to the visitor.

CHAPTER 20

CHAPTER OVERVIEW

The sediments of the ocean and the resources contained in these sediments are discussed. This chapter also contains information about the geologic history of the earth that scientists have uncovered by studying these sediments. Minerals and other materials found in or under the oceans are described, as is how fresh water may be obtained from ocean water. The plankton, nekton, and benthos that live in the ocean are described. The causes of and difficulties created by ocean pollution are explained.

Section 1: Ocean Sediments This section presents information about the ocean sediments and how minerals have been formed in part by the oceans. Resources such as iron, coal, petroleum, salt, and manganese are available either in or under the ocean. The importance of water is described, as are methods for obtaining fresh water from ocean water. The mining of deep-sea sediments and the value of sediments in informing us about the geological and climatic changes that have occurred on the earth are also discussed.

Section 2: Ocean Life This section begins with a discussion of organisms that live in the ocean. Plankton, nekton, and benthos are described. The problem of pollution is then addressed. Both its causes and consequences are considered. The pollution or potential pollution caused by sewage and garbage, industrial and agricultural chemicals, radioactive wastes, and oil spills are also discussed.

Ocean Resources

Earth's human population increases every minute. Each new person requires a share of Earth's raw materials. More fresh water is needed and more food is needed. How can the ocean help meet these needs? Humans only now are beginning to tap the vast resources of the sea. In the future these resources will be required to maintain human life on Earth.

A commercial shrimper at work

432

CHAPTER MOTIVATING ACTIVITY

Ask the students to make a list of the ways we use the world's oceans. The students will more than likely begin their list by mentioning our use of the oceans and seas for recreation, travel, and deep-sea fishing, as these categories are familiar to most students. It may not occur to many, however, that the ocean's vast reserves of fossil fuels, mineral deposits, and sediments, as well as its organic resources, provide many things that make our lives easier and more enjoyable—even if our homes are hundreds of miles from the nearest coast. Suggest that students add to their list by categorizing the ocean's vast resources according to the section headings of the chapter.

SECTION
1 Ocean Sediments

SECTION OBJECTIVES

After completing this section, you should be able to:
- **Discuss** how the world's iron, coal, and oil resources formed.
- **List** the minerals obtained from sea water and the processes used to obtain them.
- **Describe** how salt domes form.
- **Discuss** how deep-sea sediments reveal the climatic history of the earth.

NEW SCIENCE TERMS

brine
salt domes

20.1 Iron, Coal, and Petroleum

Have you ever been swimming where the bottom sediments were soft and slimy? This is not very pleasant for swimming, but bottom sediments are important because of what they may contain.

Ocean sediments, in particular, are important because they contain many of the mineral resources of the world. For example, about 3 billion years ago the ancient sea floor accumulated layers of iron oxide that alternate with layers of silicates and carbonates. The most important deposits of this type are found around Lake Superior, in eastern Canada, in the Soviet Union, in South America, and in Australia. These deposits form the major iron resources of the world today.

Fossil records show that large deposits of coal, a major energy resource, formed during a period in which the sea repeatedly invaded low, swampy lands and then retreated. Each time the sea washed in, a layer of ocean sediments was deposited on top of decaying plant matter. As these sediments accumulated, pressure and heat changed the plant matter into coal. More than a hundred coal layers, alternating with ocean sediments, are found between the Appalachian Mountains and the Mississippi River. Similar coal layers are found in England, the Soviet Union, China, South Africa, and Australia.

Figure 20–1. These deposits of iron ore (left) and coal (right) were formed millions of years ago.

BACKGROUND INFORMATION

The production of carbohydrates by land plants and plankton that eventually lead to the formation of hydrocarbon deposits or fossil fuels is the result of photosynthesis:

$$6CO_2 + 12H_2O \rightarrow C_6H_{12}O_6 + 6CO_2 + 6H_2O$$

However, only 1 percent of the total solar energy available to our planet is actually used in photosynthesis. Still, approximately 1.0×10^{22} joules per year is converted to biomass, continually resupplying a reserve of energy that is nearly 30 times greater than the total energy demand of modern society (3.0×10^{20} joules per year). Since reserves of fossil fuels are being depleted rapidly, biomass (that is, biological fuel stock) may one day provide as much as a third of the world's energy needs.

DEMONSTRATION

For a demonstration of how offshore oil rigs work, see page 431c preceding this chapter.

Section **1**: Ocean Sediments

SECTION FOCUS

Section Overview This section describes resources found in the ocean. It includes a description of the iron, coal, and oil available in the oceans. The meaning of *desalination*, as well as specific ways of obtaining both fresh water and salt from the oceans, are explored. Also discussed is the mining of manganese nodules from the ocean floor. Manganese nodules contain iron, nickel, manganese, cobalt, and phosphorous.

Section Objectives For a list of section objectives, see pupil's edition page.

New Science Terms For a list of new science terms in this section, see pupil's edition page. You may wish to familiarize students with these terms before they read the section. You may also wish to read the terms aloud so the students are familiar with the pronunciations.

SECTION DEVELOPMENT

20.1 Iron, Coal, and Petroleum

DISCUSSION Describe for the students the contents of the soft and slimy bottom sediments of the ocean. *Thinking Critically:* Ask the students whether the ocean sediment is the same near California, India, Greece, and Japan. (No, since the mineral makeup of the earth is not the same everywhere and not exactly the same species of plants grow on all the continents of the earth, the contents of the sediment will probably not be exactly the same either.)

DISCOVER:
Researching Ancient Climates

Thinking Skill (*Communicating*)

Be sure the students know where to find and how to use current encyclopedias. You may want to suggest that they consult more than one source. Have the students note particularly the existence not only of the warmer periods but also of the ice ages. Point out some alternative approaches to preparing the poster:

○ using bar or line graphs
○ using a time line
○ using a color key: warm colors and cool colors to denote these temperatures

20.2 Fresh Water and Salt

DISCUSSION Explain to the students why salt was so valuable to ancient civilizations. Use this discussion to examine the techniques described in the text for extracting salt from water. Then ask the students to name the techniques that could have been employed 100 years ago. Tell them that at different points in the development of technology, certain processes, but not others, could have been employed. ***Thinking Critically:*** Ask the students if people can utilize salt water. (Yes. For example, the students may mention that such water may be employed in industry as a cooling agent.)

THINKING SKILL (*Problem Solving*) As a rule, plants cannot tolerate salty water. Ask the students to name an approach to this problem, other than taking all the salt out of the water. (Plants can be bred with a greater tolerance for salt. Such research is currently being done and has, to some degree, been successful.)

434

DISCOVER

Researching Ancient Climates

Coal is mined in areas that were once covered with swamps. Scientists believe that millions of years ago the climate of the earth was hot and steamy.

Do some research in an encyclopedia, and then make a poster showing the progressive change of climate over the past 250 million years.

Figure 20–2. Oil is often found under a dome of rock salt.

Many ocean sediments are rich in the remains of once-living organisms. Sediments washed into the sea from the land contain many nutrients. Marine algae use these nutrients to produce food for themselves and other organisms. When these marine organisms die, their remains are added to the debris that falls to the ocean floor. Where the ocean bottom is *anaerobic*—that is, without oxygen—organic matter does not decay. Instead, it accumulates with the sediments. As more sediments accumulate, the buried organic matter becomes petroleum.

The largest known petroleum deposits on Earth formed when the African continent drifted northward and pushed organic-rich sediments against Asia. These sediments form the large petroleum reserves of the Middle East. Other deposits of petroleum are found on the north shore of Alaska, beneath the North Sea, and along the coast of the Gulf of Mexico.

○ *Where are Earth's major deposits of iron ore, coal, and petroleum?*
○ *How are the formation of iron, coal, and oil similar?*

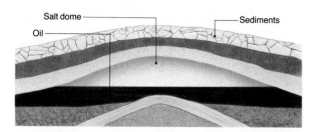

20.2 Fresh Water and Salt

Two of the ocean's most valuable resources are also the most common: water and salt. The problem is that they must be separated before they can be used. The mineral most valuable to ancient civilizations was salt. Salt, which was used primarily to preserve food, was traded like money. Salt is the major mineral resource of the oceans. Sea salt is obtained by evaporating sea water or by mining salt deposits. About one-third of the salt used in the United States is obtained by evaporation; the rest is obtained by mining.

Figure 20–3. In the United States, some salt is obtained by the evaporation of salt water, but most is mined from huge deposits such as this.

434   Chapter 20

Answers–20.1

○ Major deposits of iron are found around Lake Superior, eastern Canada, South America, and Australia. Coal has accumulated in the Appalachian Mountains, England, South Africa, and Australia. Major petroleum deposits are found in the Middle East, Alaska, the North Sea, and the Gulf of Mexico.

○ All are the products of sedimentation.

BACKGROUND INFORMATION

Salt deposits are formed by the evaporation of saline water, leaving behind deposits of varying composition. The deposits then undergo sedimentation and are shaped as all sedimentary layers are shaped. (by folding and faulting) Rock salts, called *evaporites,* are classified by their particular chemical composition.

The evaporation process uses the energy of the sun to remove the water from brine. **Brine** is the extremely salty water that accumulates in areas that are totally or partially isolated from the open ocean. As the water evaporates, salt beds form. These beds may be as thick as 1000 m.

If salt beds are buried, the deposits may be squeezed by the weight of other sediments, forming salt domes. **Salt domes** are mounds of salt, 2 km to 3 km across and 5 km to 10 km high. Most salt domes remain buried under other deposits. However, in some places, such as the floor of the Gulf of Mexico, salt domes rise above the surrounding sediments.

Fresh water is also obtained from sea water by evaporation. Although evaporation is an inexpensive method of *desalination,* or removing the salt from sea water, it is practical only in areas that receive plenty of solar energy.

The most direct way to get fresh water from sea water is to squeeze out the fresh water. In this process, called *reverse osmosis,* sea water is placed in a special container that allows the water to pass through but not the salt. Unfortunately, this method of making fresh water is expensive because of the high pressure needed and because the walls of the container must be cleaned frequently.

Desalination is an important area of continuing research and development. As world population increases, water resources for agriculture and other human uses dwindle. Desalination could provide enough water for humans all over the world if a practical, inexpensive method could be developed.

○ *How is salt obtained?*
○ *Other than salt, what else is obtained from evaporated sea water?*
○ *Why could fresh water from sea water be important to the world?*

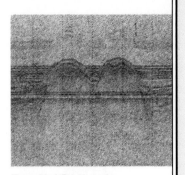

Figure 20–4. This illustration shows a sonar tracing of the Sigsbee Knowles—large salt domes—in the Gulf of Mexico.

A MATTER OF FACT

The sun is the major source of energy for life on Earth, through photosynthesis. Scientists now have photographs of deep-sea communities that inhabit the areas around crustal vents. In this warm water, organisms have evolved using heat and minerals to make food by a process called *chemosynthesis.*

Figure 20–5. Fresh water can be obtained by removing the salt from ocean water in a desalination plant.

Section 1   Ocean Sediments   **435**

DISCUSSION   Describe to the students some of the characteristics of brine. For example, brine may contain different salts; often several salts are combined. These salts include sodium bromide, sodium chloride, sodium iodide, and calcium chloride. ***Thinking Critically:*** Ask the students to name some of the uses of brine. (Brine is used in food processing and meat packing. Foods such as pickles and olives are packed in brine solutions. Brine is also used in refrigeration, water softening, and dyeing. Much of this brine is manufactured. Natural brines are a source of not only table salt, but also potash, calcium chloride, magnesium, bromine, and iodine. Some brine is pumped out of oil wells. Once separated from the oil, the brine obtained in this way is a particularly important source of iodine.)

EXTENSION   Much of the salt in salt domes is recovered by ordinary mining. However, some is obtained through solution mining. In this process, fresh water is pumped down ''wells,'' and brine is forced up. Like other brine, this must be purified and foreign particles removed before it can be used.

EXTENSION   For the location of two major salt mines, have the students look at the maps of Louisiana and Michigan in the Science Sites booklet.

Answers—20.2

○ Salt is obtained by evaporating salt water or mining salt deposits.

○ Water is the other product obtained from evaporated sea water.

○ As the world population increases, water resources available for agriculture are diminishing.

20.3 Deep-Sea Sediments

DISCUSSION   Direct the students' attention to Table 20–1. Describe a manganese nodule to the students, including its composition and general shape. Tell the students that more than half a manganese nodule is made of clay, carbonate, and quartz. None of these materials are very valuable. ***Thinking Critically:*** Ask the students why it is profitable to mine the nodules. (Iron, manganese, nickel, cobalt, and phosphorus are valuable.)

EXTENSION   (Tie-in/History)   During the Phanerozoic Eon that lasted approximately 570 million years, sediments were deposited at a rate of about 240 m per million years. The maximum recorded thickness of the complete deposit is about 138 km.

20.3 Deep-Sea Sediments

Most ocean resources are obtained from shallow water since the difficulty of recovering deep-sea resources adds to their cost. The only major mineral resources that can be gathered from the deep ocean floor are manganese nodules, which are lumps such as the ones shown in Figure 20–6.

Figure 20–6. Iron, manganese, nickel, and other elements can be obtained from manganese nodules.

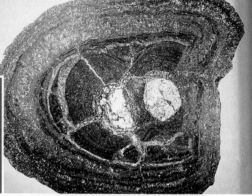

Manganese nodules, which average 5 cm to 10 cm in diameter, form slowly, adding only about 1 mm every million years. In these nodules, layers of manganese and iron oxides are mixed with sedimentary particles. Scientists estimate that in addition to manganese and iron, the manganese nodules in the Pacific contain 15 billion tons of nickel and 5 billion tons of cobalt. Plans are being made to mine these resources; however, the manganese nodules are in international waters, and mining can be done only by international agreement. The average composition of manganese nodules is shown in Table 20–1.

TABLE 20–1: AVERAGE COMPOSITION OF MANGANESE NODULES	
Material	**Percentage**
Iron	16.0
Manganese	14.0
Nickel	4.0
Cobalt	3.0
Phosphorus	1.5
Clay, carbonate, quartz	61.5

Answers–20.3 (p.437)

○ Iron, manganese, nickel, cobalt, phosphorous, clay, carbonates, and quartz can be mined from manganese nodules.

○ They contain information about the geologic and meteorologic history of the earth.

Figure 20–7. These scientists (left) are taking a core sample of the sediments on the ocean floor. The core sample is then taken to a laboratory for study (right).

In addition to being a source of valuable resources, deep-sea sediments contain a precious record of the geological and climatic changes that have occurred on Earth. Scientists collect many samples from the deep ocean floor and study them in great detail. Information gathered about shifts in populations of marine organisms and changes in seashell chemistry make it possible to reconstruct the climatic history of the earth. For example, scientists have discovered that 60 million years ago the temperature in the deep sea was 20°C. Since then the temperature has dropped to about 0°C.

○ *What minerals can be mined from manganese nodules?*
○ *What information do deep-sea sediments contain?*

A MATTER OF FACT

The oceans supply humans with over 65 million metric tons of fish, plants, and shellfish every year.

Section Review

READING CRITICALLY

1. How did iron, coal, and petroleum deposits form?
2. How are salt deposits used?
3. Why are deep-sea sediments important?

THINKING CRITICALLY

4. Much of the fresh water on Earth comes from the oceans without humans having to use desalination. Explain.
5. How might information about past climates be useful in predicting future climatic changes?

Section 1   Ocean Sediments   **437**

DISCUSSION Tell the students that scientists require many samples collected from a variety of sites in each ocean. By comparing and contrasting different segments of a large number of samples, scientists are able to draw valid conclusions. *Thinking Critically:* Ask the students to discuss the relationship between erosion of land and the formation of ocean sediments. (Soil and other substances that have been washed away from the land pass through the rivers and into the oceans, where they form sediments.)

EXTENSION (Tie-in/Social Studies) The students should realize that the history of the earth covers a far greater expanse of time than the history of humans. For example, 325 million years ago much of interior North America was covered by seas. In human history, dramatic changes can take place in only a few decades.

SECTION REVIEW

Summary   Have the students describe to each other how iron, coal, and oil formed. Ask them to discuss the significance of brine, salt domes, and desalination. Then have the students list five important minerals found in manganese nodules and discuss how they are being made available for humans.

Reinforcement   Have the students make a list of the main ideas in each paragraph for each numbered section. Then have them compare their ideas with those of other students.

Answers to Section Review

Reading Critically

1. Coal and oil deposits were formed from the remains of organic debris. Iron deposits were formed by the sedimentation of inorganic iron oxides.
2. Salt is a major mineral resource, necessary to the diet of many animals including humans. Before refrigeration was available to keep foods fresh, salt was used to preserve food.
3. Deep-sea sediments contain valuable minerals used in industry. Sediments are also used by scientists to learn more about the earth's geologic past.

Thinking Critically

4. Ocean water is desalinated naturally during the water cycle.
5. The intervals between drastic changes in climate indicated by the fossil record may help to predict when such climate changes may occur again.

Discussion After the students have read the article, hold a class discussion on the different ways to explore the oceans. Discuss the advantages of each method. (For example, tethered vehicles can go deeper than a scuba diver, but are less flexible. If unpiloted, they lack the benefit of having a human intelligence to make on-the-spot decisions.) Also discuss some of the drawbacks of underwater research, such as the expense incurred and the possibility that such research might lead to a misuse of the oceans.

Extension Some students may be interested in doing additional research on this topic. These students might look into the history and current technology available for any of the submersibles mentioned.

TECHNOLOGY: Submersibles

The ocean depths not only present many hazards but also provide many treasures. In order to claim the resources in the ocean, humans must explore these deep regions. This exploration has been made possible by the invention of the submersible. Submersibles are small vessels capable of diving to great depths. There are several types of submersibles, including tethered vehicles, roving submersibles, observation bells, and diving gear.

Tethered vehicles usually carry a single passenger, or they may be unpiloted and operated by remote control. Tethered vehicles receive their power and directions from surface craft. They are mostly used for photographing, manipulating, and repairing undersea structures or installations. The first tethered, remotely operated vehicles were used to bury underwater communication and power cables in the 1950s.

Roving vehicles make it possible for an explorer, or group of explorers, to venture far from the surface. They are self-contained vessels that provide internal air pressure equal to that at sea level, despite the enormous pressure being exerted on the craft by the water. These vehicles are battery powered and built to make long, deep journeys to the sea floor. This type of submersible has been used to explore the deepest trenches of the oceans. The Mariana Trench, however, was explored by a tethered submersible. A submersible can also be used to locate sunken vessels, as was done by oceanographer Robert Ballard in finding the lost wreck of the luxury liner *Titanic*. Sometimes, these vehicles can also be used to retrieve sunken treasure.

Another type of submersible is the observation bell. Although some observation bells are capable of independent movement, most are designed to stay at a particular station. There, the bell's mechanical arms and hands are used for the delicate manipulation of underwater objects. The passengers in an

A personal submersible vessel

observation bell are able to view a wide panorama of the sea-floor landscape through wide, pressure-resistant windows.

A different type of submersible consists of a wide range of diving gear. Simple scuba gear allows divers to explore shallow waters. The term *scuba* is actually an acronym that stands for *Self-Contained Underwater Breathing Apparatus*.

More elaborate diving suits, called *atmospheric diving suits* (ADS), provide a diver with a pressurized environment for deep-water exploration. An ADS may contain a power unit, or it may be tethered to a support vessel. The diver maneuvers by walking or by using thrusters attached to the suit. These pressurized suits are operational to a maximum depth of about 600 m.

Without submersibles, many of the mysteries and treasures of the deep would forever remain unknown.

A scuba diver and submersible (inset)

438   Chapter 20   Ocean Resources

438

SECTION

2 Ocean Life

SECTION OBJECTIVES

After completing this section, you should be able to:

■ **Identify** the categories of sea life and **give** some **examples** of each.

■ **Discuss** some uses of algae.

■ **List** the major sources of ocean pollution and **describe** the effects of ocean pollution on human and sea life.

NEW SCIENCE TERMS

plankton
nekton
benthos

Section Overview This section describes the various living organisms in the ocean. These organisms can be grouped into three categories: plankton, nekton, and benthos. Also discussed are the sources of ocean pollution—in particular, sewage and garbage, the dumping of heated water, the use of the oceans to store radioactive wastes, and oil spills.

Section Objectives For a list of section objectives, see pupil's edition page.

New Science Terms For a list of new science terms in this section, see pupil's edition page. You may wish to familiarize the students with these terms before they read the section. You may also wish to read the terms aloud so the students are familiar with the pronunciation of each.

20.4 Living Resources

If you have ever gone snorkeling or diving, or if you have visited a marine-life park, you know that the oceans and seas of the world are full of living organisms. These organisms can be grouped into three categories: *plankton, nekton,* and *benthos.*

Plankton Microscopic plantlike and animal-like organisms that drift with the tides and currents are **plankton.** The tiny, plantlike organisms are called *phytoplankton;* the animal-like organisms are called *zooplankton.*

Phytoplankton are the beginning of most food chains in the ocean. Phytoplankton also produce much of the earth's free oxygen. Zooplankton feed on phytoplankton. Small fish live on plankton and in turn are food for larger fish. A typical ocean food chain is shown as a pyramid of mass in Figure 20–8.

Phytoplankton can only grow where sunlight is strong enough to support photosynthesis. This area, known as the *euphotic zone,* reaches a depth of about 100 m. If the water is not clear, this zone may be even shallower.

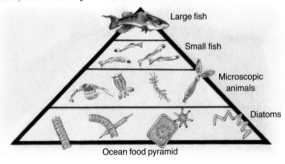

Large fish

Small fish

Microscopic animals

Diatoms

Ocean food pyramid

Figure 20–8. These diatoms (above) are at the bottom of most of the food chains in the ocean (left).

Section 2 Ocean Life **439**

SECTION DEVELOPMENT

20.4 Living Resources

DISCUSSION Direct the students' attention to Figure 20–8. Explain the function of the ocean food chain. *Thinking Critically:* Ask the students whether ocean pollution might restrict the area in which phytoplankton can grow. Ask them to explain their answer. (Yes, it may do this if the depth to which sunlight can reach is limited.)

Skill Activity:
Organizing Information

Objectives

- Compile a list of products that contain ocean resources.
- Draw conclusions from information gathered.

Preparation of Materials When telling the students how to complete the application portion of this activity, you might ask different groups of students to focus on different kinds of products. For example, one group could focus on medicines, another on ethnic foods, another on foods containing meat or fish, and another on cosmetics.

Discussion Provide the students with some background information on algae. You might mention that algae are members of a primitive group of eukaryotic organisms. They are predominantly aquatic and have no leaves, vascular system, or flowers. Seaweeds and diatoms are members of this group of organisms. The word *algae* comes from the Latin word *alga*, meaning seaweed. It refers to any green, brown, or red algae. Members of these groups have a wide diversity of form, and many (like the giant kelps) form colonies of cells that reach lengths of 15 meters or more. The extracts, alginic acid, agar, and carrageen, are used widely in foods and cosmetics. ***Thinking Critically:*** Ask the students why it would be useful if other functional extracts could be derived from seaweed. (Seaweed is plentiful, so useful chemicals and other substances might be obtained without great cost.)

440

SKILL ACTIVITY: Organizing Information

BACKGROUND

Many food products are available that contain seaweed as an ingredient. One gelatinous mixture, called *alginic acid* or *algin,* is extracted from brown kelp and used in the making of jellies and medicines. *Agar* is another extraction of kelp used in making jellies and for growing bacterial cultures for medical and other scientific purposes. A red alga, *carrageen,* is also used in foods, as well as in hand and face lotions.

PROCEDURE

In this activity you will compile a list of products, such as those shown, that contain these ocean resources: algin (or alginic acid), agar (or agar-agar), and carrageen. Make a copy of the table and fill it in as you check each product.

TABLE 1: ALGAE PRODUCTS

Item	Algin	Agar	Carrageen

INGREDIENTS: SOYBEAN OIL; WATER; HIGH FRUCTOSE CORN SYRUP; VINEGAR; TOMATO PASTE; SALT; GARLIC*; ONION*; SPICE; XANTHAN GUM AND ALGIN DERIVATIVE FOR CONSISTENCY; LEMON JUICE*; OLEORESIN PAPRIKA FOR COLOR; SUGAR; CALCIUM DISODIUM EDTA TO PRESERVE FRESHNESS.
*DEHYDRATED

APPLICATION

Visit your local supermarket and see which popular products contain these ingredients. Use a table like the one shown to record your observations.

USING WHAT YOU HAVE LEARNED

What do all of the products containing algae have in common? What role do you think the algae have in producing this characteristic?

INGREDIENTS: PARTIALLY SKIMMED MILK, MALTODEXTRIN, SUGAR, NONFAT DRY MILK, SODIUM CASEINATE, CORN OIL, ARTIFICIAL FLAVOR, MONO AND DIGLYCERIDES, ASCORBIC ACID, MAGNESIUM OXIDE, CARRAGEENAN, SOY LECITHIN, ALPHA TOCOPHERYL ACETATE, FERRIC ORTHOPHOSPHATE, ZINC SULFATE, NIACINAMIDE, D-CALCIUM PANTOTHENATE, VITAMIN A PALMITATE, COPPER GLUCONATE, MANGANESE SULFATE, ARTIFICIAL COLOR, VITAMIN B_{12}, VITAMIN D, PYRIDOXINE HYDROCHLORIDE, RIBOFLAVIN, THIAMINE HYDROCHLORIDE, FOLIC ACID, D-BIOTIN.

Nekton   While plankton can only drift, **nekton** are sea animals that swim about freely. These include all the fish, such as cod, tuna, and sharks, as well as whales, seals, octopuses, and others.

Fish and shellfish have long been food for humans, especially for people living near the coasts. Today, it is not necessary to live near the ocean to enjoy seafood. Fish can be shipped fresh, dried, canned, or frozen. Seafood is an important source of protein, unsaturated fats, and many other nutrients.

Fish can live almost anywhere there is water, from the Arctic to the tropics. In the North Atlantic, fish such as cod, herring, and haddock are harvested. Grouper and pompano are taken off the coasts of the southeastern and Gulf states, while tuna and salmon are caught in the eastern Pacific.

A MATTER OF FACT

Three of the most desirable ocean resources are crabs, shrimps, and lobsters. These creatures are scavengers of the deep. They feed on the waste and decaying materials of other organisms.

As the population of the world continues to grow, fish may become more and more important as a protein source for many people. With such modern technology as sonar, radar, and powerful winches for pulling in large nets, more fish can be caught. The increase in technology raises the danger of overfishing in some areas. Restrictions have been placed on the size of the catches of some species so that the populations can rebuild. Rare species must be protected so they do not become extinct.

Figure 20–9. Redfish were almost eliminated from the Gulf of Mexico because of the popularity of these fish in cooking (above). Now, more plentiful species of schooling fish are used (left).

Figure 20–10. Oysters (left) and shrimp (right) are shellfish harvested from the Atlantic Ocean.

Section 2   Ocean Life   **441**

DISCUSSION   Restricting the size of catches of some species of fish permits the fish population to rebuild. ***Thinking Critically:*** Ask the students how the dependence of people on products from the ocean, even if they do not live near the ocean, may be of benefit to the future of the ocean. (It is in the interest of everyone on Earth to preserve the ocean's resources and to ensure that they are not spoiled. As more people realize their dependence on the oceans, governments should be encouraged to stop practices that harm ocean life and to foster more scientific investigation.)

Thinking Skill   *(Drawing conclusions)* Ask the students in which part of the oceans they think most plankton live. (They are more commonly in offshore waters than in the open ocean. Waters near the shore contain more materials that the plankton can use as food.) Then ask whether the students think that plankton live only in the ocean or if they suspect there are freshwater plankton as well. (There are freshwater plankton in ponds, lakes, and rivers.)

EXTENSION All living organisms are related by a complex food chain. The three categories of sea life are not only dependent on the other ocean forms, but are also influenced by living organisms that inhabit the land. Both sea and land organisms are part of the same worldwide ecosystem.

EXTENSION Half a million zooplankton and more than half a billion phytoplankton may live in just 1 L of water.

20.5 Ocean Pollution

DISCUSSION Explain to the students that only recently has ocean pollution become an acute problem. Humans have been dumping things into the ocean for hundreds of years. Most of this material was either organic material that easily decomposed, or some other material that could be diluted by natural processes. *Thinking Critically:* Ask the students to define the phrase ''natural chemical processes in the ocean.'' Ask them to name other kinds of chemical processes. (The students may mention the natural breakdown of organic material by bacteria or chemical weathering of objects by the salt water. Other kinds of chemical processes require scientific equipment.)

A MATTER OF FACT

In clear ocean water, red and violet light are absorbed before reaching a depth of about 10 m. Blue light, however, reaches depths of over 100 m.

Figure 20–11. Commercial harvests from the Pacific Ocean include kelp (right) and snow crabs (left).

Benthos Organisms that live in or on the bottom of the ocean are called **benthos.** These include the invertebrates, such as worms, sponges, corals, and starfish. Crabs, lobsters, clams, and oysters are important sources of food.

Seaweeds, the red and brown algae anchored to the ocean floor, are an important resource in its own right. A brown alga called *kelp* is harvested and used to make fertilizer. Substances extracted from algae, such as algin, agar, and carrageen, are used as stabilizers in foods like ice cream, cheese spreads, and salad dressings. Agar is also used as the base on which bacteria are grown in laboratories. Seaweed is popular as a food in many countries.

○ *What are the three categories of sea life?*
○ *Which type of sea life is the most important to humans? Explain.*

20.5 Ocean Pollution

Humans have been dumping waste into lakes, rivers, and oceans for hundreds of years. In addition, natural pollutants, such as dead animals and fish and decaying plants, are also being added to the waters of the world. Natural chemical processes in the ocean can decompose small amounts of some types of pollution. However, large amounts of waste upset natural processes, killing ocean life.

The chief sources of ocean pollution are sewage and garbage, industrial waste, and agricultural chemicals and wastes. Sewage is untreated human waste and household water. Most large cities have sewage treatment plants, but this treatment does not remove some types of chemicals. When treated sewage

442   Chapter 20   Ocean Resources

Answers—20.4

○ The three categories of sea life are plankton, nekton, and benthos.

○ All sea life is important to humans because all living things are part of the same food web.

is dumped into a body of water, these chemicals go with it. The body of water may be a river or a bay or a lake, but the water eventually reaches the ocean. Many coastal areas have become unsuitable for recreation and fishing because of pollution.

Industrial chemicals kill ocean life or make seafood unfit to be eaten. Many harmful chemicals have been found in the water, in fish, and in shellfish. These chemicals become more concentrated as they move up the food chain. The effects of long-term exposure to these chemicals are not yet known.

Garbage has been polluting the oceans for many years. Most food wastes decompose, but rusting auto parts and soft drink cans litter the ocean floor. Many fish and sea mammals have died from swallowing such things as plastic bags, cans, and the plastic carriers from drink cans.

In the summer of 1987, hospital trash, including used hypodermic syringes, washed up on the beaches of New Jersey. At the same time, dead dolphins were washing up on the shore, and swimmers were complaining of skin rashes and respiratory infections. Could all these things be connected?

Many industries use water for cooling and then dump the heated water into the ocean. This causes thermal pollution. The heated water causes some fish and plants to die and some plankton to multiply rapidly, upsetting the ocean's natural balance.

Radioactive wastes from nuclear industries, such as nuclear power plants, uranium mining, and weapons manufacture, are very dangerous to ocean life. The nuclear wastes must be safely stored until they lose their radioactivity. For some wastes,

Figure 20–12. Garbage from some cities is dumped directly into the ocean (above). Some of it returns as litter on public beaches (left).

DISCOVER

Producing a Newsletter

Smog, acid rain, oil spills, contaminated wells, closed beaches, and contaminated shellfish beds are all examples of the effects of pollution. Look through several newspapers and magazines and clip articles and pictures on pollution. Also clip articles on attempts to clean up and control pollution. Then fold a large sheet of paper in half and paste your articles onto it to make a newsletter or small newspaper. Give it a title and share your newsletter with your classmates, or make a class display.

Section 2   Ocean Life   **443**

DISCUSSION   Have the students name the different sources and types of industrial pollution. *Thinking Critically:* Ask the students to explain how population growth affects pollution. (In addition to the increase in waste, there is increased demand for industrial products, fossil fuels, and goods carried on the oceans.)

EXTENSION (Tie-in/Social Studies) Because the oceans are owned by no single country, the prevention of ocean pollution as well as the clean-up of the pollution that does occur, is an international problem. This means, however, that there must be some cooperation among nations if the problem of ocean pollution is to be solved. In areas where several countries border a body of water (such as the Gulf of Mexico or the Mediterranean Sea), all the countries will have to work together. For areas such as the middle of the Atlantic and Pacific oceans, a more general answer must be found. Such areas are used by ships from many nations. Thus, international agreements involving many nations are needed if significant improvement is to be made.

DISCOVER:
Producing a Newsletter

Skill *(Communicating)*

If the students cannot obtain all the articles they need, they can substitute photocopies of articles from magazines in the school or public libraries. The students might discuss how different types of magazines cover articles on pollution. For example, a news magazine might have an article on a recent oil spill or a news analysis of attempts to deal with the problems of pollution through legislation. A science and technology magazine, on the other hand, would be more likely to contain an article on how factories can be designed to generate less pollution, or an article on certain chemical processes involved in breaking down pollutants.

DISCUSSION Draw the students' attention to the problems caused by oil tankers and ask them to name some of the ways that these problems can be handled. *Thinking Critically:* Ask the students to name some measures not directly related to the regulation of oil tankers that government can take to reduce oil pollution in the ocean. (Governments could encourage alternative energy sources and reduce oil imports.)

Thinking Skill *(Applying a Concept)* Ask the students to list 20 items in the classroom. Then have them indicate which items would decompose readily in the ocean and which would not. (Examples of materials that would readily decompose include wood and animal and plant products. Examples of materials that will not decompose easily include plastics and some manufactured fabrics. Many metals will decompose, but not readily.)

this might be 500 years; for others, 50 000 years. Experts are unsure about the safest way to store waste tanks for such long periods. The bottom of the ocean is one place where radioactive wastes are being stored; there, the tanks are covered with concrete. Would the tanks survive an earthquake without polluting the ocean?1

The oil industry has been responsible for much polluting of the oceans. Each year, huge tankers carrying hundreds of thousands of tons of crude oil sail the oceans. Spills from tanker accidents have produced massive oil slicks, which have coated beaches with oil and killed many sea birds.

In recent years the industry has changed loading and unloading procedures to eliminate the discharging of oil into the sea. Greater safety measures make accidents less likely, and new

Figure 20–13. These drums of radioactive wastes were recovered from a capsized ship. Other containers of hazardous wastes are purposely stored on the ocean floor.

Figure 20–14. Oil spills present one of the greatest problems for ocean cleanup (right). Sometimes birds and other animals are killed by the toxic oil (below).

1 The tanks might break open, especially if they are rusted.

Answers—20.5 (p.445)

○ Although large cities have sewage treatment plants that remove much harmful material from sewage, it cannot be entirely purified. So, dangerous chemicals continue to run off into the oceans.

○ Industrial chemicals can make drinking water toxic and can also kill living organisms that inhabit and thrive in the sea.

clean-up techniques make spills less disastrous, but pollution from oil is still a danger.

As the human population grows, the problem of pollution grows. And as Earth becomes more crowded, humans may look to the oceans for more resources. Pollution must be controlled to protect these resources and our future.

○ *How are the oceans being polluted by sewage?*
○ *What are the hazards of industrial waste?*

Section Review

Figure 20–15. San Diego Harbor is one of the world's busiest harbors, yet the water is fairly clean.

READING CRITICALLY

1. How has modern technology changed the fishing industry?
2. Give an example of an ocean food chain.
3. How is ocean pollution dangerous to humans?

THINKING CRITICALLY

4. Why could seafood not provide the world population with much food 100 years ago?
5. What could governments do to protect the oceans from pollution?

ACTIVITY: Recycling Water

How can you demonstrate the way nature returns fresh water to the earth through the water cycle?

MATERIALS (per group of 3 or 4)

safety goggles, glass tubing, rubber tubing, plastic bag, string, ring stand and clamp, Bunsen burner, crushed ice

PROCEDURE

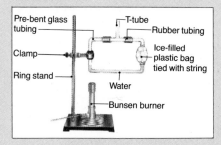

Pre-bent glass tubing — T-tube — Rubber tubing — Ice-filled plastic bag tied with string — Clamp — Ring stand — Water — Bunsen burner

CAUTION: Safety goggles must be worn at all times during this activity.

1. Set up the apparatus as shown, using the preshaped glass tubing, and then fill the bottom of the tube with water.
2. Make sure the plastic bag of ice is secured to the glass tubing with string.
3. Gently heat the water-filled part of the tube with a low Bunsen burner flame, and observe the results.

CONCLUSIONS/APPLICATIONS

1. Describe what you observed. How does this activity demonstrate the phase changes of water—solid to liquid to gas?
2. Explain how these phase changes take place in nature.
3. In what ways do machines take advantage of the phase changes of water or other fluids?

Section 2   Ocean Life   **445**

SECTION REVIEW

Summary   Have the students summarize in their own words the major concepts in this section. Various forms of plankton, nekton, and benthos live in the ocean. Pollution of the ocean is caused by sewage, industrial and agricultural chemicals, and radioactive wastes.

Reinforcement   Have the students quiz each other, using the questions at the end of each numbered subsection

ACTIVITY: Recycling Water

Thinking Skill *(Interpreting Data)*

Discussion   Mixtures of water, minerals, organic debris, and many undesirable impurities percolate through soil at varying rates. The porosity and quality of the soil or other filtering materials determine the rate at which the water settles and how well the variety of solutes is filtered from the solution.

Answers to Conclusions/Applications

1. Water boils and evaporates above the Bunsen burner and rises through the glass tube. As the vapor circulates through the tube, it condenses back into a liquid at the ice-filled bag and returns as "rain" to the water supply. The water has changed phase from liquid to vapor and vice versa.
2. Solar radiation warms the surface of the waters in oceans, lakes, and streams and causes some of the water to evaporate and rise into the atmosphere. The vapor rises and cools, condensing into clouds. If water droplets get too heavy, they fall as rain or snow and are eventually drained off the land into the ocean.
3. The pressure of water vapor created in a boiler is used to drive the flywheel of a steam engine. The water vapor is then condensed and used again. A low-boiling fluid, called freon, is used to absorb heat energy from foods placed in a refrigerator. The freon is vaporized as it absorbs the heat and then condensed back into a liquid by a compressor. Therefore the cold, liquid freon is recycled.

Answers to Section Review

Reading Critically

1. Instruments helpful in locating and catching fish have increased the number of fish being caught.
2. Answers will vary. For example, plankton are eaten by small fish, which are eaten by larger fish that die and sink to the bottom to be eaten by crabs.
3. Many pollutants are absorbed by fish that are eaten by humans.

Thinking Critically

4. At the time, fisheries did not have the technology to locate, harvest, and preserve much of the seafood available today. Seafood spoiled quickly and could not be delivered fresh to inland areas.
5. The government could continue to monitor industries that need to dispose of some of their wastes into rivers, streams, and seas. Government could limit the amount of waste material being disposed of in this manner and find new ways to recycle it.

Skill *(Filtering Water)*

Discussion   Nature recycles water in the environment by continuous evaporation, condensation, precipitation, and runoff. This activity is a simple analogy to the water cycle. Point out, however, that whereas water is distilled and cleansed of many natural products by this process, many modern chemical pollutants are not extracted from the cycle and continue to present a danger to consumers (both plant and animal) as they are transported through the environment.

Answers to Analyses and Conclusions

1. The gravel filtered out some of the soil, but by no means all of it. The water is still cloudy.
2. Yes; more mud was removed from the water.

Answers to Application

1. Organic debris accumulates in fish tanks and must be removed or the fish will die.
2. Large charcoal filters could be used to remove organic material.

INVESTIGATION 20: Water Filtration

PURPOSE

To show how water can be purified by filtration

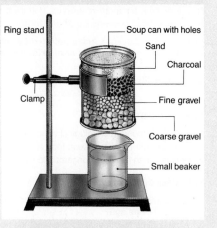

MATERIALS (per group of 3 or 4)

Hammer
Finishing nail
Empty soup can
Ring stand and clamp
Small beaker (150 mL)
Large beaker (500 mL)
Coarse gravel
Fine gravel
Muddy water
Crushed charcoal
Sand

PROCEDURE

1. Using the hammer and finishing nail, drive about one dozen small holes into the bottom of a clean soup can.
2. Secure the can above the small beaker as shown.
3. Pour about 2 cm of coarse gravel into the bottom of the can. Then pour a 2-cm layer of fine gravel on top of the coarse gravel.
4. Pour about 100 mL of muddy water into the top of the can and collect the filtered water in the small beaker below the apparatus.
5. Describe the characteristics of the filtered water as clear, cloudy, murky, or muddy. Compare your results with the results of other groups.
6. Pour out the water in the small beaker and clean the can.
7. Repeat step 3. Then pour about 2 cm of crushed charcoal on top of the fine gravel.

Pour about 2 cm of fine sand on top of the charcoal.
8. Repeat steps 4 and 5.

ANALYSES AND CONCLUSIONS

1. Was the gravel effective in filtering the muddy water? To what extent?
2. Was the filtration process improved by the presence of charcoal and sand? To what extent?

APPLICATION

1. Why do you think fish aquariums have filters with charcoal in them?
2. How could this system be used on a large scale to clean a city's water supply?

SUMMARY

- The oceans help provide many mineral resources, such as iron, coal, and petroleum. (20.1)

- Coal was formed as the sea invaded swamps and buried organic matter under heavy sediments. (20.1)

- Ocean sediments may form deposits of petroleum. (20.1)

- Fresh water and salt may be obtained from sea water by evaporation. Salt may be mined from large deposits known as salt domes. (20.2)

- Except for manganese nodules, the mining of deep-sea sediments is not cost-effective. (20.3)

- Manganese nodules may provide an economical source of iron, manganese, nickel, and cobalt. (20.3)

- Deep-sea sediments provide clues to past geologic and climatic changes. (20.3)

- Plankton are microscopic organisms that are the basis for ocean food chains. (20.4)

- Phytoplankton are plantlike organisms that photosynthesize, while zooplankton are animal-like organisms that feed on phytoplankton. (20.4)

- Nekton are free-swimming animals. They provide much of the seafood consumed by humans. (20.4)

- Benthos, or bottom-dwelling organisms, provide many products for human use. Seaweed and many shellfish are benthos. (20.4)

- Organic waste, such as sewage, may be harmful to ocean life. (20.5)

- Industrial chemicals and nuclear wastes are dangerous to ocean life and humans. (20.5)

- Oil spills have killed many sea birds and fouled many beaches. (20.5)

Write all answers on a separate sheet of paper.

SCIENCE TERMS

Correctly use each of the following terms in a sentence.

benthos (442)
brine (435)
nekton (441)
plankton (439)
salt domes (435)

SCIENCE QUIZ

Modified True-False

Mark each statement *true* or *false*. If a statement is false, change the underlined word to make the statement true.

1. <u>Salt domes</u> are important for mining commercial salt.

2. Organic matter does not decay in an <u>aerobic</u> environment.

3. <u>Manganese</u> is a major mineral resource mined from the floor of the oceans.

4. Photosynthesis takes place in the <u>euphotic</u> zone.

5. Ice cream may include stabilizers extracted from <u>seaweed</u>.

Multiple Choice

Write the letter of the choice that best answers the question or completes the statement.

6. Sharks are part of the
   a) benthos.
   b) nekton.
   c) plankton.
   d) euphotic.

7. The chief sources of ocean pollution are
   a) sewage and garbage.
   b) industrial wastes.
   c) agricultural chemicals.
   d) All of the choices are correct.

continues

Chapter 20 Review   **447**

CHAPTER REVIEW

SUMMARY

The students may review the major concepts in the chapter by reading the summary statements. The statements are cross-referenced to the chapter to facilitate reinforcement of any concepts of which the students feel unsure. Encourage the students to work in groups to quiz one another.

SCIENCE TERMS

The sentence in which the science term is used should reflect an understanding of the definition of each term. You may wish to stage a game of "Jeopardy" during which the definition is provided and the "contestants" must provide the correct term.

SCIENCE QUIZ

Answers to Modified True-False

1. true
2. false, anaerobic
3. true
4. true
5. true

Answers to Multiple Choice

6. b
7. d

continues

Answers to Multiple Choice (continued)

8. c
9. d
10. b

Answers to Completion

11. manganese nodules
12. 3 billion
13. 50 000
14. reverse osmosis
15. phytoplankton
16. coal
17. brine
18. manganese nodules
19. nekton
20. oil

Answers to Short Answers

21. Deep-sea sediments contain valuable industrial minerals such as iron and manganese. In addition, core samples containing organic and inorganic layers help geologists determine the history of climate and ocean geography.
22. Iron, manganese, nickel, cobalt, phosphorous, clay, carbonate, and quartz are obtained from manganese nodules. They are important resources, and they are records of geological changes on Earth.
23. Nektonic organisms, such as fish, swim about freely in the oceans. Benthic organisms, like crabs, are bottom dwellers. Plankton are microscopic organisms that drift with the tides and currents.

Answers to Writing Critically

24. When marine organisms die, they settle to the bottom. In that anaerobic environment the organisms accumulate rather than decay. They are buried by other sediments, put under enormous pressure, and over millions of years are chemically transformed to petroleum.
25. Answers will vary but should include the fact that they may cause trouble in later years.

SCIENCE QUIZ continued

8. The evidence of past ages is found in
   a) records of fossil fuel consumption.
   b) ocean salinity.
   c) deep-sea sediments.
   d) salt domes.

9. Ocean sediments may be mined commercially for
   a) coal.
   b) salt.
   c) petroleum.
   d) All of the choices are correct.

10. Manganese nodules also contain
    a) gold.
    b) clay, carbonate, and quartz.
    c) salt.
    d) water.

Completion

Complete each statement by supplying the correct term or phrase.

11. Commercial amounts of iron and nickel are found in _____.
12. The iron oxide deposits of the world were formed about _____ years ago.
13. Radioactive wastes may continue to pollute the ocean for as long as _____ years.
14. The most direct way to get fresh water from sea water is by _____.
15. In a euphotic food chain, zooplankton eat _____.
16. When the sea repeatedly invaded low, swampy areas, _____ formed.
17. The extremely salty water that forms in isolated parts of the ocean is _____.
18. The only minerals, besides salt, that are mined from the oceans are found in _____.
19. Sea animals that swim about freely are _____.
20. The _____ industry is responsible for much polluting of the oceans.

Short Answer

21. Explain why marine and deep-sea sediments are useful to humans.
22. List the minerals obtained from manganese nodules. Why are these minerals important?
23. Compare and contrast plankton, nekton, and benthos.

Writing Critically

24. Explain why petroleum deposits are found on land as well as under the ocean.
25. Why are containers of radioactive waste stored on the ocean floor sometimes called "time bombs" for future generations?

EXTENSION

1. Find out how an electronic fish finder works. Then explain how it can be used to locate schools of fish.
2. Research the methods of catching tuna. Explain why porpoises are sometimes caught with tuna.
3. Locate the nearest body of water to your school and find out if it is polluted. If it is, try to locate the source and the type of pollution. Suggest ways in which this source could be eliminated or treated.
4. Read and report to the class on the productivity of one of the major estuaries along the North American coast.

APPLICATIONS/CRITICAL THINKING

1. Why do you think that deep, clear areas of ocean waters are sometimes referred to as deserts?
2. Explain this statement: The earth could probably support 20 to 30 times its present population if the ocean could be farmed as extensively as the land.

ANSWERS TO EXTENSION

1. Fish finders work on a type of sonar, bouncing sound waves off schools of fish.
2. The large nets used to catch tuna also entrap porpoises feeding on the tuna.
3. Answers will vary depending on the type of pollution found.
4. Reports will vary.

ANSWERS TO APPLICATION/ CRITICAL THINKING

1. Deep, clear areas have few nutrients to support the plankton at the bottom of the food chain.
2. There is a tremendous amount of food energy in the sea; however, ways must be found to exploit it.

continues

3. If you were diving with a scuba tank containing 30 m³ of air, and you were using 2.5 m³ of air per minute at 30 m depth and 1.5 m³ at 15 m depth, how long could you dive at 20 m or 25 m?

FOR FURTHER READING

Ballard, R. "How We Found the Titanic." *National Geographic* 168(December 1985): 696. This article gives an exciting minute-by-minute account of the discovery of the lost luxury liner *Titanic*.

Golden, F. "Scientist of the Year." *Discover* 8(January 1987): 50. This article is a biography of Robert Duane Ballard, the person who discovered the lost wreckage of the luxury liner *Titanic*.

Nelson, C. and K. R. Johnson. "Whales and Walruses as Tillers of the Sea Floor." *Scientific American* (February 1987) 256:112. This article describes side-scan sonar studies of the Bering Sea that resulted in some surprising discoveries.

Sibbald, J. *Homes in the Sea: From the Shore to the Deep.* Minneapolis: Dillion, 1985. This text describes the different ocean habitats and how organisms adapt to these environments.

Whitehead, H. "The Unknown Giants." *National Geographic* 166(December 1984): 774. For centuries the great sperm and blue whales were hunted for their valuable oils. Today, they are in danger of becoming extinct. This article explains how human relationships with these gentle giants are proving to be a lesson in conservation of the ocean's varied resources.

Challenge Your Thinking

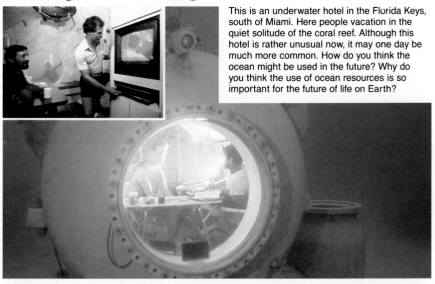

This is an underwater hotel in the Florida Keys, south of Miami. Here people vacation in the quiet solitude of the coral reef. Although this hotel is rather unusual now, it may one day be much more common. How do you think the ocean might be used in the future? Why do you think the use of ocean resources is so important for the future of life on Earth?

Chapter 20 Review   **449**

---

3. About 15 minutes at 20 m and about 10 minutes at 25 m.

The ocean floor could be used for research, for vacationing, and to supply resources not found on land.

Discussion This *Science Connection* is about tidal power. North America's first demonstration tidal-power plant at Annapolis Royal, Nova Scotia, was designed to test whether the plant's equipment will work using salt water. It has a rim generator with four stainless steel propeller blades that spin at 50 rpm to generate a steady current. Huge vanes, like a giant venetian blind, regulate the flow of water. It cost $55 million to construct; the proposed tidal power plant on the Bay of Fundy would cost approximately $23 billion to construct. Ask the students how the tidal plant uses the kinetic energy of the tides to produce electricity that can be used in homes and other buildings located hundreds of kilometers from the Bay of Fundy.

Discussion Some people are concerned because if the proposed Bay of Fundy plant caused high tides to be raised by about 15 cm, as many as 1680 hectares of land would be lost and another 4000 hectares of salt marshes would be flooded. They also point out that because the temperature of the surface water would be lower, there would be more fog and onshore winds. Salty ocean water might seep into drinking wells. Fish and clam populations might be injured. Ask the students why such a project is being considered when it might have such a harmful impact on the environment. Ask them if they think that the possible benefits would outweigh any possible harmful effects.

Extension After completing the Unit, and reading and discussing the *Science Connection*, have the students turn back to the Unit Opener on page 389 and answer the questions. When the class discussion has ended, you may wish to have interested students write a report on the tidal power plants operating in France and the Soviet Union.

UNIT 6 THE SCIENCE CONNECTION:

Power from the Tides

Soon a 15-story dam will be built across Canada's Bay of Fundy. Inside the dam there will be 100 turbines, turned by the flow of the world's highest tides.

This giant tidal power plant will take advantage of the supertides that flow in and out of the Bay of Fundy. You may recall that the gravity of the sun and the moon cause the surface of the ocean to rise and fall. A supertide occurs where the tidal flow enters a long, narrow, funnel-shaped inlet. The tidal wave sloshes up to the neck of the funnel and back. If the bay is the right shape, the tidewater arrives back at the entrance just in time to get a push from the next high tide. The wave is thus intensified, resulting in higher high tides and lower low tides. Inside the Bay of Fundy, tides rise and fall by as much as 15 meters.

This intense tidal flow makes the bay ideal for developing tidal energy. The Bay of Fundy project, across the Cobequid Bay inlet, is the largest attempt yet at harnessing the power of tides. Tidal plants are already generating electricity in France and the Soviet Union. On the Bay of Fundy, a small pilot project across the Annapolis River has been operating since 1983. However, none of these power plants matches the scale of the giant Cobequid project. When completed and operating at full capacity, the plant will generate 4800 MW (megawatts) of power. By comparison, the Indian Point nuclear plant in New York produces only 873 MW.

The engineering principles of tidal power have been known for many years. Early tidal mills trapped the inflowing tide behind a dam, then allowed the water to flow out over a waterwheel. Today's tidal plant has turbines that are turned by the tidal flow.

The Rance River tidal power plant

Tidal power turbine

Tidal power plant control room

The French tidal plant has reversible turbines, so it operates on both incoming and outgoing tides. The Cobequid turbines will turn only when the tide is flowing out. Tidal waters will flow through the turbine at a rate of 13 000 m³/s, about three times the flow rate of Canada's largest river, the St. Lawrence. The water's speed will be controlled by a set of moveable vanes, which will open and close like venetian blinds. The turbines will spin at 50 revolutions per minute.

Tidal power plants create clean electric power without consuming fossil fuels. There is no air pollution, as there is with fossil-fuel plants, and there are no safety and waste disposal problems, as there are with nuclear plants. However, tidal power does have problems. Tidal plants can work only when the tide is rising and falling. Those times may not be the same as the times when the demand for electricity is highest. Few people need to take a hot shower or bake a cake at 3 A.M. Also, tidal plants may be located hundreds of miles from the large cities that use the most electricity.

Scientists also warn that the Bay of Fundy plant may cause environmental damage. The giant dam might affect tides as far south as Cape Cod in Massachusetts. If the dam causes tides to change by only 15 cm, it could disturb plants and animals living in tidal waters. A 15 cm rise in the tide would mean that up to 2 m of coast would be covered by ocean water. That change could endanger beachfront property.

The Bay of Fundy tidal power plant is an attempt to harness one of Earth's greatest untapped resources. Similar projects could save oil, coal, and other non-renewable resources. Yet, it also raises questions about how to balance the need for electrical power with the need to protect the natural environment.

Low tide on the Bay of Fundy

UNIT OVERVIEW

This unit describes the motions of the earth and the moon, as well as the location, composition, and motions of the planets of the solar system. Also discussed are the different kinds of energy sources in the universe, optical and radio astronomy, and a brief history of space science.

Chapter 21: The Solar System, page 454

This chapter presents a general introduction to the study of the solar system. The earth's position relative to the sun, moon, and planets is described. Also discussed are the inner planets, the outer planets, asteroids, comets, and satellites of the solar system.

Chapter 22: The Universe, page 480

This chapter discusses various types of stars and groups of stars and galaxies in the universe. The sun's location and structure are explained. The light-year is defined and used to measure distances in the universe. One theory for the origin of the universe is discussed.

Chapter 23: Observing the Universe, page 502

This chapter describes the instruments astronomers use to observe the universe. The two types of optical telescopes—refracting and reflecting—are discussed. Also presented are the advantages of radiotelescopes over optical telescopes. Important discoveries made with both optical telescopes and radiotelescopes are described.

Chapter 24: Exploring Space, page 520

This chapter presents a history of space travel, from the rockets of the 1920s to the sophisticated space probes now used to explore the solar system. The principles of rocket propulsion are discussed. A survey of artificial satellites and their many uses is included in this chapter.

The Science Connection: Life on Mars, page 544

The Science Connection describes the landing of the *Viking* space probes on the surface of Mars. Also described are the experiments conducted by the landers to determine the possible existence of life on Mars.

ADVANCE PREPARATION

Chapter 22

For the Discover on page 483 and the Skill Activity on page 489, the students will need to look for constellations on a clear night. For more information, see pages 483 and 489.

For Investigation 22 on page 498, you will need a table of tangents. For more information, see page 498.

Chapter 23

For the Discover on page 507, the students will need to make night observations of the moon. For more information, see page 507.

For the Discover on page 514, the students will need to check for sources of radio waves in their homes. For more information, see page 514.

Chapter 24

For the Skill Activity on page 525, the students will need to observe the sky on a clear night. For more information, see page 525.

For the Activity on page 526 and Investigation 24 on page 540, you will need balloons. For more information, see pages 526 and 540.

BULLETIN BOARD SUGGESTIONS

Chapter 21

Display colored illustrations of the nine planets (and largest asteroids) with 3″ × 5″ index cards listing basic facts about each, such as diameter, distance from the sun, number of moons, and composition. Include a drawing of the Greek or Roman god for which the planet was named.

Chapter 22

Create displays illustrating the evolution of a star as shown in a Hertzsprung-Russell diagram, listing facts such as size, temperature, and fuel source about each type of star.

Chapter 23

Display drawings, tracings, and photographs of the "Tools of Astronomy" mentioned in this chapter. Have the students write on 3″ × 5″ index cards to explain the function of each type of tool.

Chapter 24

Display drawings, tracings, photographs, and models of rockets and satellites developed and deployed since the beginning of the space age. On 3″ × 5″ index cards, include a brief history of each space vehicle.

Building giant supercolliders, in which atoms and particles are smashed into each other at tremendous speeds, may help scientists deduce the origins of the cosmos. However, these huge amounts of money might be better spent on more direct forms of space exploration. Have the students research the scientific importance of supercolliders, then debate the pros and cons of spending the money to build them.

The United States, and probably other countries as well, is researching the feasibility of putting defensive weapons into space. Have the students debate the issue of keeping space free of all weapons, even defensive ones.

SUGGESTED PROJECTS

Over a period of several weeks, the students could determine the changing position of the sun. In a south-facing window, fasten a 3 cm square of cardboard with a hole about 1 cm in its center. At the same hour each day, the students should locate the spot of sunlight that falls on the floor or a desk. Then have the students describe or diagram the relative changes in the position of the sun.

As a special project, a group of students could prepare a script tracing the historical development of the space program. The script could be read by members of the class, or if time permits, the script could be presented to the school as a play.

TEACHER RESOURCES

Readings

Chapter 21

Van Allen, James A. "Space Science, Space Technology and the Space Station." *Scientific American* 254 (January 1986) p. 32. The author professes that a space station will do less than robot probes to further human research efforts in space.

Horowitz, N.H. "The Search for Life on Mars." *Scientific American* 237 (November 1977) p. 52. The author reports on the results of experiments performed on Mars by *Viking* probes.

Chapter 22

The Universe of Galaxies. A Scientific American Book, 1984. Compilation of articles that have appeared in *Scientific American* that reveal research about cosmic phenomena from tidal forces between galaxies to cosmic jets.

To the Edge of the Universe: The Exploration of Outer Space with NASA. New York: Bison Books Corporation, 1986. Beautiful photographs and artists' conceptions illustrating the discoveries made by NASA in its search to unravel the mysteries of the cosmos.

Gamow, George. *One Two Three . . . Infinity.* New York: The Viking Press, 1961. The author presents facts and speculations in science from the "micro" to the "macro."

Chapter 23

Hoskin, Michael. "William Herschel and the Making of Modern Astronomy." *Scientific American* (February 1986). The author discusses the career of one of astronomy's most illustrious personages.

Thorne, Kip S. "The Search for Black Holes." *Scientific American* (December 1984). The author examines the possibility that a black hole exists around the massive star Cygnus X-1.

Bennet, J.A. "Herschel's Scientific Apprenticeship and the Discovery of Uranus." In *Uranus and the Other Planets,* edited by Garry E. Hunt, Cambridge University Press, 1982. How the telescopes of Herschel led to the discovery of Uranus.

Chapter 24

Yenne, Bill. *The Encyclopedia of US Spacecraft.* New York: Exeter Books, 1987. Thoroughly illustrated catalog of United States' spacecraft, produced in cooperation with NASA.

Audiovisual/Software

Chapter 21

Computer programs, *The Solar System,* and *Time and Seasons,* are available for Apple II computers from Educational Images.

A filmstrip series, *Comet Halley: Once in a Lifetime,* is also available from Educational Images.

Chapter 22

A computer program, *Star Gazer's Guide,* is available for Apple II computers from Educational Images.

A filmstrip series, *Astronomical Wonders,* is also available from Educational Images.

Chapter 23

A computer program, *Telescopes,* is available for Apple II computers from Educational Images.

Chapter 24

A videocassette, *Space,* is available from Educational Images.

Other Materials

Chapter 22

Various celestial globes and star charts are available from Geoscience Resources.

Chapter 21: THE SOLAR SYSTEM

PLANNING THE CHAPTER

Chapter Sections	Page	Chapter Features	Page	Program Resources	Page
Section 1: Motions of Earth and the Moon	455			Investigation 21.1 *The Orbit and Phases of the Moon* **(A)**	**TRB 93** **LM 95**
21.1 Rotation of Earth **(A)**	455				
21.2 Revolution of Earth **(A)**	456	**Discover:** Rotating and Revolving **(A)**	456		
		Activity: Making an Astrolabe **(A)**	458		
21.3 Motions of the Moon **(A)**	458	Section Review	460		
Section 2: The Inner Planets and the Asteroids	461			Concept Extension: *Plotting Orbits*	**TRB 43** **LM 265**
21.4 The Solar System **(B)**	461	A Matter of Fact	462		
		Biographies: Then and Now Johannes Kepler, Irene Duhart Long	463		
21.5 Mercury and Venus **(B)**	465	A Matter of Fact	465		
21.6 Earth and the Moon **(B)**	466				
21.7 Mars and the Asteroids **(B)**	467	**Discover:** Traveling to Mars **(B)**	467		
		Section Review	469		
		Skill Activity: Measuring the Diameter of the Sun **(B)**	470		
		Investigation 21: Drawing Planet Distances to Scale **(B)**	476	Student Record Book: Textbook Investigations **(B)**	**TRB 43**
Section 3: The Outer Planets and the Comets	471			Critical Thinking **(H)** Cross-Discipline: Science and Mathematics, *Comparing the Other Planets to Earth* **(A)**	**TRB 40** **TRB 39**
21.8 Jupiter **(B)**	471			Reading for Content: *Using Charts to Summarize Information* **(A)**	**TRB 36** **LM 213**
21.9 Saturn, Uranus, Neptune, and Pluto **(B)**	473	A Matter of Fact	473		
21.10 The Comets **(B)**	475	**Discover:** Looking for Halley **(B)**	475		
		Section Review	475	Investigation 21.2: *The Orbits of Comets and Asteroids* **(A)**	**TRB 97** **LM 99**
Chapter 21 Review	477			Vocabulary **(A)**	**TRB 21** **LM 163**
				Tests **(A)** Computer Test Bank 🖥	**TRB 93**

(LM) Laboratory Manual/Study Guide, **(TRB)** Teacher's ResourceBank™

B = Basic **A** = Average **H** = Honors

The coding Basic, Average, and Honors indicates sections or subsections that might be appropriate for different levels of learners. For additional suggestions regarding choice of topic and depth of coverage, see the Pacing Chart on pages T16–T20.

CHAPTER CONCEPTS, OBJECTIVES, AND TERMS

Section	Concepts	Objectives	Science Terms
Section 1: Motions of Earth and the Moon	■ Earth rotates counterclockwise on its axis once every 24 hours. **(21.1)** ■ Earth revolves around the sun in a specific orbit every 365¼ days. The tilt of the axis from a line perpendicular to the plane of its orbit is about 23.5°. **(21.2)** ■ Phases of the moon are caused by the varying illumination of the moon's surface. **(21.3)**	■ **Describe** the earth's rotation in terms of direction of motion and time. ■ **Define** the term *revolution* as it relates to the earth's path through space. ■ **Explain** how the tilt of the earth's axis affects the seasons. ■ **Relate** the cause of solar and lunar eclipses.	axis rotation revolution orbit lunar eclipse umbra penumbra solar eclipse
Section 2: The Inner Planets and Asteroids	■ The solar system contains nine planets, each of which revolves around the sun. **(21.4)** ■ Mercury, the planet closest to the sun, has a cratered surface and practically no atmosphere. The atmosphere of Venus, the second planet from the sun, is rich in carbon dioxide. Venus is sometimes referred to as Earth's sister planet because of its similar size, mass, and density. **(21.5)** ■ Earth is the third planet from the sun, the first planet with a satellite, and the only planet known to have life. **(21.6)** ■ Mars, the fourth planet from the sun, is called the red planet. Although it may have had water at one time, none presently exists. Polar ice caps of frozen carbon dioxide can be seen on its surface. Asteroids are planetary fragments that did not join together to form a planet. Asteroids orbit between Mars and Jupiter. **(21.7)**	■ **Describe** the arrangement and orbits of the inner planets in the planetary system. ■ **Explain** why the surfaces of Mercury and the moon are heavily marked with craters while those of Venus and Earth are not. ■ **Discuss** the ways in which Venus differs from Earth. ■ **Tell** why Mars is red.	planets elliptical astronomical unit satellite asteroids meteorites
Section 3: The Outer Planets and the Comets	■ Jupiter is the largest planet in the solar system. It emits twice as much energy as it receives from the sun. **(21.8)** ■ The rings of Saturn consist of fragments of rock and ice from less than 1 cm to more than 10 m in diameter. Uranus is unusual because of its nearly horizontal tilt and retrograde rotation. **(21.9)** ■ Comets are metal and silicate particles in ice. **(21.10)**	■ **Compare** the structure of the outer planets with that of the inner planets. ■ **List** the peculiarities of the Galilean satellites of Jupiter. ■ **Discuss** the structure and composition of comets.	comets

Title	Page	Materials
Discover: Rotating and Revolving	456	*(per pair of students)* no materials needed
Activity: Making an Astrolabe	458	*(per student)* large index card; protractor; string, 30 cm; large metal washer; tape; plastic drinking straw
Discover: Traveling to Mars	467	*(per student)* paper, pencil
Skill Activity: Measuring the Diameter of the Sun	470	*(per student)* small index card (2), scissors, T-pin or ballpoint pen, tape, meter stick, paper, pencil
Discover: Looking for Halley	475	*(per student)* paper, pencil
Investigation 21: Drawing Planet Distances to Scale	476	*(per group of 3 or 4)* string, 30 m; metric ruler; construction paper; scissors; tape; markers

TEACHING SUGGESTIONS

Section 1: Motions of Earth and the Moon

Demonstration: The Seasons

Purpose
To illustrate by using models why the earth has seasons and the moon has phases

Background
The students will find it easier to conceptualize how and why the earth has seasons and the moon has phases by witnessing the motion of these celestial bodies.

Materials
Globe
Softball (or similarly sized sphere)
Lamp
Extension cord

Procedure
1. Remove the lamp shade from the lamp and set it on a desk or table in the center of the classroom.
2. Have one student handle the globe revolving around "the sun." Be sure to emphasize that the earth rotates on a tilted axis as it revolves about the sun.
3. Have another student manipulate the moon as it revolves around the earth. Be sure to emphasize that the moon (1) revolves around the earth in a plane angled 5° from the ecliptic, and (2) that it rotates only once in one complete revolution. Turn on the lamp.
4. Beginning on "January 1st," take the class through a journey lasting "one year," demonstrating the position of earth with respect to the sun during equinoxes, solstices, and seasons. Show how the moon changes phase.

Questions to Ask the Students
1. Why do we never see the "far side" of the moon? (*The rotation of the moon always keeps the same side facing Earth.*)
2. What is the inclination of the moon from the ecliptic? (*five degrees*)

Class Activity

A globe, a light source, and a ball of clay can be used to demonstrate the phases of the moon, the penumbra and umbra, and solar and lunar eclipses. You will probably need a stronger light source than a flashlight. A slide or filmstrip projector works well. Use a small ball of clay (4 cm to 5 cm in diameter) to represent the moon. Attach the ball to a pencil to make it easier to handle. As you move the moon around the globe, the students will be able to see that from the earth varying portions of the lighted surface of the moon are visible as it completes its orbit. Have the students name the phases as you demonstrate them.

To demonstrate a solar eclipse, pass the clay ball between the light and the globe. The dark umbra should be visible with the lighter penumbra around it. The students should note that the path of the shadow, as it crosses the globe, does not cover a very large area.

To demonstrate a lunar eclipse, pass the clay ball into the shadow of the earth. Move it into the shadow slowly so the students can see how the moon appears to change shape as larger and larger "bites" are taken out of it.

Class Activity

Have several students assist the local or school librarian in compiling books and magazines containing information about the planets, *but be sure that these references are at least forty years old!* By examining these older materials, students will realize that our knowledge and understanding of the solar system has changed drastically over the decades (not to mention the centuries), and that in the future this information and understanding will continually be revised.

Field Trip

Take your classes to the school library or the public library to research references to the moon in mythology, poetry, and stories. Challenge them to find a reference to each of the moon's phases in literature.

Section 2: The Inner Planets and the Asteroids

Class Activity

Suspend a model of the solar system, made of light, safe materials, from the ceiling. With the model of the sun at the center of the classroom, position the planets so that their correct order from the sun and their correct arrangement with respect to one another, is maintained. Do this by using lightly colored thread and tape to layout the orbit of each planet on the ceiling. Second, suspend plastic foam models of the planets and their satellites from the ceiling. The sizes of the planets need not be to exact scale nor do their actual distances from the sun need to be to scale. Change the positions of the planets daily (especially the position of the earth and its moon), to represent their actual movement through the solar system. The outer planets will hardly move at all in this limited span of time.

Section 3: The Outer Planets and the Comets

Outside Speaker

Invite an astronomer to class to talk about the recent planetary discoveries of the *Voyager* space probe.

7

UNIT

ASTRONOMY

For centuries scientists have wondered if life exists on other planets. Seeing the planet Mars dimly through telescopes, some astronomers thought that intelligent Martians lived in a world full of water and vegetation, and that they had built a system of canals. Some of the first interplanetary spacecraft traveled to Mars to investigate the possibility of life on the red planet. However, by 1976 scientists could more confidently answer the question, "Are we alone in the solar system?"

- What other bodies are there in the solar system?

- What instruments are used to study the universe?

- What are the theories concerning the origin of the universe?

- Before the Viking space probes, what did scientists know about Mars?

By reading the chapters in this unit, you will learn the answers to these questions. You will also develop an understanding of concepts that will allow you to answer many of your own questions about the universe.

"The view from Mars"

453

UNIT 7

INTRODUCING THE UNIT

Have the students look at the Unit Opener photograph on these pages. Read the text and use the questions to guide a class discussion. Do not expect the students to answer the questions now. The questions are motivational and the students should be aware that they will be able to answer them after completing the Unit.

Discussion Most students are interested in space and should enjoy the photograph of the Martian surface. They should agree that the surface resembles that of the moon, not Earth. Space fiction for many years invented Martians that came to Earth. You may wish to display a picture of Earth as taken from space and discuss the differences between the appearance of Earth and that of Mars.

Extension You can extend the discussion by bringing in other Unit themes. What kind of star is the sun? What is the Milky Way Galaxy? How does a star develop? When the class has completed the Unit, ask them the questions again. This follow-up discussion may help the students realize how much they have learned.

CHAPTER 21

CHAPTER OVERVIEW

This chapter presents a general introduction to the study of the solar system. The earth's position relative to the sun, moon, and planets is described. Also discussed are the inner planets, the outer planets, asteroids, comets, and satellites of the solar system.

Section 1: Motions of the Earth and the Moon The direction and time span of the earth's rotation and revolution are described. The relationship of the tilt of the earth's axis to the seasons is explained. The motions of the moon around the earth are discussed, as well as the phases of the moon. Both solar and lunar eclipses are described.

Section 2: The Inner Planets and the Asteroids This section begins with a brief overview of the components of our solar system. The surface features, atmospheres, rotation, revolution, and satellites of the inner planets are described. The location of asteroids in the solar system is presented. Similarities and differences among the inner planets are then discussed.

Section 3: The Outer Planets and the Comets In this section a comparison of the structure of the outer planets and the inner planets is made. Jupiter, Saturn, Uranus, Neptune, and Pluto are described, as well as the specific features of the four Galilean satellites. The structure and composition of comets and their place in the solar system are discussed.

The Solar System

Earth and its moon are part of a system of bodies in orbit around a rather ordinary star. This solar system, named for its central feature, the sun, is probably only one of many billions of similar systems in the universe. However, much scientific study is confined to these neighbors in space.

454

A solar eclipse

CHAPTER MOTIVATING ACTIVITY

To spark the students' interest in outer space, ask them if they think there is life on other planets. If any students answer no, ask them to explain how they arrived at that conclusion. If they answer yes, ask them to describe the characteristics that life forms on other planets might have. Encourage the students to express differing opinions.

After examining samples taken from Mars by Viking landings, scientists have deduced that living organisms would have great difficulty surviving the environment they exam-

ined, but they admit the possibility that some form of organism might exist somewhere else on the planet. Scientists are quite sure that there is no life on any of the giant planets because of their atmospheres, but living organisms may have evolved on one or more of the satellites circling those planets (for example, Titan or Ganymede).

1 Motions of Earth and the Moon

SECTION OBJECTIVES

After completing this section, you should be able to:

- **Describe** the earth's rotation in terms of direction of motion and time.
- **Define** the term *revolution* as it relates to the earth's path through space.
- **Explain** how the tilt of the earth's axis affects the seasons.
- **Relate** the cause of solar and lunar eclipses.

NEW SCIENCE TERMS

axis
rotation
revolution
orbit
lunar eclipse
umbra
penumbra
solar eclipse

21.1 Rotation of Earth

Imagine that you are an astronaut looking at Earth from a stationary space station. You would see Earth turning around its axis. An **axis** is an imaginary line that passes through the center of a planet, from pole to pole. The turning of Earth on its axis is called **rotation** (roh TAY shuhn). The earth rotates much like a spinning toy top, making one complete rotation every 24 hours. This time is called the *period of rotation*. The direction of Earth's rotation is counterclockwise as seen from above the North Pole.

If your space station were positioned so that you could see North America as Earth turned, you could watch the sun's light move across the continent. The first part of the continent to receive the sun's light is the east coast of Canada. At this time, all of the United States is still in darkness. As the earth rotates, the area of darkness moves westward, until all of North America is in sunlight. This process takes about five hours.

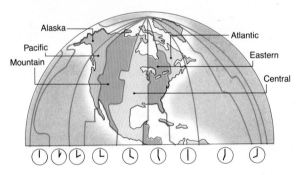

Figure 21–1. The rotation of the earth is responsible for the formation of different time zones. The time zones of North America are shown here.

DEMONSTRATION

For a demonstration on the earth's seasons and the moon's phases, see page 451e preceding this chapter.

SECTION FOCUS

Section Overview   In this section, the rotation of the earth and its revolution around the sun are explained. The position of the earth during each season is also discussed. The motions of the moon—its rotation and its revolution around the earth—are described. The relationship of the moon's position to phases and the eclipses of both the sun and the moon are discussed.

Section Objectives   For a list of section objectives, see pupil's edition page.

New Science Terms   For a list of new science terms in this section, see the pupil's edition page. This section contains many unfamiliar terms. You may wish to read the terms aloud so that the students will recognize them in the text.

SECTION DEVELOPMENT

21.1 Rotation of Earth

DISCUSSION   Direct the students' attention to the illustration in Figure 21–1. Emphasize that a complete rotation of Earth takes about 24 hours. Explain that one period of rotation of any planet, including Earth, is called a *day*. Thus Earth's day is 24 hours, while a day on Jupiter is 11 Earth hours.

THINKING   SKILL *(Interpreting Maps)* Have the students study the time-zone map of North America. Help them find the time zone in which they live. Ask the students what time it would be in Atlanta, Georgia, if it were 2:30 PM in their town. (Answers will vary.) Then ask the students to tell you the time in Washington, D.C., if it is 11:15 AM in Seattle, Washington. (2:15 PM)

DISCUSSION North America has eight time zones: Newfoundland, Atlantic, Eastern, Central, Mountain, Pacific, Alaska, and Aleutian. *Thinking Critically:* Have the students discuss why North American time zones were not necessary in 1700s. (Travel and communication were slow. The fact that the sunrise on the west coast occurred about four hours after the sunrise on the east coast had no effect on people's daily lives.)

EXTENSION As the earth rotates on its axis, the axis wobbles in a motion called *precession.* In precession, the earth's axis traces a circle in space once every twenty-two thousand years. The direction in which the axis points changes slowly over time. Today the North Pole leans toward the sun when the earth is at its farthest point from the sun. Eleven thousand years from now the South Pole will lean toward the sun when the earth is at its farthest point from the sun.

21.2 Revolution of Earth

DISCUSSION Explain to the students that the earth makes a complete revolution every 365 days, 6 hours, and 9 minutes (slightly more than one-quarter day extra). To compensate for the quarter-day, every four years an extra day is added to the year.

EXTENSION Because the supposed quarter-day is actually 9 minutes too long, century years (for example, 1900, 2000) are leap years only if they can be divided by 400. For example, 1600 is a leap year; 1900 is not one. This prevents the calendar year from lagging behind the solar year.

DISCOVER:
Rotating and Revolving

Thinking Skill *(Drawing Conclusions)*

Assign nine students to represent the nine planets and have them stand in the correct order around the sun. Ask the students if it would be possible for them to place themselves at distances from the desk that would correspond to the distances of the planets from the sun. (No. Even if Mercury were to stand only 0.5 m away from the desk, there would not be enough room in an ordinary classroom for Pluto.)

456

In order for the time of sunrise to be approximately the same all across North America, five different time zones have been established. These time zones change roughly every 15 degrees of longitude. However, time zones are not straight lines as are lines of longitude. The zones are established so that they do not divide cities. Imagine living in one time zone and going to school in another. You would have to get up an hour earlier or later than the other students just to arrive at school on time.

○ *What is an axis of a planet?*
○ *What is the earth's period of rotation?*

21.2 Revolution of Earth

From your space station, you would also observe a second motion of the earth. In addition to rotating on its axis, Earth moves around the sun. Movement of Earth around the sun is known as **revolution** (rehv uh LOO shuhn). Since it takes Earth one year, or $365\frac{1}{4}$ days, to complete one revolution, its period of revolution is one year. The path that the earth follows as it revolves around the sun is known as its **orbit.** The orbit of the earth does not change; Earth's speed of revolution remains constant as well.

Figure 21–2. The earth revolves around the sun every 365¼ days. This period of revolution determines the length of an earth year.

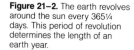

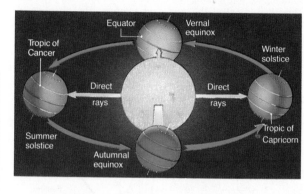

DISCOVER

Rotating and Revolving

You can demonstrate the motions of the earth and the moon with a friend. Pick an object, such as a desk, to be the sun. You, the earth, will rotate 365 times as you orbit the sun. Your moon will rotate once, and orbit you once, every 28 days. You can make this more interesting by adding other planets and their satellites.

As the earth revolves around the sun, its axis remains in a fixed position, but it is tilted. The earth's tilt, or *inclination*, is always 23.5° in relation to a line perpendicular to the plane of its orbit. Imagine a table tennis ball with a toothpick as its axis. If it is tilted about 23° from vertical, its inclination would be similar to Earth's. Because this inclination is fixed, first the Northern and then the Southern Hemisphere leans toward the sun. This creates seasonal changes in climate over most of the earth.

Answers—21.1

○ An axis is an imaginary line that passes through the center of a planet from pole to pole.

○ The earth's period of rotation is 24 hours.

When the Northern or Southern Hemisphere is tilted toward the sun, that hemisphere receives very direct rays from the sun. Also, the amount of time sunlight is received during the day is longer. This is summer. When the same hemisphere is tilted away from the sun, that hemisphere receives less direct sunlight. There are also fewer hours of sunlight during the day. It is winter.

On June 20 or 21 the tilt of the Northern Hemisphere toward the sun is greatest. The sun is directly over 23.5°N parallel, which is called the *Tropic of Cancer.* In the Northern Hemisphere this date is the *summer solstice.* The summer solstice is the day of the year with the most hours of daylight.

As the earth's revolution around the sun continues, the sun is directly over the equator on September 22 or 23. In the Northern Hemisphere this is the *autumnal equinox.* On this date the periods of daylight and darkness are about equal.

As the earth continues its revolution, the daylight hours in the Northern Hemisphere grow shorter until the *winter solstice* on December 21 or 22. On this date the sun is directly over the *Tropic of Capricorn* at latitude 23.5°S. This is the day with the least daylight in the Northern Hemisphere.

On March 20 or 21, the sun is directly over the equator again, and the hours of daylight and darkness are nearly equal. In the Northern Hemisphere this is known as the *vernal equinox.* As the earth completes its revolution around the sun, the hours of daylight continue to lengthen in the Northern Hemisphere until the summer solstice. In the Southern Hemisphere the summer and winter solstices are opposite to those in the Northern Hemisphere. Explain why this is so.

○ *What is the earth's period of revolution?*
○ *How does the earth's revolution around the sun affect the seasons in the Northern Hemisphere?*

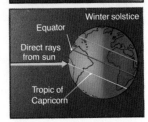

Figure 21–3. The summer solstice (top) occurs when the sun is directly over the Tropic of Cancer. The winter solstice (bottom) occurs when the sun is directly over the Tropic of Capricorn.

Figure 21–4. The seasons of the year are determined by the position of the earth with respect to the sun. In temperate climates, such as the one shown here, four seasons occur every year.

457

DISCUSSION Direct the students' attention to Figure 21–3. Point out that the direct rays of the sun, as shown by the arrow, strike the earth's surface at a 90° angle and do not spread out. Less direct rays, striking the earth at an angle other than 90°, spread out and cover a greater area. The area receiving less direct sunrays will be cooler. *Thinking Critically:* Ask the students why it is never hot at the poles. (The poles never get any direct rays of the sun.)

THINKING SKILL *(Inferring Relationships)* Have the students study Figure 21–4. Ask them whether they would observe these same seasonal changes on an island near the equator. (No; the temperatures at the equator vary little throughout the year, and thus there is almost no perceptible seasonal change.)

Answers—21.2

○ The period of revolution for the earth is 365¼ days.

○ Because the earth's axis is tilted as it revolves around the sun, sunlight strikes the surface of the planet at different angles. The Northern Hemisphere is cooler in the winter because it receives less direct sunlight, and it is warmer in the summer because it receives more direct sunlight.

ACTIVITY:
Making an Astrolabe

Skill (Measuring)

Preparation of Materials You may wish to draw the line, with its center marked, before giving the index cards to the students. You will then be sure that everyone begins the labeling in the same place. You may also want to label the angles on the index cards before handing them out.

Discussion Supervise steps 1–3 of the Procedure to be sure that the labeling is correct and that the straw is taped on straight.

CAUTION: Students should not look at the sun because of possible eye injury.

Answers to Conclusions/Applications

1. No. One must also know the distance of the structure from the observer.
2. The sextant has the advantage of utilizing the sun to determine latitude during the day.

21.3 Motions of the Moon

DISCUSSION Remind the students of the demonstration that illustrated the moon's movements around the earth. Refer them to Figure 21–5 at the top of the next page.

EXTENSION The period of the moon's revolution, 27.3 days, is called a *synodic month*. After 27.3 days, the moon is above the same spot on the earth as it was 27.3 days earlier. The moon, however, requires 2.2 days more to arrive at a position located exactly between the earth and sun. This is because the earth is revolving around the sun. Thus, the period from new moon to new moon, called the *sidereal month*, is 29.5 days long.

How can you make an instrument that can determine the distance of an object above the horizon?

MATERIALS (per person)

large index card; protractor; string, 30 cm; large metal weight; tape; plastic drinking straw

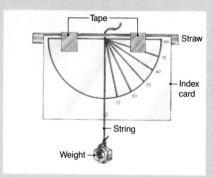

1. With a protractor, draw a line and semicircle across the index card and label it as shown.
2. Tie the weight to one end of the string. Make a small hole in the index card at the center of the straight line and thread the loose end of the string through the hole.
3. Tie the loose end of the string around the plastic straw. Tape the straw to the index card.
4. Look through the straw at a tall object or something in the sky, but do not look at the sun. Let the weight hang freely and observe where the string crosses the degree markings. That is the altitude of the object above the horizon.

CONCLUSIONS/APPLICATIONS

1. Do you think it is possible to calculate the height of a structure in meters just by knowing its angle above the horizon? Explain.
2. What other scientific instrument is similar to the astrolabe? Describe the similarities and differences.

21.3 Motions of the Moon

Studying the moon from Earth is like studying the earth from a stationary space station. The earth and the moon form a system that jointly revolves around the sun. The earth appears to be stationary in respect to any other movements of the moon.

The moon has movements that are similar to the rotation and revolution of the earth. The moon rotates on its axis as it revolves around the earth. However, since the moon's periods of rotation and revolution are the same, the same side of the moon always faces the earth.

Although the moon appears to shine on its own, moonlight is really reflected sunlight. When the entire side of the moon that faces the earth is lighted, you see a *full moon*. When the side of the moon facing the earth is not lighted, it is called a *new moon*. The series of changes in the appearance of the moon as it revolves around the earth, shown in Figure 21–5, are called the *phases of the moon*.

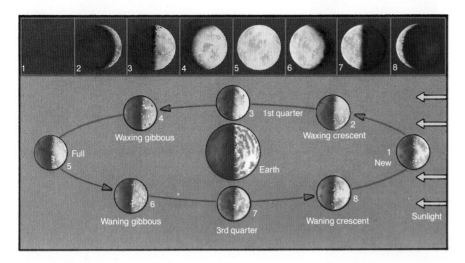

Occasionally, the moon moves into a position where the earth is directly between the moon and the sun. Then the sun's rays that would ordinarily fall on the moon's surface are temporarily blocked by the earth. When the earth's shadow falls on all or part of the moon's surface, a **lunar eclipse** (ih KLIHPS) occurs. Because of the relative sizes of the earth and the sun, two shadow zones form during a lunar eclipse. The zone of complete shadow is called the **umbra** (UHM bruh). The umbra is surrounded by a zone of partial shadow called the **penumbra** (pih NUHM bruh). During a total lunar eclipse, the moon passes through the penumbra and into the umbra. You can watch the moon pass through a region of dim light until it crosses a distinct boundary into darkness. The moon lies entirely within the umbra for almost two hours before it finally reappears in the penumbra. In a partial lunar eclipse, part of the moon passes through the penumbra but not the umbra, so the moon never completely disappears.

Figure 21–5. The moon goes through a series of phases every month. These phases, shown here, are due to the position of the moon and the earth in relation to the sun.

Figure 21–6. A lunar eclipse (below) occurs when the earth casts a shadow on the surface of the moon (left).

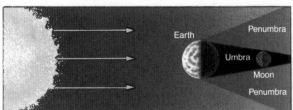

Section 1   Motions of Earth and the Moon   **459**

DISCUSSION   Be sure that the students are familiar with the technical terms for the shadows formed during an eclipse. Demonstrate the eclipse and refer the students to Figure 21–6. *Thinking Critically:* Ask the students why seeing a lunar eclipse is an unusual event. (You must be on the proper spot on the earth to see it and it must occur on a cloudless night.)

EXTENSION   Although the natural satellite of the earth is often called "the moon," its proper name is "Luna." The satellites of Mars are Phobos and Deimos.

EXTENSION   Inform the class that as the amount of the lighted surface of the moon visible from Earth increases, the moon is said to be waxing. As the amount decreases, the moon is said to be waning. Refer the students to the photos in Figure 21–5. Explain that when half the moon appears lighted, the phase is called *quarter first*. *Thinking Critically:* Ask the students why this term is appropriate. (Looking at the figure, the students should be able to note that the moon is one-quarter through its orbit.)

THINKING SKILL   (*Expressing Viewpoints*) Have half the class argue the case for continuing the space program and the other half argue the case against it. Have the students debate considerations such as the information to be gained from space travel, the many applications on Earth for space technology, the allocation of billions of dollars that could otherwise be used to improve the quality of life on Earth, and any other topics you may want to add.

THINKING SKILL   (*Making Inferences*) It is easier for the students to conceptualize how and why the earth has seasons and the moon has phases if they try to envision their motion from the perspective of future astronauts. Ask the students to imagine themselves on a future space mission to Mars or Jupiter. Ask them to consider the navigational problems of returning to Earth. They should take into account the movement of the earth in its orbit and the possibility of their colliding with other bodies that will have changed position while they are gone.

DEMONSTRATION

Demonstrate a lunar eclipse with a flashlight representing the sun and two students representing the moon and the earth, respectively. Shine the light, from about 2 m away, at the front of the student who represents the moon. Have the second student, who represents the earth, slowly move between the flashlight and the first student, until the second student blocks all the light from the flashlight. This simulates a lunar eclipse.

459

DISCUSSION   You may wish to repeat the Demonstration on the previous page, this time having the student representing the moon pass between the flashlight and the student representing the earth. Have the students study the various photographs of eclipses. **CAUTION: The students should never to look directly at a solar eclipse.**

EXTENSION   The distance between the earth and moon varies from 350 000 km to about 400 000 km. The point in the moon's orbit where it is closest to Earth is called *perigee;* the farthest point from Earth is called *apogee.* For a total solar eclipse to occur, the moon must pass very close to the earth in order to totally block the sphere of the sun. Sometimes the moon is so far from the earth when it is between the earth and the sun that it blocks only the center of the sun. The resulting eclipse, called an *annular solar eclipse,* appears as a ring of light around the moon, which is visible from only a very small area on Earth.

SECTION REVIEW

Summary   Review with the students the motions of the earth and the moon, including the earth's rotation; the earth's revolution around the sun; the position of the earth during the equinoxes and solstices; the position of the earth, moon, and sun during lunar and solar eclipses; and the moon's position during its first quarter, last quarter, full, and new phases.

Reinforcement   Have the students quiz one another using the Section Objectives at the beginning of this section.

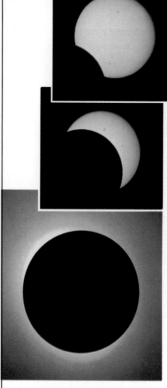
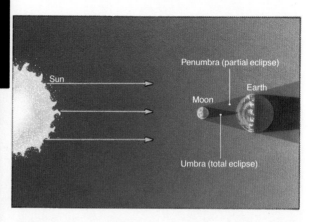

Figure 21–7. A solar eclipse (above) occurs when the moon passes between the earth and the sun (right).

Sometimes the shadow of the moon passes across the lighted side of the earth, forming a **solar eclipse.** Since the moon is much smaller than the sun, the moon forms a narrow shadow, about 258 km wide, as it moves across the face of the earth. A partial eclipse, where only part of the sun is blocked by the moon, is visible on either side of the area experiencing total solar eclipse. The moon's shadow moves rapidly across the surface of the earth. In fact, a total solar eclipse can be seen from any one position on Earth for only about 7.5 minutes. **CAUTION: Never look directly at the sun, even during an eclipse. The sun's rays may cause permanent damage to your eyes.**

○ *What are the phases of the moon?*
○ *What is a lunar eclipse? What is a solar eclipse?*

Section Review

READING CRITICALLY

1. Compare and contrast the revolution and rotation of Earth.
2. Describe the relative positions of the sun, the moon, and the earth during a full moon.
3. Explain why a solar eclipse is not observed at all points on the earth at the same time.

THINKING CRITICALLY

4. How would the seasons be different if the earth's axis were not inclined?
5. Why does a solar eclipse only last about 7.5 minutes?
6. Why is the far side of the moon never seen from Earth?

Answers–21.3

○ The phases of the moon are a series of changes in the moon's appearance as observed from Earth.

○ During a lunar eclipse, the moon's surface darkens when the earth blocks sunlight from reaching the moon. During a solar eclipse, the earth's surface darkens when the moon blocks sunlight from reaching the earth.

Answers to Section Review

Reading Critically

1. The term *revolution* describes the earth's movement around the sun. The term *rotation* describes the motion of the earth while spinning on its axis.
2. During a full moon, the earth is between the sun and moon. The full face of the moon is able to reflect the light from the sun because it revolves around the earth well above or below the plane of Earth's revolution around the sun.

continues

SECTION
2 The Inner Planets and the Asteroids

SECTION OBJECTIVES

After completing this section, you should be able to:

- **Describe** the arrangement and orbits of the inner planets in the planetary system.
- **Explain** why the surfaces of Mercury and the moon are heavily marked with craters while those of Venus and Earth are not.
- **Discuss** the ways in which Venus differs from Earth.
- **Tell** why Mars is red.

NEW SCIENCE TERMS

planets
elliptical
astronomical unit
satellite
asteroids
meteorites

21.4 **The Solar System**

The earth, moon, and sun are only part of a complex system of orbiting bodies that make up the solar system. The solar system, named for the sun, is comprised of nine planets. **Planets** are the main bodies that revolve around the sun. The nine planets of the solar system are: Mercury, Venus, Earth, Mars, Jupiter, Saturn, Uranus, Neptune, and Pluto. All the planets revolve around the sun in the same direction. Viewed from the sun's north pole, this direction, called *prograde*, is counterclockwise.

Figure 21–8. The nine planets in our solar system are shown here in their relative positions.

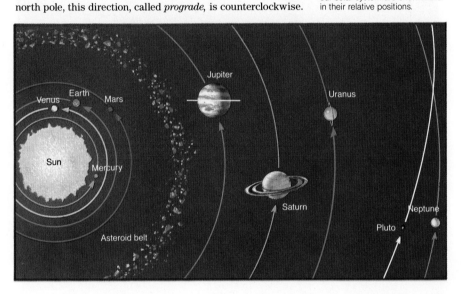

Section 2 The Inner Planets and the Asteroids **461**

Section **2**: **The Inner Planets and the Asteroids**

SECTION FOCUS

Section Overview This section begins with a general description of our solar system. Each of the inner planets, Mercury, Venus, Earth, and Mars, is described in detail. Also discussed are features of the moon and the asteroid belt between Mars and Jupiter.

Section Objectives For a list of section objectives, see pupil's edition page.

New Science Terms For a list of new science terms in this section, see pupil's edition page. Some of the terms in this section are familiar and some are unfamiliar to the students. Explain to the students that they will learn the scientific definitions of these terms.

SECTION DEVELOPMENT

21.4 **The Solar System**

DISCUSSION Explain to the students that, although the planets would appear to move counterclockwise if viewed from the sun's north pole, the planets do not appear to move counterclockwise when viewed from Earth. Their motions as viewed from Earth are apparent, not actual, and thus they appear to slow down, speed up, and even move backward in their orbits.

3. The shadow of the moon on the earth is only 258 km wide during a solar eclipse.

Thinking Critically

4. There would be no seasons, because sunlight would hit each latitude with the same amount of energy all year round.
5. Because the moon's orbit around the earth is at an angle to the earth's orbit around the sun.

6. The periods of rotation and revolution of the moon are the same, so the same side of the moon always faces the earth.

DISCUSSION Direct the students' attention to Figure 21–9. Explain that a planet travels faster as it nears perihelion and slower as it moves away from the sun. *Thinking Critically:* Ask the students if the earth will travel farther during the time from January to February or from July to August. (January to February because the earth is closest to the sun during the Northern Hemisphere's winter and would thus travel faster, covering more distance in an equal amount of time.)

THINKING SKILL *(Synthesizing Ideas)* Explain that two forces acting on the planets keep them in orbit around the sun. One of the forces, the sun's gravity, pulls the planets toward the sun. The other force, inertia, makes any object already in motion tend to travel in a straight line. The two forces working perpendicular to each other result in the elliptical orbits of the planets. Ask the students to explain what would happen if the force of the sun's gravity overcame the force of inertia. (The planets would be pulled into the sun.) Then ask them to explain what would happen if a planet gained enough inertia to overcome the sun's gravity. (The planet would shoot off into space.)

Figure 21–9. The orbit of Earth is not circular. Rather, the path that Earth takes around the sun forms an ellipse.

Accurate mathematical observations about the planets were made in the late 1500s by a Danish astronomer, Tycho Brahe (TEE koh BRAH hee). For many years Brahe made careful observations about the position of the planets. Upon his death he left his notes to his assistant, Johannes Kepler. Kepler continued to study the planets, using Brahe's data, and plotted the orbits of the planets.

A MATTER OF FACT

The size of the earth has been known since ancient times. The Greek mathematician Eratosthenes calculated the planet's circumference in about 230 B.C. He observed the difference in the length of noon shadows in Alexandria and Syene (now Aswan), Egypt, and used geometry to determine the size of the earth.

The orbits of the planets as they revolve around the sun are not exactly circular. Instead, the orbits are **elliptical,** or flattened circles with two fixed points called *foci.* The sun is at one focus while the other is empty. The point along an orbit where the planet is closest to the sun is called *perihelion* (pehr uh HEEL yuhn); the point in the orbit that is farthest from the sun is called *aphelion* (uh FEE lee uhn). Planets with large orbits take longer to revolve around the sun than planets with smaller orbits. The period is the time it takes a planet to make one revolution around the sun.

A planet's distance from the sun is measured in astronomical units. An **astronomical unit** (AU) is the average distance between the earth and the sun, or about 150 000 000 km. Table 21–1 shows the most recent photograph of each planet in the solar system and gives current information about each planet.

○ *What is an elliptical orbit?*
○ *What are astronomical units?*

462   Chapter 21   The Solar System

Answers—21.4

○ An elliptical orbit is one that is shaped like a flattened circle.

○ An astronomical unit is the average distance between the earth and the sun, or about 150 000 000 km.

BIOGRAPHIES: Then and Now

JOHANNES KEPLER (1571–1630)

Johannes Kepler was born on December 27, 1571. In his youth, he studied theology, mathematics, and philosophy at the University of Tübingen, in Germany. In the sixteenth century many people believed that the earth was the center of the solar system. In that model, the sun and all the planets revolved around the earth. Kepler became a strong defender of another theory, which had been proposed by Nicolas Copernicus in 1540. That theory states that the sun, and not the earth, is the center of the solar system. In his mid-twenties Kepler published a work entitled *Mysterium Cosmographicum*. In the book, he strongly put forth this theory.

He later worked with the accomplished astronomer Tycho Brahe who invited him to become an assistant. When Tycho Brahe died a year later, Kepler took Brahe's position at court and inherited all of the famed astronomer's notes and records. Brahe's notes included much important information about the position and movement of the planets.

Kepler made many of his own observations and also used Brahe's work to discover several laws about the motion of the planets. In 1609 he wrote *Astronomia Nova* (New Astronomy). In that book he proposed that the planets did not move in circles as was previously thought but instead moved in elliptical orbits around the sun. Kepler's laws of planetary motion served as the basis for Sir Isaac Newton's laws of gravity.

IRENE DUHART LONG (1951–)

Irene Duhart Long was born in Cleveland, Ohio, in 1951. She became interested in flight at a very young age, when she accompanied her father on his flying lessons. By the age of nine, she had decided that she wanted to work for NASA.

She began her college education at Northwestern University, in Evanston, Illinois, and received her medical degree from the St. Louis University School of Medicine, specializing in aerospace medicine. She completed her medical residency at the Kennedy Space Center.

In 1982 Dr. Long was named the Chief of the Medical and Environmental Health Office at NASA. She is currently Chief of Medical Operations. Dr. Long and her team of scientists study how the free-fall environment in outer space affects the health of astronauts. Her group provides medical care for the astronauts in the event of an emergency and collects data on the reactions of the human body to space travel. The group is particularly concerned with the effects of space travel on the heart and on the endocrine system. The information gathered is being used to develop methods to minimize the impact of space travel on the human body.

One of the highest-ranking women at the Kennedy Space Center, Dr. Long is the first black to be named Chief of Medical Operations. Her next goal is to travel in space herself, as the medical officer on a spaceflight. Then she could collect in-flight information about the effects of space travel first hand.

BIOGRAPHIES: Then and Now

Discussion   Have the students read the two biographies and discuss how the contributions of each person helped to further our understanding of space and space travel.

Thinking Skill   *(Synthesizing Information)* After the students have read the two biographies, ask them to compare the background, training, and work of scientists in Kepler's time with that required by those involved in today's space program.

Extension   Interested students might be encouraged to research further the life and works of Kepler or Long or one of the scientists listed below.

Nicholas Copernicus

Aristarchus of Samos

Ptolemy

Galileo

Sally Ride

William Herschel

Anna Fisher

DISCUSSION Explain to the students that the temperatures given in Table 21–1 are the temperatures at the visible surface of each planet. The visible surface of Mercury, Venus, Earth, and Mars is solid. In the cases of Jupiter, Saturn, Uranus, and Neptune the surface is gaseous. The temperature range given for Jupiter is its surface temperature (−95°C) and the temperature of its core (30 000°C). The temperature ranges given for the other planets are the nighttime and daytime temperatures. *Thinking Critically:* Ask the students to name the planet that has the coldest temperature and the planet that has the warmest temperature. (The coldest one is Pluto; the hottest one is Jupiter.)

TABLE 21–1: THE SOLAR SYSTEM

Mercury, with a diameter of about 4900 km, is an average of 58 million km (0.39 AU) from the sun. A year on Mercury is 88 Earth days and a day is 58.67 Earth days. Mercury has practically no atmosphere. The surface of Mercury is rocky, with many craters and steep cliffs. Temperatures on Mercury range from −170°C to over 400°C.

Venus, with a diameter of over 12 000 km, is an average of 108 million km (0.72 AU) from the sun. A year on Venus is 225 Earth days, and a day is 243 Earth days. Venus' thick atmosphere is mostly carbon dioxide. The surface of Venus has vast plains and high mountains. Temperatures on Venus average about 500°C.

Earth, with a diameter of about 12 750 km, is an average of 150 million km (1.00 AU) from the sun. A year on Earth is 365.25 days and a day is 24 hours. Earth's atmosphere is mostly nitrogen and oxygen. The surface of Earth is nearly covered with water. Temperatures on Earth range from −90°C to about 60°C.

Mars, with a diameter of nearly 6800 km, is an average of 228 million km (1.52 AU) from the sun. A year on Mars is 1.88 Earth years and a day is 24.5 hours. Mars' thin atmosphere is mostly carbon dioxide and nitrogen. The surface of Mars has polar ice caps and dry river beds. Temperatures on Mars range from −100°C to about −30°C.

Jupiter, with a diameter of nearly 143 000 km, is an average of 778 million km (5.19 AU) from the sun. A year on Jupiter is 11.86 Earth years and a day is 11 Earth hours. Jupiter's atmosphere is mostly hydrogen and helium. Jupiter has no solid surface. Temperatures range from −125°C in the atmosphere to nearly 30 000°C in the core.

Saturn, with a diameter of about 120 000 km, is an average of 1427 million km (9.51 AU) from the sun. A year on Saturn is 29.45 Earth years and a day is 10 Earth hours. Saturn's atmosphere is mostly hydrogen and helium. Saturn has no solid surface. Temperatures in the atmosphere average about −180°C.

Uranus, with a diameter of nearly 51 000 km, is an average of 2869 million km (19.13 AU) from the sun. A year on Uranus is 84.01 Earth years and a day is 16.8 Earth hours. Uranus' atmosphere is mostly hydrogen and helium. Uranus has no solid surface. Temperatures in the atmosphere average about −225°C.

Neptune, with a diameter of about 49 000 km, is an average of 4505 million km (29.90 AU) from the sun. A year on Neptune is 164.8 Earth years and a day is 18.5 Earth hours. Neptune's atmosphere is mostly hydrogen and helium. Neptune has no solid surface. Temperatures in the atmosphere average about −220°C.

Pluto, with a diameter of about 2300 km, is an average of 5890 million km (39.27 AU) from the sun. A year on Pluto is 247.7 Earth years and a day is 6.4 Earth days. Pluto's atmosphere is mostly methane. The surface of Pluto is solid with unknown features. Temperatures on Pluto average about −230°C.

21.5 Mercury and Venus

Mercury, the planet closest to the sun, is also the smallest of the inner planets. Even through a telescope Mercury appears as an indistinct object just above the horizon in the morning and evening sky.

Pictures of Mercury's surface show it to be similar to that of Earth's moon. There are many impact craters and plains of dark rock, which could be basalt. Mercury's surface has long ridges that seem to have originated as the crust and the interior cooled and contracted.

Mercury has little gravity because it is so small. Therefore, Mercury has virtually no atmosphere. Most of the gases that reach the planet's surface from its interior escape into space.

The density of Mercury has been calculated to be similar to that of Earth. This indicates that it may have an iron-nickel core as Earth has. However, the magnetic field of Mercury is very weak. Scientists believe this may be because the planet rotates too slowly for electrical currents to develop in its core.

A MATTER OF FACT

On Mercury, temperatures at noon can reach as high as 500°C, while at midnight temperatures are as cold as −180°C.

Figure 21-10. Mercury is the closest planet to the sun. The surface of Mercury is similar to that of the moon. This planet was named after the fleet-footed messenger of the Roman gods. The name was chosen because Mercury has such a rapid period of revolution.

The second planet from the sun is Venus. Venus is often thought of as Earth's sister planet because of their similar masses, densities, and sizes. Often visible in the evening or in the morning, Venus appears to have phases similar to those of the moon. These phases occur because from Earth Venus is seen with different angles of solar light.

Venus rotates very slowly, taking 243.01 Earth days to rotate once on its axis. Unlike Earth and most of the other planets, Venus rotates east to west, or *retrograde*. Venus takes 224.7 Earth days to orbit the sun, so a day on Venus is longer than a Venusian year.

Section 2   The Inner Planets and the Asteroids   **465**

21.5 Mercury and Venus

DISCUSSION   Have the students study the surface of Mercury shown in Figure 21-10. Have them indicate which features are similar to those of the moon. (They should point out the craters and large, dark plains.)

EXTENSION   You may want to provide additional data on Mercury and Venus. Mercury orbits the sun at 47.9 km/s; Venus orbits at 35 km/s. Point out that the farther the planet is from the sun, the slower is its orbital velocity. Mercury is 5.4 times denser than water; Venus is 5.2 times denser; and Earth is 5.5 times denser. The high comparative density of Mercury has indicated to scientists that Mercury may have a massive iron core.

DISCUSSION Have the students study Figure 21–11. Explain that Venus has weather, including huge wind storms, lightning, and sulfuric acid rain. ***Thinking Critically:*** Ask the students to describe the processes that may have caused the Venusian surface to appear as it does. (The students may suggest weathering by abrasion, chemical breakdown, and other processes. Erosion by wind and gravity may also be suggested. Accept any reasonable answer.)

21.6 Earth and the Moon

EXTENSION Explain to the students that many scientists agree that an interesting sequence of events led to the evolution of life on Earth. Many elements, including carbon dioxide, water vapor, carbon monoxide, and methane, made up a primitive atmosphere very different from today's. Water condensed from this early atmosphere and created the oceans. Amino acids and nucleic acids in the ocean water developed into primitive organisms that took in carbon dioxide and exhaled oxygen. Thus the amount of CO_2 decreased while the amount of O_2 increased. Eventually, more complex, oxygen-inhaling organisms developed.

Space probes have determined that the atmosphere of Venus is very dense. The Venusian atmosphere consists of 96.0 percent carbon dioxide, 3.5 percent nitrogen, and 0.5 percent sulfur dioxide, argon, and water vapor. This dense atmosphere produces a strong greenhouse effect, which allows the surface temperature to reach 500°C.

Venus has a calculated density of 5.24 g/cm³, probably with an iron-nickel core almost as large as Earth's. Venus has no magnetic field. The reason may be that Venus, like Mercury, rotates too slowly for electrical currents to develop in the core.

○ ***Describe the surface of Mercury.***
○ ***Why is Venus called Earth's sister planet?***

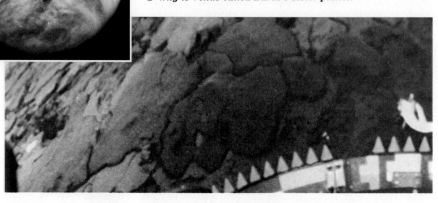

Figure 21–11. Venus is sometimes called the "cloud planet" because its atmosphere looks like the clouds of Earth. However, the atmosphere of Venus is made up of carbon dioxide rather than of nitrogen and oxygen as our atmosphere is. Venus was named after the Roman goddess of love and beauty.

21.6 Earth and the Moon

Earth is the third planet from the sun. Because of its location in the solar system, Earth's characteristics are very different from the other inner planets. Earth has many deep oceans, the atmosphere is moderately dense, and the surface temperature is moderate. These conditions probably contributed to the evolution of life on Earth.

The earth, with its iron-nickel core extending almost halfway to the surface, generates a magnetic field. The core is surrounded by a thick mantle of molten rock, which is covered with a solid crust.

Earth is the first planet from the sun to have a satellite. A **satellite** is a small body, or moon, orbiting a planet. The moon is nearly 385 000 km from Earth. Density calculations show that the moon may have a small iron-nickel core; however, it has no magnetic field. Data collected from instruments left on the moon by Apollo astronauts indicates the presence of a mantle and a 60-km-thick crust.

Answers–21.5

○ The surface of Mercury is heavily cratered.

○ Venus is similar to Earth in mass, density, and size.

Figure 21–12. The earth-moon system is shown here (left). The surface of the moon is barren due to the lack of an atmosphere (right).

The moon has two major terrains: the older highlands composed of feldspar rocks, and the younger maria, or seas, formed of basalt. These seas are not filled with water; they only resemble seas when viewed from Earth. The surface of the moon is marked with craters and covered with fine rock debris up to 20 m deep.

There are two major theories about the origin of the moon. One is that the moon was a separate body revolving around the sun, and it was captured by the earth's gravity. A second theory is that the moon is a piece of the earth that was broken off by a collision with another body. Neither theory has gained general support.

○ *What conditions are unique to the earth?*
○ *What is a satellite?*

21.7 Mars and the Asteroids

Mars, the fourth planet from the sun, orbits the sun in 687 Earth days. Mars is often called the red planet because its surface is reddish in color when viewed from Earth. The red color is probably the result of the oxidation of iron in the surface rocks. Mars has a radius a little more than half that of Earth. It has a significant, but not large, iron-nickel core and a weak magnetic field.

The surface of Mars includes deserts, dry riverbeds and flood plains, mountains, and the highest volcano in the solar system, Olympus Mons. Olympus Mons, which is now inactive, is 600 km across and rises to a height of 26 km. There are many surface craters, especially in the southern hemisphere.

> **DISCOVER**
>
> *Traveling to Mars*
> Using the table of distances on page 464, determine how long it would take to get to Mars, traveling by car at a safe 85 km per hour.

DISCUSSION You may want to explain a third theory about the moon's formation. Some scientists believe that both the earth and moon were formed in their current positions billions of years ago when the entire solar system formed. However, scientists are not certain how two large bodies could have formed so close together without combining into one body. ***Thinking Critically:*** Ask the students to explain why they agree or disagree with any of the three theories of the moon's formation. (Answers will vary. Students may suggest that they do not believe that there is evidence of another body striking the earth that would split off the moon. Some may not think that an object as large as the moon could be captured by the earth's gravity because the moon would either have crashed into the earth or hurtled past it.)

EXTENSION The examination of rocks brought back by astronauts from the moon has indicated that the dark rocks of the maria are 3.1 to 3.8 billion years old, while the light-colored rocks of the highlands are as old as 4.6 billion years. These ages, along with other information, have led scientists to conclude that until 3.8 billion years ago, huge rock debris bombarded the moon, blasting out large basins.

21.7 Mars and the Asteroids

DISCUSSION Review with the students why Mars is often called the red planet. (Its surface is reddish in color.) Ask the students how the surface of Mars is similar to Earth's. (Mars, like Earth, has streambeds, flood plains, mountains, and even volcanoes.)

DISCOVER: Traveling to Mars

Skill *(Calculating)*

When calculating the answer, have the students assume that the planets are aligned. It would take about 105 years to travel the 78 million km between the two planets.

Answers–21.6

○ Because of its distance from the sun, Earth has deep oceans and an atmosphere with moderate temperatures.

○ A satellite is a small body orbiting a larger body.

EXTENSION Begin a class discussion on what would be necessary in order to build and maintain a human colony on Mars. Explain that because it would take a very long time to get from the earth to Mars, the colony must be self-sustaining. Ask the students to list what would be needed. (The students may suggest the obvious, such as food, water, air, and shelter. They may also suggest such things as a government and health care. Accept all reasonable answers.) Divide the class into groups and assign each the task of designing one of the necessary systems. For example, one group could be responsible for food production, another for energy. Have the students research their projects, and then present a report to the class. Ask the students to take notes on one another's reports. After all the reports have been given, lead the class in a discussion of the pros and cons of each system designed.

THINKING SKILL *(Interpreting Photographs)* Have the students compare the satellites Phobos and Deimos in Figure 21–14 to the asteroid Ceres in Figure 21–15. Ask the students to identify the common features of the three. (Possible answers are that all are shaped irregularly and have deep craters, but the satellites have some areas with smoother surfaces.)

EXTENSION In addition to the asteroids in the belt between Mars and Jupiter, two other groups of asteroids exist. The Trojan asteroids are in the same orbit as Jupiter and travel in front of and in back of the planet. The Apollo asteroids travel in an orbit that brings them very near the earth.

EXTENSION You may wish to have the students look at the maps of Arizona, California, Hawaii, and New Mexico in the Science Site Booklet. These maps show the names and locations of some well-known observatories.

Figure 21–13. Mars is often called the "red planet" because it looks red as viewed from Earth. The surface of Mars is covered with red dust (right). Ice caps of frozen carbon dioxide are found at the poles.

The existence of dry riverbeds and flood plains may indicate that water was once present on Mars. Today, however, Mars is essentially dry, the water having been lost to space. Mars has two thin polar caps that expand and contract as the seasons change, but they are made of frozen carbon dioxide, not water. Space probes have shown that the atmosphere of Mars is very thin, consisting of 95.7 percent carbon dioxide, 2.7 percent nitrogen, and traces of argon, oxygen, carbon monoxide, and water.

Mars has two small satellites, Phobos and Deimos. Both have irregular shapes and may be captured asteroids. **Asteroids** are small planetary bodies that circle the sun in a belt between the orbits of Mars and Jupiter. The largest asteroid is Ceres, with a diameter of 1025 km. Large asteroids have a spherical shape; smaller asteroids have irregular shapes, like the two satellites of Mars.

Figure 21–14. Mars has two satellites: Phobos and Deimos. This planet was named after the Roman god of war. He had two attendants: Phobos, the bringer of fear, and Deimos, the bringer of panic.

468 Chapter 21 The Solar System

Answers—21.7 (p.469)

○ The surface of Mars is red because of the oxidation of its surface iron.

○ Asteroids are small planetary bodies orbiting the sun between the orbits of Mars and Jupiter.

Some scientists believe that the asteroids represent bodies that failed to combine into a planet. The gravity of the neighboring planet Jupiter was probably too strong for the asteroids to form a single body. The asteroids revolve around the sun in individual orbits, often colliding with each other. These collisions break up the asteroids and disturb their orbits.

Meteorites are asteroid fragments that leave orbit. Meteorites captured by Earth's gravity fall to Earth, occasionally forming a crater. Traces of these craters are visible on Earth. One of the largest ones, Meteor Crater, is in Arizona. Most of the craters on the moon are from meteorites.

○ *Why is Mars called the red planet?*
○ *What is an asteroid?*

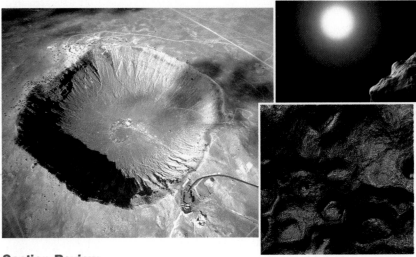

Section Review

READING CRITICALLY

1. What is the probable reason why the asteroids did not form a single planet?
2. How might the moon have formed?

THINKING CRITICALLY

3. Why are the surfaces of Mercury and the moon heavily marked with impact craters while those of Venus and Earth are not?
4. What are the possibilities that there might have been life on Mars at one time? Explain.

Figure 21–15. Ceres (top) is the largest asteroid in the asteroid belt. Meteorites (bottom) are formed when pieces of asteroids break off and leave orbit. Occasionally, a meteorite survives entry into Earth's atmosphere and crashes onto the surface of the planet (left).

Section 2   The Inner Planets and the Asteroids   **469**

THINKING SKILL *(Identifying Relationships)* Students are often confused by the difference between meteorites, meteors, and meteoroids. Meteoroids are any of the natural rock and metal fragments that revolve around the sun. Although most are asteroids, some are the remains of dead comets—either tail debris or the nucleus itself. When one of these meteoroids enters the earth's atmosphere, it burns up from friction, forming a tail of burning gases called a *meteor*. Have students discuss meteors, or "shooting stars" as they are commonly called, that they have seen. Ask the students to give examples of heat generated by friction. (Answers may include striking a match or rubbing hands together to warm them.) Explain that when a meteoroid does not burn up completely, it strikes the earth and shatters. The fragments are called *meteorites*.

SECTION REVIEW

Summary   Review with the students the main ideas of this section, including the positions of the inner planets relative to the other planets and the sun; the reason for the existence of many craters on Mercury but not on Earth or Venus; and the differences between Earth and Venus and between Earth and Mars.

Reinforcement   Have the students use the questions at the end of each numbered subsection to test one another on the major concepts of this section.

Answers to Section Review

Reading Critically

1. The gravity of the huge neighboring planet, Jupiter, probably kept the accumulating cosmic debris apart.
2. The moon may once have been a part of the earth that was broken off by a collision with another celestial body. Or it may have been a separate body captured by earth's gravitational pull.

Thinking Critically

3. Venus and Earth have atmospheres that disintegrate meteors as they plummet to their surfaces. Mercury and the moon do not.
4. It is possible that life once existed on Mars because water was probably free flowing and abundant at one time on that planet.

Objectives

- Calculate the diameter of the sun, using geometric relationships.
- Apply the geometric relationship to the measurement of other objects.

Discussion The measurement of the sun's diameter is based on the geometric rule that equilateral triangles with the same angles have proportional sides. Draw an elongated "X" on the board. Erase two-thirds of the two arms of the X on the left. Connect the end of the lines with a vertical line; also connect the two longer arms on the right with a vertical line. Bisect the two resulting triangles with a horizontal line. Explain to the class that the apex of the two triangles corresponds to the holes in their cards, the large base (D_s) corresponds to the sun's diameter, and the small base (D_i) to the diameter of the sun's image. The calculation is possible, using the ratio of the altitude of the large triangle (C_s) and small triangle (C_i) to the bases of each triangle, or $C_s/D_s = C_i/D_i$.

Answers to Application

Answers will vary but should be approximately 1 392 000 km.

Answer to Using What You Have Learned

The relationship of altitude to base would still hold true, although the moon would be represented by the smaller triangle.

SKILL ACTIVITY: Measuring the Diameter of the Sun

BACKGROUND

The diameter of the sun has been known since ancient times. All you need to make the same calculations that ancient astronomers made are a little mathematics and a simple knowledge of how light behaves.

PROCEDURE

In order to make a measurement of this sort, you must first construct the necessary instrument.

1. Cut a small index card as shown. Then, using a T-pin or the tip of a ballpoint pen, make a hole about 2 cm from the top of the middle of the cut card.
2. Tape another index card to the end of a meter stick as shown.
3. Slip the card with the hole onto the meter stick, so that you can slide it to any position along the stick.
4. With your back to the sun, place the meter stick over your shoulder so that you are looking at the card. Aim the back of the stick at the sun.
5. Beginning at the end closest to the sun, move the sliding card toward the end card. An image of the sun will appear on the end card.

APPLICATION

Once you have mastered the use of the instrument, you can begin to make your measurement.

1. Position the sliding card so that the image of the sun is in focus.
2. Circle the sun's image with a pencil as accurately as you can and measure the diameter in millimeters.
3. Note the position of the sliding card (in millimeters) from the end card.
4. Calculate the diameter of the sun according to the following formula:

$$D_s = \frac{D_i \times C_s}{C_i}$$

D_s is the diameter of the sun; D_i is the diameter of the sun's image in millimeters; C_s is the distance from the earth to the sun (use 1.49×10^{14} millimeters); and C_i is the distance in millimeters from the end card to the sliding card when the sun's image is in focus.

5. Record your results on a sheet of paper and compare them with your classmates' results. Explain any differences.

USING WHAT YOU HAVE LEARNED

Explain how this procedure could be used to measure the diameter of the moon.

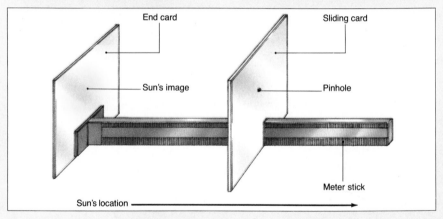

End card • Sun's image • Sliding card • Pinhole • Meter stick • Sun's location

SECTION OBJECTIVES

After completing this section, you should be able to:

- **Compare** the structure of the outer planets with that of the inner planets.
- **List** the peculiarities of the Galilean satellites of Jupiter.
- **Discuss** the structure and composition of comets.

NEW SCIENCE TERM

comets

21.8 Jupiter

The outer planets include the giant planets Jupiter, Saturn, Uranus, and Neptune, and the smallest planet, Pluto. The giant planets have many features in common. They all consist mainly of frozen gases and have atmospheres that are 90 percent hydrogen and 10 percent helium. Jupiter, Saturn, and Uranus have orbiting ring systems and numerous satellites. Neptune has only two satellites that are visible from Earth and no apparent rings.

With a mass 2.5 times that of all the other planets combined, Jupiter is by far the largest planet in the solar system. Jupiter is called a *gaseous giant* because it has no solid crust as Earth has. Beneath the thick atmosphere lies Jupiter's mantle, which extends all the way to its core. The core may have a radius of 10 000 km, larger than Earth's radius.

Jupiter's atmosphere is 17 000 km thick, and although it is not warmed much by the sun, there is convection. This may be because Jupiter itself is very warm—in fact Jupiter radiates about twice as much energy into space as it receives from the sun. This energy probably comes from radioactivity or convection in the core and mantle.

Figure 21–16. Jupiter (right), the largest planet, was named for the Roman king of the gods. This planet is characterized by a huge red spot (left), which is a swirling mass of hot gases.

471

SECTION FOCUS

Section Overview This section concentrates on the major similarities and differences among the outer planets: Jupiter, Saturn, Uranus, and Neptune. The ways in which the outer planet Pluto differs dramatically from the other planets are discussed. Comets are described and their orbits are explained.

Section Objectives For a list of section objectives, see pupil's edition page.

New Science Term For the new science term in this section, see pupil's edition page. You may wish to call the students' attention to the term before beginning the section.

SECTION DEVELOPMENT

21.8 Jupiter

DISCUSSION Direct the students' attention to Figure 21–16. Have them locate the Great Red Spot in the photograph on the right. Explain that the red spot is a swirling mass of gases that move like a giant hurricane across the surface of the planet. Ask the students to describe other distinguishing features of Jupiter. (bright white, yellow, and red bands, many of which are parallel) Tell the students that they will learn what causes the bands and the red spot on page 472.

THINKING SKILL *(Observing Relationships)* Supply the students with sheets of paper, nine different-colored pencils, and protractors. Have them draw a small point in the center of the paper to represent the sun's position. Have them use a red pencil and a ruler to draw a horizontal line through the point, extending the line 7 cm on both sides. This represents the earth's orbital plane. The following figures are the numbers of degrees from which the orbital planes of the other planets differ from Earth's: Mercury, 7°; Venus, 3.4°; Mars, 1.9°; Jupiter, 1.3°; Saturn, 2.5°; Uranus, 0.8°; Neptune, 1.8°; and Pluto, 17.2°. Have the students draw the orbital planes of the planets, using a different colored pencil for each planet. Ask the students to explain how the point of intersection in the orbital planes of two aligned planets is related to the distance between the two. (The farther the planets are from the points where their orbital planes intersect, the farther apart the two planets will be.)

DISCUSSION Have the students study the photos of Jupiter's four largest satellites in Figure 21–17. Ask them to describe the similarities and differences among the four. Then have them compare the appearance of the Galilean satellites to that of the earth's moon. Ask the students to make inferences about the events that have caused the different appearances of these satellites.

The light-colored bands of Jupiter's atmosphere are hot, rising gases; the dark-colored ones are cooler, sinking gases. These colored bands rotate around the planet at different speeds. A prominent feature of Jupiter's atmosphere is the Great Red Spot south of the equator. The red spot is a swirling mass of gases, with winds probably in excess of 1000 km per hour. Circulation in the red spot is counterclockwise because of Jupiter's Coriolis effect. How do you think the Coriolis effects of Jupiter and Earth compare?[1]

The four largest of Jupiter's 16 satellites were discovered in 1610 by the Italian astronomer Galileo. In his honor they are called *Galilean satellites.* In order of distance from the planet, the satellites observed and named by Galileo are Io, Europa, Ganymede, and Callisto.

Io, which is slightly larger than Earth's moon, has more volcanic activity than any other body in the solar system. Unlike Earth, however, the energy for the volcanic activity of Io is tidal, caused by the gravity of neighboring Jupiter.

Europa is smaller than Earth's moon, and its surface is covered with a thick layer of ice. There are very few impact craters on Europa, suggesting that the surface reseals itself after a meteorite impact.

Ganymede is larger than Mercury. In fact, it is the largest satellite in the solar system. Ganymede's surface has some impact craters, but it also has features, such as surface grooves, that may indicate more recent structural changes. These grooves may be caused by tension in the crust, like the fractures caused by tension in Earth's crust. Callisto, on the other hand, is heavily cratered, indicating that its surface has not changed much since its formation.

Figure 21–17. The four largest satellites of Jupiter are shown here: Io (top), Europa (bottom left), Gannymede (bottom right), and Callisto (bottom center).

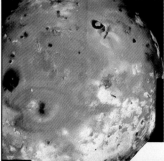

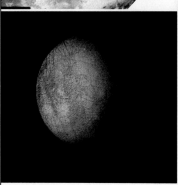

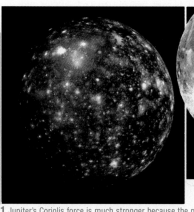

1 Jupiter's Coriolis force is much stronger because the planet is larger.

Figure 21–18. The rings of Jupiter can be seen in this Voyager photograph.

Jupiter has a ring system. The rings consist of micrometer-sized particles of metals, silicates, and ice. The ring system is not visible from Earth; it was discovered in 1976 when the United States spacecraft *Pioneer 11* photographed Jupiter.

○ *Why is Jupiter called a gaseous planet?*
○ *Name and give several characteristics of Jupiter's large satellites.*

21.9 Saturn, Uranus, Neptune, and Pluto

Saturn, shown in Figure 21–19, has the lowest density of all the planets because it consists mainly of ice. The rocky core of Saturn is small, and the structure of its mantle and atmosphere is similar to that of Jupiter. Like Jupiter, Saturn gives off more energy than it receives from the sun.

Titan, Saturn's largest satellite, is larger than Mercury but smaller than Ganymede. Titan consists of a rocky core surrounded by a thick mantle of ice. The atmosphere of Titan consists mostly of nitrogen, and there is evidence of active volcanoes on Titan. The remaining satellites of Saturn are small, heavily cratered bodies of silicate rocks and ice.

A MATTER OF FACT

Both Pluto (2300 km in diameter) and Mercury (4880 km in diameter) are smaller than Saturn's satellite Titan (5150 km in diameter) and Jupiter's satellite Ganymede (5276 km in diameter).

Figure 21–19. Saturn can be identified by its massive ring system (left). This planet was named for the Roman god of agriculture.

473

EXTENSION Supply interested students with old astronomy or science books or old encyclopedia volumes that contain information about the planets. Be sure that these references are at least 40 years old. Have the students write a report on the changes in our knowledge and understanding of the solar system within the last four decades. You may want to assign different planets to small groups of students and then have them give oral reports on their findings to the class. Point out that because new data is constantly being gathered, our knowledge of the planets needs to be continuously updated.

21.9 Saturn, Uranus, Neptune, and Pluto

DISCUSSION Point out that Titan is the second largest moon in the solar system and is the only moon having a thick atmosphere, which is about 85 percent nitrogen. There are also traces of other gases, including about 12 percent oxygen and 3 percent methane and hydrocarbons. This combination of elements, although deadly to humans, is very similar to the primitive atmosphere of Earth. *Thinking Critically:* Ask the students what hypothesis they might make concerning life on Titan, Saturn's moon. (Titan has the same essential elements as early Earth, and thus the possibility that some life forms exist on Titan is greater than in most other places in the solar system.)

EXTENSION (Tie-in/Physics) Inform the students that gravity differs greatly from planet to planet. For example, if the surface gravity on Earth is given the value of 1, the gravity of the moon would be 0.17 and that of Saturn would be 1.15. Using the formula *weight = mass × gravity*, or $w = mg$, have the students calculate what they would weigh if they were on the moon or Saturn.

Answers—21.8

○ Jupiter is called a gaseous planet because it has no solid crust.

○ Io is larger than Earth's moon and is volcanically active. Europa is smaller than Earth's moon and is covered with ice. Ganymede is larger than the planet Mercury and is covered with grooves and some impact craters. Callisto is also larger than Earth's moon and is heavily cratered.

BACKGROUND INFORMATION

In the summer of 1929, twenty-two-year-old Clyde Tombaugh became an assistant at the Lowell Observatory in Arizona. His assignment was to find the ninth planet, "Planet X." Working over a period of time, Tombaugh photographed a section of the night sky several days apart. The stars remained in place in each photo; the positions of moving planets, however, changed from one photo to the next. By painstakingly comparing minute portions of these photos, he discovered a planet—Pluto.

DISCUSSION   At this point, you may
want to lead a class discussion about the
findings of *Pioneer 10* and *11* and *Voyager 1* and *2*, space probes that were sent
to examine the environs of the outer planets. Have the students research these
probes beforehand so that they will be able
to participate actively in the discussion,
which should include information about
when the probe was launched, when each
reached or will reach the various planets,
and what data the probes sent or are sending back. ***Thinking Critically:*** Ask the
students to describe the difficulties of
space travel that specialists had to consider. (Students may suggest getting
through the asteroid belt or mechanical
malfunctions.)

THINKING SKILL   *(Recognizing Relationships)* Write the formula *density =
mass/volume* on the board. Explain that
Uranus is 15 percent more voluminous
than Neptune, but Neptune is more massive than Uranus. Ask the students to
determine which planet is denser and to
explain their answer. (Neptune must be
denser. According to the density formula,
the numerator [mass] is smaller than the
denominator [volume] for Uranus, and the
numerator is larger for Neptune. Thus, no
matter what the actual numbers are, Neptune's density is a larger ratio.)

Figure 21–20. The ring system
of Saturn (right) has been
studied by many scientists since
the time of Galileo. The study
continues today at places such
as the Jet Propulsion Laboratory
in Pasadena, California (left).

Saturn has a complex ring structure, first observed by
Galileo in 1610. The ring system lies on Saturn's equatorial plane
and consists of hundreds of concentric rings separated by gaps
of varying width. The rings are made of fragments of rock and
ice ranging in size from less than 1 cm to over 10 m.

Uranus and Neptune are gaseous planets with rocky cores.
Uranus has at least 15 satellites, while Neptune has only 2 that
are visible from Earth. Uranus is unusual in that its axis is
inclined 98° from the plane of its orbit, and it rotates retrograde.
The ring system of Uranus is thin, with particle sizes similar to
those of Saturn's rings. Neptune has at least a partial ring; scientists will know more after analyzing photographs taken by the
Voyager 2 spacecraft.

Figure 21–21. Of the three outer
planets, Uranus (left) and
Neptune (center left) are
gaseous; Pluto (right) is solid.
Scientists are just now learning
about these planets from space
probes that are passing the
planets and sending back
computer images and
information.

Pluto is the outermost planet. Pluto is also the smallest
planet in the solar system and the only outer planet without a
thick atmosphere. Little is known about Pluto except that it has
one known satellite, Charon, discovered in 1978.

○ ***What characteristics do Saturn and Uranus share with
Jupiter?***
○ ***What is unusual about Uranus?***

474   Chapter 21   The Solar System

Answers–21.9

○ They are all large, gaseous planets.

○ Uranus is tilted on an axis that is 98° from
the plane of its orbit, and it has retrograde
rotation.

Answer–21.10 (p. 475)

○ Comets are clusters of silicate rock embedded in ice that revolve around the sun
in large elliptical orbits.

21.10 The Comets

Beyond Pluto are bodies that orbit the sun over extremely long periods of time. **Comets** consist of silicate rock and metal particles embedded in ices. The ices are frozen gases such as carbon dioxide, methane, ammonia, and water. Comets may be condensations of planetary matter, formed when the solar system was beginning to take shape.

Comets are frequently captured by the giant planets—mainly Jupiter—and forced into orbits that take them close to the sun. When a comet approaches the sun, gases escaping from the comet are blown by the solar wind. These gases form a tail, as much as 1 million km thick and 10 million km long, pointing away from the sun. Comets form an envelope around the sun called the *Oört cloud.*

0:14:57   61500 km

The diameter of a typical comet is about 1 km. The body of the comet loses a layer several meters thick each time the comet moves around the sun. Scientists believe that most comets do not last more than a dozen passages around the sun. Halley's comet, however, has been observed every 76 years since 240 B.C. How many passages of Halley's comet have been seen from Earth?[1]

○ **What are comets?**

Figure 21–22. Perhaps the most famous comet is Halley's comet (left), which comes near the earth every 76 years. The core of this comet (right) is composed of rocks and ice.

DISCOVER

Looking for Halley

From the information on the period of revolution of Halley's comet, determine how old you will be the next time Halley's comet returns.

Section Review

READING CRITICALLY

1. How do the four Galilean satellites differ from each other?
2. Of what materials are the rings of Saturn made?
3. How are comets' tails formed?

THINKING CRITICALLY

4. Why are scientists so interested in the composition of comets?
5. Do you think that the ring systems of the outer planets are like unformed moons? Explain.

1 29

Section 3   The Outer Planets and the Comets   **475**

21.10 Comets

DISCUSSION Explain that the Oört cloud is spherical and that it is 15 trillion kilometers from the sun. Most comets orbit the sun over a period of several thousand years. Some return once every few years. Ask the students if they can name any comets that have recently passed the earth. (Comet Kahoutek or Halley's comet) *Thinking Critically:* Ask the students to explain in what direction the tail of a comet will point when approaching and when moving away from the sun. (The tail will always stream away from the sun because the solar wind blows away from the sun in all directions and also blows particles and gases away from comets.)

DISCOVER:
Looking for Halley

Thinking Skill *(Interpreting Data)*

Explain that the period of Halley's comet occasionally varies between 76 to 79 years. Thus in the two thousand years since the comet was first observed, it has arrived an average of one year later each 100 years. Taking this data into account, the students must first add together all the years that have passed since the comet was spotted for the first time. That amount is divided by 76. The remainder of that result is the number of years ago that the comet last passed by. However, remind students to subtract 1 year for every 100 years that have passed since 240 B.C. The students can then determine what age they will be the next time the comet appears.

SECTION REVIEW

Summary Have the students write the name of each planet on an index card. On the other side of the card, have them list each basic fact about the planet's diameter, surface features, and number of moons. Have them test each other, using these cards.

Reinforcement Have the students summarize the main idea of each paragraph in the numbered subsections of this section.

Answers to Section Review

Reading Critically

1. They are different distances from Jupiter and have different surface features. Io is volcanic. Europa is covered with ice. Ganymede has impact craters and grooves. Callisto is the most heavily cratered.
2. The rings of Saturn are made of rock and ice.
3. As the comet approaches the sun, its core material is vaporized and blown away by the solar wind.

Thinking Critically

4. Comets provide evidence of what materials may have made up the solar system during its formation.
5. Yes. The planets Jupiter, Saturn, and Uranus are known to have more than several satellites. All three planets are also known to have ring systems. The gravitational fields of their satellites could have interfered with the formation of other satellites, leaving the cosmic debris to form rings instead.

475

INVESTIGATION 21:
Drawing Planet Distances to Scale

Skill (Communicating)

Preparation of Materials

A regular $8\frac{1}{2} \times 11$ sheet of construction paper should be large enough for this investigation. Before beginning the investigation, you may want to review the use of scientific notation and exponents with the students so that the conversions can be done more quickly. You may also want students to do the scale conversions on paper before trying to draw the planets to scale.

Answers to Analyses and Conclusions

1. The diameter of the sun is more than one hundred times that of the inner planets.
2. The pieces of construction paper representing the planets are extremely small in comparison to the distances between the model planets. For example, the circle for the planet Mercury is only half a millimeter. That is barely larger than the period on a page of text. Yet at this scale, the same planet is nearly 6 m from the sun.
3. 3.84

Answer to Application

Each AU is 1 500 000 km, so each distance would need to be converted.

PURPOSE

To make a scale model of the sun and inner planets

MATERIALS (per group of 3 or 4)

String, 30 m
Metric ruler
Construction paper
Scissors
Tape
Markers

PROCEDURE

1. Table 1 shows the diameter of the sun and each planet as well as each planet's mean distance from the sun.
2. Using a scale in which 1 mm = 10^4 km, draw the sun and planets on a piece of construction paper. The size of each planet does not have to be exact. At this scale, you will find that they are all small and that exact dimensions are difficult to represent.
3. Draw a square around the sun and each planet, making sure to label each square for identification purposes.
4. Cut out the squares.
5. Tape the square of the sun to the end of the string.
6. According to the same scale used in step 2, determine the distance of each planet from the sun and tape it to the string.
7. Stretch out the string.

ANALYSES AND CONCLUSIONS

1. How do the sizes of the planets compare to that of the sun?
2. What difficulty did you experience in trying to draw the planets to scale?
3. Using the same scale of measurement, how far from Earth is Earth's moon (384.4×10^3 km) in centimeters?

APPLICATION

Explain how you would use astronomical units instead of kilometers in this investigation.

TABLE 1: THE SUN AND THE INNER PLANETS					
	Sun	**Mercury**	**Venus**	**Earth**	**Mars**
Diameter (km)	1.4×10^6	4900	12 000	12 750	6800
Distance from the Sun (km)		5.8×10^7	1.08×10^8	1.50×10^8	2.28×10^8

SUMMARY

- Earth rotates counterclockwise on its axis once every 24 hours. (21.1)

- Earth revolves around the sun in a specific orbit every 365-1/4 days. (21.2)

- The tilt of the axis from a line perpendicular to the plane of its orbit is about 23.5°. (21.2)

- Phases of the moon are caused by the varying illumination of the moon's surface. (21.3)

- The solar system contains nine planets, each of which revolves around the sun. (21.4)

- Mercury, the planet closest to the sun, has a cratered surface and practically no atmosphere. (21.5)

- The atmosphere of Venus, the second planet from the sun, is rich in carbon dioxide. Venus is sometimes referred to as Earth's sister planet because of its similar size, mass, and density. (21.5)

- Earth is the third planet from the sun, the first planet with a satellite, and the only planet known to have life. (21.6)

- Mars, the fourth planet from the sun, is called the red planet. Although it may have had water at one time, none presently exists. Polar ice caps of frozen carbon dioxide can be seen on its surface. (21.7)

- Asteroids are planetary fragments that did not join together to form a planet. Asteroids orbit between Mars and Jupiter. (21.7)

- Jupiter is the largest planet in the solar system. It emits twice as much energy as it receives from the sun. (21.8)

- The rings of Saturn consist of fragments of rock and ice from less than 1 cm to more than 10 m in diameter. (21.9)

- Uranus is unusual because of its nearly horizontal tilt and retrograde rotation. (21.9)

- Comets consist of metal and silicate particles embedded in ices. (21.10)

Write all answers on a separate sheet of paper.

SCIENCE TERMS

Correctly use each of the following terms in a sentence.

asteroids **(468)**
astronomical unit **(462)**
axis **(455)**
comets **(475)**
elliptical **(462)**
lunar eclipse **(459)**
meteorites **(469)**
orbit **(456)**
penumbra **(459)**
planets **(461)**
revolution **(456)**
rotation **(455)**
satellite **(466)**
solar eclipse **(460)**
umbra **(459)**

SCIENCE QUIZ

Modified True-False

Mark each statement *true* or *false*. If a statement is false, change the underlined term to make the statement true.

1. Jupiter gives off <u>less</u> energy than it receives.

2. The thick atmosphere of Venus makes its surface <u>cooler</u> than the surface of Mercury.

3. <u>Asteroids</u> falling to Earth usually burn up in Earth's atmosphere before they hit the ground.

4. The four largest moons of Jupiter were named by <u>Kepler</u>.

5. Between the orbits of Mars and Jupiter lies a belt of <u>comets</u>.

continues

Chapter 21 Review   **477**

CHAPTER REVIEW

SUMMARY

Students may review the major concepts in the chapter by reading the summary statements. The statements are cross-referenced to the chapter to facilitate reinforcement of any concepts of which the students feel unsure. Encourage the students to work in groups to quiz one another.

SCIENCE TERMS

The sentence in which the science term is used should reflect an understanding of the definition of the term. You may wish to stage a game of "Jeopardy" during which the definition of the term is provided and the "contestants" must guess the correct term.

SCIENCE QUIZ

Answers to Modified True-False

1. false, more
2. false, warmer
3. true
4. false, Galileo
5. false, asteroids

continues

Answers to Multiple Choice

6. d
7. c
8. a
9. b
10. d

Answers to Completion

11. Mercury
12. Jupiter
13. Jupiter
14. Pluto
15. Venus

Answers to Short Answer

16. Mercury, Venus, Earth, Mars, and Pluto
17. Jupiter's large moon, Io, has active volcanoes. Space probes have taken photographs of volcanic eruptions on this moon.
18. None of the orbits of the nine planets are perfectly circular. They are elliptical, meaning that each approaches the sun at a nearest point, called *perihelion,* and swings out to a farthest point in its orbit, called *aphelion.*
19. They were first observed by Galileo.
20. All planets rotate prograde except Venus and Uranus, which rotate retrograde.

Answers to Writing Critically

21. The material that vaporizes off a comet is pushed away from the sun by the solar wind.
22. Mercury, like the moon, does not have an atmosphere to disintegrate meteoroids as they fall to its surface. Venus, Earth, and Mars have atmospheres.
23. Pluto may have been a satellite that escaped from another planet, such as Neptune.
24. This is one of the regular motions of the universe, possibly due to movement resulting from formation of the solar system.

ANSWERS TO EXTENSION

1. Answers will vary. Howver, students may find that Dr. James A. Van Allen, the physicist who discovered the radiation belts surrounding Earth, is one of the better known scientists who has

argued that the nation should place more emphasis on robot spacecraft and less on peopled missions.
2. Answers will vary, although students should recount the difficulties scientists faced in deciding, in the first place, what to look for. They chose to test for the existence of several chemical reactions known to take place in living organisms. The Martian soil was treated with a nutrient

solution and the evolved gases were tested for the presence of organic products. The conclusions of the experiments were negative.
3. Answers will vary according to the planet chosen.

SCIENCE QUIZ **continued**

Multiple Choice

Write the letter of the choice that best answers the question or completes the statement.

6. Which of the following is NOT an inner planet?
   a) Mercury   b) Venus
   c) Mars   d) Jupiter

7. The only outer planet that is NOT a gas giant is
   a) Jupiter.   b) Saturn.
   c) Pluto.   d) Neptune.

8. The planet without rings is
   a) Pluto.   b) Jupiter.
   c) Saturn.   d) Uranus.

9. What happens to the length of a planet's revolution the nearer it is to the sun?
   a) It increases.
   b) It decreases.
   c) It remains the same.
   d) It becomes retrograde.

10. Which of the following planets is solid?
    a) Jupiter   b) Saturn
    c) Uranus   d) Mars

Completion

Complete each statement by supplying the correct term or phrase.

11. The order of the inner planets from the sun is _____, Venus, Earth, and Mars.
12. The order of the outer planets from the sun is _____, Saturn, Uranus, Neptune, and Pluto.
13. The atmosphere of _____ is heated from the planet itself.
14. The planet with the shortest day is _____.
15. Earth's sister planet is _____.

Short Answer

16. List the planets that have solid surfaces.

478   Chapter 21 Review

17. What bodies besides Earth seem to have active volcanoes? What is the evidence?
18. Describe the shape of the orbits of the nine planets.
19. Why are the four largest satellites of Jupiter known as Galilean satellites?
20. Which planets rotate prograde and which rotate retrograde?

Writing Critically

21. Explain why a comet's tail always points away from the sun.
22. Explain why the surfaces of Mercury, Mars, and Earth's moon are pockmarked with craters, while the surfaces of the inner planets, Venus and Earth, are not.
23. All of the outer planets except Pluto are gas giants. What might be a possible reason why Pluto is not gaseous?
24. Why do all the planets revolve around the sun in the same direction?

EXTENSION

1. Some scientists believe that unpiloted space probes can accomplish more than space missions with humans. Collect and report information about the current "piloted versus unpiloted" spacecraft controversy.
2. Review the information sent back from the United States' Viking probes to Mars. Report to the class on what scientists did to test for the presence of life on that planet and what the results of those experiments showed.
3. Review the conditions that exist on another planet. According to the conditions of the planet, describe an imaginary creature that might live there. For example, a Venusian would have to breathe carbon dioxide and be nearly impervious to scalding heat and acid rain.

APPLICATION/CRITICAL THINKING

1. The mean distance of the planet Jupiter from the sun is 7.783×10^8 km. If Earth and Jupiter are aligned on the same side of the sun, how long will it take a radio transmission from a space probe in orbit around the giant planet to reach Earth?

2. Explain possible reasons why the asteroids did not form a planet.

3. As the engineers working with the United States space program were preparing to send space probes to the moon, they would estimate where the moon would be several days after launch and then try to hit that spot with the probe. Why could they not aim for the location of the moon on the day of the probe's launch?

FOR FURTHER READING

Gallant, R. A. *National Geographic Picture Atlas of Our Universe.* Washington, D.C.: National Geographic Society, 1986. This beautifully done volume summarizes the wonders of the solar system. The work includes photographs taken by space probes as well as illustrations of bodies that share the neighborhood of the sun.

Weinhouse, B. "Leading Ladies of the '80s: Space Age Mother." *Ladies Home Journal* 102 (May 1985): 139. This brief article describes the life of astronaut Anna L. Fisher.

Challenge Your Thinking

You may recall from reading the chapter that Jupiter gives off about twice as much energy as it receives. Scientists believe that this energy comes from nuclear reactions within the planet. With this information, hypothesize about the possibility of Jupiter being a second sun with its own solar system.

ANSWERS TO APPLICATION/ CRITICAL THINKING

1. about 35 minutes
2. The effects of Jupiter's gravity may have disrupted the formation of a new planet.
3. The moon and Earth are both moving through space, and the location of the moon at the time the probe arrives must be determined.

ANSWER TO CHALLENGE YOUR THINKING

Jupiter gives off more energy than it receives from the sun, and the satellites of Jupiter are similar to small planets because of their size. This means that Jupiter may be a small "solar system" within the larger system of the sun.

Chapter 22: THE UNIVERSE

PLANNING THE CHAPTER

Chapter Sections	Page	Chapter Features	Page	Program Resources	Page
Section 1: Stars	481			Cross-Discipline: Science and Music, *Analyzing a Song* **(B)**	**TRB 41**
22.1 Stars **(B)** 22.2 Constellations **(B)**	481 483	A Matter of Fact **Discover:** Observing Constellations **(B)** **Activity:** Making a Chart of the Horizon **(B)**	482 483 484	Reading for Content: *Cycles and Sequencing* **(B)** Concept Extension: *Constellations* **(B)** Investigation 22.1: *The Hertzsprung-Russell Diagram* **(H)**	**TRB 38** **LM 215** **TRB 45** **LM 267** **TRB 101** **LM 103**
22.3 Red Giants, White Dwarfs, and Novas **(A)**	484	A Matter of Fact A Matter of Fact Section Review	486 488 488		
22.4 Neutron Stars and Pulsars **(A)**	487	**Skill Activity:** Using a Star Chart **(B)**	489		
Section 2: The Sun	490				
22.5 The Composition and Structure of the Sun **(A)**	490				
22.6 Motions and Activity of the Sun **(A)**	491	**Discover:** Calculating Time **(A)** Section Review	491 492		
Section 3: Galaxies	493			Critical Thinking **(H)** Investigation 22.2: *Understanding Parallax* **(A)**	**TRB 42** **TRB 105** **LM 107**
22.7 Measuring Distances in the Universe **(H)**	493	**Discover:** Traveling Fast **(H)**	493		
22.8 The Milky Way **(H)** 22.9 Galaxies, Quasars, and the Big Bang **(H)**	494 495	 A Matter of Fact	 495		
		A Matter of Fact Section Review **Careers:** Astronomer, Astrophysicist, Astronaut **Investigation 22:** Determining Distance by Parallax **(H)**	496 496 497 498	Student Record Book: Textbook Investigations **(H)**	**TRB 45**
Chapter 22 Review	499			Vocabulary **(A)** Tests **(A)** Computer Test Bank	**TRB 22** **LM 165** **TRB 96**

(LM) Laboratory Manual/Study Guide,   **(TRB)** Teacher's ResourceBank™

B = Basic   **A** = Average   **H** = Honors

The coding Basic, Average, and Honors indicates sections or subsections that might be appropriate for different levels of learners. For additional suggestions regarding choice of topic and depth of coverage, see the Pacing Chart on pages T16–T20.

CHAPTER CONCEPTS, OBJECTIVES, AND TERMS

Section	Concepts	Objectives	Science Terms
Section 1: Stars	■ A star is a huge mass of glowing gases, mainly hydrogen and helium, which formed from a swirling cloud of dust and gases. **(22.1)** ■ The magnitude of a star is a number indicating its apparent brightness as seen from Earth. **(22.1)** ■ A constellation is a group of stars in a pattern that seems to form an image. **(22.2)** ■ The Hertzsprung-Russell diagram shows how the energy emitted by a star is related to its color. **(22.3)** ■ A nova is a star that becomes suddenly brighter due to an explosion and then returns to its original brightness. **(22.3)** ■ A neutron star is the small, extremely dense star core that may be the result of a supernova. **(22.4)** ■ A black hole is the result of a supernova core that completely collapses into itself. **(22.4)**	■ **Describe** how a star forms. ■ **Discuss** the meaning of magnitude for a star. ■ **Identify** the different types of variable stars. **Tell** how the color and the temperature of a star are related.	universe magnitude variable stars constellations zodiac neutron star pulsar black holes
Section 2: The Sun	■ The sun has a layered structure consisting of the core, the radiative layer, the convective layer, and the atmosphere. The atmosphere has three layers: the photosphere, the chromosphere, and the corona. **(22.5)** ■ Sunspots are cooler areas on the sun's surface that appear dark from Earth. **(22.6)** ■ Solar flares are eruptions of the photosphere near sunspots. Solar prominences are streams of gases that explode from the photosphere and are pulled back to the sun. **(22.6)**	■ **Name** the different layers of the sun. ■ **Describe** the activity on the sun.	solar prominences
Section 3: Galaxies	■ A light-year is the distance light travels in a year—it is equal to 9.6 billion km. **(22.7)** ■ A galaxy is a network of stars. The sun is in the Milky Way Galaxy. **(22.8)** ■ A quasar is a starlike source of radio waves billions of light-years away from Earth. **(22.9)** ■ The most widely accepted theory of the beginning of the universe is the big-bang theory. **(22.9)**	■ **Explain** what a light-year is. ■ **Describe** the Milky Way Galaxy. ■ **Name** the different kinds of galaxies. ■ **Identify** nebulae and quasars. ■ **Explain** the big-bang theory.	light-year galaxy nebula quasars big-bang theory

CHAPTER MATERIALS

Title	Page	Materials
Discover: Observing Constellations	483	*(per student)* paper, pencil
Activity: Making a Chart of the Horizon	484	*(per student)* magnetic compass, drawing compass, construction paper, colored pencils, protractor, metric ruler
Skill Activity: Using a Star Chart	489	*(per student)* compass, construction paper, scissors, tracing paper, ruler, cardboard, pencil
Discover: Calculating Time	491	*(per student)* paper, pencil
Discover: Traveling Fast	493	*(per student)* paper, pencil
Investigation 22: Determining Distance by Parallax	498	*(per student)* drawing compass, construction paper, scissors, tape, insect pin, modeling clay, metric ruler, protractor, table of tangents

TEACHING SUGGESTIONS

Section 1: Stars

Demonstration: The Constellations

Purpose

To familiarize students with some of the most commonly known constellations

Materials

Aluminum foil or black construction paper
Scissors
Pin
Small piece of cardboard
Filmstrip projector
Drawings of constellations

Procedure

1. Measure and cut a strip of aluminum foil or black paper so that it is the same width as a filmstrip.
2. Using the strip of foil or paper, make a "filmstrip" of the constellations by placing the "film" on top of the cardboard and poking holes with the pin to locate each star.
3. Roll the strip and feed it into the filmstrip projector.
4. Project the "filmstrip" onto a screen and try to identify constellations.

Questions to Ask Students

1. Why was black paper or foil needed? (*To eliminate all background light.*)

2. Name the constellations you recognize. (*Answers will vary.*)

Field Trip

Write to or visit a local observatory and obtain information about how astronomers record and catalog the positions of celestial objects.

Section 2: The Sun

Class Activity

Make a "one-person planetarium" by constructing a geodesic dome out of small, hexagonal cardboard panels. Wallpaper the interior of the dome with black construction paper. Then, use a star chart to puncture small holes in the celestial sphere at the stars' correct locations. Light the stars by placing several lamps on the outside of the dome.

Field Trip

Visit a local planetarium and view a "star show" about constellations and other lights in the universe.

Section 3: Galaxies

Outside Speaker

Have an amateur astronomer visit the class and talk about his or her experiences of viewing the night sky. The person could also talk about the scientific discoveries made by amateur astronomers.

CHAPTER 22

CHAPTER OVERVIEW

This chapter discusses various types of stars and groups of stars, or galaxies, in the universe. The sun's location and structure are explained. The light-year is defined and used to measure distances in the universe. One theory for the origin of the universe is discussed.

Section 1: Stars The origin of stars is discussed. The classification of stars by magnitude, or brightness, is explained. The kinds of variable stars are described, and constellations are examined. The further classification of stars by temperature and color is discussed and related to the Hertzsprung-Russell diagram.

Section 2: The Sun The layered structure of the sun is described and each layer, from its core to its outer atmosphere, is defined. The motion of the sun is explained, and sunspots, solar flares, and solar prominences are discussed.

Section 3: Galaxies The vastness of the universe is discussed, and the light-year is defined. A galaxy is described, with emphasis on the Milky Way Galaxy. Galaxies are classified as spiral, irregular, and elliptical, and a nebula is defined. The big-bang theory of the origin of the universe is briefly discussed.

CHAPTER

The Universe

On any clear night you can see thousands of stars in the sky. People have been looking at and studying stars for thousands of years. How many stars are there? How far away are they? Could you travel to one? Are stars just lights in the sky? Many of these questions can be answered by studying the night sky.

The Orion Nebula

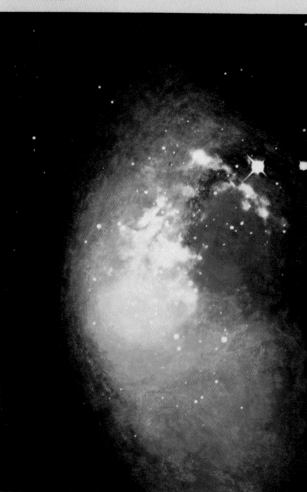

480

CHAPTER MOTIVATING ACTIVITY

Draw a diagram of the constellation Orion on the chalkboard. Point out that Orion, the mighty hunter of Greek mythology, is visible mostly during the winter months in the Northern Hemisphere. This constellation is most easily identified by the belt: three second-magnitude stars that descend from right to left across the night sky.

Have a picture, poster, or projected slide of that portion of the night sky and point out the following: The red giant Betelgeuse, which dwarfs the sun by a million times,

marks Orion's right shoulder. Betelgeuse is 520 light-years away. Bellatrix, 470 light-years away, forms Orion's left shoulder. One of the brightest, blue supergiants in the sky, Rigel, marks Orion's left leg. It is 900 light-years away. Rigel has a companion star, Rigel B, that is a double star. The Great Nebula forms Orion's sword. Just below Orion's belt lies another nebula, the Horsehead Nebula. Both nebulae are at least 1500 light-years from Earth.

SECTION OBJECTIVES

After completing this section, you should be able to:

- **Describe** how a star forms.
- **Discuss** the meaning of magnitude for a star.
- **Identify** the different types of variable stars.
- **Tell** how the color and the temperature of a star are related.

NEW SCIENCE TERMS

universe
magnitude
variable stars
constellations
zodiac
neutron star
pulsar
black holes

22.1 **Stars**

When you observe the sky on a clear night, you can see many, many stars. What are stars, and where did they come from? Stars are huge masses of glowing gases, mainly hydrogen and helium.

Scientists believe stars form from clouds of dust and gases in the universe. The **universe** is all the matter and energy that exists: all the stars, planets, dust, gases, and energy in space. Gravity causes the clouds in space to contract. The pressure at the center increases, and the temperature of the center becomes hotter and hotter. When the temperature reaches about 1 million°C, nuclear fusion begins. Hydrogen atoms fuse, creating helium and vast amounts of energy. The mass begins to glow, and the star shines.

The brightness of a star can be measured by an instrument called a *photometer*. The photometer gives a number that is the star's magnitude. **Magnitude** indicates the star's brightness as it appears from Earth—not its actual, or absolute, brightness. The magnitude of a star depends on two things: how much light it emits and how close it is to Earth. A dim star that is nearer Earth may appear more brilliant than a bright star that is farther away.

Figure 22–1. On a clear night, many stars in the Milky Way Galaxy can be seen from Earth.

Section 1 Stars **481**

SECTION FOCUS

Section Overview This section discusses the origin of stars and the grouping of stars into constellations. The characteristic of magnitude is explained, and a star's temperature and color are related by use of the Hertzsprung-Russell diagram. Novas and supernovas are described, and the resultant neutron stars, pulsars, and black holes are described.

Section Objectives For a list of section objectives, see pupil's edition page.

New Science Terms For a list of new science terms in this section, see pupil's edition page. You may wish to familiarize the students with these terms before they read the section.

SECTION DEVELOPMENT

22.1 **Stars**

DISCUSSION Call the students' attention to Figure 22–1. Have the students compare these stars to the ones they might see if they looked up into a clear night sky. (They will probably say that they have never seen that many stars.) *Thinking Critically:* Ask the students why they would never see that many stars with their unaided eyes. (The students should mention light-blocking factors such as air pollution, light pollution, and the atmosphere—even if it is not polluted.) Discuss the life cycle of stars and the relative aspect of magnitude with the students.

BACKGROUND INFORMATION

The stars in the night sky have played many roles in human history. People wish on the first star. Shooting stars were thought to foretell disasters. For centuries the stars were the only means travelers had of plotting locations at sea. They were also helpful guideposts on land. And they are just nice to look at.

Stars begin as clouds of gas and dust. A star's history depends on its mass. If its mass is less than one-tenth that of the sun, a star may become and stay a brown dwarf like Jupiter. A star like the sun "burns" hydrogen for billions of years, after which it becomes a red giant and may finally collapse into a white dwarf. If the mass of a star is ten times greater than the sun, it may become a nova, which explodes and collapses into a neutron star. If a star's mass were 40 times that of the sun, it could become a supernova that explodes and collapses into a black hole.

DISCUSSION Have the students explain why telescopes have made it possible to extend the Greeks' magnitude classifications list from 6 to 25. Then have the students look at the variable stars in Figure 22–3. The idea of a pulsating variable should be easy for the students to understand. Eclipsing variables can be illustrated with flashlights. *Thinking Critically:* Ask the students to discuss the difference between a pulsating variable and a nova. (The pulsating variable has a regular, repeating cycle of increasing and decreasing brightness. The nova becomes brighter when there is an explosion of the star. This does not happen at regular intervals.)

EXTENSION Stars are created in cloudlike nebulae composed primarily of hydrogen and helium. Dark nebulae are frequently silhouetted by the light of stars positioned behind them. The Horsehead Nebula in the constellation Orion is one such nebula. Reflecting nebulae, like the Pleiades star cluster located between the constellations Perseus and Taurus, reflect the light of stars situated within it. The gases of emission nebulae, like the Great Nebula in Orion's sword, are heated to more than 10 000 Kelvin by the ultraviolet light of the stars within, causing them to fluoresce.

EXTENSION You may wish to have the students look at the map of your state in the Science Sites booklet. If there is a planetarium in your area, have the students add it to the map if it is not already shown.

Figure 22–2. The magnitude, or brightness, of stars varies. The bright star shown here has a lower magnitude than the stars around it.

A MATTER OF FACT

The universe contains about 100 billion galaxies. If a galaxy contains an average of 100 billion stars, then the total number of stars in the universe would be around 10 000 billion billion.

Figure 22–3. Shown here are a variable star (left) and a binary star system (right).

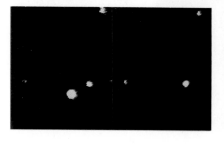

The ancient Greeks classified stars into six magnitudes. First-magnitude stars appear the brightest. The second, third, fourth, and fifth magnitudes include stars of decreasing brightness. Stars of the sixth magnitude are barely visible to the unaided eye. Modern telescopes have made it possible to see stars even as faint as those of the twenty-fifth magnitude. Negative magnitudes are used for very bright objects. The sun, for example, has a magnitude of –27.

Some stars are called **variable stars** because they change magnitudes. One type of variable star is the *pulsating variable*. Energy from below the surface of the star heats the gases of the visible, shining surface. The surface becomes hotter and brighter, and it expands. As the surface expands, the gases cool and the star becomes dimmer. The cooler gases contract, and a new cycle begins as the gases again become hotter and the surface expands.

The first pulsating variable discovered was *Mira*, whose magnitude changes every 331 days. The time from bright to dim and back to bright again is the star's period. Mira is a long-period variable. The period of most of the thousands of other pulsating variables that have been discovered is 100 days or less. Astronomers call these short-period variables *cepheid* (SEF ee ihd) *variables*. The North Star, for instance, is a cepheid variable with a period of about 4 days.

Other variable stars are *eclipsing binaries* and *exploding stars*. Eclipsing binaries are pairs of stars that move around each other. Their brightness appears to vary as one star periodically blocks the light from the other star. Exploding variables are stars that have bursts of energy that make them appear many thousands of times brighter for days or even years. An exploding star is called a *nova* or a *supernova* because in ancient times it was thought to be a new star; *nova* is Latin for "new." A nova eventually returns to its original brightness, but a supernova is such a huge explosion that the star blows itself apart.

○ *What is a star?*
○ *What is the system that measures a star's apparent brightness as seen from Earth?*
○ *Name three kinds of variable stars.*

Answers–22.1

○ A star is a huge mass of glowing gases.

○ The apparent brightness of a star is measured by a system of relative brightness called magnitude.

○ The three kinds of variable stars are pulsating variables, eclipsing variables, and exploding variables.

DEMONSTRATION

For a demonstration of the constellations see page 479c preceding this chapter.

22.2 Constellations

Have you ever looked at clouds floating in a blue sky and imagined you saw pictures in the cloud shapes? Ancient astronomers looked at the stars scattered across the sky and saw pictures of animals, humans, and other objects in certain groups of stars. Rather like the way you might make a dot-to-dot picture, ancient astronomers filled in shapes around groups of stars. The groups of stars that form images are the **constellations.** The ancient Greeks recognized 48 constellations. Today 88 are recognized, and different areas of the sky are named for the constellation located in that area.

Imagine that the constellations and their stars are painted on a glass sphere that surrounds the earth. This is called a celestial sphere. Just as a globe shows Earth with an equator midway between its poles, the sphere has a midline called the *celestial equator.* The sphere also has a line that represents the earth's orbit, called the *ecliptic.* The twelve constellations along the ecliptic form the signs of the **zodiac.** These twelve constellations roughly parallel the twelve months of the year.

DISCOVER

Observing Constellations

Using the star chart on page 584 of the Reference Section, locate as many constellations of the zodiac as you can. Draw the constellation that corresponds to your zodiac sign and write a short report on its features.

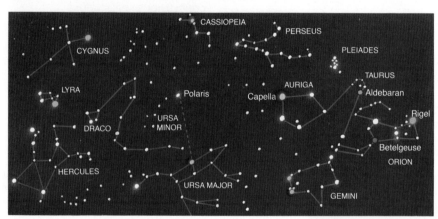

Figure 22–4. Some groups of stars form identifiable figures in the sky.

Figure 22–4 shows several of the constellations of the zodiac. You can use it to identify the major stars of the night sky. Star color can also aid in identification. Some stars, like Betelgeuse, are reddish in color, and others, like Sirius and Rigel, are bluish. You can locate these stars on the star chart and verify their colors by observing the night sky.

○ *What is a constellation?*
○ *What is the celestial equator?*
○ *What is the zodiac?*

Section 1 Stars **483**

BACKGROUND INFORMATION

The constellations will not remain fixed forever. As all nebulae, stars, and galaxies continue to move through the universe, their apparent positions, as seen from Earth, will change. Of course this change will occur slowly, over many thousands of years. Someday the Big Dipper (Ursa Major) may more closely resemble a "big skillet."

Answers–22.2

○ A constellation is an arrangement of stars that appear to form an image on the celestial sphere.

○ The celestial equator is an imaginary line drawn on the celestial sphere, corresponding to an extension of the earth's equator.

○ The zodiac consists of 12 constellations lying on the ecliptic.

DISCOVER:
Observing Constellations

Skills *(Identifying, Diagramming)*

The students may need star charts to help them identify the constellations. A star chart is located in the Reference Section on page 590. You may wish to point out the features that make some of the constellations easy to identify. Reports will vary with the zodiac signs of the students in the class.

22.2 Constellations

DISCUSSION Call the students' attention to the constellations shown in Figure 22–4. You may wish to have star diagrams for some of the constellations not shown in the figure. Ask the students to guess what the constellation might be or to make up their own constellations. Remind the students that the star patterns might not look exactly as they did when the Greeks named the various constellations because of movement of the stars. To provide a historical perspective, you may wish to have pictures of the constellations of the zodiac and the dates for each one available for the students. Wire circles suspended around a globe may help the students visualize the celestial equator and the ecliptic.

EXTENSION Just as the Big Dipper identifies Ursa Major (the Big Bear) and the Little Dipper identifies Ursa Minor (the Little Bear), there are features that help to identify other constellations. The most prominent feature of Cygnus (the swan) is the five bright stars that form the Northern Cross. Cassiopeia (the queen) is identified by the five stars that form an irregular, sideways W. Five stars of Cepheus (the king) form a peaked, roofed house. Andromeda (the princess) is identified by the three bright stars in a nearly straight line. The three stars in the belt of Orion point to the star Sirius, second only in magnitude to the sun. Sirius is a star in the collar of the dog constellation Canis Major.

ACTIVITY:
Making a Chart of the Horizon

Skills (*Observing, Recording Data*)

Preparation of Materials If enough magnetic compasses are not available for each student to have one, two or three students may be able to share one for the activity.

Hints Noting positions on the horizon will prove useful in locating constellations from one evening to the next. Using the astrolabe constructed in Chapter 21 will aid the students in estimating the elevation of the constellation above the horizon, where the center of the chart represents the zenith of the celestial sphere directly above the observer.

Answers to Conclusions/Applications

1. It is extremely important to note the date and time that each chart was made, since the celestial sphere rotates and tilts constantly as the earth rotates on its axis and revolves around the sun. As a result, the position of constellations with respect to landmarks on the horizon change hourly and, of course, daily.

2. By making photocopies of the horizon landmark chart and designating the position of constellations from one evening to the next, one will readily observe the shift in the celestial sphere.

22.3 Red Giants, White Dwarfs, and Novas

DISCUSSION The students may have seen "red hot" coals in a fire or "red hot" coils on a stove. Something that is "white hot" is hotter than something that is "red hot." If the students can observe stars on a very clear night, they may be able to see the different colors in the stars. Discuss with the class how stars are plotted on the Hertzsprung-Russell. (by spectral class, or color, and by temperature) *Thinking Critically:* Ask the students why higher temperatures and higher energy emissions are related. (Atoms move faster at higher temperatures, so more energy, or light, is given off.)

ACTIVITY: Making a Chart of the Horizon

How can you make a chart of the local horizon to aid in locating the constellations?

MATERIALS (per student)

magnetic compass, drawing compass, construction paper, colored pencils, protractor, metric ruler

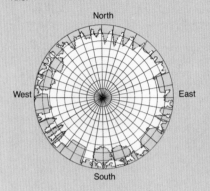

North

West

East

South

PROCEDURE

1. Use the magnetic compass to locate north, south, east, and west.
2. Use the drawing compass to draw a dark circle 18 cm in diameter on the construction paper and label the circle as shown.
3. Using a pencil, sketch the horizon so that natural landmarks correspond with their true locations on the horizon.
4. In the evening, locate familiar stars and constellations and mark their positions on the chart. You may use the astrolabe you constructed for Chapter 21, or a protractor, to help you determine the position of the constellations above the horizon.
5. Note the date and time that you made your chart.

CONCLUSIONS/APPLICATIONS

1. Why is it important to note the date and time that you made your chart?
2. What information could you gather from charts drawn at the same time every evening during the course of a year?

22.3 Red Giants, White Dwarfs, and Novas

The energy output of stars varies tremendously. Some stars emit only one one-thousandth as much energy as the sun; other stars emit 100 000 times more energy than the sun. The more energy a star emits, the hotter it is. The color of a star and its temperature are closely related. The hottest stars are bluish in color, and the cooler ones are reddish.

In 1913 American astronomer Henry Norris Russell and Danish astronomer Ejnar Hertzsprung published a graph showing that the energy emitted by a star is related to its color. This graph is known as the *Hertzsprung-Russell diagram.*

Most of the stars that have been plotted so far form a diagonal line on the diagram called the *main sequence.* Hot, blue stars are at the top left; cooler, red stars are at the bottom right. Yellow stars, like the sun, are plotted near the middle.

Figure 22–5. A star's color is based on the amount of energy the star emits.

484 Chapter 22 The Universe

BACKGROUND INFORMATION

A star begins as a swirling mass of gases. As gravity causes the core of the star to contract, its temperature rises until the star finally begins to glow. As the fuel is consumed, the core contracts more and the increased pressure causes an increase in temperature. When the hydrogen of a main sequence star is virtually spent, the increased temperature causes the surface of the star to expand and cool. The star is transformed into a red giant. The fusion of heavier elements in the red giant star releases voluminous energy and pressure that swells the star to its enormous size. As the dense core of the red giant exhausts that fuel and collapses even further, it blows off its outer layers and eventually becomes a white dwarf. The white dwarf continues to shine until it is completely burned out. The black, burned-out star is called a black dwarf.

Besides color differences, the Hertzsprung-Russell diagram shows star masses. Star mass increases from the bottom to the top of the main sequence. The stars at the bottom right have masses about one-tenth that of the sun. At the top left, there are stars with masses more than ten times greater than that of the sun. The largest stars are called *supergiants*. The smallest stars are called *dwarfs*. Although small in size, dwarf stars are extremely dense. Supergiants are large, but they may be even less dense than the earth's outer atmosphere.

Figure 22–6. This graph, called a *Hertzsprung-Russell diagram,* shows the relationship between a star's color and its energy.

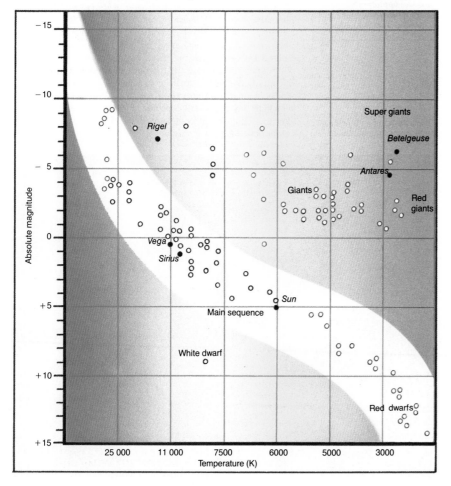

DISCUSSION Be sure the students understand that the lower a star's magnitude number, the brighter the star is. By looking at the Hertzsprung-Russell diagram, the students should be able to identify the sun as a medium bright, yellowish star, whose surface temperature is about 5550°C. Have the students similarly describe other stars in the diagram.

SKILL *(Classifying)* Have the students tell where, on the Hertzsprung-Russell diagram, each of the following would be plotted.

Absolute Magnitude/Surface Temperature

Star A	+10	10 000°C
Star B	−2	3500°C
Star C	−6	6000°C
Star D	+3	7000°C

485

DISCUSSION Have the students share examples of fuel being used up. (Some examples would be a car running out of gas, a lantern going out, a fire burning out.) The students should understand that stars have a limited supply of fuel (hydrogen) although a supply that lasts billions of years is certainly large. Direct the students' attention to Figure 22–7 and trace the life cycle of a main-sequence star. *Thinking Critically:* Have the students discuss why a star of absolute magnitude +7 would last longer than a star of -7, if the stars were of the same mass. (The fuel in the less bright star would not be used up as quickly.)

THINKING SKILL (*Inferring*) Ask the students what might happen to the solar system as the sun becomes a red giant in about 5 billion years. (Some answers might be that Earth will become too hot for life as it now exists. Conditions on other planets may become more like Earth and humans could live there. As the sun expands, maybe the solar system will expand and Earth would not change very much.)

A MATTER OF FACT

The red supergiant Betelgeuse, which marks the right shoulder of the constellation Orion, is 520 light-years from Earth. At that distance, photographing the star is like capturing the image of a flea from more than 20 km away.

The Hertzsprung-Russell diagram can also be used to show the life cycle of stars. All stars produce energy by nuclear reactions. A star with a mass similar to the mass of the sun produces energy for about 10 billion years. The nuclear reaction in smaller stars is slower, so smaller stars last longer than larger stars.

Scientists believe that the sun is about 4.6 billion years old. It probably has enough hydrogen left in its core to continue shining as it does now for at least another 4.6 billion years. According to the most accepted theory, when the sun is about 10 billion years old, all the hydrogen in its core will be gone. Then the core will contract, its temperature will rise, and its surface will expand outward. After another 100 million years, the sun will become a red giant, its surface reaching halfway to Venus. Then the center of the sun will get hotter, its outer layer will blow off, and the sun will become a white dwarf. Eventually, the nuclear reactions in the center will stop. The sun will be a small star, no more than 10 000 km in radius, with a very high density. It will continue shining for billions of years, slowly cooling. When completely burned out, it will be a cold cinder called a *black dwarf*.

Figure 22–7. Shown here is the life cycle of a main-sequence star.

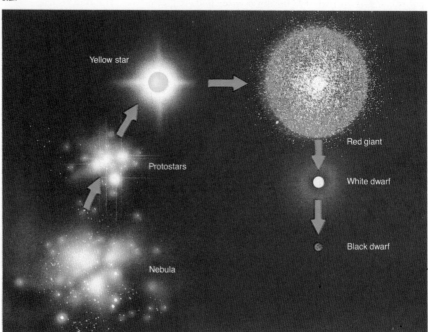

486   Chapter 22   The Universe

If a white dwarf is part of a binary system, it may undergo a different change. Instead of slowly cooling, it may capture hydrogen from its companion star. When the temperature of the captured hydrogen reaches 10 million°C, a nuclear explosion occurs. The star increases its brightness about 10 000 times and appears as a nova. A nova reaches its maximum brightness in a day or two and then gradually returns to its original intensity.

○ *What does the Hertzsprung-Russell diagram show?*
○ *How will a yellow star, like the sun, change when its supply of hydrogen is gone?*

22.4 Neutron Stars and Pulsars

A star that has a large mass has a different life cycle from that of a smaller star like the sun. When the hydrogen in the core of a large star is depleted, the core can no longer support the weight of the outer layers. It contracts with such violence that the entire star blows up. Actually it is a two-way explosion; the core implodes, or is blown inward, and the rest of the star explodes outward. An exploding star is called a *supernova*, and it produces more energy than the sun will produce in its entire lifetime. Astronomers can see part of this energy as light that is as bright as 500 million suns.

Figure 22–8. Novas occur when white dwarfs explode. Novas are like supernovas, except that the total mass of a nova is much less than that of a supernova.

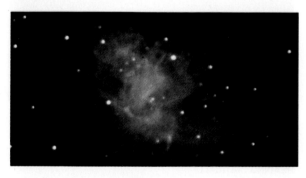

Figure 22–9. A nebula, such as the Crab Nebula shown here, can be created by the explosion of a supernova.

During a supernova explosion, 90 percent of the star's mass scatters into space, becoming the matter from which new stars may be born. The other 10 percent, the core of the star, is blown inward, becoming a neutron star. A **neutron star** is a very small star, as small as 15 km in radius. Neutron stars rotate at high speed and have strong magnetic fields. They are also very dense. If the Great Pyramid of Cheops in Egypt, which weighs 3 million tons, had the density of a neutron star, it would be the size of a pinhead.

Section 1 Stars **487**

DISCUSSION Have the students study Figure 22–8. The phenomenon known as a nova could be compared to the flare-up of a match – after the flare, it settles into its normal flame.

22.4 Neutron Stars and Pulsars

EXTENSION If a variable star has a mass several times that of the sun, its core contracts with such a release of energy that the outer layers of the star blow apart. The supernova is a rare phenomenon in the universe. The Crab Nebula is the result of a supernova that occurred in 1054. A supernova in February 1987 in the Large Magellanic Cloud, one of the two galaxies closest to the Milky Way Galaxy, allowed astronomers to expand their knowledge of the galaxy while they wait for the next supernova in the Milky Way Galaxy. The last was in 1604; they occur about every 300 years.

DISCUSSION Direct the students' attention to Figure 22–9. *Thinking Critically:* Ask the students why the Crab Nebula looks like the result of an explosion. (Some answers might be that it looks like an irregular cloud of smoke, it seems to have fragments in it, it looks like matter is being blown out from a center.) The contracted core of the supernova is a star made of neutrons. It is nearly impossible to imagine the density of a neutron star, but the Pyramid example may help.

Answers–22.3

○ The Hertzsprung-Russell diagram relates the energy emission of a star to its absolute magnitude and mass. It can also be interpreted to represent the life cycle of a star.

○ It will become a red giant and, eventually, a white dwarf.

BACKGROUND INFORMATION

Whether or not a star will evolve into a white dwarf depends upon its mass. If the star's mass is between 1.2 and 3.2 solar masses, it cannot become a white dwarf but instead becomes either a neutron star or pulsar. This limit of mass is called the Chandrasekhar Limit after the Indian astronomer, Subrahmanyan Chandrasekhar (1910–). If the star's mass exceeds 3.2 solar masses, it is destined to become a black hole.

487

EXTENSION As a neutron star rapidly spins, the star may emit energy, perhaps radio waves, from its magnetic poles. Since the star is spinning, the energy is received on Earth in pulses of energy, like the intermittent light from a lighthouse beacon.

DISCUSSION Have the students study the photographs of the pulsar. *Thinking Critically:* Ask the students why they think the first pulsar was not discovered until 1967. (Modern astronomical instruments and computers have made new discoveries possible.) You may wish to have the students compare the life cycle of the giant stars with the life cycle of the main sequence stars. The concept of a black hole is hard to understand. Explain to the students that it is matter that is so dense and has such a strong gravitational field that not even light can escape its force. It is not visible and is not a hole that things will "fall into" in space. However, things may be attracted by the gravitational force.

SECTION REVIEW

Summary Have the students make a list of the various types of stars in the universe and the characteristics of each.

Reinforcement Have the students pair off and review the section by quizzing each other on the questions at the end of each subsection.

Figure 22–10. Pulsars (above) are stars that seem to blink on and off. Pulsars form during the life cycle of giant stars (right).

A MATTER OF FACT

Some scientists thought that energy bursts from newly discovered pulsars were messages from alien civilizations. They labeled the stars "LGMs," meaning "little green men."

A neutron star may capture gas from space, or from a companion star, or nearby star. Perhaps the captured gas becomes ionized and, as the neutron star rotates, pulses of light are emitted, similar to the beacon of a lighthouse. The first rapidly pulsating neutron star, or **pulsar,** was discovered in 1967. Scientists are not certain what causes pulsars. Over 350 pulsars have been discovered since then.

If the core mass of a supernova is more than three times greater than the sun's mass, the force of the implosion collapses the core into itself so that no energy, including light, can be emitted. Collapsed stars that emit no energy are called **black holes.**

○ *What is a supernova?*
○ *What is a neutron star?*
○ *What is a black hole?*

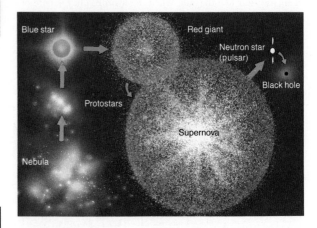

Section Review

READING CRITICALLY

1. How does a star form?
2. What are some of the differences among stars?
3. How does a star of great mass die?

THINKING CRITICALLY

4. What does it mean to say that "looking at stars is looking at the past"?
5. What can astronomers learn about the sun from studying other stars?

Answers–22.4

○ A supernova is an exploding star that is blown apart.
○ A neutron star is a densely packed star, as small as 15 km in diameter, that can result from a supernova.
○ A black hole is a collapsed, dense star that emits no energy because of its intense gravitational field.

Answers to Section Review

Reading Critically

1. Stars are formed by the accumulation and compression of large masses of hydrogen and helium.
2. Stars vary in size, brightness, composition, and color.
3. A star of great mass dies in a violent explosion called a supernova.

continues

SKILL ACTIVITY: Using a Star Chart

BACKGROUND

The star chart below displays a map of the most familiar constellations visible in the Northern Hemisphere. However, not all of these constellations can be seen on a single night, since the view of the celestial sphere from Earth changes from day to day as our planet revolves around the sun. In this activity, you can create your own star chart to help you become familiar with the relative positions of these constellations.

PROCEDURE

1. Measure the radius of the star chart with a compass and draw a circle of the same size on construction paper. Cut out the circle with scissors.
2. Trace the star chart on tracing paper and tape the tracing paper neatly to the circular piece of construction paper.
3. Draw another, slightly larger circle (with a radius that is larger by 1 cm) and cut it out. Drawing through the center of that circle, use a ruler to divide the circle evenly into 12 slices. At the rim of the circle, label the slices, numbering 1 through 12 to represent the months in a year.
4. Place the tracing of the star chart on the large circle and pin both pieces of construction paper, at their centers, to a piece of cardboard.
5. Study the positions of the constellations by comparing their locations relative to the North Star, or polestar.

APPLICATION

Use the star chart to find some of the constellations by positioning your chart in front of you so that the constellations on the celestial sphere match the direction you are facing. Locate the constellations Hercules and Orion.

1. What are the positions of the constellations Hercules and Orion relative to the *North Star?*
2. Explain why not all constellations in the northern celestial sphere are visible at once.

USING WHAT YOU HAVE LEARNED

Use your star chart to locate as many of the constellations as you can.

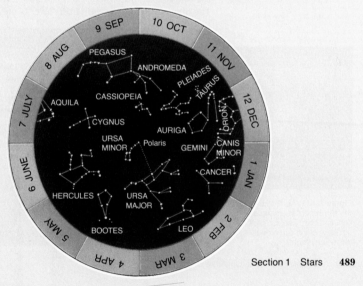

Section 1   Stars   **489**

Thinking Critically

4. It takes light time to travel through space. So seeing a star means seeing the light that left the star perhaps millions of years before.
5. They learn about the past, present, and future behavior of the sun.

SKILL ACTIVITY: Using a Star Chart

Objectives

- Become familiar with the constellations visible in the Northern Hemisphere.
- Observe the constellations visible on a particular night.

Preparation of Materials   Each student will need a letter-size piece of tracing paper and half a sheet of construction paper. The cardboard can be any size larger than the larger-sized circle of tracing paper. A brass paper fastener would be useful in

Discussion   Using the star chart at various times during the year will help the students appreciate the earth's yearly revolution around the sun. The students in rural areas should be able to see more stars than those in urban areas due to less light pollution. ***Thinking Critically:*** Ask the students if they think the stars would be in exactly the same positions in the constellations if they looked at them again in 10 years and again in 50 years. (The stars would not be in the same positions in 10 years because the universe is constantly moving, but the difference might not be great enough to observe until 50 years or more have passed.)

Answers to Application

1. Hercules and Orion are on opposite sides of the North Star.
2. The constellations seem to rise and set, just as the sun does.

Section 2: The Sun

SECTION FOCUS

Section Overview This section focuses on a particular star, the sun. The sun's layered structure and its atmosphere are studied. Sunspots, solar flares, and solar prominences are described.

Section Objectives For a list of section objectives, see pupil's edition page.

New Science Term For the new science term in this section, see pupil's edition page. You may wish to call this term to the students' attention before they begin reading the section.

SECTION DEVELOPMENT

22.5 The Composition and Structure of the Sun

DISCUSSION Mechanical energy depends on fuel, and so does animal energy. *Thinking Critically:* Have the students discuss why all energy on Earth depends on the sun. (Solar energy comes directly from the sun. Wind is the result of air currents being heated and cooled by the sun. The fossil fuels, oil, gas, and coal contain the stored energy of the sun from millions of years ago. People and animals get their energy from food, which contains the energy of the sun that was changed into chemical energy by photosynthesis.) Have the students look at Figure 22–11 and discuss the structure of the sun.

NEW SCIENCE TERM
solar prominences

SECTION OBJECTIVES
After completing this section, you should be able to:
- **Name** the different layers of the sun.
- **Describe** the activity on the sun.

22.5 The Composition and Structure of the Sun

Astronomers say that the sun is just an average star. However, to the inhabitants of Earth it is the most important star in the universe, for life on Earth depends on the sun's energy.

The sun's radius is 696 000 km, 109 times greater than the radius of the earth, while its mass is over 330 000 times greater than the mass of Earth. In fact, the sun contains about 99.85 percent of all the mass of the solar system.

At the center of the sun is a dense core of hydrogen and helium, which is at a temperature of about 15 million°C. Above the core is a thick *radiative layer.* The energy produced in the core warms this layer just as heat from a radiator warms a room. In this layer, the temperature averages about 3 million°C. Next is the *convective layer,* where energy is transferred by convection. The temperature at the top of this layer is only 8000°C.

Figure 22–11. You may think of the sun simply as a source of light and heat. The complex structure of the sun produces this energy (right).

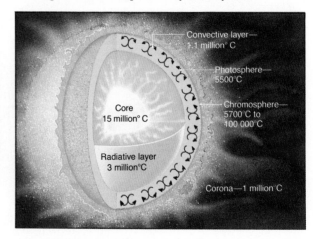

Convective layer—1.1 million° C

Photosphere—5500°C

Chromosphere—5700°C to 100 000°C

Core 15 million° C

Radiative layer 3 million°C

Corona—1 million° C

490   Chapter 22   The Universe

BACKGROUND INFORMATION

Core matter inside the sun has a density about 10 times greater than the density of iron (approximately 80 g/cm³). The nuclear fusion reaction of the core uses hydrogen as fuel. Two protons fuse to form deuterium (one proton and one neutron) and neutrinos are released. Neutrinos are so infinitesimally small that they pass though the layers of the sun, through surrounding space, and even through the planets without hitting anything. Deuterium nuclei then fuse with other free protons to form helium. As they do this, the fusion reaction gives off gamma rays. The resulting isotopic helium nuclei continue to collide, breaking off two protons and, thus, leaving the final helium product. At each collision, light and heat energy are released.

490

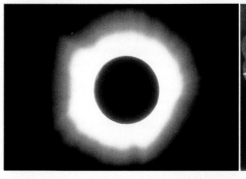 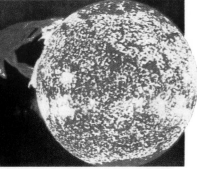

Above the convective layer is the sun's atmosphere, which has three layers. The *photosphere* is the inner layer of the sun's atmosphere, which forms the visible surface of the sun. The photosphere is similar to the surface of a pan of sauce; it boils and bubbles in much the same way.

The temperature of the photosphere is about 5500°C. The surface of the photosphere has many small, bright areas called *granules.* The granules, shown in Figure 22–12, are actually the tops of rising columns of hot gases. The dark areas between granules are sinking areas of cooler gases.

Above the photosphere is the *chromosphere,* a 2500-km-thick layer, where the temperature increases to 100 000°C. Above the chromosphere is the *corona,* a layer of thin solar gases that merge with outer space. The temperature in the corona is approximately 1 million°C.

○ *Name the layers of the sun and its atmosphere, starting from the core.*
○ *What are granules?*

Figure 22–12. During a solar eclipse (left), the corona of the sun is sometimes visible. At other times, it is possible to observe the surface of the sun (right), using a telescope equipped with a special filter.

22.6 Motions and Activity of the Sun

The sun rotates on its axis in the same direction as the planets revolve, that is, counterclockwise as seen from above the sun's north pole. Since the sun is a gaseous body, not all parts rotate at the same speed; the rotational speed decreases toward the sun's north and south poles. At the solar equator, one sun day is equal to about 25 Earth days.

Solar energy is produced by the fusion of hydrogen nuclei into helium nuclei. The helium nucleus that is the end product of the fusion reaction has two protons and two neutrons. Scientists calculate that it takes about 2 million years for energy from the sun's core to reach its surface and 8.3 minutes to reach Earth.

DISCOVER

Calculating Time

Using the table on page 464, calculate how long it takes for the sun's energy to reach each of the planets.

DISCUSSION Emphasize to the students that they should learn everything they want to know about the sun from photographs and diagrams. **CAUTION: Students should never, even with dark glasses, look directly at the sun because of the risk of damage to the retina of the eye.** Tell the students that they can get a good idea of solar granules by observing a pot of any boiling, thick liquid. Steam comes from the bubbles as they burst at the surface and the spot does appear darker. ***Thinking Critically:*** After the students have finished reading about the areas of the sun's atmosphere, ask them what unusual property exists. (The temperature of the atmosphere increases out to the corona. You would expect the temperature of the atmosphere to decrease as it moves away from the heat source, but the opposite is true. Scientists are not sure of the reason for this.)

22.6 Motions and Activity of the Sun

DISCUSSION Have the students compare the sun's rotation to the earth's rotation. You may also wish to tell the students that just as the earth revolves around the sun, so does the sun revolve around the center of the Milky Way Galaxy. One revolution takes about 250 million years. Have the students discuss what they know about nuclear reactions. (Nuclei combine to form nuclei of a new element, and in the process energy is given off. Some students may mention nuclear bombs, nuclear power plants, or nuclear submarines.)

EXTENSION About three-fourths of the matter of the sun is hydrogen. Scientists estimate that there is enough hydrogen remaining in the sun to keep the sun's fusion reaction going for about 5 billion years more.

DISCOVER:
Calculating Time

Skill (*Calculating*)

From the table, find the AU distance of each planet from the sun, then multiply each value by 8.3 minutes. (the time it takes for sunlight to reach the earth = 1 AU)

Answers–22.5

○ The core, radiative layer, convective layer, photosphere, chromosphere, and corona are the layers of the sun.

○ Granules are dark areas on the surface of the photosphere.

BACKGROUND INFORMATION

Sunspots represent cooler areas on the surface of the photosphere. They occur in 11-year cycles and may affect the earth's weather; years of lower sunspot activity cause the atmosphere of the earth to be cooler. It is believed that sunspots form along magnetic lines of force flowing from the north to south solar poles. But since the equator rotates faster than the poles, the lines of forces are drawn out like a rubber band, twisted and kinked, until they wind about the sun, spreading poleward.

DISCUSSION   Have the students study the activity features of the sun as shown in the three photographs. Scientists are not sure what causes sunspots, but the activity does follow the 11-year cycle, or a 22-year cycle if considered from minimum activity to maximum and back to minimum again. Solar flares and solar prominences occur more frequently during a sunspot maximum.

EXTENSION   The first sunspots to appear in a new sunspot cycle appear in higher solar latitudes. As the cycle progresses, more and more sunspots appear near the solar equator. They never seem to appear near the solar poles.

SECTION REVIEW

Summary   Have the students draw a circle to represent a cutaway view of the sun. They should mark a point for the center, and then label the layers of the sun out to the corona. Ask them to sketch the solar activity features discussed.

Reinforcement   You may wish to have the students each make up three questions for the subsections. They can form groups of three or four students, pool their questions, and quiz each other.

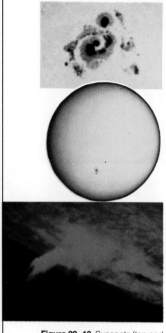

Figure 22–13. Sunspots (top and center) are the result of magnetic activity in the sun. Another form of solar activity is a solar prominence (bottom).

The convective layer and the photosphere of the sun consist of ionized atoms and free electrons. Turbulence in these layers moves the charged particles around and creates local magnetic fields. These fields tend to combine with each other as magnets do, forming larger fields. When a magnetic field grows to about 2000 km across, convection stops. The top of the field radiates energy away and becomes cooler. These cooler areas appear dark from Earth. They combine, forming larger dark areas called *sunspots*.

Sunspot formation follows an 11-year cycle. During a sunspot minimum, the surface of the sun may be completely free of sunspots, while at a sunspot maximum, several major groups may dot the surface of the sun at any given time. In sunspots, charged particles such as protons, electrons, and ions speed up until they escape from the sun's atmosphere. These particles cause an increase in strength of the solar wind. *Solar flares*, spectacular eruptions that occur near sunspots, cause periodic increases in the solar wind as well.

You may recall from Chapter 8 that the solar wind particles, which reach Earth in about 2.5 days, are responsible for disturbances in the earth's ionosphere. A particularly strong solar wind can disturb the earth's magnetic field, causing interference in radio and television transmissions and in microwave communications.

Another impressive display of solar activity is the solar prominence, shown in Figure 22–13. **Solar prominences** are streams of gases that explode above the sun's surface. They eventually fall back to the sun's surface under the pull of gravity.

○ *What are sunspots?*
○ *What is the solar wind?*
○ *What are solar flares and solar prominences?*

Section Review

READING CRITICALLY

1. How does sunlight reach the earth?
2. How do activities on the sun affect the earth?

THINKING CRITICALLY

3. Why is Earth the only planet in the solar system known to have life?
4. Why would it take 2 million years for energy from the sun's core to reach its surface but only 8.3 minutes for it to reach the earth?

492   Chapter 22   The Universe

Answers–22.6

○ Sunspots are cooler areas of the sun's surface.
○ Solar wind is the sum total of solar energy being blown out into space.
○ Solar flares and solar prominences are spectacular eruptions on the sun's surface.

Answers to Section Review

Reading Critically

1. Energy from the sun's core radiates to the surface where visible sunlight from the photosphere radiates out into space. Some of the radiation spreads through space in the direction of the earth.
2. Sun spots, solar flares, and solar prominences influence the radiation output of the sun. This radiation can interfere with the earth's magnetic field and affect weather, radio transmissions, and compass needles.

Thinking Critically

3. Because it is the only planet with the proper conditions for life.
4. There is very little matter to impede the flow of radiation through space after it has left the surface of the sun.

SECTION OBJECTIVES

After completing this section, you should be able to:

- **Explain** what a light-year is.
- **Describe** the Milky Way Galaxy.
- **Name** the different kinds of galaxies.
- **Identify** nebulae and quasars.
- **Explain** the big-bang theory.

NEW SCIENCE TERMS

light-year
galaxy
nebula
quasars
big-bang theory

22.7 Measuring Distances in the Universe

Stars generally are very far from each other. If the sun were a marble 1 cm in diameter, the nearest marble, or star, would be over 300 km away. As modern instruments make it possible to discover stars at greater distances from Earth, a unit was needed to measure these distances. The unit scientists have chosen to use is called the **light-year**—the distance light travels in one year. At a speed of over 300 000 km/sec, this is about 9.5×10^{12} km.

In 1838 German astronomer Wilhelm Bessel determined that the star known as 61 Cygni was 11 light-years away from the earth. This means that it has taken 11 years for the light from 61 Cygni to reach Earth. There are stars closer to Earth than 61 Cygni. The closest star, Proxima Centauri, is 4.26 light-years away.

○ *What is a light-year?*
○ *What star is closest to the earth?*

DISCOVER

Traveling Fast

Determine the relationship between astronomical units and the speed of light. Then determine how long it would take to travel from Earth to Pluto. Use the table on page 464 to find the astronomical units that you need to use.

Figure 22–14. Proxima Centauri, part of the bright star system shown here, is the closest star (other than the sun) to Earth.

Section 3: Galaxies

SECTION FOCUS

Section Overview   In this section, a galaxy is described, with emphasis on the Milky Way Galaxy. A light-year is defined as the unit used to measure great distances in the universe. Galaxies are classified as spiral, irregular, or elliptical. The discovery of quasars and the big-bang theory for the origin of the universe are described.

Section Objectives   For a list of section objectives, see pupil's edition page.

New Science Terms   For a list of new science terms in this section, see pupil's edition page. You may wish to read the terms aloud for the students before beginning the section.

SECTION DEVELOPMENT

22.7 Measuring Distances in the Universe

DISCUSSION   Have a small marble in class and pick a location about 300 km away. This should help the students appreciate the distance between the stars. The students may understand the immense distance defined by a light-year if you write it on the chalkboard as 95 followed by 11 zeroes. ***Thinking Critically:*** Ask the students why they think the radiation of light through the universe was chosen as the basis for the unit of measuring great distances. (Students may mention that light travels very fast or that light shines from all the stars.)

DISCOVER:
Traveling Fast

Skill *(Calculating)*

Subtract 1 AU from the distance (in AU) that Pluto is from the sun. Next multiply that by 8.3 minutes to determine the time it would take to travel from Earth to Pluto at the speed of light.

BACKGROUND INFORMATION

Models of the universe have changed continually ever since people began looking at the universe. Conceptions of the universe are largely influenced by the technological innovations that allow a closer examination of the sky. From Galileo's first refracting telescope to NASA's incredible orbiting Infrared Astronomical Satellite (IRAs), perceptions of celestial objects have been constantly refined.

The earliest people thought the earth was the center of the universe. Nicolaus Coper-

nicus created a model of the solar system with the planets revolving around the sun.

Answers—22.7

○ A light-year is the distance light travels in one year, about 9.5×10^{12} km/year.

○ Proxima Centauri is the closest star to Earth.

22.8 The Milky Way

DISCUSSION Have the students look at the Milky Way Galaxy in Figure 22–15. Have them approximate the location of the sun. (It is in the central plane of the galaxy, about three-fifths of the way from the center.) Discuss the other photographs with the class. ***Thinking Critically:*** Ask the students why they think the inhabitants of Earth would have named their own galaxy the Milky Way Galaxy. (People in a galaxy cannot see the whole galaxy. With the unaided eye, the arm of the spiral that can be seen from Earth appears to be a milky streak across the sky.)

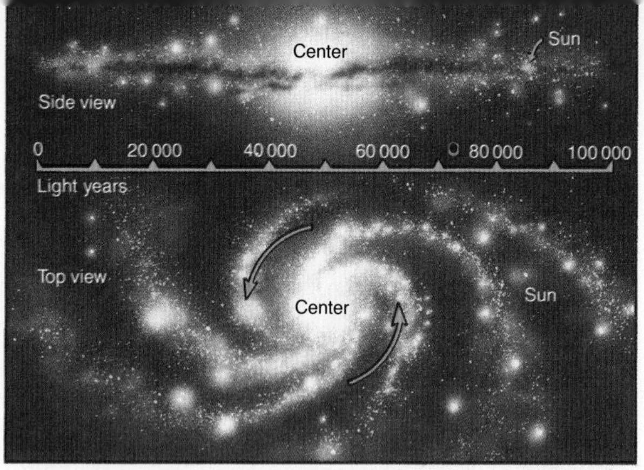

Figure 22–15. The earth is part of the Milky Way Galaxy, shown here in relation to the sun.

Figure 22–16. There are many types of star clusters. Shown here are two examples, the globular cluster (top) and the open cluster (bottom).

22.8 The Milky Way

Have you ever noticed, on a very clear night, that there is a ribbon of stars so dense that it looks like a starlit cloud? This ribbon, called the *Milky Way Galaxy*, consists of billions of stars. A **galaxy** is a large grouping of stars. All the stars you see with your unaided eye belong to the Milky Way Galaxy. The Milky Way Galaxy contains about 180 billion stars and a large amount of interstellar gases and dust. The Milky Way Galaxy is a spiral galaxy; it is shaped like a disk that bulges in the middle and has curved arms.

The Milky Way Galaxy is 100 000 light-years across and 10 000 light-years thick at the center, decreasing to 1000 light-years at the edge. The sun is on the inner rim of one arm of the galaxy, about 30 000 light-years from the center and midway between the upper and lower edges of the galaxy. As the Milky Way Galaxy rotates, the sun travels around the center of the galaxy at a speed of 250 km/s, making one complete turn every 250 million years.

The average distance between stars in the Milky Way Galaxy is several light-years. Near the sun, the average distance between stars is four to five light-years. All around the Milky Way Galaxy, there are groups of stars that are close together. These stars form star clusters. The stars in a cluster seem to be arranged in a spherical mass. They are probably the galaxy's oldest stars, estimated to be about 10 to 15 billion years old. Each cluster contains over 100 000 stars and has little or no interstellar dust.

○ ***What is a galaxy?***
○ ***In what galaxy is the sun?***
○ ***What is a star cluster?***

494   Chapter 22   The Universe

Answers–22.8

○ A galaxy is a large grouping of stars.

○ The sun is in the Milky Way Galaxy.

○ A star cluster is a group of stars that seem unusually close together.

22.9 Galaxies, Quasars, and the Big Bang

The Milky Way Galaxy is just one of billions of galaxies that exist in the universe. One of these galaxies is visible without a telescope and appears as a hazy star or cloud, called a *nebula*. A **nebula** is a cloud of dust and gases in space, which sometimes glows by reflected light. When astronomers began using telescopes, they found that some of these clouds were actually galaxies. Other nebulas are not galaxies, but the remains of exploded stars. For instance, the Crab Nebula, in the constellation Taurus, is the result of the explosion of a supernova in 1054.

A MATTER OF FACT

The sword of the constellation Orion is marked by a great nebula, not a star.

Figure 22–17. The Magellanic Clouds are the closest galaxies to the Milky Way Galaxy.

The two galaxies closest to Earth, known as the Large Magellanic Cloud and the Small Magellanic Cloud, are about 170 000 light-years away. The Great Spiral Galaxy, visible in the constellation Andromeda, is about 2 million light-years away.

Galaxies are named for their shapes as they appear from Earth. Spiral and irregular galaxies contain a lot of dust and gases, in addition to stars. Elliptical galaxies contain very little dust and gas. Some scientists believe that spiral and irregular galaxies have a black hole at their center around which the galaxy rotates. Elliptical galaxies either do not rotate or rotate very slowly.

Figure 22–18. Galaxies have several shapes (left). The Southern Pinwheel Galaxy (right), in the constellation Hydra, is an example of a barred galaxy.

Elliptical	Spiral	Barred	Irregular

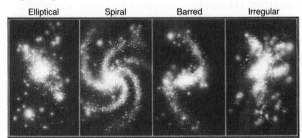

Section 3   Galaxies   **495**

22.9 Galaxies, Quasars, and the Big Bang

DISCUSSION Emphasize the difference between a nebula and a galaxy. (A nebula is a cloud of dust and gases. A distant galaxy may appear to be a nebula but is actually a grouping of billions of stars.) You may wish to refer the students to the photograph of the Crab Nebula in the chapter opening. Discuss the three classifications of galaxies. Ask the students to classify the Large and Small Magellanic Clouds. (They are irregular galaxies.)

EXTENSION A small group of 17 galaxies is called the Local Group. The Milky Way Galaxy belongs to the Local Group. Some of the other galaxies in the Local Group are the Large and Small Magellanic Clouds and the Great Spiral, which is in the constellation Andromeda and is about 2 million light-years away from Earth.

DISCUSSION Have the students discuss why it was not until the 1960s that quasars were discovered. (More sophisticated instruments were needed to discover them.) Direct the students' attention to Figure 22–20 and discuss the big-bang theory. Mention that this is just one theory for the beginning of the universe. ***Thinking Critically:*** Scientists have observed that the universe is still expanding. Ask the students how this would substantiate the big-bang theory. (If there were a tremendous explosion, the matter released in the explosion would travel outward from it.)

SECTION REVIEW

Summary Have the students write two- or three-sentence summaries of each subsection. They may use their books for help.

Reinforcement Have the students complete the Section Review. Allow any students who need assistance to look up the answers of which they are unsure.

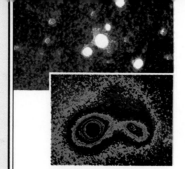

Figure 22–19. Quasars are most easily observed using radiotelescopes. The "pictures" that astronomers receive look like those above.

A MATTER OF FACT

If time since the big bang were compressed into 24 hours, Earth would not have formed until sunset, and humans would have been on Earth for only one second.

Figure 22–20. The big-bang theory explains one way in which the universe may have originated. Shown here is an artist's conception of how the initial explosion may have looked.

In the 1960s astronomers located what appear to be the largest energy sources in the universe. **Quasars** emit radio waves and light. Although they are several billion light-years away, they may be the brightest objects in the universe, emitting the energy of hundreds of galaxies. However, scientists still do not fully understand the nature of quasars.

How did the universe come to be? Physicists claim that the universe began about 15 billion years ago. Currently, the most accepted theory for the beginning of the universe is the big-bang theory. According to the **big-bang theory,** a tremendous explosion created the universe.

If, as American astronomer Edwin Hubble observed in 1929, the universe is expanding, it must have been smaller at one time. If the universe were run in reverse, it would contract to a densely packed mass. Probably about 15 billion years ago, the mass exploded into a giant cloud of gases and dust that has been expanding and cooling ever since.

○ *What are the three kinds of galaxies?*
○ *What is a quasar?*
○ *What is the big-bang theory of the origin of the universe?*

Section Review

READING CRITICALLY

1. Why are some galaxies originally thought to be nebulas?
2. Why is the universe expanding?

THINKING CRITICALLY

3. Why has the study of astronomy changed so much over the years?
4. Why would the density of the universe relate to its future?

Answers—22.9

○ Spiral galaxies, irregular galaxies, and elliptical galaxies are the three kinds of galaxies.

○ A quasar is a distant object emitting enormous amounts of energy primarily in the radio range.

○ The big-bang theory suggests that the universe was at one time condensed into a single object that exploded and eventually evolved into the present universe.

Answers to Section Review

Reading Critically

1. They were viewed without the aid of a good telescope, and the individual stars could not be seen.

2. At one time all matter was condensed into a single object that exploded in a violent "big bang." All matter has been moving away from the center of that explosion ever since.

continues

CAREERS

ASTRONOMER

The study of the movements of stars, planets, and galaxies is called *astronomy*. An *astronomer* is trained to observe, catalog, and draw conclusions about the patterns of movement of stars and other heavenly bodies.

Professional astronomers attend colleges and universities and usually have a master's degree or a doctorate. However, many amateur astronomers with only a high-school education have also made important contributions to the study of the universe.

The universe is vast, and the catalog of objects continues to grow with the work of both professionals and amateurs.

For Additional Information
Amateur Astronomers, Inc.
Sperry Observatory
1033 Springfield Avenue
Cranford, NJ 07016

ASTROPHYSICIST

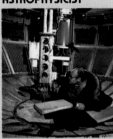

An *astrophysicist* studies the physical characteristics of the universe. Such a scientist needs a great deal of training in the physical sciences, especially physics and mathematics. Most astrophysicists have doctorates, and they teach and conduct research at universities. They must have a clear understanding of current theories regarding the relationship between the matter and energy that comprise the universe.

The major divisions of astrophysics include the following studies: solar system, stellar atmospheres, nebulae and interstellar matter, and stellar evolution.

For Additional Information
American Astronomical Society
c/o Dr. Peter B. Boyce
2000 Florida Ave., N.W. #300
Washington, DC 20009

ASTRONAUT

The word *astronaut* means "star voyager." Since exploration of the heavens is both challenging and dangerous, astronauts must be highly trained before undertaking missions above Earth's atmosphere. They must be well educated, confident, and resourceful.

The first American astronauts were military test pilots. However, since that time research scientists have joined the space program.

In the 1960s only 43 men were assigned as astronauts on the Mercury, Gemini, Apollo, and Skylab projects. Since 1978 many candidates, both males and females, have been selected for assignments aboard the space shuttle.

For Additional Information
Young Astronauts Council
P.O. Box 65432
1211 Connecticut Ave., N.W.
Washington, DC 20036

Section 3   Galaxies   **497**

CAREERS

Discussion Have interested students write to one or all of the addresses provided for additional information. Have the students discuss the opportunities offered by the various careers.

Extension You may wish to have a "Career Day" in the class during which students report on the information they have obtained. You may also wish to have the students assume the role of a person in each career and make a presentation to the class covering why he or she chose the career, what training was necessary, and what a typical workday involves. The students may be interested in researching information on the following additional careers:

Radio Astronomer

Planetarium Lecturer

Astronomy Professor

Thinking Critically

3. Astronomy has been drastically changed by the invention of new and more powerful instruments with which to examine the universe.
4. The force of gravity throughout the universe is determined by the amount of its matter. Whether or not the universe will continue to expand, or begin contracting sometime in the distant future, depends upon the amount of that matter.

Skills (Observing, Calculating)

Preparation of Materials

You may wish to have the half-circles and the long strip of paper pre-cut for the students. Each student needs only a small lump of clay to act as a base for the star and pins.

Parallax is the apparent shift in the position of an object with respect to background due to a shift in the position of the observer. The parallax method used to measure the distance to faraway objects is similar to the method of triangulation used by surveyors.

Answers to Analyses and Conclusions

1. The tangent of an angle decreases as the length of the side of the triangle opposite the tangent angle decreases. Since that side is analogous to the distance D^s, the tangent angle decreases as the star is moved closer and closer.

2. The base of the triangle used to derive stellar distances is the limiting measure. Even using the diameter of Earth's orbit (300 000 000 km) during a six-month period as the base of the triangle used to determine parallax angles, only the distances of stars less than 300 light-years away can be accurately determined.

INVESTIGATION 22: Determining Distance by Parallax

PURPOSE

To learn how parallax is used to determine the distance of stars

MATERIALS (per student)

Drawing compass
Construction paper
Scissors
Tape
Insect pin
Modeling clay
Metric ruler
Protractor
Table of Tangents

PROCEDURE

1. Use the drawing compass to make a half-circle 25 to 30 cm in diameter on the construction paper.
2. Cut out a long strip of construction paper that will represent the "celestial sphere" as shown. Perpendicular to the diameter of the half-circle, draw a line to the middle of the celestial sphere.
3. Tape the celestial sphere to the half-circle and place it at the edge of a flat surface.
4. Draw a small image of a star, and attach it to the insect pin. Stand it up, using the modeling clay as a base. Place the star anywhere along the midline of the celestial sphere.
5. Kneel down and, with the bridge of your nose at the position indicated, sight the distant star with one eye. Mark the position of your eye on the diameter line and draw a line from that point to the star.
6. Measure the distance from your eye position to the position of the bridge of your nose when you made the sighting.
7. Using the protractor, measure the angle made by the line to the star and the line between the sighting eye and nose positions.
8. Use a Table of Tangents to find the tangent of the angle.
9. Calculate the distance to the star, using the formula

$$D_S = T_A \times D_E$$

where D_S is the distance to the star, T_A is the tangent of the angle, and D_E is the distance from the sighting eye to the bridge of the nose.
10. Repeat steps 5 to 9 several times, moving your eye position each time.

ANALYSES AND CONCLUSIONS

1. After taking several measurements, what did you discover about the tangent of the angle as the distance of the star from your nose decreased?
2. If you try closing first one eye and then the other during a sighting, the position of the star shifts back and forth. This shifting is called *parallax* and is useful in measuring the distance to objects. What factors would limit this type of procedure in measuring the distance to many stars?

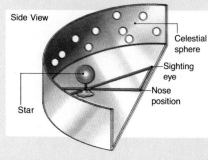

Side View

Celestial sphere

Sighting eye

Nose position

Star

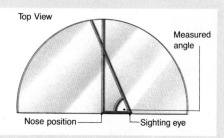

Top View

Measured angle

Nose position

Sighting eye

SUMMARY

- A star is a huge mass of glowing gases, mainly hydrogen and helium, which formed from a swirling cloud of dust and gases. (22.1)

- The magnitude of a star is a number indicating its apparent brightness as seen from the earth. (22.1)

- A constellation is a group of stars in a pattern that seems to form an image. (22.2)

- The Hertzsprung-Russell diagram shows how the energy emitted by a star is related to its color. (22.3)

- A nova is a star that becomes suddenly brighter due to an explosion and then returns to its original brightness. (22.3)

- A neutron star is the small, extremely dense core of a supernova. (22.4)

- A black hole is the core of a supernova that completely collapses into itself. (22.4)

- The sun has a layered structure consisting of the core, the radiative layer, the convective layer, and the atmosphere. The atmosphere has three layers: the photosphere, the chromosphere, and the corona. (22.5)

- Sunspots are cooler areas on the sun's surface that appear dark from Earth. (22.6)

- Solar flares are eruptions of the photosphere near sunspots. Solar prominences are streams of gases that form in the corona and are pulled back to the sun. (22.6)

- A light-year is the distance light travels in a year—it is equal to 9.5 trillion km. (22.7)

- A galaxy is a grouping of stars. The sun is in the Milky Way Galaxy. (22.8)

- A quasar is a starlike source of radio waves billions of light-years away from Earth. (22.9)

- The most widely accepted theory of the beginning of the universe is the big-bang theory. (22.9)

Write all answers on a separate sheet of paper.

SCIENCE TERMS

Correctly use each of the following terms in a sentence.

big-bang theory **(496)**
black holes **(488)**
constellations **(483)**
galaxy **(494)**
light-year **(493)**
magnitude **(481)**
nebula **(495)**
neutron star **(487)**
pulsar **(488)**
quasars **(496)**
solar prominences **(492)**
universe **(481)**
variable stars **(482)**
zodiac **(483)**

SCIENCE QUIZ

Modified True-False

Mark each statement *true* or *false*. If the statement is false, change the underlined word to make the statement true.

1. Most galaxies contain <u>thousands</u> of stars.

2. Star groupings that form a pattern as seen from Earth are called <u>galaxies</u>.

3. Cepheid-variable stars change magnitude in periods of 100 days or <u>less</u>.

4. According to the big-bang theory, the universe may continue to <u>expand</u>.

5. Rapidly rotating neutron stars are called <u>quasars</u>.

6. The theory that explains how the universe may have formed is called the <u>big-bang</u> theory.

continues

CHAPTER REVIEW

SUMMARY

The students may review the major concepts in the chapter by reading the summary statements. The statements are cross-referenced to the chapter to facilitate reinforcement of any concepts of which the students feel unsure. Encourage the students to work in groups to quiz one another.

SCIENCE TERMS

The sentence in which the science term is used should reflect an understanding of the definition of the term. You may wish to have the students make flash cards for the terms and definitions with which they can quiz each other.

SCIENCE QUIZ

Answers to Modified True-False

1. false, millions
2. false, constellations
3. true
4. true
5. false, pulsars
6. true

continues

Answers to Multiple Choice

7. b
8. c
9. a
10. d
11. c
12. b

Answers to Completion

13. big-bang
14. first
15. Hertzsprung – Russell
16. black hole
17. the big-bang

Answers to Short Answer

18. spiral galaxies, irregular galaxies, and circular galaxies
19. Energy radiates by conduction until it reaches the sun's atmosphere. There it radiates by convection.
20. open and closed star clusters
21. Red giants, white dwarfs, neutron stars, and pulsars are all types of stars.
22. Stars formed in gaseous nebulae eventually become main sequence stars, red giants, and then white dwarfs. Depending on the mass of the star, it may then become either a black dwarf, neutron star, or a black hole.
23. Drawings will vary.

Answers to Writing Critically

24. The sun's energy is created at the core and is radiated through a thick radiative layer to a convective layer and finally to the surface photosphere.
25. If the mean density of the universe is great enough, the universe will eventually stop expanding and begin to contract. If the density is not great enough, the universe will expand forever.

ANSWERS TO EXTENSION

1. Answers will vary, but the students should summarize the steady-state theorists' suggestion that as the universe expands, new matter is continually being created.
2. Answers will vary.
3. Answers will vary.

ANSWERS TO APPLICATION/ CRITICAL THINKING

1. At a speed of 313 600 km/sec, light

Multiple Choice

Write the letter of the choice that best answers the question or completes the statement.

7. What elements fuse in the reactions of the sun?
   a) hydrogen and oxygen
   b) hydrogen
   c) hydrogen and carbon
   d) all heavy elements

8. How far away is the nearest star?
   a) 4 260 000 kilometers
   b) 4.26 astronomical units
   c) 4.26 light-years
   d) 10 light-years

9. The most distant objects in the sky are known as
   a) quasars.
   b) galaxies.
   c) supernovas.
   d) spiral galaxies.

10. How old is the universe according to the big-bang theory?
    a) 4.5 million years
    b) 4.5 billion years
    c) 15 million years
    d) 15 billion years

11. The leftover core of a supernova may be any of the following except a
    a) black hole.   b) pulsar.
    c) quasar.   d) neutron star.

12. Most stars of the main sequence, such as the sun, will eventually end up as
    a) black holes.
    b) black dwarfs.
    c) white dwarfs.
    d) supernovas.

Completion

Complete each statement by supplying the correct term or phrase.

13. The _____ theory states that an explosion created the universe.

14. A _____ magnitude star is the brightest.

500   Chapter 22 Review

---

15. The life cycle of stars is represented on the _____ diagram.

16. The center of a rotating spiral galaxy may contain a _____.

17. The universe will continue to expand if the theory of _____ is correct.

Short Answer

18. Describe several types of galaxies.

19. Explain how energy from the sun's core reaches its surface.

20. Describe the differences between the types of star clusters.

21. Name four different types of stars.

22. List the phases of the sun's evolution from its birth to its death.

23. Draw the arrangement of stars in five constellations.

Writing Critically

24. Explain why the energy of the sun's core takes so long to reach the sun's surface.

25. Compare and contrast opposing theories about the future of the universe.

EXTENSION

1. Research the steady-state theory of the universe; then write a report contrasting this theory with the currently accepted big-bang theory.

2. Research and report on the methods astronomers use to approximate the distance to the stars. Be sure to include the advantages and limitations of each method.

3. Make a table showing the distances to several close stars in kilometers, astronomical units, and light-years.

APPLICATION/CRITICAL THINKING

1. Calculate the distance to Proxima Centauri in kilometers.

---

travels 9.6×10^{12} km per year. Multiply the distance from Earth to Proxima Centauri by the length of a light-year in km. Answer: 4.26 light-years $\times 9.6 \times 10^{12}$ km per light-year $= 4.09 \times 10^{13}$ km.

continues

2. Describe how variable stars might be used to calculate stellar distances.

3. Since black holes give off no energy, how can they be detected?

4. Most scientists believe that the universe is still expanding from the big bang. Some scientists believe the universe will eventually start to contract, and that it will continue contracting until another big bang occurs. Describe the events that might take place during this "big crunch."

FOR FURTHER READING

Davies, O. "Comet Collectors." *Omni* 8 (July 1986): 24. This article is a brief essay describing the successful attempts by scientists to use existing satellites for gathering information on the comets Giacobini-Zinner and Halley.

Heidmann, J. *Extragalactic Adventure: Our Strange Universe.* New York: Cambridge University Press, 1982. This book describes the evolution of the universe, space, time, black holes, and the possible existence of extraterrestrials.

Mitton, S. and J. Mitton. *Invitation to Astronomy.* New York: Basil Blackwell, 1986. The authors explain what astronomy is all about and what astronomers do. It is an excellent book for anyone who is considering a career in astronomy.

Wolkomir, R. "SS-433, What Are You?" *Omni* 8 (September 1986): 30. This article describes the star designated SS-433, which is now known to be a binary star. Before an in-depth analysis was performed on the object, no one was quite sure what it was.

Challenge Your Thinking

This photograph is an artist's idea of a black hole. In some movies the crew of spacecraft travel through a black hole to return home safely. Using your knowledge of black holes, explain why this would not be possible.

ANSWER TO CHALLENGE YOUR THINKING

A black hole is not a hole into which things fall, instead, it is an area of extremely dense matter.

2. Variable stars are stars whose brightness varies periodically. By comparing the absolute brightness of a close variable star to the apparent brightness of a distant variable star with a similar period, it is possible to calculate the distance to the far star. This calculation is possible because the rate at which the intensity of light diminishes with distance is known (intensity of light decreases as the square of the distance from the light source).

3. Black holes are detected by the way they influence near-by objects.

4. The expanding universe slows down, stops, and then begins to come together due to gravity.

Chapter 23: OBSERVING THE UNIVERSE

PLANNING THE CHAPTER

Chapter Sections	Page	Chapter Features	Page	Program Resources	Page
Section 1: Optical Astronomy 23.1 Lenses and Light **(A)** 23.2 Refracting Telescopes **(A)** 23.3 Reflecting Telescopes **(A)**	503 503 506 508	**Discover:** Observing the Moon **(A)** **Activity:** Refracting Light **(A)** A Matter of Fact Section Review **Skill Activity:** Using a Telescope **(A)** **Investigation 23:** Measuring Focal Length **(H)**	507 508 510 510 511 516	Cross-Discipline: Science and Mathematics, *Measuring Distances in Outer Space* **(A)** Reading for Content: *Examining Reflection and Refraction* **(A)** Investigation 23.1: *Refraction of Light by Glass* **(H)** Investigation 23.2: *Refracting Lenses* **(A)** Student Record Book: Textbook Investigations **(H)**	TRB 43 TRB 40 LM 217 TRB 107 LM 109 TRB 111 LM 113 TRB 47
Section 2: Radio Astronomy 23.4 Radio Waves **(H)** 23.5 Radiotelescopes **(H)**	512 512 513	A Matter of Fact **Discover:** Identifying Sources of Radio Waves **(H)** Section Review **Technology:** Radiotelescopes	513 514 514 515	Critical Thinking **(H)** Concept Extension: *Space Probe* **(A)**	TRB 44 TRB 47 LM 269
Chapter 23 Review	517			Vocabulary **(A)** Tests **(A)** Computer Test Bank	TRB 23 LM 167 TRB 99

(LM) Laboratory Manual/Study Guide, **(TRB)** Teacher's ResourceBank™

B = Basic   **A** = Average   **H** = Honors

The coding Basic, Average, and Honors indicates sections or subsections that might be appropriate for different levels of learners. For additional suggestions regarding choice of topic and depth of coverage, see the Pacing Chart on pages T16–T20.

CHAPTER CONCEPTS, OBJECTIVES, AND TERMS

Section	Concepts	Objectives	Science Terms
Section 1: Optical Astronomy	■ Refraction occurs when a beam of light hits the surface of a substance through which light travels at a slower speed. **(23.1)** ■ A lens that causes light to converge on one point is called a convex lens. A lens that causes light to diverge is called a concave lens. **(23.1)** ■ Focal length is the distance between the center of a lens and its focus. **(23.1)** ■ Refracting telescopes have two convex lenses that focus the light and produce an upside-down image. **(23.2)** ■ The size of refractors is limited by the problems associated with making large lenses. **(23.2)** ■ The use of photographic plates can greatly extend the range of telescopes. **(23.2)** ■ Reflectors are telescopes that focus light with mirrors. **(23.3)** ■ Reflecting telescopes have no distortion, absorb less light than lenses, and can be made larger than refractors. **(23.3)**	■ **Discuss** the importance of telescopes. ■ **Explain** what refraction and reflection are. ■ **Define** what is meant by *focus* and *focal length*. ■ **Describe** the difference between refractors and reflectors.	refraction reflection lens focal length telescope magnification spectrograph
Section 2: Radio Astronomy	■ Radio waves are long-wavelength energy. Many objects in space produce radio waves. **(23.4)** ■ Radiotelescopes are used to detect long-wavelength radiation from outer space. **(23.4)** ■ Radiotelescopes consist of movable dishes that focus radio waves on an antenna. **(23.5)** ■ Many important discoveries about the universe have been made by radio astronomers. **(23.5)**	■ **Explain** the difference between radio waves and visible light. ■ **Describe** how radiotelescopes work. ■ **List** the major discoveries made with radiotelescopes.	radio waves radiotelescope

CHAPTER MATERIALS

TEACHING SUGGESTIONS

Section 1: Optical Astronomy

Demonstration: Night Photography

Purpose

To show how the quality of photographic images depends upon film characteristics and exposure time, as well as other factors

Background

The quality of the images will vary, depending upon the ASA values (the sensitivity) of the film, the shutter speed, and aperture setting. The image of the candle will be brighter as exposure time and aperture opening increases (a longer time for more light to expose the film).

However, the quality of the image may also increase with exposure time. Discuss with the class how telescopic images recorded using cameras mounted at the eyepiece are also dependent upon these factors.

Materials

Camera with several shutter speeds and aperture openings
Film (two rolls with different ASA values)
Candle

Procedure

1. Darken the room and light the candle in one corner of the classroom.
2. Load one of the rolls of film and set up the camera on the other side of the room.
3. Take a series of photographs of the candle at a fixed aperture setting, systematically changing time exposures (i.e., 1/500 second to 1 second). Be sure to record the shutter speed for each photo.
4. Take another series at a single shutter speed, this time varying the aperture settings (f-stop 1 to 16). Be sure to record the settings for each photo.
5. Remove the film.
6. Repeat Steps 2 through 5 with the other roll.
7. Have the film developed and display them to the class, making sure to indicate the film used, the aperture setting, and the exposure time for each photograph.

Question to Ask Students

What is meant by film speed? *(Faster films can take photographs in less light or with faster shutter speeds.)*

Class Activity

Make some nighttime observations of the stars or moon with reflecting and refracting telescopes.

Section 2: Radio Astronomy

Class Activity

Refer to the classified or advertisement sections of magazines like *Popular Science* or *Popular Mechanics*. Write companies that manufacture, sell, or install television or radio antennas and request free catalogs describing their products.

Field Trip

Write or visit a local television station or cable company and obtain information about how radio antennas are used to receive programming. If there is a company nearby that sells and installs backyard radio antennas, pay them a visit as well.

Outside Speaker

Have a radio astronomer talk to the class about the use of radiotelescopes in explaining the universe.

CHAPTER 23

CHAPTER OVERVIEW

This chapter describes the instruments astronomers use to observe the universe. The two types of optical telescopes, refracting and reflecting, are discussed. Also presented are the advantages of radiotelescopes over optical telescopes. Important discoveries made with both optical telescopes and radiotelescopes are then described.

Section 1: Optical Astronomy This section begins with a discussion of visible light. The properties of refraction and reflection are defined. Then the application of these properties is used to describe how refracting and reflecting telescopes work. These two types of telescopes are compared and contrasted.

Section 2: Radio Astronomy In this section, the difference between radio waves and visible light is explained. Radiotelescopes are described, and some of the discoveries that have been made with them are highlighted.

Observing the Universe

Stars, gas clouds, novae, and neutron stars all emit energy. Cosmic rays, gamma rays, X rays, ultraviolet rays, visible light rays, infrared rays, and radio waves all travel through space at the speed of light. How are these rays detected? What instruments do astronomers use?

Kitt Peak Observatory, in Arizona

502

CHAPTER MOTIVATING ACTIVITY

Obtain books from the library that show photographs of telescopes, observatories, radiotelescope arrays, and orbiting telescopes. As you show them to the class, explain how these instruments have become larger and more complex as a consequence of people's desire to see farther and farther into outer space. Begin with illustrations of simple refracting telescopes and list some of the discoveries made with them such as the the discovery of Saturn's rings and mountains on the moon. Then move on to the larger reflecting telescopes that Sir William Herschel (1738–1822) used to categorize "nebulosities" and galaxies that were previously identified as "fuzzy stars."

SECTION
1 Optical Astronomy

SECTION OBJECTIVES

After completing this section, you should be able to:

- **Discuss** the importance of telescopes.
- **Explain** what refraction and reflection are.
- **Define** what is meant by *focus* and *focal length*.
- **Describe** the difference between refractors and reflectors.

NEW SCIENCE TERMS

refraction
reflection
lens
focal length
telescope
magnification
spectrograph

23.1 Lenses and Light

Have you ever looked at the millions of stars in the night sky? The stars shine because of the energy they give off. The energy you see is in the form of light. You may recall from Chapter 8 that light travels through space as particles called *photons*.

Visible light is a very small part of all the energy in the universe. Some other forms of energy include radio waves, X rays, infrared radiation, and ultraviolet radiation. Visible light, however, is the easiest for people to detect. Light travels through space at the constant speed of 299 792 458 m/s (3.0×10^8 m/s). Light also travels through transparent matter but at slower rates of speed. Table 23–1 shows the speed of light in various substances. Notice that the speed of light in air is similar to the speed of light in space.

TABLE 23–1: THE SPEED OF LIGHT IN DIFFERENT MEDIA	
Substance	**Speed of Light (m/s)**
Empty space	299 792 458 (3.0×10^8)
Air	299 702 547 (3.0×10^8)
Water	225 407 863 (2.2×10^8)
Glass	197 231 880 (2.0×10^8)
Diamond	123 881 181 (1.2×10^8)

Refraction Light behaves in certain predictable ways. A beam of light traveling through air will slow down if it strikes a glass pane. If the beam hits the glass at a right angle, even though it slows down, it will come out of the glass going in the same direction as when it went in. If a light beam strikes a glass pane at an angle other than 90 degrees, it not only slows down but also changes direction.

Figure 23–1. Observing the stars is not simply a matter of looking through a telescope. Modern astronomical telescopes are controlled by computers.

Section 1   Optical Astronomy   **503**

Section 1: Optical Astronomy

SECTION FOCUS

Section Overview This section describes the principles of optical astronomy. Two important properties of light—refraction and reflection—are discussed. The differences between convex and concave lenses are explained.

Section Objectives For a list of section objectives, see pupil's edition page.

New Science Terms For a list of new science terms in this section, see pupil's edition page. Point out to the students the new terms they will be learning in this section. Help the students make sure that they can distinguish between *refraction* and *reflection* as they read the section.

SECTION DEVELOPMENT

23.1 Lenses and Light

DISCUSSION Remind the students that the movement of light through space was discussed in Chapter 8. Point out that all electromagnetic waves travel at the same speed. (the speed of light)

DISCUSSION Refer the students to Table 23–1. The table shows that the speed of light in air is slightly less than the speed of light in space. *Thinking Critically:* Ask the students to explain the reason for this difference. (Although air looks empty to us, it is a mixture of molecules of various gases. These gas molecules slow down electromagnetic waves passing through.)

BACKGROUND INFORMATION

At one time it was thought that light travels with infinite speed instantaneously from its source to an observer. Olaus Romer (Danish, 1644–1710) first estimated the speed of light by observing the eclipses of various satellites of Jupiter. More accurate measurements were made during the nineteenth century. The theory of relativity predicts that the speed of light in a vacuum is the limiting velocity for material particles. A particle can be accelerated to a speed closely approaching the speed of light, but can never reach the speed of light.

DEMONSTRATION

For a demonstration of night photography see page 501c preceding this chapter.

503

DISCUSSION Make sure the students realize that refraction occurs virtually instantaneously. They may mistakenly think that a light ray requires an appreciable amount of time to bend.

DISCUSSION Point out that sound waves are reflected just as light waves are because both normally travel in straight lines. *Thinking Critically:* Ask the students how reflection is used to measure distances. (Since the speed of light is known, measuring the amount of time it takes for light to travel from a source, bounce off a distant object, and return to the source enables the observer to calculate the distance to the object. By pointing a laser at a mirror placed on the moon by astronauts, scientists have been able to measure the distance to the moon with great precision.)

Consider a drill team marching in formation. As they come to a corner, the person on the end of a line turns the corner first and then marches in place. The rest of the people in line continue marching until they reach the corner, and then they also turn and march in place. When the entire line has turned, they again begin marching in a straight line.

Figure 23–2. The sun's energy easily passes through clear glass into this room, and is trapped inside as heat.

If the person who turned the corner first did not march in place, then the line would bend as a beam of light does. One side of the light beam hits the glass first and is slowed before the other side even reaches the glass. This causes the beam to bend. As the beam emerges from the other side of the glass, the side of the beam that reached the glass first leaves the glass first and moves faster than the other side. The beam then changes direction again. The change in direction of a light beam as it passes from one type of matter to another is called **refraction.**

Figure 23–3. Light striking a shiny surface is reflected (left). Light that strikes a pane of glass at a 90° angle passes through the glass without refracting (center). If light strikes a pane of glass at a lesser angle, the light is refracted (right).

Reflection   Light is bounced back from a smooth, shiny surface. This bouncing back of light is called **reflection.** If light hits a shiny surface at any angle other than 90°, it bounces off at an angle. The angle at which the light is reflected is equal to the angle at which the light hit the surface. This property of reflected light enables anyone with a mirror to bounce light in practically any direction.

504   Chapter 23   Observing the Universe

DEMONSTRATION

Have the students observe the shape of the point of a sharpened pencil. Then place the pencil, point down, in a glass of water and have the students observe the shape of the pencil point again. The pencil point will appear shorter when viewed through the water.

BACKGROUND INFORMATION

The ancient Greeks and Romans used mirrors made of highly polished metal. Glass mirrors, which have a back coated with a mixture of tin and mercury, came into use during the Middle Ages. Since 1840, the backing on most high-quality mirrors has been silver.

Lenses The properties of refraction and reflection of light are used in astronomy. Understanding refraction allows scientists to use a series of lenses to focus the light given off by distant stars. A piece of curved glass that can be used to focus light is called a **lens.** If the surfaces are curved outward as shown in Figure 23–4, all light rays will converge, or focus, at a single point. Lenses that cause light rays to converge are called *convex lenses*. Lenses that cause light rays to diverge, or separate, are called *concave lenses*. These lenses curve inward as shown in Figure 23–4. The distance between the center of the lens and the point of focus is called the **focal length.**

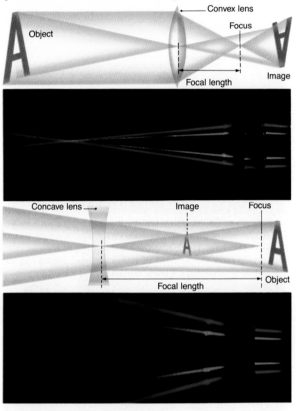

Figure 23–4. Light rays passing through a convex lens will converge at a focus (top). Light rays passing through a concave lens will diverge (bottom).

○ *What is refraction of light?*
○ *What is reflection of light?*
○ *What is a lens?*

Section 1   Optical Astronomy   **505**

DISCUSSION Ask the students what characteristics a material used to make lenses must have. (It must be transparent, able to be shaped and smoothed, free from imperfections, and not easily scratched.) Point out that plastic instead of glass is now commonly used for many eyeglass lenses.

EXTENSION The images made by convex lenses may be larger than, smaller than, or the same size as the object. For example, a camera uses a convex lens to make a small image of an object on the film. A slide projector uses a convex lens to make a large image on a screen. Images made by concave lenses are always smaller than the actual object. *Thinking Critically:* Ask the students if they can think of other uses of concave lenses. (Concave lenses are used in doorway peepholes to give the viewer an image of a relatively large area; they are used in eyeglasses to correct nearsightedness.)

EXTENSION Lenses are categorized according to their aperture and focal length. The aperture of a lens is the diameter of the clear material through which light is able to pass. The aperture is usually the figure noted when referring to the size of a telescope. (such as 500 cm) The focal length is the distance from the center of the lens to the point where all the rays converge. The ratio of the focal length to the aperture is the focal number of the lens. All lenses tend to distort images by dispersing colors of different wavelengths. This problem, called chromatic aberration, is an inherent defect of lenses. It can be overcome by some compound lenses.

Answers–23.1

○ Refraction is the change in direction of light waves as they pass from one medium into another.

○ Reflection is the bouncing of light waves off a surface.

○ A lens is a piece of curved glass that can be used to focus light.

23.2 Refracting Telescopes

DISCUSSION Direct the students' attention to Figure 23–5 which shows how a refracting telescope works. Remind the students that the image produced by a simple refracting telescope is upside down.

SKILL (*Making Mathematical Calculations*) The amount of light gathered by a telescope (the brightness of the image) depends on the area of the objective lens. Ask the students to calculate what happens to the light-gathering power of a circular lens if its radius is doubled or tripled. (Area of a circle = πr^2; if the radius is doubled the light-gathering power is four times greater. If the radius is tripled, the light-gathering power is nine times greater.)

EXTENSION All refracting telescopes house an objective lens and an ocular lens (the eyepiece). Light from an object reaches the objective lens in parallel rays that are refracted to converge at a focal point behind the lens. A first image of the object is produced behind the focal point in the focal plane, which is a plane perpendicular to the axis of the lens. The eyepiece is positioned in front of the focal point in order to magnify the rays refracted by the objective lens. Almost all refracting telescopes produce chromatic aberration.

Figure 23–5. A refracting telescope, such as the one that Galileo used (above), has convex lenses (right).

23.2 Refracting Telescopes

A **telescope** is an instrument used to enlarge the image of a distant object. In 1609 Galileo Galilei, an Italian astronomer and physicist, made a telescope based on the principle of refraction. Although this instrument was primitive, Galileo made many discoveries, such as the craters and mountains on the moon, the phases of Venus, and the four large satellites of Jupiter.

A *refracting telescope* is usually made with two tubes that fit together snugly. Light from the object being viewed passes through the *objective lens*. The light comes to focus inside the tube and then passes through a second lens, called the *ocular lens*, to the observer's eye. The tubes can be moved to get the best focus for the observer.

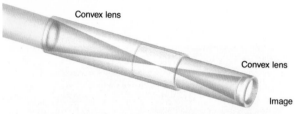

Convex lens

Convex lens

Image

The **magnification** of a telescope is equal to the focal length of the objective lens divided by the focal length of the ocular lens. The combination of a long objective-lens focal length and a short ocular-lens focal length produces a telescope with great magnification. Galileo's first telescope had a magnification of 3. The brightness of the image depends on the size of the surface of the objective lens. The larger the surface area of the objective lens, the greater the amount of light captured. Increasing the amount of light entering a telescope increases the brightness of the image.

Figure 23–6. A modern refracting telescope has basically the same structure as the one that Galileo used in the early 1600s.

506

In order to improve the quality of refracting telescopes, larger lenses were made. The first large lenses were 38 cm in diameter and were made in Germany in the 1830s. The Yerkes Observatory, founded in 1895 in Williams Bay, Wisconsin, has the largest refracting telescope ever made. The objective lens of the Yerkes telescope is 102 cm in diameter. The glass used to make refracting lenses must be perfectly smooth. For this reason it is very difficult to make large lenses. Lenses larger than that of the Yerkes refractor will probably not be made because the glass in lenses larger than 100 cm in diameter tends to change shape, causing distortion.

Until the end of the last century, most astronomical observations were made with the human eye. Now most observations are made using photographic plates, which astronomers then study. Computers align the telescope, and motors on the telescope keep it focused on an object for hours or even days. This tracking system allows enough time for even very faint objects to leave an image on the photographic plates.

Photographic plates taken of the same area of the universe at different times can be studied in order to see changes in the stars. When a supernova was observed in 1987, studying photographic plates made it possible to identify which star had exploded, since the star simply was no longer there.

Telescopes have also been connected to spectrographs to study the composition of stars. A **spectrograph** is an optical instrument used for observing the spectrum of light given off by a star. A spectrograph can be used to determine the composition of stars because light from each element gives off a unique spectrum. When the spectrum from a star is compared with the spectrum of elements known on Earth, astronomers can identify the elements present in the star.

○ *What is a refracting telescope?*
○ *What is a spectrograph?*

DISCOVER

Observing the Moon

On a clear night, observe the moon and draw what you see, in as much detail as possible. Then observe the moon with a telescope or binoculars. Again draw as much detail as you can. Describe the features that were not visible with the unaided eye.

Figure 23–7. Many astronomical observations are made by studying film exposed through a large telescope (left). In some cases, film has been replaced by an electronic plate that records light more precisely than film does (below).

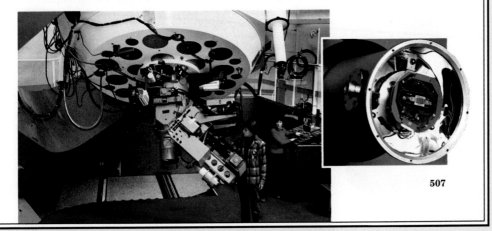

507

DISCOVER:
Observing the Moon

Thinking Skill *(Comparing)*

Have the students compare their two sketches of the moon. They should have observed many more craters and mountains with binoculars or with a telescope than they did with the unaided eye. A student who uses a telescope should produce a sketch that is upside down relative to the unaided eye sketch.

EXTENSION Review the wavelength and frequency range of the visible portion of the electromagnetic spectrum. Show the spectra of the sun and several elements. Explain how spectral analysis of stars, including absorption lines, is used to determine the composition of stars. Remind the students that only a limited variety of objects can be recorded using the visible spectrum.

Answers–23.2

○ A refracting telescope is a tool that uses lenses to magnify distant objects.

○ A spectrograph is an optical instrument used to examine the colors of light emitted by a star.

BACKGROUND INFORMATION

With many astronomers competing for telescope time, the amount of time available for obtaining spectra is limited. To help meet this need, a telescope dedicated to spectroscopic surveys is being constructed at the MacDonald Observatory in Fort Davis, Texas. This telescope will enable astronomers to obtain the large number of spectra needed to study and classify stars and galaxies.

ACTIVITY:
Refracting Light

Skill *(Experimenting)*

Preparation of Materials Another method for doing this activity is to fill a large glass container with water. Place part of a long ruler into the water and have the students observe and describe what they see. (The students should observe that the part of the ruler in the water seems bent.) Point out to the students that the bending of a light beam as it passes from one medium (air) into a different one (water) is called refraction.

Answers to Conclusions/Applications

1. The light rays are also traveling in a different direction than the white beam that entered the glass.
2. The beam is again refracted to its orignal path.

23.3 Reflecting Telescopes

DISCUSSION Direct the students' attention to Figure 23–9 on page 509, which shows how a reflecting telescope works. Then compare it with Figure 23–5 on page 506, which shows the refracting telescope.

How can you show that white light is refracted as it passes through glass?

MATERIALS (per group of 3 or 4)

glass blocks (2), flashlight, white construction paper

PROCEDURE

1. On a flat surface, stand one glass block on end, point the flashlight at the block, and turn the flashlight on.
2. Hold the white construction paper as shown, and move it around the glass block until the paper reflects an image of the light coming through the glass block.
3. Position the second glass block between the paper and the first block, and note the angle of the light as it passes through both blocks.

CONCLUSIONS/APPLICATIONS

1. How is the white light changed as it passes through the first block? Give a specific description of the nature of the change.
2. What happens to the light passed through the first block when it is intercepted and passed through the second block?

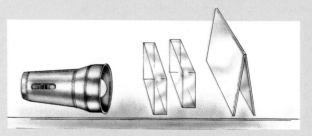

Figure 23–8. This telescope is a replica of the reflecting telescope made by Isaac Newton.

508

23.3 Reflecting Telescopes

Sir Isaac Newton, an English scientist, constructed the first *reflecting telescope*. Newton realized that concave mirrors have three advantages over lenses. First, they reflect light instead of refracting it and therefore produce no distortion. Second, a good mirror absorbs less light than even the most transparent glass, so the image will be brighter. Third, a mirror is not as heavy as a lens of the same size. Therefore, a mirror is easier to support, and much larger telescopes can be constructed.

Reflecting telescopes have two mirrors. The primary mirror is a large concave mirror that receives the image and reflects it to a secondary mirror. The secondary mirror, a flat mirror placed in front of the focus, reflects the light through an ocular lens to the observer. The secondary mirror is so small in comparison to the primary mirror that the light it intercepts does not affect the quality of the image.

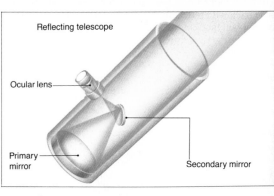
Reflecting telescope

Ocular lens

Primary mirror

Secondary mirror

The largest telescopes used today are reflectors. The largest reflector in the world, located on Mt. Semirodriki in the Soviet Union, has a primary mirror 6 m across. The second-largest reflector is located at the Palomar Observatory, near San Diego, California.

Even though the atmosphere lets in visible light, the light is distorted by the atmosphere and seems to flicker. This is the reason why stars seem to twinkle. As a result of this flickering, images produced by ground-based telescopes tend to be a bit fuzzy. There are plans to place into Earth orbit a large reflector for the observation of distant space. A telescope operating outside the atmosphere will produce sharper images and allow astronomers to study objects in space in greater detail.

Figure 23–9. Reflecting telescopes have a pair of mirrors instead of lenses to focus the light (left). This student is making observations with a modern reflecting telescope (right).

Figure 23–10. The primary mirror of a reflecting telescope can be quite large (right). Astronomical observatories are often placed on mountain tops to eliminate much of the distortion caused by the earth's atmosphere (left).

Section 1   Optical Astronomy   **509**

THINKING SKILL *(Comparing)* Ask the students how the size of the largest reflector compares with the size of the largest refractor. (six m diameter for the reflector, compared with 102 cm or just over 1 m for the largest refraction—diameter six times larger; area nine times larger)

DISCUSSION   Point out to the students that the size of some huge refracting telescopes requires the observer to sit in a special cage inside the tube in order to view the image. ***Thinking Critically:*** Photographs of astronomers at work often show them wearing coats or jackets because observatories are not heated. How might heating affect astronomers' observations? (Hot air causes distortion of light rays and could fog mirrors and lenses.)

EXTENSION   Because reflecting telescopes use mirrors to gather and converge light rays, they are free of chromatic aberration. They also produce brighter images, since they reflect rather than absorb light. Reflecting telescopes do not require tubes of as great a length as refracting telescopes of similar magnification. In addition, the surface of a reflecting mirror need not be as perfect as a refracting lens. This makes reflecting telescopes simpler to manufacture and maintain. In addition, a mirror can be supported more firmly than a lens. A lens can be supported only at its edges.

BACKGROUND INFORMATION

A relatively new type of reflecting telescope is the multiple-mirror telescope (MMT), in which a number of smaller mirrors are arranged so as to throw their reflections into a single image. The first MMT was completed in 1968. Since then, new techniques of casting very large mirrors have been developed, and the MMT is now being replaced by a single mirror. Other MMTs are in the process of being developed.

THINKING SKILL (*Drawing Conclusions*) Ask the students to name the advantages an infrared telescope might have over a light telescope. (It can pick up very low-energy infrared radiations, such as radiations emitted by the dust and gas between the stars. Many of these emit too little energy to be seen by reflecting telescopes.)

Infrared radiation, or heat energy, from stars is studied, as is visible light, with reflecting telescopes. One problem is that surrounding buildings and the telescope are warm and, therefore, produce their own infrared radiation. A photographic plate at the focus of an infrared reflector receives more infrared radiation from the mirror itself than it does from space. To minimize this problem, infrared telescopes are placed on top of remote mountains and cooled with liquid nitrogen or liquid helium.

○ **How does a reflecting telescope focus light?**

Figure 23–11. Shown here are infrared photographs of a nova (above) and of the edge of the Milky Way Galaxy (right).

A MATTER OF FACT

Launched in early 1983, NASA's Infrared Astronomical Satellite gathered information on more than 200 000 celestial objects emitting low-energy infrared radiation. In addition to identifying whole new galaxies, the satellite discovered thousands of previously unseen asteroids.

Section Review

READING CRITICALLY

1. What is the difference between a reflecting telescope and a refracting telescope?
2. What advantages do reflecting telescopes have over refracting telescopes?
3. Why is the spectrograph a useful instrument in astronomy?

THINKING CRITICALLY

4. If the city of San Diego expands, what will happen to the effectiveness of the Palomar Observatory?
5. How has the telescope advanced the knowledge of astronomy?

510   Chapter 23   Observing the Universe

SECTION REVIEW

Summary Have the students write a short paragraph that compares and contrasts refraction and reflection. Then have them list advantages and disadvantages of refracting and reflecting telescopes.

Reinforcement Have the students quiz each other, using the questions at the end of each numbered subsection.

Answer–23.2

○ A reflecting telescope uses a concave mirror to focus light.

Answers to Section Review

Reading Critically

1. Reflecting telescopes use mirrors and lenses to magnify distant objects. Refracting telescopes use only lenses to do this.
2. Reflecting telescopes can be built much larger because mirrors are easier to construct, handle, and maintain than lenses.
3. A spectrograph makes it possible for astronomers to determine the composition of stars.

continues

SKILL ACTIVITY: Using a Telescope

BACKGROUND

The word *telescope* comes from the Greek word *teleskopos,* meaning "to see from a distance." A telescope is a device used to make distant objects, such as stars, appear brighter than the unaided eye perceives them. The first refracting telescopes combined a convex and concave lens as shown. In this activity, you will construct and use a telescope similar to the one used by the great astronomer Galileo Galilei.

PROCEDURE

1. Fit together two short paper tubes made of rolled construction paper. One tube should fit snugly into the other and slide back and forth with ease.
2. Into the end of the smaller-diameter tube (the eyepiece), insert a double concave lens. Hold the lens in place with modeling clay.
3. Into the end of the larger-diameter tube (the objective), insert a double convex lens, using clay to secure it.
4. Slide the tubes back and forth to focus and enlarge distant objects.

APPLICATION

1. Use the telescope to view objects in your classroom.
2. Draw to scale several objects as they appear both with the unaided eye and through the telescope.

USING WHAT YOU HAVE LEARNED

1. Use your telescope to view the moon. What features can you see with the telescope that you could not see before?
2. When you have mastered the use of this telescope, read about reflecting telescopes and try making and using a simple one.

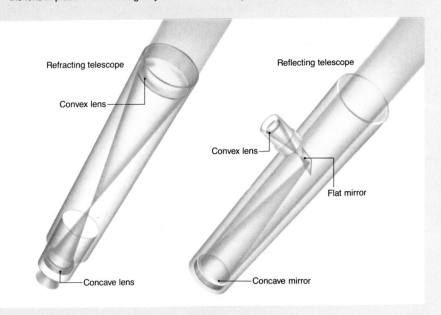

Refracting telescope

Convex lens

Concave lens

Reflecting telescope

Convex lens

Flat mirror

Concave mirror

SKILL ACTIVITY:
Using a Telescope

Objectives

- Construct a simple refracting telescope.
- Practice using the telescope in order to learn how a refracting telescope works.

Discussion   Objects viewed by the students with their telescopes will appear upside down. ***Thinking Critically:*** Ask the students what limitations are inherent in their telescopes. (The tube is short and the small lens can't gather very much light, so only fairly bright objects can be seen.)

Answer to Application

Drawings will vary.

Answer to Using What You Have Learned

1. Answers will vary. Students should be able to distinguish mountains and craters that they could not see without the telescope.

Thinking Critically

4. The additional city lights will obscure the images photographed by the observatory telescope.
5. The telescope has allowed astronomers to increase the information available to them. They were able to see more of the universe and therefore were able to document many more celestial objects.

SECTION FOCUS

Section Overview In this section, radio waves are defined as long-wavelength electromagnetic waves. The operation of radiotelescopes is described, and some important discoveries that have been made with radiotelescopes are discussed.

Section Objectives For a list of section objectives, see pupil's edition page.

New Science Terms For a list of new science terms in this section, see pupil's edition page. You may wish to familiarize students with these terms before they begin the section.

SECTION DEVELOPMENT

23.4 Radio Waves

DISCUSSION Stress the point that because radio waves are not absorbed by matter, they can carry information from distant parts of the sky where no light is received. Radio waves pass through Earth's atmosphere during the day as well as at night.

SECTION
2 **Radio Astronomy**

NEW SCIENCE TERMS
radio waves
radiotelescope

SECTION OBJECTIVES
After completing this section, you should be able to:
- **Explain** the difference between radio waves and visible light.
- **Describe** how radiotelescopes work.
- **List** the major discoveries made with radiotelescopes.

23.4 Radio Waves

Objects in the universe that cannot be seen can still be studied. Other forms of energy, especially radio waves, can be detected from Earth by special telescopes. **Radio waves** are the long-wavelength electromagnetic waves. Light rays can have a wavelength as short as 0.000 000 5 m. Radio waves, on the other hand, can have wavelengths that range from 10 m to several kilometers. Any time you listen to a radio, you are hearing sounds that are transmitted as radio waves. Your radio is tuned to receive a particular frequency, or wavelength, of radio waves. In 1931 an American engineer named Karl Jansky wondered why he was getting static on his radio receiver. He hypothesized that the static was radio waves coming from space.

Figure 23–12. Radio waves have very long wavelengths.

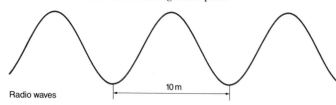

.000001m
Light waves

Radio waves

10 m

Now scientists know that radio waves of many wavelengths are produced in outer space. Stars, planets as big as Jupiter, and the cores of many galaxies emit radio waves. Quasars emit powerful radio waves. Radio waves are not absorbed by matter as visible light waves are. Radio waves from outer space may cross the entire universe with little absorption.

Astronomers capture radio waves from outer space with a special instrument called a **radiotelescope.** These waves provide astronomers with much information about the structure of the universe and the distribution of matter.

○ *What are radio waves?*
○ *What is a radiotelescope?*

512   Chapter 23   Observing the Universe

Answers–23.4

○ Radio waves are long-wavelength electromagnetic waves with wavelengths from 10 m to several kilometers long.

○ A radiotelescope is an instrument that gathers radio waves from space.

23.5 Radiotelescopes

Since the 1940s, radiotelescopes have been used to study space. Some radiotelescopes are like large TV antennas. Most, however, are metallic dishes like the one shown in Figure 23–13.

Radio waves are reflected by metallic surfaces just as light is reflected by a mirror. The wavelength of radio waves is very long; therefore, it is possible to build very large dishes for receiving distant signals. Like optical telescopes, the dishes can be rotated in different directions. The largest movable radiotelescope is 100 m across.

The largest radiotelescope is near Arecibo, Puerto Rico. A wire mesh acts as a reflecting surface. This telescope cannot be moved because it is built in a natural depression, but the rotation of the earth enables the telescope to point in different directions.

At the focus of a radiotelescope is a radio antenna. Electrons in the metal of the antenna are excited by the radio waves focused onto it. The resulting current is fed through a tuner and an amplifier, much as in a common radio. The radio astronomer uses the tuner to select the radio frequency he or she wishes to study. The strength and wavelengths of the signals are recorded, and the source of their direction is plotted.

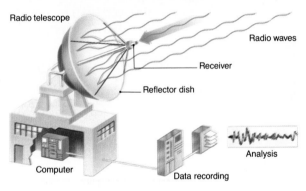

Radio telescope

Radio waves

Receiver

Reflector dish

Computer

Data recording

Analysis

A single radiotelescope provides an image of the sky that is less precise than the image provided by an optical telescope. Radiotelescopes can be linked together, however, so that the image received by one can be compared and mixed with the image received by another. Astronomers can use this type of information to measure astronomical distances and wavelengths of energy from space. Combining images from several radiotelescopes can produce a much clearer radio image of the sky. Near Socorro, New Mexico, an array of 27 dishes has been

Figure 23–13. The shape of this large radiotelescope antenna in Puerto Rico focuses radio waves to the receiver above the dish.

A MATTER OF FACT

The first radiotelescope was built in the 1930s.

Figure 23–14. This diagram illustrates the operation of a radiotelescope.

Section 2   Radio Astronomy   **513**

23.5 Radiotelescopes

DISCUSSION Direct the students' attention to Figure 23–14, which shows how a radiotelescope works. ***Thinking Critically:*** Ask the students to compare optical telescopes and radiotelescopes. How are they similar? How are they different? (Similar: Both gather and focus electromagnetic radiation; optical telescopes magnify images, while radiotelescopes magnify radio waves; both enable astronomers to study objects they cannot actually see. Different: Radiotelescopes do not provide a visual image; radiotelescopes can detect objects that do not give off visible light.)

SKILL (*Making Mathematical Calculations*) The largest radiotelescope, at Arecibo, is over 300 m in diameter. Have the students calculate how much larger it is than the largest reflecting telescope. (300 m divided by 6 m = diameter 50 times greater; area is 25^2, or 625 times greater.)

EXTENSION The radiotelescope at Arecibo is set into a natural depression in the earth.

EXTENSION Radio waves that penetrate the atmosphere from outer space have wavelengths from 10^{-2} to 30 meters. Although the earth intercepts radio energy on the order of approximately 100 watts, only 10^{-14} watt has been captured by radiotelescopes since their use began. Radio signals must be amplified 10^{12} times before they can be recorded.

DISCOVER:
Identifying Sources of Radio Waves

Thinking Skill (Relating)

You may have to help the students think about household items. They may name such items as remote controls for television and stereo equipment, electronic garage-door openers, and telephones. Some hobbyists may have remote-controlled model airplanes or boats.

DISCUSSION Point out to the students that all types of telescopes have disadvantages. The use of optical telescopes may be hindered by atmospheric conditions. An array of radiotelescopes, on the other hand, takes up a great deal of space on the ground.

SECTION REVIEW

Summary Review the electromagnetic spectrum with the students. Ask them to write a sentence in which they differentiate between visible light and radio waves.

Reinforcement Have the students quiz each other, using the questions at the end of each numbered subsection.

DISCOVER

Identifying Sources of Radio Waves

In your home, locate all the objects that produce or receive radio waves. After you have identified several sources, do some library research to determine the frequency that each produces or receives.

built in the desert. This group of radiotelescope dishes provides an image of the sky that compares favorably with the image provided by the best optical telescopes.

Radio astronomers have made major discoveries about the universe. The first discovery, made in 1931, was that the center of the Milky Way Galaxy emits radio waves. It was later discovered that all galaxies emit radio waves and that some galaxies emit more intense waves than others. Even stronger is the radio-wave emission of quasars.

The most important discovery made by radio astronomers was the detection of background radiation resulting from the big bang. The detection of this radiation in 1964 provided strong evidence to support the big-bang theory.

○ *How does a radiotelescope collect information?*
○ *What are the advantages of radiotelescopes over optical telescopes?*

Figure 23–15. This array of 27 radiotelescopes provides a much clearer image of the universe than a single dish does.

Section Review

READING CRITICALLY

1. How are radio waves different from visible light?
2. Why are radio waves from outer space important in astronomical studies?

THINKING CRITICALLY

3. Why are radiotelescopes curved?
4. Why is the big bang still considered a theory, even with the information provided by radio astronomy?
5. Do you think the spectrograph could help explain the beginnings of the solar system? Why?

514   Chapter 23   Observing the Universe

Answers–23.5

○ Radio waves are reflected by a concave mirror to a radio antenna at the focal point of the reflection.

○ Radiotelescopes need not be constructed with the same precision as optical telescopes; they can be made larger.

Answers to Section Review

Reading Critically

1. Radio waves have a longer wavelength and slower frequency than visible light waves.
2. Radio waves supply additional support for the big-bang theory of the origin of the universe.

continues

TECHNOLOGY: Radiotelescopes

Modern optical telescopes are capable of revealing many startling cosmic objects that would not ordinarily be visible to the unaided eye. These large refracting, and even larger reflecting, telescopes allow astronomers to view in detail the wonders of objects never seen before. Nevertheless, despite their advanced capabilities, optical telescopes are "blind" to many other structures that exist in our universe. They cannot "see" many of the events that alter and transform the size and shape of the cosmos.

The visible spectrum is only a small fraction of the entire range of electromagnetic radiation that floods the vacuum of outer space. Gamma rays, X rays, ultraviolet and infrared rays, radar, microwaves, shortwaves, and radio waves are given off by cosmic bodies. Telescopes of many different designs have been engineered to capture these other important forms of energy.

Radiotelescopes must have very powerful amplifiers be-cause the radio waves that reach Earth are extremely weak.

A radiotelescope consists of three components: (1) a large reflecting surface that collects and focuses the incident radiation, (2) an electronic receiver that detects and amplifies cosmic signals, and (3) a data display device used by astronomy laboratory technicians to view and record the information.

The principle of operation of the radiotelescope is very similar to that of the reflecting telescope. The electromagnetic energy strikes a reflecting, concave surface and is focused on the receiver, just as visible light is reflected by a concave mirror and focused through the eyepiece of an optical telescope.

However, the reflecting surface of a radiotelescope must be precisely measured to take into account the frequency and wavelength of the energy it is designed to capture. Radiotelescopes have a parabolic shape that must focus all radiation to the same, exact point; otherwise, the image will be blurred.

A Radiotelescope at Stanford, California

A recent innovation in radiotelescope technology has revolutionized the study of radio waves from space. Astronomers prefer to construct large radiotelescopes that can detect large fields of radiation. However, because of the accuracy that is required in the construction of these giants, larger and larger structures are more difficult to engineer. Therefore, astronomers have worked out a method of combining the efforts of a series of individual radiotelescopes to form what is called a very large array, or VLA. One such array in New Mexico includes a series of 27 radio dishes that can achieve a resolution equal to that of a single dish 40 km wide.

A VLA requires a large complex of computers to integrate the information coming in from the separate radiotelescopes. However, astrophysicists believe that the effort is worth it. With such an array, they can study the most distant objects of the cosmos.

515

Radio picture of the Andromeda Galaxy

TECHNOLOGY: Radiotelescopes

After the students have read this article, hold a class discussion on the advantages and disadvantages of the three major types of telescopes discussed in this chapter: refracting telescopes, reflecting telescopes, and radiotelescopes. You might want to write or visit a local television station or cable company and obtain information about how radio antennae are used to receive programming from satellites. If there is a company nearby that sells and installs home radio antennae, pay them a visit as well.

Thinking Critically

3. A curved radio dish can reflect and focus incident radiation to a single point.
4. The big-bang theory explains most of the phenomena that astronomers have observed and is a good model for explaining how the universe could have formed. However, some of the predictions of the model have not yet been demonstrated.
5. Yes. The spectrograph is helpful in determining the composition of distant objects. It will aid astronomers in comparing and contrasting the matter that makes up the planets and their moons. This will help them to decide if the bodies of the solar system had a common origin.

Skill (Measuring)

Preparation of Materials

If your classroom has enough outlets, you can use electric lights (in sockets) instead of candles. Put a triangular piece of insulating tape on each light bulb; the tape will form an image on the paper.

Answers to Analyses and Conclusions

1. As thickness increases, focal length decreases. The image is upside down and seems to be edged by a thin rainbow.
2. Answers will vary according to the way the materials were set up, but differences in shape and colors will be mentioned.
3. The focal length gets shorter.

Answer to Application

The focal length could be determined from the drawings made.

INVESTIGATION 23: Measuring Focal Length

PURPOSE

To measure the focal length of different convex lenses

MATERIALS (per group of 3 or 4)

Short candle
Matches
Petri dish
Meter stick
Scissors
White construction paper
Modeling clay
Convex lenses of varying thickness

PROCEDURE

1. **CAUTION: An open flame is used in this investigation; be careful of hair and clothing.** Carefully warm the bottom of the candle with a lighted match and allow the melting wax to drip into the center of the petri dish. Blow out the match and put the candle in the hot wax so that as the wax cools the candle is held in place.
2. Place the meter stick several meters behind the dish.
3. Cut the white construction paper as shown so that it can stand vertically without support. Then place the paper near the end of the meter stick.
4. Mold the modeling clay as a base for the lenses and mount each lens in the clay as shown.
5. Light the candle and put one of the lenses between the candle and the construction paper. Darken the room. Focus the image of the candle on the paper.
6. When the image is sharp, measure and record the distance from the lens to the construction paper. This is the focal length of the lens.
7. Repeat the procedure for the other lenses, keeping a record of each focal length.

ANALYSES AND CONCLUSIONS

1. Prepare a chart to compare and contrast the focal lengths of the lenses. Can you see any relationships between the two quantities? Explain.
2. Does the actual candle flame differ from its image on the construction paper? If yes, how? Be specific.
3. What happens to the focal length of a lens as its thickness increases?

APPLICATION

How could this procedure be used to find the focal length of telescopes and other optical instruments?

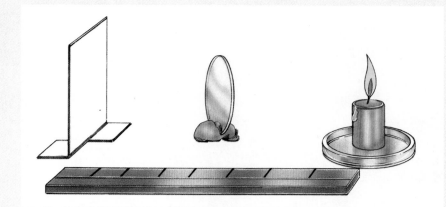

SUMMARY

- Refraction occurs when a beam of light hits the surface of a substance through which light travels at a slower speed. (23.1)

- A lens that causes light to converge on one point is called a convex lens. A lens that causes light to diverge is called a concave lens. (23.1)

- Focal length is the distance between the center of a lens and its focus. (23.1)

- Refracting telescopes have two lenses that focus the light and produce an upside-down image. (23.2)

- The size of refractors is limited by the problems associated with making large lenses. (23.2)

- The use of photographic plates can greatly extend the range of telescopes. (23.2)

- Reflectors are telescopes that focus light with mirrors. (23.3)

- Reflecting telescopes have no distortion, absorb less light than lenses, and can be made larger than refractors. (23.3)

- Radio waves are long-wavelength energy. Many objects in space produce radio waves. (23.4)

- Radiotelescopes are used to detect long-wavelength radiation from outer space. (23.4)

- Radiotelescopes consist of movable dishes that focus radio waves on an antenna. (23.5)

- Many important discoveries about the universe have been made by radio astronomers. (23.5)

Write all answers on a separate sheet of paper.

SCIENCE TERMS

Correctly use each of the following terms in a sentence.

focal length **(505)**
lens **(505)**
magnification **(506)**
radio waves **(512)**
radiotelescope **(512)**
reflection **(504)**
refraction **(504)**
spectrograph **(507)**
telescope **(506)**

SCIENCE QUIZ

Modified True-False

Mark each statement *true* or *false*. If a statement is false, change the underlined term to make the statement true.

1. Energy is emitted from quasars as <u>light</u> waves.

2. The speed of light in air is nearly the same as it is in <u>space</u>.

3. Light bouncing off an object is <u>reflected</u>.

4. Distortions occur in <u>reflecting</u> telescopes.

5. The refracting telescope was used by the Italian astronomer <u>Galileo</u>.

Multiple Choice

Write the letter of the choice that best answers the question or completes the statement.

6. Which of the following emit radio waves?
   a) stars
   b) the moon
   c) asteroids
   d) All of the choices are correct.

continues

CHAPTER REVIEW

SUMMARY

Students may review the major concepts in the chapter by reading the summary statements. The statements are cross-referenced to the chapter to facilitate reinforcement of any concepts of which the students feel unsure. Encourage the students to work in groups to quiz one another.

SCIENCE TERMS

The sentence in which the science term is used should reflect an understanding of the definition of the term. You may wish to stage a game of "Jeopardy" during which the definition is provided and the contestants must provide the correct term.

SCIENCE QUIZ

Answers to Modified True-False

1. false, radio
2. true
3. true
4. false, refracting
5. true

Answers to Multiple Choice

6. a

continues

Answers to Multiple Choice (continued)

7. a
8. a
9. d
10. b

Answers to Completion

11. refract
12. refracting
13. focal length
14. inverted
15. composition

Answers to Short Answer

16. Refracting telescopes use lenses to magnify distant objects. Reflecting telescopes use mirrors and lenses to magnify distant objects.
17. Large mirrors are much easier to manufacture, handle, and maintain than lenses.
18. The wavelength of light absorbed by a gas is the same as the wavelength of light emitted by a gas when it is excited. Therefore, absorption lines on a spectrograph reveal the elements in a star.

Answers to Writing Critically

19. The radiotelescope has allowed astronomers to catalogue objects that radiate energy other than visible light.
20. Answers will vary, but should include the fact that the earth's atmosphere will not interfere with the images recorded by such telescopes.

ANSWERS TO EXTENSION

1. Answers will vary.
2. Answers will vary, but should mention Landsat images.
3. Answers will vary.
4. Answers will vary, but should include information about the Hubble telescope's construction.

ANSWERS TO APPLICATION/ CRITICAL THINKING

1. There is no interference with the earth's atmosphere.
2. The magnification of a telescope is equal to the focal length of the objective lens divided by the focal length of the ocular lens. For this telescope the magnification is 25 cm divided by 2.5 cm, or 10X.

518

---

7. What will increasing the surface area of a lens in a refracting telescope do?
   a) brighten the image
   b) darken the image
   c) sharpen the image
   d) nothing

8. Which of the following have longer wavelengths than light?
   a) radio waves
   b) X rays
   c) ultraviolet rays
   d) gamma rays

9. What happens to a light wave as it hits glass at a 90° angle?
   a) It bends.
   b) It increases in wavelength.
   c) It speeds up.
   d) It slows down.

10. Who first used a reflecting telescope?
    a) Galileo
    b) Newton
    c) Kepler
    d) Brahe

Completion

Complete each statement by supplying the correct term or phrase.

11. Lenses _____ light.

12. A telescope with an objective lens and an ocular lens is called a _____ telescope.

13. The magnification of a refracting telescope is related to the _____ of the lenses.

14. Refracting telescopes usually produce an _____ image.

15. Spectrographs are used to determine the _____ of a star.

Short Answer

16. Compare and contrast the operation of refracting and reflecting telescopes.

17. Why can reflecting telescopes be larger than refracting telescopes?

18. Explain how a spectrograph can be used to determine the composition of a star.

Writing Critically

19. Explain how the radiotelescope has added to the catalog of known objects in the universe.

20. Discuss the reasons why astronomers would like to launch optical telescopes, infrared telescopes, and radiotelescopes into orbit around the earth.

EXTENSION

1. Write a report on the efforts of NASA to design, launch, and maintain a fleet of observatories in space.

2. Research and report on the uses of orbiting optical and infrared telescopes that are currently pointed at Earth.

3. Find out how a body, such as Pluto, appears through different telescopes. Then make a drawing of the different views and describe them to your classmates.

4. Research and write a report about the Hubble space telescope.

APPLICATION/CRITICAL THINKING

1. The Hubble space telescope is to be placed into orbit above the earth's atmosphere by the space shuttle. What advantage is there to placing a telescope above the earth's atmosphere? Why are scientists so interested in viewing the distant parts of the universe?

2. What is the magnification of a telescope with an ocular focal length of 2.5 cm and an objective focal length of 25 cm?

3. If the frequency of an electromagnetic wave traveling through a vacuum is about 24 000 vibrations per second, what is its frequency in air? Explain.

---

3. Nearly the same, because the frequency does not change much from a vacuum to air.

FOR FURTHER READING

Bartusiak, M. "Megascope." *Omni* 8(July 1986): 16. This article is a discussion of how improved technology may once again make arrays of reflecting telescopes valuable to astronomers.

Bartusiak, M. "Ultimate Catalog." *Omni* 8(February 1986): 26. This article is a summary of how scientists use complex computers to catalog over 20 million celestial objects.

Berry, R. *Build Your Own Telescope: Complete Plans for Five High-Quality Telescopes That Anyone Can Build.* New York: Scribner's, 1985. This is a step-by-step manual that shows anyone with basic carpentry skills how to build a telescope.

Kerrod, R. *The All Color Book of Space.* New York: Arco, 1985. This is a series of essays that relates astronomy to space exploration. The author describes early telescopes, space exploration technology, and space pioneers.

Renner-Smith, S. "New Satellite Antennas." *Popular Science* 227(December 1985): 69. This article gives pointers on how to select inexpensive, state-of-the-art television receivers.

Challenge Your Thinking

This array of radiotelescopes is "eavesdropping" on the universe, listening for signals that might indicate the presence of other life forms. If other life forms are listening to us, what kinds of things do you think they have learned about humans in the last 10 years?

Answers will vary, but should include ideas from radio and television programs, microwave communications, and satellite transmissions.

PLANNING THE CHAPTER

Chapter Sections	Page	Chapter Features	Page	Program Resources	Page
Section 1: The Road to Space	521			Investigation 24.2: *Observing Spectra* **(A)** Concept Extension: *Life in Space* **(H)**	TRB 115 LM 117 TRB 49 LM 271
24.1 Rocket Propulsion **(B)**	521	**Discover:** Calculating Speed **(B)**	522		
24.2 Rocket Pioneers **(B)**	522	A Matter of Fact	523		
24.3 Humans in Space **(B)**	524	**Skill Activity:** Making a Star Chart for the Seasons **(B)**	525		
		Section Review	526		
		Activity: Observing Newton's Third Law **(A)**	526		
		Investigation 24: Measuring Balloon Rocket Thrust **(A)**	540	Student Record Book: Textbook Investigations **(A)**	TRB 48
Section 2: Space Travel	527			Critical Thinking **(H)** Reading for Content: *Making Time Lines* **(A)** Investigation 24.1: *Craters on the Moon* **(B)**	TRB 46 TRB 41 LM 219 TRB 113 LM 115
24.4 Humans on the Moon **(B)**	527	A Matter of Fact	528		
24.5 Space Stations **(B)**	529				
24.6 The Space Shuttle **(B)**	530	A Matter of Fact	530		
		A Matter of Fact	531		
		Discover: Reading About *Challenger* **(B)**	532		
		Section Review	532		
		Technology: The Space Shuttle	533		
Section 3: Space Science	534			Cross-Discipline: Science and Social Studies, *Writing a World Space Treaty* **(H)**	TRB 45
24.7 Communications Satellites **(A)**	534	A Matter of Fact	535		
24.8 Weather Satellites **(A)**	536				
24.9 Observing the Earth **(A)**	537				
24.10 Exploring the Solar System **(A)**	538	**Discover:** Learning About Mars **(A)**	538		
		Section Review	539		
Chapter 24 Review	541			Vocabulary **(A)**	TRB 24 LM 169
				Tests **(A)** Computer Test Bank	TRB 103

(LM) Laboratory Manual/Study Guide, **(TRB)** Teacher's ResourceBank™

B = Basic **A** = Average **H** = Honors

The coding Basic, Average, and Honors indicates sections or subsections that might be appropriate for different levels of learners. For additional suggestions regarding choice of topic and depth of coverage, see the Pacing Chart on pages T16–T20.

CHAPTER CONCEPTS, OBJECTIVES, AND TERMS

Section	Concepts	Objectives	Science Terms
Section 1: The Road to Space	■ The momentum of burned gases expelled from a rocket propels the rocket forward. **(24.1)** ■ Solid rocket fuels consist of mixtures of powdered metals, plastics, and rubber. **(24.1)** ■ The liquid fuel used by Robert Goddard for his first rocket was gasoline. **(24.2)** ■ The V-2 was a rocket adapted to warfare by the Germans. **(24.2)** ■ The Soviets put *Sputnik 1,* the first artificial satellite, into orbit in 1957. **(24.3)** ■ The first American in orbit was Lieutenant Colonel John Glenn. Glenn orbited the earth three times in 1962. **(24.3)**	■ **State** the principle on which rocket propulsion is based. ■ **Discuss** the contributions to rocketry of Tsiolkovsky and Goddard. ■ **Name** the first Russian and the first American to orbit the earth.	momentum
Section 2: Space Travel	■ *Apollo 11,* launched in 1969, put the first humans on the moon. **(24.4)** ■ The Soviets have orbited a series of seven space stations called *Salyut.* **(24.5)** ■ *Skylab* was a large American space station that operated during 1973–1974. **(24.5)** ■ The space shuttle is a spacecraft that is launched by rockets but is capable of landing on an airstrip as an airplane does. **(24.6)**	■ **Discuss** why it was important for humans to go to the moon. ■ **Explain** the importance of space stations. ■ **Describe** what the space shuttle can do.	There are no new terms in this section.
Section 3: Space Science	■ Geostationary orbits pass above the equator, while polar orbits pass above the poles. **(24.7–24.8)** ■ Weather satellites continuously monitor the earth and allow the accurate forecasting of weather. **(24.8)** ■ Landsats are designed to observe the earth's surface. **(24.9)** ■ The sun and the solar system have been studied by a series of space probes launched by the United States and the Soviet Union. **(24.10)** ■ *Voyager 1* and *Voyager 2* discovered a total of 20 new satellites orbiting Jupiter, Saturn, and Uranus. **(24.10)**	■ **Explain** what communications satellites do. ■ **Tell** why weather satellites are important. ■ **Summarize** the major scientific advances produced by space exploration.	geostationary space probes

CHAPTER MATERIALS

Title	Page	Materials
Discover: Calculating Speed	522	*(per student)* paper, pencil
Skill Activity: Making a Star Chart for the Seasons	525	*(per student)* compass, construction paper, tracing paper, ruler, pin
Activity: Observing Newton's Third Law	526	*(per group of 3 or 4)* string (10 m), balloon, plastic straw, tape, ring stands (2)
Discover: Reading About the Challenger	532	*(per student)* reference books
Discover: Learning About Mars	538	*(per student)* reference books
Investigation 24: Measuring Balloon Rocket Thrust	540	*(per group of 3 or 4)* spring balance, balloons (2), string (10 m), plastic straw, tape, ring stands (2), metric ruler, stopwatch

TEACHING SUGGESTIONS

Section 1: The Road to Space

Demonstration: Inertia

Purpose

To investigate properties of inertia

Background

Because of a property called *inertia*, a moving object will continue to move in a straight line until acted upon by a force. But planets revolve around the sun and moons revolve around the planets. They do not move in straight lines. Why? They are acted upon by a force—the force of gravity. Gravitation between bodies in space keeps the planets and satellites in their orbits.

Materials

Lined paper
Smooth, circular hoop (an embroidery hoop)
Marble

Procedure

1. Place a sheet of lined paper on the flat surface of a table. Mark an X between two lines about one third of the way from the top of the paper.
2. Place the hoop on the paper. Place the hoop so that the X is at the top. Label the X "1." Place a second X at the bottom of the hoop and label this X "2." Place two more X's at the left and at the right of the hoop and label these X's "3" and "4."
3. Place the marble inside the hoop and swirl the hoop around the paper so that the marble rolls rapidly around the inside of the hoop. Stop the hoop with the inside upper edge touching X 1 on the paper. Race the marble

slowly enough so that you can lift the hoop just as the marble reaches X 1.
4. Watch the path of the marble as it leaves the hoop. Mark its path with a pencil. Try this procedure several times.
5. Now move the marble in the opposite direction. Repeat the activity several times, again making a record of the path of the marble with the pencil.
6. Repeat the entire procedure. This time let the marble roll away when it reaches X 2.
7. Then repeat the demonstration letting the marble roll away at X 3 and again at X 4.
8. On the chalkboard show how the marble leaves the hoop at the four places indicated by X 1, X 2, X 3, and X 4.

Questions to Ask the Students

1. Describe what happens when the marble is released from the hoop. (*The marble rolls away from each X in a straight line.*)
2. What law does the motion of the marble illustrate? (*The Law of Inertia— moving objects continue to move in a straight line.*)

Section 2: Space Travel

Class Activity

Obtain information at the library about space capsules used in the Mercury, Gemini, and Apollo Programs and construct cardboard mock-ups (life sized or box sized) of their interiors. Construct a model of the Space Shuttle cockpit and compare and contrast it to the other smaller craft.

Class Activity

Obtain information about the NASA sponsored space camp and the adventures they have planned for vacationing space cadets. Write to:

The Space & Rocket Center
One Tranquility Base
Huntsville, Alabama 35807–0680

Field Trip

If you live near one of NASA's facilities, arrange for a visit and a conducted tour.

Section 3: Space Science

Outside Speaker

Have a representative from NASA visit the class and talk about various satellites and space probes.

CHAPTER 24

CHAPTER OVERVIEW

This chapter presents a history of space travel, from the first rockets invented in the 1920s to the sophisticated space probes now used to explore the solar system. The principles of rocket propulsion are discussed. A survey of satellites and their many uses is included in this chapter.

Section 1: The Road to Space In this section, the principles of rocket propulsion are explained. A history of early rocketry is provided, along with a description of the earliest manned space flights.

Section 2: Space Travel This section describes the first moon flights and moon landings. A survey of the space stations used in the early development of space exploration is provided, with special emphasis on the achievements of the space-shuttle missions.

Section 3: Space Science This section discusses the scientific and communication uses of satellites. Emphasis is placed on communication, weather, and space research conducted via satellites that orbit the earth or that venture out into the solar system.

Exploring Space

"Space, the final frontier": These are familiar words to even the most casual "Trekkie," but they are a reality in your lifetime. Humans first went into space nearly 30 years ago, and it is likely that there are humans in space as you read this. Humans have been to the moon and are now ready to explore the planets in person. Are you ready for the adventure?

A space-shuttle launch

CHAPTER MOTIVATING ACTIVITY

Ask the students if they have ever considered working on a space station or living in a colony on the moon or Mars. What type of vehicle might transport them there? What difficulties or dangers might they encounter in such an environment? How would they have to prepare themselves, academically, emotionally, and physically for the voyage? You may wish to have the class brainstorm ideas. Then have the students use these ideas to write a description of a day in a space station.

Many filmstrips and videos are available through public and private sources, documenting the personal experiences of men and women in space. Keep several of these audio-visuals on hand and uses them to help the students answer some of the questions put forth above. Students may also find dramatized accounts of astronauts in space (such as the movie, *The Right Stuff*) very informative. Films that are purely fictional, but realistic (such as the movie, *Marooned* or segments of *2001*), can help students to visualize some of the technical problems of living and working in space.

Section **1**: The Road to Space

SECTION OBJECTIVES

After completing this section, you should be able to:

- **State** the principle on which rocket propulsion is based.
- **Discuss** the contributions to rocketry of Tsiolkovsky and Goddard.
- **Name** the first Russian and the first American to orbit the earth.

NEW SCIENCE TERM

momentum

24.1 Rocket Propulsion

Have you ever blown up a toy balloon and then released it to fly around a room? If so, you have demonstrated the basic principle of rocket propulsion. Rocket propulsion is based on Newton's third law of motion; that is, for each action there is an equal and opposite reaction. The air rushes out of the back of the balloon, and the balloon is pushed forward.

This law establishes the principle of conservation of momentum. Mathematically, **momentum** is equal to the mass of an object times its velocity, but a simple example provides a more useful description. A loaded cannon has no momentum. When the cannon is fired, however, the forward momentum of the cannonball is equal to the backward momentum of the cannon. The sum of the forward momentum and backward momentum is zero because they are equal and in opposite directions. The original momentum, which was zero, is conserved.

Fireworks, jet engines, and rockets all work on the same principle. Fuel is burned, and the gases produced by the burning are expelled through a nozzle facing the rear. The momentum of the gases toward the rear propels the object forward.

Figure 24–1. This jet has momentum because of the exhaust gases. The momentum for the runner is provided by leg muscles.

Section 1   The Road to Space   **521**

BACKGROUND INFORMATION

The Chinese were actually the first to experiment with rocket propulsion. They first developed rockets to launch fireworks and by the 1200s were using rockets to propel "arrows of flying fire" in warfare. By the 1800s, the British were using rockets to propel explosives used in warfare. Remind the students that our own national anthem, "The Star Spangled Banner," written in 1814, includes the phrase: "the rockets' red glare."

DEMONSTRATION

For a demonstration of inertia, see page 519c preceding this chapter.

SECTION FOCUS

Section Overview   In this section, the principles of rocket propulsion are explained. The designs and uses of early rockets are described. This section also presents a history of the first piloted flights into space.

Section Objectives   For a list of section objectives, see pupil's edition page.

New Science Term   For the new science term in this section, see pupil's edition page. You may wish to call the students' attention to the new term.

SECTION DEVELOPMENT

24.1 Rocket Propulsion

DISCUSSION   You may wish to use a balloon to demonstrate Newton's third law of motion, on which the principle of rocket propulsion is based. ***Thinking Critically:*** Ask the students to describe what happens to the air pressure inside the balloon before and after the balloon is released. (When the balloon is closed, air pressure is equal on all sides. After release, pressure on the inside of the balloon is released as air escapes. The escaping air on one end of the balloon thrusts the balloon in the opposite direction.) Another example of conservation of momentum would be the jumping of objects on a table when a person's hand slams the table.

THINKING SKILL   *(Applying Concepts)* Ask the students to explain how the conservation of momentum applies to fuel-powered rockets. (A momentum that is equal but opposite to the momentum of the gases toward the rear of the rocket propels the rocket forward.)

521

DISCOVER: Calculating Speed

Skills (*Calculating, Using a Formula*)

The students should write the formula and then rewrite it with the numerals substituted for the letters. Then they can perform the calculations. (*v = 9.9 m/s*)

EXTENSION   Point out to the students that some rockets use liquid oxygen. These rockets are different from other jet-propelled systems because they do not require atmospheric oxygen for combustion to take place. Instead, they carry a supply of liquid oxygen in a separate tank for oxidation of another fuel.

24.2 Rocket Pioneers

DISCUSSION   The accomplishments of early pioneers were possible because of the mathematical, scientific, and literary works of others who preceded them. The principles set forth by Copernicus, Kepler, Galileo, Newton, and even Jules Verne, who wrote science fiction and fantasy, provided the foundation for rocket experimentation. ***Thinking Critically:*** Ask the students to suggest how problems in the early development of rocketry may have been solved by using principles developed earlier. (Some students may suggest Newton's laws of motion and theories about gravity.)

Figure 24–2. Momentum for launching a rocket is provided by the burning of a fuel. The momentum of the balloon is from escaping air.

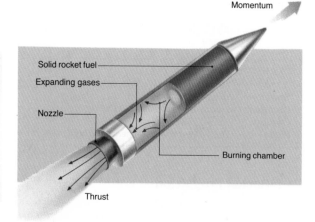

Momentum

Solid rocket fuel
Expanding gases
Nozzle
Burning chamber
Thrust

DISCOVER

Calculating Speed

Calculate the racing speed of a sprinter who has a mass of 90 kg and a momentum of 891 kg-m/s. Substitute the data into the formula: $p = m \times v$, where p = momentum, m = mass, and v = velocity.

Fireworks, jets, and rockets are launched by igniting fuel. Air, oxygen, or a chemical oxidant allows the fuel to burn. Fireworks fuel consists of gunpowder; solid fuel for rockets contains mixtures of powdered metals and plastics or rubber. Jet engines and some rockets use kerosene as fuel. The fuel is compressed, ignited, and expelled.

○ *What is Newton's third law of motion?*
○ *What force propels jet engines and rockets forward?*

24.2 Rocket Pioneers

The first person to scientifically study the use of rockets for space travel was Konstantin Tsiolkovsky (KAHN stuhn teen tseel KOHF skee), a Russian inventor. In the 1920s, Tsiolkovsky designed a spaceship with a bullet-shaped passenger cabin in

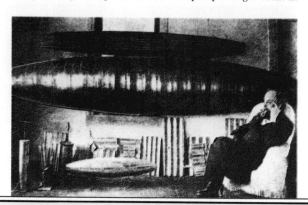

Figure 24–3. Konstantin Tsiolkovsky is shown here with a model rocket.

522

Answers–24.1

○ Newton's third law of motion states that "for every action, there is an equal and opposite reaction."

○ The momentum of exhaust gases rushing backward out of the nozzle forces the rocket forward.

BACKGROUND INFORMATION

Time after time, great advances are made in science when one scientist learns from and expands on the work of another scientist. Albert Einstein once put his accomplishments in perspective for his generation by observing that he was able to see as far as he did because he "stood on the shoulders of giants."

the nose and fuel tanks filled with liquid hydrogen and liquid oxygen in the tail. He planned for the fuel and oxygen to burn in a cone-shaped chamber that faced the rear. Tsiolkovsky designed support systems for his astronauts and pressurized suits to enable them to leave their cabins and construct space stations. Although Tsiolkovsky never launched a rocket himself, his thinking was fundamental to the development of rocket flight.

At about the same time, an American physicist, Dr. Robert Goddard, was actually experimenting with rockets. Goddard contributed to the development of liquid fuels for rockets. In 1926 Goddard launched the world's first liquid-fueled rocket, shown in Figure 24–4. The fuel was gasoline, and the oxidant was liquid oxygen. The rocket was 1.2 m long, and it rose to a height of 56 m at an average speed of 103 km/h.

During World War II, the Germans constructed many rockets for delivering bombs. On October 3, 1944, a rocket called the *V-2* was used to bomb London. The rocket was self-steering and was launched from a land-based platform.

A MATTER OF FACT

In 1930 Tsiolkovsky published a book, *The Jet Plane*, in which he predicted that jet planes would someday replace propeller-driven planes.

Figure 24–4. Robert Goddard (right) launched a gasoline-fueled rocket in 1926. After World War II, the United States conducted experiments with captured German *V-2* rockets (left).

At the end of World War II, scientists in the United States used captured German rockets to scientifically explore the upper atmosphere. An American two-stage rocket, with a captured German *V-2* rocket as its first stage, reached a record altitude of 393 km in 1949.

○ *Who was the first person to study the use of rockets for space travel?*
○ *For what purpose did the Americans use captured V-2 rockets?*

Section 1   The Road to Space   **523**

EXTENSION Charles Yaeger will always be remembered as the first to break the sound barrier. On October 14, 1947, he flew a rocket-powered plane, the Bell X-1, to an altitude of 10 668 m at a speed faster than the speed of sound. Sound travels at a speed of about 3353 m per second. This historic flight paved the way for future space travel.

THINKING SKILL (*Making Judgments*) The earliest uses of rockets were primarily military. Ask the students to suggest possible non-military uses for rockets. (Answers will vary, but will probably deal with space travel and space exploration.)

Answers–24.2

○ Konstantin Tsiolkovsky was the first to study the use of rockets for space travel.

○ They used the *V-2* to study the upper atmosphere.

24.3 Humans in Space

DISCUSSION Point out to the students that they probably take space travel for granted. Ask the students to imagine what it may have been like for their parents' generation to be the first people in history to witness human beings traveling in space. ***Thinking Critically:*** John Glenn was treated as a national hero. Have the students discuss why his flight of just three orbits around the earth was a major accomplishment. (Some answers might be that it was the first time it had been done by an American; John Glenn had done something extremely dangerous; and he had made the first step on a long list of space accomplishments.)

THINKING SKILL *(Making Judgments)* Remind the students that because there is no atmosphere in space, special equipment must be used to protect astronauts from possible harm. Ask the students to name some of the conditions that exist and the equipment and techniques that might be used to enable people to ''walk'' in space. (Oxygen must be carried; astronauts must learn to move in an environment without gravity; spacesuits must protect against radiation and temperature extremes.)

EXTENSION (Tie-in/Life Science) Workers in the field of space medicine have conducted research to learn about the long-term effects of space travel on humans. Long-term weightlessness can cause muscles, including heart muscles, to atrophy. Another possible effect is the leaching of calcium from bone cells. Some of these symptoms persist for long periods after astronauts return to Earth. Engineers are working on new spacesuit designs and other technologies that might alleviate some of the problems of space travel.

Figure 24–5. *Sputnik* (right) was the first satellite placed into Earth orbit. *Explorer* (left) was the first satellite placed into orbit by the United States.

Figure 24–6. Yuri Gagarin (right) was the first human to be launched into space. The United States' first astronaut to orbit the earth was John Glenn (left).

24.3 Humans in Space

The development and use of rockets proceeded slowly until October 4, 1957, when the Soviet Union surprised the world by placing into orbit an artificial satellite weighing 83.6 kg. Called *Sputnik 1*, the satellite carried instruments to measure the density and temperature of the upper atmosphere. A month later, on November 3, 1957, the Soviet Union put another satellite, weighing 508 kg, into orbit. This second satellite carried a passenger—a dog named Laika. The purpose for orbiting the animal was to study the effects of space travel on a living organism.

The first United States satellite, *Explorer 1*, was launched on January 31, 1958. The rocket was 2.05 m long and carried a scientific package for measuring temperature, cosmic rays, and meteoroids. *Explorer 1* discovered the Van Allen belts of radiation described in Chapter 8.

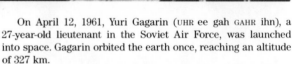

On April 12, 1961, Yuri Gagarin (UHR ee gah GAHR ihn), a 27-year-old lieutenant in the Soviet Air Force, was launched into space. Gagarin orbited the earth once, reaching an altitude of 327 km.

Lieutenant Colonel John Glenn of the United States Marine Corps was launched into space on February 20, 1962, and made three Earth orbits. Glenn's flight followed two suborbital flights, each of which lasted about 15 minutes.

SKILL ACTIVITY: Making a Star Chart for the Seasons

BACKGROUND

The earth's equator is inclined 23.5° from its orbital plane. Because of this, our view of the sky, seen from either the Northern or Southern Hemisphere, alters during the course of the year. In the Skill Activity in Chapter 22, you made a star chart that helped you become familiar with the most visible constellations. In this activity, you will complete your star chart so that it gives a more accurate view of these constellations as they appear during the changing seasons.

PROCEDURE

1. Use a compass to measure the radius of your star chart and draw a circle of equal size on a piece of construction paper.
2. On tracing paper, trace the circle shown in Figure A, cut it out (leaving the tab intact), and use it as a guide to draw a similar ellipse onto the circle made of construction paper.
3. Use a ruler to divide the circle into 24 sections, representing the hours in a day.
4. Put a pin through the tab and attach the circle to the star chart as shown in Figure B.

APPLICATION

When viewing stars at night, align the time and approximate date before viewing the window of the celestial sphere.

1. Why would you need to make an adjustment for the time of night?
2. Why is the date important in using the sphere correctly?

USING WHAT YOU HAVE LEARNED

What modifications would have to be made to the star chart if you lived near the equator or near the Arctic Circle? Explain.

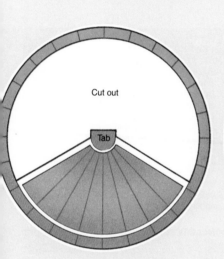

Figure A

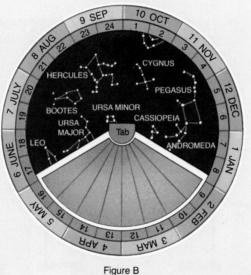

Figure B

SKILL ACTIVITY:
Making a Star Chart for the Seasons

Objectives

- Complete a star chart.
- Observe constellations in their constantly changing positions.

Discussion   Tell the students that in 1543 Copernicus published the book *Concerning the Revolutions of the Celestial Spheres* in which he described the movements of bodies in our solar system. Copernicus' ideas provided the foundation for understanding the visible movements of stars.

Procedure

The rim of the window of the star chart should coincide, approximately, with the horizon on a given night. As mentioned in Chapter 23, the entire celestial sphere is not visible every night because mid-latitude constellations are positioned on the daylight side of Earth on given days and are, therefore, obscured by the sun's light. Polaris should be found at the center of the star chart.

Answers to Application

1. During the course of a single night, the celestial sphere seems to rotate above the observer. To accurately track the stars from one evening to the next, the observer must consider this apparent rotation and record the time that all observations are made.
2. As the earth rotates on its axis, it also revolves around the sun. Due to the earth's progression along the plane of its orbit, counterclockwise as viewed from above the ecliptic, stars will migrate from day to day and arrive 3 minutes 56 seconds earlier at a given position in the sky than they did on the preceding evening. This makes noting the date of an observation extremely important.

Answer to Using What You Have Learned

Very near the equator the North Star would not be visible, and a different reference star would have to be used. At the North Pole, the visible constellations would not rise and set but would instead make a circle around the pole.

EXTENSION (Tie-in/Social Studies) The 1960s saw many new developments in space flight. One reason for this was the competition between the United States and the Soviet Union. ***Thinking Critically:*** Ask the students to suggest reasons for the competition between the two countries. (Answers will vary but will probably involve matters of power and politics.)

SECTION REVIEW

Summary Have the students create a chart listing the major developments in rocketry. The chart should include primitive rocket mechanisms used in the nineteenth century and major events in twentieth-century space travel.

Reinforcement Have the students complete the Section Review. Allow the students to look up answers of which they are unsure.

ACTIVITY:
Observing Newton's Third Law

Thinking Skill *(Drawing Conclusions)*

Preparation of Materials The straw is a slide on which the balloon travels. It is probably easiest to hold the inflated balloon against the underside of the straw and, using a 5–cm piece of tape, tape the straw to the balloon.

Discussion Remind the students that Newton's third law states that for each action there is an equal and opposite reaction. (Please note: This apparatus can be used again in Investigation 24 on p. 540.)

Answers to Conclusions/ Applications

1. It demonstrated that every action produces an equal and opposite reaction.
2. Newton's third law is derived from the Law of Conservation of Momentum. Since the original momentum of the balloon was zero, the product of the mass and velocity of air released at one end was cancelled by the product of the mass and velocity of the moving balloon.
3. Friction slowed the balloon.

Figure 24–7. Former space-shuttle astronaut Sally Ride was the first woman astronaut for the United States.

The first woman to go into space was Valentina Tereshkova (tehr uhsh KOH vah), launched in June 1963. In 1965 Soviet Lieutenant Colonel Alexei Leonov (LEE uhn uhf) became the first human to "walk" in space. Leonov, dressed in a protective spacesuit, left his pressurized spacecraft cabin and floated for 10 minutes in space. All these space firsts had been accomplished in less than 10 years after the launch of *Sputnik 1;* the next 10 years would be even more dramatic.

○ ***When did the United States put its first satellite into orbit? What was it called?***
○ ***Who was the first person to "walk" in space?***

Section Review

READING CRITICALLY

1. What kinds of fuel did Goddard use to launch his early rockets?
2. Who were the first Soviet and American humans launched into Earth orbit?

THINKING CRITICALLY

3. Why do you think that the Soviet Union is so far ahead of the United States in some areas of space exploration?
4. What effects of space travel did the Soviets wish to study by sending Laika into orbit?

ACTIVITY: Observing Newton's Third Law

How can you apply Newton's third law of Motion?

MATERIALS (per group of 3 or 4)

string (10 m), balloon, plastic straw, tape, ring stands (2)

PROCEDURE

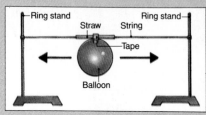

1. Place the ring stands on desks on opposite sides of the room.
2. Tie one end of the string to the top of one of the ring stands.
3. Thread the other end of the string through the plastic straw and tie that end to the other ring stand. Make sure that the string is tightly stretched.
4. Inflate the balloon but do not tie it closed. Hold the balloon closed and tape it to the straw.
5. On the count of "5, 4, 3, 2, 1 launch!" release the balloon at one end of the string and observe it as it flies across the room.

CONCLUSIONS/APPLICATIONS

1. Why is this a demonstration of Newton's third law of motion?
2. What forces drove the balloon forward?
3. What forces tended to slow the balloon down?

526 Chapter 24 Exploring Space

Answers–24.3

○ The United States launched *Explorer* on January 31, 1958.
○ Alexei Leonov was the first person to "walk" in space.

Answers to Section Review

Reading Critically

1. Both used liquid oxygen as an oxidant. Tsiolkovsky used liquid hydrogen, while Goddard used gasoline.

2. Yuri Gagarin orbited the earth on April 12, 1961. John Glenn orbited the earth on February 20, 1962.

Thinking Critically

3. The Soviet Union made "being number one in space" a national goal, and provided the scientists and resources necessary to achieve that goal.
4. They wanted to study the effects of space travel on living organisms.

Section 2: Space Travel

SECTION OBJECTIVES

After completing this section, you should be able to:

- **Discuss** why it was important for humans to go to the moon.
- **Explain** the importance of space stations.
- **Describe** what the space shuttle can do.

NEW SCIENCE TERMS

There are no new terms in this section.

24.4 Humans on the Moon

As a challenge to the Soviet Union's leadership in space, President John F. Kennedy committed the United States to placing an astronaut on the moon before the end of the 1960s. In preparation for this journey to the moon, many additional Earth-orbiting missions, called *Mercury* and *Gemini*, were launched. American astronauts learned to live and work in space.

In October 1968, the Apollo system, which was designed to send three astronauts to the moon, was tested in Earth orbit. *Apollo 7* consisted of a three-stage rocket, called *Saturn V*, and a command module for the flight crew. *Apollo 8*, led by Colonel Frank Borman, made the first flight to the moon in December 1968, but it did not land. The spacecraft orbited the moon 10 times and sent televised views of the lunar surface back to Earth.

Landing humans on the lunar surface involved launching a spaceship with three separate units: a command module, a service module, and a lunar module. The command module was the capsule in which the astronauts traveled to the moon and

Figure 24–8. The *Saturn V* rocket (right) is the most powerful rocket ever built. It was used to launch the Apollo missions to the moon (left).

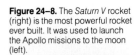

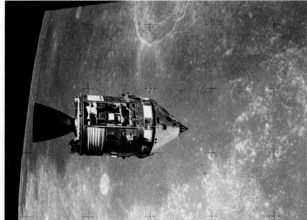

BACKGROUND INFORMATION

In April 1970, *Apollo 13* nearly met with disaster. En route to the moon, an explosion ruptured oxygen tanks in the service module, severely crippling the spacecraft. Trained to act quickly and calmly in the face of impending disaster, the crew of *Apollo 13* constructed a makeshift air purifier to prevent suffocation as the level of carbon dioxide increased. For days, they took refuge in the lunar module, *Aquarius,* and returned to Earth cold and exhausted, but alive.

SECTION FOCUS

Section Overview This section describes the lunar missions that took place in the 1960s and 1970s. It also describes space stations that orbit the earth to gather scientific data. Special emphasis is placed on the space-shuttle missions.

Section Objectives For a list of section objectives, see pupil's edition page.

New Science Terms There are no new science terms in this section.

SECTION DEVELOPMENT

24.4 Humans on the Moon

DISCUSSION Point out to the students that space travel developed rapidly during the 1960s as a result of competition between the United States and the Soviet Union. *Thinking Critically:* Ask the students to suggest what prompted the competition between the two world powers to develop space technology. (Development of sophisticated space technology provides advantages in military, industrial, computer, and other technologies, as well as scientific information that can be applied to many branches of science such as meteorology and astronomy.)

DISCUSSION Take-apart models of the Apollo spacecraft are available from many sources, as are videotapes of the flights. The students should enjoy seeing these. Many science programs on public television deal with these topics; have the students watch for them in TV listings. *Thinking Critically:* Ask the students how Newton's third law of motion was used in the functioning of the LEM. (An engine was fired toward the surface of the moon to provide a sort of cushion as the LEM landed. Without the engine thrust pushing it away from the moon's surface, the LEM would have crashed.)

EXTENSION Point out to the students that samples of rocks from the moon collected by members of the Apollo crew provide valuable information about the geologic history of the moon. This information was used by scientists to further their knowledge of the earth's geologic history, as well as to provide clues to the formation of our solar system.

Figure 24–9. The LEM *Eagle* (left) landed on the moon July 20, 1969. Neil Armstrong and "Buzz" Aldrin (right) spent 2-1/2 hours walking on the moon.

A MATTER OF FACT

Astronauts Eugene Cernan and Harrison Schmitt hold the record for the farthest moon safari aboard the incredible lunar rover: 35 km. The lunar buggy traveled at a top speed of 16 km/h.

returned to Earth. The service module supplied oxygen, power, and fuel to the command module. The lunar excursion module (LEM) landed two astronauts on the moon's surface. The LEM included an engine to slow the craft for descent and four legs for a landing platform. The lunar-module engine was used for the liftoff from the moon and the return to the command module.

On July 16, 1969, the crew of *Apollo 11* were launched from Cape Canaveral, Florida. Neil Armstrong and Edwin "Buzz" Aldrin were to land on the moon's surface. The third astronaut, Michael Collins, was to remain in orbit around the moon.

The lunar module landed on the surface of the moon on July 20 at 8:17 P.M. Greenwich Mean Time (GMT). Armstrong and Aldrin left the lunar module at 2:56 A.M. GMT and spent 2-1/2 hours on the surface of the moon. They collected rocks, set up experiments, and took many pictures. Their mission accomplished, the astronauts left the moon and returned to Earth, landing in the Pacific Ocean 195 hours and 18 minutes after launch.

The last lunar mission, *Apollo 17*, was completed on December 19, 1972. No humans have been to the moon since then, but there are preliminary plans for a permanent station on the moon early in the twenty-first century.

○ *Describe the three parts of the spaceship used to land on the lunar surface.*
○ *Name the astronauts aboard the first mission to land on the moon.*

Figure 24–10. The "Lunar Rover" was used by astronauts to explore wide areas of the moon's surface.

528

Answers–24.4

○ The command module transported the astronauts to the moon. The service module provided oxygen, power, and fuel to the command module. The lunar-excursion module (LEM) was used to fly astronauts to and from the lunar surface.

○ Armstrong, Aldrin, and Collins were aboard the first moon mission.

24.5 Space Stations

In addition to the information obtained from the Apollo moon missions, astronaut-scientists have conducted many experiments from space stations. The first space station was the Soviet Union's *Salyut 1*, launched in April 1971. *Salyut 1* was cylindrical in shape; it was 15 m long, over 4 m in diameter, and weighed 18.9 metric tons. Three Soviet cosmonauts made a variety of scientific observations from this station.

In May 1973 the United States launched *Skylab*, shown in Figure 24–11. *Skylab* was much bigger than *Salyut 1;* it had a length of 25.7 m and a diameter of 6 m. *Skylab* was joined three times by Apollo-type spacecraft carrying teams of three astronauts. The three *Skylab* missions, all in 1973, lasted 28, 59, and 84 days, respectively.

Figure 24–11. The *Skylab* space station (left) was used three times in 1973. Inside *Skylab* (center and right), astronauts conducted many experiments in the reduced gravity of space.

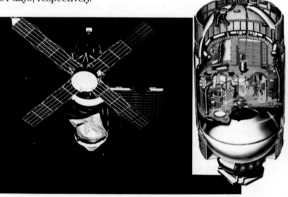

On July 15, 1975, an American Apollo spacecraft joined with a Soviet Soyuz spacecraft 225 km above Earth. The three American astronauts and two Soviet cosmonauts visited each other through a passage connecting the two spacecraft and jointly conducted several small experiments.

Figure 24–12. Astronauts from the United States and the Soviet Union (left) met in space for the first time on July 15, 1975. The Soyuz and Apollo spacecraft were linked by a common passage (right).

529

DISCUSSION Emphasize to the students the importance of international cooperation in the development of space technology. The joining of the American and Soviet space crafts in 1975 was only one instance in which countries cooperated. The world scientific community cooperates in many other ways such as international scientific conferences and the publication of articles in scientific journals that are read all over the world. *Thinking Critically:* Ask the students what the advantages are to sharing scientific information among the nations of the world. (Sharing information helps develop technologies that can be used to better the standard of living of people around the world.)

DISCUSSION Astronauts are generally trained in a variety of skills to enable them to perform the myriad tests and experiments planned for a space mission. They have to know a lot more than simply how to operate the controls of their spacecraft. Experiments in space include gathering medical data on their own bodies' responses to conditions in space, gathering meteorological data, and gathering data on how the spacecraft responds to various conditions in space. *Thinking Critically:* Ask the students what other kinds of data might be collected on a space mission. (Students' answers will vary, but may include data about radiation and zero-gravity conditions.)

EXTENSION In November 1966, the United States Air Force launched an unpiloted mock-up of a Manned Orbital Laboratory (MOL). The station was about 12.2 meters long and about 3 meters in diameter and was designed to dock with the two-seat Gemini capsule. However, efforts to successfully journey to the moon drained funds from this project and it was cancelled in 1969.

EXTENSION Encourage any students who may be visiting Washington, D.C., to visit the Air and Space Museum of the Smithsonian Institution. They can see models of rockets, the LEM, and *Skylab*. They should pay special attention to the ingenious gadgets in *Skylab* developed to make life in close quarters possible.

24.6 The Space Shuttle

DISCUSSION The space shuttle orbiter has crew quarters and a large cargo area. *Thinking Critically:* Ask the students why a space shuttle was deemed necessary for construction of a space station. (A space station would be in space for very long periods of time. A shuttle could be used to resupply the space station with materials, supplies, and crew members.) Have the students look in a dictionary for the meaning of *shuttle* ("to move back and forth") and discuss why it is an appropriate name for the spacecraft.

 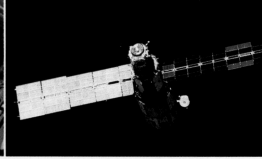

Figure 24–13. Scientists from the Soviet Union have occupied various space stations almost continuously since 1971.

In September 1977 a new space station, *Salyut 6*, was launched by the Soviet Union. *Salyut 6* had two docking ports. On December 11, 1977, *Soyuz 26* docked at the rear port, and on January 11, 1978, *Soyuz 27* docked at the front port. For the first time, three spacecraft were linked in space. There was a total of six people on board.

On January 20, 1978, the first unpiloted ferry craft was launched by the Soviets to resupply *Salyut 6*. Resupplying by crewless flights meant that the crews on *Salyut 6* could stay in space indefinitely. *Salyut 6* was in operation until the summer of 1982, when it was replaced by *Salyut 7*, or *Mir*. In 1984 a two-person crew remained aboard *Mir* for 238 days. The Soviet Union is now developing a shuttle to ferry crews and supplies to and from future space stations.

The Soviet Union is clearly ahead of the United States in establishing a permanent space station. The only American space station to date was *Skylab*, which operated from May 1973 to February 1974. Although plans are being considered for an American space station, it will not become operational for several more years.

○ *Name the Soviet and American space stations.*
○ *Why did the Soviets wish to resupply space stations by unpiloted flights?*

24.6 The Space Shuttle

A space vehicle that could be lifted into orbit by rockets and returned to Earth like an airplane seemed necessary for the construction of an American space station. The final design was the space shuttle, a system consisting of four parts: two rocket boosters with solid fuel, one disposable liquid-fuel tank, and an orbiter.

A MATTER OF FACT

Although a single one of the space shuttle's main rocket engines has a mass of only 3000 kg, it has a power output equal to seven Hoover Dams.

Answers-24.5

○ *Salyut* and *Skylab* are the Soviet and American space stations.

○ Resupply by unpiloted crafts meant that their cosmonauts could spend more time in space.

BACKGROUND INFORMATION

Some Facts About the Shuttle Fleet

Orbiting Vehicle Name	Weight Unloaded (dry)	First Flight
Columbia	69005 kg	Apr. 12, 1981
Discovery	67123 kg	Aug. 30, 1984
Atlantis	67123 kg	Sept. 20, 1985

Columbia, the first orbiter to be used in space, looks like a snub-nosed airplane with short wings. The orbiter is 37.24 m long, 17.27 m high, and has a wingspan of 23.79 m. *Columbia* has three main engines, each with a thrust of nearly 170 metric tons. The two smaller maneuvering engines have a thrust of nearly 3 tons. The cargo deck, which is 18 m long, is used for scientific equipment or to carry satellites into orbit.

The solid rocket boosters contain rubber and aluminum powder mixed with an oxidant as fuel. The thrust of each booster engine is over 1200 metric tons. The external fuel tank contains liquid oxygen and liquid hydrogen, which are mixed and burned in the orbiter's engines.

Figure 24-14. The United States' space shuttles are very complex spacecraft.

A MATTER OF FACT

Like astronauts of television screen and comic strip fame, Astronaut Bruce McCandless made the first untethered walk in space in February 1984. Strapped into a jet-thruster backpack called the *manned maneuvering unit* (MMU), McCandless ventured more than 90 m away from space shuttle *Challenger.* He was, in fact, a human satellite, cruising high above Earth at an orbital speed of 28 000 km/h.

Figure 24–15. Space shuttles are prepared in the huge Vehicle Assembly Building (left) at the Kennedy Space Center in Florida. After landing in California (below), space shuttles are flown back to Florida on top of a "jumbo jet."

Section 2   Space Travel   **531**

DISCUSSION   You may wish to have a plastic scale model of the space shuttle for the students to examine. Remind the students about what they learned about rocket propulsion in Section 1 of this chapter. *Thinking Critically:* Ask the students to explain the purpose of the liquid hydrogen and liquid oxygen that is carried in the space shuttle's fuel tanks. (Liquid hydrogen is used as fuel; liquid oxygen allows the fuel to burn.) Direct the students' attention to the photographs. Have them notice the size of objects around the shuttle so they can appreciate its size.

EXTENSION   Point out to the students that while astronaut Bruce McCandless was orbiting the earth at a speed of 28 000 km/h, people on Earth were actually hurtling through space at a speed of 41 358 km/h—the speed at which the earth orbits the sun. At the same time, the earth follows the sun and its solar system through space at a speed of about 27 000 km/h. A third motion is the earth's spinning on its own axis at the speed of about 620 km/h. Just as people on Earth do not sense their hurtling through space, astronauts in space have no sense of the enormous speed at which they are traveling.

DISCUSSION The space shuttle was designed as a reusable craft. The solid-rocket boosters are recovered from the sea to be reused. ***Thinking Critically:*** Ask the students to name some advantages of reusable parts. (Reusable parts cut the cost of each space flight and eliminate some spaceflight trash.)

THINKING SKILL *(Applying Concepts)* Ask the students why the orbiter does not require separate fuel tanks to power it during orbit around the earth. (The earth's gravitational pull helps maintain the orbiter's path around the earth.)

DISCOVER:
Reading About *Challenger*

Skill *(Researching Data)*

As a result of this tragedy, engineers and scientists have increased their research on the rocket engine. They have also built into the space shuttle new means of emergency exit in case of engine malfunctions.

SECTION REVIEW

Summary Have the students summarize in their own words the major design features of the space-shuttle fleet.

Reinforcement Have the students complete the Section Review. Allow the students to look up any answers of which they are unsure. You may wish to allow the students to work in pairs to check each others' answers.

Figure 24–16. The crew of a space shuttle can launch satellites (left) or conduct scientific experiments inside the Spacelab module (right).

The shuttle system is lifted from its launch pad at Kennedy Space Center in Florida by the rocket engines of the boosters and the orbiter. At an altitude of 45 km, the boosters are detached and returned to Earth by parachutes for recovery and reuse. The external tank, however, is dropped into the Atlantic Ocean. The orbiter is placed into orbit by its maneuvering engines, which are also fired to slow the craft for reentry and landing. The orbiter lands at about 350 km/h on a long aircraft runway.

Three similar orbiters, *Challenger, Discovery,* and *Atlantis,* were added to the space shuttle fleet between 1980 and 1985. However, the January 1986 flight of *Challenger* ended in tragedy as one of the two boosters developed a leak and the assembly exploded 75 seconds into the flight. The seven astronauts died.

A consortium of 11 European nations has built a small laboratory, called *Spacelab,* that fits into the cargo bay of a shuttle orbiter. The first Spacelab mission was in November 1983. Since that time, there have been additional Spacelab missions, all of which produced much scientific information.

○ ***What was the name of the first American space shuttle to be used in space?***

○ ***Which parts of the space shuttle system are recovered and reused, and which parts are not reused?***

DISCOVER

Reading About Challenger

Using the resources of a library, learn about the final flight of the space shuttle *Challenger.* How has the emphasis on human space exploration changed as a result of this tragedy?

Section Review

READING CRITICALLY

1. Of what components did the Apollo launch system consist?
2. What dimensions did *Skylab* have?
3. How is the space shuttle launched, and how does it return to Earth?

THINKING CRITICALLY

4. Why did the United States decide to send astronauts to the moon?
5. Explain the importance of Americans and Soviets meeting in space in 1975.

Answers–24.6

○ *Columbia* was the first American space shuttle to be launched.

○ The solid-rocket boosters are recovered, and the shuttle itself is reused. The large fuel tank is not yet reused.

Answers to Section Review

Reading Critically

1. The *Apollo* command module, service module, and LEM were launched into orbit aboard a three-stage *Saturn V* rocket.

2. *Skylab* was 25.7 m long and 6 m in diameter.
3. The shuttle is launched into orbit using the fuel in a large liquid fuel tank, two solid-fuel rocket boosters, and its own engines. It returns to Earth as a glider.

Thinking Critically

4. President John F. Kennedy committed the nation to a program of space exploration. A journey to the moon focused these efforts toward a single goal.

continues

TABLE 24–1: SPACE PROBES

Pioneer The United States launched eight Pioneer probes between 1959 and 1973. Studies of the sun were conducted by many of the Pioneers. *Pioneer 5,* launched in 1960, was the first probe sent to explore interplanetary space. *Pioneer 10* and *Pioneer 11* sent back to Earth the first close-up photographs of Jupiter in 1972 and 1973. *Pioneer 11* sent back close-up photographs of Saturn in 1974.

Mariner The United States launched a series of Mariner probes beginning in 1962. *Mariner 2* passed close to Venus in 1962, and *Mariner 4* passed close to Mars in 1964. Both made many important new discoveries. *Mariner 9* orbited Mars in 1971, taking many photographs, and *Mariner 10* passed close to Mercury in 1973. The photographs of Mercury showed it to be heavily cratered like the moon.

Venera A series of Venera probes was launched by the Soviet Union. *Venera 9* and *Venera 10* landed on Venus in 1975, taking one photograph of the surface before being destroyed. *Venera 13* and *Venera 14* landed on Venus in 1982. Together they took four photographs, including the first color photograph of the surface of Venus. Many other measurements of the planet were made as well.

Viking *Viking 1* and *Viking 2,* launched by the United States, orbited Mars in 1976, taking over 50 000 photographs. Landers were sent to the surface to make direct observations of the surface. The Viking landers took photographs of the surface, recorded quakes and weather conditions, analyzed Martian soil, and conducted experiments to search for life on Mars.

Voyager Two Voyager probes were launched by the United States to explore the outer planets. In 1979 and 1980 *Voyager 1* and *Voyager 2* passed close to Jupiter, discovering many new satellites and a system of rings. In 1980 and 1981 the Voyager probes passed close to Saturn, again making many discoveries. Voyager continued on to explore Uranus in 1986 and Neptune in 1989.

Section Review

READING CRITICALLY

1. What is a geostationary satellite?
2. What is the advantage of a polar orbit?

THINKING CRITICALLY

3. Describe three ways in which satellites affect your life.
4. Why do you think certain nations have spent so much money and effort exploring other planets in the solar system?

DISCUSSION Many of the probes in Table 24–1 were operating at the same time. You may wish to have the students make an analysis of space-probe history by year, starting with 1959 and going through the present.

SKILL *(Interpreting Tables)* Have the students list the information scientists now have as a result of the data sent back to Earth by the probes listed in Table 24–1.

THINKING SKILL *(Relating Information)* Have the students imagine that they are preparing a space probe to be sent beyond our solar system. Tell them that in hope of their probe contacting intelligent life in another solar system, they should include objects that tell something about life on Earth. Have the students work in small groups to list objects they would place in their space probe. Have the students explain the reasons for their choices.

SECTION REVIEW

Summary Have the students summarize the uses of satellites and space probes. Encourage them to think of additional uses that are not discussed in this section. Allow them to work together in small groups to facilitate brainstorming of original ideas.

Reinforcement Have the students complete the Section Review. Encourage them to share their answers with the class.

Answers to Section Review

Reading Critically

1. A geostationary satellite remains in orbit above one spot on Earth.
2. Polar orbits allow the satellite to observe the earth as it rotates.
3. They allow meteorologists to observe weather patterns on a global scale. This makes weather predictions more accurate and gives scientists a better understanding of the mechanisms that determine the weather.

Thinking Critically

4. Answers will vary, but should mention our seemingly insatiable curiosity about the universe around us. They might mention that the development of space programs has advanced technological expertise in other fields, including geology, medicine, oceanography, and physics.

Skill *(Calculating)*

Hint

Be sure the students realize that this time they are calculating the force they demonstrated in the Activity.

Discussion Point out that the students' calculations are merely estimations of the balloon rocket's force of thrust. In this investigation, the students used a measure of average velocity when, in fact, the balloon obviously accelerated at "blast off" and may have decelerated as it approached the end of its flight. Nevertheless, its change in momentum from zero to the measured quantity, $P = m \times v$, represents the force pushing the balloon. Recall that Newton's Second Law of Motion can be summarized as follows:
$$F = m_b \times a = m_b \times v2 - m_b \times v1$$
where F is force, m^b is the mass of the balloon, a is the acceleration of the balloon, v_1 is the velocity of the balloon prior to launch (zero), and v_2 is the average velocity of the balloon.

Answers to Analyses and Conclusions

1. Numerical answers will vary; however, the momentum of the balloon just prior to colliding with the ringstand is the product of its mass and velocity at that instant.
2. Numerical answers will vary; however, the frictional force of the air and string combined with the impeding force of the ringstand required to stop the balloon must be equal to the thrust force required to set the balloon in motion.

INVESTIGATION 24: Measuring Balloon Rocket Thrust

PURPOSE

To measure the force required to accelerate a balloon rocket

MATERIALS (per group of 3 or 4)

Laboratory balance
Balloons (2)
String (10 m)
Plastic straw
Tape
Ring stands (2)
Metric ruler
Stopwatch

PROCEDURE

1. Use a balance to find the mass in grams of an inflated balloon. Record that mass.
2. Turn to the Procedure of the Activity on page 526, and follow steps 1 through 4. This time, measure and record the distance between the ring stands to the nearest centimeter.
3. Set the stopwatch to zero and begin the countdown.
4. Measure the time it takes for the balloon to race to the end of the string.
5. Calculate the balloon's average velocity during its brief trip, using the formula

$$v = \frac{d}{t}$$

   where v is the average velocity, d is the distance between the ring stands, and t is the time.
6. Calculate the momentum of the balloon when it reached the end of the string, using the formula

$$p = m \times v$$

   where p is the momentum, m is the mass of the similarly inflated balloon on the balance, and v is the average velocity of the balloon.
7. Repeat the procedure several times, taking an average measure for momentum. Since the velocity, and therefore the momentum, of the balloon became zero at the end of the string, p represents a measure of the *force* required to stop the balloon's forward motion.

540   Chapter 24   Exploring Space

ANALYSES AND CONCLUSIONS

1. What was the momentum of the balloon at the instant prior to its collision with the ring stand?
2. How much force was required to stop the balloon? What is the unit of measure for this force?

APPLICATION

How could this procedure be used to measure the momentum of a small rocket motor that could be purchased at a hobby shop?

CHAPTER 24 REVIEW

SUMMARY

- The momentum of burned gases expelled from a rocket propels the rocket forward. (24.1)
- Solid rocket fuels consist of mixtures of powdered metals, plastics, or rubber. (24.1)
- The liquid fuel used by Robert Goddard for his first rocket was gasoline. (24.2)
- The *V-2* was a rocket adapted to warfare by the Germans. (24.2)
- The Soviets put *Sputnik 1,* the first artificial satellite, into orbit in 1957. (24.3)
- The first American in orbit was Lieutenant Colonel John Glenn. Glenn orbited the earth three times in 1962. (24.3)
- *Apollo 11,* launched in 1969, put the first humans on the moon. (24.4)
- The Soviets have orbited a series of seven space stations called *Salyut.* (24.5)
- *Skylab* was a large American space station that operated during 1973–1974. (24.5)
- The space shuttle is a spacecraft that is launched by rockets but is capable of landing on an airstrip as an airplane does. (24.6)
- Geostationary orbits are high above the equator, while polar orbits pass above the poles. (24.7–24.8)
- Weather satellites continuously monitor the earth and allow the forecasting of weather. (24.8)
- Landsats are designed to observe the earth's surface. (24.9)
- The sun and the solar system have been studied by a series of space probes launched by the United States and the Soviet Union. (24.10)
- *Voyager 1* and *Voyager 2* discovered a total of 20 new moons orbiting Jupiter, Saturn, and Uranus. (24.10)

Write all answers on a separate sheet of paper.

SCIENCE TERMS

Correctly use each of the following terms in a sentence.

geostationary **(534)**
momentum **(521)**
space probes **(538)**

SCIENCE QUIZ

Modified True-False

Mark each statement *true* or *false.* If the statement is false, change the underlined word to make the statement true.

1. Rocketry is an application of Newton's <u>first</u> law of motion.
2. The first human to orbit Earth was <u>John Glenn.</u>
3. The fact that <u>momentum</u> is conserved makes rocket travel possible.
4. <u>John F. Kennedy</u> committed the United States to landing humans on the moon.
5. Geostationary orbits are also called <u>Clarke</u> orbits.

Multiple Choice

Write the letter of the choice that best answers the question or completes the statement.

6. One of the first American rocket pioneers was
   a) Konstantin Tsiolkovsky.
   b) Robert Goddard.
   c) John Glenn.
   d) Yuri Gagarin.

7. In order to burn their fuel, most rockets must carry a chemical
   a) propellant.
   b) catalyst.
   c) oxidant.
   d) repellent.

continues

Chapter 24 Review **541**

CHAPTER REVIEW

SUMMARY

The students may review the major concepts in this chapter by reading the summary statements. The statements are cross-referenced to the chapter to facilitate reinforcement of any concepts of which the students feel unsure. Encourage the students to work in groups to quiz one another.

SCIENCE TERMS

The sentence in which the science term is used should reflect an understanding of the definition of the the term. You may wish to allow the students to draw diagrams to illustrate the meanings of the terms.

SCIENCE QUIZ

Answers to Modified True-False

1. false, third
2. false, Yuri Gagarin
3. true
4. true
5. true

Answers to Multiple Choice

6. b
7. c

continues

Answers to Multiple Choice (continued)

8. a
9. d
10. c
11. a
12. c
13. b
14. a
15. d

Answers to Completion

16. live and work in space
17. Mars
18. Voyager
19. geostationary
20. V-2

Answers to Short Answer

21. Konstantin Tsiolkovsky was the first person to scientifically study how rockets could be used in space travel. Dr. Robert Goddard launched the first liquid-fueled rocket.
22. Pioneer space probes were the first to explore the near reaches of the solar system. Viking spacecraft have landed on the planet Mars. Voyager spacecraft have flown by several of the outer planets.
23. Neil Armstrong, Edwin Aldrin, and Michael Collins journeyed to the moon in *Apollo 11*. Armstrong and Aldrin descended to the lunar surface, set up experiments, collected samples of moon rock, and took photographs. They returned to the orbiting command module piloted by Collins and then returned to Earth.

Answers to Writing Critically

24. Many Americans feared that the Soviet Union might dominate outer space if the United States did not become actively involved in space exploration.
25. Construction of a space station will require a considerable amount of supplies and equipment. The space shuttle is the most efficient means of putting those materials into orbit.

SCIENCE QUIZ continued

8. Who was the first person to set foot on the moon?
   a) Neil Armstrong   b) John Glenn
   c) Yuri Gagarin   d) Frank Borman
9. Which of the following is a good reason to launch satellites?
   a) monitor weather
   b) scientific research
   c) communications
   d) All of the choices are correct.
10. Which of the following is a function of *Landsat 1?*
   a) monitor ice sheets
   b) monitor forest fires
   c) Both choices are correct.
   d) Both choices are incorrect.
11. Which space probe was the first to land on Mars?
   a) Viking
   b) Venera
   c) Pioneer
   d) Mariner
12. What was the United States space program that landed the first humans on the moon?
   a) Mercury
   b) Gemini
   c) Apollo
   d) Columbia
13. Which space shuttle was destroyed shortly after launch on January 28, 1986?
   a) Columbia
   b) Challenger
   c) Discovery
   d) Atlantis
14. What was the first United States space station called?
   a) *Skylab*
   b) *Mir*
   c) *Salyut*
   d) *Spacelab*
15. Which of the following is not a part of a space shuttle?
   a) orbiter
   b) external fuel tank
   c) solid rocket boosters
   d) LEM

542   Chapter 24 Review

Completion

Complete each statement by supplying the correct term or phrase.

16. The Mercury and Gemini space missions increased the ability of humans to _____.
17. *Viking 1* soft-landed on the planet _____.
18. The outer planets, Jupiter, Saturn, and Uranus, have been photographed in detail by the space probes _____.
19. A satellite that remains over a single spot on Earth is in a _____ orbit.
20. During World War II, Germany used a rocket called the _____ to bomb London, England.

Short Answer

21. List the names of two early rocket pioneers and describe their accomplishments.
22. Describe three early probes into space and tell what each did.
23. Describe the mission of *Apollo 11*.

Writing Critically

24. Why did the United States speed up its efforts to develop a productive space program after the launch of *Sputnik 1*?
25. Why is an operational space shuttle necessary to the development of a permanant United States space station?

EXTENSION

1. Do library research and write a report on the Mercury, Gemini, Apollo, Skylab, or Apollo-Soyuz space projects.
2. Debate the need for the United States to have a permanant space station.
3. Read about the causes of the *Challenger* explosion, and report on the corrections that have been made to the space shuttle system to avoid a similar tragedy in the future.
4. Read about the history of NASA, the National Aeronautics and Space Administration, and write a summary on the accomplishments of the United States space program.

ANSWERS TO EXTENSION

1. Reports will vary in style and content, but should include the major purposes and achievements of the space projects that students research.
2. Students should present facts to support their opinions.
3. Accounts should report the failure of the rocket booster seam rings.
4. Summary should include information about the early launch failures, as well as later successes in launching satellites and space probes.

APPLICATION/CRITICAL THINKING

1. The space shuttle orbiters are covered with hundreds of heat-absorbing tiles that protect the craft and the crew during reentry into Earth's atmosphere. During the first few shuttle flights, several of the tiles fell off. What kinds of problems might occur if a large number of these tiles were lost during a mission?

2. Why was there such a long delay in launching a space shuttle after the *Challenger* explosion?

FOR FURTHER READING

Black, R. "Shuttle Hybrid." *Omni* 8(September 1986): 20. Because of the *Challenger* explosion, the author of this article considers alternative methods of getting large payloads into space without the shuttle.

Joels, K. *The Mars One Crew Manual.* NY: Ballantine Books, 1985. The author shows how modern technology will be used to make the first trip to Mars. The book is your crew member's manual, describing ship systems, crew's duties, flight time, and experiments to be conducted.

Kerrod, R. *The Illustrated History of NASA.* New York: Gallery Books, 1986. This book is a photographic history of the National Aeronautics and Space Administration.

Wolkomir, R. "Alien Worlds: The Search Heats Up." *Discovery* 8(October 1987): 66. This article explains that wobbling stars provide evidence that this is not the only solar system in the Milky Way Galaxy.

Challenge Your Thinking

In January 1986, seven American astronauts died in the fiery explosion of the space shuttle *Challenger.* Yet, the families of every one of the astronauts have encouraged the continued human exploration of space. Why is it necessary for humans to fly into space when robot probes, such as Viking, have sent so much data back from space?

Chapter 24 Review   **543**

ANSWERS TO APPLICATION/ CRITICAL THINKING

1. Students should recognize the dangers of the tremendous heat generated by friction during re-entry.
2. Many design features of the shuttle had to be re-examined and, in some cases, redesigned to assure safety in future missions.

ANSWERS TO CHALLENGE YOUR THINKING

Humans have the ability to adjust to new or changing situations. For example, on the *Apollo 13* mission, if no crew had been on board, the spacecraft would probably have been lost.

THE SCIENCE CONNECTION:
Life on Mars

Discussion This *Science Connection* is about the possibility of life on Mars. Discuss why fiction writers so often invented life on Mars, and not Venus or Jupiter. The students should understand that the surface of Venus would have been too hot since it was much closer to the sun and the surface of Jupiter would have been too cold for the opposite reason. Scientists kept supposing life might exist on Mars, given the surface changes they could view through a telescope.

Discussion Have the students study the far right photograph of the surface of Mars. Ask the students why such a surface would not give any appearance of having life. Have them discuss whether or not places on Earth look that barren.

Discussion Until the Viking landers performed their experiments, scientists still hoped that life might exist or might have existed on Mars. Discuss why it would have been so interesting to find intelligent life on another planet. Students might suggest that, if Martian civilization were more advanced than that of Earth, the ''Martians'' might have shared their technological advances or perhaps their knowledge about the rest of the Universe.

Extension After completing the Unit and reading and discussing the *Science Connection,* have the students turn back to the Unit Opener on page 453 and answer the questions. When the class discussion has ended, have the students make a list of the reasons why water and an atmosphere are critical to life on a planet. Then have them write a report on how conditions on Earth would change if the atmosphere gradually evaporated. Have them also discuss why it is so important to protect Earth's atmosphere.

UNIT 7 THE SCIENCE CONNECTION:
Life on Mars

In the summer of 1976, two Viking spacecraft landed on Mars and sampled its soil and atmosphere. Some scientists hoped that the Mars landing would yield clues about the existence of life on the red planet. However, the Viking landers' tests demonstrated that no life, as we know it, presently exists on Mars. With the Viking landings, science fiction's long fantasy of great civilizations beyond Earth finally ended.

Launch of a Viking aboard a Titan rocket

544

The fantasy began in the early 1900s with an amateur astronomer named Percival Lowell. Lowell built an observatory in Arizona just to look at Mars. He claimed that an intelligent race of Martians had built irrigation canals on the dry planet. Lowell also believed that polar ice caps proved the existence of water on Mars. Color changes on the planet's surface seemed to indicate the seasonal growth of plants.

The hopes for finding life on Mars were dashed in 1965 when a Mariner spacecraft sent back pictures of the Martian surface. The photos revealed a dry, cratered planet, more like the moon than like the earth. Mariner also revealed that Mars had virtually no atmosphere. The atmospheric pressure

A Viking lander

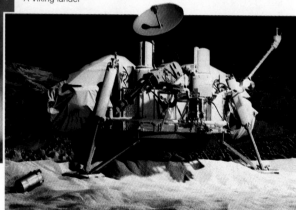

on Mars was only one two-hundredth that of Earth. The pressure was too low to hold liquid water; it would boil away into space.

A second Mariner renewed hopes for finding life on Mars. It photographed dry channels that might have been the result of water runoff. Perhaps, scientists said, life on Mars had begun during a wet period and had evolved to survive the recent dry conditions.

Enough scientists still believed in the possibility of life on Mars that biological experiments were built into the Viking landers. Two Viking spacecraft landed on the Martian surface. Each lander analyzed Martian soil. One biological experiment involved mixing Martian soil with a nutrient-rich liquid nicknamed

A Viking lander on Mars

The Viking probes provided much evidence as to why Mars has no life. Mars has almost no water vapor in its atmosphere. The polar ice caps are made of frozen carbon dioxide, or dry ice, rather than water. Earth has oceans full of life-sustaining water; Mars has only a few bucketfuls at most. Besides having no water, Mars has virtually no atmosphere. This allows its surface to be bombarded by ultraviolet rays. These rays, which destroy many known life-forms, are filtered out by the ozone in Earth's atmosphere. Finally, the surface of Mars is covered with metal oxides, similar to rust. Oxides are poisonous to most primitive forms of life.

Even though no life was found on Mars, scientists did learn something about life on Earth. Life as we know it can survive only with very special conditions. Those conditions exist on Earth but nowhere else in the solar system. Therefore, Earth's life-forms are all closely related. All life shares a basic chemical and genetic makeup.

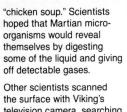

"chicken soup." Scientists hoped that Martian micro-organisms would reveal themselves by digesting some of the liquid and giving off detectable gases.

Other scientists scanned the surface with Viking's television camera, searching for signs of movement. After days of watching the television screen, project scientists admitted that "no rock has gotten up and moved away." A reaction had taken place in the "chicken soup"; however, it was almost certainly a chemical process, since it provided no evidence of any life form. The surface of Mars, it seems, was entirely dead.

The surface of Mars

REFERENCE SECTION

Safety Guidelines

Participating in laboratory investigations should be an enjoyable experience as well as a learning experience. You can ensure both learning and enjoyment from the experience by making the laboratory a safe place in which to work. Carelessness, lack of attention, and showing off are the major causes of laboratory accidents. It is, therefore, important that you follow safety guidelines at all times. If an accident should occur, you should know exactly where to locate emergency equipment. Good safety practice means being responsible for your fellow students' safety as well as your own.

You will be expected to practice the following safety guidelines whenever you are in the laboratory.

1. **Preparation** Study your laboratory assignment in advance. Before beginning your investigation, ask your teacher to explain any procedures you do not understand.

2. **Eye Safety** Wear goggles when handling acids or bases, using an open flame, or performing any activity that could harm the eyes. If a solution is splashed into the eyes, wash the eyes with plenty of water and notify your teacher at once. Never use reflected sunlight to illuminate a microscope. This practice is dangerous to the eyes.

3. **Safety Equipment** Know the location of all safety equipment. This includes fire extinguishers, fire blankets, first-aid kits, eyewash fountains, and emergency showers. Note the location of the nearest telephone. Take responsibility for your fellow students and report all accidents and emergencies to your teacher immediately.

4. **Neatness** Keep work areas free of all unnecessary books and papers. Tie back long, loose hair and button or roll up loose sleeves when working with chemicals or near an open fire.

5. **Chemicals and Other Dangerous Substances** Chemicals can be dangerous if they are handled carelessly.

   Never taste chemicals or place them near your eyes. Do not use mouth suction when using a pipette to transfer chemicals. Use a suction bulb instead.

   Never pour water into a strong acid or base. The mixture produces heat. Sometimes the heat causes splattering. To keep the mixture cool, pour the acid or base slowly into the water.

   If any solution is spilled on a work surface, wash it off at once with plenty of water. When noting the odor of chemical substances, wave the fumes toward your nose with your hand rather than putting your nose close to the source of the odor.

   Never eat in the laboratory. Counters and glassware may contain substances that can contaminate food. Handle toxic substances in a well-ventilated area or under a ventilation hood.

   Do not use flammable substances near flame. Do not place flammable chemicals in a refrigerator. Sparks from the refrigeration unit can ignite these substances or their fumes.

6. **Heat** Whenever possible, use an electric hot plate instead of an open flame. If you must use an open flame, shield it with a wire screen that has a ceramic center. When heating chemicals in a test tube, do not point the test tube toward anyone.

Keep combustible materials away from heat sources.

7. **Electricity** Be cautious around electrical wiring. When using a microscope with a lamp, do not place its cord where it can cause someone to trip and fall. Do not let cords hang loose over a table edge in a way that permits equipment to fall if the cord is tugged. Do not use equipment with frayed cords.

8. **Knives** Use knives, razor blades, and other sharp instruments with extreme care. Do not use double-edged razor blades in the laboratory.

9. **Glassware** Examine all glassware before heating. Glass containers for heating should be made of boro-silicate glass or some other heat-resistant material. Never use cracked or chipped glassware. Never force glass tubing into rubber stoppers. Broken glassware should be swept up immediately, never picked up with the fingers. Broken glassware should be discarded in a special container, never into a sink.

10. **First Aid** In case of severe bleeding, apply pressure or a compress directly to the wounded area and see that the injured student reports immediately to the school nurse or a physician.

    Minor burns caused by heat should be treated by applying ice. Immerse the burn in cold water if ice is not available. Treat acid burns by applying sodium bicarbonate (baking soda). Use boric acid to treat burns caused by bases. Any burn, regardless of cause, should be reported to your teacher and referred to the school nurse or a physician.

In case of fainting, place the person's head lower than the rest of the body and see that the person has fresh air. Report to your teacher immediately.

In case of poisoning, report to your teacher at once. Try to determine the poisoning agent if possible.

11. **Unauthorized Experiments** Do not perform any experiment that has not been assigned by your teacher. Never work alone in the laboratory.

12. **Cleanup** Wash your hands immediately after handling hazardous materials. Before leaving the laboratory, clean up all work areas. Put away all equipment and supplies. Make sure water, gas, burners, and electric hot plates are turned off.

Remember at all times that a laboratory is a safe place only if you regard laboratory work as serious work.

The instructions for your laboratory investigations will include cautionary statements where necessary. In addition, you will find that the following safety symbols appear whenever a procedure requires extra caution:

Wear safety goggles   Flame/heat

Wear laboratory apron   Dangerous chemical/poison

Sharp/pointed object   Electrical hazard

Biohazard/disease-causing organisms   Radioactive material

Laboratory Procedures

Reading a Metric Ruler

1. Examine your metric ruler. The numbers on it represent lengths in centimeters. The usual metric ruler is about 30 cm long. There are 10 marked spaces within each centimeter, which represent tenths of centimeters (0.1 cm).

2. To measure the width of a piece of paper, place the ruler on the paper. The zero end of the ruler must line up exactly with one edge of the paper. Look at the other edge of the paper to see which of the marks on the ruler is closest to that edge. In Figure A, for example, the edge of the paper is nearest to the second line beyond the 7. Therefore, the width of the paper is 7.2 cm.

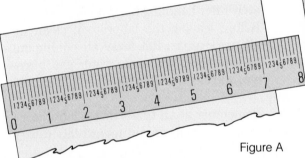

Figure A

3. The edge of the paper might fall exactly on one of the centimeter marks. In Figure B, the edge is just on the 5-cm mark. The width of this paper is 5.0 cm. You must write in the .0 to indicate that the measurement is accurate to the nearest tenth of a centimeter; that is, it is more than 4.95 cm and less than 5.05 cm.

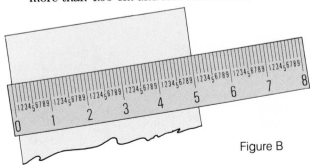

Figure B

4. Sometimes you may want to make a reading with more accuracy. It is possible to estimate readings to the nearest hundredth of a cen- timeter, but you must be very careful. Look at Figure A again. You can guess the number of tenths in the distance between the marks. The edge of the paper is about 3 tenths of the space between 7.2 and 7.3. The best estimate, then, is that the width of the paper is 7.23 cm.

5. In Figure C, the edge of the paper falls exactly on the 8.6 mark. If you are taking careful read- ings, accurate to the nearest hundredth of a centimeter, you must record the width as 8.60 cm.

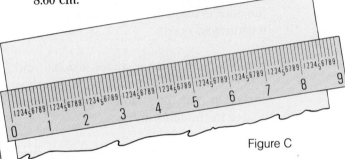

Figure C

6. Note the general rule: You can estimate scale readings to the nearest tenth of a scale divi- sion. If the scale is marked in tenths, you can estimate the hundredths place, but never more than that.

Metric Conversion Table

In SI, it is easy to convert from unit to unit. To convert from a larger unit to a smaller unit, move the decimal to the left. To convert from a smaller unit to a larger unit, move the decimal to the right. Figure D shows you how to move the deci- mals to convert in SI.

Figure D

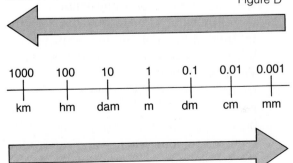

METRIC CONVERSION TABLE

Metric Units		Converting Metric to English	Converting English to Metric
Length			
kilometer (km)	= 1000 m	1 km = 0.62 mile	1 mile = 1.609 km
meter (m)	= 100 cm	1 m = 1.09 yards	1 yard = 0.914 m
		= 3.28 feet	1 foot = 0.305 m
centimeter (cm)	= 0.01 m	1 cm = 0.394 inch	= 30.5 cm
millimeter (mm)	= 0.001 m	1 mm = 0.039 inch	1 inch = 2.54 cm
micrometer (μm)	= 0.000 001 m		
nanometer (nm)	= 0.000 000 001 m		
Area			
square kilometer (km²)	= 100 hectares	1 km² = 0.3861 square mile	1 square mile = 2.590 km²
hectare (ha)	= 10 000 m²	1 ha = 2.471 acres	1 acre = 0.4047 ha
square meter (m²)	= 10 000 cm²	1 m² = 1.1960 square yards	1 square yard = 0.8361 m²
			1 square foot = 0.0929 m²
square centimeter (cm²)	= 100 mm²	1 cm² = 0.155 square inch	1 square inch = 6.4516 cm²
Mass			
kilogram (kg)	= 1000 g	1 kg = 2.205 pounds	1 pound = 0.4536 kg
gram (g)	= 1000 mg	1 g = 0.0353 ounce	1 ounce = 28.35 g
milligram (mg)	= 0.001 g		
microgram (μg)	= 0.000 001 g		
Volume of Solids			
1 cubic meter (m³)	= 1 000 000 cm³	1 m³ = 1.3080 cubic yards	1 cubic yard = 0.7646 m³
		= 35.315 cubic feet	1 cubic foot = 0.0283 m³
1 cubic centimeter (cm³)	= 1000 mm³	1 cm³ = 0.0610 cubic inch	1 cubic inch = 16.387 cm³
Volume of Liquids			
kiloliter (kL)	= 1000 L	1 kL = 264.17 gallons	1 gallon = 3.785 L
liter (L)	= 1000 mL	1 L = 1.06 quarts	1 quart = 0.94 L
milliliter (mL)	= 0.001 L	1 mL = 0.034 fluid ounce	1 pint = 0.47 L
microliter (μL)	= 0.000 001 L		1 fluid ounce = 29.57 mL

Reading a Graduate

1. Examine the graduate and note how the scale is marked. The units are milliliters (mL). A milliliter is a thousandth of a liter, and is equal to a cubic centimeter. Note carefully how many milliliters are represented by each scale division on the graduate.

2. Pour some liquid into the graduate and set the cylinder on a level surface. Notice that the upper surface of the liquid is flat in the center, and curved at the edges. This curve is called the *meniscus* and may be either upward or downward. In reading the volume, you must ignore the curvature and read the scale at the flat part of the surface.

3. Bring your eye to the level of the surface and read the scale at the level of the flat surface of the liquid.

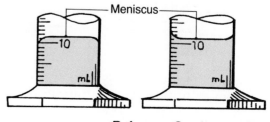

Reference Section   551

Laboratory Procedures

Using a Laboratory Balance

1. Make sure the balance is on a level surface. Use the leveling screws at the bottom of the balance to make any necessary adjustments.
2. Place all the countermasses at zero. The pointer should be at zero. If it is not, adjust the balancing knob until the pointer rests at zero.
3. Place the object you wish to measure on the pan. **CAUTION: Do not place hot objects or chemicals directly on the balance pan as they can damage its surface.**
4. Move the largest countermass along the beam to the right until it is at the last notch that does not tip the balance. Follow the same procedure with the next largest countermass. Then move the smallest countermass until the pointer rests at zero.
5. Determine the readings on all beams and add them together to determine the mass of the object.
6. When massing crystals or powders, use a piece of filter paper. First, mass the paper; then add the crystals or powders and remass. The actual mass is the total minus the mass of the paper. When massing liquids, first mass the empty container, then mass the liquid and container. Finally, subtract the mass of the container from the mass of the liquid and the container to get the mass of the liquid.

Triple-beam balance

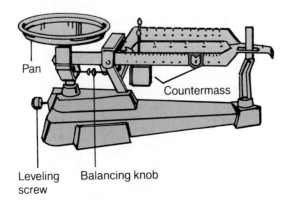

Pan

Countermass

Leveling screw   Balancing knob

Using a Bunsen Burner

1. Before lighting the burner, observe the locations of fire extinguishers, fire blankets, and sand buckets. Wear safety goggles, gloves, and an apron. Tie back long hair and roll up long sleeves.
2. **CAUTION: If the burner is not operating properly, the flame may burn inside the base of the barrel. Carbon monoxide, an odorless gas, is released from this type of flame. Should this situation occur, immediately turn off the gas at the laboratory gas valve. Do not touch the barrel of the burner.** After the barrel has cooled, partially close the air ports before relighting the burner.
3. Close the air ports of the burner and turn the gas full on by using the valve at the laboratory outlet.

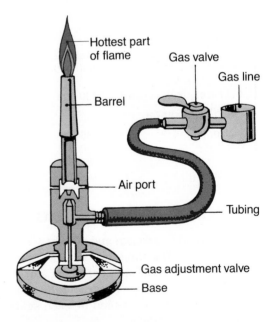

Hottest part of flame

Gas valve

Gas line

Barrel

Air port

Tubing

Gas adjustment valve

Base

4. Hold the striker in such a position that the spark will be placed just above the rim of the burner. Strike a spark.
5. Open the air ports until you can see a blue cone inside the flame. If you hear a roaring sound, the ports are open too wide.

6. Adjust the gas-flow valve and the air ports on the burner until you get a flame of the desired size—not roaring—with a blue cone. The hottest part of the flame is just above the tip of the blue cone.

Filtering Techniques

1. To separate a precipitate from a solution, pass the mixture through filter paper. To do this, first obtain a glass funnel and a piece of filter paper.

2. Fold the filter paper in fourths. Then open up one fourth of the folded paper, as shown in Figure A. Put the paper, pointed end down, into the funnel.

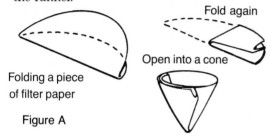

Fold again

Open into a cone

Folding a piece of filter paper

Figure A

3. Support the funnel in a ring stand, with its stem inserted into a beaker.

4. Stir the mixture well and pour it quickly into the filter paper within the funnel. Wait until all the liquid has flowed through.

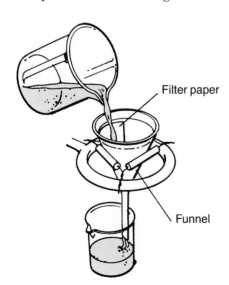

Filter paper

Funnel

5. You may wish to wash the solid that is left in the filter paper. If this is the case, pour some distilled water into the filter paper. Your teacher will tell you how much water to use.

Using Reagents

1. For safety reasons, it is important to learn how to pour a reagent from a bottle into a flask or beaker. Begin with the reagent bottle on the table.

2. While holding the bottle steady with your left hand, grasp the stopper of the bottle between the first and second fingers of your right hand. Remove the stopper from the bottle. DO NOT put the stopper down on the table top.

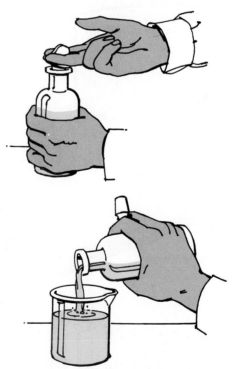

3. While still holding the stopper, lift the bottle with your right hand and pour the reagent into your container.

4. Replace the bottle on the table top and replace the stopper.
(If you are left-handed, these instructions may be reversed.)

Common Laboratory Equipment

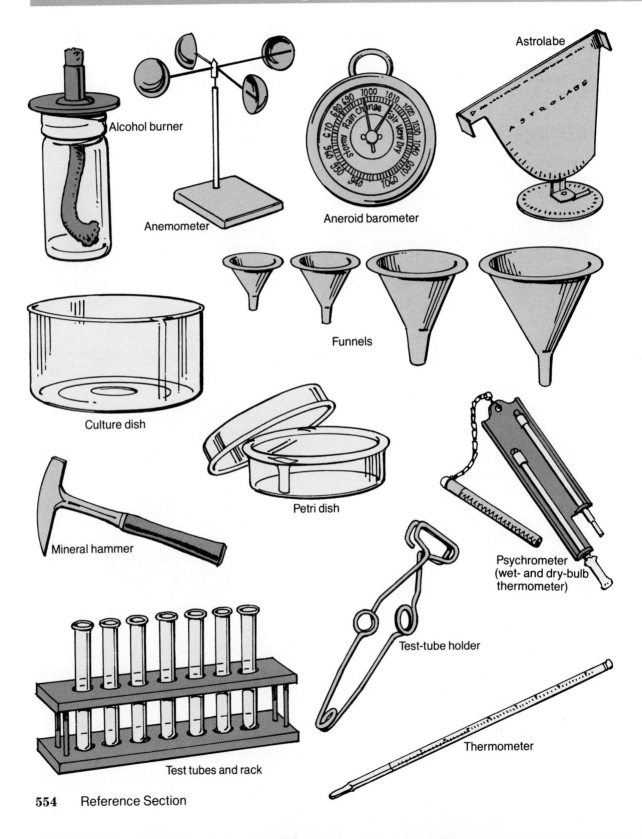

Alcohol burner

Anemometer

Aneroid barometer

Astrolabe

Funnels

Culture dish

Petri dish

Mineral hammer

Psychrometer
(wet- and dry-bulb
thermometer)

Test-tube holder

Test tubes and rack

Thermometer

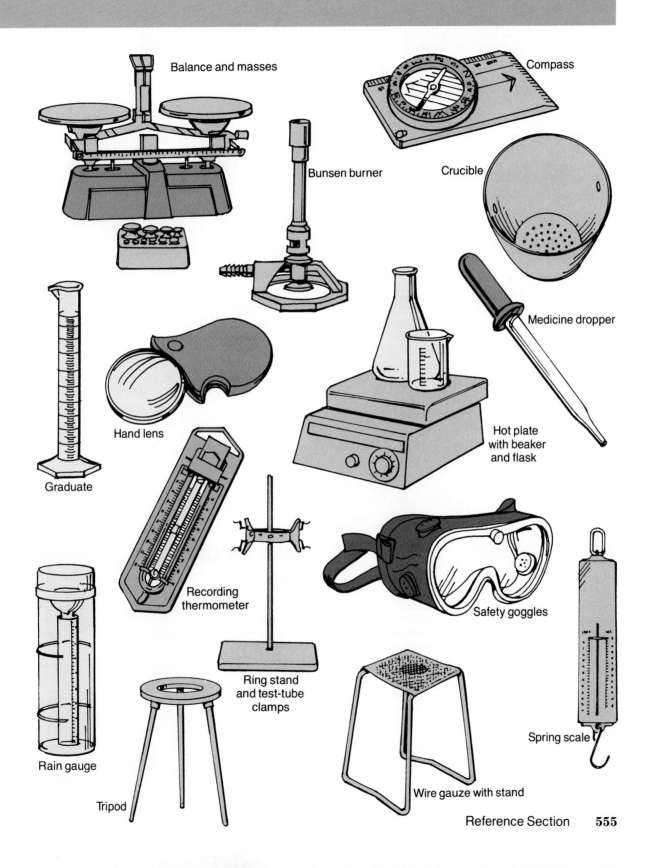

Balance and masses

Compass

Bunsen burner

Crucible

Medicine dropper

Graduate

Hand lens

Hot plate
with beaker
and flask

Recording
thermometer

Safety goggles

Rain gauge

Ring stand
and test-tube
clamps

Spring scale

Tripod

Wire gauze with stand

Key Discoveries in Earth Science

3000 B.C.-501 B.C.

3000 B.C. The Sumerians of the Tigris-Euphrates valley reach high levels in the use of bronze. They are aware that copper can be obtained from mineral ores by using fire, mixed with tin, and cast into bronze instruments.

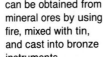

Bronze Age tool

500 B.C.-799 A.D.

500 B.C. Greek philosophers Thales and Anaximander recognize that fossil fish are preserved evidence of once-living fish.

Fossil fish

1600s

1608 Hans Lippershey invents the first telescope by accidentally combining two lenses. In 1609 Galileo refines the telescope and discovers four satellites orbiting Jupiter, rings around Saturn, and mountains and craters on the moon.

Galileo's telescope

1643 Torricelli, an Italian inventor, designs the first barometer, using a column of water in a tube to measure air pressure.

1700s

1785 James Hutton publishes *The Theory of the Earth,* which stresses the geological activity of the internal heat of the earth.

1799 William Smith publishes a usable method of rock classification.

1900s-1930s

1908 Anschutz-Kampfe, a German engineer, develops a gyrocompass. A gyrocompass points to true north rather than to magnetic north as a magnetic compass does.

1940s

1947 John Bardeen, Walter Brattain, and William Shockley invent a transistor out of germanium. These tiny transistors replace vacuum tubes in many common electronic instruments.

1950s

1957 *Sputnik* is launched into an orbit around the earth by USSR. This is followed in 1958 by the launch of *Explorer,* the first United States satellite placed into orbit.

1961 Yuri Gagarin of the Soviet Union becomes the first human to see the earth from space.

Explorer

The key discoveries given here are only a limited selection over 2500 years and are not meant to be a definitive list. The number of women scientists has increased dramatically over the past 10 years.

A.D. 829 Al-Mamum founds an astronomical observatory in Baghdad. Observations of the precession of the equinox and the orbit of the earth are made.

Early cartographer

1250 Roger Bacon of Oxford University states that the true student should learn natural science by experimentation, not from scholars who base their opinions on fallible authority or the weight of custom.

1569 Gerardus Mercator publishes a new map, based on a projection that shows meridians and parallels as straight lines that cross each other at right angles. This is a great aid to navigation.

1824 Joseph Aspdin of Leeds, England, invents cement by pulverizing limestone, mixing it with clay and water, and then heating it in a kiln.

1830 Charles Lyell publishes *The Principles of Geology* as an attempt to explain that changes in the earth's surface were caused by processes still in operation. This is basically a summary of the earlier theories of James Hutton.

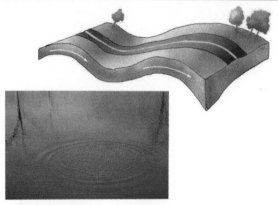

Surface waves

1969 Neil Armstrong of the United States becomes the first human to walk on the moon.

"One small step"

1976 The United States places two *Viking* spacecraft in orbit around Mars. Each orbiter sends a lander to the Martian surface, where they take photographs and conduct experiments on the Martian soil, searching for signs of life.

1986 Archaeological dig near Titusville, Florida unearth preserved brain tissue of humans that lived over 5000 years ago.

Archaeological dig

Reference Section   **557**

Periodic Table

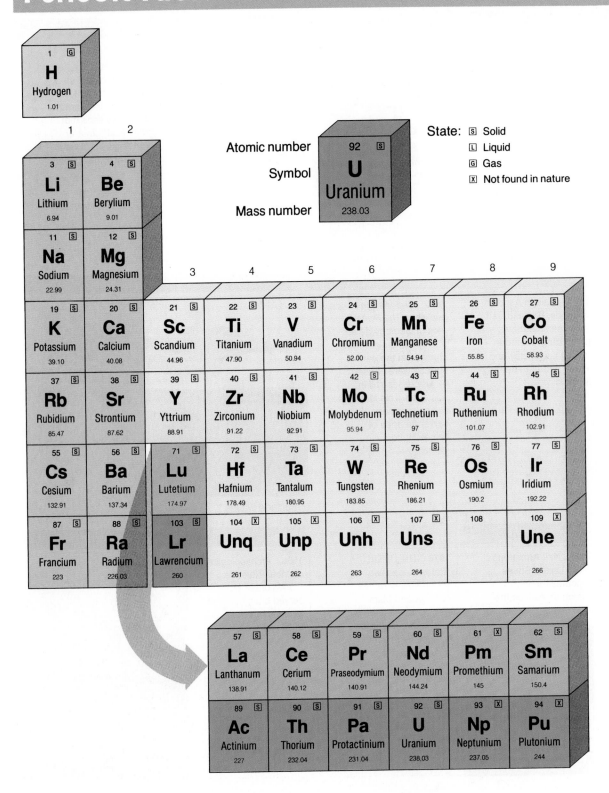

State:
- Ⓢ Solid
- Ⓛ Liquid
- Ⓖ Gas
- Ⓧ Not found in nature

Atomic number — 92 Ⓢ
Symbol — U
Uranium
Mass number — 238.03

Metals

Transition metals

Nonmetals

Noble gases

Lanthanide series

Actinide series

			13	14	15	16	17	18
								2 [G] **He** Helium 4.00
			5 [S] **B** Boron 10.81	6 [S] **C** Carbon 12.01	7 [G] **N** Nitrogen 14.01	8 [G] **O** Oxygen 16.00	9 [G] **F** Fluorine 19.00	10 [G] **Ne** Neon 20.18
10	11	12	13 [S] **Al** Aluminum 26.98	14 [S] **Si** Silicon 28.09	15 [S] **P** Phosphorus 30.97	16 [S] **S** Sulfur 32.06	17 [G] **Cl** Chlorine 35.45	18 [G] **Ar** Argon 39.95
28 [S] **Ni** Nickel 58.71	29 [S] **Cu** Copper 63.55	30 [S] **Zn** Zinc 65.38	31 [S] **Ga** Gallium 69.72	32 [S] **Ge** Germanium 72.59	33 [S] **As** Arsenic 74.92	34 [S] **Se** Selenium 78.96	35 [L] **Br** Bromine 79.90	36 [G] **Kr** Krypton 83.80
46 [S] **Pd** Palladium 106.4	47 [S] **Ag** Silver 107.87	48 [S] **Cd** Cadmium 112.40	49 [S] **In** Indium 114.82	50 [S] **Sn** Tin 118.69	51 [S] **Sb** Antimony 121.75	52 [S] **Te** Tellurium 127.60	53 [S] **I** Iodine 126.90	54 [G] **Xe** Xenon 131.30
78 [S] **Pt** Platinum 195.09	79 [S] **Au** Gold 196.97	80 [L] **Hg** Mercury 200.59	81 [S] **Tl** Thallium 204.37	82 [S] **Pb** Lead 207.2	83 [S] **Bi** Bismuth 208.98	84 [S] **Po** Polonium 209	85 [S] **At** Astatine 210	86 [G] **Rn** Radon 222

63 [S] **Eu** Europium 151.96	64 [S] **Gd** Gadolinium 157.25	65 [S] **Tb** Terbium 158.93	66 [S] **Dy** Dysprosium 162.50	67 [S] **Ho** Holmium 164.93	68 [S] **Er** Erbium 167.26	69 [S] **Tm** Thulium 168.93	70 [S] **Yb** Ytterbium 173.04
95 [X] **Am** Americium 243	96 [S] **Cm** Curium 247	97 [X] **Bk** Berkelium 247	98 [X] **Cf** Californium 251	99 [X] **Es** Einsteinium 254	100 [X] **Fm** Fermium 257	101 [X] **Md** Mendelevium 258	102 [X] **No** Nobelium 259

Scientific Notation

The distance from the earth to the sun is 150 000 000 000 m. The diameter of a hydrogen atom is 0.000 000 000 053 m. You could get dizzy counting all those zeroes, so scientists have developed a much neater method, called *scientific notation,* to express very large and very small numbers.

People often give approximate values in round numbers, such as "a thousand" or "about a million." In scientific notation, this sort of approximation is stated by writing the number 10 raised to some power. For example:

$100 = 10^2$

a hundred thousand $= 100\ 000 = 10^5$

a billion $= 1\ 000\ 000\ 000 = 10^9$

In each case, *the number of zeroes after the 1 is the exponent of 10 in scientific notation.*

To write a more exact number in scientific notation, use a technique like this:

$450\ 000 = 4.5 \times 100\ 000 = 4.5 \times 10^5$

The trick is to place the decimal point just after the first digit. Then count the number of places from there to the end of the number. This gives you the power of 10.

To express the distance from the earth to the sun (150 000 000 000 m) in scientific notation, put the decimal point after the 1. From that point, there are 11 digits to the end of the number. Therefore, the distance from the earth to the sun is 1.5×10^{11} m.

For numbers less than 1, the exponent of 10 is negative. Place the decimal point after the first nonzero digit. Then count the number of places from there, going left to the decimal point. That number is the negative exponent of 10.

To express the diameter of a hydrogen atom (0.000 000 000 053 m) in scientific notation, place the decimal point after the 5. Counting left from the decimal point, there are 11 digits. Thus, the diameter of a hydrogen atom is 5.3×10^{-11} m.

Significant Digits

In measuring, there is an important difference between an answer of 9.2 cm and 9.20 cm. If you write 9.2 cm, you are saying that you know the value to the nearest 0.1 cm. That is, the correct value lies between 9.15 cm and 9.25 cm. A measurement of 9.20 cm says that you know the value correct to the nearest hundredth of a cm. The correct value lies between 9.195 cm and 9.205 cm.

The accuracy of a measurement is expressed by the number of significant digits it contains. A reading of 9.20 cm has three digits. It is a more accurate measurement than is expressed by the two digits of 9.2 cm.

Accuracy has nothing to do with the location of the decimal point. Thus, if the length of a road is given as 4935 m, the reading has four significant digits. It is a more accurate measurement than a statement that the width of the road is 20.7 m, which has only three digits.

The zeroes in a number are not always significant. Thus, a measurement made in meters might come out to be 0.058 m. This is no different from 5.8 cm—both numbers have only two significant

digits. Zeroes before the first nonzero digit are needed only to give the location of a decimal point, and are never significant.

Final zeroes may or may not be significant. If you guess the length of a field as "about 350 m," the final zero is not even a guess. It is put in only to locate the decimal point. The 5 is a guess, so the number has two significant digits. However, if you measure the field carefully and it comes to 350 m, the zero is significant. The reading has three significant digits.

Sometimes, a decimal point is put after a zero to indicate that the final zero is significant. Using this system, 350 m has two significant digits, and 350. m has three significant digits.

Zeroes after the decimal point are not needed to locate the decimal point and are never written unless they are significant. A measurement of 9.300 has four significant digits.

In scientific notation, all digits are significant. The guess about the length of the field would be written 3.5×10^2 m, and the measurement would be 3.50×10^2 m.

Suggested Science Projects

1. Make a table showing the metric measurement of some common objects. For example, a doorknob is about 1 m off the floor.

2. Make a topographic map of your neighborhood, showing all the houses and streets; also, use contour lines to show changes in elevation.

3. Make a collection of mineral crystals. Start your collection with some large crystals of salt.

4. Make a list of some different kinds of rock and ways in which they are used in your community. For example, your school may be built of limestone or sandstone.

5. Under the supervision of your science teacher, try smelting a piece of copper ore to recover the copper metal.

6. Make a model of several layers of sedimentary rocks with different colors of modeling clay; then try folding and faulting the layers to produce various landforms.

7. Make a model of a cinder-cone volcano out of plaster or papier-mâché. With the supervision of your science teacher, create an eruption in the volcano by igniting potassium permanganate.

8. Make a model showing the Coriolis effect, using a marble and a record turntable. Vary the turntable speed to see what it does to the Coriolis effect.

9. Try making a cloud in a laboratory flask by first adding some condensation nuclei in the form of dust or smoke, and then reducing the air pressure in the flask with a laboratory pump.

10. Set up a weather station with a barometer, hygrometer, recording thermometer, and anemometer, and try making weather forecasts. Start by predicting the weather for the next 12 hours. Then try forecasting for longer periods with the aid of weather maps and satellite photographs from a newspaper.

11. Try making several different types of soil from varying percentages of sand, humus, and clay. Plant some seeds in each of the soils to determine the best soil mixture.

12. Make a model of stream erosion, using a flat pan, some soil, and a garden hose. Try creating different stream patterns and different types of landforms to be eroded.

13. Using topographic maps of Wisconsin, Pennsylvania, or New York, try to find the southern limit of the last glaciation of these areas by looking for glaciated landforms, such as drumlins, kettle lakes, and moraines.

14. Visit a national, state, or local park near your home. Ask the attendants for as much information as they can provide for you on the topography, geology, and hydrology of the area.

15. Make a collection of fossils from the area in which you live, or from some place with which you are familiar. Look for fossils in layers of sedimentary rock, particularly shale.

16. Make a diorama of one of the periods of Earth's history. Paint the background scenery with typical plants and animals of the period, then make models of the organisms for the center of the display.

17. Compare the heat value of various fossil fuels by burning equal amounts of each in a simple calorimeter. A calorimeter can be made from a soda can, a test tube, a one-holed stopper, a thermometer, and a large beaker.

18. Make a working model of one of the locks that raise and lower ships along the St. Lawrence Seaway.

19. If you live near the ocean or a bay, determine the salinity of the streams or rivers that flow into the bay or ocean. Determine the salinity at 1 km intervals from the ocean or bay.

20. Make a list of the products that you use daily that come directly or indirectly from the ocean. For example, shrimp come directly from the ocean, while oil comes indirectly from the fossils of organisms that once lived in the ocean.

21. Make a model of the solar system, showing all of the planets and their known satellites. Try making each planet and satellite its correct relative size. As an alternative, try making a model of the orbits of the inner planets, making the orbits of each the proper relative size.

22. Determine what constellations are visible, and their location on different nights, or at different times during the same night.

23. Make a telescope out of two paper-towel rolls and a pair of convex lenses.

24. Make a model of the Apollo/Saturn system that carried astronauts to the moon. Using the correct modules for each phase, model a typical mission.

25. Make a relief model of the surface of the earth, showing mountains, lakes, oceans, and seas. Show the elevations of the land and the depths of the oceans as close as possible to relative scale.

Building a Science Vocabulary

Are you confused and intimidated by all those big words used in science books? Some students are so overwhelmed by them that they just give up, which is a pity, because scientific words are not really hard once you know a trick about them.

The thing to realize about words used in science is that they are made up of little "pieces," or word elements, from Greek and Latin. Those elements are parts of words that you probably already know. For example, *helicopter* is made up of the word elements *helix* and *pter*. *Helix* means "spiral," and *pter* means "wing." Do you recall that a pterodactyl is a "wing-fingered" dinosaur? Now, look at the science word *orthoptera*. You know that *pter* means "wing," and you know that an orthodontist straightens crooked teeth, so orthoptera means "something with straight wings." It is the scientific way to say *dragonfly*. When you see a difficult word, look for those parts whose meanings you know; then try to figure out the rest.

Abrasion is one of the first words you encounter in the glossary of this book. You probably already know the word, but here is what the elements mean. *Ab* means "away" (as in *absence*), and *raso* means "rub" (as in *erase*). Therefore, *abrasion* means "rubbing away."

Alluvial is a word that might cause some difficulty unless you know that the first part, *al*, means "to" or "at" (as in *ally*) and the last part, *luvia*, means "wash" (as in *lavatory*). *Alluvial* comes from "to wash," and an alluvial fan is a fan-shaped pile of debris that has been washed over the ground by a river's flow.

Anticline is easy if you know that *anti* means "against" and *cline* means "to lean" (as in "That wall is inclining so much that it may soon recline on the floor"). An anticline is a place where layers of rocks have been pushed upward and lean against each other.

Asthenosphere means "a weak part of a sphere." The word element *a* (or *an*) at the beginning of a science word means "no" or "not" (as in *asexual*, or *asymmetrical*). The Greek element *sthenos* means "strength," and *sphere* means "ball." Of course, you know that the earth is a sphere, but the outer part of that sphere is the only part that is solid. The inside is still in the molten state because it has not cooled off enough to be solid. That is why we call the interior part of the earth (the lower mantle, the inner core, and the outer core) the *asthenosphere*, or the "no-strength part of the ball."

Atmosphere is another word that incorporates the word *sphere*. The Greek word for vapor is *atmos*, and the atmosphere is that part of the spherical earth that contains air, water, and other vapors.

Barometer can better be remembered if you know what its word elements mean. *Baros* means "heavy," and *metr* means "to measure." Just as a speedometer measures the speed of a car, a barometer measures how heavy the atmosphere is.

Benthos is a very deep subject—deep in the sea, that is, because *benthic* creatures live deep in the sea. *Benthos* is close to the word *bathos*, which really means "deep." A bathysphere is a round vehicle that people travel in to the bottom of the sea.

Caldera is an interesting word because we all know that a caldron is a large, open-topped kettle. If you get too close to a boiling caldron, it might scald you. The center of a volcanic cone, the caldera, looks rather like a giant boiling kettle.

Cartography might seem, at first glance, to be a funny name for mapmaking. However, the Greek word *charte* means "a piece of paper," and you may have read about sailing uncharted seas. The word *card* also means "paper," and *graphy* means "description," so *cartography* refers to descriptions on pieces of paper.

Convection is one of the ways in which heat is transferred, but you might have difficulty remembering that. The first element, *con*, means "with" (as in *convoy* and *conjunction*). The word element *vec* means "to carry" (as in *convey*). It should now be easy to understand that when heat is moved or transferred by convection, it is carried with something, such as a liquid or a gas.

Delta is not part of a word; it is a letter in the Greek alphabet, and it is shaped like a triangle. *Delta* has two meanings in science. The symbol means "change"; the word means "the triangular-shaped pile of silt deposited at the mouth of a muddy river."

Exfoliation will be easy to remember if you know that *ex* means "off." Many small leaves make up foliage, and *exfoliation* means "the falling away of small flakes of rock from boulders," similar to the falling of autumn leaves from trees.

Foliated rocks are rocks that are in layers. This textbook has many layers as well, but we call them *leaves*, or in Latin, *foli*.

Galaxy In the sky at night, you can see a galaxy that is called the *Milky Way*. The Greeks called it that also, because *galax* means "milk." They thought that the ribbon of stars they saw looked as white as milk. At parties, the Greeks would wear their whitest clothes. That is why, even now, a dress-up party is called a *gala*.

Geologic and **geomorphic** both contain the word element *geo*, which means "earth." The other elements in these words are *logos*, which means "a study of," and *morphic*, which means "form" or "shape."

Igneous is a hot item in earth science classes because it means "fire" (as in *ignite*). Whenever you think of igneous rocks, think about the fact that they form in the earth's hot mantle.

Lithosphere *Lithos* is a Greek word that means "stone." *Lithosphere* means "rock sphere" and refers to the rock-solid outer layer of the earth.

Magma is a strange name for the molten rock inside the earth. It becomes more understandable when you think about the consistency of molten lava—about the same as that of the dough from which bread is made. *Magma* is a Latin word that means "dough."

Metamorphic contains the word element *morphic*, which means "form." The element *meta* means "among" or "changing," so metamorphic rocks are rocks that have changed to form new rocks.

Nekton is a Greek word meaning "swim near the surface of the sea." This word is a little like the word *nata*, which just means "to swim." However, Nekto was a mortal who angered a god and was forced to swim at the surface of the sea forever.

Nucleus is a nutty-sounding word—and it should be—because when scientists first looked inside a cell through a microscope, they saw what looked like a nut (a walnut, to be exact). The Latin name for walnut is *nucleus*. We use the same term for the inner core of an atom.

Ozone is the term used for a molecule of three oxygen atoms. The word element *ozo* means "smell." Pure ozone has a smell that is hard to forget. It smells a little like acetylene gas or the odor from a spark generator. Its toxic fumes will quickly give you a headache.

Penumbra refers to a kind of shadow. The first part of the word, *penu*, means "almost" (as in *penultimate*, or "next to last"). *Umbra* means "shade" (as in *umbrella*), so *penumbra* means "an area of partial eclipse."

Permeable has the word element *per*, which means "through" (as in *perjury* and *perspective*). Think of permeable things as being able to let fluids pass though with ease.

Playa is neither Greek nor Latin. It is a Spanish word that means "beach." The word came from the early Spanish conquistadors who saw shallow pools of water in the desert. The area around the pools looked like a beach, so the name stuck.

Refraction is usually a hard word to remember. Think of the word element *fract*, which is part of *fracture*. *Fract* means "broken." Refracted light is light that is broken, or bent. A fraction is a "broken" whole number.

Saltation is not a salty term. It is more of a bouncy, dancing little word because *salta* actually means "to bounce" or "to dance." Think of little pieces of sand and gravel being blown so hard that they bounce when they strike a surface.

A **seismograph** is a device for measuring earthquakes. You know what a graph is; would it surprise you to learn that *seism* means "to shake"?

Smog is a fairly new word. It was first used in the early 1950s. The word is a combination of *smoke* and *fog*, describing what smog looks like.

A **syncline** is the opposite of an anticline. Since *cline* means "to lean," and *syn* means "within," you can figure out that a syncline must be a formation in which the rocks lean toward one another.

Tectonics (as in "the theory of plate tectonics") has the word element *techton. Techton* is a Greek word meaning "carpenter." Plate tectonics is a theory that describes how huge continental plates drift into different patterns over the ages, much as large boards can be put together by a carpenter into different shapes.

Zodiac is a term that is not merely a part of astrology. The word element *zo* means "animal" (as in *zoo* and *protozoa*). The term *zodiac* arose because ancient astronomers thought they could see the shapes of animals in certain patterns of stars. They called those patterns the *zodiac*.

TABLE 1: WORD ELEMENTS

Element	Meaning	Examples
A	No, Not	Atheist, Anarchy, Asthenosphere
AB	From, Away	Abrasion, Absorb, Abscond
ANTI	Against	Antidote, Antiseptic, Anticline
AQUA	Water	Aquatic, Aqueduct, Aquifer
BARI	Heavy	Barometer, Barium, Baritone
BENTHO	Depths (of the sea)	Benthic, Benthal, Benthos
CALD	Hot	Caldera, Caldron, Scald
CART	Piece of paper	Cartography, Chart, Magna Charta
CLAST	To break	Iconoclast, Clastic, Orthoclase
CLIMA	Region, Slope	Climate, Climax
CLINO	Lean	Anticline, Incline, Recline
CO	With	Cooperate, Covalent bond, Collaborate
COM	Together	Composite, Complex, Complicate
CON	With, Together	Convection, Conclude, Convey
DELTA	Triangle shape	River delta, Deltoid, Deltoidal
EPI	Upon	Epicenter, Epidemic, Epitaph
EX	Out of	Extrusive, Exodus, Exclude
FER	To bear, Ferry	Coniferous, Ferry, Aquifer
FLATU	Wind	Deflation, Inflate
FOLI	Plant leaves	Foliage, Folio, Foliated
FRACT	Broken	Fraction, Refraction, Fracture
GALA	Milky white	Galaxy, Galactic, Gala party
GEO	Earth	Geology, Geomorphic, Geographical
GRAPHY	Written description	Cartography, Geography, Photography
HYDRO	Water	Hydrophobia, Hydraulic, Hydrogen
HYPO	Under	Hypothesis, Hypodermic, Hypotonic
IGNIS	Fire	Ignite, Igneous rock, Ignition
ION	Violet color	Cation, Ionize, Iodine
ISO	Equal	Isobar, Isosceles, Isomer
LAVO	To wash	Alluvial, Lavatory, Laundry
LITH	Stone, Rock	Lithosphere, Monolith, Neolithic
LOGO	Speech, Thought	Analog, Dialogue, Logical
MAGMA	Dough, To knead	Magma, Magmatic
META	Changing	Metaphase, Metastasis, Metamorphic
METR	To measure	Barometer, Centimeter, Thermometer
MORPH	Form, Shape	Mesomorph, Metamorphic, Morphology
NEKTON	Swim	Nekton, Nekteric, Nektonic
NUCLE	Nut, Walnut	Nuclear, Nucleoplasm, Nucleus
OZO	Smell	Ozone, Ozonosphere
PENU	Almost	Penultimate, Penumbra, Penumbrous
PER	Through	Perjury, Perspective, Permeable
POSI	Place	Position, Composite, Deposit
RAS	Scrape, Rub away	Rasp, Abrasion, Razor
SALTA	Jump, Dance	Saltation, Saltatory, Salticidae
SEDI	Sit	Sediment, Sedentary, Sedimentary rock
SEISM	Shake	Seismic, Seismograph, Seismology
STHENOS	Strength	Asthenosphere, Asthenia, Asthenolith
TECTON	Carpenter	Plate tectonics, Tectology, Tectological
TOMO	Cut	Atom, Appendectomy, Epitome
TOPO	Surface	Topography, Topectomy, Topographical
TRUDO	Thrust	Intrusive, Protrude, Extrude
UMBRA	Shade	Umbrella, Penumbra, Umbrage
VALE	Value, Worth	Covalent bond, Equivalent, Valence
VEC	To carry	Convection, Vector, Convey
VERG	To incline	Converge, Diverge, Divergent
ZO	Animal	Zodiac, Zoo, Protozoa

Mineral Classification Key

DIRECTIONS: Examine your mineral sample. First, determine whether the sample has a metallic luster, which places it in Category I, or a nonmetallic luster, which places it in Category II. If the luster is metallic, determine the mineral's streak. If the streak is black (Choice A), determine the mineral's hardness. If the streak is yellow or red, move to the Key of Mineral Groups.

If the luster is nonmetallic, determine whether the mineral is light colored (Choice A) or dark colored (Choice B). With each choice, determine the mineral's hardness, then move to the proper group in the Key of Mineral Groups.

Key for Determining Mineral Groups

I.   Metallic luster
  A.   Streak black, greenish-black, or dark gray
      1.   Hardness 3 or less . . . . .Group 1
      2.   Hardness between 3 and 6 . . . . .Group 2
      3.   Hardness 6 or over . . . . .Group 3
  B.   Streak yellow to brown . . . . .Group 4
  C.   Streak red . . . . .Group 5

II.   Nonmetallic Luster
  A.   Mineral white or light colored
      1.   Hardness 3 or less . . . . .Group 6
      2.   Hardness between 3 and 6 . . . . .Group 7
      3.   Hardness 6 or over . . . . .Group 8
  B.   Mineral dark colored, black, or green
      1.   Hardness 3 or less . . . . .Group 9
      2.   Hardness between 3 and 6 . . . . .Group 10
      3.   Hardness 6 or over . . . . .Group 11

Key of Mineral Groups

DEFINITIONS: The following terms are used in the Key of Mineral Groups.
  Cleavage—
        basal—one direction
        cubic—three directions, 90° angles
        diamond—two directions, less than 90° angles
        rhombohedral—three directions, less than 90° angles
        square—two directions, 90° angles
        octahedral—four directions
        dodecahedron—six directions
  Concretionary—looks like chunks of different-colored minerals glued together; luster is earthy or dull, color of samples
  Saline—tastes like salt
  Malleable—will bend without breaking

Group 1
Galena—cubic cleavage, very high specific gravity

Graphite—basal cleavage, greasy feeling

Group 2
Chalcopyrite—dark brassy color

Group 3
Magnetite—magnetic, black color

Pyrite—light yellow color

Group 4
Limonite—no cleavage, rarely has metallic luster

Group 5
Copper—malleable, reddish color

Hematite—brittle, black to reddish in color, can also have appearances of glitter

Group 6
Bauxite—no cleavage, earthy, concretionary, tan, white color

Calcite—rhombohedral cleavage, hardness 3

Gypsum—nonelastic, hardness 2, basal cleavage not often evident

Halite—cubic cleavage, saline taste

Kaolinite—no cleavage, white color, earthy, smells like clay when moist

Muscovite—perfect basal cleavage

Serpentine—variety of asbestos, separates into silky fibers, although this may not always be evident

Talc—may show basal cleavage, nonelastic, greasy feeling

Group 7
Amphibole—hornblende, variety of tremolite, diamond-shaped cleavage, white to gray color

Apatite—no cleavage, brown or green color

Barite—three cleavage planes, two at right angles, high specific gravity

Dolomite—rhombohedral cleavage

Fluorite—octahedral cleavage, hardness 4

Malachite—no cleavage, green color

Sphalerite—dodecahedral cleavage, resinous luster

Group 8
Corundum—no cleavage, may show parting, hardness 9

Orthoclase—feldspar, two cleavages at right angles, hardness 5, tan or pink color

Plagioclase—feldspar, two cleavages at 86° and 94°, striations, blue-gray to white color

Quartz—no cleavage, hardness 7, conchoidal fracture

Tourmaline—no cleavage, pink, blue, or green color

Group 9
Bauxite—no cleavage, earthy, concretionary

Biotite—perfect basal cleavage, elastic, black color

Chlorite—perfect cleavage, nonelastic, green color

Graphite—basal cleavage, greasy feeling

Hematite—red streak, high specific gravity

Limonite—no cleavage, brown streak

Muscovite—perfect basal cleavage, elastic

Talc—basal cleavage, nonelastic, greasy feeling

Group 10
Amphibole—hornblende, diamond-shaped cleavage, two planes not equal to 90°

Apatite—no cleavage, brown or green color, hardness 5

Azurite—no cleavage, blue color

Fluorite—octahedral cleavage, hardness 4

Hematite—red streak, high specific gravity

Limonite—no cleavage, brown streak

Malachite—no cleavage, green color, green streak

Pyroxene—resembles hornblende, square cleavage, two planes at 92°

Serpentine—no cleavage, green, waxy luster

Sphalerite—dodecahedral cleavage, resinous luster

Group 11
Corundum—no cleavage, hardness 9

Garnet—no cleavage, red color

Olivine—olive-green color, never occurs with quartz

Orthoclase—feldspar, two cleavages at right angles, hardness 6, white, tan, or pink color

Plagioclase—feldspar, two cleavages at 86° and 94°, striations, white to blue-gray color, hardness 6

Quartz—no cleavage, hardness 7

Tourmaline—no cleavage, usually black color

Rock Classification Key

DIRECTIONS: Examine your rock sample. First, determine if it belongs in Category I by asking yourself the question, "Does the rock have layers or clastic particles?" If you decide the answer is yes, then determine whether choice A or B applies to your rock sample.

If you decide that the rock does not belong in Category I, then go to Category II. Continue through the Categories until you find one that fits your rock sample. Then read the detailed choices to determine the name of your rock sample.

I. Does the sample have layers, or clastic particles glued together? If it does, go to choice A. If it does not, go to Category II.

    A. Does the rock react to HCl? If it does, go to Example 1. If it does not, go to Choice B.
- 1. Limestone—not shiny; whole surface reacts
- 2. Calcareous sandstone—sand grains present
- 3. Limestone conglomerate—pebble-sized particles; HCl reacts with matrix
- 4. Marble—shiny

    B. If the rock does not react to HCl and it is not foliated, go to Example 1. If the rock is foliated, go to Category II.
- 1. Conglomerate—rounded pebbles
- 2. Breccia—sharp-edged pebbles
- 3. Sandstone—sand-sized grains
- 4. Shale—clay-sized grains; frequently shows thin layers
- 5. Bituminous coal—black color which rubs off on your hands

II. Is the rock foliated or banded? If the sample is foliated or banded, go to Choice A. If the sample is not foliated or banded, go to Category III.

    A. Banded
- 1. Gneiss—shows layers of different-colored minerals

    B. Foliated
- 1. Schist—distinct mica-type layering with wavy surfaces
- 2. Phyllite—shows some "mica shine" but looks less foliated than schist

III. Does the rock contain holes? If it does, go to Choice A. If it does not, go to Category IV.

    A. Scoria—dark color, dense and heavy

    B. Pumice—light color, tan, brown, or red, lightweight

IV. Does the rock have luster like glass? If it does, it is obsidian or anthracite. If it does not, go to Category V.

    A. Obsidian—hard, shows conchoidal fracture, black to gray color

    B. Anthracite coal—very lightweight, black color, glassy luster

V. Can mineral crystals be identified? If yes, the rock is granite. If not, go to Choice A.

    A. Rhyolite—light color, dull luster

    B. Basalt—dark, dull luster

    C. Obsidian—shiny luster like glass

Granite

Basalt

Pumice

Rhyolite

Obsidian

Scoria

Limestone

Conglomerate

Breccia

Sandstone

Shale

Siltstone

Coquina

Chert

Rock salt

Marble

Quartzite

Slate

Wet-Bulb Depression (°C)
(Dry-bulb temperature minus wet-bulb temperature)

Dry-Bulb Temperature (°C)	1	2	3	4	5	6	7	8	9	10	11	12	13	14	15	16	17	18	19	20
-10	-15	-22																		
-8	-12	-18	-30																	
-6	-9	-14	-23																	
-4	-7	-11	-17	-30																
-2	-5	-8	-13	-20																
0	-2.5	-6	-10	-15	-25															
2	-0.5	-3	-7	-11	-18	-30														
4	2	-1	-4	-7.5	-12	-17														
6	4	1.5	-1	-4	-8	-14	-22													
8	6	4	1	-1.7	-4.5	-9	-15	-20												
10	8	6	4	1.7	-1.5	-5	-9.5	-15	-28											
12	10	9	6	4	1	-2	-5.5	-10	-16	-30										
14	12	11	8	6	4	1	-2	-6	-10	-17.5										
16	14	12.5	10.7	8.5	6	4	1	-2	-6	-10	-18									
18	16	14.5	13	11	9	6.5	4	1	-2	-4.5	-10	-18								
20	18	16.7	15	13	10.5	9.5	7	4.5	2	-1	-5	-10	-18							
22	20	18.7	17	16	13.5	11.5	10	7.5	5	2	-1.5	-5	-10	-18						
24	22	20.7	19	17.5	16	14	12	10	8	5	2.5	-1	-5	-10	-18					
26	24	22.7	21	19.5	18	16.5	15	13	10.5	8	6	3	1	-1	-5	-10	-18			
28	26	24.7	23	22	20	19	17	15	13	11	9	6	3	1	-1	-5	-10	-18		
30	28	26.7	25	24	22	21.5	20	18	16	14	12	10	6	3	1	-1	-5	-10	-18	
32	30	28.7	27	26	24	23	22	20	18	17	15	13	10	6	3	1	-1	-5	-10	-18
34	32	30.7	29	28	26	25	24	22	20	19	17	15	13	10	6	3	1	-5	-10	-18
36	34	32.7	31	30	28	27	26	24	22	21	19	17	15	13	10	6	3	1	-5	-10
38	36	34.7	33	32.5	30	29	28	26	24	23	21	19	17	15	13	10	6	3	1	-5
40	38	36.9	35	34	32	31	30	28	26	25	23	21	19	17	15	13	10	6	3	1

Wet-Bulb Depression (°C)
(Dry-bulb temperature minus wet-bulb temperature)

Dry-Bulb Temperature (°C)	1	2	3	4	5	6	7	8	9	10	11	12	13	14	15	16	17	18	19	20
-10	67	35																		
-8	71	43	15																	
-6	74	49	25																	
-4	77	55	33	12																
-2	79	60	40	22																
0	81	64	46	29	13															
2	84	68	52	37	22	7														
4	85	71	57	43	29	16														
6	86	73	60	48	35	24	11													
8	87	75	63	51	40	29	19	8												
10	88	77	66	55	44	34	24	15	6											
12	89	78	68	58	48	39	29	21	12											
14	90	79	70	60	51	42	34	26	18	10										
16	90	81	71	63	54	46	38	30	23	15	8									
18	91	82	73	65	57	49	41	34	27	20	14	7								
20	91	83	74	66	59	51	44	37	31	24	18	12	6							
22	92	83	76	68	61	54	47	40	34	28	22	17	11	6						
24	92	84	77	69	62	56	49	43	37	31	26	20	15	10	5					
26	92	85	78	71	64	58	51	46	40	34	29	24	19	14	10	5				
28	93	85	78	72	65	59	53	48	42	37	32	27	22	18	13	9	5			
30	93	86	79	73	67	61	55	50	44	39	35	30	25	21	17	13	9	5		
32	93	86	80	74	68	62	57	51	46	41	37	32	28	24	20	16	12	9	5	
34	93	87	81	75	69	63	58	53	48	43	39	35	31	27	23	19	15	12	8	5
36	94	87	81	75	70	64	59	54	50	45	41	37	33	29	25	21	18	15	11	8
38	94	88	82	76	71	66	61	56	51	47	43	39	35	31	27	24	20	17	14	11
40	94	88	82	77	72	67	62	57	53	48	44	40	36	33	29	26	23	20	16	14

HEAT INDEX CHART (APPARENT TEMPERATURE)
RELATIVE HUMIDITY (%)

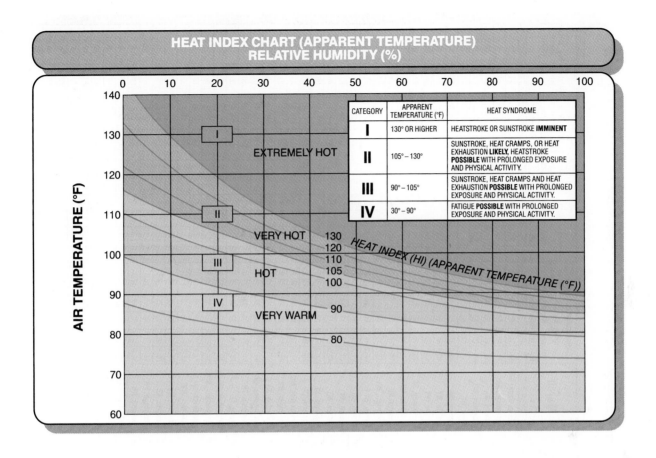

CATEGORY	APPARENT TEMPERATURE (°F)	HEAT SYNDROME
I	130° OR HIGHER	HEATSTROKE OR SUNSTROKE **IMMINENT**
II	105° – 130°	SUNSTROKE, HEAT CRAMPS, OR HEAT EXHAUSTION **LIKELY**, HEATSTROKE **POSSIBLE** WITH PROLONGED EXPOSURE AND PHYSICAL ACTIVITY.
III	90° – 105°	SUNSTROKE, HEAT CRAMPS AND HEAT EXHAUSTION **POSSIBLE** WITH PROLONGED EXPOSURE AND PHYSICAL ACTIVITY.
IV	30° – 90°	FATIGUE **POSSIBLE** WITH PROLONGED EXPOSURE AND PHYSICAL ACTIVITY.

WIND CHILL CHART (APPARENT TEMPERATURE)

Wind Speed		Cooling Power of Wind Expressed as "Equivalent Chill Temperature"																			
Knots	MPH	Temperature (°F)																			
Calm	Calm	40	35	30	25	20	15	10	5	0	−5	−10	−15	−20	−25	−30	−35	−40	−45	−50	−55	−60
		Equivalent Chill Temperature																			
3	6	5	35	30	25	20	15	10	5	0	−5	−10	−15	−20	−25	−30	−35	−40	−45	−50	−55	−65	−70
7	10	10	30	20	15	10	5	0	−10	−15	−20	−25	−35	−40	−45	−50	−60	−65	−70	−75	−80	−90	−95
11	15	15	25	15	10	0	−5	−10	−20	−25	−30	−40	−45	−50	−60	−65	−70	−80	−85	−90	−100	−105	−110
16	19	20	20	10	5	0	−10	−15	−25	−30	−35	−45	−50	−60	−65	−75	−80	−85	−95	−100	−110	−115	−120
20	23	25	15	10	0	−5	−15	−20	−30	−35	−45	−50	−60	−65	−75	−80	−90	−95	−105	−110	−120	−125	−135
24	28	30	10	5	0	−10	−20	−25	−30	−40	−50	−55	−65	−70	−80	−85	−95	−100	−110	−115	−125	−130	−140
29	32	35	10	5	−5	−10	−20	−30	−35	−40	−50	−60	−65	−75	−80	−90	−100	−105	−115	−120	−130	−135	−145
33	36	40	10	0	−5	−15	−20	−30	−35	−45	−55	−60	−70	−75	−85	−95	−100	−110	−115	−125	−130	−140	−150

Winds above 40 have little additional effect

Little Danger

Increasing Danger (Flesh may freeze within 1 min.)

Great Danger (Flesh may freeze within 30 sec.)

Danger of freezing exposed flesh for properly clothed persons

January

City	Temperatures (Deg. F) Average Max	Min	Record Max	Min	Mean Days of Precip.	Amount of Precip. Total Avg.	Max	Snowfall Avg.	Max	% of Possible Sunshine	Mean Clear Days	Mean Cloudy Days	Relative Humidity (%)	Avg. Wind Speed (MPH)
Albuquerque, NM	47.2	22.3	69	-17	3.8	.41	1.32	2.5	9.5	72	13.0	10.2	40	8.1
Anchorage, AK	20.0	6.0	50	-34	7.0	.80	2.09	10.1	21.1	40	8.3	18.2	70	6.1
Atlanta, GA	51.2	32.6	79	-5	11.6	4.91	10.82	0.8	8.3	49	8.3	16.4	60	10.6
Bismarck, ND	17.5	-4.2	62	-44	7.9	.51	1.29	7.2	25.0	54	6.8	16.5	67	10.0
Boise, ID	37.1	22.6	63	-17	12.4	1.64	3.87	7.2	21.4	39	4.4	21.8	70	8.2
Boston, MA	36.4	22.8	63	-12	11.7	3.99	10.55	12.6	35.9	53	9.3	15.0	57	14.0
Buffalo, NY	30.0	17.0	72	-16	20.2	3.02	6.88	23.9	68.3	32	1.4	23.4	73	14.3
Casper, WY	32.5	11.9	60	-40	7.4	.50	1.19	10.1	22.1	NA	6.5	16.8	60	16.5
Charleston, SC	58.8	36.9	83	10	9.8	3.33	6.68	0.1	1.0	58	8.9	15.6	55	9.2
Charlotte, NC	50.3	30.7	78	-3	10.2	3.80	7.44	2.0	11.7	55	9.1	15.5	56	7.9
Chicago, IL	29.2	13.6	61	-26	11.3	1.60	4.11	11.1	34.3	42	6.9	18.2	67	11.6
Cincinnati, OH	37.3	20.4	69	-25	12.1	3.13	9.43	7.5	31.5	41	5.1	19.7	68	10.7
Dallas, TX	54.0	33.9	88	4	7.0	1.65	3.60	1.4	12.1	53	9.8	15.5	59	11.1
Denver, CO	43.1	15.9	73	-25	5.8	.51	1.44	7.7	23.7	71	10.0	11.8	46	8.8
Detroit, MI	30.6	16.1	62	-21	13.0	1.86	3.63	9.6	29.6	40	4.3	20.0	70	11.7
Helena, MT	28.1	8.1	62	-42	8.2	.66	2.78	9.1	35.6	46	4.6	20.5	61	6.9
Honolulu, HI	79.9	65.3	87	53	10.1	3.79	14.74	0.0	—	62	9.1	8.9	63	9.8
Houston, TX	61.9	40.8	84	12	10.3	3.21	7.68	0.0	0.2	43	7.3	18.3	65	8.2
Kansas City, MO	34.5	17.2	68	-17	8.2	1.08	2.66	5.8	14.2	59	9.6	15.4	64	11.0
Las Vegas, NV	56.0	33.0	77	8	3.0	.50	2.41	1.2	16.7	77	13.9	10.9	30	7.4
Lexington, KY	39.8	23.1	76	-21	12.6	3.57	6.65	6.2	21.9	NA	5.8	19.3	69	11.2
Los Angeles, CA	64.6	47.3	88	23	6.1	3.06	9.60	T	T	69	12.0	10.8	56	6.7
Memphis, TN	48.3	30.9	78	-4	10.1	4.61	12.21	2.3	12.2	50	7.9	17.2	63	10.3
Miami, FL	75.0	59.2	88	31	6.5	2.08	6.66	0.0	—	67	9.6	8.5	60	9.4
Milwaukee, WI	26.0	11.3	62	-26	11.1	1.64	4.04	13.0	33.6	44	7.2	17.7	68	12.8
Minneapolis, MN	19.9	2.4	58	-34	8.8	.82	1.93	9.9	46.4	52	8.4	15.3	66	10.4
New Orleans, LA	61.8	43.0	83	14	10.1	4.97	13.63	0.0	0.1	48	6.7	16.9	66	9.4
New York, NY	38.0	25.6	72	-6	11.2	3.21	10.52	7.6	27.4	50	8.1	13.7	60	10.7
Philadelphia, PA	38.6	23.8	74	-7	11.0	3.18	8.86	6.4	23.4	50	7.4	16.1	59	10.3
Phoenix, AZ	65.2	39.4	88	17	3.8	.73	2.41	T	T	78	13.8	10.3	32	5.3
Pittsburgh, PA	34.1	19.2	68	-18	16.4	2.86	6.25	12.1	40.2	33	3.0	21.9	65	10.7
Portland, OR	44.3	33.5	62	-2	18.6	6.16	12.83	3.9	41.4	27	2.8	25.0	76	8.8
Sacramento, CA	52.6	37.9	70	23	10.2	4.03	9.14	T	T	45	6.3	18.7	71	7.6
Salt Lake City, UT	37.4	19.7	62	-22	9.9	1.35	3.14	13.2	32.3	46	5.5	18.9	68	7.7
San Antonio, TX	61.7	39.0	89	0	8.1	1.55	8.52	0.2	4.7	48	8.9	15.9	57	9.2
San Francisco, CA	55.5	41.5	72	24	10.8	4.65	10.43	0.0	1.5	56	8.6	14.7	66	7.1
San Diego, CA	65.2	48.4	88	29	6.9	2.11	6.26	T	T	72	12.4	11.2	56	5.8
Seattle, WA	43.9	34.3	64	0	19.0	6.04	12.92	6.0	57.2	24	2.6	24.5	74	9.9
Tampa, FL	70.0	49.5	84	22	6.3	2.17	8.02	0.0	0.2	64	9.5	11.6	59	8.8
Washington, D.C.	42.9	27.5	79	-5	10.3	2.76	7.11	5.1	21.3	48	7.6	16.2	54	10.0

March

City	Temperatures (Deg. F) Average Max	Min	Record Max	Min	Mean Days of Precip.	Amount of Precip. Total Avg.	Max	Snowfall Avg.	Max	% of Possible Sunshine	Mean Clear Days	Mean Cloudy Days	Relative Humidity (%)	Avg. Wind Speed (MPH)
Albuquerque, NM	60.7	31.7	85	8	4.5	.52	2.18	1.9	13.9	73	11.3	9.6	24	10.2
Anchorage, AK	31.7	15.7	51	-24	7.7	.69	2.76	9.1	31.0	53	7.4	17.9	56	6.7
Atlanta, GA	63.2	41.7	85	10	11.7	5.91	11.66	0.4	7.9	58	8.9	14.9	51	10.9
Bismarck, ND	36.4	15.6	81	-31	8.2	.70	3.19	8.4	31.1	59	5.4	17.2	63	11.0
Boise, ID	51.8	41.4	81	6	9.7	1.03	2.76	1.9	11.9	62	6.0	17.8	45	10.1
Boston, MA	45.0	31.8	81	6	11.9	4.13	11.0	7.8	31.2	57	7.8	15.3	57	13.8
Buffalo, NY	40.4	25.6	81	-7	16.2	2.97	5.59	11.8	29.2	45	3.6	19.9	67	13.4
Casper, WY	43.4	20.2	73	-21	9.6	.99	2.43	15.2	36.2	NA	5.8	16.5	48	14.0
Charleston, SC	68.0	45.3	90	15	10.4	4.38	11.11	0.1	2.0	67	9.1	13.7	51	10.1
Charlotte, NC	61.6	39.1	90	4	11.4	4.83	8.76	1.5	19.3	63	9.1	13.8	51	8.9
Chicago, IL	44.3	27.6	87	-8	12.7	2.59	5.91	7.6	24.7	50	4.7	17.6	61	11.8
Cincinnati, OH	51.5	32.0	83	-11	13.2	3.95	12.48	4.5	13.0	28	5.0	19.3	61	11.1
Dallas, TX	67.2	44.9	96	15	7.3	2.42	6.39	0.2	2.5	59	9.6	13.7	51	13.0
Denver, CO	51.2	24.7	84	-11	8.7	1.21	4.56	13.1	30.5	70	7.8	12.9	41	9.8
Detroit, MI	43.4	26.5	77	-4	13.2	2.54	4.48	6.8	16.1	51	5.3	18.3	62	11.7
Helena, MT	42.5	20.6	77	-30	8.6	.69	1.62	7.4	21.6	60	3.6	19.1	46	8.5
Honolulu, HI	81.6	67.3	88	55	9.0	3.48	20.79	0.0	—	68	7.3	9.6	58	11.5
Houston, TX	72.1	49.8	90	22	9.8	2.68	8.52	0.0	—	47	6.5	17.9	60	9.5
Kansas City, MO	51.3	31.7	85	12	10.7	2.41	9.08	3.9	11.4	58	6.3	16.9	59	12.5
Las Vegas, NV	68.3	42.3	91	23	2.9	.41	1.83	0.0	0.1	83	13.8	8.2	22	10.1
Lexington, KY	53.7	34.1	83	-2	13.1	4.83	10.38	2.8	17.7	NA	5.5	18.0	58	11.5
Los Angeles, CA	65.1	49.7	93	34	5.8	1.76	6.37	0.0	—	73	11.6	10.7	61	8.1
Memphis, TN	61.4	41.9	85	12	11.0	5.44	12.08	1.0	17.3	56	8.1	16.3	56	11.1
Miami, FL	79.3	64.1	92	32	5.7	1.89	7.22	0.0	—	77	8.7	8.1	56	10.4
Milwaukee, WI	39.2	24.9	81	-10	11.9	2.58	6.93	9.2	26.7	50	6.0	17.2	65	13.0
Minneapolis, MN	37.5	20.8	83	-32	9.0	1.47	4.75	10.6	40.0	55	6.9	16.4	62	11.3
New Orleans, LA	71.2	51.6	89	25	9.0	4.73	19.09	T	—	57	7.7	15.1	60	9.9
New York, NY	48.6	34.1	86	3	11.5	4.22	10.41	5.1	30.5	56	8.8	12.1	55	11.0
Philadelphia, PA	50.5	33.1	87	7	11.1	3.86	7.01	4.0	13.4	55	7.5	15.3	53	11.4
Phoenix, AZ	74.5	46.7	95	25	3.5	.81	4.16	T	T	83	14.6	8.6	24	6.7
Pittsburgh, PA	47.6	29.4	80	-1	16.1	3.58	6.10	8.7	21.3	44	4.0	20.1	58	10.9
Portland, OR	54.5	37.4	80	19	17.2	3.61	7.52	0.5	12.9	46	3.1	23.6	60	8.3
Sacramento, CA	64.1	42.4	86	26	8.4	2.06	7.12	T	T	72	10.4	12.5	53	8.8
Salt Lake City, UT	51.5	29.9	78	2	9.9	1.72	3.97	10.4	41.9	63	7.0	15.7	47	9.3
San Antonio, TX	73.7	49.8	100	19	7.0	1.33	4.19	T	T	57	8.5	15.2	47	10.6
San Francisco, CA	60.6	44.9	85	30	9.7	2.64	9.01	T	T	69	9.7	12.6	63	10.4
San Diego, CA	65.9	52.1	93	39	7.2	1.60	6.57	0.0	—	71	11.2	10.5	59	7.3
Seattle, WA	51.1	37.2	72	11	17.1	3.59	8.40	1.4	18.2	49	3.1	22.1	62	9.9
Tampa, FL	76.2	56.1	91	29	6.8	3.46	12.64	T	T	71	10.2	10.8	55	9.6
Washington, D.C.	55.0	36.6	89	11	11.0	3.46	7.43	2.3	17.1	55	7.3	15.1	50	10.9

July

July	Temperatures (Deg. F.) Average Max	Average Min	Record Max	Record Min	Mean Days of Precip.	Amount of Precip. Total Avg.	Total Max	Snowfall Avg.	Snowfall Max	% of Possible Sunshine	Mean Clear Days	Mean Cloudy Days	Relative Humidity (%)	Avg. Wind Speed (MPH)
Albuquerque, NM	92.8	64.7	105	54	8.8	1.30	3.33	0.0	-	76	12.1	4.7	27	9.1
Anchorage, AK	65.1	51.1	81	38	11.6	1.97	4.44	0.0	-	43	3.4	22.0	61	7.1
Atlanta, GA	87.9	69.2	105	53	12.0	4.73	11.26	0.0	-	63	5.6	12.0	61	7.5
Bismarck, ND	84.4	56.4	109	35	8.9	2.05	5.24	0.0	-	75	11.5	6.7	42	9.2
Boise, ID	90.6	58.5	111	40	2.3	.26	1.62	T		87	20.9	3.1	22	8.4
Boston, MA	81.8	65.1	102	54	9.1	2.68	8.12	0.0	-	67	7.0	11.6	55	10.9
Buffalo, NY	80.2	61.2	94	43	9.8	2.96	6.43	0.0	-	68	7.0	10.9	54	10.4
Casper, WY	87.1	54.7	104	30	7.8	1.06	3.05	0.0	-	NA	14.1	5.9	26	10.1
Charleston, SC	89.4	71.6	101	58	13.7	7.33	18.46	0.0	-	68	4.9	13.8	63	7.9
Charlotte, NC	88.3	68.7	103	53	11.3	3.92	9.12	0.0	-	68	6.5	13.0	58	6.6
Chicago, IL	83.3	62.7	102	40	9.8	3.63	8.33	0.0	-	69	8.8	9.2	57	8.1
Cincinnati, OH	85.8	64.9	101	47	10.0	4.28	8.36	0.0	-	70	7.4	11.9	57	7.1
Dallas, TX	97.8	74.7	110	59	4.9	2.59	11.13	0.0	-	81	15.3	6.4	44	9.5
Denver, CO	88.0	58.7	104	47	9.1	1.93	6.41	0.0	-	72	9.2	6.0	34	8.4
Detroit, MI	83.1	60.7	102	41	9.1	3.10	6.02	0.0	-	70	9.3	9.5	53	8.3
Helena, MT	83.6	52.2	102	36	7.4	1.04	3.89	T		78	14.8	5.3	29	7.8
Honolulu, HI	87.1	73.1	91	67	7.5	.54	2.01	0.0	-	73	7.5	5.4	52	13.5
Houston, TX	93.6	72.5	104	62	9.7	3.33	8.10	0.0	-	66	7.7	8.1	58	6.8
Kansas City, MO	88.5	68.5	107	52	7.1	4.35	8.71	0.0	-	74	14.5	7.3	53	8.9
Las Vegas, NV	104.5	75.9	116	61	2.6	.45	2.48	0.0	-	87	19.6	3.4	15	10.1
Lexington, KY	85.9	65.9	103	47	11.2	4.95	10.64	0.0	-	NA	7.8	10.5	59	7.4
Los Angeles, CA	75.3	62.6	97	49	0.5	.01	.15	0.0	-	82	12.7	5.4	67	7.7
Memphis, TN	91.5	72.6	108	52	8.7	4.03	8.84	0.0	-	74	10.4	8.8	57	7.5
Miami, FL	88.7	76.2	98	69	15.8	5.98	13.51	0.0	-	79	2.5	11.3	63	7.8
Milwaukee, WI	79.8	61.1	101	40	9.5	3.54	7.66	0.0	-	70	10.3	9.5	61	9.6
Minneapolis, MN	83.4	62.7	104	43	9.7	3.51	7.10	0.0	-	71	10.4	9.1	53	9.3
New Orleans, LA	90.7	73.5	101	60	14.8	6.73	13.07	0.0	-	83	5.1	11.5	66	6.1
New York, NY	85.3	68.2	106	52	10.4	3.77	11.89	0.0	-	65	8.5	9.5	55	7.6
Philadelphia, PA	86.1	66.8	104	51	9.1	3.88	8.33	0.0	-	62	7.1	12.0	54	8.0
Phoenix, AZ	105.0	79.5	118	61	4.3	.74	5.15	0.0	-	85	16.3	4.4	20	7.1
Pittsburgh, PA	82.7	61.3	99	42	10.8	3.83	7.43	0.0	-	59	5.3	12.5	53	7.3
Portland, OR	79.5	55.8	107	43	3.8	.46	2.68	0.0	-	70	13.0	9.3	45	7.6
Sacramento, CA	93.3	57.9	114	48	0.3	.05	.79	0.0	-	97	27.1	0.9	28	9.0
Salt Lake City, UT	93.2	61.8	107	40	4.5	.72	2.57	0.0	-	83	16.9	4.3	29	9.5
San Antonio, TX	94.9	74.3	106	62	4.3	1.92	8.19	0.0	-	74	9.2	6.6	45	9.3
San Francisco, CA	71.0	53.3	104	43	0.3	.03	.35	0.0	-	66	20.6	2.9	59	13.5
San Diego, CA	75.6	64.9	93	55	0.3	.01	.19	0.0	-	69	13.3	4.8	65	7.3
Seattle, WA	75.2	54.3	98	43	5.1	.74	2.39	T		65	10.4	10.4	49	8.3
Tampa, FL	90.0	74.2	97	63	15.7	7.35	20.59	0.0	-	62	2.2	12.8	63	7.4
Washington, D.C.	87.9	69.9	103	55	9.7	3.88	11.06	0.0	-	64	7.8	11.4	53	8.2

September

September	Temperatures (Deg. F.) Average Max	Average Min	Record Max	Record Min	Mean Days of Precip.	Amount of Precip. Total Avg.	Total Max	Snowfall Avg.	Snowfall Max	% of Possible Sunshine	Mean Clear Days	Mean Cloudy Days	Relative Humidity (%)	Avg. Wind Speed (MPH)
Albuquerque, NM	83.0	54.9	100	37	5.7	.85	1.99	T		79	16.9	5.3	31	8.6
Anchorage, AK	55.2	41.1	73	20	13.8	2.45	5.43	0.2	4.6	41	4.0	20.7	63	6.2
Atlanta, GA	82.3	63.6	98	36	7.5	3.17	7.52	0.0	-	64	9.7	10.5	60	8.0
Bismarck, ND	71.4	43.2	105	11	7.0	1.38	6.93	0.3	5.0	66	10.4	10.8	44	10.0
Boise, ID	77.6	48.7	102	23	3.7	.58	2.54	0.0	-	81	16.9	6.0	30	8.3
Boston, MA	72.3	56.9	100	38	8.7	3.41	8.31	0.0	-	64	10.5	11.3	60	11.1
Buffalo, NY	71.4	52.7	98	32	10.8	3.37	8.99	T		59	6.5	13.6	60	10.4
Casper, WY	74.2	42.5	96	16	6.3	.76	3.40	1.2	11.5	NA	13.4	8.0	30	11.0
Charleston, SC	84.6	66.7	99	42	9.4	4.94	17.31	0.0	-	63	6.6	12.8	62	7.9
Charlotte, NC	81.7	62.3	104	39	7.2	3.59	10.89	0.0	-	67	9.2	11.4	57	6.7
Chicago, IL	75.5	53.9	99	28	9.6	3.35	11.44	T		58	9.0	11.3	58	8.8
Cincinnati, OH	78.7	56.3	98	33	8.0	2.91	8.61	0.0	-	64	9.5	11.6	58	7.4
Dallas, TX	89.7	67.5	105	43	6.8	3.31	9.52	0.0	-	74	13.1	8.4	54	9.4
Denver, CO	77.5	47.7	97	20	6.1	1.23	4.67	1.6	21.3	75	13.5	7.5	34	8.0
Detroit, MI	74.4	52.2	98	29	9.7	2.25	5.83	0.0	-	61	8.3	12.2	57	8.7
Helena, MT	70.3	40.8	99	18	6.6	.83	3.37	1.6	13.7	67	10.2	11.1	35	7.4
Honolulu, HI	88.2	72.9	93	66	7.0	.62	2.74	0.0	-	75	8.2	5.9	52	11.6
Houston, TX	88.7	68.1	100	48	9.5	4.93	11.35	0.0	-	62	8.9	10.6	62	6.8
Kansas City, MO	78.6	58.1	98	33	7.6	4.14	11.34	0.0	-	65	11.8	10.0	59	9.3
Las Vegas, NV	94.7	65.6	113	46	1.8	.32	1.58	0.0	-	91	22.7	2.3	17	8.8
Lexington, KY	79.3	58.1	103	35	8.1	3.28	9.69	0.0	-	NA	10.6	10.9	59	7.7
Los Angeles, CA	76.4	62.5	110	47	1.0	.15	4.39	0.0	-	79	12.9	6.6	65	7.2
Memphis, TN	84.3	64.1	103	36	7.0	3.62	7.61	0.0	-	70	12.4	10.0	56	7.5
Miami, FL	87.8	75.7	95	68	17.6	8.07	24.40	0.0	-	72	2.3	12.9	67	8.2
Milwaukee, WI	71.2	52.5	98	28	9.0	2.88	9.87	T		59	9.7	11.1	63	10.5
Minneapolis, MN	71.0	50.2	98	26	9.4	2.50	7.53	0.0	1.7	61	10.1	11.3	59	9.9
New Orleans, LA	86.8	70.1	101	42	9.8	5.87	16.74	0.0	-	63	9.9	9.8	65	7.3
New York, NY	76.4	60.1	102	39	8.3	3.66	16.85	0.0	-	63	10.6	9.4	57	8.1
Philadelphia, PA	77.8	58.6	100	35	7.9	3.42	8.78	0.0	-	60	9.8	11.3	55	8.2
Phoenix, AZ	98.2	70.9	118	47	3.0	.64	4.23	0.0	-	89	22.0	3.0	23	6.3
Pittsburgh, PA	74.8	53.3	97	31	9.3	2.80	5.42	0.0	-	58	7.6	12.1	56	7.4
Portland, OR	74.2	51.1	101	34	7.8	1.61	3.98	T		61	10.1	11.9	49	6.5
Sacramento, CA	87.6	55.8	108	43	1.2	.27	1.81	0.0	-	93	23.7	2.2	31	7.7
Salt Lake City, UT	80.0	50.0	100	27	5.3	.89	7.04	0.1	4.0	83	16.9	5.1	29	9.1
San Antonio, TX	89.3	69.4	102	41	7.1	3.75	15.78	0.0	-	67	9.6	9.4	52	8.6
San Francisco, CA	73.4	54.3	103	38	0.9	.19	2.30	0.0	-	72	18.4	3.6	57	11.0
San Diego, CA	76.8	65.1	111	51	1.1	.19	1.90	0.0	-	68	15.0	5.8	65	6.9
Seattle, WA	68.7	51.2	94	35	9.4	2.02	5.95	T		59	7.6	13.6	58	8.1
Tampa, FL	88.9	72.8	96	57	13.1	6.23	13.98	0.0	-	61	4.5	12.0	62	8.0
Washington, D.C.	80.1	62.0	101	39	7.6	3.22	12.36	0.0	-	62	10.1	11.7	55	8.2

Reference Section 573

Geologic Time Line

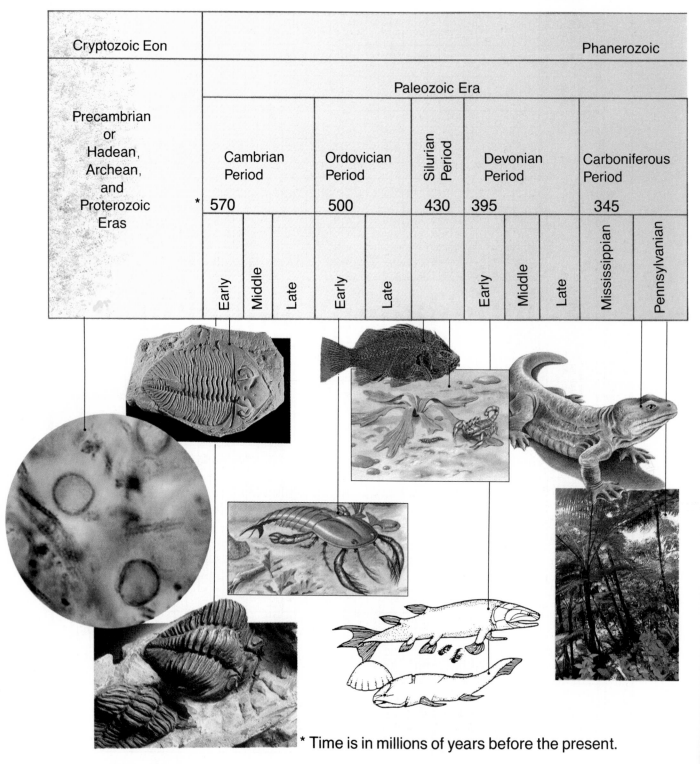

Cryptozoic Eon	Phanerozoic									
	Paleozoic Era									
Precambrian or Hadean, Archean, and Proterozoic Eras	Cambrian Period *570			Ordovician Period 500		Silurian Period 430	Devonian Period 395			Carboniferous Period 345	
	Early	Middle	Late	Early	Late		Early	Middle	Late	Mississippian	Pennsylvanian

* Time is in millions of years before the present.

Eon															
		Mesozoic Era								Cenozoic Era					
Permian Period		Triassic Period			Jurassic Period			Cretaceous Period		Tertiary Period					Quaternary Period
280		225			190			136		65					2.5
Early	Late	Early	Middle	Late	Early	Middle	Late	Early	Late	Paleocene	Eocene	Oligocene	Miocene	Pliocene	Pleistocene	Holocene

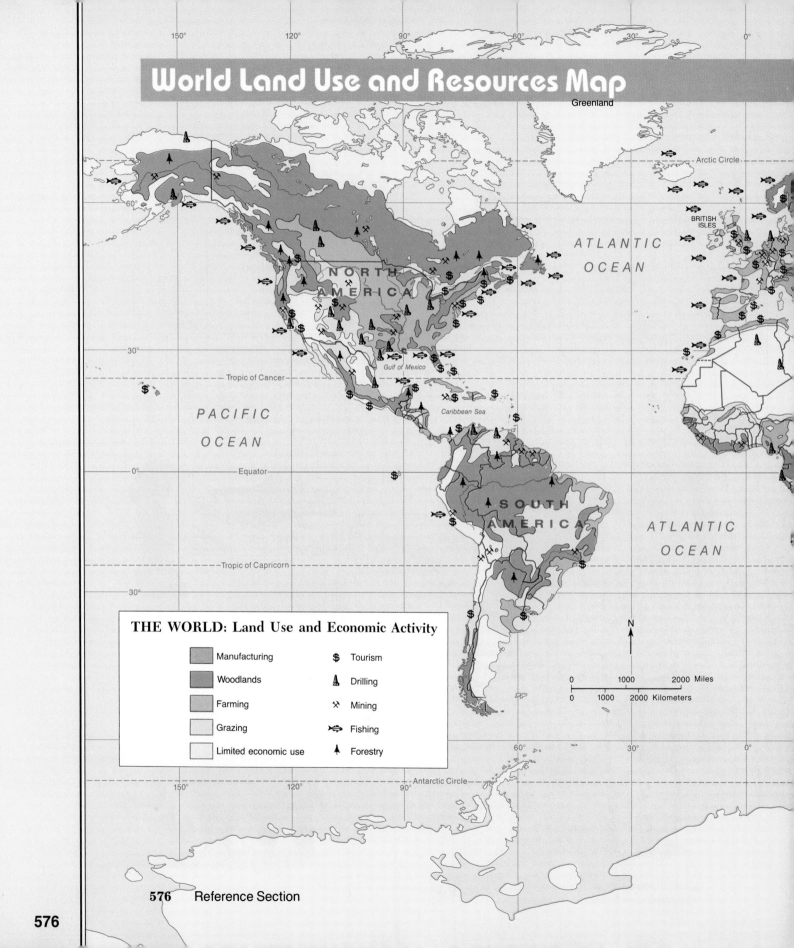

World Land Use and Resources Map

Greenland

ATLANTIC OCEAN

Arctic Circle

60°

BRITISH ISLES

NORTH AMERICA

PACIFIC OCEAN

30°

Tropic of Cancer

Gulf of Mexico

Caribbean Sea

Equator

0°

SOUTH AMERICA

ATLANTIC OCEAN

Tropic of Capricorn

30°

THE WORLD: Land Use and Economic Activity

▨ Manufacturing	$	Tourism
▨ Woodlands	⚑	Drilling
▨ Farming	⚒	Mining
▨ Grazing	🐟	Fishing
▢ Limited economic use	♠	Forestry

N

0	1000	2000 Miles
0	1000	2000 Kilometers

60°   30°   0°

Antarctic Circle

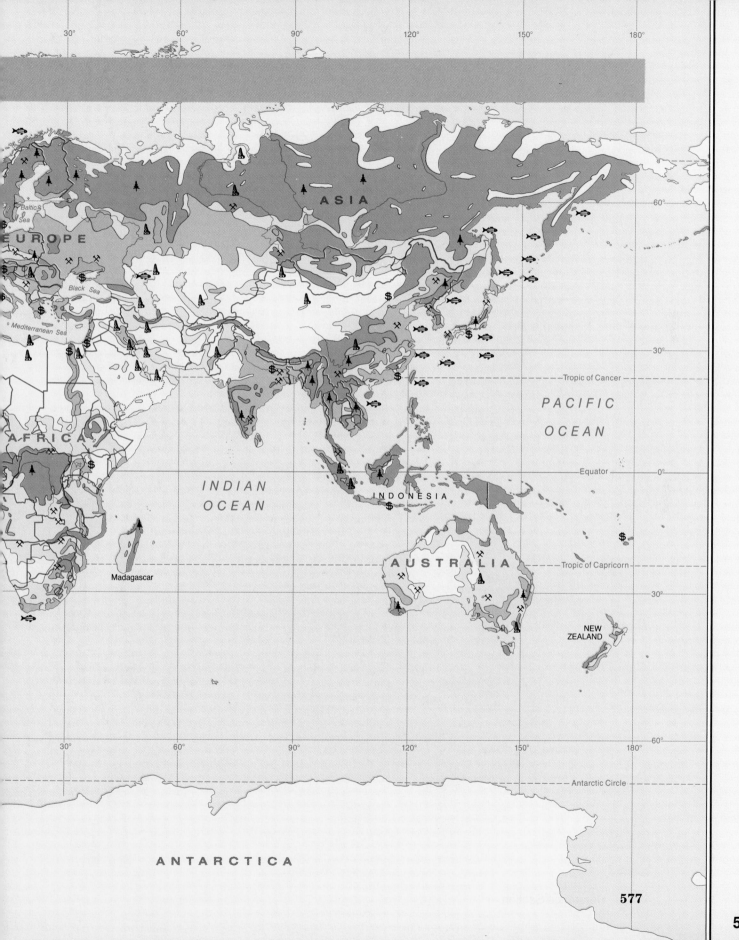

EUROPE

Baltic Sea

Black Sea

Mediterranean Sea

AFRICA

ASIA

Madagascar

INDIAN OCEAN

INDONESIA

AUSTRALIA

NEW ZEALAND

PACIFIC OCEAN

Tropic of Cancer

Equator

Tropic of Capricorn

Antarctic Circle

ANTARCTICA

30° 60° 90° 120° 150° 180°

60°

30°

0°

30°

60°

30° 60° 90° 120° 150° 180°

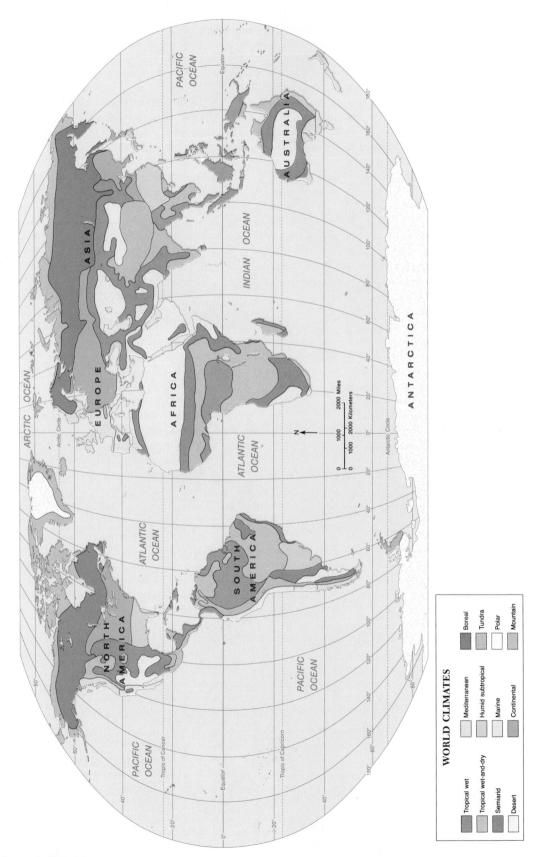

Map of World Climates

WORLD CLIMATES

- Tropical wet
- Tropical wet-and-dry
- Semiarid
- Desert
- Mediterranean
- Humid subtropical
- Marine
- Continental
- Boreal
- Tundra
- Polar
- Mountain

Map of Ocean Currents

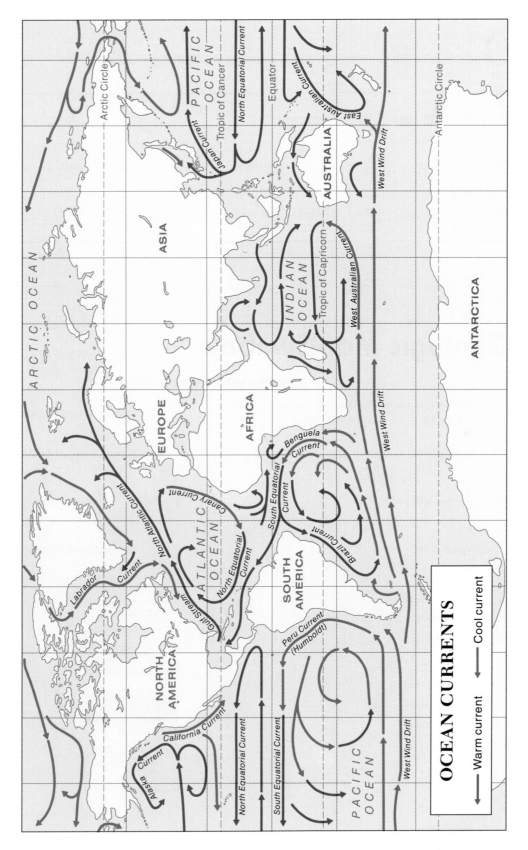

ARCTIC OCEAN

ASIA

EUROPE

AFRICA

NORTH AMERICA

SOUTH AMERICA

AUSTRALIA

ANTARCTICA

PACIFIC OCEAN

ATLANTIC OCEAN

INDIAN OCEAN

Arctic Circle

Tropic of Cancer

Equator

Tropic of Capricorn

Antarctic Circle

North Equatorial Current

Japan Current

East Australian Current

West Wind Drift

West Australian Current

North Atlantic Current

Canary Current

Labrador Current

Gulf Stream

North Equatorial Current

South Equatorial Current

Benguela Current

Brazil Current

West Wind Drift

Peru Current (Humboldt)

California Current

Alaska Current

North Equatorial Current

South Equatorial Current

West Wind Drift

OCEAN CURRENTS

→ Warm current   → Cool current

Topographic Maps and Map Symbols

Symbol	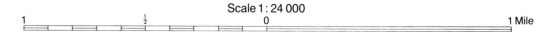								
Color	Black	Black	Red or black	Black	Blue	Blue	Blue	Brown	Blue or green
Meaning	Buildings	School	Road	Railroad	Stream	Intermittent Stream	Lake or pond	Depression	Marsh or swamp

Scale 1 : 24 000

1   ½   0   1 Mile

Geologic Map Symbols

Symbols indicating structure

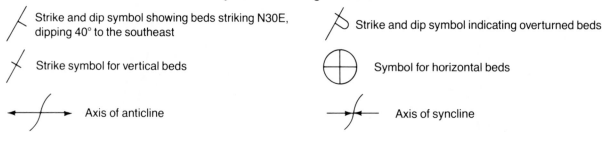

Strike and dip symbol showing beds striking N30E, dipping 40° to the southeast

Strike and dip symbol indicating overturned beds

Strike symbol for vertical beds

Symbol for horizontal beds

Axis of anticline

Axis of syncline

Geological ages

Q   Quaternary

T   Tertiary

K   Cretaceous

J   Jurassic

TR   Triassic

P   Permian

P   Pennsylvanian

M   Mississippian

D   Devonian

S   Silurian

O   Ordovician

Є   Cambrian

PЄ   Precambrian

Symbols indicating rock types

Conglomerate

Basic lava flows

Bedded sandstone

Other lava flows

Shale

Granitic rock

Massive limestone

Folded schist

UNITED STATES
DEPARTMENT OF THE INTERIOR
GEOLOGICAL SURVEY

BOOTHBAY QUADRANGLE
MAINE
15 MINUTE SERIES (TOPOGRAPHIC)

SCALE 1:62500

CONTOUR INTERVAL 20 FEET

Use with Investigation 2 on page 44.

Reference Section   581

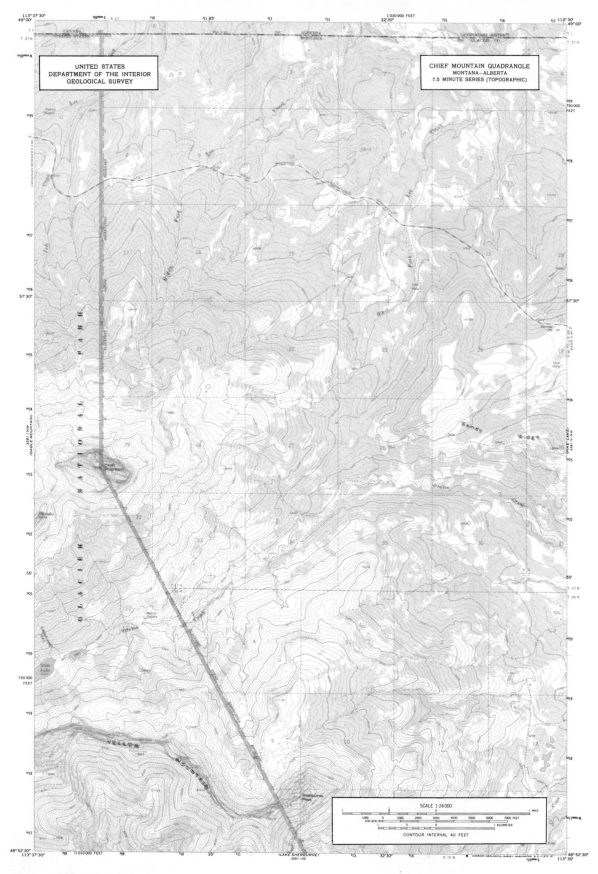

582 Reference Section

Use with Investigation 13 on page 294.

Weather Map Symbols

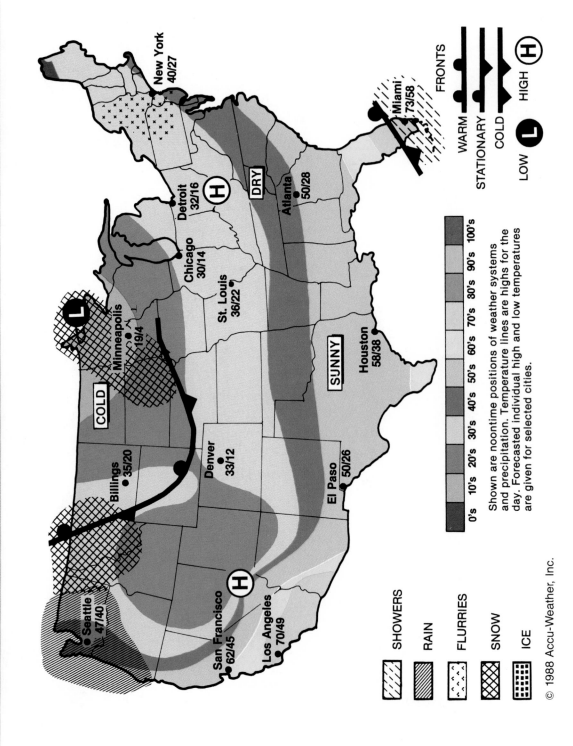

New York
40/27

Miami
73/58

Detroit
32/16

Atlanta
50/28

DRY

Chicago
30/14

St. Louis
36/22

L

Minneapolis
19/4

Houston
58/38

COLD

SUNNY

Billings
35/20

Denver
33/12

El Paso
50/26

H

Seattle
47/40

San Francisco
62/45

Los Angeles
70/49

H

FRONTS

WARM

STATIONARY

COLD

HIGH (H)

LOW (L)

0's 10's 20's 30's 40's 50's 60's 70's 80's 90's 100's

Shown are noontime positions of weather systems
and precipitation. Temperature lines are highs for the
day. Forecasted individual high and low temperatures
are given for selected cities.

SHOWERS

RAIN

FLURRIES

SNOW

ICE

© 1988 Accu-Weather, Inc.

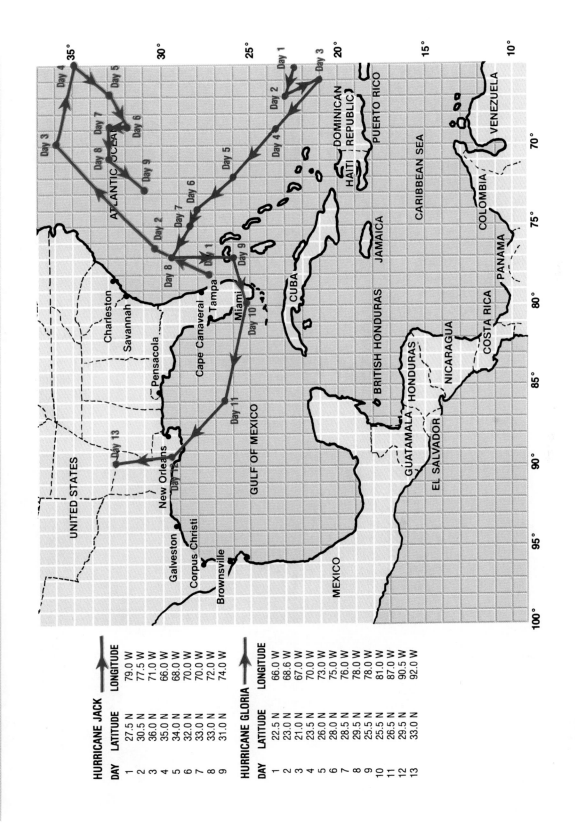

Hurricane Tracking Map

HURRICANE JACK →

DAY	LATITUDE	LONGITUDE
1	27.5 N	79.0 W
2	30.5 N	77.5 W
3	36.0 N	71.0 W
4	35.0 N	66.0 W
5	34.0 N	68.0 W
6	32.0 N	70.0 W
7	33.0 N	70.0 W
8	33.0 N	72.0 W
9	31.0 N	74.0 W

HURRICANE GLORIA →

DAY	LATITUDE	LONGITUDE
1	22.5 N	66.0 W
2	23.0 N	68.6 W
3	21.0 N	67.0 W
4	23.5 N	70.0 W
5	26.0 N	73.0 W
6	28.0 N	75.0 W
7	28.5 N	76.0 W
8	29.5 N	78.0 W
9	25.5 N	78.0 W
10	25.5 N	81.0 W
11	26.5 N	87.0 W
12	29.5 N	90.5 W
13	33.0 N	92.0 W

Weather Maps with Satellite Photos

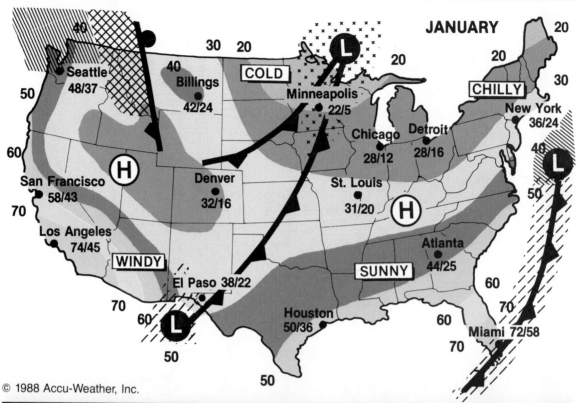

JANUARY

COLD

CHILLY

WINDY

SUNNY

Seattle 48/37

Billings 42/24

Minneapolis 22/5

Chicago 28/12

Detroit 28/16

New York 36/24

San Francisco 58/43

Denver 32/16

St. Louis 31/20

Los Angeles 74/45

El Paso 38/22

Atlanta 44/25

Houston 50/36

Miami 72/58

© 1988 Accu-Weather, Inc.

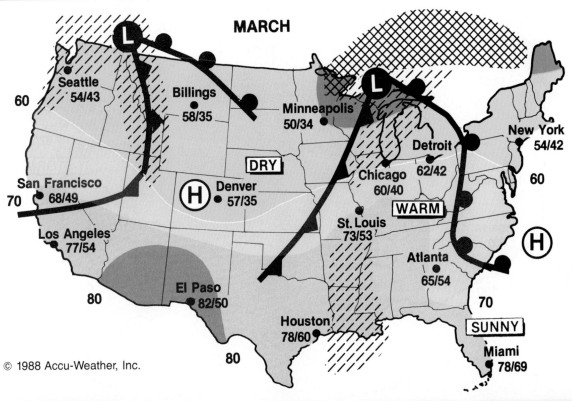

MARCH

Seattle 54/43
60
Billings 58/35
Minneapolis 50/34
New York 54/42
60
DRY
Detroit 62/42
San Francisco 68/49
70
Chicago 60/40
H Denver 57/35
WARM
H
Los Angeles 77/54
St. Louis 73/53
Atlanta 65/54
El Paso 82/50
80
70
Houston 78/60
SUNNY
80
Miami 78/69

© 1988 Accu-Weather, Inc.

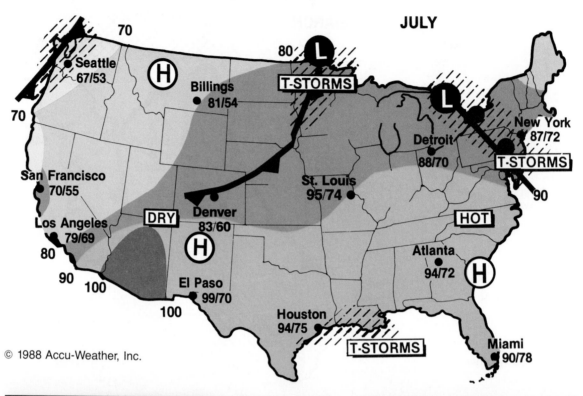

JULY

Seattle 67/53
Billings 81/54
San Francisco 70/55
Los Angeles 79/69
Denver 83/60
El Paso 99/70
Houston 94/75
St. Louis 95/74
Detroit 88/70
New York 87/72
Atlanta 94/72
Miami 90/78

DRY
HOT
T-STORMS

© 1988 Accu-Weather, Inc.

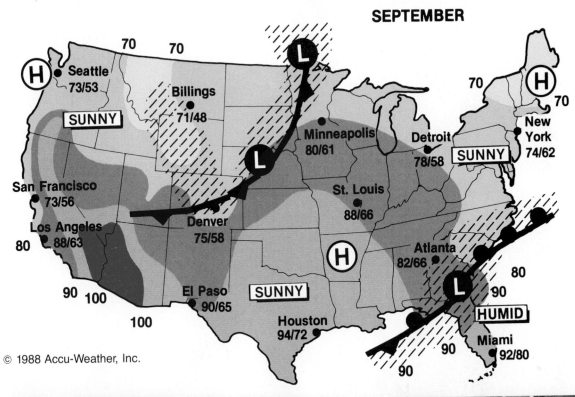

SEPTEMBER

Seattle 73/53

Billings 71/48

SUNNY

San Francisco 73/56

Los Angeles 88/63

El Paso 90/65

Denver 75/58

Houston 94/72

Minneapolis 80/61

St. Louis 88/66

Atlanta 82/66

Detroit 78/58

SUNNY

New York 74/62

SUNNY

HUMID

Miami 92/80

70   70

70   70

80

90   100

100

80

90

90   90

© 1988 Accu-Weather, Inc.

Star Chart

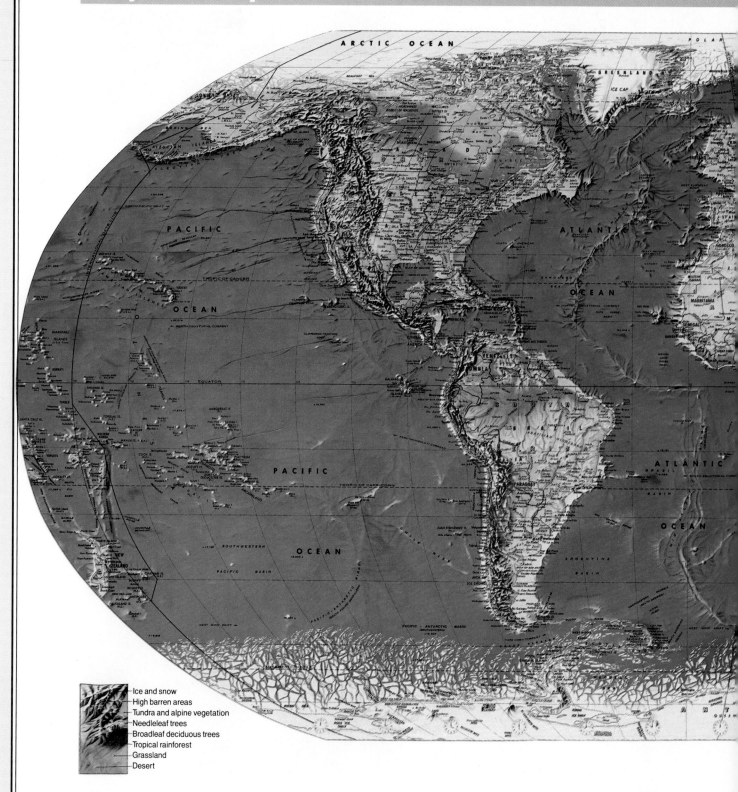

Ice and snow
High barren areas
Tundra and alpine vegetation
Needleleaf trees
Broadleaf deciduous trees
Tropical rainforest
Grassland
Desert

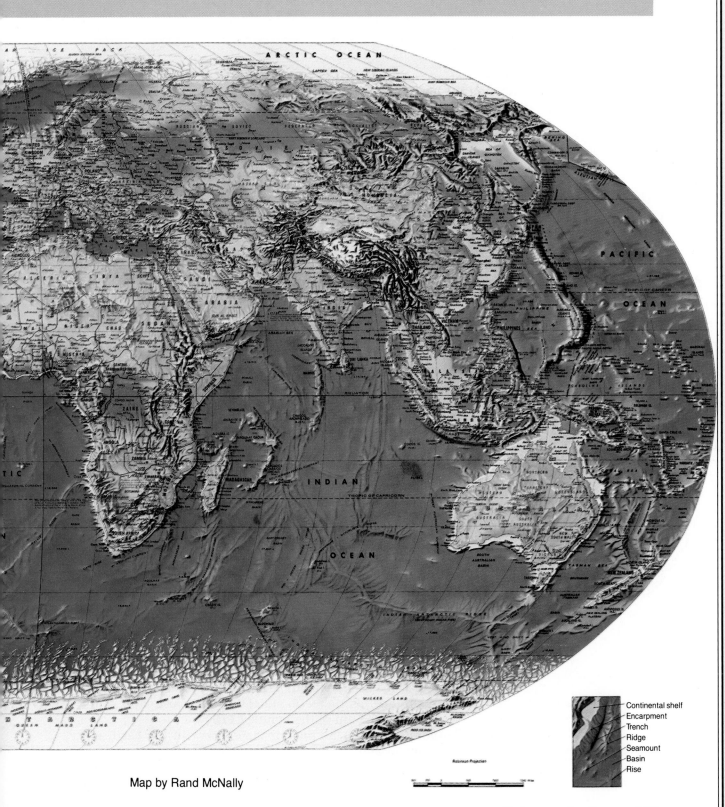

	Continental shelf
	Encarpment
	Trench
	Ridge
	Seamount
	Basin
	Rise

Map by Rand McNally

GLOSSARY

PRONUNCIATION KEY

Symbol	As In	Phonetic Respelling		Symbol	As In	Phonetic Respelling	
a	bat	a	(bat)	ô	dog	aw	(dawg)
ā	face	ay	(fays)	oi	foil	oy	(foyl)
ā	careful	ai	(CAIR fuhl)	ou	mountain	ow	(MOWN tuhn)
ä	argue	ah	(AHR gyoo)	s	sit	s	(siht)
ch	chapel	ch	(CHAP uhl)	sh	sheep	sh	(sheep)
e	test	eh	(tehst)	u	love	uh	(luhv)
ē	eat	ee	(eet)	u̇	pull	u	(pull)
	ski	ee	(skee)	ü	mule	oo	(myool)
ėr	fern	ur	(furn)	zh	treasure	zh	(TREH zhuhr)
i	bit	ih	(biht)	ə	medal	uh	(MEHD uhl)
ī	ripe	y	(ryp)		effect	uh	(uh FEHKT)
	idea	eye	(eye DEE uh)		serious	uh	(SIHR ee uhs)
k	card	k	(kahrd)		onion	uh	(UHN yuhn)
o	lock	ah	(lahk)		talent	uh	(TAL uhnt)
ō	over	oh	(OH vuhr)				

alluvial fan

abrasion   weathering by physical contact, causing the rounding of rocks   **(234)**

absolute dating   process of determining the age of a sample by using the radioactive isotopes in the sample   **(329)**

aerosol   solid in the air that acts as a nucleus for condensation   **(195)**

air mass   large body of air with uniform temperature and humidity   **(207)**

alloy   mixture of two or more metals, or a mixture of metals and nonmetals   **(100)**

alluvial fan   fan-shaped deposit at the base of a mountain   **(263)**

anticline   structure formed when rock layers are folded upward   **(123)**

aquifer   permeable rock layer that is between two nonpermeable layers and transports water   **(267)**

arid   dry; usually regarding climate   **(275)**

asteroid   small planetary body that circles the sun in a belt between the orbits of Mars and Jupiter   **(468)**

asthenosphere (as THEHN uh sfihr)   100-km-thick layer of the earth, near the top of the mantle **(119)**

astronomical unit   average distance between the earth and the sun, or about 150 000 000 km   **(462)**

astronomy   study of the characteristics of the planets, the sun, and the stars   **(5)**

atmosphere layer of gases that surrounds the earth **(165)**

atmospheric pressure weight of the atmosphere **(175)**

atom building block of matter **(51)**

atomic number number of protons an atom contains **(56)**

axis imaginary line that passes through the center of a planet from pole to pole **(455)**

barchan (BAHR kahn) **dune** crescent-shaped dune with the rounded side facing into the wind **(280)**

barometer (buh RAHM uh tuhr) instrument used to measure atmospheric pressure **(175)**

base level low point of a stream or river **(259)**

batholith body of magma that cools beneath the surface, forming an intrusion **(143)**

bay indentation of the sea, partially enclosed by land **(402)**

bench mark bronze marker indicating location and elevation of a specific surveyed place in the United States **(27)**

benthos organisms that live in or on the bottom of the ocean **(442)**

big-bang theory theory stating that a tremendous explosion created the universe; the most accepted theory for the beginning of the universe **(496)**

biomass once-living matter **(379)**

black hole collapsed star that emits no energy **(488)**

brine extremely salty water that collects in areas totally or partially isolated from the open ocean **(435)**

bench mark

caldera wide basin in the center of a volcanic cone **(146)**

carbonation chemical weathering process in which carbonic acid reacts with the minerals in rocks such as limestone, dissolving them **(236)**

cartography (kahr TAHG ruh fee) science of mapmaking **(26)**

cellular respiration breakdown of carbohydrates in the presence of oxygen **(346)**

cementation process in which water that carries natural cements with it soaks through the earth, coating sediments and binding them together **(84)**

chemical sediment sediment that comes from minerals dissolved in water **(85)**

cinder-cone volcano steep-sided volcano that consists of alternating layers of ash, tuff, and lava **(145)**

cirrus cloud highest cloud; made of ice crystals **(197)**

clastic sediments pieces of other rocks **(85)**

cleavage tendency of minerals to break along flat surfaces **(66)**

climate all the weather that characterizes a region **(211)**

cinder-cone volcano

comet

cloud   collection of water droplets in the atmosphere **(195)**

comet   object consisting of silicate rock and metal particles embedded in ices **(475)**

composite volcano   volcano that consists of layers produced by alternating eruptions of different lavas **(145)**

compound   combination of two or more different kinds of elements **(54)**

compression   pushing together of the earth's crust **(126)**

conclusion   statement about an initial hypothesis, based on all the information that was gathered **(8)**

condensation   change from a vapor to a liquid **(194)**

conservation   wise use of materials **(372)**

constellation   group of stars that forms an image **(483)**

continental drift   slow movement of large portions of the earth's crust **(129)**

continental shelf   extension of the continents, forming a gently sloping ocean floor **(394)**

continental slope   steep ocean floor farther offshore than the continental shelf **(394)**

control   in an experiment, the part that does not change **(8)**

convection   transfer of heat by the circulation or movement of solids, liquids, or gases **(120)**

convergent boundary   boundary where two plates of land press together **(140)**

core   innermost layer of the earth **(119)**

Coriolis (cawr ee OH lihs) **effect**   tendency of objects to move to the right in the Northern Hemisphere **(181)**

covalent bond   bond formed when atoms share electrons **(55)**

creep   slow, downhill movement of rock materials saturated with water **(237)**

crossbedding   design in rocks that shows directional changes in wind patterns **(325)**

crust   thin, rocky layer on the surface of the earth **(117)**

crystal   three-dimensional structure with each face, or surface, having a definite shape and orientation **(69)**

cumulus cloud

cumulus (KYOO myoo luhs) **cloud**   cloud that forms as hot, humid air rises **(197)**

current   stream of water that moves like a river through the ocean **(419)**

cyclone   any weather system that circulates around a low-pressure area **(209)**

D

data   information gathered by observation or investigation **(7)**

deflation   removal of all the soil of an area by wind erosion **(279)**

delta   area of sand and soil deposits, formed by a stream that slows down as it enters a large body of water **(262)**

density   measure of the mass contained in a certain volume of matter **(18)**

dew point   temperature at which dew forms **(194)**

differential weathering   difference in weathering of rocks that are in the same formation and are exposed to the same environment **(240)**

dip   angle at which a rock layer is tilted from the horizontal **(123)**

divergent boundary   area of the earth where the crust is spreading apart **(139)**

drainage basin   area drained by a river and its tributaries **(255)**

dip

earthquake   violent shaking of the earth's interior **(139)**

electromagnetic wave   energy wave that travels at the speed of light **(170)**

electron   subatomic particle that has a negative electrical charge **(52)**

element   substance made of only one kind of atom **(51)**

elliptical   describes a flattened circle **(462)**

epicenter   area on the earth's surface directly above the focus of an earthquake **(151)**

erosion   process that moves soil and reshapes landscapes **(254)**

evaporation   process of changing a liquid to a vapor **(191)**

evaporite (ih VAP uh ryte)   chemical sedimentary rock that forms when minerals dissolved in water are left as the water evaporates **(87)**

exfoliation (ehks foh lee AY shuhn)   weathering process in which the surface of rocks flakes off in sheets **(233)**

experiment   test designed to give a scientist information under carefully controlled conditions **(8)**

extrusive (ehk STROOS ihv) **rocks**   fine-grained igneous rocks formed from lava **(80)**

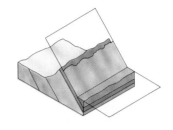

exfoliation

fault   rock fracture along which there is movement **(125)**

fermentation   partial breakdown of sugar to release energy **(344)**

fiord (fee AWRD)   long, deep, narrow, U-shaped sea channel that was originally formed by ice **(402)**

firn   snow pile in which half of the air spaces are filled with ice **(283)**

fission   breakdown of atomic nuclei, with the release of energy as a result **(379)**

flood plain   area over which a stream spreads in times of high-volume runoff **(260)**

flow   process in which water-saturated rock material moves downhill at a noticeable rate **(238)**

focus

focal length   distance between the center of a lens and the focal point   **(505)**

focus   source of an earthquake   **(151)**

fold   in a rock layer, a bend caused by force   **(122)**

foliated rock   metamorphic rock having obvious layers, formed from more than one mineral type   **(92)**

forge   furnace through which a stream of air is forced   **(102)**

fossil   remains or traces of a once-living organism, preserved in the earth   **(5)**

fracture   break in minerals that occurs along curved or irregular surfaces   **(66)**

front   surface between different air masses   **(207)**

frost action   weathering process that involves the freezing and cracking of rocks   **(233)**

fusion energy   energy produced when small atomic nuclei fuse to form new atoms   **(379)**

galaxy   large grouping of stars   **(494)**

gem   mineral that has beauty, durability, and value   **(69)**

geologic map   map that shows different layers and types of rock by the use of different colors   **(29)**

geology   study of the solid earth   **(4)**

geomorphic province   area with a distinctive pattern of landforms   **(301)**

geomorphology   study of landforms of the earth   **(299)**

geostationary   describes a satellite that is in a Clarke orbit and does not appear to move from its position relative to the earth   **(534)**

geostrophic current   ocean current that flows from a high water level to a lower level   **(421)**

geothermal power   steam power from the earth   **(377)**

glaciation   period in which great continental ice sheets advance and retreat repeatedly   **(354)**

glacier   sheet of ice that covers mountains or a large part of a continent   **(282)**

greenhouse effect   temperature rise caused by the addition of carbon dioxide to the atmosphere   **(174)**

half-life   time it takes for half of the atoms of a radioactive isotope to decay   **(328)**

headward erosion   process by which streams increase their length   **(258)**

horizon   development of soil in horizontal layers   **(246)**

hot spot   area of concentrated heat deep within the earth   **(147)**

humidity   amount of water vapor in the air   **(193)**

humus (HYOO muhs)   organic part of the soil   **(244)**

hurricane   large storm that develops over a tropical ocean   **(219)**

horizon

hydration   chemical weathering process in which water is combined with certain minerals to form new compounds   **(235)**

hydrocarbon   compound containing only hydrogen and carbon   **(107)**

hydroelectric power   power produced from the energy of falling water   **(375)**

hydrologic map   map that shows where surface water and underground water are located, as well as special landforms produced by water   **(29)**

hydrology   study of all the water on land   **(4)**

hypothesis (hy PAHTH uh sihs)   educated guess made by scientists about what they think will occur   **(8)**

iceberg   great mass of freshwater ice freed from a glacier by a process called *calving*   **(283)**

igneous (IHG nee uhs) **rock**   rock that forms from magma or lava   **(76)**

intrusive (ihn TROO sihv) **rock**   igneous rock produced when magma cools slowly and forms large crystals   **(79)**

ion   atom that has gained or lost one or more electrons and therefore has an electrical charge   **(53)**

ionic bond   bond that forms when electrons move from one atom to another   **(54)**

isotope   atom of an element that has a different number of neutrons   **(53)**

iceberg

jet stream   high-speed wind circling both polar areas at an altitude of 12 km   **(183)**

lake   large body of fresh or salt water completely surrounded by land   **(398)**

landslide   mass movement that occurs when dry soil or rock moves rapidly and unexpectedly down a steep slope   **(238)**

latitude (LAT uh tood)   distance from the earth's equator measured by a parallel imaginary line that runs east and west   **(30)**

lava   molten rock that reaches the surface of the earth   **(76)**

law   description of what happens in a given situation   **(8)**

leaching   movement of minerals from an upper horizon to a lower horizon   **(247)**

lens   piece of curved glass that can be used to focus light   **(505)**

lightning   stroke of electrons traveling at speeds of about 1000 km/s   **(214)**

light-year   distance that light travels in one year   **(493)**

lithosphere (LIHTH uh sfihr)   solid layer of the earth, composed of the crust and upper mantle   **(119)**

load   amount of eroded material carried by a stream or a river   **(254)**

lightning

lunar eclipse

loess (LEHS)   thick layers of fine glacial till   **(287)**
longitude (LAHN juh tood)   distance measured by a vertical imaginary line that runs around the earth through the North and South Poles   **(30)**
lunar eclipse (ih KLIHPS)   occurs when Earth's shadow falls on all or part of the moon's surface   **(459)**
luster   appearance of a mineral's surface in reflected light   **(63)**

M

magma   molten rock within the earth   **(76)**
magnification   mathematically, the focal length of the objective lens of a telescope divided by the focal length of the ocular lens   **(506)**
magnitude   star's brightness as it appears from Earth—not its actual, or absolute, brightness   **(481)**
mantle   layer of the earth immediately beneath the Moho   **(118)**
mass   amount of matter an object contains   **(17)**
mass movement   downhill movement of rocks and soil due to the force of gravity   **(237)**
meander   broad curve formed by the erosion of a mature stream   **(261)**
metamorphic (meht uh MAWR fihk) **rock**   rock produced from an existing rock as heat and pressure cause the minerals to form new combinations or new crystals   **(77)**
meteorite   asteroid fragment that leaves orbit and enters the earth's atmosphere   **(469)**
meteorology   study of the earth's atmosphere   **(4)**
mineral   naturally occurring substance consisting of a single element or compound   **(59)**
model   something scientists create to help them understand how things work   **(11)**
molecule   group of two or more atoms   **(54)**
momentum   mathematically, the mass of an object multiplied by the object's velocity   **(521)**
moraine   mound of unsorted rock materials that builds up along the edge of a glacier   **(286)**

meander

N

nebula   cloud of dust and gas in space; sometimes glows by reflected light   **(495)**
nekton   sea animals that swim about freely   **(441)**
neutron   subatomic particle that has no electrical charge   **(52)**
neutron star   very small, dense star, as small as 15 km in radius   **(487)**
nonfoliated rock   metamorphic rock having no visible layers   **(92)**
nucleus   core of an atom   **(52)**

ocean   major expanse of salt water with a surface area larger than 10 million km^2 **(391)**

oceanography   study of the waters of the oceans   **(4)**

orbit   path that the earth follows as it revolves around the sun   **(456)**

ore   economically important mineral compound of a metal   **(99)**

organic sediment   hard remains of a once-living organism   **(85)**

orogeny (oh RAHJ uh nee)   process of mountain building   **(122)**

outwash plain   thick blanket of sediments formed as meltwater deposits material in front of a stationary glacier   **(287)**

oxidation   chemical weathering process in which oxygen combines directly with an element   **(235)**

ozone   gas molecule consisting of three oxygen atoms   **(168)**

penumbra   in a lunar eclipse, a zone of partial shadow surrounding the umbra   **(459)**

permeability   ability of rocks to allow water to pass through   **(266)**

pesticide   chemical used to kill crop-destroying insects   **(370)**

photon (FOH tahn)   particle of light, which can travel from the sun to the earth in 8.3 minutes   **(170)**

photosynthesis (foht oh SIHN thuh sihs)   process of making food from water and carbon dioxide   **(166)**

planet   main body that revolves around the sun   **(461)**

planetesimal (plan uh TEHS uh muhl)   small body of gases and solids that formed a planet   **(342)**

plankton   microscopic plantlike and animal-like organisms that drift with the tides and currents   **(439)**

playa   desert plain where runoff collects temporarily in muddy pools   **(263)**

plucking   process in which glacial ice freezes around rocks and plucks them out of the ground as the glacier moves   **(284)**

pollution   unwanted dirt and waste that fouls water, air, and land   **(364)**

porosity   ratio of the volume of air space in a rock to the total volume of the rock   **(266)**

precipitate (pruh SIHP uh tayt)   chemical sedimentary rock that forms when dissolved minerals precipitate, or fall out of solution   **(87)**

precipitation   solid or liquid water that falls to the earth's surface   **(199)**

primary coast   coast resulting from erosional processes of the land   **(402)**

outwash plain

playa

quasar

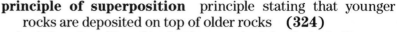

principle of superposition principle stating that younger rocks are deposited on top of older rocks **(324)**

principle of uniformitarianism principle that similar processes produce similar results in different times **(323)**

proton subatomic particle that has a positive electrical charge **(52)**

pulsar rapidly pulsating neutron star **(488)**

P wave primary wave; longitudinal wave produced during an earthquake **(151)**

quasar body that emits radio waves and light in a starlike fashion; largest energy source in the universe **(496)**

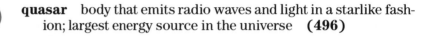

radiation balance balance between solar radiation coming into Earth's atmosphere and radiation going out from Earth **(172)**

radioactive decay process in which particles and energy are given off by unstable nuclei **(328)**

radiotelescope instrument for capturing radio waves from outer space **(512)**

radio wave long-wavelength electromagnetic wave **(512)**

rain shadow area shielded from precipitation by a mountain or other topographic feature **(276)**

recycle to use something again **(371)**

reflection light bounced back from a smooth, shiny surface **(504)**

refraction change in direction of a light beam as it passes from one type of matter to another **(504)**

relative dating dating an igneous formation absolutely, and then estimating the age of the neighboring formation by its position relative to the igneous formation **(332)**

revolution movement of Earth around the sun **(456)**

Richter (RIHK tur) **scale** special scale used to report the size, or magnitude, of an earthquake **(154)**

rip current returning water flowing at right angles to the beach **(425)**

rock naturally occurring combination of minerals **(75)**

rock cycle continuous process of change in which new rocks are formed from old rock material **(78)**

rotation (roh TAY shuhn) turning of the earth on its axis **(455)**

runoff movement of water on the surface of the earth toward a lower level **(253)**

rip current

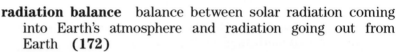

salinity measure of the amount of salts dissolved in ocean water **(412)**

saltation bouncing of sand that is blown by the wind **(279)**

salt dome mound of salt 2 km to 3 km across and 5 km to 10 km high **(435)**

satellite   small body orbiting a larger body or planet   **(466)**

saturated air   air that is filled with as much water as it can hold   **(192)**

scale   relationship between a distance on a map and a distance on the earth's surface   **(36)**

science   organized body of knowledge that has developed through observation and experimentation   **(5)**

scientific method   steps that scientists use to answer questions   **(9)**

sea   expanse of salt water less than 3 million km^2 in area   **(392)**

seamount   underwater volcano   **(147)**

secondary coast   coast shaped by erosion caused by the sea   **(403)**

secondary coast

sedimentary (sehd uh MEHN tuhr ee) **rock**   rock that forms from compressed sediments   **(76)**

seismograph (SYZ muh graf)   instrument used to record earthquake waves   **(152)**

shear   action of pieces of the earth's crust dragging against each other in opposite directions   **(127)**

shield volcano   volcano with a gentle slope and a wide base formed by quiet eruptions of lava   **(144)**

smelting   process of separating metals from their ores by using heat   **(99)**

smog   chemical fog produced by the reaction between sunlight and pollutants   **(367)**

soil   combination of small rock fragments and organic material in which plants can grow   **(231)**

solar cell   device that produces electricity from sunlight   **(378)**

solar eclipse   occurs when the moon's shadow passes across the lighted side of the earth   **(460)**

solar power   power produced by energy from the sun   **(377)**

solar prominence   stream of gases that explodes above the sun's surface   **(492)**

solar wind   high-energy radiation from the sun   **(165)**

sonar   device that bounces sound waves off a solid surface to determine its distance   **(394)**

space probe

space probe   spacecraft that is sent out into the solar system but is not sent into Earth orbit   **(538)**

spectrograph   optical instrument used for observing the colors of light given off by a star   **(507)**

stratus cloud   extended cloud layer that may cover the entire sky   **(197)**

streak   color of the powder that is left when a mineral is rubbed on a rough surface   **(63)**

strike   straight line along the earth's surface, formed where a rock layer dips   **(124)**

subsidence (suhb SY duhns)   sinking of rock material to a lower level without its sliding down a hillside   **(238)**

sunspot   enormous whorl of gases on the surface of the sun   **(168)**

surface wave   earthquake wave that is produced by the interaction of P waves and S waves and that causes the earth's surface to roll   **(152)**

S wave   secondary wave; transverse wave produced during an earthquake   **(151)**

syncline   structure formed when rock layers are folded downward   **(123)**

telescope

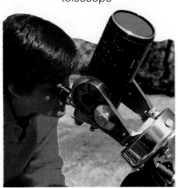

telescope   instrument used to enlarge the image of a distant object   **(506)**

tension   force that causes the earth's crust to pull apart   **(125)**

tetrahedron   network of atoms that forms a four-sided structure   **(59)**

texture   size of the crystals in igneous rocks   **(79)**

theory   statement that explains why things happen the way they do   **(8)**

theory of plate tectonics   theory stating that the lithosphere is divided into plates, each moving independently of the others **(138)**

thermocline   separation between warm surface water and colder deep water   **(415)**

thermohaline   ocean circulation due to temperature and salinity differences   **(415)**

thunderstorm   violent storm that develops from cumulonimbus clouds   **(214)**

tidal power   energy of changing tides   **(376)**

tide   crest of ocean water pulled up by the gravity of the moon as the earth turns on its axis   **(425)**

till   unsorted rock debris deposited directly by glaciers **(285)**

topographic map

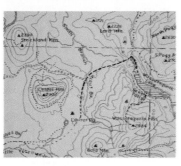

topographic map   map that shows the shape of the land, using fine lines drawn in patterns to indicate features such as mountains, valleys, and rivers   **(29)**

topography   contour or shape of the land surface   **(39)**

tornado   small, extremely violent storm   **(216)**

transform fault   type of lateral fault that forms when two plates slide past each other in parallel but opposite directions   **(142)**

umbra (UHM bruh)   zone of complete shadow in a lunar eclipse   **(459)**

universe   all the matter and energy that exists: all the stars, planets, dust, gas, and energy in space   **(481)**

Van Allen belts two atmospheric belts that trap incoming protons and electrons and protect the earth from high-energy particles; discovered in 1958 by an American physicist, James Van Allen **(180)**

variable in a scientific experiment, something that can be changed **(8)**

variable star star that changes magnitude **(482)**

viscosity liquid's resistance to flow **(144)**

volcano opening in the earth's crust that allows magma to reach the earth's surface **(139)**

volume total space that an object occupies **(15)**

waterspout vortex that occurs over open water **(218)**

water table zone of saturation, where all the air spaces in the soil are filled with water **(265)**

wave surface motion of water, produced by a local wind **(423)**

weathering processes that break rocks into smaller fragments, eventually producing soil **(231)**

weight measure of the force, or pull, of gravity on the mass of an object **(17)**

zodiac signs of the 12 constellations along the ecliptic **(483)**

waterspout

INDEX

Boldface numbers refers to an illustration on that page.

Aa lava, 144, **144**
Abrasion (weathering), 234
Absolute dating, 329
Absolute humidity, 193
Acadia National Park, 313
Accelerated mass spectrometry, 335
Acid rain, 167, 368
  damage of, **368**
  effects of, **236**
Aeration, **365**
Aerosols, 195
Agar, 442
Air, 367–368
Air mass, 207
Air pollution, **367, 368**
Air scrubbers, 368
Alaskan Province, 307–308, **308**
Alaska Range, 308, **308**
Aldrin, Edwin "Buzz," 528, **528**
Aleutians, 307, 396
Algin, 442
Allahabad, 195
Alloy, 100
Alluvial fans, 263, **263**
Alpine glaciers, 289–290
Aluminum, 371
Alvarez, Luis, biography of, 19, **19**
Alvarez, Walter, 321, 386–387, **387**
Amazon River, 419
Amber, 332, **332**
Amino acids, 344, **357**
Amphibians
  of Devonian Period, 351
Amplitude (wave), 170
Anchorage, **211**
Ancient map, **25**
Andes Mountains, 141
Andromeda, 495
Aneroid barometer, 175, **175**
Angel Falls, 259
Annapolis River (Canada), 450
Antarctica, 282
Antarctic Circumpolar Current, 420
Antarctic Ocean, 392
Anthracite, 92, 93, 110
Anticlines, 123, **123, 135**
Anticyclones, 209, **209**

Apatite, **65**
Aphelion, 462
Apollo/Soyuz linkup, 529
Apollo spacecraft, **529**
Apollo spaceflights, 527, 528
Appalachian Mountain range, 129
Appalachian Plateau, 312, **312**
Appalachian provinces, 310–313
Apprenticeships, 58
Aquifers, 267, **267**
Arabian Sea, **140**
Aral Sea, 398
Archean Era, 344–345
Arctic Ocean, 229, 308, **414,** 431, **431**
Arctic Slope, 308, **308**
Arecibo, 513
Arid environments, 275–277
Armstrong, Neil, 528, **528**
Artesian wells, 267, **267**
Asbestos, **66**
Asia, central, **212**
Asian plate, 141
Asphalts, 108
Asteroids, 468–469
Asthenosphere, 119, **120,** 138
Astronauts, 17. *See also*
  Individual astronauts
  career as, 497, **497**
Astronomer, **5**
  career as, 497, **497**
Astronomia Nova (Kepler), 463
Astronomical unit, 462
Astronomy
  defined, 5
  optical, 503–511
  radio, 512–515
Astrophysicist, career as, 497, **497**
Aswan Dam, 260
Atacama Desert, 199, **199**
Atlantic Ocean, 229, 394–395
Atlantis, 532
Atlas, earth science, 576–593
Atmosphere (earth's), **178**
  composition of, 165–168
  defined, 165
  gases of, 165–166
  layers of, 177–180
  motions of, 181–185
  present, 167–168
  primitive, 165–167
  primitive and present
    compared, *table* 166
  structure of, 175–180
  temperature of, **179**

Atmosphere chemist, career as, 176, **176**
Atmospheric diving suits (ADS), 438
Atmospheric pressure, 175–177
Atolls, 308
Atomic number, 56
Atomic structure, 51–53
  atoms of, 51–52
  and ions, 53
  and isotopes, 53
Atoms, 51–52
  computer model of, **52**
  and radioactive decay, 328
Aurora, 179
Aurora australis, 164
Aurora borealis, 164, **164**
Automotive technician, career as, 176, **176**
Autumnal equinox, 457
Axis
  of crystal shapes, *table* 68
  of the earth, 455
  of the moon, 458
  of Uranus, 474

Bacteria, 166
Bagnold, R.A., 296, 297
Balance, **17, 18**
Bald eagle, **370**
Ballard, Robert Duane, biography of, 418, **418;** 438
Baltic Sea, 393
Baraboo Dome, 309
Barchan dune, 280, 281
Barometer, 175
Barrier islands, 404, **404**
Barrow, **225**
Basalt, 79, **79,** 80
Base level, **259**
  of streams, **259**
  temporary, **259**
Basin (geologic feature), **135**
Basin and Range Province, 305
Batholith, 143
Bay, 402
Bay of Fundy, 389, 426, 450–451, **451**
Bench mark, 27, **27,** 35
Benioff, Victor Hugo, 155
Benthos, 442
Beryl, 69

Credits

PHOTO CREDITS
Key: (t) top, (b) bottom, (l) left, (r) right, (c), center.

COVER: Bill Ross/Westlight.

FRONTISPIECE: David W. Hamilton/The Image Bank.

Contents pages: vi(t), E. R. Degginger; vi(b), Breck P. Kent/Earth Scenes; vii(t), Weinberg/Clark/The Image Bank; vii(c), Townsend P. Dickinson/Photo Researchers; vii(bl), The Granger Collection; vii(bc), E. R. Degginger; vii(br), Paolo Koch/Photo Researchers; viii(t), J. L. Lepore/Photo Researchers; viii(bl), Science Source/Photo Researchers; viii(bc), Runk-Schoenberger/Grant Heilman Photography; viii(br), Michael Schneps/The Image Bank; ix(t), John Eastcott/Wa Momatiak/ Woodfin Camp & Assoc.; ix(c), Mike Andrews/Earth Scenes; ix(bl), Ralph A. Reinhold/Earth Scenes; ix(br), Miriam Reinhart/Photo Researchers; x(t), Farrell Grehan/Photo Researchers; x(bl), John Lemker/Earth Scenes; x(bc), Georg Gerster/Photo Researchers; x(br), Bill Bachman/Photo Researchers; xi(l), Holt Confer/ Grant Heilman Photography; xi(c), Kees Van Den Berg/Photo Researchers; xi(r), M. Paternostro/Science Photo Library/Photo Researchers; xii(t), Chip Henderson/ Woodfin Camp & Assoc.; xii(c), Official U. S. Navy Photograph; xii(bl), Woods Hole Oceanographic Institution; xii(bc), Alex Bartel/Science Photo Library/Photo Researchers; xii(br), Jeff Rotman/Peter Arnold, Inc.; xiii(t), A. S. P./Science Source/ Photo Researchers; xiii(c), Photri; xiii(b), NASA/Science Source/Photo Researchers; xiv, Paul Silverman/Fundamental Photographs; xv(t), Townsend P. Dickinson/Photo Researchers; xv(b), Thomas R. Taylor/Photo Researchers; xvii(t), The Granger Collection; xvii(c), Pam Taylor/Bruce Coleman, Inc; xvii(b), Hank Morgan/ Photo Researchers; xviii(t), Bob Evans/Peter Arnold, Inc.; xviii(b), Hank Morgan/ Photo Researchers;

ESSAY: xx(tl), Finley Holiday Film Corp.; xx(tr), NASA; xx(c), John Bova/Photo Researchers; xx(bl), Gary Guisinger/Photo Researchers; xx(bc), Richard Kolar/ Earth Scenes; xx(br), E. R. Degginger; xxi(t), HBJ Photo/Ted Flagg; xxi(b), Shambroom/Photo Researchers; xxii(t), Grant Heilman Photography; xxii(bl), Frank Fisher/Gamma-Liaison; xxii(br), U. S. Geological Survey, Department of the Interior; xxiii(t), Eugene A. Shinn; xxiii(c), TASS/Sovfoto; xxiii(b), Lawrence Gould/Oxford Scientific Films/Earth Scenes;

UNIT 1: Pages; xxv—1, Michel Tcherevkoff/The Image Bank; 2, NASA/Science Source/Photo Researchers; 3, NASA/Science Source/Photo Researchers; 4(t), S. J. Krasemann/Photo Researchers; 4(c), Lawrence Gould/Oxford Scientific Films/ Earth Scenes; 4(b), Paul Shambroom/Photo Researchers; 5(l), Gary Guisinger/ Photo Researchers; 5(r), Paul Shambroom/Photo Researchers; 6(tl), E. R. Degginger; 6(tc), Breck P. Kent/Earth Scenes; 6(tr), E. R. Degginger; 6(bl), Mickey Gibson/Earth Scenes; 6(bc), E. R. Degginger; 6(br), E. R. Degginger; 7, Gregory Heisler/The Image Bank; 8(t), HBJ Photo; 8(b), HBJ Photo; 11, Guy Gillette/Photo Researchers; 13(t), NASA/Science Source/Photo Researchers; 13(b), E. R. Degginger; 14(l), HBJ Photo/Earl Kogler; 14(r), HBJ Photo/Earl Kogler; 15, HBJ Photo/ Ted Flagg; 16(t), HBJ Photo/Ted Flagg; 16(b), HBJ Photo/Ted Flagg; 17(t), HBJ Photo/Sam Joosten; 17(b), NASA/Science Source/Photo Researchers; 18, HBJ Photo/Sam Joosten; 19(l), The Granger Collection; 19(r), Lawrence Berkeley Laboratory/Science Photo Library/Photo Researchers; 20(l), E. R. Degginger; 20(r), E. R. Degginger; 23, Peress/Magnum Photos; 24, Georg Gerster/Photo Researchers; 25, E. R. Degginger; 26, Ray Fairbanks/Photo Researchers; 27(t), Photo Researchers; 27(b), Photri; 28(t), E. R. Degginger; 28(c), HBJ Library; 28(b), HBJ Library; 29, Precision Visuals International/Science Photo Library/Photo Researchers; 31, Phil Degginger; 35(t), Pam Taylor/Bruce Coleman, Inc.; 35(c), GECO UK/Science Photo Library/Photo Researchers; 35(b), Hank Morgan/Photo Researchers; 36(l), from Atlanta Street Map, ©1988 Rand McNally & Co.; 36(r), from Atlanta Street Map, ©1988 Rand McNally & Co.; 37, U. S. Geological Survey, Department of the Interior; 39(t), U. S. Geological Survey, Department of the Interior; 39(b), U. S. Geological Survey, Department of the Interior; 40(l), U. S. Geological Survey, Department of the Interior; 40(r), U. S. Geological Survey, Department of the Interior; 41, U. S. Geological Survey, Department of the Interior; 42, U. S. Geological Survey, Department of the Interior; 44, U. S. Geological Survey, Department of the Interior; 45(t), Bruce Coleman, Inc.; 45(b), Kees Van Den Berg/Photo Researchers; 46(t), Photri; 46(b), Photri; 47(t), Photo Researchers; 47(b), NASA/Science Source/Photo Researchers;

UNIT 2: 48—49, Comstock; 50, Paul Silverman/Fundamental Photographs; 51, E. R. Degginger; 52, Photo Researchers; 53(t), E. R. Degginger; 53(b), J. L. Weber/Peter Arnold, Inc.; 54, Breck P. Kent/Earth Scenes; 55(t), Chemical Design Ltd.; 55(bl), Chemical Design Ltd.; 55(br), Chemical Design Ltd.; 57(l), E. R. Degginger; 57(lc), E. R. Degginger; 57(rc), E. R. Degginger; 57(r), E. R. Degginger; 58(t), Anne Van Der Vaeren/The Image Bank; 58(c), Geco UK/Science Photo Library/Photo Researchers; 58(b), Richard Hutchings/Photo Researchers; 59(t), Breck P. Kent/Earth Scenes; 59(b), E. R. Degginger; 60(l), Breck P. Kent/Earth Scenes; 60(r), HBJ Photo/Earl Kogler; 61(t), Paul Silverman/Fundamental Photographs; 61(b), Paul Silverman/Fundamental Photographs; 62, Paul Silverman/Fundamental Photographs; 63(tl), Breck P. Kent/Earth Scenes; 63(tr), Fundamental Photographs; 63(b), Paul Silverman/Fundamental Photographs; 65(#1), Breck P. Kent/Earth Scenes; 65(#2), E. R. Degginger; 65(#3), Breck P. Kent/Earth Scenes; 65 (#4), E. R. Degginger; 65(#5), Breck P. Kent/Earth Scenes; 65(#6), E. R. Degginger; 65(#7), E. R. Degginger; 65(#8), Breck P. Kent/Earth Scenes; 65(#9), E. R. Degginger; 65(#10), Breck P. Kent/Earth Scenes; 66(tl), E. R. Degginger; 66(tc), E. R. Degginger; 66(tr), Paul Silverman/Fundamental Photographs; 66(tc), Breck P. Kent/Earth Scenes; 66(bc), Paul Silverman/Fundamental Photographs; 66(b), Breck P. Kent/Earth Scenes; 68(t-b), E. R. Degginger; 68, Breck P. Kent/Earth Scenes; 68, E. R. Degginger; 68, E. R. Degginger; 68, E. R. Degginger; 68, E. R. Degginger; 69, Harold & Erica Van Pelt; 73(t), Kelly Farris; 73(b), HBJ Photo; 74, J. H. Robinson/Earth Scenes; 75(l), E. R. Degginger; 75(r), Breck P. Kent/Earth Scenes; 76(t), Gianni Tortoli/Photo Researchers; 76(b), E. R. Degginger; 77(tl), E. R. Degginger; 77(tlc), E. R. Degginger; 77(trc), E. R. Degginger; 77(tr), E. R. Degginger; 77(bl), E. R. Degginger; 77(br), E. R. Degginger; 78(l), E. R. Degginger; 78(c), E. R. Degginger; 78(r), E. R. Degginger; 79(t), E. R. Degginger; 79(b), E. R. Degginger; 80(t), E. R. Degginger; 80(bl), Michael & Barbara Reed/Earth Scenes; 80(br), Roger Evans/Photo Researchers; 81(t), Breck P. Kent/Earth Scenes; 81(bl), E. R. Degginger; 81(br), Breck P. Kent/Earth Scenes; 82(l), E. R. Degginger; 82(r), Dana Hyde/ Photo Researchers; 83(l), Tom McHugh/Photo Researchers; 83(r), Charles R. Belinky/Photo Researchers; 84(t), E. R. Degginger; 84(b), Richard Choy/Peter Arnold, Inc.; 85(t), E. R. Degginger; 85(c), L. L. T. Rhodes/Earth Scenes; 85(b), Nadine Markova/The Image Bank; 86(t), Breck P. Kent/Earth Scenes; 86(cl), Breck P. Kent/Earth Scenes; 86(c), E. R. Degginger; 86(cr), E. R. Degginger; 86(bl), E. R. Degginger; 86(br), Breck P. Kent/Earth Scenes; 87(t), E. R. Degginger; 87(c), E. R. Degginger; 87(bl), E. R. Degginger; 87(br), Breck P. Kent/Earth Scenes; 88(t), E. R. Degginger; 88(b), E. R. Degginger; 89(t), Stuart Cohen/The Stock Market; 89(b), E. R. Degginger; 90(l), The Granger Collection; 90(r), Ronald B. Parker, PhD; 92(t), Breck P. Kent/Earth Scenes; 92(bl), Breck P. Kent/Earth Scenes; 92(br), Breck P. Kent/Earth Scenes; 93(t), Art Stein/Photo Researchers; 93(b), E. R. Degginger; 97, Earth & Mineral Sciences Museum, The Pennsylvania State University; 98, Honeywell Inc.; 99(l), Joe Bereswill; 99(tr), E. R. Degginger; 99(br), E. R. Degginger; 100(tl), Guy Gillette/Photo Researchers; 100(tr), Day Williams/Photo Researchers; 100(bl), The Granger Collection; 100(br), The Granger Collection; 101(l), The Granger Collection; 101(r), The Granger Collection; 102(tl), The Granger Collection; 102(tc), Art Resource; 102(tr), Art Resource; 102(b), M. E. Warren/Photo Researchers; 103(t), Korody/Sygma; 103(bl), M. Serraillier/Photo Researchers; 103(bcl), C. Falco/Photo Researchers; 103(bcr), Dailloux/Rapho/Photo Researchers; 103(br), George Haling/Photo Researchers; 104(t), E. R. Degginger; 104(bl), Corning Glass Works; 104(br), Bill Nation/Sygma; 105(t), SIV/Photo Researchers; 105(bl), Jay Freis/The Image Bank; 105(bc), Frank Wing/The Image Bank; 105(br), Weinberg/Clark/The Image Bank; 106(l), Tom Mareschal/The Image Bank; 106(c), C. Kuhn/The Image Bank; 106(r), Michael L. Abramson/Woodfin Camp & Assoc.; 107(l), Joseph Nettis/Photo Researchers; 107(tr), Tom McHugh/ Photo Researchers; 107(br), Larry Voigt/Photo Researchers; 108(t), David Halpern/ Photo Researchers; 108(b), Wide World Photos; 109(l), David Hurn/Magnum Photos; 109(c), Louisa Preston/Photo Researchers; 109(r), E. R. Degginger; 110(l), Zig Leszczynski/Earth Scenes; 110(r), Frank Fisher/Gamma-Liaison; 111(l), Hank Morgan/Photo Researchers; 111(r), Erich Hartman/Magnum Photos; 114, Amy Ellis Photography; 115, AP/Wide World Photos; 116, Yoshikazu Shirakawa/The Image Bank; 121(tl), Solarfilma; 121(tr), The Image Bank 121(b), Georg Gerster/Photo Researchers; 122, Charlie Ott/Photo Researchers; 123, Kent & Donna Dannen/ Photo Researchers; 124, James P. Rowan/Click/Chicago 125(t), Dick Rowan/Photo Researchers; 125(b), E. R. Degginger; 126(t), Breck P. Kent/Earth Scenes; 126(b), National Park Service; 127(t), Francois Gohier/Photo Researchers; 127(b), Map 1, Geologic Map of Pennsylvania, 1980. Compiled by T. M. Berg, copyright The Commonwealth of Pennsylvania; 130, Townsend P. Dickinson/Photo Researchers; 131(t), Doug Wechsler/Earth Scenes; 131(c), HBJ Photo/Sam Joosten; 131(b), C. C. Lockwood/Earth Scenes; 135(t), John Gerlach/Earth Scenes; 135(b), Bruce Coleman, Inc.; 136, Dick Rowan/Science Source/Photo Researchers; 137(l), Dr. Steve Gull, Dr. John Fielden & Dr. Alan Smith/Science Photo Library/Photo Researchers; 137(c), Dr. Steve Gull, Dr. John Fielden & Dr. Alan Smith/Science Photo Library/Photo Researchers; 137(r), Dr. Steve Gull, Dr. John Fielden & Dr. Alan Smith/Science Photo Library/Photo Researchers; 140(t), NASA; 140(b), Francois Gohier/Photo Researchers; 141(l), The Image Bank; 141(r), Earl Roberge/Photo Researchers; 142, Field Museum, Chicago/Tom McHugh/Photo Researchers; 143, Phil Degginger; 144(tl), E. R. Degginger/Earth Scenes; 144(tc), E. R. Degginger; 144(tr), E. R. Degginger; 144(b), Mickey Gibson/Earth Scenes; 145(tl), Krafft-Explorer/Photo Researchers; 145(tr), Vulcain-Explorer/Photo Researchers; 145(bl), Paolo Koch/Photo Researchers; 145(br), David R. Frazier/Photo Researchers; 146(t), Georg Gerster/Photo Researchers; 146(b), Photri; 147(l), Georg Gerster/Photo Researchers; 147(r), Leonard Lee Rue III/Photo Researchers; 148, Georg Gerster/Photo Researchers; 150, The Granger Collection; 152, E. R. Degginger; 153(l), Breck P. Kent/Earth Scenes; 153(r), Tom McHugh/Photo Researchers; 155(tl), Jay Dickman/Gamma-Liaison; 155(tr), Hank Morgan/Photo Researchers; 155(b), Krafft-Explorer/Photo Researchers; 159, Wesley Bocxe/Photo Researchers; 160(l), DeBeers, Ltd.; 160(r), Ray Ellis/Photo Researchers; 161(t), Daniel Zirinsky/Photo Researchers; 161(bl), Gerhard Gscheidle/The Image Bank; 161(br), Paul Katz/The Image Bank;

UNIT 3: 162—163; Cliff Feulner/The Image Bank; 164, E. R. Degginger; 165, Catherine Ursillo/Photo Researchers; 166, Michael Abbey/Photo Researchers; 167(l), E. R. Degginger; 167(r), E. R. Degginger; 168(t), G. C. Fuller/Science Photo Library/ Photo Researchers; 168(b), Michael Gilbert/Science Photo Library/Photo Researchers; 170, John Deeks/Photo Researchers; 171, Dr. G. J. Chafaris/E. R. Degginger; 173(t), Jules Bucher/Photo Researchers; 173(b), E. R. Degginger; 174, J. L. Lepove/Photo Researchers; 175, Runk-Schoenberger/Grant Heilman Photography; 176(t), Dick Luria/Science Source/Photo Researchers; 176(c), Lawrence

Kabilinsky/Peter Arnold, Inc.; 176(b), Grapes Michaud/Photo Researchers; 177, Phil Degginger; 179, Yoshikazu Shirakawa/The Image Bank; 180, NASA; 184, Paolo Koch/Photo Researchers; 185(l), Stephen Krasemann/Photo Researchers; 185(r), Jules Bucher/Photo Researchers; 189, AP/Wide World Photos; 190, Eric Meola/The Image Bank; 191(l), Harvey Lloyd/The Stock Market; 192, HBJ Photo/Earl Kogler; 193(t), Thermometer Corporation of America; 193(b), HBJ Photo/Rodney Jones; 194(t), Nicholas Foster/The Image Bank; 194(bl), Bob Silverman/The Stock Market; 194(br), Chris Jones/The Stock Market; 195(l), Michael Schneps/The Image Bank; 195(r), Frank Whitney/The Image Bank; 196(t to b), Leonard Lee Rue III/Photo Researchers; 196, Gregory K. Scott/Photo Researchers; 196, Henry Lansford/Photo Researchers; 196, Joyce Photographics/Photo Researchers; 196, Mary Fuller/Taurus Photos; 197(t to b), Richard Kolar/Earth Scenes; 197, Joe Devenney/The Image Bank; 197, H. Wendler/The Image Bank; 197, Edna Bennett/Photo Researchers; 197, Eric Meola/The Image Bank; 197, Howie Bluestein/Photo Researchers; 199(tl), Frans Lanting/Photo Researchres; 199(tr), Francois Gohier/Photo Researchers; 199(b), Science Source/Photo Researchers; 200(l), NOAA; 200(r), E. R. Degginger; 201(l), Universitetsbiblioteket 1 Oslo, Hovedbiblioteket; 202, HBJ Photo/Earl Kogler; 204, Larry Dale Gordon/The Image Bank; 205(l), John Gerlach/Earth Scenes; 205(r), NCAR; 206, Bruce Coleman, Inc.; 207, Walter Bibikow/The Image Bank; 211(l), Carleton Ray/Photo Researchers; 211(c), Margot Granitsas/Photo Researchers; 211(r), Mathias Oppersdorff/Photo Researchers; 212(tl), G. Whiteley/Photo Researchers; 212(bl), William Bacon III/Photo Researchers; 212(r), E. R. Degginger; 213(tl), Hank Morgan/Photo Researchers; 213(tr), Hank Morgan/Photo Researchers; 213(b), Hank Morgan/Photo Researchers; 214, Larry Dale Gordon/The Image Bank; 215(l), Lionel Brown/The Image Bank; 215(r), E. R. Degginger; 216(tl), Howard B. Bluestein/Science Source/Photo Researchers; 216(tc), Howard B. Bluestein/Science Source/Photo Researchers; 216(tr), Howard B. Bluestein/Science Source/Photo Researchers; 216(bl), Howard B. Bluestein/Science Source/Photo Researchers; 216(bc), Howard B. Bluestein/Science Source/Photo Researchers; 217, E. R. Degginger; 218, C. C. Lockwood/Earth Scenes; 220, NOAA/Science Photo Library/Photo Researchers; 221(l), UPI/Bettmann Newsphotos; 221(r), AP/Wide World Photos; 225, Jim Cartier/Photo Researchers; 226(l), National Severe Storms Lab, Norman, OK.; 226(r), Photri; 227(t), Howard Bluestein/Photo Researchers; 227(b), E. R. Degginger;

UNIT 4: 228—229, National Park Service; 230, E. R. Degginger; 231(l), Sepp Seitz/Woodfin Camp & Assoc.; 231(r), HBJ Photo/Sam Joosten; 232(l), Townsend P. Dickinson/Photo Researchers; 232(r), Townsend P. Dickinson/Photo Researchers; 233, John Eastcott/Yva Momatiuk/Woodfin Camp & Assoc.; 234(t), Miriam Reinhart/Photo Researchers; 234(bl), E. R. Degginger; 234(br), Townsend P. Dickinson/Photo Researchers; 235(t), Franke Keating/Photo Researchers; 235(c), C. C. Lockwood/Earth Scenes; 235(bl), Joy Glenn; 235(br), HBJ Photo/Sam Joosten; 236(t), Mike Andrews/Earth Scenes; 236(b), Gregory D. Dimijian/Photo Researchers; 237, Georg Gerster/Photo Researchers; 238(tl), R. Call/Gamma-Liaison; 238(tr), E. R. Degginger/Earth Scenes; 238(b), Leif Skoogfors/Woodfin Camp & Assoc.; 239(l), E. R. Degginger; 239(r), E. R. Degginger; 240(t), Bill Ross/Woodfin Camp & Assoc.; 240(bl), E. R. Degginger; 240(br), U. S. Geological Survey, Department of the Interior; 241, Photri; 243, E. R. Degginger; 244(l), Pat & Tom Leeson/Photo Researchers; 244(r), E. R. Degginger; 245(t), Betsy Blass/Photo Researchers; 245(b), Phil Degginger; 245(inset), Lowell Georgia/Photo Researchers; 246, USDA; 247(tl), Owen Franken/Stock, Boston; 247(tr), E. R. Degginger; 247(b), E. R. Degginger; 251(l), UPI/Bettmann Newsphotos; 251(r), The Bettmann Archive; 252, E. R. Degginger; 253, Mike Maple/Woodfin Camp & Assoc.; 254(l), E. R. Degginger; 254(r), Charles R. Belinky/Photo Researchers; 255, NASA; 257(l), The Bancroft Library/University of California, Berkeley; 257(r), Fred Seidman; 258(l), Ty & Julie Hotchkiss/Photo Researchers; 258(r), E. R. Degginger; 259, Ralph A. Reinhold/Earth Scenes; 260(l), John Bova/Photo Researchers; 260(r), John Lemker/Earth Scenes; 261(tl), Dan Getsug/Photo Researchers; 261(tr), C. C. Lockwood/Earth Scenes; 261(cr), Dan Guravich/Photo Researchers; 261(br), Earth Scenes; 262(l), NASA/Photri; 262(r), Dan Guravich/Photo Researchers; 265(t), William M. Partington/Photo Researchers; 265(b), Francois Gohier/Photo Researchers; 266(l), E. R. Degginger; 266(r), E. R. Degginger; 267, E. R. Degginger; 268, U. S. Geological Survey, Department of the Interior; 269(t), Breck P. Kent/Earth Scenes; 269(bl), Bruce Roberts/Photo Researchers; 269(br), Robert Knowles/Photo Researchers; 272, R. R. Donnelley & Sons Company Cartographic Services; 273, Breck P. Kent/Earth Scenes; 274, Richard Weymouth Brooks/Photo Researchers; 276(t), Zig Leszczynski/Animals Animals; 276(bl), George Holton/Photo Researchers; 276(br), Kummerly & Frey, Bern edition 1984; 277, Mike Yamashita/Woodfin Camp & Assoc.; 278(t), Larry Miller/Photo Researchers; 278(bl), Bill Bachman/Photo Researchers; 278(br), Joseph Gerster/Photo Researchers; 279(tl), Charlie Ott/Photo Researchers; 279(tr), E. R. Degginger; 279(b), E. R. Degginger; 280(t), Phil Degginger; 280(b), E. R. Degginger; 282, W. Curtsinger/Photo Researchers; 283(tl), E. R. Degginger; 283(tc), E. R. Degginger; 283(tr), Farrell Grehan/Photo Researchers; 283(bl), Nuridsany et Perennou/Photo Researchers; 283(bc), Thomas W. Martin/Photo Researchers; 283(br), Nuridsany et Perennou/Photo Researchers; 284(t), Paolo Koch/Photo Researchers; 284(c), Francois Gohier/Science Source/Photo Researchers; 284(b), Leonard Lee Rue III/Earth Scenes; 285(t), Paolo Koch/Photo Researchers; 285(b), Steve Coombs/Photo Researchers; 286(tl), Jim Brandenburg/Woodfin Camp & Assoc.; 286(tr), John Lemker/Earth Scenes; 286(bl), E. R. Degginger; 286(br), Charles R. Belinky/Photo Researchers; 287(tl), E. R. Degginger; 287(tr), D. Pulleston/Photo Researchers; 287(b), Ira Block/The Image Bank; 288(t), Bob Daemmrich/Stock, Boston; 288(c), Carleton Ray/Photo Researchers; 288(b), Bonnie Freer/Photo Researchers; 289, E. R. Degginger/Bruce Coleman, Inc.; 290(tl), Breck P. Kent/Earth Scenes; 290(tr), Leo Touchet/Woodfin Camp & Assoc.; 290(bl), Sven Olof Lindblad/Photo Researchers; 290(bc), E. R. Degginger; 290(br), NASA; 291, Michael Giannechini/Photo Researchers; 292(l), The Granger Collection; 292(r), R.

R. Donnelley & Sons Company Cartographic Services; 293(l), E. R. Degginger; 293(r), William W. Bacon III/Photo Researchers; 294(l), U. S. Geological Survey, Department of the Interior; 294(r), U. S. Geological Survey, Department of the Interior; 297, Dick Rowan/Photo Researchers; 298, Pat & Tom Leeson/Photo Researchers; 299(l), E. R. Degginger; 299(r), Mickey Gibson/Earth Scenes; 301(t), R. R. Donnelley & Sons Company Cartographic Services; 301(bl), Richard Kolar/Earth Scenes; 301(bc), John Bova/Photo Researchers; 301(br), E. R. Degginger; 302(l), The Granger Collection; 302(r), AP/Wide World Photos; 303(l), Doug Wechsler/Earth Scenes; 303(r), Dan Guravich/Photo Researchers; 304(t), R. R. Donnelley & Sons Company Cartographic Services; 304(bl), Kent & Donna Dannen/Photo Researchers; 304(br), E. R. Degginger; 305(l), Stouffer Enterprises/Earth Scenes; 305(r), Lillian N. Bolstad/Photo Researchers; 306(l), Steve Van Matre/National Audubon Society/Photo Researchers; 306(r), E. R. Degginger; 307(l), E. R. Degginger; 307(r), Roger Archibald/Earth Scenes; 308(t), R. R. Donnelley & Sons Company Cartographic Services; 308(c), Johnny Johnson/Earth Scenes; 308(bl), Stephen J. Krasemann/Photo Researchers; 308(br), Stephen J. Krasemann/Photo Researchers; 309(tl), E. R. Degginger; 309(tc), E. R. Degginger; 309(tr), R. R. Donnelley & Sons Company Cartographic Services; 309(b), Thomas Hollyman/Photo Researchers; 310(l), E. R. Degginger; 310(r), R. R. Donnelley & Sons Company Cartographic Services; 311(t), Bill Curtsinger/Photo Researchers; 311(bl), E. R. Degginger; 311(bc), J. H. Robinson/Photo Researchers; 311(br), John Lemker/Earth Scenes; 312(tl), Wendell D. Metzen/Bruce Coleman, Inc.; 312(tr), Photo Researchers; 312(bl), Michael Habicht/Earth Scenes; 312(br), Kent & Donna Dannen/Photo Researchers; 313(t), Stephen J. Krasemann/Photo Researchers; 313(b), E. R. Degginger; 314(l), U. S. Geological Survey, Department of the Interior; 314(r), Map courtesy of the Commonwealth of Virginia; 316, Photri; 317, McCarthy/The Image Bank; 318(l), National Park Service; 318(r), Phil Degginger/Bruce Coleman, Inc.; 319(t), National Park Service; 319(c), National Park Service; 319(b), National Park Service;

UNIT 5: 320—321, Tom McHugh/Photo Researchers; 322, Michael & Barbara Reed/Earth Scenes; 323, Paolo Koch/Photo Researchers; 324(t), E. R. Degginger/Earth Scenes; 324(bl), E. R. Degginger; 324(br), C. C. Lockwood/Earth Scenes; 325(tl), E. R. Degginger/Earth Scenes; 325(tr), E. R. Degginger; 325(b), Breck P. Kent/Earth Scenes; 326, E. R. Degginger; 327(tl), Townsend P. Dickinson/Photo Researchers; 327(tc), Townsend P. Dickinson/Photo Researchers; 327(tr), Townsend P. Dickinson/Photo Researchers; 327(c), Louise K. Broman/Photo Researchers; 327(bc), Townsend P. Dickinson/Photo Researchers; 327(br), Townsend P. Dickinson/Photo Researchers; 328, Campbell/Sygma; 331(l), Charles Naeser/U. S. Geological Survey; 331(tr), Paolo Koch/Photo Researchers; 331(br), Milner/Sygma; 332(l), E. R. Degginger; 332(bl), J. Koivula/Photo Researchers; 332(br), Novosti from Sovfoto; 333, Richard Weiss/Peter Arnold, Inc.; 334, Michael Collier/Stock, Boston; 335(tl), Photo Researchers; 335(tr), The Stock Market; 335(b), John Zoiner/Peter Arnold, Inc.; 336, Louise K. Broman/Photo Researchers; 339, E. R. Degginger; 340, Paul Fusco/Magnum Photos; 341(l), Lauren Freudmann/Woodfin Camp & Assoc.; 341(r), Bill Bachman/Photo Researchers; 342(t), Thomas Hopker/Woodfin Camp & Assoc.; 342(b), NASA/Photo Researchers; 343(l), Paolo Koch/Photo Researchers; 343(r), Chris Alan Wilton/The Image Bank; 344(l), E. R. Degginger; 344(r), UPI/Bettmann Newsphotos; 345(tl), Michael Abbey/Photo Researchers; 345(tr), Eric Grave/Science Source/Photo Researchers; 345(b), Eugene A. Shinn; 346(t), M. I. Walker/Photo Researchers; 346(bl), Photo Researchers; 346(bc), Biophoto Assoc./Photo Researchers; 346(br), Michael Abbey/Photo Researchers; 347(l), Nancy Sefton/Photo Researchers; 347(r), Bill Curtsinger/Photo Researchers; 349, J. L. Lepore/Photo Researchers; 350(t), Tom McHugh/Photo Researchers; 350(b), E. R. Degginger; 351(t), Porterfield/Chickering/Photo Researchers; 351(b), Fred Whitehead/Animals Animals; 353(l), John Zoiner/Peter Arnold, Inc.; 353(r), Stephen J. Krasemann/Peter Arnold, Inc.; 354(l), Michael Giannechini/Photo Researchers; 354(r), Emil Muench/Photo Researchers; 355(l), Chicago Zoological Park/Tom McHugh/Photo Researchers; 355(r), Helen Williams/Photo Researchers; 356, Will & Deni McIntyre/Photo Researchers; 357(l), AP/Wide World Photos; 357(r), Boston University Photo Services; 361, Jeff Foott/Bruce Coleman, Inc.; 362, F. Stuart Westmorland/Tom Stack & Assoc.; 363(l), Grant Heilman Photography; 363(r), Grant Heilman Photography; 364(l), Bruce Coleman, Inc.; 364(r), Kees Van Den Berg/Photo Researchers; 365(tl), William Felger/Grant Heilman Photography; 365(tc), Runk-Schoenberger/Grant Heilman Photography; 365(tr), Runk-Schoenberger/Grant Heilman Photography; 365(cr), R. Perron/Nawrocki Stock Photo; 365(bl), Steve Peterson/Tom Stack & Assoc.; 265(br), Doug Lee/Tom Stack & Assoc.; 366(tl), Holt Confer/Grant Heilman Photography; 366(tr),Larry Lefever/Grant Heilman Photography; 366(b), HBJ Photo/Beverly Brosius; 367(t), Grant Heilman Photography; 367(bl), Rick McIntyre/Tom Stack & Assoc.; 367(br), Steve Solum/Bruce Coleman, Inc.; 368(tl), Runk-Schoenberger/Grant Heilman Photography; 368(tr), John Shaw/Tom Stack & Assoc.; 368(b), Jake Rajs/The Image Bank; 369(tl), Betsy Blass/Photo Researchers; 369(tr), John L. Stage/The Image Bank; 369(bl), Richard Hoffman/Sygma; 369(br), Grant Heilman Photography; 370(tl), Charles G. Summers, Jr./Tom Stack & Assoc.; 370(tr), Grant Heilman Photography; 370(bl), Grant Heilman Photography; 370(br), Adrienne T. Gibson/Tom Stack & Assoc.; 371, The Image Bank; 372(tl), Grant Heilman Photography; 372(tr), E. R. Degginger/Bruce Coleman, Inc.; 372(b), Jan Halaska/Photo Researchers; 374(t), Lowell Georgia/Photo Researchers; 374(b), Dick Durrance/Woodfin Camp & Assoc.; 375(tl), R. L. Carlton/Photo Researchers; 375(tr), Grant Heilman Photography; 375(b), Grant Heilman Photography; 376(t), William Felger/Grant Heilman Photography; 376(tr), Georg Gerster/Photo Researchers; 376(b), A. Bollinger/The Image Bank; 377(tl), Gerald A. Corsi/Tom Stack & Assoc.; 377(tr), J. Glab/The Stock Shop; 377(bl), Tom McHugh/Photo Researchers; 377(br), John Colwell/Grant Heilman Photography; 378(tl), Nancy A. Potter/Bruce Coleman, Inc.; 378(tc), Jean Guichard/Photo Researchers; 378(tr), Jim Brandenburg/Bruce Coleman, Inc.; 378(bl), Tom McHugh/